Lecture Notes in Artificial Intelligence   3632
Edited by J. G. Carbonell and J. Siekmann

Subseries of Lecture Notes in Computer Science

W0111338

Robert Nieuwenhuis (Ed.)

Automated Deduction – CADE-20

20th International Conference on Automated Deduction
Tallinn, Estonia, July 22-27, 2005
Proceedings

 Springer

Series Editors

Jaime G. Carbonell, Carnegie Mellon University, Pittsburgh, PA, USA
Jörg Siekmann, University of Saarland, Saarbrücken, Germany

Volume Editor

Robert Nieuwenhuis
Technical University of Catalonia (UPC)
Dpt. Lenguajes y Sistema Informaticos (LSI)
Jordi Girona 1, 08034 Barcelona, Spain
E-mail: roberto@lsi.upc.es

Library of Congress Control Number: 2005929197

CR Subject Classification (1998): I.2.3, F.4.1, F.3, F.4, D.2.4

ISSN   0302-9743
ISBN-10   3-540-28005-7 Springer Berlin Heidelberg New York
ISBN-13   978-3-540-28005-7 Springer Berlin Heidelberg New York

Springer is a part of Springer Science+Business Media

springeronline.com

© Springer-Verlag Berlin Heidelberg 2005
Printed in Germany

Typesetting: Camera-ready by author, data conversion by Scientific Publishing Services, Chennai, India
Printed on acid-free paper   SPIN: 11532231   06/3142   5 4 3 2 1 0

Preface

This volume contains the proceedings of the *20th International Conference on Automated Deduction* (CADE-20). It was held July 22–27, 2005 in Tallinn, Estonia, together with the Workshop on Constraints in Formal Verification (CFV'05), the Workshop on Empirically Successful Classical Automated Reasoning (ES-CAR), the Workshop on Non-Theorems, Non-Validity, Non-Provability (DIS-PROVING), and the yearly CADE ATP System Competition (CASC).

CADE is the major forum for the presentation of research in all aspects of automated deduction. The first CADE conference was held in 1974. Early CADEs were mostly biennial, and annual conferences started in 1996.

Logics of interest include propositional, first-order, equational, higher-order, classical, intuitionistic, constructive, modal, temporal, many-valued, substructural, description, and meta-logics, logical frameworks, type theory and set theory.

Methods of interest include saturation, resolution, tableaux, sequent calculi, term rewriting, induction, unification, constraint solving, decision procedures, model generation, model checking, natural deduction, proof planning, proof presentation, proof checking, and explanation.

Applications of interest include hardware and software development, systems analysis and verification, deductive databases, functional and logic programming, computer mathematics, natural language processing, computational linguistics, robotics, planning, knowledge representation, and other areas of AI.

This year, there were 78 submissions, of which 9 system descriptions. Each submission was assigned to at least four program committee members, who carefully reviewed the papers, in many cases with the help of one or more of a total number of 115 external referees. For each submission at least four reviews were produced and forwarded to the authors. The merits of the submissions were discussed by the program committee for ten days through the Internet by means of the *EasyChair* system. Finally, the program committee selected for publication 25 regular research papers and 5 system descriptions.

Also included in this volume are three invited papers, by Randal Bryant (Decision Procedures Customized for Formal Verification), Gilles Dowek (What Do We Know When We Know That A Theory Is Consistent?), and by Frank Wolter (Temporal Logics over Transitive States). Not contained in this volume are the materials of two invited tutorials, given by Bruno Blanchet (An Automatic Security Protocol Verifier Based on Resolution Theorem Proving) and by Enrico Giunchiglia (Beyond SAT: QSAT, and SAT-Based Decision Procedures), and the tutorial on Integrating Object-Oriented Design and Deductive Verification of Software, by Wolfgang Ahrend and others.

At CADE-20 the 2005 Herbrand Award for Distinguished Contributions to Automated Reasoning was delivered to Martin Davis, in recognition of his role as

- a founding father of the field of automated reasoning;
- coauthor of both papers that introduced what is now called the Davis-Putnam or Davis-Putnam-Logemann-Loveland procedure, one of the most outstanding and useful proof procedures known today;
- historian regarding the early history of the field of automated deduction;

and for his numerous other contributions to the field.

Many people helped to make CADE-20 a success. I am of course grateful to the members of the program committee and to the external reviewers, as well as to the local organizers and the sponsors.

Special thanks go to Andrei Voronkov for providing the *EasyChair* software, which includes many features which improved the quality of the reviewing process, like the extremely good interface for discussion, or the complete hiding of information to PC members with conflicts of interest; for me, it also eliminated most of the administrative work related to paper assignment, author notification, etc.

Finally, I thank the organizers of all co-located events at CADE-20, which made it even more interesting and attractive to a larger audience, and among them, of course, the Conference Chair, Tanel Tammet.

July 2005   Robert Nieuwenhuis

Organization

Conference Chair

Tanel Tammet (Tallinn University of Technology)

Local Organization

Jüri Vain (Tallinn University of Technology)
Jaan Penjam (Institute of Cybernetics, Tallinn)

Program Chair

Robert Nieuwenhuis (Universidad Politécnica de Cataluña)

Workshop and Tutorial Chair

Frank Pfenning (Carnegie Mellon University)

Publicity Chair

Brigitte Pientka (McGill University)

System Competition

Geoff Sutcliffe (University of Miami)

Program Committee

Franz Baader (Technische Universität Dresden)
Peter Baumgartner (Max-Planck Institute)
Amy Felty (University of Ottawa)
Ian Horrocks (University of Manchester)
Deepak Kapur (University of New Mexico)
Chris Lynch (Clarkson University)
Fabio Massacci (University of Trento)
Ilkka Niemela (Helsinki University of Technology)
Robert Nieuwenhuis (Universidad Politécnica de Cataluña)
Dale Miller (INRIA/Ecole Polytechnique)
Tobias Nipkow (Technische Universität München)

Frank Pfenning (Carnegie Mellon University)
Andreas Podelski (Max-Planck Institute)
Manfred Schmidt-Schauss (Universität Frankfurt)
Peter Schmitt (Universität Karlsruhe)
Stephan Schulz (Technische Universität München)
Carsten Schürmann (Yale University)
Aaron Stump (Washington University)
Geoff Sutcliffe (University of Miami)
Tanel Tammet (Tallinn University of Technology)
Cesare Tinelli (University of Iowa)
Ashish Tiwari (SRI)
Moshe Vardi (Rice University)
Miroslav Velev (Reservoir Labs)
Andrei Voronkov (University of Manchester)
Toby Walsh (University of New South Wales)

External Reviewers

Roger Antonsen
Thomas Baar
Massimo Benerecetti
Sergey Berezin
Armin Biere
Amine Chaieb
Thomas Colcombet
Byron Cook
Joerg Denzinger
William Farmer
Alexander Fuchs
Malay Ganai
Silvio Ghilardi
Juergen Giesl
Rajeev Gore
Jean Goubault-Larrecq
Nadeem Abdul Hamid
Pat Hayes
Miki Hermann
Dieter Hutter
Tomi Janhunen
Victor Jauregui
Lydia Kavraki
Vladimir Klebanov
Alexander Koller
Konstantin Korovin
Viktor Kuncak

Jeremy Avigad
Bernhard Beckert
Christoph Benzmueller
Stefan Berghofer
Alexander Bolotov
Koen Claessen
Véronique Cortier
Michael De Coster
Ho Ngoc Duc
Alan Frisch
Didier Galmiche
Vijay Ganesh
Martin Giese
Birte Glimm
Shuvendu Lahiri
Bernhard Gramlich
James Harland
Keijo Heljanko
Doug Howe
Radu Iosif
Matti Jarvisalo
Tommi Junttila
Nils Klarlund
Michael Kohlhase
Boris Konev
Mathis Kretz
Ralf Küsters

Oliver Kutz
Gary T. Leavens
Lei Li
Salvador Lucas
Ines Lynce
Matthias Mann
Marius Minea
Barbara Morawska
Leonardo de Moura
Madanlal Musuvathi
Joachim Niehren
Alexander Nittka
Andreas Nonnengart
Nicolas Peltier
Alexander Pretschner
Jochen Renz
Jussi Rintanen
Andrey Rybalchenko
Andrei Sabelfeld
Christelle Scharff
Konrad Slind
Volker Sorge
Mark-Oliver Stehr
Monika Sturm
Cesare Tinelli
Ralf Treinen
Anni-Yasmin Turhan
Mathieu Turuani
Kumar Neeraj Verma
Luca Vigano
Heike Wehrheim
Mary-Anne Williams
Hantao Zhang
Evgeny Zolin

Timo Latvala
Reinhold Letz
Bernd Löchner
Carsten Lutz
Patrick Maier
Thomas Meyer
Ralf Möller
Boris Motik
Erik T. Müller
Aleks Nanevski
Peter Nightingale
Hans de Nivelle
Albert Oliveras
Brigitte Pientka
Claude-Guy Quimper
Alexandre Riazanov
Albert Rubio
David Sabel
Jun Sawada
Steffen Schlager
Viorica Sofronie-Stokkermans
Bruce Spencer
Gernot Stenz
Nobu-Yuki Suzuki
Alwen Tiu
Dmitry Tsarkov
Daniele Turi
Rakesh Verma
Alexander Veynberg
Arild Waaler
Thomas Wies
Calogero Zarba
Lintao Zhang

Table of Contents

What Do We Know When We Know That a Theory Is Consistent?
Gilles Dowek 1

Reflecting Proofs in First-Order Logic with Equality
Evelyne Contejean, Pierre Corbineau 7

Reasoning in Extensional Type Theory with Equality
Chad E. Brown 23

Nominal Techniques in Isabelle/HOL
Christian Urban, Christine Tasson 38

Tabling for Higher-Order Logic Programming
Brigitte Pientka 54

A Focusing Inverse Method Theorem Prover for First-Order Linear Logic
Kaustuv Chaudhuri, Frank Pfenning 69

The CORE Calculus
Serge Autexier 84

Simulating Reachability Using First-Order Logic with Applications to
Verification of Linked Data Structures
*T. Lev-Ami, N. Immerman, T. Reps, M. Sagiv, S. Srivastava,
G. Yorsh* 99

Privacy-Sensitive Information Flow with JML
Guillaume Dufay, Amy Felty, Stan Matwin 116

The Decidability of the First-Order Theory of Knuth-Bendix Order
Ting Zhang, Henny B. Sipma, Zohar Manna 131

Well-Nested Context Unification
Jordi Levy, Joachim Niehren, Mateu Villaret 149

Termination of Rewrite Systems with Shallow Right-Linear, Collapsing,
and Right-Ground Rules
Guillem Godoy, Ashish Tiwari 164

The OWL Instance Store: System Description
Sean Bechhofer, Ian Horrocks, Daniele Turi 177

Temporal Logics over Transitive States
Boris Konev, Frank Wolter, Michael Zakharyaschev 182

Deciding Monodic Fragments by Temporal Resolution
Ullrich Hustadt, Boris Konev, Renate A. Schmidt 204

Hierarchic Reasoning in Local Theory Extensions
Viorica Sofronie-Stokkermans 219

Proof Planning for First-Order Temporal Logic
Claudio Castellini, Alan Smaill 235

System Description: MULTI A Multi-strategy Proof Planner
Andreas Meier, Erica Melis 250

Decision Procedures Customized for Formal Verification
Randal E. Bryant, Sanjit A. Seshia 255

An Algorithm for Deciding BAPA: Boolean Algebra with Presburger
Arithmetic
Viktor Kuncak, Huu Hai Nguyen, Martin Rinard 260

Connecting Many-Sorted Theories
Franz Baader, Silvio Ghilardi 278

A Proof-Producing Decision Procedure for Real Arithmetic
Sean McLaughlin, John Harrison 295

The MathSAT 3 System
Marco Bozzano, Roberto Bruttomesso, Alessandro Cimatti,
Tommi Junttila, Peter van Rossum, Stephan Schulz,
Roberto Sebastiani 315

Deduction with XOR Constraints in Security API Modelling
Graham Steel 322

On the Complexity of Equational Horn Clauses
Kumar Neeraj Verma, Helmut Seidl, Thomas Schwentick 337

A Combination Method for Generating Interpolants
Greta Yorsh, Madanlal Musuvathi 353

sKizzo: A Suite to Evaluate and Certify QBFs
Marco Benedetti 369

Regular Protocols and Attacks with Regular Knowledge
   Tomasz Truderung 377

The Model Evolution Calculus with Equality
   Peter Baumgartner, Cesare Tinelli 392

Model Representation via Contexts and Implicit Generalizations
   Christian G. Fermüller, Reinhard Pichler 409

Proving Properties of Incremental Merkle Trees
   Mizuhito Ogawa, Eiichi Horita, Satoshi Ono 424

Computer Search for Counterexamples to Wilkie's Identity
   Jian Zhang 441

KRHyper - In Your Pocket
   Alex Sinner, Thomas Kleemann 452

Author Index 459

What Do We Know
When We Know That a Theory Is Consistent?

Gilles Dowek

École polytechnique and INRIA,
LIX, École polytechnique, 91128 Palaiseau Cedex, France
Gilles.Dowek@polytechnique.fr
http://www.lix.polytechnique.fr/~dowek

Abstract. Given a first-order theory and a proof that it is consistent, can we design a proof-search method for this theory that fails in finite time when it attempts to prove the formula $\perp$?

1 Searching for Proofs in a Theory

1.1 Who Knows That Higher-Order Logic Is Consistent?

It is well known that higher-order logic can be presented as a first-order theory, *i.e.* that there exists a first-order theory H and a function Φ translating closed formulas of higher-order logic to closed formulas of the language of H such that the sequent $\vdash A$ is provable in higher-order logic if and only if the sequent $\vdash \Phi A$ is provable in first-order logic (see, for instance, [4]). Thus, instead of using a proof-search method specially designed for higher-order logic, such as higher-order resolution [8,9], it is possible to use a first-order method, such as resolution, to search for proofs in higher-order logic.

However, this reduction is inefficient. Indeed, if we attempt to prove the formula $\perp$ with higher-order resolution, the clausal form of the formula $\perp$ is the empty set of clauses, from these clauses, we can apply neither the higher-order resolution rule that requires two clauses to be applied, nor any other rule of higher-order resolution, that all require at least one clause. Thus, this attempt to prove the formula $\perp$ fails immediately. In contrast, the axioms of H give an infinite number of opportunities to apply the resolution rule and thus when searching for a proof of $H \vdash \perp$, the search space in infinite.

Thus, we can say that higher-order resolution "knows" that higher-order logic is consistent, because an attempt to prove the formula $\perp$ fails in finite time, while first-order resolution does not.

1.2 A Proof-Search Method for the Theory H

There is an link between higher-order logic and higher-order resolution and another link between higher-order logic and the first-order theory H. But can we establish a direct link between higher-order resolution and the theory H, without referring to higher-order logic?

R. Nieuwenhuis (Ed.): CADE 2005, LNAI 3632, pp. 1–6, 2005.

The answer is positive because the translation Φ can be inverted: there exists a function Ψ translating closed formulas of the language of H to closed formulas of higher-order logic such that the sequent $\mathsf{H} \vdash B$ is provable in first-order logic if and only if the sequent $\vdash \Psi B$ is provable in higher-order logic. Thus, a way to search for a proof of a sequent $\mathsf{H} \vdash B$ is to apply higher-order resolution to the sequent $\vdash \Psi B$. Thus, independently of higher-order logic, higher-order resolution can be seen as special proof-search method for a the first-order theory H.

As $\Psi \bot = \bot$ this first-order proof-search method immediately fails when attempting to prove the sequent $\mathsf{H} \vdash \bot$. Thus, this method is much more efficient than applying first-order resolution to the sequent $\mathsf{H} \vdash \bot$ as it "knows" that the theory H is consistent.

1.3   A Proof-Search Method for a Theory T

Can we generalize this to other theories than H? Given an arbitrary first-order theory T and a proof that T is consistent, can we always *build in* the theory T, *i.e.* exploit the consistency of T to design a proof-search method that fails in finite time when required to prove the formula $\bot$?

Of course, as we are not interested in the trivial solution that first tests if the formula to be proven is $\bot$ and then applies any method when it is not, we have to restrict to proof-search methods that do not mention the formula $\bot$.

It is clear that the consistency of the theory T is a necessary condition for such a method to exist: if T is inconsistent, a complete proof-search method should succeed, and not fail, when attempting to prove the formula $\bot$. The main problem is to know if this hypothesis is sufficient.

2   Resolution Modulo

2.1   Resolution Modulo

Resolution modulo is a proof-search method for first-order logic that generalizes higher-order resolution to other theories than the theory H.

Some axioms of the theory H are equational axioms. How to build in equational axioms is well-known: we drop equational axioms and we replace unification by equational unification modulo these axioms (see, for instance, [13,12]). Equational unification modulo the equational axioms of H is called *higher-order unification.*

From a proof-theoretical point of view, this amounts to define a congruence on formulas generated by the equational axioms and to identify congruent formulas in proofs. For instance, if we identify the terms $2 + 2$ and 4, we do not need the axiom $2+2 = 4$ that is congruent to $4 = 4$, but when we substitute the term 2 for the variable x in the term $x + 2$, we obtain the term 4. We have called *deduction modulo* the system obtained by identifying congruent formulas in proofs.

But not all axioms can be expressed as equational axioms. For instance, if the axiom of arithmetic $S(x) = S(y) \Rightarrow x = y$ can be replaced by the equivalent equational axiom $Pred(S(x)) = x$, the axiom $\neg 0 = S(x)$, that has no one-point model, cannot be replaced by an equational axiom.

Thus, we have extended deduction modulo by identifying some atomic formulas with not atomic ones. For instance, identifying formulas with the congruence generated by the rewrite rules $Null(0) \longrightarrow \top$ and $Null(S(x)) \longrightarrow \bot$ is equivalent to having the axiom $\neg 0 = S(x)$.

When we have such rewrite rules operating directly on formulas, equational resolution has to be extended. Besides the resolution rule, we need to add another rule called *Extended narrowing*. For instance, if we identify the formula $P(1)$ with $\neg P(0)$, we can refute the set of clauses $\{\neg P(x)\}$, but to do so, we have to be able to substitute the term 1 for the variable x in the clause $\neg P(x)$, deduce the clause $P(0)$ and conclude with the resolution rule. More generally, the *Extended narrowing* rule allows to narrow any atom in a clause with a propositional rewrite rule. The proposition obtained this way must then be put back in clausal form. Equational resolution extended with this rule is called ENAR — *Extended Narrowing and Resolution* — or *resolution modulo* for short.

When we orient the axioms of H as rewrite rules and use resolution modulo, we obtain exactly higher-order resolution.

2.2 Proving Completeness

Proving the completeness of higher-order resolution, and more generally of resolution modulo, is not very easy. Indeed higher-order resolution knows that higher-order logic is consistent, *i.e.* it fails in finite time when attempting to prove the formula $\bot$. Thus, a finitary argument shows that the completeness of higher-order resolution implies the consistency of higher-order logic, and by Gödel's second incompleteness theorem, the completeness of higher-order resolution cannot be proved in higher-order logic itself. This explains that some strong proof-theoretical results are needed to prove the completeness of higher-order resolution, at least the consistency of higher-order logic. The completeness proof given by Andrews and Huet [1,8,9] uses a result stronger than consistency: the cut elimination theorem for higher-order logic.

In the same way, the completeness of resolution modulo rests upon the fact that deduction modulo the considered congruence has the cut elimination property. Indeed, when the congruence is defined by rules rewriting atomic formulas to non-atomic ones, deduction modulo this congruence may have the cut elimination property or not. For instance, deduction modulo the rule $P \longrightarrow Q \wedge R$ has the cut elimination property, but not deduction modulo the rule $P \longrightarrow Q \wedge \neg P$ [6] and resolution modulo this second rule is incomplete.

Is it possible to weaken this cut elimination hypothesis and require, for instance only consistency? The answer is negative: the rule $P \longrightarrow Q \wedge \neg P$ is consistent, but resolution modulo this rule is incomplete. More generally, Hermant [7] has proved that the completeness of resolution modulo a congruence implies cut elimination for deduction modulo this congruence.

2.3   A Resolution Strategy

At least in the propositional case, resolution modulo can be seen as a strategy of resolution [2].

For instance, consider the rule $P \longrightarrow Q \wedge R$. The *Extended narrowing* rule allows to replace an atom P by $Q \wedge R$ and to put the formula obtained this way back in clausal form. With this rule, from a clause of the form $C \vee P$ we can derive the clauses $C \vee Q$ and $C \vee R$ and from a clause of the form $C \vee \neg P$ we can derive the clause $C \vee \neg Q \vee \neg R$.

We can mimic this rule by adding three clauses $\neg \underline{P} \vee Q$, $\neg \underline{P} \vee R$, $\underline{P} \vee \neg Q \vee \neg R$ and restricting the application of the resolution rules as follows: (1) we cannot apply the resolution rule using two clauses of the this set (2) when we apply the resolution rule using one clause of this set the eliminated atom must be the underlined atom. Notice that this set of clauses is exactly the clausal form of the formula $\underline{P} \Leftrightarrow (Q \wedge R)$. This strategy is in the same spirit as hyper-resolution, but the details are different.

If we apply the same method with the formula $\underline{P} \Leftrightarrow (Q \wedge \neg P)$, we obtain the three clauses $\neg \underline{P} \vee Q$, $\neg \underline{P} \vee \neg P$, $\underline{P} \vee \neg Q \vee P$ with the same restriction and, like resolution modulo, this strategy is incomplete: it does not refute the formula Q.

The fact that this strategy is complete for one system but not for the other is a consequence of the fact that deduction modulo the rule $P \longrightarrow Q \wedge R$ has the cut elimination property, but not deduction modulo the rule $P \longrightarrow Q \wedge \neg P$.

Understanding resolution modulo as a resolution strategy seems to be more difficult when we have quantifiers. Indeed, after narrowing an atom with a rewrite rule, we have to put the formula back in clausal form and this involves skolemization.

3   From Consistency to Cut Elimination

We have seen in section 2 that the theory $\mathsf{T} = \{P \Leftrightarrow (Q \wedge \neg P)\}$ is consistent, but that resolution modulo the rule $P \longrightarrow (Q \wedge \neg P)$ is incomplete.

Thus, it seems that the consistency hypothesis is not sufficient to design a complete proof-search method that knows that the theory is consistent. However the rule $P \longrightarrow (Q \wedge \neg P)$ is only one among the many rewrite systems that allow to express the theory T in deduction modulo. Indeed, the formula $P \Leftrightarrow (Q \wedge \neg P)$ is equivalent to $\neg P \wedge \neg Q$ and another solution is to take the rules $P \longrightarrow \bot$ and $Q \longrightarrow \bot$. Deduction modulo this rewrite system has the cut elimination property and hence resolution modulo this rewrite system is complete. In other words, the resolution strategy above with the clauses $\neg \underline{P}$, $\neg \underline{Q}$ is complete and knows that the theory is consistent.

Thus, the goal should not be to prove that if deduction modulo a congruence is consistent then it has the cut elimination property, because this is obviously false, but to prove that a consistent set of axioms can be transformed into a congruence in such a way that deduction modulo this congruence has the cut elimination property. To stress the link with the project of Knuth and Bendix [10], we call this transformation an *orientation* of the axioms.

A first step in this direction has been made in [3] following an idea of [11]. Any consistent theory in propositional logic can be transformed into a polarized rewrite system such that deduction modulo this rewrite system has the cut elimination property.

To do so, we first put the theory T in clausal form and consider a model ν of this theory (*i.e.* a line of a truth table).

We pick a clause. In this clause there is either a literal of the form P such that $\nu(P) = 1$ or a literal of the form $\neg Q$ such that $\nu(Q) = 0$.

In the first case, we pick all the clauses where P occurs positively $P \vee A_1, ..., P \vee A_n$ and replace these clauses by the formula $(\neg A_1 \vee ... \vee \neg A_n) \Rightarrow P$. In the second, we pick all the clauses where Q occurs negatively $\neg Q \vee B_1, ..., \neg Q \vee B_n$ and replace these clauses by the formula $Q \Rightarrow (B_1 \wedge ... \wedge B_n)$. We repeat this process until there are no clauses left. We obtain this way a set of axioms of the form $A_i \Rightarrow P_i$ and $Q_j \Rightarrow B_j$ such that the atomic formulas P_i's and Q_j's are disjoint.

The next step is to transform these formulas into rewrite rules and this is difficult because they are implications and not equivalences. But, this is possible if we extend deduction modulo allowing some rules to apply only to positive atoms and others to apply only to negative atoms. This extension of deduction modulo is called *polarized deduction modulo*. We get the rules $P_i \longrightarrow_+ A_i$ and $Q_j \longrightarrow_- B_j$. Using the fact that the P_i's and the Q_j's are disjoint, it is not difficult to prove cut elimination for deduction modulo these rules [3].

So, this result is only a partial success because resolution modulo is defined for non-polarized rewrite systems and orientation yields a polarized rewrite system. There may be two ways to bridge the gap: the first is to extend resolution modulo to polarized rewrite systems. There is no reason why this should not be possible, but this requires some work. A more ambitious goal is to produce a non-polarized rewrite system when orienting the axioms. Indeed, the axiom $P \Rightarrow A$ can be oriented either as the polarized rewrite rule $P \longrightarrow_- A$ or as the non-polarized rule $P \longrightarrow (P \wedge A)$, and similarly the axiom $A \Rightarrow P$ can be oriented as the rule $P \longrightarrow (P \vee A)$. But the difficulty here is to prove that deduction modulo the rewrite system obtained this way has the cut elimination property.

Bridging this gap would solve our initial problem for the propositional case. Starting from a consistent theory, we would build a model of this theory, orient it using this model, *i.e.* define a congruence and resolution modulo this congruence would be a complete proof search method for this theory that knows that the theory is consistent.

But, this would solve only the propositional case and for full first-order logic, everything remains to be done.

We have started this note with a problem in automated deduction: given a theory T and a proof that it is consistent, can we design a complete proof-search method for T that knows that T is consistent? We have seen that this problem boils down to a problem in proof theory: given a theory T and a proof that it is consistent, can we orient the theory into a congruence such that deduction modulo this congruence has the cut elimination property?

We seem to be quite far from a full solution to this problem, although the solution for the propositional case seems to be quite close.

Some arguments however lead to conjecture a positive answer to this problem: first the fact that the problem seems almost solved for propositional logic, then the fact that several theories such as arithmetic, higher-order logic, and some version of set theory have been oriented. Finally, we do not have examples of theories that can be proved to be non orientable (although some counter examples exist for intuitionistic logic). However, some theories still resist to being oriented, for instance higher-order logic with extensionality or set theory with the replacement scheme.

A positive answer to this problem could have some impact on automated theorem proving, as in automated theorem proving, like in logic in general, axioms are often a burden.

References

1. P.B. Andrews, Resolution in type theory, *The Journal of Symbolic Logic*, 36, 1971, pp. 414-432.
2. G. Dowek, Axioms vs. rewrite rules: from completeness to cut elimination, H. Kirchner and Ch. Ringeissen (Eds.) *Frontiers of Combining Systems*, Lecture Notes in Artificial Intelligence 1794, Springer-Verlag, 2000, pp. 62-72.
3. G. Dowek, What is a theory?, H. Alt, A. Ferreira (Eds.), *Symposium on Theoretical Aspects of Computer Science*, Lecture Notes in Computer Science 2285, Springer-Verlag, 2002, pp. 50-64.
4. G. Dowek, Th. Hardin, and C. Kirchner, HOL-lambda-sigma: an intentional first-order expression of higher-order logic, *Mathematical Structures in Computer Science*, 11, 2001, pp. 1-25.
5. G. Dowek, Th. Hardin, and C. Kirchner, Theorem proving modulo, *Journal of Automated Reasoning*, 31, 2003, pp. 33-72.
6. G. Dowek and B. Werner, Proof normalization modulo, *The Journal of Symbolic Logic*, 68, 4, 2003, pp. 1289-1316.
7. O. Hermant, Déduction Modulo et Élimination des Coupures : Une approche syntaxique, *Mémoire de DEA*, 2002.
8. G. Huet, Constrained Resolution A Complete Method for Higher Order Logic, *Ph.D Thesis*, Case Western Reserve University, 1972.
9. G. Huet, A Mechanisation of Type Theory, *Third International Joint Conference on Artificial Inteligenge*, 1973, pp. 139-146.
10. D.E. Knuth and P.B. Bendix. Simple word problems in universal algebras. J. Leech (Ed.), *Computational Problems in Abstract Algebra*, Pergamon Press, 1970, pp. 263-297.
11. S. Negri and J. Von Plato. Cut elimination in the presence of axioms. *The Bulletin of Symbolic Logic*, 4, 4, 1998, pp. 418-435.
12. G. Peterson, and M. E. Stickel, Complete Sets of Reductions for Some Equational Theories, *Journal of the ACM* 28, 1981, pp. 233-264.
13. G. Plotkin, Building-in equational theories, *Machine Intelligence* 7, 1972, pp. 73-90.

Reflecting Proofs in First-Order Logic with Equality[*]

Evelyne Contejean and Pierre Corbineau

PCRI — LRI (CNRS UMR 8623) & INRIA Futurs,
Bât. 490, Université Paris-Sud, Centre d'Orsay,
91405 Orsay Cedex, France

Abstract. Our general goal is to provide better automation in interactive proof
assistants such as Coq. We present an interpreter of proof traces in first-order
multi-sorted logic with equality. Thanks to the reflection ability of Coq, this interpreter is both implemented and formally proved sound — with respect to a reflective interpretation of formulae as Coq properties — inside Coq's type theory.
Our generic framework allows to interpret proofs traces computed by any automated theorem prover, as long as they are precise enough: we illustrate that on
traces produced by the CiME tool when solving unifiability problems by ordered
completion. We discuss some benchmark results obtained on the TPTP library.

The aim of this paper is twofold: first we want to validate a reflective approach for
proofs in interactive proof assistants, and second show how to provide a better automation for such assistants. Both aspects can be achieved by using external provers designed
to automatically solve some problems of interest: these provers can "feed" the assistant
with large proofs, and help to compare the direct and the reflective approaches, and they
can also release the user from (parts of) the proof.

The proof assistant doesn't rely on the soundness of the external tool, but keeps
a skeptical attitude towards the external traces by rechecking them. Moreover incompleteness of this tool is not an issue either, since when it fails to produce an answer,
the user simply has to find another way to do his proof. But a key point is that it has to
produce a trace which can be turned into a proof.

Proof checkers usually have a very fine grained proof notion, whereas automated
theorem provers tend to do complex inferences such as term normalization and paramodulation in one single step. Reflection techniques [10] provide a good intermediate
layer to turn traces missing a lot of implicit information into fully explicit proofs. They
rely on the computation abilities of the proof assistant for trivial parts of proofs, leaving
the hard but interesting work of finding proofs to automated tools. Bezem *et al.* [5] use
reflection techniques to handle the clausification part of a proof but the derivation of the
empty clause is provided by an external tool.

Our approach extends the reflection technique to the proof itself, turning the proof
assistant into a *skeptical trace interpreter* from an intermediate language of proof traces
to its own native format. We have implemented this technique inside the Coq proof
assistant [17] using a sequent calculus for multi-sorted intuitionistic first-order logic

---
[*] Supported by the EU coordinated action TYPES and the French RNTL project AVER-
ROES, `www.cs.chalmers.se/Cs/Research/Logic/Types/` `www-verimag.imag.fr/AVERROES/`

R. Nieuwenhuis (Ed.): CADE 2005, LNAI 3632, pp. 7–22, 2005.

with equality as a semantics for the intermediate language. To validate the reflective approach, we used the CiME tool [4] to produce traces when solving word and unifiability problems by ordered completion and had these traces checked by Coq.

Other works integrate either reflection and/or rewriting inside Coq. Nguyen [14] explains how to produce term rewriting proofs, but does not use reflection, whereas Alvarado [1] provides a reflection framework dedicated to proofs of equality of terms. Both of them consider that the rewriting system is fixed *a priori*. Our approach is close to the work of Crégut [6] who also interprets proof traces thanks to reflection, but for quantifier-free formulae in Peano's Arithmetic.

In our work, the rewriting system changes during the completion process, and in order to address the problem of the trace, we add some information on the usual rules of ordered completion. Thanks to this extra information, only the useful rules are extracted from the completion process, and the resulting formal proof is significantly shorter than the completion process. But this is far from a formal proof. We explain how to turn the annotations into useful lemmata, that is universally quantified equalities together with their proofs as sequences of equational steps.

Section 1 presents a general framework for reflection of first-order logic with equality in type theory. In Section 2, we give key details of the implementation. In Section 3, we briefly recall ordered completion, give an annotated version of the completion rules and show how to extract a CiME proof from a successful completion run. In Section 4, we explain the translation of CiME proofs into Coq reified proofs. In Section 5, we comment some benchmarks obtained on the TPTP library with CiME and Coq.

1   Type Theory and the Reflection Principle

1.1   Type Theory and the Conversion Rule

Coq's type theory is an extension of dependently-typed λ-calculus with inductive types, pattern matching and primitive recursion. Coq acts as an interpreter/type-checker for this language. The use of the *proofs as programs* paradigm allows its use as an interactive proof assistant: the typing relation $\Gamma \vdash t : T$ can be seen either as *"according to the types of variables in Γ, t is an object in set T"* or as *"supposing the hypotheses in Γ, t is a proof of the assertion T"*. Together with the λ function binder, Coq's most important construction is the $\forall$ binding operator, which builds either a dependent function type or a universally quantified logical proposition. The term $\forall x : A.B$ is written $A {\rightarrow} B$ as long as x does not occur free in B; in this case it represents the function space from type A to type B or the implication between propositions A and B. To establish a clear distinction between informative objects (datatypes and functions) and non-informative objects (proofs of propositions), types intended as datatypes are in the sort Set and logic propositions are in the sort Prop.

A main feature of Coq's type theory, which will be essential for the reflection mechanism, is the *conversion* rule: if $\Gamma \vdash t : T$ then $\Gamma \vdash t : T'$ for any type T' equivalent to T, i.e. having the same normal form with respect to the reduction of Coq terms. In particular, any term t can be proved equal to its normal form $\mathrm{nf}(t)$ inside Coq's type theory : we can use the reflexivity axiom $\texttt{refl}_=$ to build the proof $(\texttt{refl}_= \ t) : t = t$ and thanks to the conversion rule this is also a proof of $t = \mathrm{nf}(t)$ *e.g.* take the term 2×2

which reduces to 4, then $(\mathtt{refl}_= 4)$ is a proof of $4 = 4$ and a proof of $2 \times 2 = 4$ (both propositions have $4 = 4$ as normal form).

1.2 The Reflection Principle

The power of type theory appears when building types or propositions by case analysis on informative objects: suppose there is a type $\mathtt{form}$ of sort Set intended as a concrete representation of logical formulas, and an interpretation function $[\![\_]\!]$: $\mathtt{form}{\rightarrow}\mathsf{Prop}$, then one can define a function $\mathtt{decide}$: $\mathtt{form}{\rightarrow}\mathtt{bool}$ deciding whether an object in $\mathtt{form}$ represents a tautology, and prove a correctness theorem:

$$\mathtt{decide\_correct} : \forall F : \mathtt{form}, ((\mathtt{decide}\ F) = \mathtt{true}){\rightarrow} [\![F]\!]$$

This gives a useful way of proving that some proposition A in the range of $[\![\_]\!]$ is valid. First, by some external means, A is *reified*, i.e. some representation $\dot{A}$ such that $[\![\dot{A}]\!]$ is convertible to A is computed. Then the term $(\mathtt{decide\_correct}\,\dot{A}\,(\mathtt{refl}_=\mathtt{true}))$ is built. This term is well typed if, and only if, $(\mathtt{decide}\ \dot{A})$ is convertible to $\mathtt{true}$, and then its type is $[\![\dot{A}]\!]$ so this term is a proof of A. Otherwise, the term is not typable. The advantage of this approach is that it is left to the internal reduction procedure of Coq to check if A is provable. Moreover, careful implementation of the $\mathtt{decide}$ function allows some improvement in the time and space required to check the proof.

The distinction between properties and their representations is necessary since there is no way to actually compute something from a property (for example there is no function of type $\mathsf{Prop}{\rightarrow}\mathsf{Prop}{\rightarrow}\mathtt{bool}$ deciding the syntactic equality) since there is no such thing as pattern-matching or primitive recursion in Prop, whereas the representation of properties can be defined as an inductive type such as $\mathtt{form}$ on which functions can be defined by case analysis.

As tempting as this approach may seem, there are several pitfalls to avoid in order to be powerful enough. First, the propositions that can be proven thanks to this approach are the interpretations of the representations F : $\mathtt{form}$ such that $(\mathtt{decide}\ F) = \mathtt{true}$. This set is limited by the range of the interpretation function and by the power of the decision function: the function that always yields $\mathtt{false}$ is easily proved correct but is useless. Besides, we need to take into account space and time limitations induced by the cost of the reduction of complex Coq terms inside the kernel.

1.3 Reflecting First-Order Proofs

Since validity in first-order logic is undecidable, we have two options if we want to prove first-order assertions by reflection: either restrict the $\mathtt{decide}$ function to a decidable fragment of the logic, or change the $\mathtt{decide}$ function into a proof-checking function taking a proof trace (of type $\mathtt{proof}$: Set) as an extra argument. The correctness theorem for this function $\mathtt{check}$ reads:

$$\mathtt{check\_correct} : \forall \pi : \mathtt{proof}.\forall F : \mathtt{form}.((\mathtt{check}\ F\ \pi) = \mathtt{true}){\rightarrow} [\![F]\!]$$

The proof process using this proof trace approach is shown in Figure 1, it shows how an external tool can be called by Coq to compute a proof trace for a given formula. The

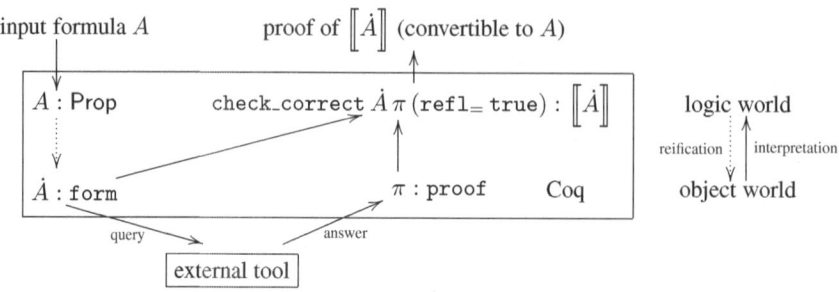

Fig. 1. Reflection scheme with proof traces

reification phase is done inside the Coq system at the ML level and not by a type-theory function (remember there is no case analysis on Prop).

The proof traces can have very different forms, the key point being the amount of implicit information that check will have to recompute when the kernel will type-check the proof. Moreover, if too many things have to be recalculated, the correctness theorem may become trickier to prove. What we propose is to give a derivation tree for a sequent calculus as a proof trace, and this is what the next section is about.

2   Proof Reflection for Multi-sorted First-Order Logic

2.1   Representation for Indexed Collections

The first attempt to represent collections of indexable objects with basic lists indexed by unary integers as in [3] was most inefficient, consuming uselessly time and space resources. For efficiency purposes, lists were replaced by binary trees indexed by binary integers $\mathbb{B}$. These binary trees $\mathbb{T}(\tau)$ containing objects of type τ are equipped with access and insertion functions get : $\mathbb{T}(\tau) \to \mathbb{B} \to \tau^?$ and add : $\mathbb{T}(\tau) \to \mathbb{B} \to \tau \to \mathbb{T}(\tau)$.

The question mark in the get function stands for the basic option type constructor adding a dummy element to any type. Indeed any Coq function must be total, which means that even the interpretation of badly formed objects needs a value. Most of the time the option types will be the solution to represent partiality.

Trees will sometimes be used as indexable stacks by constructing a pair in type $\mathbb{B} \times \mathbb{T}(\tau)$, the first member being the next free index in the second member. The empty stack and the empty tree will be denoted $\emptyset$. As a shortcut, if $S = (i, T)$, the notation $S; x$ stands for $(i + 1, \mathrm{add}(T, i, x))$ (pushing x on top of S). The i index in $S; i : x$ means that i is the stack pointer of S, i.e. the index pointing to x in $S; i : x$ (see Example 1). The notation S_j stands for get(T, j), assuming j is a valid index, i.e. $j < i$.

2.2   Representation and Interpretation of Formulae

Since Coq's quantifiers are typed, in order to be able to work with problems involving several types that are unknown when proving the reflection theorems, our representation will be *parameterized* by a *domain signature*. This enables the representation of multi-sorted problems, and not only problems expressed with a domain chosen *a priori*. We represent the quantification domains by our binary integers, referring to a binary tree Dom : $\mathbb{T}(\mathsf{Set})$ which we call *domain signature*. The interpretation function, given a domain signature, is a Coq function $[\![\_]\!]_{\mathrm{Dom}} : \mathbb{B} \to \mathsf{Set}$, which maps undefined indices to $\mathbb{1} : \mathsf{Set}$, the set with one element.

Given a domain signature, function symbols are defined by a dependent record: its first field is a (possibly empty) list of binary integers $d_1, \ldots, d_n$, the second field is an integer r for the range of the function, and the third field is the function itself, of type $[\![d_1]\!]_{\mathrm{Dom}} \to \ldots \to [\![d_n]\!]_{\mathrm{Dom}} \to [\![r]\!]_{\mathrm{Dom}}$ (its type is computed from Dom and the two first fields). Predicate symbols are defined in a similar way, except there is no range field and the last field has type $[\![d_1]\!]_{\mathrm{Dom}} \to \ldots \to [\![d_n]\!]_{\mathrm{Dom}} \to \mathsf{Prop}$. The *function signature* Fn is defined as a tree of function symbols, and the *predicate signature* Pred is defined as a tree of predicate symbols.

Example 1. In order to express properties of integers in $\mathbb{N}$ and arrays of integers (in $\mathcal{A}_{\mathbb{N}}$), we define the domain signature as $\mathrm{Dom} = 1 : \mathbb{N} ; 2 : \mathcal{A}_{\mathbb{N}}$.

In the signature, we put the $\mathbf{0}$ constant as well as two relations $<$ and $\geq$ over the integers. We also add the get and set functions over arrays and the sorted predicate. The valid predicate states that an integer is a valid index for a given array. The function and predicate signatures corresponding to this theory will be:

$$\mathrm{Fn} = 1 : \{\emptyset, 1, \mathbf{0}\}; 2 : \{[2; 1], 1, \mathsf{get}\}; 3 : \{[2; 1; 1], 2, \mathsf{set}\}$$

$$\mathrm{Pred} = 1 : \{[1; 1], <\}; 2 : \{[1; 1], \geq\}; 3 : \{[2], \mathsf{sorted}\}; 4 : \{[1; 2], \mathsf{valid}\}$$

Definition 1 (term, formula). *Terms and formulae are recursively defined as follows:*

```
term := FvB | BvN | App B args                args := ∅ | term, args
form := Atom B args | term ≐B term | ⊥ | ∀̇B form | ∃̇B form
      | form →̇ form | form ∧̇ form | form ∨̇ form
```

In terms, the Fv constructor represents free variables and their indices will refer to the slot they use in the sequent context (see below). The Bv constructor represents bound variables under quantifiers, using the deBruijn notation [7]: the indices are unary integers $\mathbb{N}$, 0 standing for the variable bound by the innermost quantifier over the position of the variable, 1 for the next innermost, etc. The indices in the App and Atom constructors refer to symbols in the signature, whereas those in the equality and quantifiers refer to domains in Dom. There is no variable in the quantifiers since the deBruijn notation takes care of which quantifier binds which variable without having to name the variable. The logical negation of a formula F can be expressed by $F \dot{\to} \bot$.

A term is closed if it contains no Bv constructor, and a formula is closed if all Bv constructors have indices less than their quantifier depth.

Example 2. Using the signature of the previous example, the property:

$\forall a : \mathcal{A}_{\mathbb{N}}, \mathsf{sorted}(a) \to \exists i : \mathbb{N}, \forall : j : \mathbb{N}, \mathsf{valid}(j, a) \to$
$\qquad ((j < i) \to (\mathsf{get}(a, j) < \mathbf{0})) \wedge ((j \geq i) \to (\mathsf{get}(a, j) \geq \mathbf{0}))$ is represented by:

$\dot{\forall}_2 (\mathsf{Atom}\ 3\ \mathsf{Bv}_0) \dot{\to} \dot{\exists}_1 \dot{\forall}_1 (\mathsf{Atom}\ 4\ (\mathsf{Bv}_0, \mathsf{Bv}_2)) \dot{\to}$
$\qquad (\mathsf{Atom}\ 1\ (\mathsf{Bv}_0, \mathsf{Bv}_1)) \dot{\to} (\mathsf{Atom}\ 1\ (\mathsf{App}\ 2\ (\mathsf{Bv}_2, \mathsf{Bv}_0), \mathsf{App}\ 1\ \emptyset)) \dot{\wedge}$
$\qquad (\mathsf{Atom}\ 2\ (\mathsf{Bv}_0, \mathsf{Bv}_1)) \dot{\to} (\mathsf{Atom}\ 2\ (\mathsf{App}\ 2\ (\mathsf{Bv}_2, \mathsf{Bv}_0), \mathsf{App}\ 1\ \emptyset))$

Terms	$[\![\mathrm{Bv}_n]\!]^r_{\gamma,\delta} = x$ if $\delta_n = \{x \in r'\}$ and $r = r'$
	$[\![\mathrm{Fv}_i]\!]^r_{\gamma,\delta} = x$ if $\gamma_i = \{x \in r'\}$ and $r = r'$
	$[\![\mathrm{App}\ i\ a]\!]^r_{\gamma,\delta} = [\![f\,|\,a]\!]^d_{\gamma,\delta}$ if $\mathrm{Fn}_i = (d, r', f)$ and $r = r'$
Arguments	$[\![\varPhi\,|\,\emptyset]\!]^0_{\gamma,\delta} = \varPhi \qquad [\![\varPhi\,|\,t :: a]\!]^{r::d}_{\gamma,\delta} = \left[\!\!\left[\varPhi([\![t]\!]^r_{\gamma,\delta})\,\middle|\,a\right]\!\!\right]^d_{\gamma,\delta}$
Formulae	$[\![\mathrm{Atom}\ i\ a]\!]_{\gamma,\delta} = [\![P\,|\,a]\!]^d_{\gamma,\delta}$ if $\mathrm{Pred}_i = (d, P)$
	$[\![t_1 \doteq_d t_2]\!]_{\gamma,\delta} = [\![t_1]\!]^d_{\gamma,\delta} \underset{[d]_{\mathrm{Dom}}}{=} [\![t_2]\!]^d_{\gamma,\delta} \qquad \left[\!\!\left[\dot{\bot}\right]\!\!\right]_{\gamma,\delta} = \bot$
	$\left[\!\!\left[\dot{\forall}_d F\right]\!\!\right]_{\gamma,\delta} = \forall x : [d]_{\mathrm{Dom}}, [\![F]\!]_{\gamma,\{x \in d\}::\delta}$
	$\left[\!\!\left[\dot{\exists}_d F\right]\!\!\right]_{\gamma,\delta} = \exists x : [d]_{\mathrm{Dom}}, [\![F]\!]_{\gamma,\{x \in d\}::\delta}$
	$[\![F_1 \dot{\rightarrow} F_2]\!]_{\gamma,\delta} = [\![F_1]\!]_{\gamma,\delta} \rightarrow [\![F_2]\!]_{\gamma,\delta}$
	$[\![F_1 \dot{\wedge} F_2]\!]_{\gamma,\delta} = [\![F_1]\!]_{\gamma,\delta} \wedge [\![F_2]\!]_{\gamma,\delta}$
	$[\![F_1 \dot{\vee} F_2]\!]_{\gamma,\delta} = [\![F_1]\!]_{\gamma,\delta} \vee [\![F_2]\!]_{\gamma,\delta}$
Contexts	$[\![\emptyset \,\|\, \varPsi]\!] = \varPsi(\emptyset) \qquad [\![\varGamma; F \,\|\, \varPsi]\!] = \left[\!\!\left[\varGamma \,\middle\|\, \gamma \mapsto [\![F]\!]_{\gamma,\emptyset} \rightarrow \varPsi(\gamma)\right]\!\!\right]$
	$[\![\varGamma; id : \,\|\, \varPsi]\!] = \left[\!\!\left[\varGamma \,\middle\|\, \gamma \mapsto \forall x : [d]_{\mathrm{Dom}}, \varPsi(\mathrm{add}\ \gamma\ i\ \{x \in d\})\right]\!\!\right]$
Sequents	$[\![\varGamma \vdash G]\!] = \left[\!\!\left[\varGamma \,\middle\|\, \gamma \mapsto [\![G]\!]_{\gamma,\emptyset}\right]\!\!\right]$

Fig. 2. Interpretation of terms, formulae, sequents

Definition 2 (Sequent). *A sequent is a pair written* $\varGamma \vdash G$*, where* $\varGamma$ *is the context, of type* $\mathbb{T}(\mathbb{B} + \mathtt{form})$ *containing objects in* $\mathbb{B}$ *or in* $\mathtt{form}$*. Objects in* $\mathtt{form}$ *represent logical hypotheses whereas objects in* $\mathbb{B}$ *represent the domain of assumed variables (referred to using the* Fv *constructor).* $G : \mathtt{form}$ *is called the goal of the sequent.*

On Figure 2, we explain how to interpret reified objects as Coq propositions. The interpretation functions use *global and local valuations*, these valuations contain dependent pairs $\{v \in d\}$ in type $\mathtt{val} : \mathtt{Set}$. They contain a domain index d and an object $v : [d]_{\mathrm{Dom}}$. The letter γ will usually denote global valuations, in $\mathbb{T}(\mathtt{val})$. We say γ is an *instantiation* for a context $\varGamma$ if it maps indices of global variables in $\varGamma$ to pairs in the same domain (it gives values to global variables in $\varGamma$). The local context usually written δ, is a list of the above pairs (addition of x in the list L is written $(x :: L)$).

The interpretation of terms takes an extra argument which is the intended domain r of the result, and returns an optional value in $[r]^?_{\mathrm{Dom}}$. If the represented term is not well typed, the dummy optional value is returned. The interpreter for arguments uses an accumulator $\varPhi$ which is successively applied to the interpretation of arguments. This accumulator trick is necessary to build well typed terms, as well as the use of a Coq proof of $r = r'$ to coerce objects of $[r']_{\mathrm{Dom}}$ to $[r]_{\mathrm{Dom}}$. The interpretation of atomic formulae uses the arguments interpreter with a different kind of accumulator. The interpretation of badly formed formulae is $\top$, the trivial formula. The interpretation of contexts is presented inside out so it is easier to reason about, it builds a function $\varPsi$ from global valuations to Prop by adding hypotheses recursively.

$$\frac{}{(\text{Ax } i) : \Gamma \vdash G}\ \Gamma_i = G \qquad \frac{\pi_1 : \Gamma \vdash A \quad \pi_2 : \Gamma; A \vdash G}{(\text{Cut } A\ \pi_1\ \pi_2) : \Gamma \vdash G} \qquad \frac{}{(\bot_E\ i) : \Gamma \vdash G}\ \Gamma_i = \dot{\bot}$$

$$\frac{\pi : \Gamma; A \vdash B}{(\rightarrow_I\ \pi) : \Gamma \vdash A \dot{\rightarrow} B} \qquad \frac{\pi : \Gamma; B \vdash G}{(\rightarrow_E\ i\ j\ \pi) : \Gamma \vdash G} \quad \begin{cases} \Gamma_i = A \dot{\rightarrow} B \\ \Gamma_j = A \end{cases}$$

$$\frac{\pi_1 : \Gamma; A; B \dot{\rightarrow} C \vdash B \quad \pi_2 : \Gamma; C \vdash G}{(\rightarrow_D\ i\ \pi_1\ \pi_2) : \Gamma \vdash G} \quad \Gamma_i = (A \dot{\rightarrow} B) \dot{\rightarrow} C$$

$$\frac{\pi_1 : \Gamma \vdash A \quad \pi_2 : \Gamma \vdash B}{(\wedge_I\ \pi_1\ \pi_2) : \Gamma \vdash A \dot{\wedge} B} \qquad \frac{\pi : \Gamma; A; B \vdash G}{(\wedge_E\ i\ \pi) : \Gamma \vdash G} \quad \Gamma_i = A \dot{\wedge} B$$

$$\frac{\pi : \Gamma; A \dot{\rightarrow} B \dot{\rightarrow} C \vdash G}{(\wedge_D\ i\ \pi) : \Gamma \vdash G} \quad \Gamma_i = (A \dot{\wedge} B) \dot{\rightarrow} C$$

$$\frac{\pi : \Gamma \vdash A}{(\vee_{I_1}\ \pi) : \Gamma \vdash A \dot{\vee} B} \qquad \frac{\pi : \Gamma \vdash B}{(\vee_{I_2}\ \pi) : \Gamma \vdash A \dot{\vee} B} \qquad \frac{\pi_1 : \Gamma; A \vdash G \quad \pi_2 : \Gamma; B \vdash G}{(\vee_E\ i\ \pi_1\ \pi_2) : \Gamma \vdash G} \quad \Gamma_i = A \dot{\vee} B$$

$$\frac{\pi : \Gamma; A \dot{\rightarrow} C; B \dot{\rightarrow} C \vdash G}{(\vee_D\ i\ \pi) : \Gamma \vdash G} \quad \Gamma_i = (A \dot{\vee} B) \dot{\rightarrow} C$$

$$\frac{\pi : \Gamma;_j d \vdash (\text{inst } A\ \text{Fv}_j)}{(\forall_I\ \pi) : \Gamma \vdash \dot{\forall}_d A} \qquad \frac{\pi : \Gamma; (\text{inst } A\ t) \vdash G}{(\forall_E\ i\ t\ \pi) : \Gamma \vdash G} \quad \Gamma_i = \dot{\forall}_d A$$

$$\frac{\pi_1 : \Gamma \vdash \dot{\forall}_d A \quad \pi_2 : \Gamma; B \vdash G}{(\forall_D\ i\ \pi_1\ \pi_2) : \Gamma \vdash G} \quad \Gamma_i = (\dot{\forall}_d A) \dot{\rightarrow} B$$

$$\frac{\pi : \Gamma \vdash (\text{inst } A\ t)}{(\exists_I\ t\ \pi) : \Gamma \vdash \dot{\exists}_d A} \qquad \frac{\pi : \Gamma;_j d; (\text{inst } A\ \text{Fv}_j) \vdash G}{(\exists_E\ i\ \pi) : \Gamma \vdash G} \quad \Gamma_i = \dot{\exists}_d A$$

$$\frac{\pi : \Gamma; \dot{\forall}_d (A \dot{\rightarrow} B) \vdash G}{(\exists_D\ i\ \pi) : \Gamma \vdash G} \quad \Gamma_i = (\dot{\exists}_d A) \dot{\rightarrow} B$$

$$\frac{}{(=_I\ t) : \Gamma \vdash t \dot{=} t} \qquad \frac{\pi : \Gamma; A \vdash G}{(=_D\ i\ \pi) : \Gamma \vdash G} \quad \Gamma_i = (t \overset{\tau}{\dot{=}} t) \dot{\rightarrow} A$$

$$\frac{\pi : \Gamma \vdash (\text{rewrite } s\ t\ p\ G)}{(=_{E_1}\ i\ \rightarrow\ p\ \pi) : \Gamma \vdash G}\ \Gamma_i = s \overset{d}{\dot{=}} t \qquad \frac{\pi : \Gamma \vdash (\text{rewrite } t\ s\ p\ G)}{(=_{E_1}\ i\ \leftarrow\ p\ \pi) : \Gamma \vdash G}\ \Gamma_i = s \overset{d}{\dot{=}} t$$

$$\frac{\pi : \Gamma; (\text{rewrite } s\ t\ p\ A) \vdash G}{(=_{E_2}\ i\ j\ \rightarrow\ p\ \pi) : \Gamma \vdash G} \begin{cases} \Gamma_i = s \overset{d}{\dot{=}} t \\ \Gamma_j = A \end{cases} \qquad \frac{\pi : \Gamma; (\text{rewrite } t\ s\ p\ A) \vdash G}{(=_{E_2}\ i\ j\ \leftarrow\ p\ \pi) : \Gamma \vdash G} \begin{cases} \Gamma_i = s \overset{d}{\dot{=}} t \\ \Gamma_j = A \end{cases}$$

Fig. 3. Well-formedness of proofs

2.3 Proof Traces

We define the proof traces and their meaning in Figure 3. The judgement $\pi : \Gamma \vdash G$ means *"π is a correct proof trace for the sequent $\Gamma \vdash G$"*. We adapted Roy Dychkoff's contraction free sequent calculus for intuitionistic logic [8], in order to allow multiple sorts and possibly empty domains. Even though the choice of this sequent calculus may seem exotic, the proof technique is quite generic and our proofs may be easily adapted to other kinds of mono-succedent sequent calculi such as classical ones with excluded middle axiom or implicit non-empty domains.

In order to reify first-order reasoning, it is necessary to decide syntactic equality between terms and between formulae, which is done by the boolean functions $=_{\texttt{term}}$ and $=_{\texttt{form}}$. The inst function implements the substitution of a given closed term (no lift operator needed) to the first bound variable in a formula, which is used in the definition of proof steps for quantifiers. Finally, the rewrite function replaces a closed term by another at a given set of occurrences, checking that the first term is the same as the subterms at the rewrite positions.

2.4 Correctness Proof

For every object defined in the last paragraphs (terms, arguments, formulae, contexts, sequents), there is an implicit notion of well-formedness contained in the definition of the interpretation functions. These well-formedness properties are implemented inside Coq both as inductive predicates in Prop named WF_* (with $* = \texttt{term}, \texttt{args}, \texttt{form}\ldots$) and as boolean functions check_*. For every object a correctness lemma is proved:

$$\texttt{WF\_checked\_*} : \forall x : *, (\texttt{check\_*}\ x) = \texttt{true} \rightarrow (\texttt{WF\_*}\ x)$$

The next step is the definition of WF_proof which is a Coq inductive predicate representing the ":" in the well-formedness of proofs. Then, a check_proof boolean function is implemented and the corresponding WF_checked_proof lemma can be proved. Using these, the fundamental theorem can be stated.

Theorem 1 (Logical soundness). *There are Coq proof terms of these two theorems:*

$$\forall \pi : \texttt{proof}.\forall \Gamma, \forall G, (\texttt{WF\_sequent}\ \Gamma \vdash G) \rightarrow (\texttt{WF\_proof}\ \Gamma \vdash G\ \pi) \rightarrow [\![\Gamma \vdash G]\!] \quad (1)$$

$$\forall \pi : \texttt{proof}.\forall \Gamma, \forall G, (\texttt{check\_sequent}\ \Gamma \vdash G) = \texttt{true} \rightarrow$$
$$(\texttt{check\_proof}\ \Gamma \vdash G\ \pi) = \texttt{true} \rightarrow [\![\Gamma \vdash G]\!] \quad (2)$$

Coq proofs for these theorems, along with all the definitions above are available at http://www.lri.fr/~corbinea/ftp/programs/rfo.tar.gz.

Sketch of the proof. (2) is a consequence of (1) by composition with the WF_checked_* lemmata. Formula (1) is proved by induction on the proof trace, so it amounts to prove that in each step the interpretation of the conclusion is a consequence of the interpretation of the premises. Many intermediate steps are needed, the key ones being lemmata about the semantics of inst and rewrite, weakening lemmata, and lemmata about the stability of the interpretation functions (an interpretation is preserved by the addition of variables in the context).

3 Rewriting Traces for Ordered Completion

3.1 Completion Rules

The purpose of ordered completion [15,13,2] is to build a convergent rewriting system from a set of equations in order to decide the word problem. Some provers such as CiME [4] or Waldmeister [12] use an enhanced version of ordered completion in order to solve unifiability problems instead of word problems.

We adopt the classical presentation of the completion process as a sequence of applications of inference rules. The input is a pair $(E_0, s = t)$, where E_0 is a set of (implicitly universally quantified) equalities defining an equational theory, and $s = t$ is a conjecture, that is an implicitly existentially quantified equation. The output is True when there exists a substitution σ such that $s\sigma$ and $t\sigma$ are equal modulo E_0, False otherwise. Of course, this semi-decision procedure may not terminate. The whole procedure is parameterized by a reduction ordering $>$. The completion rules are working on a triple (E, R, C), where E and C are sets of equations (unordered pairs of terms), and R is a set of rules (ordered pairs of terms).

The set of completion rules may be divided into several subsets. First the rule **Init** initializes the process, by building the initial triple from the initial set of axioms and the conjecture. Then there are the rules **Orient** and **Orient'** which create rewrite rules for R from unordered pairs of terms $u = v$ in E. If u and v are comparable for $>$, a single rewrite rule is created, otherwise, two rules are created since it may be the case that $u\sigma > v\sigma$ or $v\sigma > u\sigma$, depending on σ. The rewrite rules are used by the **Rewrite, Rewrite', Collapse** and **Compose** for rewriting respectively in an equation, in a conjecture, in the left-hand side of a rule and in the right-hand side of a rule. Some new facts are computed by deduction, either between two rewrite rules (**Critical pair**) or between a rewrite rule and a conjecture (**Narrow**).

A recursive application of the rules may run forever, or stop with an application of the **Success** rule. Due to lack of space, we do not recall the rules, they can be read from Figure 4 by erasing the annotations.

3.2   Completion Rules with Traces

In order to build a trace for a formal proof management system as Coq, the inference rules have to be annotated with some additional information. From this respect, there are two kinds of rules, those which *simplify* equations or rules by rewriting the inside terms, and those which *create* some new facts. The application of the simplification rules is recorded in the rewritten term itself as a part of its history whereas the new facts contain the step by which they were created as a trace. Moreover, one has to precisely identify the left- and right-hand sides of equations, which leads to duplicate some of the rules. We shall only write down the left version of them and mention that there is a right version when there is no ambiguity. Hence the data structures are enriched as follows:

- An equation is a pair of terms with a trace. It will sometimes be denoted by $u = v$, u and v being the two terms, when the trace is not relevant.
- A term u has a current version u^*, an original version u^0, and an history, that is sequence of rewriting steps, which enables to rewrite u^0 into u^*.
- A rewriting step is given by a position, a substitution and a rewrite rule.
- A rewrite rule is an oriented ($+$ or $-$) equation $u = v$ denoted by $u \overset{+}{\to} v$ or $v \overset{-}{\to} u$.
- A trace is either an axiom (ω), or a peak corresponding with a critical pair (κ) or with a narrowing (either in a left-hand side (ν_L) or in a right-hand side (ν_R)).
- A peak is given by its top term t, two rewrite rules rl_1, rl_2, and the position p where the lowest rewrite rule rl_2 has to be applied to t: $\overset{rl_2}{\underset{p}{\leftarrow}} t \to^{rl_1}_{\Lambda}$.

An axiom, that is a pair of usual terms $s = t$, is lifted into an equation $\widehat{s = t}$ by turning s and t into general terms with an empty history and by adding the trace $\omega(s = t)$.

Init $\dfrac{}{\widehat{E_0, \emptyset, \{s = t\}}}$ if E_0 is the set defining the equational theory and $s = t$ is the conjecture.

Orient$_L$ $\dfrac{\{u = v\} \cup E, R, C}{E, \{u \xrightarrow{\pm} v\} \cup R, C}$ if $u^* > v^*$ **Orient$_R$** $\dfrac{\{u = v\} \cup E, R, C}{E, \{v \xrightarrow{\pm} u\} \cup R, C}$ if $v^* > u^*$

Orient' $\dfrac{\{u = v\} \cup E, R, C}{E, \{u \xrightarrow{\pm} v; v \xrightarrow{\pm} u\} \cup R, C}$ if u^* and v^* are not comparable w.r.t. to $>$.

Rewrite$_L$ $\dfrac{\{u = v\} \cup E, \{l \xrightarrow{\pm} r\} \cup R, C}{\{Rew(u, \underset{p,\sigma}{\xrightarrow{l \xrightarrow{\pm} r}}) = v\} \cup E, \{l \xrightarrow{\pm} r\} \cup R, C}$ if $u^*|_p = l^*\sigma$ and $l^*\sigma > r^*\sigma$.

Rewrite'$_L$ $\dfrac{E, \{l \xrightarrow{\pm} r\} \cup R, \{s = t\} \cup C}{E, \{l \xrightarrow{\pm} r\} \cup R, \{Rew(s, \underset{p,\sigma}{\xrightarrow{l \xrightarrow{\pm} r}}) = t\} \cup C}$ if $s^*|_p = l^*\sigma$ and $l^*\sigma > r^*\sigma$.

Rewrite$_R$ **Rewrite'$_R$**

Collapse$_+$ $\dfrac{E, \{l \xrightarrow{+} r; g \xrightarrow{\pm} d\} \cup R, C}{\{Rew(l, \underset{p,\sigma}{\xrightarrow{g \xrightarrow{\pm} d}}) = r\} \cup E, \{g \xrightarrow{\pm} d\} \cup R, C}$ if $l^*|_p = g^*\sigma$.

Collapse$_-$ $\dfrac{E, \{l \xrightarrow{-} r; g \xrightarrow{\pm} d\} \cup R, C}{\{r = Rew(l, \underset{p,\sigma}{\xrightarrow{g \xrightarrow{\pm} d}})\} \cup E, \{g \xrightarrow{\pm} d\} \cup R, C}$ if $l^*|_p = g^*\sigma$.

Compose $\dfrac{E, \{l \xrightarrow{\pm} r; g \xrightarrow{\pm} d\} \cup R, C}{E, \{l \xrightarrow{\pm} Rew(r, \underset{p,\sigma}{\xrightarrow{g \xrightarrow{\pm} d}}); g \xrightarrow{\pm} d\} \cup R, C}$ if $r^*|_p = g^*\sigma$.

Critical pair $\dfrac{E, \{l \xrightarrow{\pm} r; g \xrightarrow{\pm} d\} \cup R, C}{\left\{\begin{array}{l} l^*\rho_1[d^*\rho_2]_p\sigma = r^*\rho_1\sigma \\ \text{by } \kappa(\underset{p}{\xleftarrow{g \xrightarrow{\pm} d}} l\rho_1\sigma \underset{\Lambda}{\xrightarrow{l \xrightarrow{\pm} r}}) \end{array}\right\} \cup E, \{l \xrightarrow{\pm} r; g \xrightarrow{\pm} d\} \cup R, C}$ if $l^*\rho_1|_p\sigma = g^*\rho_2\sigma$.

Narrow$_L$ $\dfrac{E, \{g \xrightarrow{\pm} d\} \cup R, \{s = t\} \cup C}{E, \{g \xrightarrow{\pm} d\} \cup R, \left\{\begin{array}{l} s^*\rho_1[d^*\rho_2]_p\sigma = t^*\rho_1\sigma \\ \text{by } \nu_L(\underset{p}{\xleftarrow{g \xrightarrow{\pm} d}} s\rho_1\sigma \underset{\Lambda}{\xrightarrow{s \xrightarrow{\pm} t}}) \end{array}\right\} \cup \{s = t\} \cup C}$ if $s^*\rho_1|_p\sigma = g^*\rho_2\sigma$.

Narrow$_R$ $\dfrac{E, \{g \xrightarrow{\pm} d\} \cup R, \{s = t\} \cup C}{E, \{g \xrightarrow{\pm} d\} \cup R, \left\{\begin{array}{l} s^*\rho_1\sigma = t^*\rho_1[d^*\rho_2]_p\sigma \\ \text{by } \nu_R(\underset{p}{\xleftarrow{g \xrightarrow{\pm} d}} t\rho_1\sigma \underset{\Lambda}{\xrightarrow{t \to s}}) \end{array}\right\} \cup \{s = t\} \cup C}$ if $t^*\rho_1|_p\sigma = g^*\rho_2\sigma$.

Success $\dfrac{E, R, \{s = t\} \cup C}{True}$ if s^* and t^* are unifiable.

Fig. 4. Annotated rules for ordered completion

Given a term u, a rule $l \to r$, a position p and a substitution σ, when $u^*|_p = l^*\sigma$ the term u can be rewritten into a new term $Rew(u, \xrightarrow{l \to r}_{p,\sigma})$ where the original version keeps the same, the current value is equal to $u^*[r^*\sigma]_p$ and the new history is equal to $h@(p, \sigma, l \to r)$, h being the history of u. The annotated completion rules are given in Figure 4.

3.3   Constructing a Formal Proof from a Trace

When a triple (E, R, C) can trigger the completion rule **Success** all the needed information for building a formal proof can be extracted from the trace of the conjecture $s = t$ such that s^* and t^* are unifiable. Building the formal proof is done in two steps, first the information (as sequences of equational steps) is extracted from the trace, then the proof is built (either as a reified proof object or as a Coq script).

Word Problem. We shall first discuss the easiest case, when the original conjecture is an equation between two closed terms. The narrowing rules never apply, the set of conjectures is always a singleton, and the **Success** rule is triggered when both current sides of the conjecture are syntactically equal.

There are two possibilities for building a proof. The first one is to give a sequence of equational steps with respect to E_0 between both sides of the conjecture (cut-free proof). Since extracting the information from the trace in order to build such a proof is quite involved (the critical pairs have to be recursively unfolded), we start by the other alternative which is more simpler, where the proof is given by a list of lemmata mirroring the computation of critical pairs.

Critical Pairs as Cuts. First, it is worth noticing that all the applications of completion rules made so far do not have to be considered, but only the useful ones, which can be extracted and sorted by a dependency analysis, starting from the rewrite rules which occur in the history of s and t.

Now each critical pair can be seen as a lemma stating a universally quantified equality between two terms, and can be proven by using either the original equalities in E_0 or the previous lemmata.

When the trace of $l \xrightarrow{\pm} r$ is $\omega(u = v)$, this means that the rule is obtained from the original equation $u = v$ of E_0 by possibly rewriting both left and right hand sides. l^* and r^* can be proven equal by the sequence of equational steps obtained by the concatenation of **1.** the reverted history of l, **2.** $u = v$ at the top with the identity substitution either forward when the rule has the same orientation as it parent equation (+ case) or backward when the orientation is different (− case), **3.** the history of r.

When the trace of a rule $l \xrightarrow{\pm} r$ is $\kappa(\xleftarrow[p]{l_2 \xrightarrow{\pm} r_2} l_1\sigma \xrightarrow[\Lambda]{l_1 \xrightarrow{\pm} r_1})$, the rule is obtained from a critical pair between two rules and possibly a change of orientation. l^* and r^* can be proven equal by the sequence of equational steps obtained from the concatenation of **1.** the reverted history of l, **2.** a proof between l^0 and r^0 depending on the orientation; when there is no change of orientation, **2.a** $l_2 \xrightarrow{\pm} r_2$ applied backward at position p with the substitution σ_2 such that $l_2^*\sigma_2 = l_1^*\sigma|_p$ and $r_2^*\sigma_2 = l^0|_p$, **2.b** $l_1 \xrightarrow{\pm} r_1$ applied forward at the top with the substitution σ_1 such that $l_1^*\sigma_1 = l_1^*\sigma$ and $r_1^*\sigma_1 = r^0$, (when there is a change of orientation, **2.a'** $l_1 \overset{\pm}{\rightarrow} r_1$ applied backward at the top with the substitution σ_1 such that $l_1^* \sigma_1 = l_1^* \sigma$ and $r_1^* \sigma_1 = l^0$, **2.b'** $l_2 \overset{\pm}{\rightarrow} r_2$ applied forward at position p with the substitution σ_2 such that $l_2^* \sigma_2 = l_1^* \sigma|_p$ and $r_2^* \sigma_2 = r^0|_p$), **3.** the history of r.

The proof of the original conjecture $s^0 = t^0$ is then given by the concatenation of the history of s and the reverted history of t.

Unfolding the Proof. The main proof between s^0 and t^0 can be unfolded: as soon as a rule is used between u and v, at position p with the substitution σ, this rule is recursively replaced by its proof, plugged into the context $u[\_]_p$ when used forward or the context $v[\_]_p$ when used backward, and the whole sequence being instantiated by σ.

Unifiability Problem. In this case the narrowing rules may apply and the proof can be given as a sequence of lemmata, or as a substitution and a sequence of equational steps using only E_0.

There are two kinds of lemmata, those coming from critical pairs as above, and those coming from narrowing. Again, the useful ones can be extracted from the conjecture $s = t$ which triggered the **Success** rule.

Critical Pairs and Narrowing Steps as Cuts. The computation of the set of useful lemmata is similar as in the word case when the trace of $s = t$ is of the form $\omega(\_)$, but the starting set has to be extended with $l_1 \overset{\pm}{\rightarrow} r_1$ and $l_2 \overset{\pm}{\rightarrow} r_2$ when the trace is of the form $\nu_{L/R}(\overset{l_2 \overset{\pm}{\rightarrow} r_2}{\underset{p}{\leftarrow}} l_1\sigma \overset{l_1 \overset{\pm}{\rightarrow} r_1}{\underset{\Lambda}{\rightarrow}})$. This set can be sorted as above, and the proof of a critical pair lemma is exactly the same. The proof of a narrowing lemma is different: let $s' = t'$ be a conjecture obtained by narrowing from the rule $g \overset{\pm}{\rightarrow} d$ and the conjecture $s = t$. From the formal proof point of view, this means that the goal $s = t$ has been replaced by $s' = t'$; one has to demonstrate that the replacement is sound, that is $s' = t'$ implies $s = t$. Hence the proof of $s = t$ is actually a sequence of rewriting steps between s^* and t^* instantiated by the appropriate substitution, using $s' = t'$ and some smaller critical pair lemmata. In the case of left narrowing, for example, the substitution is equal to σ_1 such that $s^* \sigma_1 = s^* \sigma$ and $t^* \sigma_1 = t'^0$ and the sequence of rewriting steps is obtained by the concatenation of **1.** the forward application of the rule $g \overset{\pm}{\rightarrow} d$ at position p with the substitution σ_2 such that $g^* \sigma_2 = s^* \sigma|_p$ and $d^* \sigma_2 = s'^0|_p$, **2.** the history of s', **3.** the forward application of $s' = t'$ at the top with the identity substitution, **4.** the reverted history of t'. The case of right narrowing is similar, and the case of the original conjecture has already been described in the word problem case.

The proof of the last goal, that is the conjecture $s = t$ which actually triggered the **Success** rule is the substitution σ which unifies s^* and t^*, and the sequence of rewriting steps obtained by the concatenation of the history of s and the reverted history of t, every step being then instantiated by σ.

Unfolding the Proof. As in the word problem case, the main proof can also be unfolded, but one has also to propagate the substitution introduced by each narrowing lemma.

4   Translation of CiME Proofs into Reified Proofs

First we describe the way we model unifiability problems inside Coq, then in Subsection 4.2 we explain how to reify a single sequence of rewrite steps in an suitable context, and finally in Subsections 4.3 and 4.4 how to obtain a global proof from several lemmata by building the needed contexts, and by combining the small proofs together.

4.1   Modelling Unifiability Problems Inside Coq

A natural way to represent unifiability problems inside Coq is to represent algebraic terms with Coq terms (deep embedding), but since Coq has a typed language we first need to suppose we have a type `domain` in the Coq sort `Set`.

This encoding is not harmless at all since Coq types are not inhabited by default, which contradicts the semantics of unifiability problems: with an empty domain, the provability of the conjecture $f(x) = f(x)$ (which is implicitly existentially quantified) depends on the existence of a constant symbol in the signature. Therefore we assume we have a `dummy` constant in our `domain` type to model the non-emptiness assumption.

We need to introduce a Coq object of type $\overbrace{\texttt{domain} \to \cdots \to \texttt{domain}}^{n \text{ times}} \to \texttt{domain}$ for every n-ary function symbol in our signature. To represent the equality predicate we use Coq's standard polymorphic equality which is the interpretation of the $\doteq$ construction. Since we always use the `domain` type, with index 1 in our domain signature, 1 will be the default subscript for $\doteq$, $\dot\forall$ and $\dot\exists$. For each equality $s = t$ we choose an arbitrary total ordering on variables. We suppose rules will always be quantified in increasing order with respect to this ordering.

A unifiability problem is formed by a list of universally quantified equalities $R_1, \ldots, R_n$ which are declared as Coq hypotheses and an existentially quantified equality G which is declared as the goal we want to prove.

4.2   Sequence of Rewrite Steps

In a tool like CiME, the equations and the rewrite rules are implicitly universally quantified and some new variables are created when needed, whereas in the intermediate sequent calculus, any variable has to come from a context. The main difficulty of the subsection is to fill this gap by explaining in which contexts we are able to reify a sequence of rewrite steps.

Let $R_1, \ldots, R_n$ be rules, suppose CiME has given a rewriting trace between s_1 and t as follows: $s_1 \xrightarrow[\sigma_1, p_1]{\pm R_1} s_2 \xrightarrow[\sigma_2, p_2]{\pm R_2} \cdots \xrightarrow[\sigma_n, p_n]{\pm R_n} s_{n+1} = t$. We say Γ is an *adapted context* for this trace if it contains a representation of $y_1, \ldots, y_m$, the free variables in s_1 and t, a representation of dummy, and the closed hypotheses $\dot R_1, \ldots, \dot R_n$. In such a context Γ, the reification of an open term t is defined by if $t = (f\ a_1 \ldots a_p)$ then $\dot t = (\dot f\ \dot a_1 \ldots \dot a_p)$, otherwise if $t = y_i$ then $\dot t$ is the corresponding variable in Γ, and if t is an unknown variable then $\dot t = $ dummy. These unknown variables appear for example when the conjecture $f(a, a) = f(b, b)$ is proven using the hypothesis $\forall xy. f(x, x) = y$.

Theorem 2. *There exists a proof-trace of $\Gamma \vdash \dot{s}_1 \doteq t$ for any adapted context Γ.*

Proof. We prove by downwards induction on i that for any adapted context Γ there exist π such that $\pi : \Gamma \vdash \dot{s}_i \doteq t$:

If $i = n + 1$ then s_i is t, so we have $=_I \; :\Gamma \vdash (t \doteq t)$ for any Γ.

Otherwise, let Γ be an adapted context of length h and k be the index of $\dot{R}_i$ in Γ, let $z_1, \ldots, z_l$ be the free variables in R_i. We can build the following proof tree:

$$
\begin{array}{c}
\text{Induction hypothesis with } \Gamma; h{+}1 : R_i\{z_1 \xmapsto{} z_1\sigma_i\} \ldots; h{+}l : \dot{R}_i\sigma_i \\
\vdots \\
\hline
\pi : \Gamma; \ldots; h{+}l : \dot{R}_i\sigma_i \vdash \dot{s}_{i+1} \doteq t \\
\hline
(=_{E_1} (h{+}l) \leftrightarrow \dot{p}_i\, \pi) : \Gamma; \ldots; h{+}l : \dot{R}_i\sigma_i \vdash \dot{s}_i \doteq t \\
\vdots \; l \; \forall_E \text{ steps} \\
\hline
(\forall_E \, k \, z_1\dot{\sigma}_i(\forall_E \, (h{+}1)\, z_2\dot{\sigma}_i \ldots (\forall_E \,(h{+}l{-}1)\, z_l\dot{\sigma}_i(=_{E_1} (h{+}l) \leftrightarrow p_i\, \pi)) \ldots) : \Gamma \vdash \dot{s}_i \doteq t
\end{array}
$$

The induction step is valid since any extension of an adapted context stays adapted. $\square$

4.3   Closed Goals and Critical Pairs

Theorem 3. *Let G be a closed goal, $O_1, \ldots, O_n$ some original rules and $C_1, \ldots, C_m$ an ordered list of critical pairs used in the CiME trace of G. Let Γ^k be the context $\square; \text{dummy}; \dot{O}_1; \ldots; \dot{O}_n; \dot{C}_1; \ldots; \dot{C}_k$ with $0 \leq k \leq m$. There is a proof trace π for $\Gamma^0 \vdash \dot{G}$.*

Proof. We build inductively a proof trace π for the sequent $\Gamma^i \vdash \dot{G}$, starting from $i = m$ and going backwards to $i = 0$.

- We can build a proof trace π such that $\pi : \Gamma^m \vdash \dot{G}$ using the Theorem 2 and the trace given by CiME for G.
- Suppose we have a proof trace π such that $\pi : \Gamma^i \vdash \dot{G}$, and suppose C_i is of the form $\forall x_1 \ldots x_p.s = t$, we build the following tree to obtain a proof of $\Gamma^{i-1} \vdash G$:

$$
\begin{array}{cc}
\begin{array}{c}
\text{Theorem 2} \\
\vdots \\
\pi' : \Gamma^{i-1}; \dot{x}_1; \ldots; \dot{x}_p \vdash \dot{s} \doteq t \\
\vdots \; p \; \forall_I \text{ steps} \\
\hline
(\forall_I \ldots (\forall_I \, \pi') \ldots) : \Gamma \vdash \dot{\forall}_1 \ldots \dot{\forall}_1 \dot{s} \doteq t
\end{array}
&
\begin{array}{c}
\text{Induction hypothesis} \\
\vdots \\
\pi : \Gamma^i \vdash \dot{G}
\end{array}
\\
\hline
\multicolumn{2}{c}{(\text{Cut } \dot{C}_i \, (\forall_I \ldots (\forall_I \, \pi') \ldots) \; \pi) : \Gamma^{i-1} \vdash \dot{G}}
\end{array}
$$

$\square$

4.4   Open Goals and Narrowings

Open Goals. When the goal is of the form $\exists x_1, \ldots, x_n.s = t$, CiME provides a substitution σ and a rewriting trace for $s\sigma = t\sigma$. Using Theorem 2 we build a trace π for $\dot{s}\sigma \doteq t\dot{\sigma}$. The proof trace for the quantified goal is $(\exists_I \, x_1\sigma \ldots (\exists_I \, x_n\sigma\, \pi) \ldots)$.

Narrowings. Assume the current goal is an equality $\exists x_1, \ldots, x_n.s = t$ and CiME gives a proof trace with a narrowing $N = \exists y_1, \ldots, y_m.s' = t'$. We do a cut on $\dot{N}$.

In the left premise, the goal becomes $\dot{N}$ and we build here the trace for the rest of the lemmata. In the right premise, we apply m times the $\exists_E$ rule to obtain the hypotheses $\dot{y}_1, \ldots, \dot{y}_m, \dot{s}' \doteq \dot{t}'$, and we are in the case of the open goal described above.

5   Benchmarks

We run successfully CiME and Coq together on 230 problems coming from the TPTP library [16]. These problems are a subset of the 778 unifiability and word problems of TPTP. Some of the 778 TPTP problems are discarded because they are not solved within the given time (298), some others because they do not have a positive answer (11) or because they involve AC symbols (239), not handled by our framework yet.

The experiments were made on a 1.8GHz Pentium PC with 1Gb RAM, and a time-out of 600s on the completion process. For each of the 230 completion successes, 4 proofs were automatically generated, a short reified proof, a short proof with tactics, a cut-free reified proof and a cut-free proof with tactics. We used the current CVS versions of CiME3 and Coq, with the virtual machine turned on, which helps the Coq kernel reduce terms (see [9]).

Coq has been run on each of the 4*230 generated proofs, again with a timeout of 600s. We have observed that the short proofs are always checked in less that 1 second, for reified proofs as well as for tactics proofs, whatever the completion time. A short reified proof takes less time than the corresponding short tactics proof, but this is on very short times, so not very significant. The cut-free reified proofs take less time than the cut-free tactics, and the factor varies between 1 and 30. There is even an example (GRP614-1) where the reified proof is checked in 2 seconds and Coq gives up on the tactics proof. Some of the cut-free proofs are actually huge (several millions of lines) and cannot be handled by Coq (14 reified proofs and 16 script proofs).

6   Conclusion

We have described how to use reflection for proofs in Coq, how to annotate the usual ordered completion rules in order to build a trace, and how to turn the obtained trace into a reflective proof. The experiments made so far have validated the reflective approach and shown that some automation may be introduced in Coq and release the user from a part of the proof.

Previous works on reflection either aimed at proving meta-properties of proof trees in a very general framework [11] or at actually solving domain specific problems and at providing some automation for interactive provers [3,6]. We claim that our work belongs to the second trend but without loss of generality since our development is parameterized by the signature. Special care has been devoted to efficiency of proof-checking functions written in the Coq language.

We plan to work in several directions. First finalize the existing implementation as a Coq tactic by adding glue code, then extend the reflection mechanism to other calculi; for example $LJT I$ which adds arbitrary non-recursive connectives to first-order logic with a contraction-free presentation [5], or classical multi-succedent calculi to handle more general traces produced *e.g.* by classical tableaux provers. Finally handle AC function symbols by reflecting AC-steps.

References

1. C. Alvarado. *Réflexion pour la réécriture dans le calcul des constructions inductives*. PhD thesis, Université Paris-Sud, Dec. 2002.
2. L. Bachmair, N. Dershowitz, and D. A. Plaisted. Completion without failure. In H. Aït-Kaci and M. Nivat, editors, *Resolution of Equations in Algebraic Structures 2: Rewriting Techniques*, chapter 1, pages 1–30. Academic Press, New York, 1989.
3. M. Bezem, D. Hendriks, and H. de Nivelle. Automated proof construction in type theory using resolution. *Journal of Automated Reasoning*, 29(3):253–275, 2002.
4. E. Contejean, C. Marché, and X. Urbain. CiME3, 2004. http://cime.lri.fr/.
5. P. Corbineau. First-order reasoning in the Calculus of Inductive Constructions. In S. Berardi, M. Coppo, and F. Damiani, editors, *TYPES 2003 : Types for Proofs and Programs*, volume 3085 of *LNCS*, pages 162–177, Torino, Italy, Apr. 2004. Springer-Verlag.
6. P. Crégut. Une procédure de décision réflexive pour l'arithmétique de Presburger en Coq. Deliverable, Projet RNRT Calife, 2001.
7. N. G. de Bruijn. Lambda calculus with nameless dummies, a tool for automatic formula manipulation, with application to the Church-Rosser theorem. *Proc. of the Koninklijke Nederlands Akademie*, 75(5):380–392, 1972.
8. R. Dyckhoff. Contraction-free sequent calculi for intuitionistic logic. *The Journal of Symbolic Logic*, 57(3):795–807, 1992.
9. B. Grégoire and X. Leroy. A compiled implementation of strong reduction. In *International Conference on Functional Programming 2002*, pages 235–246. ACM Press, 2002.
10. J. Harrison. Metatheory and reflection in theorem proving: A survey and critique. Technical Report CRC-053, SRI Cambridge, Millers Yard, Cambridge, UK, 1995.
11. D. Hendriks. Proof Reflection in Coq. *Journal of Automated Reasoning*, 29(3-4):277–307, 2002.
12. T. Hillenbrand and B. Löchner. The next waldmeister loop. In A. Voronkov, editor, *Proceedings of CADE 18*, LNAI, pages 486–500. Springer-Verlag, 2002.
13. J. Hsiang and M. Rusinowitch. On word problems in equational theories. In T. Ottmann, editor, *14th International Colloquium on Automata, Languages and Programming*, volume 267 of *LNCS*, pages 54–71, Karlsruhe, Germany, July 1987. Springer-Verlag.
14. Q.-H. Nguyen. Certifying Term Rewriting Proofs in ELAN. In M. van den Brand and R. Verma, editors, *Proceedings of the International Workshop RULE'01*, volume 59 of *Electronic Notes in Theoretical Computer Science*. Elsevier Science Publishers, 2001.
15. G. E. Peterson. A technique for establishing completeness results in theorem proving with equality. *SIAM Journal on Computing*, 12(1):82–100, Feb. 1983.
16. G. Sutcliffe and C. Suttner. The TPTP Problem Library: CNF Release v1.2.1. *Journal of Automated Reasoning*, 21(2):177–203, 1998.
17. The Coq Development Team. *The Coq Proof Assistant Reference Manual – Version V8.0*, Apr. 2004. http://coq.inria.fr.

Reasoning in Extensional Type Theory with Equality

Chad E. Brown

Universität des Saarlandes, Saarbrücken, Germany
cebrown@ags.uni-sb.de

Abstract. We describe methods for automated theorem proving in extensional type theory with primitive equality. We discuss a complete, cut-free sequent calculus as well as a compact representation of cut-free (ground) proofs as extensional expansion dags. Automated proof search can be realized using a few operations to manipulate extensional expansion dags with variables. These search operations form a basis for complete search procedures. Procedures based on these ideas are implemented in the higher-order theorem prover TPS.

1 Introduction

Church's type theory [12] is a form of higher-order logic which is sufficiently powerful to represent much of traditional mathematics. The original Hilbert-style proof theory in [12] does not provide a convenient calculus for automated deduction. In an effort to study automated deduction for higher-order logic, fragments of Church's type theory have been considered. In particular, the higher-order theorem proving system TPS has traditionally searched for proofs in elementary type theory. Elementary type theory is Church's type theory without axioms of extensionality, descriptions, choice, or infinity. Andrews introduced a Hilbert-calculus for elementary type theory in [2]. Three important steps regarding the development of automated reasoning in elementary type theory can be sketched as follows:

1. In [2] Andrews proved a cut-free sequent calculus complete relative to $\mathscr{T}$.
2. Miller [15] demonstrated that every theorem of elementary type theory has an *expansion proof* as described in [15,16,6,4].
3. Procedures were developed to search for expansion proofs by manipulating expansion trees (with progressively instantiated variables). The implementation of such search procedures in TPS are described in [3,6].

We can generalize the three steps above and add another to outline a method for studying automated reasoning in a logic.

1. Develop a cut-free ground calculus for the logic.
2. Develop a compact representation of cut-free ground proofs.
3. Design a set of search operations for approximating compact ground proofs (using variables).
4. Establish completeness of search by verifying that some selection of the search operations will lead to a proof, if a proof exists.

R. Nieuwenhuis (Ed.): CADE 2005, LNAI 3632, pp. 23–37, 2005.

A ground expansion proof can be approximated by an expansion tree with expansion variables. First-order methods can be naturally generalized to search for expansion proofs. In particular, TPS combines mating search with Huet's higher-order pre-unification. For completeness, one must also consider *primitive substitutions* for set variables. A primitive substitution introduces a logical constant or projection for a variable at the head of a literal. Primitive (and general) substitutions in TPS are discussed in [6] on page 331. Such enumeration techniques place practical limitations on the search space. Intuitively, TPS attempts to converge towards a ground expansion proof by adding appropriate connections, performing higher-order unification steps and performing primitive substitutions.

To complete the analysis of automated reasoning in elementary type theory, such a search procedure would need to be proven complete. That is, one must show that some sequence of search operations applied to expansion trees successfully terminates if an expansion proof exists. This has never been carried out for expansion proofs. One reason is TPS traditionally has performed primitive substitutions (and most quantifier duplications) in a pre-processing step. Since new quantifiers can be introduced during search (via primitive substitutions), one cannot expect restricting primitive substitution applications to pre-processing leads to a complete search procedure.

Another interesting fragment of Church's type theory is extensional type theory. Extensional type theory adds principles of Boolean and functional extensionality to elementary type theory. The system LEO [8] searches using a resolution calculus for extensional type theory. TPS can now also prove theorems of extensional type theory using an appropriately modified notion of an expansion proof. We also extend the notion of expansion proof to include reasoning with primitive equality.

One advantage of working with extensional type theory is that the theory is closer to mathematical reasoning. For example, in mathematics one does not distinguish between $A \cup B$ and $B \cup A$. These sets could be different in models of elementary type theory, but not in models of extensional type theory.

Another advantage of working with extensional type theory is that one can simplify the search for instantiations of set variables. (This was, in fact, the motivation for adding extensionality reasoning to TPS .) While primitive substitutions are still necessary, one can obtain certain restrictions on which logical constants must be available for primitive substitutions. Also, one can represent some theorems in a natural manner which avoids introducing certain set variables. In particular, the use of primitive equality instead of Leibniz equality completely eliminates the set variables introduced by Leibniz equality.

A lifting lemma is proven for the resolution calculus in [8] in order to show completeness of search. However, this lifting lemma required a *Flex-Flex rule* to apply a substitution to a flex-flex pair. Since such a rule is notoriously branching, LEO does not actually apply such a rule during search. One of the goals of the development of extensional expansion proofs and corresponding search procedures was to show completeness of search without requiring operations on flex-flex pairs. In the end, we not only show completeness of search without any *Flex-Flex rule*, but also without considering connections between two flexible nodes.

We begin by describing extensional type theory relative to a signature of logical constants. By formulating extensional type theory in this way, we make precise the possible restrictions on set instantiations and hence primitive substitutions. We proceed by describing each of the four steps outlined above relative to (fragments of) extensional type theory. We describe a cut-free sequent calculus (complete with respect to an appropriate semantics). We describe extensional expansion proofs as extensional expansion dags (generalizing expansion trees) which satisfy certain properties. We indicate completeness of ground extensional expansion proofs relative to extensional type theory. When searching for extensional expansion proofs, we use variables which are progressively instantiated. We give a set of operations one can perform during search on extensional expansion dags (with variables). A lifting argument shows completeness of search once one has completeness of the ground case. Finally, we indicate some theorems which can be proven automatically in TPS using the new methods. This paper describes the work contained in [11], focusing on the results for automated proof search in extensional type theory.

2   Terms and Propositions

In first-order logic, one first defines terms (inductively) then atomic propositions (using terms and relations) and finally propositions (inductively). In higher-order logic, one can inductively define the terms of type α and let propositions be the terms of a particular type o of truth values. However, in order to express propositions as a term of type o, one must use logical constants such as $\neg_{oo}$, $\vee_{ooo}$ and $\Pi^\alpha_{o(o\alpha)}$. For completeness of proof search, one must consider primitive substitutions for each such logical constant. Since one of our purposes is to restrict primitive substitutions, it is worthwhile to define propositions at one level higher than terms (as is done in the first-order case).

First, we define the set of (simple) types inductively. o is the type of truth values, ι is a type of individuals and $(\alpha\beta)$ is a type of functions from β to α whenever α and β are types. Suppose $\mathcal{S}$ is a set of typed logical constants, $\mathcal{V}$ is a set of typed variables and $\mathcal{P}$ is a set of typed parameters. For each type α, we assume the set $\mathcal{V}_\alpha$ (variables of type α) and the set $\mathcal{P}_\alpha$ (parameters of type α) are infinite. Next, we can inductively define the set $wff_\alpha(\mathcal{S})$ of terms of type α using logical constants in $\mathcal{S}$, variables, parameters, application and λ-abstraction. We make the dependence on $\mathcal{S}$ explicit in order to consider different sets of logical constants.

To formulate higher-order logic, Church [12] assumed the signature included logical constants $\neg_{oo}$, $\vee_{ooo}$ and $\Pi^\alpha_{o(o\alpha)}$ for each type α. From these, one can define the other logical operators as in [12]. Equality at type α can be defined by Leibniz equality. The alternative pursued in [5] is to have primitive equality $=^\alpha_{o\alpha\alpha}$ in the signature for each type α. The logical connectives and quantifiers can be defined from primitive equality (assuming full extensionality). In the general case, we will not assume any logical constants to be in the signature $\mathcal{S}$. We will always assume, however, that $\mathcal{S}$ is a subset of the collection

$$\{\top_o, \bot_o, \neg_{oo}, \wedge_{ooo}, \vee_{ooo}, \supset_{ooo}, \equiv_{ooo}\}$$
$$\cup \{\Pi^\alpha_{o(o\alpha)} \mid \alpha \in \mathcal{T}\} \cup \{\Sigma^\alpha_{o(o\alpha)} \mid \alpha \in \mathcal{T}\} \cup \{=^\alpha_{o\alpha\alpha} \mid \alpha \in \mathcal{T}\}.$$

Without (enough) logical constants we should not consider the type theory to be "higher-order logic" since one cannot define the same sets using the restricted language as one can define using higher-order logic. Each collection of logical constants yields a fragment of higher-order logic.

The set $prop(\mathcal{S})$ of $propositions$ over a signature $\mathcal{S}$ is defined inductively.

- If $\mathbf{A} \in \mathit{wff}_o(\mathcal{S})$, then $\mathbf{A} \in prop(\mathcal{S})$.
- If α is a type and $\mathbf{A}, \mathbf{B} \in \mathit{wff}_\alpha(\mathcal{S})$, then $[\mathbf{A} \doteq^\alpha \mathbf{B}] \in prop(\mathcal{S})$.
- $\dot{\top} \in prop(\mathcal{S})$.
- If $\mathbf{M} \in prop(\mathcal{S})$, then $[\dot{\neg}\mathbf{M}] \in prop(\mathcal{S})$.
- If $\mathbf{M}, \mathbf{N} \in prop(\mathcal{S})$, then $[\mathbf{M} \dot{\vee} \mathbf{N}] \in prop(\mathcal{S})$.
- If $\mathbf{M} \in prop(\mathcal{S})$, then $[\dot{\forall}x_\alpha \mathbf{M}] \in prop(\mathcal{S})$.

We use the notation $\dot{\top}$, $\dot{\neg}$, $\dot{\vee}$ and $\dot{\forall}$ to distinguish these constructors (at the level of propositions) from the corresponding logical constants $\top$, $\neg$, $\vee$ and Π^α.

Suppose $\mathcal{S}$ is a signature with $\neg \in \mathcal{S}$ and $\mathbf{A} \in \mathit{wff}_o(\mathcal{S})$. Then $\neg\mathbf{A} \in \mathit{wff}_o(\mathcal{S})$ is both a term of type o and a proposition. Also, $\mathbf{A}$ is a term of type o and a proposition. Hence $\dot{\neg}\mathbf{A}$ is a proposition, but is not a term of type o.

We assume the usual notions of α, β and η conversion for terms. These can be easily extended (by induction) to propositions along with notions of substitution, β-normalization, $\beta\eta$-normalization, etc. We use $\mathbf{M}^\downarrow$ to denote the $\beta\eta$-normal form of $\mathbf{M}$ (whether $\mathbf{M}$ is a term or a proposition). Likewise, the notions of free and bound variables is defined as usual for terms and propositions. A term or proposition is $closed$ if it contains no free variables. A $sentence$ is a closed proposition.

An advantage of adding this extra level of propositions is that many theorems can be stated as propositions without assuming any logical constants are in $\mathcal{S}$. For example, the surjective form of Cantor's Theorem can be stated as the sentence

$$\dot{\neg} \dot{\exists} g_{o\iota\iota} \dot{\forall} f_{o\iota} \dot{\exists} j_\iota. \, g \, j \doteq^{o\iota} f \tag{1}$$

which is in $prop(\emptyset)$. (We use $\dot{\exists}$ as shorthand for $\dot{\neg}\dot{\forall}\dot{\neg}$.) If we insisted on representing Cantor's Theorem as a term of type o, then we would need several logical constants. By separating the levels of terms and propositions, we can distinguish between the expressive power of a fragment of type theory from the proof strength of the fragment.

3   Sequent Calculus

For any signature $\mathcal{S}$ of logical constants, a sequent calculus $\mathcal{G}^\mathcal{S}_{\beta f b}$ is defined in [11] (where sequents are multisets of sentences). Using this proof theory, we define the

$\mathcal{S}$-fragment of extensional type theory by the set of theorems of $\mathcal{G}^{\mathcal{S}}_{\beta\mathfrak{f}\mathfrak{b}}$ (provable sequents containing only one sentence). The rules of $\mathcal{G}^{\mathcal{S}}_{\beta\mathfrak{f}\mathfrak{b}}$ are shown in Figure 1. The rules $\mathcal{G}(\forall^{W})$ and $\mathcal{G}(\mathfrak{f}^{W})$ require the usual condition that the parameter W be new with respect to the sequent in the conclusion of the rule. The rules $\mathcal{G}(\neg\forall, \mathcal{S})$ and $\mathcal{G}(\neg =^{\rightarrow}, \mathcal{S})$ depend directly on the signature $\mathcal{S}$. (The more logical constants are in $\mathcal{S}$, the more terms can be used in these rules.) The rules $\mathcal{G}(Init^{=})$ and $\mathcal{G}(Dec)$ apply for any $n \geq 0$ and parameter H (of the appropriate type).

$$\frac{\Gamma, \mathbf{M}, \mathbf{M}}{\Gamma, \mathbf{M}} \mathcal{G}(Contr) \qquad \frac{\Gamma, \mathbf{M}^{\downarrow}}{\Gamma, \mathbf{M}} \mathcal{G}(\beta\eta) \qquad \frac{}{\Gamma, \top} \mathcal{G}(\top) \qquad \frac{\Gamma, \mathbf{M}}{\Gamma, \neg\neg\mathbf{M}} \mathcal{G}(\neg\neg)$$

$$\frac{\Gamma, \mathbf{M}, \mathbf{N}}{\Gamma, [\mathbf{M} \vee \mathbf{N}]} \mathcal{G}(\vee) \qquad \frac{\Gamma, \neg\mathbf{M} \quad \Gamma, \neg\mathbf{N}}{\Gamma, \neg[\mathbf{M} \vee \mathbf{N}]} \mathcal{G}(\neg\vee)$$

$$\frac{\Gamma, [[W/x]\mathbf{M}]}{\Gamma, [\forall x_{\alpha}\mathbf{M}]} \mathcal{G}(\forall^{W}) \qquad \frac{\Gamma, \neg[[C/x]\mathbf{M}] \quad C \in cwff_{\alpha}(\mathcal{S})}{\Gamma, \neg[\forall x_{\alpha}\mathbf{M}]} \mathcal{G}(\neg\forall, \mathcal{S})$$

$$\frac{\Gamma, \mathbf{A}^{\sharp} \quad \mathbf{A} \in cwff_{o}(\mathcal{S})}{\Gamma, \mathbf{A}} \mathcal{G}(\sharp) \qquad \frac{\Gamma, \neg(\mathbf{A}^{\sharp}) \quad \mathbf{A} \in cwff_{o}(\mathcal{S})}{\Gamma, \neg\mathbf{A}} \mathcal{G}(\neg\sharp)$$

$$\frac{\Gamma, [[\mathbf{G}\,W] =^{\alpha} [\mathbf{H}\,W]]}{\Gamma, [\mathbf{G} =^{\alpha\beta} \mathbf{H}]} \mathcal{G}(\mathfrak{f}^{W}) \qquad \frac{\Gamma, \neg[[\mathbf{G}\,\mathbf{B}] =^{\alpha} [\mathbf{H}\,\mathbf{B}]] \quad \mathbf{B} \in cwff_{\beta}(\mathcal{S})}{\Gamma, \neg[\mathbf{G} =^{\alpha\beta} \mathbf{H}]} \mathcal{G}(\neg =^{\rightarrow}, \mathcal{S})$$

$$\frac{\Gamma, \neg\mathbf{A}, \mathbf{B} \quad \Gamma, \neg\mathbf{B}, \mathbf{A}}{\Gamma, [\mathbf{A} =^{o} \mathbf{B}]} \mathcal{G}(\mathfrak{b}) \qquad \frac{\Gamma, \mathbf{A}, \mathbf{B} \quad \Gamma, \neg\mathbf{A}, \neg\mathbf{B}}{\Gamma, \neg[\mathbf{A} =^{o} \mathbf{B}]} \mathcal{G}(\neg =^{o})$$

$$\frac{\Gamma, [\mathbf{A} =^{\iota} \mathbf{C}] \quad \Gamma, [\mathbf{B} =^{\iota} \mathbf{D}]}{\Gamma, \neg[\mathbf{A} =^{\iota} \mathbf{B}], [\mathbf{C} =^{\iota} \mathbf{D}]} \mathcal{G}(EUnif_{1}) \qquad \frac{\Gamma, [\mathbf{A} =^{\iota} \mathbf{D}] \quad \Gamma, [\mathbf{B} =^{\iota} \mathbf{C}]}{\Gamma, \neg[\mathbf{A} =^{\iota} \mathbf{B}], [\mathbf{C} =^{\iota} \mathbf{D}]} \mathcal{G}(EUnif_{2})$$

$$\frac{\Gamma, [\mathbf{A}^{1} = \mathbf{B}^{1}] \quad \cdots \quad \Gamma, [\mathbf{A}^{n} = \mathbf{B}^{n}]}{\Gamma, [H\,\overline{\mathbf{A}^{n}}], \neg[H\,\overline{\mathbf{B}^{n}}]} \mathcal{G}(Init^{=}) \qquad \frac{\Gamma, [\mathbf{A}^{1} = \mathbf{B}^{1}] \quad \cdots \quad \Gamma, [\mathbf{A}^{n} = \mathbf{B}^{n}]}{\Gamma, [[H\,\overline{\mathbf{A}^{n}}] =^{\iota} [H\,\overline{\mathbf{B}^{n}}]]} \mathcal{G}(Dec)$$

Fig. 1. Sequent Rules for $\mathcal{G}^{\mathcal{S}}_{\beta\mathfrak{f}\mathfrak{b}}$

The rules $\mathcal{G}(\sharp)$ and $\mathcal{G}(\neg\sharp)$ provides the connection between the logical constants in $\mathcal{S}$ and the level of propositions using the operation taking a term $\mathbf{A} \in wff_{o}(\mathcal{S})$ to a proposition $\mathbf{A}^{\sharp}$. The definition of $\mathbf{A}^{\sharp}$ depends on the head of the term $\mathbf{A}$. In particular, $\top^{\sharp}$ is $\top$, $[\neg\mathbf{B}]^{\sharp}$ is $\neg\mathbf{B}$, $[\mathbf{B} \vee \mathbf{C}]^{\sharp}$ is $[\mathbf{B} \vee \mathbf{C}]$, $[\mathbf{D} =^{\alpha} \mathbf{E}]^{\sharp}$ is $[\mathbf{D} =^{\alpha} \mathbf{E}]$ and $[\Pi^{\alpha} \mathbf{F}]^{\sharp}$ is $[\forall x_{\alpha} . \mathbf{F}\,x]$. For other logical constants, we rely on the usual methods of representing different operations in terms of $\top$, $\vee$ and $\forall$. For example, $\perp^{\sharp}$ is $\neg\top$ and $[\mathbf{B} \supset \mathbf{C}]^{\sharp}$ is $[\neg\mathbf{B} \vee \mathbf{C}]$. If the head of $\mathbf{A}$ is not a logical constant, then $\mathbf{A}^{\sharp}$ is simply defined to be $\mathbf{A}$. (Note that the operation is not recursive. For example, $[\neg\neg\mathbf{B}]^{\sharp}$ is $\neg\neg\mathbf{B}$ as opposed to $\neg\neg\mathbf{B}$.)

In [11], a model class $\mathfrak{M}_{\beta\mathfrak{fb}}(\mathcal{S})$ is defined for the $\mathcal{S}$-fragment of extensional type theory. The sequent calculus $\mathcal{G}^{\mathcal{S}}_{\beta\mathfrak{fb}}$ is proven sound and complete with respect to $\mathfrak{M}_{\beta\mathfrak{fb}}(\mathcal{S})$ (see Theorems 3.4.14 and 5.7.18 of [11]). As a consequence of completeness, we can prove that the surjective form of Cantor's "theorem" (1) cannot be proven in the $\emptyset$-fragment of extensional type theory. (The intuitive reason for this is the need to use $\neg$ in the definition of the diagonal set.) In fact, one cannot prove (1) in the $\mathcal{S}$-fragment of extensional type theory even if $\mathcal{S}$ is $\{\top, \bot, \wedge, \vee\}$ (see Corollary 6.7.9 in [11]). There is a concrete model in $\mathfrak{M}_{\beta\mathfrak{fb}}(\{\top, \bot, \wedge, \vee\})$ in which (1) is false.

We do not include a cut rule in $\mathcal{G}^{\mathcal{S}}_{\beta\mathfrak{fb}}$ since we are interested in representing (and searching for) cut-free proofs. We know cut is admissible in $\mathcal{G}^{\mathcal{S}}_{\beta\mathfrak{fb}}$ as a consequence of completeness (see Corollary 5.7.19 of [11]).

We do not include an initial rule for deriving general sequents of the form $\Gamma, \neg\mathbf{M}, \mathbf{M}$ or a reflexivity rule for deriving general sequents of the form $\Gamma, [\mathbf{A} \doteq \mathbf{A}]$. The sequent calculus is complete without such rules (essentially because one can reduce to special cases of $\mathcal{G}(Init^{\doteq})$ and $\mathcal{G}(Dec)$). The reason we do not include a general initial rule or a general reflexivity rule is to avoid needing to perform arbitrary $\beta\eta$-unification when proving lifting results to show completeness of automated search.

While we do include a contraction rule in $\mathcal{G}^{\mathcal{S}}_{\beta\mathfrak{fb}}$, there are only a limited number of situtations in which contraction is necessary to obtain a proof. For example, the contraction rule is often used along with the rules $\mathcal{G}(\neg\forall, \mathcal{S})$ and $\mathcal{G}(\neg \doteq \rightarrow, \mathcal{S})$ in order to allow multiple instantiations. In the sequent calculus $\mathcal{G}^{\mathcal{S}}_{\beta\mathfrak{fb}}$, one may also need to make multiple uses of formulas in instances of the rules $\mathcal{G}(Init^{\doteq})$, $\mathcal{G}(EUnif_1)$ and $\mathcal{G}(EUnif_2)$. Finally, we may need to use contraction to provide a copy of an equation for an instances of $\mathcal{G}(Dec)$ as well as an instance of either $\mathcal{G}(EUnif_1)$ or $\mathcal{G}(EUnif_2)$. The next example demonstrates different essential applications of contraction in conjunction with the other rules of $\mathcal{G}^{\mathcal{S}}_{\beta\mathfrak{fb}}$.

For purposes of illustration, the main example we will consider in this paper is THM615 (discussed briefly in [11]):

$$H_{\iota o}[H \top =^{\iota} H \bot] \doteq^{\iota} H \bot$$

This is a sentence in $prop(\mathcal{S})$ if we assume $\top, \bot, =^{\iota} \in \mathcal{S}$. There is a reasonably simple proof of THM615. Either $[H \top =^{\iota} H \bot]$ is true or false. If true, then THM615 is equivalent to $[H \top =^{\iota} H \bot]$ which we have assumed true. If false, then THM615 is equivalent to $[H \bot =^{\iota} H \bot]$ which is true by reflexivity.

This short proof of THM615 can be formalized using of the cut rule with cut formula $[H \top =^{\iota} H \bot]$. Since cut is admissible, there must be a derivation of THM615 without using cut. The only possible rules of $\mathcal{G}^{\mathcal{S}}_{\beta\mathfrak{fb}}$ which can be used to conclude THM615 are contraction or the decomposition rule $\mathcal{G}(Dec)$. Since $[H \top =^{\iota} H \bot] =^o \bot$ is not a theorem, one cannot expect a derivation of THM615 to end with the decomposition rule. Instead, one can derive the sequent

$$[H_{\iota o}[H \top =^{\iota} H \bot] \doteq^{\iota} H \bot], [H \top =^{\iota} H \bot] \doteq^o \bot \qquad (2)$$

and complete the derivation of THM615 using decomposition followed by contraction.

The rule $\mathcal{G}(\mathfrak{b})$ (for Boolean extensionality) reduces deriving (2) to deriving (3) and (4):

$$[H_{\iota o}\,[H\,\top\ =^\iota\ H\,\bot]\ \dot{=}^\iota\ H\,\bot],\ \neg[H\,\top\ =^\iota\ H\,\bot],\ \bot \tag{3}$$

$$[H_{\iota o}\,[H\,\top\ =^\iota\ H\,\bot]\ \dot{=}^\iota\ H\,\bot],\ \neg\bot, [H\,\top\ =^\iota\ H\,\bot]. \tag{4}$$

The sequent (4) is derived using $\mathcal{G}(\neg\sharp)$, $\mathcal{G}(\neg\neg)$ and $\mathcal{G}(\top)$ (since $\neg(\bot^\sharp)$ is $\neg\neg\top$).

Nontrivial equality reasoning is required to derive (3). One can conclude (3) using $\mathcal{G}(EUnif_1)$ and contraction from (5) and (6):

$$[[H\,\top]\ \dot{=}\ [H.[H\,\bot] = [H\,\bot]]],\ \neg[H\,\top\ \dot{=}\ H\,\bot],\ \bot \tag{5}$$

$$[[H\,\bot]\ \dot{=}\ [H\,\bot]],\ \neg[H\,\top\ \dot{=}\ H\,\bot],\ \bot \tag{6}$$

The sequent (6) contains an instance of reflexivity and has an easy derivation (with several steps since there is no general rule for reflexivity). Using $\mathcal{G}(Dec)$ and $\mathcal{G}(\mathfrak{b})$, (5) can be reduced to deriving (7) and (8):

$$\neg\top,\ [[H\,\bot] = [H\,\bot]],\ \neg[H\,\top\ \dot{=}\ H\,\bot],\ \bot \tag{7}$$

$$\top,\ \neg[[H\,\bot] = [H\,\bot]],\ \neg[H\,\top\ \dot{=}\ H\,\bot],\ \bot \tag{8}$$

Both (7) and (8) are straightforward to derive.

4   Extensional Expansion Dags

While one can formulate automated search based on a sequent calculus, it is more common to choose a proof representation which eliminates certain redundancies. Expansion proofs provide a compact representation of cut-free proofs in elementary type theory. In particular, an expansion proof contains the essential information regarding instantiations and which atoms are used in initial sequents without recording all the information about the order of sequent rule applications. In [15] and [16] Dale Miller defined expansion proofs consisting of expansion trees with an acyclic dependence relation and a complete mating. A complete mating [3] is a set of connections which spans every vertical path.

Frank Pfenning defined a notion of extensional expansion proofs in his thesis [17] by adding a new kind of (extensional) node to expansion trees. The new extensional nodes introduce instances of extensionality axioms. However, during automated search there is no criteria for when to introduce such a node. Hence the notion of extensional expansion proof in [17] was never implemented as part of TPS or used as part of an automated search procedure.

An alternative notion of extensional expansion proof is defined and studied in [11]. An extensional expansion proof (as defined in [11]) is an extensional expansion dag (instead of an expansion tree) with an acyclic dependence relation and no *unsolved parts* (instead of no unspanned vertical paths). A formal definition of extensional expansion proofs can be found in [11] (see Definition 7.4.18). Here we describe the different elements of extensional expansion proofs by considering examples and noting the relationship to the sequent calculus $\mathcal{G}^{\mathcal{S}}_{\beta f b}$.

As in an expansion tree, each node of an extensional expansion dag has an associated *shallow formula* and *polarity*. Extensional expansion dags also distinguish between primary nodes and certain other nodes (such as connection nodes). Consider the sentence $\neg[P_{oo} \perp_o] \vee [P[\neg\top]]$. An expansion tree for this formula is shown on the left in Figure 2. We could try to connect LEAF0 to LEAF1, but we will never be able to syntactically unify $[P\perp]$ and $[P[\neg\top]]$. Extensionally, we can prove these are equal since the arguments $\perp$ and $\neg\top$ are equivalent, hence equal by Boolean extensionality. n extensional expansion dag which would provide a proof of the sentence is shown on the right in Figure 2. We make the connection between ATOM0 and ATOM1 explicit via the connection node MATE. The node MATE corresponds to the goal of showing $[P\perp]$ and $[P[\neg\top]]$ are equal. Whenever we create a mate node by connecting two atoms, the two atoms must have the same parameter in the head position. The children of the mate node correspond to showing the arguments of the two atoms are equal. The process of connecting two atoms by creating a mate node and forming the children of the new mate node corresponds to an application of the $\mathcal{G}(Init^{\pm})$ rule in the sequent calculus $\mathcal{G}^{\mathcal{S}}_{\beta f b}$. In Figure 2, the node BOOL corresponds to showing $\perp$ and $[\neg\top]$ are equal (and is related to the $\mathcal{G}(\flat)$ sequent rule). One child of BOOL corresponds to showing $[\neg\top]$ implies $\perp$ (which uses the positive node for $[\neg\top]$, but not the negative node for $\perp$). The other child of BOOL corresponds to showing $\perp$ implies $[\neg\top]$ (which uses the positive node for $\perp$, but not the negative node for $[\neg\top]$). Many of the nodes below BOOL simply pass from logical constants in terms to the level of propositions. Such nodes correspond to applications of the sequent rules $\mathcal{G}(\sharp)$ and $\mathcal{G}(\neg\sharp)$.

In Figure 3 we show an extensional expansion proof of THM615, omitting a few minor portions. (The nodes labelled by $\supset$ actually stand for a $\vee$ node and a $\neg$ node.) We can directly compare this extensional expansion proof to the sequent calculus derivation described in the previous section. Note that the root node EQN0 is a negative equation node with two children, DEC0 and EQNGOAL0, both of which correspond to the same negative equation. EQN0 behaves like an application of contraction giving two copies of the equation. The child DEC0 is a decomposition node which will correspond to the principal formula in applications of the rule $\mathcal{G}(Dec)$. The child EQNGOAL0 is an equation goal node which will correspond to a principal formula in applications of the rule $\mathcal{G}(EUnif_1)$ and $\mathcal{G}(EUnif_2)$. We obtained the subgoal sequent (2) by applying $\mathcal{G}(Contr)$ and $\mathcal{G}(Dec)$. This corresponds to passing from the root node EQN0 to the children DEC0 and EQNGOAL0 and then to the child BOOL0 of DEC0. Hence the se-

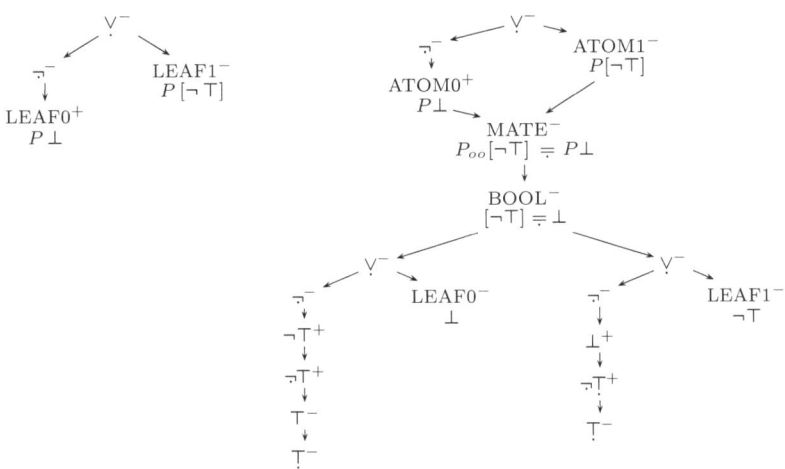

Fig. 2. Expansion Tree and Extensional Expansion Dag

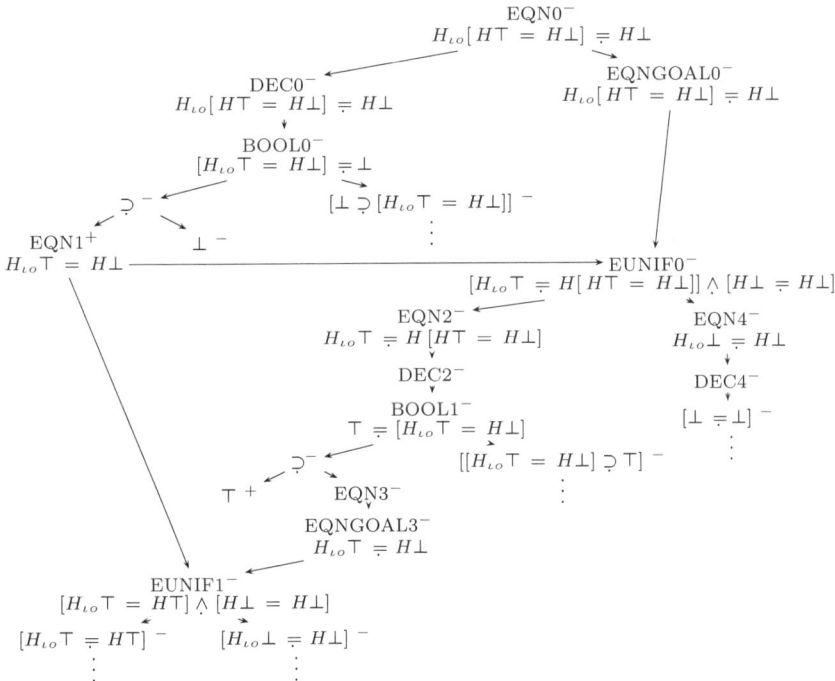

Fig. 3. Extensional Expansion Proof for THM615

quent (2) corresponds to the set {EQNGOAL0, BOOL0}. Each sequent Γ which occurs in the sequent derivation of THM615 corresponds to a set of nodes in the extensional expansion dag:

sequent	set of nodes
(3)	$\{\mathrm{EQNGOAL0, EQN0}, \perp^-\}$
(5)	$\{\mathrm{EQN2, EQN1}, \perp^-\}$
(6)	$\{\mathrm{EQN4, EQN1}, \perp^-\}$
(7)	$\{\top^+, \mathrm{EQN1, EQNGOAL3}, \perp^-\}$

This correspondence can be used to translate between sequent derivations and extensional expansion proofs (see Theorems 7.9.1 and 7.10.12 in [11]).

5   Search Operations

While searching for an extensional expansion proof, we manipulate extensional expansion dags containing (free) expansion variables. Some search operations extend the structure by adding information (e.g., nodes or edges). Other search operations partially instantiate variables using projection and imitation terms (as in Huet's pre-unification algorithm). One search operation (*Flex-Rigid Mate*) instantiates a variable (using an imitation term) and adds a connection.

First we consider a search operation which can always be applied eagerly without sacrificing completeness. We refer to this operation as development.

– **Development** The operation is technically defined by a large number of cases. We only provide a few example cases. If a is a leaf node with shallow formula $[\neg \mathbf{M}]$ and polarity p, then add a new successor node b to a with shallow formula $\mathbf{M}$ and polarity $-p$. If a is a positive leaf node with shallow formula $[\forall x_\alpha \mathbf{M}]$, then let y_α be a new variable and add a new positive node b with shallow formula $[y/x]\mathbf{M}$ and add a new expansion arc from a to b labelled with a new variable y_α.

The remaining operations are relative to a given set of nodes. A set $\mathfrak{p}$ of nodes is called an *unsolved part* (or *u-part*) of an extensional expansion dag if the set satisfies certain closure conditions (see Definition 7.4.4 in [11]). For example, we require every child of a conjunctive node in $\mathfrak{p}$ (e.g., a negative $\vee$ node) to be in $\mathfrak{p}$ and we require that every disjunctive node in $\mathfrak{p}$ has some child in $\mathfrak{p}$. These conditions are analogous to conditions defining the vertical paths of a matrix representation of a formula. We also require a condition for $\mathfrak{p}$ arising from the existence of connection nodes: If a and b are connected by a node c and $a, b \in \mathfrak{p}$, then $c \in \mathfrak{p}$.

Given any expansion node in a u-part, we can increase its multiplicity.

– **Duplication** Suppose $\mathfrak{p}$ is a u-part and $e \in \mathfrak{p}$ is a positive expansion node with shallow formula $[\forall x_\alpha \mathbf{M}]$. A *Duplication step* creates a new child of e with shallow formula $[y_\alpha/x]\mathbf{M}$ where y_α is a new (free) expansion variable.[1]

Three operations add connections between two nodes in a u-part.

---

[1] This simple form of duplication does not "copy" any information from previously existing children of e.

- **Rigid-Rigid Mate** Suppose $\mathfrak{p}$ is a u-part, $a \in \mathfrak{p}$ is a positive atomic node with shallow formula $[P\,\overline{\mathbf{A}^n}]$ and $b \in \mathfrak{p}$ is a negative atomic node with shallow formula $[P\,\overline{\mathbf{B}^n}]$ where P is a parameter. A *Rigid-Rigid Mate step* creates a new mate node with shallow formula $[[P\,\overline{\mathbf{A}^n}] \;{=}^o\; [P\,\overline{\mathbf{B}^n}]$ beneath a and b.
- **E-Unification** Suppose $\mathfrak{p}$ is a u-part, $a \in \mathfrak{p}$ is a positive equation node with shallow formula $[\mathbf{A} \;\dot{=}^\iota\; \mathbf{B}]$ and $b \in \mathfrak{p}$ is a negative equation goal node with shallow formula $[\mathbf{C} \;\dot{=}^\iota\; \mathbf{D}]$. An *E-unification step* creates a new E-unification node with shallow formula $[[\mathbf{A} \;\dot{=}^\iota\; \mathbf{C}] \wedge [\mathbf{B} \;\dot{=}^\iota\; \mathbf{D}]]^2$ beneath a and b.
- **Symmetric E-Unification** This is analogous to the E-unification step except the new node has shallow formula $[[\mathbf{A} \;\dot{=}^\iota\; \mathbf{D}] \wedge [\mathbf{B} \;\dot{=}^\iota\; \mathbf{C}]]$.

We next consider two operations which instantiate an expansion variable. The first operation corresponds to pre-unification and the second operation corresponds to applying primsubs.

- **Flex-Rigid** Suppose $\mathfrak{p}$ is a u-part and $a \in \mathfrak{p}$ is a negative equation node with shallow formula $[x\,\overline{\mathbf{A}^n} \;{=}^\iota\; F\,\overline{\mathbf{B}^m}]$ (or $[F\,\overline{\mathbf{B}^m} \;{=}^\iota\; x\,\overline{\mathbf{A}^n}]$) where x is a variable and F is a parameter. A *Flex-Rigid step* instantiates x by imitating F or by projecting onto any argument of x with an appropriate type.
- **Primitive Substitution** Suppose $\mathfrak{p}$ is a u-part and $a \in \mathfrak{p}$ is a positive flexible node with shallow formula $[p\,\overline{\mathbf{A}^n}]$ where p is a (set) variable. A *Primitive Substitution step* instantiates p by imitating any logical constant in $\mathcal{S}$ or projecting onto any argument of p with an appropriate type.

Finally, we consider an operation which mates a positive flexible node with a negative atomic node.

- **Flex-Rigid Mate** Suppose $\mathfrak{p}$ is a u-part, $a \in \mathfrak{p}$ is a positive flexible node with shallow formula $[p\,\overline{\mathbf{A}^n}]$ and $b \in \mathfrak{p}$ is a negative atomic node with shallow formula $[Q\,\overline{\mathbf{B}^m}]$ where p is a (set) variable and Q is a parameter. A *Flex-Rigid Mate step* instantiates p by imitating Q and creates a new mate node beneath a and b.

A search procedure proceeds by applying the search operations above (starting with an initial extensional expansion dag) until one obtains a pre-solved extensional expansion dag (i.e., an extensional expansion dag such that every u-part is pre-solved).

We have not introduced search operations which operate on negative flex-flex equation nodes or that connect two flexible nodes. The fact that such (highly branching) rules are not necessary for completeness is significant. The fact that flex-flex pairs (and flex-flex connections) can always be delayed corresponds to the fact that flex-flex pairs are always delayed in higher-order pre-unification [14]. In higher-order pre-unification, if one reaches a set of disagreement pairs such that every pair is flex-flex, then the set is *pre-unified* and there exist many (easy to construct) solutions to the unification problem. In the context of extensional

---

[2] Technically, this proposition is $\neg[\neg[\mathbf{A} \;\dot{=}^\iota\; \mathbf{C}] \vee \neg[\mathbf{B} \;\dot{=}^\iota\; \mathbf{D}]]$.

expansion dags, we say a u-part $\mathfrak{p}$ is *pre-solved* if there is either a negative flex-flex equation node in $\mathfrak{p}$ or a negative flexible node in $\mathfrak{p}$. Extensional expansion proofs are required to contain no u-parts. We can weaken this condition to say every unsolved part is pre-solved. Given an appropriate extensional expansion dag such that every unsolved part is pre-solved, one can easily construct an extensional expansion proof (see Theorem 8.3.3 in [11]). The idea for solving u-parts containing flex-flex pairs is the same as in higher-order pre-unification: one imitates a new parameter C_ι of base type. The idea for solving u-parts containing negative flexible nodes is analogous: one imitates the logical constant $\top_o$. Using this idea, one can solve any collection of pre-solved u-parts simultaneously.

Among the operations presented above, the least directed are primitive substitution steps. A primitive substitution can involve imitating any logical constant in $\mathcal{S}$ whenever a u-path contains a positive flexible node. In general, the set $\mathcal{S}$ is infinite (e.g., if $\Pi^\alpha \in \mathcal{S}$ for every type α). In an effort to minimize the problem of primitive substitutions, one can consider smaller signatures of logical constants. Some restrictions on the signature $\mathcal{S}$ do not result in any essential incompleteness. For example, assuming $\{\neg, \vee\} \subseteq \mathcal{S}$, there is no need to include $\wedge$, $\supset$ or $\equiv$ in the signature $\mathcal{S}$. In particular, performing primitive substitutions for $\wedge$ in addition to primitive substitutions for $\vee$ and $\neg$ does not increase the theorems one can prove. Furthermore, assuming $\{\neg, \vee\} \subseteq \mathcal{S}$, there is no need to include Π^α, Σ^α or $=^\alpha$ in $\mathcal{S}$ for propositional types α (i.e., types α which are constructed solely from type o). If one is concerned only with the theory of propositional types (as in [13,1]), then one can reduce the signature (and hence primitive substitutions) to $\{\neg, \vee\}$ (or other such minimal signatures) without sacrificing completeness. This fact does not hold if one does not assume extensionality. Consider the simple proposition involving only propositional types:

$$\exists s_{o(oo)} \forall f_{oo}. s f \equiv \exists p_o f p \qquad (9)$$

Essentially, (9) expresses the existence of the Σ^o (existence) operator for type o. Assuming extensionality, the witness $[\lambda f_{oo}. f A_o \vee [f \neg A]]$ (expressed using only $\vee$, $\neg$ and an arbitrary parameter A_o) can be used to prove the theorem. On the other hand, there is a non-extensional model which includes interpretations for $\neg$ and $\vee$ in which (9) is false.

One can also consider other restrictions (such as eliminating all logical constants by choosing $\mathcal{S}$ to be empty) which do rule out the possibility of proving certain theorems of extensional type theory (e.g., the surjective form of Cantor's theorem). On the other hand, $\mathfrak{M}_{\beta\mathfrak{f}\mathfrak{b}}(\emptyset)$ provides a model theory for the $\emptyset$-fragment of extensional type theory, $\mathcal{G}^\emptyset_{\beta\mathfrak{f}\mathfrak{b}}$ provides a cut-free proof theory and the search operations above provide for automated search without primitive substitutions (except projections).

6   Completeness of Search

In order to prove completeness of search, we assume there is a ground extensional expansion proof $\mathcal{Q}^*$ of a sentence $\mathbf{M}$. Given $\mathcal{Q}^*$ we can define the notion of a *lifting map* $f = \langle \varphi, f \rangle$ from an extensional expansion dag $\mathcal{Q}$ (with variables) to $\mathcal{Q}^*$ (where φ is a ground substitution and f maps the nodes of $\mathcal{Q}$ to the nodes of $\mathcal{Q}^*$). We can also define an ordinal $\|f : \mathcal{Q} \to \mathcal{Q}^*\|$ for each such lifting map (see Definitions 8.4.3 and 8.4.4 in [11]) which roughly measures how *close* $\mathcal{Q}$ is to $\mathcal{Q}^*$. For $f = \langle \varphi, f \rangle$, $\|f : \mathcal{Q} \to \mathcal{Q}^*\|$ is defined in terms of the nodes in $\mathcal{Q}^*$ missing from the image of the map f and the size of the long $\beta\eta$-form of the values of the ground substitution φ. Using $\mathcal{Q}^*$ we can guide the application of search operations to $\mathcal{Q}$ and use the lifting map f to maintain the relationship between the current extensional expansion dag $\mathcal{Q}$ and the known proof $\mathcal{Q}^*$.

A key step for proving completeness is to verify one can always make progress. That is, given any ground extensional expansion proof $\mathcal{Q}^*$, extensional expansion dag $\mathcal{Q}$ (where $\mathcal{Q}$ is not already pre-solved) and lifting map f from $\mathcal{Q}$ to $\mathcal{Q}^*$, there must be some search operation one can apply to $\mathcal{Q}$ to obtain $\mathcal{Q}'$ and some lifting map f' from $\mathcal{Q}'$ to $\mathcal{Q}^*$ such that $\|f' : \mathcal{Q}' \to \mathcal{Q}^*\| < \|f : \mathcal{Q} \to \mathcal{Q}^*\|$. In other words, some search operation gives a new dag which is *closer* to the goal proof $\mathcal{Q}^*$.

Once one knows progress can always be made, one can argue completeness of search as follows: There is a trivial lifting map from an initial extensional expansion dag (with one node corresponding to **M**) to a ground extensional expansion proof $\mathcal{Q}^*$ of **M**. Using the progress property, one can appropriately choose search operations until one obtains a pre-solved extensional expansion dag. Since $\|f : \mathcal{Q} \to \mathcal{Q}^*\|$ is an ordinal, the process will terminate in a pre-solved extensional expansion dag after a finite number of appropriate choices. From a pre-solved extensional expansion dag, one can obtain a ground extensional expansion proof (which may actually be different from $\mathcal{Q}^*$).

This completeness argument is further refined as in [11]. First, one can assume that no development step can be applied to the given extensional expansion proof $\mathcal{Q}^*$. Using this assumption, one can show that given any extensional expansion dag $\mathcal{Q}$ and lifting map $f : \mathcal{Q} \to \mathcal{Q}^*$, if any development step applies to $\mathcal{Q}$, then this development step results in an extensional expansion dag $\mathcal{Q}'$ and lifting map $f' : \mathcal{Q}' \to \mathcal{Q}^*$ such that $\|f' : \mathcal{Q}' \to \mathcal{Q}^*\| < \|f : \mathcal{Q} \to \mathcal{Q}^*\|$ (see Lemma 8.6.17 in [11]). Consequently, development steps are a form of *don't-care* non-determinism.

Assuming no development step can be applied to $\mathcal{Q}$, then one can show a progress property relative to u-parts (which are not pre-solved). In particular, for any u-part $\mathfrak{p}$ either $\mathfrak{p}$ is pre-solved or there is a search operation (other than development) relative to $\mathfrak{p}$ which results in an extensional expansion dag $\mathcal{Q}'$ with a lifting map $f' : \mathcal{Q}' \to \mathcal{Q}^*$ such that $\|f' : \mathcal{Q}' \to \mathcal{Q}^*\| < \|f : \mathcal{Q} \to \mathcal{Q}^*\|$ (see Lemma 8.8.4 in [11]). Consequently, choosing the (not pre-solved) u-part $\mathfrak{p}$ is also a form of *don't-care* non-determinism while choosing the operation relative to $\mathfrak{p}$ is a form of *don't-know* non-determinism. We can always restrict our attention to u-parts which are not pre-solved since search ends precisely when every u-part is pre-solved.

Finally, instead of simply maintaining a current extensional expansion dag $\mathcal{Q}$ and a search lifting map f, one can also represent information regarding quantifier duplications and set variables. Using this information one can impose certain conditions on the order of search operations. In particular, one can perform quantifier duplications (on a single expansion node), then primitive substitutions (for certain set variables) and lastly create connections and perform unification. (If new expansion nodes or set variables are created during search, then new duplications or primitive substitutions for these new objects are allowed.) Completeness of this ordered form of search follows from Theorem 8.8.6 of [11].

7   Examples

The MS04-2 search procedure implemented in TPS uses a bounded best-first search strategy with iterative deepening. Under the basic flag settings, MS04-2 only considers u-parts which are not presolved and only considers the options described in Section 5. Flags provide weights for ordering different options.

Using MS04-2, TPS can prove THM615 automatically in less than a second. Other examples TPS can prove automatically in less than a second using MS04-2 include $\mathbf{E}_1^{ext}$, $\mathbf{E}_2^{ext}$ and $\mathbf{E}_3^{ext}$ from [7].

The example $\mathbf{E}^{Dec}$ in [7] motivates a special decomposition rule in LEO for functions. The formulation in [7] is with respect to Leibniz equality:

$$\forall X_{\alpha\alpha} \forall Y_{\alpha} \, [f_{\alpha\alpha(\alpha\alpha)} \, X \, Y \doteq^{\alpha} g_{\alpha\alpha(\alpha\alpha)} \, X \, Y] \wedge \forall Z_{\alpha} \, [h_{\alpha\alpha} \, Z \doteq^{\alpha} j_{\alpha\alpha} \, Z] \supset f \, h \doteq^{\alpha\alpha} g \, j$$

In this proposition, we use the notation $[\mathbf{A} \doteq^{\alpha} \mathbf{B}]$ to denote Leibniz equality of $\mathbf{A}$ and $\mathbf{B}$ at the level of propositions: $[\forall q_{o\alpha} \cdot [q \, \mathbf{A}] \supset [q \, \mathbf{B}]]$. We can form an example EDEC2 by replacing all instances of Leibniz equality with primitive equality: The proof of $\mathbf{E}^{Dec}$ (using Leibniz equality) requires two primitive substitutions using $=^{\iota}$ followed by unification steps including 8 imitations and 4 projections. Even when flag settings are optimized for this example (while still only allowing basic search operations), TPS takes over 2 hours to find the proof. On the other hand, EDEC2 requires no primitive substitutions and very little unification (3 imitation steps and 1 projection step) and can be proven in less than a second.

Our final example is a theorem of topology we call THM616:

$$\forall G_{o(o\iota)} \, [G \subseteq OPEN_{o(o\iota)} \supset OPEN \, [\bigcup G]]$$

$$\supset \forall B_{o\iota} \cdot \forall x_{\iota} \, [B \, x \supset \exists D \, . \, OPEN \, D \wedge D \, x \wedge D \subseteq B] \supset OPEN \, B$$

where $\bigcup$ abbreviates $\lambda G_{o(o\alpha)} \lambda x_{\alpha} \exists S_{o\alpha} \, . \, G \, S \wedge S \, x$. THM616 is a modified version of BLEDSOE-FENG-SV-10 (studied in [9,10]) where closure of $OPEN$ under unions is stated in a more natural manner. However, THM616 is only a theorem if extensionality is assumed (whereas BLEDSOE-FENG-SV-10 is a theorem of elementary type theory). Consequently, no previous TPS search procedure could possibly prove THM616. By combining the extensionality reasoning described here with the set constraint reasoning described in [10], TPS can prove THM616 in about 10 seconds.

8   Conclusion

We have developed the theory of automated reasoning in extensional type theory with primitive equality by describing a complete sequent calculus, compact representations of proofs, search operations and lifting results which verify completeness of search.

References

1. Peter B. Andrews. A reduction of the axioms for the theory of propositional types. *Fundamenta Mathematicae*, 52:345–350, 1963.
2. Peter B. Andrews. Resolution in type theory. *Journal of Symbolic Logic*, 36:414–432, 1971.
3. Peter B. Andrews. On connections and higher-order logic. *Journal of Automated Reasoning*, 5:257–291, 1989.
4. Peter B. Andrews. Classical type theory. In Alan Robinson and Andrei Voronkov, editors, *Handbook of Automated Reasoning*, volume 2, chapter 15, pages 965–1007. Elsevier Science, 2001.
5. Peter B. Andrews. *An Introduction to Mathematical Logic and Type Theory: To Truth Through Proof.* Kluwer Academic Publishers, second edition, 2002.
6. Peter B. Andrews, Matthew Bishop, Sunil Issar, Dan Nesmith, Frank Pfenning, and Hongwei Xi. TPS: A theorem proving system for classical type theory. *Journal of Automated Reasoning*, 16:321–353, 1996.
7. Christoph Benzmüller. *Equality and Extensionality in Automated Higher-Order Theorem Proving.* PhD thesis, Universität des Saarlandes, 1999.
8. Christoph Benzmüller and Michael Kohlhase. System description: LEO — a higher-order theorem prover. In Claude Kirchner and Hélène Kirchner, editors, *Proceedings of the 15th International Conference on Automated Deduction*, volume 1421 of *Lecture Notes in Artificial Intelligence*, pages 139–143, Lindau, Germany, 1998. Springer-Verlag.
9. W. W. Bledsoe and Guohui Feng. Set-Var. *Journal of Automated Reasoning*, 11:293–314, 1993.
10. Chad E. Brown. Solving for set variables in higher-order theorem proving. In Andrei Voronkov, editor, *Proceedings of the 18th International Conference on Automated Deduction*, volume 2392 of *Lecture Notes in Artificial Intelligence*, pages 408–422, Copenhagen, Denmark, 2002. Springer-Verlag.
11. Chad E. Brown. *Set Comprehension in Church's Type Theory.* PhD thesis, Department of Mathematical Sciences, Carnegie Mellon University, 2004.
12. Alonzo Church. A formulation of the simple theory of types. *Journal of Symbolic Logic*, 5:56–68, 1940.
13. Leon Henkin. A theory of propositional types. *Fundamenta Mathematicae*, 52:323–344, 1963.
14. Gérard P. Huet. A unification algorithm for typed λ-calculus. *Theoretical Computer Science*, 1:27–57, 1975.
15. Dale A. Miller. *Proofs in Higher-Order Logic.* PhD thesis, Carnegie Mellon University, 1983. 81 pp.
16. Dale A. Miller. A compact representation of proofs. *Studia Logica*, 46(4):347–370, 1987.
17. Frank Pfenning. *Proof Transformations in Higher-Order Logic.* PhD thesis, Carnegie Mellon University, 1987. 156 pp.

Nominal Techniques in Isabelle/HOL

Christian Urban[1] and Christine Tasson[2]

[1] Ludwig-Maximilians-University Munich
urban@mathematik.uni-muenchen.de
[2] ENS Cachan Paris
tasson@dptmaths.ens-cachan.fr

Abstract. In this paper we define an inductive set that is bijective with the α-equated lambda-terms. Unlike de-Bruijn indices, however, our inductive definition includes names and reasoning about this definition is very similar to informal reasoning on paper. For this we provide a structural induction principle that requires to prove the lambda-case for fresh binders only. The main technical novelty of this work is that it is compatible with the axiom-of-choice (unlike earlier nominal logic work by Pitts *et al*); thus we were able to implement all results in Isabelle/HOL and use them to formalise the standard proofs for Church-Rosser and strong-normalisation.

Keywords: Lambda-calculus, nominal logic, structural induction, theorem-assistants.

1   Introduction

Whenever one wants to formalise proofs about terms involving binders, one faces a problem: how to represent such terms? The "low-level" representations use concrete names for binders (that is they represent terms as abstract syntax trees) or use de-Bruijn indices. However, a brief look in the literature shows that both representations make formal proofs rather strenuous in places (typically lemmas about substitution) that are only loosely concerned with the proof at hand. Three examples from the literature: VanInwegen wrote [19, p. 115]:

> *"Proving theorems about substitutions (and related operations such as alpha-conversion) required far more time and HOL code than any other variety of theorem."*

in her PhD-thesis, which describes a formalisation of SML's subject reduction property based on a "concrete-name" representation for SML-terms. Altenkirch formalised in LEGO a strong normalisation proof for System-F (using a de-Bruijn representation) and concluded [1, p. 26]:

> *"When doing the formalization, I discovered that the core part of the proof... is fairly straightforward and only requires a good understanding of the paper version. However, in completing the proof I observed that in certain places I had to invest much more work than expected, e.g. proving lemmas about substitution and weakening."*

R. Nieuwenhuis (Ed.): CADE 2005, LNAI 3632, pp. 38–53, 2005.
© Springer-Verlag Berlin Heidelberg 2005

Hirschkoff made a similar comment in [10, p. 167] about a formalisation of the π-calculus:

> *"Technical work, however, still represents the biggest part of our implementation, mainly due to the managing of de Bruijn indexes...Of our 800 proved lemmas, about 600 are concerned with operators on free names."*

The main point of this paper is to give a representation for α-*equated* lambda-terms that is based on names, is inductive and comes with a structural induction principle where the lambda-case needs to be proved for only fresh binders. In practice this will mean that we come quite close to the informal reasoning using Barendregt's variable convention. Our work is based on the nominal logic work by Pitts *et al* [16,6]. The main technical novelty is that our work by giving an explicit construction for α-equated lambda-terms is compatible with the axiom of choice. Thus we were able to implement all results in Isabelle/HOL and formalise the simple Church-Rosser proof of Tait and Martin-Löf described in [3], and the standard Tait-style strong normalisation proof for the simply-typed lambda-calculus given, for example, in [7,17].

The paper is organised as follow: Sec. 2 reviews α-equivalence for lambda-terms. Sec. 3 gives a construction of an inductive set that is bijective with the α-equated lambda-terms and adapts some notions of the nominal logic work for this construction. An induction principle for this set is derived in Sec. 4. Examples of Isabelle/HOL formalisations are given in Sec. 5. Related work is mentioned in Sec. 6, and Sec. 7 concludes.

2   Preliminaries

In order to motivate a design choice later on, we begin with a review of α-equivalence cast in terms of the nominal logic work. The set of lambda-terms is inductively defined by the grammar:

$$\Lambda: \quad t ::= a \mid t\, t \mid \lambda a.t$$

where a is an atom drawn from a countable infinite set, which will in what follows be denoted by $\mathbb{A}$.

The notion of α-equivalence for Λ is often defined as the least congruence of the equation $\lambda a.t =_\alpha \lambda b.t[a := b]$ involving a renaming substitution and a side-condition, namely that b does not occur freely in t. In the nominal logic work, however, atoms are manipulated not by renaming substitutions, but by permutations—bijective mappings from atoms to atoms. While permutations have some technical advantages, for example they preserve α-equivalence which substitutions do not [18], their primary reason in the nominal logic work is that one can use them to define the notion of *support*. This notion generalises what is meant by the set of free atoms of an object, which is usually clear in case the object is an abstract syntax tree, but less so if the object is a function. The generalisation of "free atoms" to functions, however, will play a crucial rôle in our construction of the bijective set.

There are several ways for defining the operation of a permutation acting on a lambda-term. One way [18] that can be easily implemented in Isabelle/HOL is to represent permutations as finite lists whose elements are swappings (i.e., pairs of atoms). We write such permutation as $(a_1\ b_1)(a_2\ b_2)\cdots(a_n\ b_n)$; the empty list $[]$ stands for the identity permutation. The permutation action, written $\pi\bullet(-)$, can then be defined on lambda-terms as:

$$[]\bullet a \overset{\text{def}}{=} a$$
$$(a_1\ a_2) :: \pi\bullet a \overset{\text{def}}{=} \begin{cases} a_2 & \text{if } \pi\bullet a = a_1 \\ a_1 & \text{if } \pi\bullet a = a_2 \\ \pi\bullet a & \text{otherwise} \end{cases} \qquad \begin{aligned} \pi\bullet(t_1\ t_2) &\overset{\text{def}}{=} (\pi\bullet t_1\ \pi\bullet t_2) \\ \pi\bullet(\lambda a.t) &\overset{\text{def}}{=} \lambda(\pi\bullet a).(\pi\bullet t) \end{aligned} \qquad (1)$$

where $(a\ b) :: \pi$ is the composition of a permutation followed by the swapping $(a\ b)$. The composition of π followed by another permutation π' is given by list-concatenation, written as $\pi'@\pi$, and the inverse of a permutation is given by list reversal, written π^{-1}.

While the representation of permutations based on lists of swappings is convenient for definitions like permutation composition and the inverse of a permutation, this list-representation is not unique; for example the permutation $(a\ a)$ is "equal" to the identity permutation. Therefore some means to identify "equal" permutations is needed.

Definition 1 (Disagreement Set and Permutation Equality). *The dis-agreement set of two permutations, say π_1 and π_2, is the set of atoms on which the permutations disagree, that is $\mathtt{ds}(\pi_1,\pi_2) \overset{\text{def}}{=} \{\, a \mid \pi_1\bullet a \neq \pi_2\bullet a \,\}$. Two permutations are equal, written $\pi_1 \sim \pi_2$, provided $\mathtt{ds}(\pi_1,\pi_2) = \varnothing$.*

Using the permutation action on lambda-terms, α-equivalence for Λ can be defined in a syntax directed fashion using the relations $(-)\approx(-)$ and $(-)\notin \mathtt{fv}(-)$; see Fig. 1. Because of the "asymmetric" rule $\approx_{\lambda 2}$, it might be surprising, but:

Proposition 1. $\approx$ *is an equivalence relation.*

The proof of this proposition is omitted: it can be found in a more general setting in [18]. (We also omit a proof showing that $\approx$ and $=_\alpha$ coincide). In the following, $[t]_\alpha$ will stand for the α-equivalence class of the lambda-term t, that is $[t]_\alpha \overset{\text{def}}{=} \{\, t' \mid t' \approx t \,\}$, and $\Lambda_{/\approx}$ for the set Λ quotient by $\approx$.

3   The Bijective Set

In this section, we will define a set Φ; inside this set we will subsequently identify (inductively) a subset, called Λ_α, that is in bijection with $\Lambda_{/\approx}$. In order to obtain the bijection, Φ needs to be defined so that it contains elements corresponding, roughly speaking, to α-equated atoms, applications and lambda-abstractions—that is to $[a]_\alpha$, $[t_1 t_2]_\alpha$ and $[\lambda a.t]_\alpha$. Whereas this is straightforward for atoms and applications, the lambda-abstractions are non-trivial: for them we shall use some

$$\frac{}{a \approx a} \approx_{\text{var}} \quad \frac{t_1 \approx s_1 \quad t_2 \approx s_2}{t_1\,t_2 \approx s_1\,s_2} \approx_{\text{app}} \quad \frac{t \approx s}{\lambda a.t \approx \lambda a.s} \approx_{\lambda 1} \quad \frac{a \neq b \quad t \approx (a\,b)\bullet s \quad a \notin \text{fv}(s)}{\lambda a.t \approx \lambda b.s} \approx_{\lambda 2}$$

$$\frac{a \neq b}{a \notin \text{fv}(b)}\,\text{fv}_{\text{var}} \quad \frac{a \notin \text{fv}(t_1) \quad a \notin \text{fv}(t_2)}{a \notin \text{fv}(t_1\,t_2)}\,\text{fv}_{\text{app}} \quad \frac{}{a \notin \text{fv}(\lambda a.t)}\,\text{fv}_{\lambda 1} \quad \frac{a \neq b \quad a \notin \text{fv}(t)}{a \notin \text{fv}(\lambda b.t)}\,\text{fv}_{\lambda 2}$$

Fig. 1. Inductive definitions for $(-) \approx (-)$ and $(-) \notin \text{fv}(-)$

specific "partial" functions from $\mathbb{A}$ to Φ (by "partial" we mean functions that return "error" for undefined values[1]). Thus the set Φ is defined by the grammar

$$\Phi: \quad t ::= \text{er} \mid \text{am}(a) \mid \text{pr}(t,t) \mid \text{se}(fn)$$

where er stands for "error", a for atoms and fn stands for functions from $\mathbb{A}$ to Φ.[2] This grammar corresponds to the inductive datatype that one might declare in Isabelle/HOL as:

```
datatype phi = er
             | am "atom"
             | pr "phi × phi"
             | se "atom ⇒ phi"
```

where it is presupposed that the type atom has been declared. The constructors am, pr and se will be used in Λ_α for representing α-equated atoms, applications and lambda-abstractions. Before the subset Λ_α can be carved out from Φ, however, some terminology from the nominal logic work needs to be adapted. For this we overload the notion of permutation action, that is $\pi\bullet(-)$, and define abstractly sets that come with a notion of permutation:

Definition 2 (PSets). *A set X equipped with a permutation action $\pi\bullet(-)$ is said to be a* pset, *if for all $x \in X$, the permutation action satisfies the following properties:*

(i) $[]\bullet x = x$
(ii) $\pi_1 @ \pi_2 \bullet x = \pi_1 \bullet (\pi_2 \bullet x)$
(iii) if $\pi_1 \sim \pi_2$ then $\pi_1 \bullet x = \pi_2 \bullet x$

The informal notation $x \in pset$ will be adopted whenever it needs to be indicated that x comes from a pset. The idea behind the permutation action, roughly speaking, is to permute all atoms in a given pset-element. For lists, tuples and sets the permutation action is therefore defined point-wise:

lists: $\pi\bullet[] \stackrel{\text{def}}{=} []$ tuples: $\pi\bullet(x_1,\ldots,x_n) \stackrel{\text{def}}{=} (\pi\bullet x_1,\ldots,\pi\bullet x_n)$

$\pi\bullet(x::t) \stackrel{\text{def}}{=} (\pi\bullet x) :: (\pi\bullet t)$ sets: $\pi\bullet X \stackrel{\text{def}}{=} \{\pi\bullet x \mid x \in X\}$

---

[1] This is one way of dealing with partial functions in Isabelle.

[2] Employing (on the meta-level) a lambda-calculus-like notation for writing such functions, one could in this grammar just as well have written $\lambda a.f$ instead of fn.

The permutation action for Φ is defined over the structure as follows:

$$\pi \bullet \text{er} \stackrel{\text{def}}{=} \text{er} \qquad\qquad \pi \bullet \text{pr}(t_1, t_2) \stackrel{\text{def}}{=} \text{pr}(\pi \bullet t_1, \pi \bullet t_2)$$

$$\pi \bullet \text{am}(a) \stackrel{\text{def}}{=} \text{am}(\pi \bullet a) \qquad \pi \bullet \text{se}(\textit{fn}) \stackrel{\text{def}}{=} \text{se}(\lambda a. \pi \bullet (\textit{fn}\ (\pi^{-1} \bullet a)))$$

where a lambda-term (on the *meta-level!*) specifies how the permutation acts on the function *fn*, namely as $\pi \bullet \textit{fn} \stackrel{\text{def}}{=} \lambda a. \pi \bullet (\textit{fn}\ (\pi^{-1} \bullet a))$.

When reasoning about Λ_α it will save us some work, if we show that certain sets are psets and then show properties (abstractly) for pset-elements.

Lemma 1. The following sets are psets: $\mathbb{A}$, Λ, Φ, and every set of lists (similarly tuples and sets) containing elements from psets.

Proof. By routine inductions.   □

The most important notion of a pset-element is that of its *support* (a set of atoms) and derived from this the notion of *freshness*[6]:

Definition 3 (Support and Freshness). *Given an $x \in pset$, its support is defined as:*[3]

$$\text{supp}(x) \stackrel{\text{def}}{=} \{a \mid \inf\{b \mid (a\,b) \bullet x \neq x\}\}\ .$$

An atom a is said to be fresh for such an x, written $a \mathbin{\#} x$, provided $a \notin \text{supp}(x)$.

Note that as soon as one fixes the permutation action for elements of a set, the notion of support is fixed as well. That means that Def. 3 defines the support for lists, sets and tuples as long as their elements come from psets. Calculating the support for terms in Λ is simple: $\text{supp}(a) = \{a\}$, $\text{supp}(t_1\ t_2) = \text{supp}(t_1) \cup \text{supp}(t_2)$ and $\text{supp}(\lambda a.t) = \text{supp}(t) \cup \{a\}$. Because of the functions in se(*fn*), the support for terms in Φ is more subtle. However, later on, we shall see that for terms of the subset Λ_α there is simple structural characterisation for their support, just like for lambda-terms.

First, some properties of support and freshness are established.

Lemma 2. For all $x \in pset$,

  (i) $\pi \bullet \text{supp}(x) = \text{supp}(\pi \bullet x)$, and
  (ii) $a \mathbin{\#} \pi \bullet x$ if and only if $\pi^{-1} \bullet a \mathbin{\#} x$.

Proof. (i) follows from the calculation:

$$\begin{aligned}
\pi \bullet \text{supp}(x) &\stackrel{\text{def}}{=} \pi \bullet \{a \mid \inf\{b \mid (a\,b) \bullet x \neq x\}\} \\
&\stackrel{\text{def}}{=} \{\pi \bullet a \mid \inf\{b \mid (a\,b) \bullet x \neq x\}\} \\
&= \{\pi \bullet a \mid \inf\{\pi \bullet b \mid (a\,b) \bullet x \neq x\}\} & (*^1) \\
&= \{a \mid \inf\{b \mid (\pi^{-1} \bullet a\ \ \pi^{-1} \bullet b) \bullet x \neq x\}\} \\
&= \{a \mid \inf\{b \mid \pi \bullet (\pi^{-1} \bullet a\ \ \pi^{-1} \bullet b) \bullet x \neq \pi \bullet x\}\} & (*^2) \\
&= \{a \mid \inf\{b \mid (a\,b) \bullet \pi \bullet x \neq \pi \bullet x\}\} \stackrel{\text{def}}{=} \text{supp}(\pi \bullet x) & (*^3)
\end{aligned}$$

---

[3] The predicate `inf` will stand for a set being infinite.

where $(*^1)$ holds because the sets $\{b|\ldots\}$ and $\{\pi\bullet b|\ldots\}$ have the same number of elements, and where $(*^2)$ holds because permutations preserve (in)equalities; $(*^3)$ holds because π commutes with the swapping, that is $\pi@(a\,b) \sim (\pi\bullet a\ \pi\bullet b)@\pi$. (ii): For all π, $a \in \text{supp}(x)$ if and only if $\pi\bullet a \in \pi\bullet\text{supp}(x)$. The property follows then from (i) and $x \in pset$.   □

Another important property is the fact that the freshness of two atoms w.r.t. an pset-element means that a permutation swapping those two atoms has no effect:

Lemma 3. For all $x \in pset$, if $a \# x$ and $b \# x$ then $(a\,b)\bullet x = x$.

Proof. The case $a = b$ is clear by Def. $2(i, iii)$. In the other case, the assumption implies that both $\{c\,|\,(c\,a)\bullet x \neq x\}$ and $\{c\,|\,(c\,b)\bullet x \neq x\}$ are finite, and therefore also their union must be finite. Hence the corresponding co-set, that is $\{c\,|\,(c\,a)\bullet x = x \wedge (c\,b)\bullet x = x\}$, is infinite (recall that $\mathbb{A}$ is infinite). If one picks from this co-set one element, which is from now on denoted by c and assumed to be different from a and b, one has $(c\,a)\bullet x = x$ and $(c\,b)\bullet x = x$. Thus $(c\,a)\bullet(c\,b)\bullet(c\,a)\bullet x = x$. The permutations $(c\,a)(c\,b)(c\,a)$ and $(a\,b)$ are equal, since they have an empty disagreement set. Therefore, by using Def. $2(ii, iii)$, one can conclude with $(a\,b)\bullet x = x$.   □

A further restriction on psets will filter out all psets containing elements with an infinite support.

Definition 4 (Fs-PSet). *A pset X is said to be an* fs-pset *if every element in X has finite support.*

Lemma 4. The following sets are fs-psets: $\mathbb{A}$, Λ, and every set of lists (similarly tuples and finite sets) containing elements from fs-psets.

Proof. The support of an atom a is $\{a\}$. The support of a lambda-term t is the set of atoms occurring in t. The support of a list is the union of the supports of its elements, and thus finite for fs-pset-elements (ditto tuples and finite sets).   □

The set Φ is *not* an fs-pset, because some functions from $\mathbb{A}$ to Φ have an infinite support. Similarly, some infinite sets have infinite support, even if all their elements have finite support. On the other hand, the infinite set $\mathbb{A}$ has *finite* support: $\text{supp}(\mathbb{A}) = \varnothing$ [6]. The main property of elements of fs-psets is that there is always a fresh atom.

Lemma 5. For all $x \in fs\text{-}pset$, there exists an atom a such that $a \# x$.

Proof. Since $\mathbb{A}$ is an infinite set and the support of x is by assumption finite, there must be an $a \notin \text{supp}(x)$.   □

We mentioned earlier that we are not going to use all functions from $\mathbb{A}$ to Φ for representing α-equated lambda-abstractions, but some specific functions.[4] The following definition states what properties these functions need to satisfy.

---

[4] This is in contrast to "weak" and "full" HOAS [15,4] which use the full function space for representing lambda-abstractions.

Definition 5 (Nominal Abstractions). *An operation, written* $[-].(-)$, *taking an atom and a pset-element is said to be a* nominal abstraction, *if it satisfies the following properties (where* $a \neq b$*):*

(i) $\pi \cdot ([a].x) = [\pi \cdot a].(\pi \cdot x)$
(ii) $[a].x_1 = [b].x_2$ *if and only if either:*

$$a = b \wedge x_1 = x_2, \text{ or}$$
$$a \neq b \wedge x_1 = (a\,b) \cdot x_2 \wedge a \# x_2$$

The first property states that the permutation action needs to commute with nominal abstractions. The second property ensures that nominal abstractions behave, roughly speaking, like lambda-abstractions. To see this reconsider the rules $\approx_{\lambda 1}$ and $\approx_{\lambda 2}$ given in Fig. 1, which can be used to decide when two lambda-terms are α-equivalent. Property *(ii)* paraphrases these rules for nominal abstractions. The similarities, however, do not end here: given a $[a].x$ with $x \in$ *fs-pset*, then freshness behaves like $(-) \notin \mathtt{fv}(-)$, as shown next:

Lemma 6. Given $a \neq b$ and $x \in$ *fs-pset*, then

(i) $a \# [b].x$ if and only if $a \# x$, and
(ii) $a \# [a].x$

Proof. (i⇒): Since $x \in$ *fs-pset*, $\mathtt{supp}([b].x) \subseteq \mathtt{supp}(x) \cup \{b\}$ and therefore the support of $[a].x$ must be finite. Hence $(a, b, x, [b].x)$ is finitely supported and by Lem. 5 there exists a c with $(*)$ $c \# (a, b, x, [b].x)$. Using the assumption $a \# [b].x$ and the fact that $c \# [b].x$ (from $*$), Lem. 3 and Def. 5(i) give $[b].x = (c\,a)[b].x = [b].(c\,a) \cdot x$. Hence by Def. 5(ii) $x = (c\,a) \cdot x$. Now $c \# x$ (from $*$) implies that $c \# (c\,a) \cdot x$; and moving the permutation to the other side by Lem. 2(ii) gives $a \# x$. (i⇐): From $(*)$, $c \# [b].x$ and therefore by Lem. 2(ii) $(a\,c) \cdot c \# (a\,c).([b].x)$, which implies by Def. 5(i) that $a \# [b].((a\,c) \cdot x)$. From $(*)$ $c \# x$ holds and from the assumption also $a \# x$; then Lem. 3 implies that $x = (a\,c) \cdot x$, and one can conclude with $a \# [b].x$.
(ii): By $c \# x$ and $c \neq a$ (both from $*$) we can use (i) to infer $c \# [a].x$. Further, from Lem. 2(ii) it holds that $(c\,a) \cdot c \# (c\,a) \cdot [a].x$. This is a $a \# [c].(c\,a) \cdot x$ using Def. 5(i). Since $c \neq a$, $c \# x$ and $(c\,a) \cdot x = (c\,a) \cdot x$, Def. 5(ii) implies that $[c].(c\,a) \cdot x = [a].x$. Therefore, $a \# [a].x$. $\qquad\square$

The functions from $\mathbb{A}$ to Φ we identify next satisfy the nominal abstraction properties. Let $[a].t$ be defined as follows

$$[a].t \overset{\text{def}}{=} \mathtt{se}(\lambda b. \,\mathtt{if}\ a = b\ \mathtt{then}\ t\ \mathtt{else}\ \mathtt{if}\ b \# t\ \mathtt{then}\ (a\,b) \cdot t\ \mathtt{else}\ \mathtt{er})\,. \quad (2)$$

This operation takes two arguments: an $a \in \mathbb{A}$ and a $t \in \Phi$. To see how this operation encodes an α-equivalence class, consider the α-equivalence class $[\lambda a.(a\,b)]_\alpha$ and the corresponding Φ-term $[a].\mathtt{pr}(a, b)$ (for the moment we ignore the term constructor $\mathtt{se}$ and only consider the function given by $[a].\mathtt{pr}(a, b)$). The graph of this function is as follows: the atom a is mapped to $\mathtt{pr}(a, b)$ since the first $\mathtt{if}$-condition is true. For b, the first $\mathtt{if}$-condition obviously fails, but also the second one fails, because $b \in \mathrm{supp}(\mathrm{pr}(a, b))$; therefore b is mapped to er. For all other atoms c, we have $a \neq c$ and $c \,\#\, \mathrm{pr}(a, b)$; so the c's are mapped by the function to $(a\,c)\bullet\mathrm{pr}(a, b)$, which is just $\mathrm{pr}(c, b)$. Clearly, the function returns er whenever the corresponding lambda-term is *not* in the α-equivalence class—in this example $\lambda b.(b\,b) \notin [\lambda a.(a\,b)]_\alpha$; in all other cases, however, it returns an appropriately "renamed" version of $\mathrm{pr}(a, b)$.

Lemma 7. The operation $[-].(-)$ given for Φ in (2) is a nominal abstraction.

Proof. Def. 5(i) follows from the calculation:

$$\pi\bullet[a].t$$
$$\stackrel{\mathrm{def}}{=} \pi\bullet\mathrm{se}(\lambda b.\, \text{if } a = b \text{ then } t \text{ else if } b \,\#\, t \text{ then } (a\,b)\bullet t \text{ else er})$$
$$\stackrel{\mathrm{def}}{=} \mathrm{se}(\lambda b.\, \pi\bullet\text{if } a = \pi^{-1}\bullet b \text{ then } t \text{ else if } \pi^{-1}\bullet b \,\#\, t \text{ then } (a\,\pi^{-1}\bullet b)\bullet t \text{ else er})$$
$$= \mathrm{se}(\lambda b.\, \text{if } a = \pi^{-1}\bullet b \text{ then } \pi\bullet t \text{ else if } b\,\#\,\pi\bullet t \text{ then } \pi\bullet(a\,\pi^{-1}\bullet b)\bullet t \text{ else er})\,(*)$$
$$= \mathrm{se}(\lambda b.\, \text{if } a = \pi^{-1}\bullet b \text{ then } \pi\bullet t \text{ else if } b \,\#\, \pi\bullet t \text{ then } (\pi\bullet a\,b)\bullet\pi\bullet t \text{ else er})$$
$$= \mathrm{se}(\lambda b.\, \text{if } \pi\bullet a = b \text{ then } \pi\bullet t \text{ else if } b \,\#\, \pi\bullet t \text{ then } (\pi\bullet a\,b)\bullet\pi\bullet t \text{ else er})$$
$$\stackrel{\mathrm{def}}{=} [\pi\bullet a].(\pi\bullet t)$$

where we use in $(*)$ the fact that $\pi\bullet\text{if...then...else...} = \text{if...then } \pi\bullet\text{...else } \pi\bullet\text{...}$ and Lem 2(ii). In case $a = b$, Def. 5(ii) is by a simple calculation using extensionality of functions. In case $a \neq b$ and Def. 5$(ii \Rightarrow)$, the following formula can be derived from the assumption by extensionality:

$$\forall c.\ \text{if } a = c \text{ then } t_1 \text{ else if } c \,\#\, t_1 \text{ then } (a\,c)\bullet t_1 \text{ else er} =$$
$$\text{if } b = c \text{ then } t_2 \text{ else if } c \,\#\, t_2 \text{ then } (b\,c)\bullet t_2 \text{ else er}$$

Instantiating this formula once with a and once with b yields the two equations

$$t_1 = \text{if } a \,\#\, t_2 \text{ then } (b\,a)\bullet t_2 \text{ else er}$$
$$t_2 = \text{if } b \,\#\, t_1 \text{ then } (a\,b)\bullet t_1 \text{ else er}$$

Next, one distinguishes two cases where $a \,\#\, t_2$ and $\neg a \,\#\, t_2$, respectively. In the first case, $t_1 = (b\,a)\bullet t_2$, which by Lem. 1 and Def. 2(iii) is equal to $(a\,b)\bullet t_2$; and obviously $a \,\#\, t_2$ by assumption. In the second case $t_1 = $ er. This substituted into the second equation gives $t_2 = \text{ if } b \,\#\, \text{ er then } (a\,b)\bullet\text{er else er}$. Since $\mathrm{supp}(\text{er}) = \varnothing$, $t_2 = (a\,b)\bullet\text{er} = \text{er}$. Now there is a contradiction with the assumption $\neg a \,\#\, t_2$, because $a \,\#\, \text{er}$. Def. 5$(ii \Leftarrow)$ for $a \neq b$ is by extensionality and a case-analysis. $\square$

Note that, in *general*, one cannot decide whether two functions from $\mathbb{A}$ to Φ are equal; however Def. 5(ii) provides means to decide whether $[a].t_1 = [b].t_2$ holds: one just has to consider whether $a = b$ and then apply the appropriate property in Def. 5(ii)—just like deciding the α-equivalence of two lambda-terms using $(-)\approx(-)$.

Now everything is in place for defining the subset Λ_α. It is defined inductively by the rules:

$$\frac{a \in \mathbb{A}}{\mathrm{am}(a) \in \Lambda_\alpha} \qquad \frac{t_1 \in \Lambda_\alpha \quad t_2 \in \Lambda_\alpha}{\mathrm{pr}(t_1, t_2) \in \Lambda_\alpha} \qquad \frac{a \in \mathbb{A} \quad t \in \Lambda_\alpha}{[a].t \in \Lambda_\alpha}$$

using in the third inference rule the operation defined in (2). For Λ_α we have:

Lemma 8. Λ_α is:

    (i) an fs-pset, and
    (ii) closed under permutations, that is if $x \in \Lambda_\alpha$ then $\pi \bullet x \in \Lambda_\alpha$.

Proof. (i): The pset-properties of Φ carry over to Λ_α. The fs-pset property follows by a routine induction on the definition of Λ_α using the fact derived from Lem. 6(i,ii) that for $x \in$ *fs-pset*, $\mathrm{supp}([a].x) = \mathrm{supp}(x) - \{a\}$. (ii) Routine induction over the definition of Λ_α. □

Taking Lem. 8(i) and Lem. 6 together gives us a simple characterisation of the support of elements in Λ_α: $\mathrm{supp}(\mathrm{am}(a)) = \{a\}$, $\mathrm{supp}(\mathrm{pr}(t_1, t_2)) = \mathrm{supp}(t_1) \cup \mathrm{supp}(t_2)$ and $\mathrm{supp}([a].t) = \mathrm{supp}(t) - \{a\}$. In other words it coincides with what one usually means by the free variables of a lambda-term.

Next, one of the main points of this paper: there is a bijection between $\Lambda_{/\approx}$ and Λ_α. This is shown by using the following mapping from Λ to Λ_α:

$$q(a) \stackrel{\mathrm{def}}{=} \mathrm{am}(a) \qquad q(t_1\, t_2) \stackrel{\mathrm{def}}{=} \mathrm{pr}(q(t_1), q(t_2)) \qquad q(\lambda a.t) \stackrel{\mathrm{def}}{=} [a].q(t)$$

and the following lemma:

Lemma 9. $t_1 \approx t_2$ if and only if $q(t_1) = q(t_2)$.

Proof. By routine induction over definition of Λ_α. □

Theorem 1. There is a bijection between $\Lambda_{/\approx}$ and Λ_α.

Proof. The mapping q needs to be lifted to α-equivalence classes (see [14]). For this define $q'([t]_\alpha)$ as follows: apply q to every element of the set $[t]_\alpha$ and build the union of the results. By Lem. 9 this must yield a singleton set. The result of $q'([t]_\alpha)$ is then the singleton. Surjectivity of q' is shown by a routine induction over the definition of Λ_α. Injectivity of q' follows from Lem. 9 since $[t_1]_\alpha = [t_2]_\alpha$ for all $t_1 \approx t_2$. □

4   Structural Induction Principle

The definition of Λ_α provides an induction principle for free. However, this induction principle is not very convenient in practice. Consider Fig. 2 showing a typical informal proof involving lambda-terms—it is Barendregt's proof of the substitution lemma taken from [3]. This informal proof considers in the lambda-case only binders z that have suitable properties (namely being fresh for x, y, N

Substitution Lemma: If $x \not\equiv y$ and $x \notin FV(L)$, then

$$M[x := N][y := L] \equiv M[y := L][x := N[y := L]].$$

Proof: By induction on the structure of M.

Case 1: M is a variable.

    Case 1.1. $M \equiv x$. Then both sides equal $N[y := L]$ since $x \not\equiv y$.

    Case 1.2. $M \equiv y$. Then both sides equal L, for $x \notin FV(L)$ implies
        $L[x := \ldots] \equiv L$.

    Case 1.3. $M \equiv z \not\equiv x, y$. Then both sides equal z.

Case 2: $M \equiv \lambda z.M_1$. By the variable convention we may assume that $z \not\equiv x, y$ and z is not free in N, L. Then by induction hypothesis

$$
\begin{aligned}
(\lambda z.M_1)[x := N][y := L] &\equiv \lambda z.(M_1[x := N][y := L]) \\
&\equiv \lambda z.(M_1[y := L][x := N[y := L]]) \\
&\equiv (\lambda z.M_1)[y := L][x := N[y := L]].
\end{aligned}
$$

Case 3: $M \equiv M_1 M_2$. The statement follows again from the induction hypothesis.

                $\square$

Fig. 2. The informal proof of the substitution lemma copied from [3]. In the lambda-case, the variable convention allows Barendregt to move the substitutions under the binder, to apply the induction hypothesis and then to pull out the substitutions

and L). If we would prove the substitution lemma by induction over the definition of Λ_α, then we would need to show the lambda-case for *all* z, not just the ones being suitably fresh. This would mean we have to rename binders and establish a number of auxiliary lemmas concerning such renamings. In this section we will derive an induction principle which allows a similar convenient reasoning as in Barendregt's informal proof.

For this we only consider induction hypotheses of the form $P\ t\ x$, where P is the property to be proved; P depends on a variable $t \in \Lambda_\alpha$ (over which the induction is done), and a variable x standing for the "other" variables or *context* of the induction. Since x is allowed to be a tuple, several variables can be encoded. In case of the substitution lemma in Fig. 2 the notation $P\ t\ x$ should be understood as follows: the induction variable t is M, the context x is the tuple (x, y, N, L) and the induction hypothesis P is

$$\lambda M.\ \lambda(x, y, N, L).\ M[x := N][y := L] \equiv M[y := L][x := N[y := L]]$$

where we use Isabelle's convenient tuple-notation for the second lambda-abstraction [11]. So by writing $P\ t\ x$ we just make explicit all the variables involved in the induction.

From the inductive definition of Λ_α we can derive a structural induction principle that requires to prove the lambda-case for binders that are fresh for the context x—this is what the variable convention assumes.

Lemma 10 (Induction Principle). Given an induction hypothesis $P\ t\ x$ with $t \in \Lambda_\alpha$ and $x \in$ *fs-pset*, then proving the following:

- $\forall x\ a.\ P\ \mathtt{am}(a)\ x$
- $\forall x\ t_1\ t_2.\ P\ t_1\ x \wedge P\ t_2\ x \Rightarrow P\ \mathtt{pr}(t_1, t_2)\ x$
- $\forall x\ a.\ a \mathbin{\#} x \Rightarrow (\forall t.\ P\ t\ x \Rightarrow P\ [a].t\ x)$

gives $\forall t\ x.\ P\ t\ x$.

Proof. By induction over the definition of Λ_α. We need to strengthen the induction hypothesis to $\forall t\ \pi\ x.\ P\ (\pi \cdot t)\ x$, that means considering t under all permutations π. Only the case for terms of the form $[a].t$ will be explained. We need to show that $P\ (\pi \cdot [a].t)\ x$, where $\pi \cdot [a].t = [\pi \cdot a].(\pi \cdot t)$ by Def. 5(*i*). By IH, $(*^1)\ \forall \pi\ x.\ P\ (\pi \cdot t)\ x$ holds. Since $x, \pi \cdot t, \pi \cdot a \in$ *fs-pset* holds, one can derive by Lem. 5 that there is a c such that $(*^2)\ c \mathbin{\#} (x, \pi \cdot t, \pi \cdot a)$. From $c \mathbin{\#} x$ and the assumption, one can further derive $(\forall t.\ P\ t\ x \Rightarrow P\ [c].t\ x)$. Given $(*^1)$ we have that $P\ ((c\ \pi \cdot a) :: \pi \cdot t)\ x$ holds and thus also $P\ ([c].((c\ \pi \cdot a) :: \pi \cdot t))\ x$. Because of $(*^2)\ c \neq \pi \cdot a$ and $c \mathbin{\#} \pi \cdot t$, and by Def. 5(*ii*) we have that $[c].((c\ \pi \cdot a) :: \pi \cdot t) = [\pi \cdot a].(\pi \cdot t)$. Therefore we can conclude with $P\ (\pi \cdot [a].t)\ x$. $\qquad\square$

With this we have achieved what we set out in the introduction: we have a representation for α-equivalent lambda-terms based on names (for example $[\lambda a.t]_\alpha$ is represented by $[a].t$) and we have an induction principle where the lambda-case needs to be proved for binders that are fresh w.r.t. the variables in the context of the induction, i.e., we can reason as if we had employed a variable convention.

5   Examples

It is reasonably straightforward to implement the results from Sec. 3 and 4 in Isabelle/HOL: the set Φ is an inductive datatype, the pset and fs-pset properties can be formulated as axiomatic type-classes [20], and the subset Λ_α can be defined using the Isabelle's `typedef`-mechanism. This section focuses on how reasoning over Λ_α pans out in practice.

The first obstacle is that so far Isabelle's datatype package is not general enough to allow a direct definition of functions over Λ_α: although Λ_α contains only terms of the form $\mathtt{am}(a)$, $\mathtt{pr}(t_1, t_2)$ and $[a].t$, pattern-matching in Isabelle requires the injectivity of term-constructors. But clearly, $[a].t$ is *not* injective. Fortunately, one can work around this obstacle by, roughly speaking, defining functions as inductive relations and then use the definite description operator *THE* of Isabelle to turn the relations into functions.

We give an example: capture-avoiding substitution can be defined as a four-place relation (the first argument contains the term into which something is being substituted, the second the variable that is substituted for, the third the term that is substituted, and the last contains the result of the substitution):

```
consts Subst :: "(Λα × 𝔸 × Λα × Λα) set"
inductive Subst
intros
s1: "(am(a),a,t',t')∈Subst"
s2: "a≠b ⟹ (am(b),a,t',am(b))∈Subst"
s3: "⟦(s₁,a,t',s₁')∈Subst; (s₂,a,t',s₂')∈Subst⟧
                          ⟹ (pr(s₁,s₂),a,t',pr(s₁',s₂'))∈Subst"
s4: "⟦b#(a,t');(s,a,t',s')∈Subst⟧ ⟹ ([b].s,a,t',[b].s')∈Subst"
```

While on first sight this relation looks as if it defined a non-total function, one should be careful! Clearly, the lambda-case (i.e. ([b].s,a,t',[b].s') ∈ Subst) holds only under the precondition b#(a,s)—roughly meaning that $a \neq b$ and b cannot occur freely in s. However, Subst *does* define a total function, because Subst is defined over α-equivalent lambda-terms (more precisely Λ_α), *not* over lambda-terms. We can indeed show "totality":

Lemma 11. *For all* t_1, a, t_2, $\exists t_3.$ $(t_1, a, t_2, t_3) \in$ Subst .

Proof. The proof in Isabelle/HOL uses the induction principle derived in Thm. 10. It is as follows:

```
proof (nominal_induct t₁)
   case (1 b) (* variable case *)
   show "∃t₃. (am(b),a,t₂,t₃)∈Subst" by (cases "b=a") (force+)
next
   case (2 s₁ s₂) (* application case *)
   thus "∃t₃. (pr(s₁,s₂),a,t₂,t₃)∈Subst" by force
next
   case (3 b s) (* lambda case *)
   thus "∃t₃. ([b].s,a,t₂,t₃)∈Subst" by force
qed
```

The induction method nominal_induct brings the induction hypothesis automatically into the form

$$\underbrace{(\lambda t_1 \, \lambda(a, t_2). \exists t_3.(t_1, a, t_2, t_3) \in \text{Subst})}_{P} \underbrace{t_1}_{t} \underbrace{(a, t_2)}_{x}$$

by collecting all free variables in the goal, and then it applies Thm. 10. This results in three cases to be proved—variable case, application case and lambda-case. The requirement that the context (a, t_2) is a *fs-pset*-element is enforced by using axiomatic type-classes and relying on Isabelle's type-system. Note that in the lambda-case it is important to know that the binder b is fresh for a and t_2. The proof obligation in this case is:

$$b \mathbin{\#} (a, t_2) \land \exists t_3.(s, a, t_2, t_3) \text{ implies } \exists t_3.([b].s, a, t_2, t_3)$$

which can be easily be shown by rule s4. As a result, the only case in which we really need to manually "interfere" is in the variable case where we have to give Isabelle the hint to distinguish the cases $b = a$ and $b \neq a$. □

```
lemma substitution_lemma:
assumes a1: "x≠y"
    and a2: "x#L"
shows "M[x:=N][y:=L] = M[y:=L][x:=N[y:=L]]"
proof (nominal_induct M)
  case (1 z) (* case 1: variables *)
  have "z=x ∨ (z≠x ∧ z=y) ∨ (z≠x ∧ z≠y)" by force
  thus "am(z)[x:=N][y:=L] = am(z)[y:=L][x:=N[y:=L]]"
    using a1 a2 forget by force
next
  case (2 z M₁) (* case 2: lambdas *)
  assume ih: "M₁[x:=N][y:=L] = M₁[y:=L][x:=N[y:=L]]"
  assume f1: "z#(L,N,x,y)"
  from f1 fresh_fact1 have f2: "z#N[y:=L]" by simp
  show "([z].M₁)[x:=N][y:=L]=([z].M₁)[y:=][x:=N[y:=L]]" (is "?LHS=?RHS")
  proof -
    have "?LHS = [z].(M₁[x:=N][y:=L])" using f1 by simp
    also have "...= [z].(M₁[y:=L][x:=N[y:=L]])" using ih by simp
    also have "...= ([z].(M₁[y:=L]))[x:=N[y:=L]]" using f1 f2 by simp
    also have "...= ?RHS" using f1 by simp
    finally show "?LHS = ?RHS" by simp
  qed
next
  case (3 M₁ M₂) (* case 3: applications *)
  thus "pr(M₁,M₂)[x:=N][y:=L]=pr(M₁,M₂)[y:=L][x:=N[y:=L]]" by simp
qed
```

Fig. 3. An Isabelle proof using the Isar language for the substitution lemma shown in Fig. 2. It uses the following auxiliary lemmas: forget which states that $x \# L$ implies L[x:=T]=L, needed in the variable case. This case proceeds by stating the three subcases to be considered and then proving them automatically using the assumptions a1 and a2. The lemma fresh_fact1 in the lambda-case shows from $z \# (L,N,x,y)$ that $z \# N[x:=L]$ holds. This lemma is not explicitly mentioned in Barendregt's informal proof, but it is necessary to pull out the substitution from under the binder z. This case proceeds as follows: the substitutions on left-hand side of the equation can be moved under the binder z; then one can apply the induction hypothesis; after this one can pull out the second substitution using $z \# N[y:=L]$ and finally move out the first substitution using $z \# (L,N,x,y)$. This gives the right-hand side of the equation

Together with a uniqueness-lemma (whose proof we omit) asserting that

$$\forall s_1 s_2.(t_1, a, t_2, s_1) \in \text{Subst} \wedge (t_1, a, t_2, s_2) \in \text{Subst} \Rightarrow s_1 = s_2 \qquad (3)$$

one can prove the stronger totality-property, namely for all t_1, a, t_2:

$$\exists! t_3. (t_1, a, t_2, t_3) \in \text{Subst} . \qquad (4)$$

Having this at our disposal, we can use Isabelle's definite description operator THE and turn capture-avoiding substitution into a function; we write this function as $(-)[(-) := (-)]$, and establish the equations:

$$
\begin{aligned}
\mathsf{am}(\mathsf{a})[\mathsf{a} := \mathsf{t}] &= \mathsf{t} \\
\mathsf{am}(\mathsf{b})[\mathsf{a} := \mathsf{t}] &= \mathsf{am}(\mathsf{b}) && \text{provided } \mathsf{a} \neq \mathsf{b} \\
\mathsf{pr}(\mathsf{s}_1, \mathsf{s}_2)[\mathsf{a} := \mathsf{t}] &= \mathsf{pr}(\mathsf{s}_1[\mathsf{a} := \mathsf{t}], \mathsf{s}_2[\mathsf{a} := \mathsf{t}]) \\
([\mathsf{a}].\mathsf{s})[\mathsf{a} := \mathsf{t}] &= [\mathsf{a}].(\mathsf{s}[\mathsf{a} := \mathsf{t}]) && \text{provided } \mathsf{b} \mathbin{\#} (\mathsf{a}, \mathsf{t})
\end{aligned}
\tag{5}
$$

These equations can be supplied to Isabelle's simplifier and one can reason about substitution "just like on paper". For this we give in Fig. 3 one *simple* example as evidence—giving the whole formalised Church-Rosser proof from [3, p. 60–62] would be beyond the space constraints of this paper. The complete formalisations of all the results, the Church-Rosser and strong normalisation proof is at http://www.mathematik.uni-muenchen.de/∼urban/nominal/ .

6   Related Work

There are many approaches to formal treatments of binders; this section describes the ones from which we have drawn inspiration.

Our work uses many ideas from the nominal logic work by Pitts *et al* [16,6]. The main difference is that by constructing, so to say, an explicit model of the α-equated lambda-terms based on functions, we have no problem with the axiom-of-choice. This is important. For consider the alternative: if the axiom-of-choice causes inconsistencies, then one cannot build a framework for binding on top of Isabelle/HOL with its rich reasoning infrastructure. One would have to interface on a lower level and has to redo the effort that has been spend to develop Isabelle/HOL. This was attempted in [5], but the attempt was later abandoned.

Closely related to our work is [9] by Gordon and Melham; it has been applied and further developed by Norrish [13]. This work states five axioms characterising α-equivalence and then shows that a model based on de-Bruijn indices satisfies the axioms. This is somewhat similar to our approach where we construct explicitly the set Λ_α. In [9] they give an induction principle that requires in the lambda-case to prove (using their notation)

$$
\forall x\, t.\, (\forall v.\, P\, (t[x := \mathit{VAR}\, v])) \implies P\, (\mathit{LAM}\, x\, t)
$$

That means they have to prove $P(\mathit{LAM}\, x\, t)$ for a variable x for which nothing can be assumed; explicit α-renamings are then necessary in order to get the proof through. This inconvenience has been alleviated by the version of structural induction given in [8] and [12], which is as follows

$$
\exists X.\, \mathrm{FINITE}\, X \wedge (\forall\, x\, t.\, x \notin X \wedge P\, t \implies P\, (\mathit{LAM}\, x\, t))
$$

For this principle one has to provide a finite set X and then has to show the lambda-case for all binders not in this set. This is very similar to our induction principle, but we claim that our version based on freshness fits better with informal practise and can make use of the infrastructure of Isabelle (namely the axiomatic type-classes enforce the finite-support property).

Like our Λ_α, HOAS uses functions to encode lambda-abstractions; it comes in two flavours: *weak* HOAS [4] and *full* HOAS [15]. The advantage of full HOAS over our work is that notions such as capture-avoiding substitution come for free. We, on the other hand, load the work of such definitions onto the user. The advantage of our work is that we have no difficulties with notions such as simultaneous-substitution (a crucial notion in the usual strong normalisation proof), which in full HOAS seem rather difficult to encode. Another advantage we see is that by inductively defining Λ_α one has induction for "free", whereas induction requires considerable effort in full HOAS. The main difference of our work with weak HOAS is that we use *some* specific functions to represent lambda-abstractions; in contrast, weak HOAS uses the *full* function space. This causes problems known by the term "exotic terms"—essentially junk in the model.

7   Conclusion

The paper [2], which sets out some challenges for automated proof assistants, claims that theorem proving technologies have almost reached the threshold where they can be used *by the masses* for formal reasoning about programming languages. We hope to have pushed with this paper the boundary of the state-of-the-art in formal reasoning closer to this threshold. We showed all our results for the lambda-calculus. But the lambda-calculus is only *one* example. We envisage no problems generalising our results to other term-calculi. In fact, there is already work by Bengtson adapting our results to the π-calculus. We also do not envisage problems with providing a general framework for reasoning about binders based on our results. The real (implementation) challenge is to integrate these results into Isabelle's datatype package so that the user does not see any of the tedious details through which we had to go. For example one would like that the subset construction from a bigger set is done completely behind the scenes. Deriving an induction principle should also be done automatically. Ideally, a user just defines an inductive datatype and indicates where binders are—the rest of the infrastructure should be provided by the theorem prover. This is future work.

Acknowledgements. The first author is very grateful to Andrew Pitts and Michael Norrish for the many discussions with them on the subject of the paper. We thank James Cheney, Aaron Bohannon, Daniel Wang and one anonymous referee for their suggestions. The first authors interest in this work was sparked by an email-discussion with Frank Pfenning and by a question from Neil Ghani at the spring school of Midland Graduate School.

References

1. T. Altenkirch. A Formalization of the Strong Normalisation Proof for System F in LEGO. In *Proc. of TLCA*, volume 664 of *LNCS*, pages 13–28, 1993.
2. B. E. Aydemir, A. Bohannon, M. Fairbairn, J. N. Foster, B. C. Pierce, P. Sewell, D. Vytiniotis, G. Washburn, S. Weirich, and S. Zdancewic. Mechanized Metatheory for the Masses: The PoplMark Challenge. accepted at tphol 05.
3. H. Barendregt. *The Lambda Calculus: Its Syntax and Semantics*, volume 103 of *Studies in Logic and the Foundations of Mathematics*. North-Holland, 1981.
4. J. Despeyroux, A. Felty, and A. Hirschowitz. Higher-Order Abstract Syntax in Coq. In *Proc. of TLCA*, volume 902 of *LNCS*, pages 124–138, 1995.
5. M. J. Gabbay. *A Theory of Inductive Definitions With α-equivalence*. PhD thesis, University of Cambridge, 2000.
6. M. J. Gabbay and A. M. Pitts. A New Approach to Abstract Syntax with Variable Binding. *Formal Aspects of Computing*, 13:341–363, 2001.
7. J.-Y. Girard, Y. Lafont, and P. Taylor. *Proofs and Types*, volume 7 of *Cambridge Tracts in Theoretical Computer Science*. Cambridge University Press, 1989.
8. A. D. Gordon. A Mechanisation of Name-Carrying Syntax up to Alpha-Conversion. In *Proc. of Higher-order logic theorem proving and its applications*, volume 780 of *LNCS*, pages 414–426, 1993.
9. A. D. Gordon and T. Melham. Five Axioms of Alpha-Conversion. In *Proc. of TPHOL*, volume 1125 of *LNCS*, pages 173–190, 1996.
10. D. Hirschkoff. A Full Formalisation of π-Calculus Theory in the Calculus of Constructions. In *Proc. of TPHOL*, volume 1275 of *LNCS*, pages 153–169, 1997.
11. T. Nipkow, L. C. Paulson, and M. Wenzel. *Isabelle HOL: A Proof Assistant for Higher-Order Logic*, volume 2283 of *LNCS*. Springer-Verlag, 2002.
12. M. Norrish. Mechanising λ-calculus using a Classical First Order Theory of Terms with Permutations, forthcoming.
13. M. Norrish. Recursive Function Definition for Types with Binders. In *Proc. of TPHOL*, volume 3223 of *LNCS*, pages 241–256, 2004.
14. L. Paulson. Defining Functions on Equivalence Classes. To appear in ACM Transactions on Computational Logic.
15. F. Pfenning and C. Elliott. Higher-Order Abstract Syntax. In *Proc. of the ACM SIGPLAN Conference PLDI*, pages 199–208. ACM Press, 1989.
16. A. M. Pitts. Nominal Logic, A First Order Theory of Names and Binding. *Information and Computation*, 186:165–193, 2003.
17. A. S. Troelstra and H. Schwichtenberg. *Basic Proof Theory*, volume 43 of *Cambridge Tracts in Theoretical Computer Science*. Cambridge University Press, 2000.
18. C. Urban, A. M. Pitts, and M. J. Gabbay. Nominal Unification. *Theoretical Computer Science*, 323(1-2):473–497, 2004.
19. M. VanInwegen. *The Machine-Assisted Proof of Programming Language Properties*. PhD thesis, University of Pennsylvania, 1996. Available as MS-CIS-96-31.
20. M. Wenzel. *Using Axiomatic Type Classes in Isabelle*. Manual in the Isabelle distribution.

Tabling for Higher-Order Logic Programming

Brigitte Pientka

School of Computer Science, McGill University
bp@cs.mcgill.ca

Abstract. We describe the design and implementation of a higher-order tabled logic programming interpreter where some redundant and infinite computation is eliminated by memoizing sub-computation and re-using its result later. In particular, we focus on the table design and table access in the higher-order setting where many common operations are undecidable in general. To achieve a space and time efficient implementation, we rely on substitution factoring and higher-order substitution tree indexing. Experimental results from a wide range of examples (propositional theorem proving, refinement type checking, small-step evaluator) demonstrate that higher-order tabled logic programming yields a more robust and more powerful proof procedure.

1 Introduction

Efficient redundancy elimination techniques such as loop detection or tabling play an important role in the success of first-order theorem proving and logic programming systems. The central idea of tabling is to eliminate infinite and redundant computation by memoizing subcomputation and reusing its results later on. Up to now, higher-order theorem proving and logic programming systems lack such memoization techniques, thereby limiting their success in many applications. This paper describes the design and implementation of tabling for the higher-order logic programming systems Twelf [16,18] and presents a broad experimental evaluation demonstrating the feasibility and benefits of tabling in the higher-order setting.

Higher-order logic programming as Twelf [16] or λProlog [12] extends first-order logic programming along two orthogonal dimensions: First, we allow dynamic assumptions to be added and used during proof search. Second, we allow a higher-order term language which contains terms defined via λ-abstraction. Moreover, execution of a query will not only produce a yes or no answer, but produce a proof term as a certificate which can be checked independently. These features make higher-order logic programming an ideal generic framework for implementing formal systems and executing them.

Most recently, higher-order logic programming has been successfully employed in several certified code projects, where programs are equipped with a certificate (proof) that asserts certain safety properties [3,5,2]. The safety policy can be represented as a higher-order logic program and the higher-order logic programming interpreter can be used to execute the specification and generate a

R. Nieuwenhuis (Ed.): CADE 2005, LNAI 3632, pp. 54–68, 2005.

certificate that a given program fulfills a specified safety policy. However, these applications also demonstrate that present technology is inadequate to permit prototyping and experimenting with safety and security policies. Many specifications are not directly executable and long response times lead to slow development of safety policies in many applications. In [19], we outline a proof-theoretic foundation for tabled proof search to overcome some of these deficiencies, by memoizing sub-computations and re-using its result later. This paper focuses on the realization and implementation of a tabled higher-order logic programming interpreter in practice and presents a broad experimental evaluation. This work is inspired by the success of memoization techniques in tabled first-order logic programming, namely the XSB system [24] where it has been applied in different problem domains such as implementing recognizers for grammars [27], representing transition systems CCS, writing model checkers [6].

In the higher-order setting, tabling introduces several complications. First, we must store intermediate goals together with dynamic assumptions which may be introduced during proof search. Second, many operations necessary to achieve efficient table access such unifiability or instance checking, are undecidable in general for higher-order terms. Our approach relies on linear higher-order patterns [21] and adapts higher-order substitution tree indexing [19] to permit lookup and possible insertion of terms to be performed in a single pass. To avoid repeatedly scanning terms when reusing answers, we adapt substitution factoring [22]. Third, since storing and reusing fully explicit proof terms to certify tabled proofs is impractical due to their large size, we propose a compact proof witness representation inspired by [13] which only keeps track of a proof footprint. As the experimental results from a wide range of examples (propositional theorem proving, refinement type checking, small-step evaluation) demonstrate, tabling leads to a more robust and more powerful higher-order proof search procedure.

The paper is organized as follows: In Sec. 2 we introduce higher-order logic programming. In Sec. 3 we describe the basic principles behind table design guided by substitution factoring and linearization of higher-order terms. This is followed by higher-order term indexing (Sec. 4), and compact proof witness generation (Sec. 5). Experimental results are discussed in Sec. 6. We conclude with a discussion of related work (Sec. 7).

2   Motivating Example: Sequent Calculus

To illustrate the proof search problems and challenges in higher-order logic programming, we introduce a sequent calculus which includes implication, conjunction, and universal quantification. This logic can be viewed as a simple example of a general safety logic. It is small, but expressive enough that it allows us to discuss the basic principles and challenges of proof search in this setting. It can also easily be extended to a richer fragment which includes the existential quantifier, disjunction and falsehood. We will focus here on the higher-order logic programming language Elf [16], which is based on the logical framework LF [9]. We will briefly discuss the representation of a first-order logic in the logical framework LF, and then illustrate how higher-order logic programming interpreter proceeds and what problems arise. We can characterize this fragment of first-order logic as follows, where P represents atomic propositions.

$$\text{Propositions } A, B, C := P \mid \text{true} \mid A \wedge B \mid A \supset B \mid \forall x.A$$
$$\text{Context} \qquad \Gamma \qquad := . \mid \Gamma, A$$

The main judgment to describe provability is: $\Gamma \Longrightarrow A$ which means proposition A is provable from the assumptions in Γ. While A, B, C denote propositions, we will use T to denote terms. The rules for the intuitionist sequent calculus are then straightforward.

$$\frac{\Gamma \Longrightarrow A \quad \Gamma \Longrightarrow B}{\Gamma \Longrightarrow A \wedge B} \text{ andR} \qquad \frac{\Gamma, A \wedge B, A \Longrightarrow C}{\Gamma, A \wedge B \Longrightarrow C} \text{ andL}_1 \qquad \frac{\Gamma, A \wedge B, B \Longrightarrow C}{\Gamma, A \wedge B \Longrightarrow C} \text{ andL}_2$$

$$\frac{\Gamma, A \Longrightarrow B}{\Gamma \Longrightarrow A \supset B} \text{ impR} \qquad \frac{\Gamma, A \supset B \Longrightarrow A \quad \Gamma, A \supset B, B \Longrightarrow C}{\Gamma, A \supset B \Longrightarrow C} \text{ impL}$$

$$\frac{}{\Gamma, A \Longrightarrow A} \text{ axiom} \qquad \frac{\Gamma \Longrightarrow [a/x]A \quad a \text{ is new}}{\Gamma \Longrightarrow \forall x.A} \text{ allR} \qquad \frac{\Gamma, \forall x.A, [T/x]A \Longrightarrow C}{\Gamma, \forall x.A \Longrightarrow C} \text{ allL}$$

The logical framework LF is ideally suited to support the representation and implementation of logical systems such as the intuitionist sequent calculus above. The representation of formulas and judgments follows [9]. We will distinguish between propositions (conc A) we need to prove and propositions (hyp A) we assume. The main judgment to show that a proposition A is provable from the assumptions $A_1, \ldots, A_n$ can be then viewed as: hyp $A_1, \ldots,$ hyp $A_n \Longrightarrow$ conc A. This will allow a direct representation within the logical framework LF and the higher-order logic program describing the inference rules is given next.

axiom : conc A
    $\leftarrow$ hyp A.

andR : conc (A and B)
    $\leftarrow$ conc A
    $\leftarrow$ conc B.

andL$_1$: conc C
    $\leftarrow$ hyp (A and B)
    $\leftarrow$ (hyp $A \rightarrow$ conc C).

andL$_2$: conc C
    $\leftarrow$ hyp (A and B)
    $\leftarrow$ (hyp $B \rightarrow$ conc C).

impR : conc (A imp B)
    $\leftarrow$ (hyp $A \rightarrow$ conc B).

impL : conc C
    $\leftarrow$ hyp (A imp B)
    $\leftarrow$ conc A
    $\leftarrow$ (hyp $B \rightarrow$ conc C).

allR : conc (forall $\lambda x.A\ x$)
    $\leftarrow \Pi x{:}i.$conc ($A\ x$).

allL : conc C
    $\leftarrow$ hyp (forall $\lambda x.A\ x$)
    $\leftarrow$ (hyp ($A\ T$) $\rightarrow$ conc C).

There are two key ideas which make the encoding of the sequent calculus elegant and direct. First, we use higher-order abstract syntax to encode the bound variables in the universal quantifier. We can read the allR clause as follows: To prove conc (forall $\lambda x.A.x$) we need to prove for all parameters x, that conc ($A\ x$) is true, where the Π-quantifier denotes the universal quantifier in the meta-language. Second, we use the power of dynamic assumptions which higher-order logic programming provides, to eliminate the need to manage assumptions in a list explicitly. To illustrate, we consider the clause impR. To prove conc $(A$ imp $B)$, we prove conc B assuming hyp A. In other words, the proof for conc B may use the dynamic assumption hyp A.

When we need to access an assumption from the context, we simply try to prove hyp A using the axiom clause axiom. All the left rules andL$_1$, andL$_2$, impL and allL follow the same pattern. The andL$_1$ rule can be read operationally as follows: we can prove conc C, if we have an assumption hyp $(A$ and $B)$ and we can prove conc C under the assumption hyp A.

For the propositional fragment of the sequent calculus, proof search is decidable. Therefore, we expect that simple examples such as conc $((A$ and $B)$ imp $B)$ should be easily be provable. Unfortunately, an execution with a depth-first search interpreter will lead to an infinite loop, as we will continue to apply the andL$_1$ rule, and generate the following subgoals.

Dynamic assumption	Goal	Justification
$A{:}o, B{:}o$	$\vdash$ conc $((A$ and $B)$ imp $B)$	
$A{:}o, B{:}o, h_1{:}$hyp $(A$ and $B)$	$\vdash$ conc B	impR
$A{:}o, B{:}o, h_1{:}$hyp $(A$ and $B), h_2{:}$hyp A	$\vdash$ conc B	andL$_1$
$A{:}o, B{:}o, h_1{:}$hyp $(A$ and $B), h_2{:}$hyp $A, h_3{:}$hyp $A \vdash$ conc B		loop

To prevent looping, we need to detect two independent problems. First, we need to prevent adding dynamic assumptions, which are already present in some form. However, this is only part of the solution, since we also need to detect that we keep trying to prove the goal conc B. In this paper, we will propose the use of tabling in higher-order logic programming to detect loops. The essential idea is to memoize subgoals together with its dynamic assumptions and re-use their results later. This will prevent that the computation will be trapped in infinite paths and can potentially improve performance by re-using the result of previous proofs. Note that although the subgoals encountered in the previous example were all ground, and did not contain any existential variables, this may in general not be the case. Consider for the slightly different version of the previous example which corresponds to $\exists y'.\forall x.\exists y.((Q\ y') \wedge (P\ x)) \supset (P\ y))$:

exists $\lambda y'$. forall λx. exists λy. $(((Q\ y')$ and $(P\ x))$ imp $(P\ y))$

We first remove the existential quantifier by introducing an existential variable Y'. Next, we eliminate the allR-rule by introducing a new parameter x. Then we remove the second existential quantifier, by introducing a second existential variable Y. Existential variables (or logic variables) such as Y' and Y are subject to higher-order unification during proof search. Parameter dependencies such as that the existential Y is allowed to depend on the parameter x, while Y' is not, is naturally enforced by allowing higher-order terms and existential variables which can be instantiated with functions. Using the impR-rule, we introduce the assumption hyp $((Q\ Y')$ and $(P\ x))$, and the sequence of subgoals we will then encounter by continuing to apply the andL$_1$-rule is:

x:i, u:hyp $((Q\ Y')$ and $(P\ x))$ $\vdash$ conc $(P\ (Y\ x))$

x:i, u:hyp $((Q\ Y')$ and $(P\ x)), u_1$:hyp $(Q\ Y')$ $\vdash$ conc $(P\ (Y\ x))$ andL$_1$

x:i, u:hyp $((Q\ Y')$ and $(P\ x)), u_1$:hyp $(Q\ Y'), u_2$:hyp $(Q\ Y') \vdash$ conc $(P\ (Y\ x))$ loop

As in the previous example, we will end up in an infinite loop, however the subgoals now contain existential variables Y' and Y.

Although it is possible to design specialized propositional sequent calculus with loop detection [10], this is often non-trivial and complicates the implementation of the proof search procedure. Moreover, proving the correctness of such a more refined propositional calculus, is non-trivial, because we need to reason explicitly about the structure of memoization. Finally, the certificates, which are produced as a result of the execution, are larger and contain references to the explicit memoization data-structure. This is especially undesirable in the context of certified code where certificates are transmitted to and checked by a consumer, as sending larger certificates takes up more bandwidth and checking them takes more time. Tabled logic programming provides generic memoization support for proof search and allows us to factor out common sub-proofs during proof search, thereby potentially obtaining smaller and more compact certificates. Since tabled logic programming terminates for programs with the bounded term-size property, we are also able to disprove certain statements. This in turn helps the user to debug the specification and implementations and increases the expressive power and usefulness of the overall system. In the case of the propositional sequent calculus, we obtain a decision procedure for free.

3   Tabling in Higher-Order Logic Programming

Tabling methods eliminate redundant and infinite computation by memoizing subgoals and their answers in a table and re-using the results later. Our search is based on the multi-stage strategy by Tamaki and Sato [25], which differs only insignificantly from SLG resolution [4] for first-order logic programs without negation. Tabled search proceeds in stages and relies on a table to keep track of all the subgoals encountered, and answers which were derived for them. When trying to prove a goal G from the dynamic assumptions Γ, we first check if there exists a variant of $\Gamma \vdash G$ in the table. If yes, then we suspend the computation and backtrack. If no, we add $\Gamma \vdash G$ to the table, and proceed proving the goal using the dynamic assumptions in Γ and the program clauses. If we derive an answer for a goal $\Gamma \vdash G$, then this answer is added to the table. This first stage terminates, once all possible search paths have been explored, and the leafs in the search tree are either failure, success, or suspended nodes. In the next stage, we will re-consider the suspended nodes in the search tree, and try to grow the tree further by re-using answers of previous stages from the table. For a more detailed description of the search we refer the reader to [19,20]. Here we will discuss the basic design principles underlying tabled search, how to manage and access the table efficiently in the higher-order setting. These principles are largely independent of the actual strategy of how to reuse answers from the table. There are three main table access operations:

Call CheckInsert. When we encounter a tabled subgoal, we need to check whether this subgoal is redundant. We check, if there exists a table entry $\Gamma' \vdash G'$ s.t. $\Gamma \vdash G$ (the current goal) is a variant (or instance) of the already existing entry $\Gamma' \vdash G'$.

Answer CheckInsert. When an answer together with its proof witness is derived for a tabled subgoal, we need to check whether this answer has been already entered into the table answer list for this particular subgoal. If it has then the search fails, otherwise the answer together with its proof witness is added to the answer list, and may be re-used in later stages.

Answer Backtracking. When a tabled subgoal is encountered, and answers for it are available from the table, we need to backtrack through all the answers.

A naive implementation can result in repeatedly rescanning terms and large table size thereby degrading performance considerably and rendering tabling impractical. This problem has been named *table access problem* in first-order logic programming [22]. In this section, we will describe design and implementation solution, which shares common structure and common operations in the higher-order setting using substitution factoring, linear higher-order patterns, higher-order substitution tree indexing, and compact proof witnesses.

3.1 Design of Memo-Table

The table records intermediate goals $\Gamma \vdash G$ together with answers and proof witnesses. As we have seen in the previous example, intermediate goals may refer to existential (or logic variables) which will be instantiated during proof search. In an implementation, existential variables are typically realized via references and destructive updates. This achieves that instantiations of existential variables are propagated immediately. On the other hand, we may need to undo these instantiations for existential variables upon backtracking. This is usually achieved by keeping a separate trail of existential variables and their corresponding instantiations. As a consequence, we must take special care in an implementation when memoizing and suspending the computation of intermediate goals. When suspending nodes, we copy the trail to re-instantiate the existential variables adapting ideas from [8]. Before storing intermediate goals in a memo-table, we must abstract over all the existential variables in a goal, to avoid pollution of the table. To illustrate recall the previous subgoal:

$$x{:}\mathsf{i}, u{:}\mathsf{hyp}\ ((Q\ Y')\ \mathsf{and}\ (P\ x)) \vdash \mathsf{conc}\ (P\ (Y\ x))$$

To store this subgoal in a table, we abstract over the existential variables Y' and Y, to obtain the following table entry:

$$\begin{array}{cccc} \Delta & ; & \Gamma & \vdash & G \\ y' : \mathsf{i}, y : \mathsf{i} \to \mathsf{i} & ; & x{:}\mathsf{i},\ u{:}\mathsf{hyp}\ ((Q\ y')\ \mathsf{and}\ (P\ x)) & \vdash \mathsf{conc}\ (P\ (y\ x)) \end{array}$$

Δ refers to a context describing existential variables, Γ describes the context for the bound variables and dynamic assumptions and G describes the goal we are attempting to prove. To allow easy comparison of goals G with dynamic assumptions Γ modulo renaming of the variables in Δ and Γ, we represent terms internally using explicit substitutions [1] and de Bruijn indices.

Once this subgoal is solved and we inferred a possible instantiation for the existential variables in Δ, we will add the answer to the table. The answer is a substitution for the existential variables in Δ. In the previous example, the correct instantiation for Y is $\lambda x.x$, while the existential variable Y' is unconstrained. As we see in this example, not all instantiations for existential variables need to be ground. To avoid pollution of the answer substitution in the table, we again must abstract over the existential variables in the computed answer, which leads to the following abstracted answer substitution:

$$y'{:}i \quad \vdash (y'/y', \ \lambda x.x/y) \quad : \quad y'{:}i, y{:}i \rightarrow i$$

In general, the invariant about table entries and answer substitutions are:

Table entry	Answer substitution
$\Delta; \Gamma \vdash G$	$\Delta' \vdash \theta : \Delta$

The design supports naturally substitution factoring based on explicit substitutions [22]. With substitution factoring the access cost is proportional to the size of the answer substitution rather than the size of the answer itself. It guarantees that we only store the answer substitutions, and create a mechanism of returning answers to active subgoals that takes time linear in the size of the answer substitution θ rather than the size of the solved query $[\theta]G$. In other words, substitution factoring ensures that answer tables contain no information that also exists in their associated call table. Operationally, this means that the constant symbols in the subgoal need not be examined again during either answer checkInsert or answer backtracking. For this setup to work cleanly in the higher-order setting, it is crucial that we distinguish between existential variables in Δ and bound variables and assumptions in Γ.

To support selective memoization, we provide a user keyword which allows the user to mark predicates to be tabled or not. If the predicate in G is not marked tabled, then nothing will change. We design the tabled search in such a way that it is completely separate from non-tabled search. The only overhead in non-tabled computation will be a check whether a given predicate is tabled.

3.2 Optimization: Linearization

A common optimization for first-order terms is linearization which enforces that every existential variable occurs only once. This means that any necessary consistency checks can be delayed and carried out in a post-processing step. In the higher-order setting, we extend this linearization step to eliminate any computationally expensive checks involving bound variables and enforce that terms fall into the linear higher-order pattern fragment, where every existential variable occurs only once and must be applied to all the bound variables. Linear higher-order patterns refine the notion of higher-order patterns[11,17] further and factor out any computationally expensive parts. As shown in [21], many terms encountered fall into this fragment and linear higher-order pattern unification performs well in practice. Consider again, the previous example:

$$\Delta \qquad ; \qquad \Gamma \qquad \vdash \qquad G$$
$$y' : i, y : i \rightarrow i \quad ; \quad x{:}i, \; u{:}\mathsf{hyp}\,(Q\,x) \quad \vdash \mathsf{conc}\,(P\,(y\,x))$$

Both occurrences of the existential variable y and y' are higher-order patterns, since they are applied to a distinct set of bound variables. However, the variable y' and y are not linear higher-order patterns, since neither is applied to all the bound variables which occur in Γ. During linearization, we will translate the goal into a linear higher-order pattern together with residual equations which factor out non-linear sub-parts. We abbreviate $y_1\,x\,u$ as $y_1\lfloor \mathsf{id} \rfloor$.

$$\Delta \qquad ; \qquad \qquad \Gamma \qquad \qquad \vdash \qquad G$$
$$y_1 : i \rightarrow i \rightarrow i,$$
$$y_2 : i \rightarrow i \rightarrow i$$
$$y' : i, y : i \rightarrow i \quad ; \quad x{:}i, \; u{:}\mathsf{hyp}\,((Q\,y_1\lfloor \mathsf{id} \rfloor)\text{ and }(P\,x)) \quad \vdash \mathsf{conc}\,(P\,(y_2\lfloor \mathsf{id} \rfloor))$$

together with the residual equations R: $y_1\lfloor \mathsf{id} \rfloor \doteq y' \wedge y_2\lfloor \mathsf{id} \rfloor \doteq y\,x$
This motivates the final table design:

Table entry	Residual Equ.	Answer substitution
$\Delta; \Gamma \vdash G$	$\Delta; \Gamma \vdash R$	$\Delta' \vdash \theta : \Delta$

where G is a linear higher-order pattern, Γ denotes the bound variables and dynamic assumptions and Δ describes the existential variables occurring in G and Γ. This linearization step can be done together with abstraction and standardization over the existential variables in goal, hence only one pass through the term is required.

3.3   Optimization: Strengthening

We have seen previously that strengthening of the dynamic assumptions is necessary to prevent some loops. We previously concentrated on strengthening by removing duplicate assumptions from the dynamic context. However, in general we use in addition two other forms of strengthening based on a type-dependency analysis called subordination [26]. First, we eliminate dynamic assumption in Γ which cannot possibly contribute to a proof of G. Second, we eliminate bound variable dependencies in existential variable. Strengthening allows us to detect more loops during proof search and eliminate more redundant computation. Furthermore, it allows us to store some information more compactly.

4 Higher-Order Term Indexing for Tabling

To achieve an efficient and successful tabled logic programming interpreter, it is crucial to support efficient indexing of terms in the table to facilitate compact storage and rapid retrieval of table entries. Although a wide range of indexing techniques exists in the first-order setting, indexing techniques for higher-order terms are almost not existent. The main problem in handling higher-order terms lies in the fact that most operations such as testing whether two terms are unifiable, computing the most specific generalization of two terms etc. are undecidable in the higher-order setting.

We propose substitution tree indexing for linear higher-order patterns. In [19,20], we give a formal description for computing the most specific generalization of two linear higher-order patterns, for inserting terms in the index and for retrieving a set of terms from the index s.t. the query is an instance of the term in the index, and show correctness. Here we will concentrate on the adaptations to support tabling. The main algorithm of building a substitution tree follows the description in [23]. To illustrate higher-order substitution tree indexing let us consider the following set of linear higher-order patterns.

Goal	Residual Equation
conc (forall $\lambda z.$ $((P\ (f\ x)\ y_1[\mathsf{id}])$ and $\quad Q\quad z\ ))$	$y_1[\mathsf{id}] \doteq y[x/x]$
conc (forall $\lambda z.$ $((P\quad z\quad y_1[\mathsf{id}])$ and $Q\ y_2[\mathsf{id}]))$	$y_1[\mathsf{id}] \doteq y' \wedge y_2[\mathsf{id}] \doteq y[x/x]$
conc (forall $\lambda z.$ $((P\quad z\quad y_1[\mathsf{id}])$ and $Q\ y_2[\mathsf{id}]))$	$y_1[\mathsf{id}] \doteq y[\mathsf{id}] \wedge y_2[\mathsf{id}] \doteq y'$
conc (forall $\lambda z.$ $((P\ (f\ z)\ y_1[\mathsf{id}])$ and $Q\ (f\ y_2[\mathsf{id}]))))$	$y_1[\mathsf{id}] \doteq y[\mathsf{id}] \wedge y_2[\mathsf{id}] \doteq y[x/x]$

For simplicity, we assume each of the goals has the same dynamic context $\Gamma = x{:}i, u{:}\mathsf{hyp}\ ((P\ y_0[\mathsf{id}]x)$ and $(Q\ x)$ and the context Δ describing the existential variables contains y_0, y_1, y_2, y, and y'.

A higher-order substitution tree is a tree whose nodes are substitutions together with a context Δ_i which describes the existential variables occurring in the node. For example, the substitution in the top-most node contains the existential variable y_1, while the node with the substitution $f(i_3[\mathsf{id}])/i_1$ does not refer to any existential variable. It is crucial that we ensure that any internal variable i which is applied to all the variables in whose scope it occurs in. However, for any operation on the index, we must treat internal existential variables i differently than globally existential variables y. Internal existential variables i will be instantiated at a later point as we traverse the tree. While existential variables defined in Δ_i are potentially subject to "global" instantiation, if we check whether the current goal is an instance of a table entry. The intention is that all the Δ_i along a given path together with the Δ_n at the leaf constitutes the full context of existential variables Δ. As there are no typing dependencies among the variables in Δ and all the variables in Δ are linear, they can be arbitrarily re-ordered. Distributing Δ along the nodes in the substitution tree, makes it easier to guarantee correctness in an implementation where variables are represented via de Bruijn indices. Below we show the substitution tree for the given set of linear higher-order patterns. Each table predicate will have its own substitution tree.

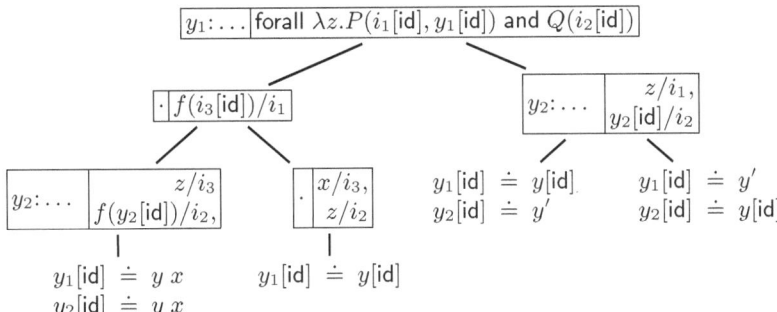

At the leafs, we will store linear residual equations, dynamic assumptions Γ, the existential variables Δ' occurring in the residual equations and in Γ, as well as a pointer to the answer list. Note we omitted the two latter parts in the figure above. By composing the substitutions along a path and collecting all the existential variables Δ_i along this path, we will obtain the table entry $\Delta; \Gamma \vdash G$ together with its residual equations. By composing the substitutions in the left-most branch, we obtain the term (4).

We distinguish between internal existential variables i which are defined in the context Σ and "global" existential variables u and v which are defined in the context Δ. A higher-order substitution tree is an ordered tree and is defined as follows:

1. A tree is a leaf node with substitution ρ such that $\Delta_n \vdash \rho : \Sigma$.
2. A tree is a node with substitution ρ such that $(\Delta_j, \Sigma) \vdash \rho : \Sigma'$ and children nodes $N_1, \ldots, N_n$ where each child Node N_i has a substitution ρ_i such that $(\Delta_i, \Sigma_i) \vdash \rho_i : \Sigma$.

For every path from the top node ρ_0 where $(\Delta_0, \Sigma_1) \vdash \rho_0 : \Sigma_0$ to the leaf node ρ_n, we have $\Delta = \Delta_0 \cup \Delta_1 \cup \ldots \cup \Delta_n$ and $\Delta \vdash \rho_n \circ \rho_{n-1} \circ \ldots \circ \rho_0 : \Sigma_0$. In other words, there are no internal existential variables left after we compose all the substitutions ρ_n up to ρ_0. As there are no typing dependencies among the variables in Σ and all the variables are linear, they can be arbitrarily reordered. At the leaf, we also store a list of answer substitutions θ, where we have $\Delta' \vdash \theta : \Delta$ and the dynamic context Γ.

5   Compact Proof Witnesses

Generating certificates as evidence of a proof is essential if we aim to use the tabled logic programming interpreter as part of a certifying code infrastructure. Moreover, it is helpful in guaranteeing correctness of the tabled search and debugging the logic programming interpreter. The naive solution to generate certificates when tabling intermediate sub-goals and their results, is to store the corresponding proof term together with the answer substitution in the table. However this may take up considerable space and results in high computational overhead, due to their large size[14]. Hence it is impractical to store the full proof term. In our implementation, we only store a footprint of the proof from which it is possible to recover the full proof term. Essentially we just keep track of the id of the applied clause thereby obtaining a string of numbers which corresponds to the actual proof. This more compact proof witness can be de-compressed and checked by building and re-running a deterministic higher-order logic programming engine. This idea to represent proof witnesses as a string of (binary) numbers is inspired by [13].

6   Experimental Results

In this section, we discuss some experimental results with different examples. All experiments are done on a machine with the following specifications: 2.4GHz Intel Pentium Processor, 512 MB RAM. We are using SML of New Jersey 110.0.7 under Linux Red Hat 9. Times are measured in seconds.

6.1   Propositional Theorem Proving

We report on experiments with an implementation of the propositional sequent calculus where we chain all invertible rules together and use focusing for the non-deterministic choices We will only memoize subgoals during focusing, thereby controlling the table size and employ strengthening. The implementation of the sequent calculus within Twelf will not be executable using a logic programming interpreter based on depth-first search, however it is possible to use the iterative deepening theorem prover which is part of the meta-theorem prover [18]. Iterative deepening will stop after finding the first solution, hence we compare it to finding the first solution using tabled search. We also include the time it takes for tabled search to terminate, and conclude that no other solution exists.

Focusing Calculus (Propositional theorem proving) – run time in sec

Name	ItDeep	Tab(1)	Tab(all)
$(A \vee B) \wedge (D \vee E) \wedge (G \vee H) \supset (A \wedge D) \vee$ $(A \wedge G) \vee (D \wedge G) \vee (B \wedge E) \vee (B \wedge H) \vee (E \wedge H)$	0.23	0.05	0.05
$((((A \leftrightarrow B) \supset (A \wedge B \wedge C)) \wedge ((B \leftrightarrow C) \supset (A \wedge B \wedge C))$ $\wedge((C \leftrightarrow A) \supset (A \wedge B \wedge C))) \supset (A \wedge B \wedge C))$	∞	0.46	0.40
$(((A \vee B \vee C) \wedge (D \vee E \vee F) \wedge (G \vee H \vee J) \wedge$ $(K \vee L \vee M)) \supset ((A \wedge D) \vee (A \wedge G) \vee (A \wedge K) \vee$ $(D \wedge G) \vee (D \wedge K) \vee (G \wedge K) \vee (B \wedge E) \vee (B \wedge H) \vee$ $(B \wedge L) \vee (E \wedge H) \vee (E \wedge L) \vee (H \wedge L) \vee (C \wedge F) \vee$ $(C \wedge J) \vee (C \wedge M) \vee (F \wedge J) \vee (F \wedge M) \vee (J \wedge M)))$	∞	4.12	4.23

Focusing Calculus (Propositional Theorem Proving) – Disproving

Formula	tab
$(((A \wedge (B \vee C)) \supset (C \vee (C \wedge D))) \supset ((\neg A) \vee ((A \vee B) \supset C)))$	0.01
$(((A \vee B \vee C) \wedge (D \vee E \vee F)) \supset ((A \wedge B) \vee (B \wedge E) \vee (C \wedge F)))$	0.02
$(((((\neg(\neg(\neg A \vee \neg B))) \supset (\neg A \vee \neg B)) \supset ((\neg(\neg(\neg A \vee \neg B)))) \vee$ $\neg(\neg A \vee \neg B))) \supset (\neg(\neg(\neg A \vee \neg B)) \vee \neg(\neg A \vee \neg B)))$	11.99

In our experiments, we consider all the propositional test-cases reported by J. Howe in [10], which he used to compare and evaluate two special purpose propositional theorem prover which employ loop-detection. Tabled higher-order logic programming is able to prove or disprove 14 of the 15 propositional examples from Howe's test-suite within 1 sec or below, and only one example took 4.12 sec thereby providing a decision procedure for propositional logic for free. Only in one example the tabled logic programming interpreter started thrashing. The table size in these examples was up to 2600 table entries and up to 22000 suspended goals, which seems to be the main limiting factor.

Not surprisingly, iterative deepening is not powerful enough to prove most of the examples from Howe's test-suite.

6.2   Refinement Type Checking

In this example, we explore refinement type checking as described by Davies and Pfenning in [7]. This is an advanced type system for a small functional language MiniML where expressions may have more than one type and there may be many ways of inferring a type. The type system is executable with a depth-first logic programming interpreter, however the redundancy may severely hamper the performance. We will compare the performance between depth-first search and tabled search, and group the examples in three categories: 1) Finding the first solution and finding all possible solutions 2) Discovering that a given program cannot be typed.

Refinement type checking – Typable examples (runtime in sec)

Name		lp(1)	tab(1)	lp(all)	tab(all)
sub :	$((\mathsf{nat} \to \mathsf{pos} \to \mathsf{nat})\&(\mathsf{pos} \to \mathsf{nat} \to \mathsf{nat})\&$ $(\mathsf{pos} \to \mathsf{pos} \to \mathsf{nat})\&\mathsf{nat} \to \mathsf{nat} \to \mathsf{nat})$	0.10	0.38	3.43	0.43
mult :	$((\mathsf{pos} \to \mathsf{nat} \to \mathsf{nat})\&(\mathsf{nat} \to \mathsf{nat} \to \mathsf{nat})\&$ $(\mathsf{nat} \to \mathsf{pos} \to \mathsf{nat})\&(\mathsf{pos} \to \mathsf{pos} \to \mathsf{pos}))$	0.06	0.66	∞	0.84
square:	$(\mathsf{pos} \to \mathsf{nat}\&\mathsf{nat} \to \mathsf{nat})$	0.02	0.70	∞	1.06
square:	$(\mathsf{pos} \to \mathsf{pos})$	0.10	0.90	∞	0.88

– time out after 1h

Refinement type checking – Untypable examples

Name		lp	tab
plus :	$((\mathsf{nat} \to \mathsf{nat} \to \mathsf{nat})\&(\mathsf{nat} \to \mathsf{pos} \to \mathsf{pos})\&$ $(\mathsf{pos} \to \mathsf{nat} \to \mathsf{pos})\&(\mathsf{pos} \to \mathsf{pos} \to \mathsf{zero}))$	8.14	0.20
mult :	$(\mathsf{nat} \to \mathsf{pos} \to \mathsf{nat})$	805.97	0.35
mult :	$((\mathsf{pos} \to \mathsf{nat} \to \mathsf{nat})\&(\mathsf{nat} \to \mathsf{nat} \to \mathsf{nat})\&$ $(\mathsf{nat} \to \mathsf{pos} \to \mathsf{nat})\&(\mathsf{pos} \to \mathsf{pos} \to \mathsf{zero}))$	∞	0.620
square :	$(\mathsf{pos} \to \mathsf{zero})\&(\mathsf{pos} \to \mathsf{nat})$	∞	0.72

– time out after 1h

As the results demonstrate, logic programming is superior, if we are only interested in finding the first solution, but is not able to disprove that a given program is in fact not well-typed. Similarly, finding all possible types for a given program is too unwieldy. The table contains up to 400 table entries and 300 suspended goals. The fact that depth-first-search is superior to tabled search is not surprising since managing the table imposes some computational overhead. Moreover, the tabled strategy delays the re-use of answers hence imposing a penalty. However, the tabled logic programming interpreter is able to solve all the examples within 1 sec. This attests to the strength and robustness of the system.

6.3   Parsing

Next, we present experiments with a parser for formulas into higher-order abstract syntax where we mix right and left recursion to model right and left associativity in the grammar. This leads to specifications which are not executable using a depth-first search. Hence we compare iterative deepening with tabled search. As the results demonstrate, tabling is clearly superior to iterative deepening, and provides a practical way of experimenting with parsers and grammars. We only compare finding the first solution with tabling and finding the first solution with iterative deepening, and report on the time depending on the number of tokens parsed. Table size ranges up to 1500 table entries, and up to 1750 suspended goals.

Parsing: Provable – runtime

Name	#tokens	ItDeep	tab(1)
1	5	0.01	0.02
2	20	0.78	0.07
3	32	79	0.28
4	60	2820.02	0.94
6	118	∞	3.22

Time limit : 1h

Parsing: Not provable – runtime

Name	#tokens	Tab
1	19	0.01
2	31	0.27
3	58	0.50
4	117	2.24

7   Conclusion

In this paper, we described the design and implementation of a tabled higher-order logic programming interpreter within the Twelf system. The system including the test-suites is available at http://www.cs.cmu.edu/~twelf as part of the Twelf distribution. Crucial ingredients in the design are substitution factoring, linear higher-order patterns, higher-order substitution tree indexing, and compact proof witnesses. These techniques are the key ingredients to enabling tabling in higher-order logic programming or theorem proving systems. They should also be applicable to systems such as λProlog [12] or Isabelle [15].

The wide range of examples we have experimented with demonstrates that tabling is a significant step towards obtaining a more robust and more powerful proof search engine in the higher-order setting. Tabling leads to improved performance and more meaningful, quicker failure behavior. This does not mean that tabling is a panacea for all the proof search problems, but rather the first step towards integrating and adapting some of the more sophisticated first-order theorem proving techniques to the higher-order setting.

Unlike most descriptions of tabling which rely on modifying the underlying WAM to enable tabling support, we have identified and implemented the essential tabling mechanisms independently of the WAM. Although we have tried to carefully design and implement tabling within the higher-order logic programming system Twelf, there is still quite a lot of room for improvements. The most severe limitation currently is due to the multi-stage strategy which re-uses answers in stages, and prevents the use of answers as soon as they are available. Different strategies have been developed in first-order tabled logic programming such as SCC scheduling (strongly connected components), which allows us to consume answers as soon as they are available and garbage collect unproductive suspended nodes [24]. In the future, we plan to adapt these techniques to the higher-order setting, and incorporate more first-order theorem proving techniques such as ordering constraints.

References

1. Martín Abadi, Luca Cardelli, Pierre-Louis Curien, and Jean-Jacques Lèvy. Explicit substitutions. In *Conference Record of the Seventeenth Annual ACM Symposium on Principles of Programming Languages, San Francisco, California*, pages 31–46. ACM, 1990.
2. Andrew W. Appel and Edward W. Felten. Proof-carrying authentication. In *ACM Conference on Computer and Communications Security*, pages 52–62, 1999.
3. W. Appel and Amy P. Felty. A semantic model of types and machine instructions for proof-carrying code. In *27th ACM SIGPLAN-SIGACT Symposium on Principles of Programming Languages (POPL '00)*, pages 243–253, Jan. 2000.
4. W. Chen and D. S. Warren. Tabled evaluation with delaying for general logic programs. *Journal of the ACM*, 43(1):20–74, January 1996.
5. Karl Crary. Toward a foundational typed assembly language. In *30th ACM Symposiumn on Principles of Programming Languages (POPL)*, pages 198–212, New Orleans, Louisisana, January 2003. ACM-Press.
6. B. Cui, Y. Dong, X. Du, K. N. Kumar, C.R. Ramakrishnan, I.V. Ramakrishnan, A. Roychoudhury, S.A. Smolka, and D.S. Warren. Logic programming and model checking. In *International Symposium on Programming Language Implementation and Logic Programming (PLILP'98)*, volume 1490 of *Lecture Notes in Computer Science*, pages 1–20. Springer, 1998.
7. Rowan Davies and Frank Pfenning. Intersection types and computational effects. In *Proceedings of the International Conference on Functional Programming (ICFP 2000), Montreal, Canada*, pages 198–208. ACM Press, 2000.
8. B. Demoen and K. Sagonas. CAT: The copying approach to tabling. In *International Symposium on Programming Language Implementation and Logic Programming (PLILP'98)*, Lecture Notes in Computer Science (LNCS), vol. 1490, pages 21–36, 1998.
9. Robert Harper, Furio Honsell, and Gordon Plotkin. A framework for defining logics. *Journal of the Association for Computing Machinery*, 40(1):143–184, January 1993.
10. Jacob M. Howe. Two loop detection mechanisms: a comparison. In *Proceedings of the 6th Workshop on Theorem Proving with Analytic Tableaux and Related Methods*, pages 188–200. Springer, 1997. LNCAI 1227.

11. Dale Miller. Unification of simply typed lambda-terms as logic programming. In *Eighth International Logic Programming Conference*, pages 255–269, Paris, France, June 1991. MIT Press.

12. Gopalan Nadathur and Dale Miller. An overview of λProlog. In *Fifth International Logic Programming Conference*, pages 810–827, Seattle, Washington, August 1988. MIT Press.

13. G. Necula and S. Rahul. Oracle-based checking of untrusted software. In *28th ACM Symposium on Principles of Programming Languages (POPL'01)*, pages 142–154, 2001.

14. G. Necula and P. Lee. Efficient representation and validation of logical proofs. In *Proceedings of the 13th Annual Symposium on Logic in Computer Science (LICS'98)*, pages 93–104, June 1998. IEEE Computer Society Press.

15. Lawrence C. Paulson. Natural deduction as higher-order resolution. *Journal of Logic Programming*, 3:237–258, 1986.

16. Frank Pfenning. Elf: A language for logic definition and verified meta-programming. In *Fourth Annual Symposium on Logic in Computer Science*, pages 313–322, Pacific Grove, California, June 1989. IEEE Computer Society Press.

17. Frank Pfenning. Unification and anti-unification in the Calculus of Constructions. In *Sixth Annual IEEE Symposium on Logic in Computer Science*, pages 74–85, Amsterdam, The Netherlands, July 1991.

18. Frank Pfenning and Carsten Schürmann. System description: Twelf — a meta-logical framework for deductive systems. In *Proceedings of the 16th International Conference on Automated Deduction (CADE-16)*, pages 202–206, Trento, Italy, July 1999. Springer Lecture Notes in Artificial Intelligence (LNAI) 1632.

19. Brigitte Pientka. A proof-theoretic foundation for tabled higher-order logic programming. In *18th International Conference on Logic Programming, Copenhagen, Denmark*, Lecture Notes in Computer Science (LNCS), 2401, pages 271–286. Springer, 2002.

20. Brigitte Pientka. *Tabled higher-order logic programming*. PhD thesis, Department of Computer Sciences, Carnegie Mellon University, December 2003. CMU-CS-03-185.

21. Brigitte Pientka and Frank Pfennning. Optimizing higher-order pattern unification. In *19th International Conference on Automated Deduction, Miami, USA*, Lecture Notes in Artificial Intelligence (LNAI) 2741, pages 473–487. Springer, July 2003.

22. I. V. Ramakrishnan, P. Rao, K. Sagonas, T. Swift, and D. Warren. Efficient access mechanisms for tabled logic programs. *Journal of Logic Programming*, 38(1):31–54, Jan 1999.

23. I. V. Ramakrishnan, R. Sekar, and A. Voronkov. Term indexing. In Alan Robinson and Andrei Voronkov, editors, *Handbook of Automated Reasoning*, volume 2, pages 1853–1962. Elsevier Science Publishers B.V., 2001.

24. K. Sagonas and T. Swift. An abstract machine for tabled execution of fixed-order stratified logic programs. *ACM Transactions on Programming Languages and Systems*, 20(3):586–634, 1998.

25. H. Tamaki and T. Sato. OLD resolution with tabulation. In *Proceedings of the 3rd International Conference on Logic Programming*, volume 225 of *Lecture Notes in Computer Science*, pages 84–98. Springer, 1986.

26. Roberto Virga. *Higher-Order Rewriting with Dependent Types*. PhD thesis, Department of Mathematical Sciences, Carnegie Mellon University, Available as Technical Report CMU-CS-99-167, Sep 1999.

27. David S. Warren. *Programming in tabled logic programming*. draft available from http://www.cs.sunysb.edu/w̃arren/xsbbook/book.html, 1999.

A Focusing Inverse Method Theorem Prover for First-Order Linear Logic

Kaustuv Chaudhuri and Frank Pfenning*

Department of Computer Science,
Carnegie Mellon University
{kaustuv, fp}@cs.cmu.edu

Abstract. We present the theory and implementation of a theorem prover for first-order intuitionistic linear logic based on the inverse method. The central proof-theoretic insights underlying the prover concern resource management and focused derivations, both of which are traditionally understood in the domain of backward reasoning systems such as logic programming. We illustrate how resource management, focusing, and other intrinsic properties of linear connectives affect the basic forward operations of rule application, contraction, and forward subsumption. We also present some preliminary experimental results obtained with our implementation.

1 Introduction

Linear logic [1] extends classical logic by internalizing *state*. This is achieved by forcing *linear assumptions* to be used exactly once during a proof. Introducing or using a linear assumption then corresponds to a change in state. This alters the fundamental character of the logic so that, for example, even the propositional fragment is undecidable. The expressive power of linear logic has been exploited to provide a logical foundation for phenomena in a number of diverse domains, such as planning, concurrency, functional programming, type systems, and logic programming.

Despite this wide range of potential applications, there has been relatively little effort devoted to theorem proving for linear logic. One class of prior work consists of fundamental proof-theoretic studies of the properties of linear logic [2,3,4]. Most of these are concerned with backward reasoning, that is, with analyzing the structure of proof search starting from a given goal sequent. The most important results, such as the discovery of focusing [2] and solutions to the problem of resource management [5,6] have made their way into logic programming languages [7,8], but not implemented theorem provers. The term *resource management* in this context refers to the problem of efficiently ensuring that the single-use semantics of linear assumptions is respected during proof search.

Another class of prior work is directly concerned with theorem proving for linear logic in the forward direction. Here, we are only familiar with Mints's theoretical study

---
* This work has been supported by the Office of Naval Research (ONR) under grant MURI N00014-04-1-0724 and by the National Science Foundation (NSF) under grant CCR-0306313.

R. Nieuwenhuis (Ed.): CADE 2005, LNAI 3632, pp. 69–83, 2005.

of resolution for linear logic [9] and Tammet's implementation of a resolution prover for classical propositional linear logic [10].

In this paper we describe a theorem prover for first-order intuitionistic linear logic based on the inverse method. The choice of the logic is motivated by two considerations. First, it includes the core of the Concurrent Logical Framework (CLF) [11], so our theorem prover can reason with specifications written in CLF; many such example CLF specifications have been investigated, including Petri nets, the π-calculus, and Concurrent ML. For many of these applications, the intuitionistic nature of the framework is essential. Second, this logic is almost a worst-case scenario for theorem proving, combining the difficulties of intuitionistic logic, and modal, linear, and first-order connectives. However, nothing specifically in our approach prevents it from being applied just as readily to classical linear logic, which might even be simpler because of the single-sided formulation and a fuller symmetry in the connectives.

Our variant of the inverse method combines the forward-reasoning techniques of Mints and Tammet with the (backward) proof-theoretic analyses by Andreoli and others by incorporating focusing into the inverse method. This allows the inverse method to proceed in bigger steps than single inferences, generating fewer sequents to be kept and tested for subsumption. In addition, we develop a new approach to resource management for reasoning in the forward direction and show how it extends to the first-order case. We further treat the delicate interactions of resource management with contraction and subsumption, two critical operations in a forward inference engine. Finally, we present some experimental results which demonstrate a significant speedup over Tammet's prover and help us quantify the effects of several internal optimizations.

Most closely related to the results reported in this paper is a recent submission [12] which presents focusing for the inverse method in intuitionistic *propositional* linear logic, but does not discuss many of the necessary implementation choices. We therefore concentrate here on the first-order and implementation aspects of the prover and only briefly sketch focusing. The propositional prover from [12] and the first-order prover here are separate implementations, since the propositional case—though also undecidable—affords a number of optimizations that do not directly apply in the first-order case.

Among related developments are recent implementations of theorem provers for the logic of bunched implications [13,14]. BI-logic and linear logic have a common core, so some techniques may be transferable, something we plan to consider in future work. Méry's prover [13] uses labeled tableaux in a goal-directed manner and is therefore quite different from ours. Donnelly et al. [14] use the inverse method, but the essential difficulty in its design concerns a particular interaction between weakening and contraction that is germane to bunched implication, but foreign to linear logic. Moreover, both provers are quite preliminary at this stage and do not incorporate techniques such as subsumption or indexing.

The remainder of the paper is organized as follows. In sect. 2 we give the briefest sketch of a cut-free sequent calculus for intuitionistic linear logic and its critical subformula property. In sect. 3 we present the ground inverse method underlying our prover and show how we solve the resource management problem. In sect. 4 we show how to lift the results and operations to the case of sequents with free variables which are necessary for a first-order prover. In sections 5 and 6 we present the main points in the design of an efficient implementation and some experimental results. In sect. 7 we conclude with remarks regarding the scope of our methods and future work.

2 Backward Sequent Calculus and the Subformula Property

We use a backward cut-free sequent calculus for propositions constructed out of the linear connectives $\{\otimes, \mathbf{1}, \multimap, \&, \top, !, \forall, \exists\}$. To simplify the presentation slightly we leave out $\oplus$ and $\mathbf{0}$, though the implementation supports them and some of the experiments in Sec. 6 use them. Propositions are written using uppercase letters A, B, C, with P standing for atomic propositions. Atomic propositions contain terms t, which can be term variables x, parameters a, b, or function applications $f(t_1, \ldots, t_n)$. As usual, term constants are treated as nullary functions. The sequent calculus is a standard fragment of JILL [15], containing dyadic two-sided sequents of the form $\Gamma ; \Delta \Longrightarrow C$: the zone Γ contains the unrestricted hypotheses, and Δ contains the linear hypotheses. Both Γ and Δ are unordered. For the rules of this calculus we refer the reader to [15, page 14]; the missing quantifier rules are below:

$$\frac{\Gamma ; \Delta \Longrightarrow [a/x]A}{\Gamma ; \Delta \Longrightarrow \forall x.\, A}\ \forall R^a \quad \frac{\Gamma ; \Delta, [t/x]A \Longrightarrow C}{\Gamma ; \Delta, \forall x.\, A \Longrightarrow C}\ \forall L \quad \frac{\Gamma ; \Delta \Longrightarrow [t/x]A}{\Gamma ; \Delta \Longrightarrow \exists x.\, A}\ \exists R \quad \frac{\Gamma ; \Delta, [a/x]A \Longrightarrow C}{\Gamma ; \Delta, \exists x.\, A \Longrightarrow C}\ \exists L^a$$

The superscript in $\forall R$ and $\exists L$ indicates the proviso that a may not occur in the conclusion. Also in [15] are the standard structural properties for the unrestricted hypotheses and admissibility of cut, which extend naturally to the first-order setting.

Definition 1 (subformulas). *A* decorated formula *is a tuple* $\langle A, s, w \rangle$ *where* A *is a proposition,* s *is a* sign *(+ or −) and* w *is a* weight *(h or l). The* subformula relation $\leq$ *is the smallest reflexive and transitive relation between decorated subformulas to satisfy the following conditions.*

$$\langle A, s, h \rangle \leq \langle !\, A, s, w \rangle \quad \langle A, \bar{s}, l \rangle \leq \langle A \multimap B, s, w \rangle \quad \langle B, s, l \rangle \leq \langle A \multimap B, s, w \rangle$$

$$\left.\begin{array}{ll}\langle [a/x]A, +, l \rangle \leq \langle \forall x.\, A, +, w \rangle & \langle [a/x]A, -, l \rangle \leq \langle \exists x.\, A, -, w \rangle \\[4pt] \langle [t/x]A, +, l \rangle \leq \langle \exists x.\, A, +, w \rangle & \langle [t/x]A, -, l \rangle \leq \langle \forall x.\, A, -, w \rangle\end{array}\right\} \cdots \begin{array}{l} \text{a } \textit{any parameter} \\ \text{t } \textit{any term}\end{array}$$

$$\langle A_i, s, l \rangle \leq \langle A_1 \star A_2, s, w \rangle \qquad \cdots \star \in \{\otimes, \&\}, i \in \{1, 2\}$$

where $\bar{s}$ *is the opposite of* s*. The notation* $*$ *can stand for either h or l, as necessary. Decorations and the subformula relation are lifted to (multi)sets in the obvious way.*

We also need the notion of free subformula *which is precisely as above, except that we use* A *instead of* $[t/x]A$ *for subformulas of positive existentials and negative universals.*

Property 2 (strong subformula property). *In any sequent* $\Gamma' ; \Delta' \Longrightarrow C'$ *used to prove* $\Gamma ; \Delta \Longrightarrow C$: $\langle \Gamma', -, h \rangle \cup \langle \Delta', -, * \rangle \cup \{\langle C', +, * \rangle\} \leq \langle \Gamma, -, h \rangle \cup \langle \Delta, -, l \rangle \cup \{\langle C, +, l \rangle\}$.

For the rest of the paper, all rules are to be understood as being restricted to decorated subformulas of the goal sequent. A right rule is only applicable if the principal formula A is a positive decorated subformula of the goal sequent (i.e., $\langle A, +, * \rangle \leq$ goal);

similarly, a left rule only applies to A if $\langle A, -, * \rangle \leq$ goal. Of the judgmental rules, init is restricted to atomic subformulas that are *both* positive and negative decorated subformulas, and the copy rule is restricted to cases where the copied formula A is a heavy negative decorated subformula, i.e., $\langle A, -, h \rangle \leq$ goal.

3 Forward Sequent Calculus and Resource Management

Following the "recipe" for the inverse method outlined in [16], we first present a forward sequent calculus in a ground form (containing no term variables), and then lift it to free subformulas containing term and parameter variables and explicit unification.

As mentioned before, backward proof-search in linear logic suffers from certain kinds of non-determinism peculiar to the nature of linear hypotheses. A binary multiplicative rule like $\otimes R$ as indicated

$$\frac{\Gamma ; \Delta_1 \Longrightarrow A \quad \Gamma ; \Delta_2 \Longrightarrow B}{\Gamma ; \Delta_1, \Delta_2 \Longrightarrow A \otimes B} \otimes R$$

is actually deceptive as it hides the fact that the split Δ_1, Δ_2 of the linear zone is not structurally obvious. Linearity forces the division to be exact, but there are an exponential number of such divisions, which is an unreasonable branching factor for proof-search. This form of resource non-determinism is typical in backward search, but is of course entirely absent in a forward reading of multiplicative rules where the parts Δ_1 and Δ_2 are inputs. The nature of resource non-determinism in forward and backward search is therefore significantly different.

To distinguish forward from backward sequents, we shall use a single arrow ($\longrightarrow$), but keep the names of the rules the same. In the forward direction, the primary context management issue concerns rules where the conclusion cannot be simply assembled from the premises. The backward $\top R$ rule has an arbitrary linear context Δ,

$$\frac{}{\Gamma ; P \Longrightarrow P} \text{ init}$$

$$\frac{}{\Gamma ; \Delta \Longrightarrow \top} \top R$$

and the unrestricted context Γ is also unknown in several rules such as init and $\top R$. For the unrestricted zone, this problem is solved in the usual (non-linear) inverse method by collecting only the needed unrestricted assumptions and remembering that they can be weakened if needed [16]. We adapt the solution to the linear zone, which may either be precisely determined (as in the case for initial sequents) or subject to weakening (as in the case for $\top R$). We therefore differentiate sequents whose linear context can be weakened and those who can not.

Definition 3 (forward sequents). *A forward sequent is of the form* $\Gamma ; \Delta \longrightarrow^w C$, *with w a Boolean (0 or 1) called the* weak-flag. *The correspondence between forward and backward sequents is governed by the following conditions:*

$$\Gamma ; \Delta \longrightarrow^0 C \quad \text{corresponds to} \quad \Gamma' ; \Delta \Longrightarrow C \quad \text{if } \Gamma \subseteq \Gamma'$$

$$\Gamma ; \Delta \longrightarrow^1 C \quad \text{corresponds to} \quad \Gamma' ; \Delta' \Longrightarrow C \quad \text{if } \Gamma \subseteq \Gamma' \text{ and } \Delta \subseteq \Delta'$$

Sequents with $w = 1$ are called weakly linear *or simply* weak, *and those with $w = 0$ are* strongly linear *or* strong.

It is easy to see that weak sequents model affine logic, which is familiar from embeddings into linear logic that translate affine implications $A \rightarrow B$ as $A \multimap (B \otimes \top)$. Initial sequents are always strong, since their linear context cannot be weakened. On the other

judgmental rules

$$\frac{}{\cdot\,;P \longrightarrow^0 P}\;\text{init} \qquad \frac{\Gamma\,;\Delta,A \longrightarrow^w C}{\Gamma,A\,;\Delta \longrightarrow^w C}\;\text{copy} \qquad \frac{\Gamma,A,A\,;\Delta \longrightarrow^w C}{\Gamma,A\,;\Delta \longrightarrow^w C}\;\text{contr}$$

multiplicative connectives

$$\frac{\Gamma\,;\Delta \longrightarrow^w A \quad \Gamma'\,;\Delta' \longrightarrow^{w'} B}{\Gamma,\Gamma'\,;\Delta,\Delta' \longrightarrow^{w\vee w'} A\otimes B}\;\otimes R \qquad \frac{\Gamma\,;\Delta,A,B \longrightarrow^w C}{\Gamma\,;\Delta,A\otimes B \longrightarrow^w C}\;\otimes L \qquad \frac{\Gamma\,;\Delta,A_i \longrightarrow^1 C \quad (A_j \notin \Delta)}{\Gamma\,;\Delta,A_1\otimes A_2 \longrightarrow^1 C}\;\otimes L_i$$
$$(i,j)\in\{(1,2),(2,1)\}$$

$$\frac{}{\cdot\,;\cdot \longrightarrow^0 1}\;1R \qquad \frac{\Gamma\,;\Delta \longrightarrow^0 C}{\Gamma\,;\Delta,1 \longrightarrow^0 C}\;1L \qquad \frac{\Gamma\,;\Delta,A \longrightarrow^w B}{\Gamma\,;\Delta \longrightarrow^w A\multimap B}\;\multimap R \qquad \frac{\Gamma\,;\Delta \longrightarrow^1 B \quad (A\notin\Delta)}{\Gamma\,;\Delta \longrightarrow^1 A\multimap B}\;\multimap R'$$

$$\frac{\Gamma\,;\Delta,B \longrightarrow^w C \quad \Gamma'\,;\Delta' \longrightarrow^{w'} A \quad (w=0\vee B\notin\Delta')}{\Gamma,\Gamma'\,;\Delta,\Delta',A\multimap B \longrightarrow^{w\vee w'} C}\;\multimap L$$

additive connectives

$$\frac{\begin{array}{c}\Gamma\,;\Delta_1 \longrightarrow^{w_1} A \\ \Gamma'\,;\Delta_2 \longrightarrow^{w_2} B \\ (\Delta_1/w_1 + \Delta_2/w_2 \rightsquigarrow \Delta)\end{array}}{\Gamma,\Gamma'\,;\Delta \longrightarrow^{w_1\wedge w_2} A\,\&\,B}\;\&R \qquad \frac{}{\cdot\,;\cdot \longrightarrow^1 \top}\;\top R \qquad \frac{\Gamma\,;\Delta,A_i \longrightarrow^w C}{\Gamma\,;\Delta,A_1\,\&\,A_2 \longrightarrow^w C}\;\&L_i$$
$$i\in\{1,2\}$$

exponentials

$$\frac{\Gamma\,;\cdot \longrightarrow^w A}{\Gamma\,;\cdot \longrightarrow^0 !A}\;!R \qquad \frac{\Gamma,A\,;\Delta \longrightarrow^w C}{\Gamma\,;\Delta,!A \longrightarrow^w C}\;!L \qquad \frac{\Gamma\,;\Delta \longrightarrow^0 C \quad (A\notin\Gamma)}{\Gamma\,;\Delta,!A \longrightarrow^0 C}\;!L'$$

quantifiers

$$\frac{\Gamma\,;\Delta \longrightarrow^w [a/x]A}{\Gamma\,;\Delta \longrightarrow^w \forall x.\,A}\;\forall R^a \qquad \frac{\Gamma\,;\Delta,[t/x]A \longrightarrow^w C}{\Gamma\,;\Delta,\forall x.\,A \longrightarrow^w C}\;\forall L$$

$$\frac{\Gamma\,;\Delta \longrightarrow^w [t/x]A}{\Gamma\,;\Delta \longrightarrow^w \exists x.\,A}\;\exists R \qquad \frac{\Gamma\,;\Delta,[a/x]A \longrightarrow^w C}{\Gamma\,;\Delta,\exists x.\,A \longrightarrow^w C}\;\exists L^a$$

Fig. 1. forward linear sequent calculus

hand, $\top R$ always produces a weak sequent. The collection of inference rules for the forward calculus is in Fig. 1.

For binary rules, the unrestricted zones are simply juxtaposed. We can achieve the effect of taking their union by applying the explicit contraction rule (which is absent, but admissible in the backward calculus). The

$$\frac{\Gamma\,;\Delta \Longrightarrow A \quad \Gamma\,;\Delta \Longrightarrow B}{\Gamma\,;\Delta \Longrightarrow A\,\&\,B}\;\&R$$

situation is not as simple for the linear zone. As shown above, in the backward direction the same linear zone is copied into both premises of the $\&R$ rule: This rule is easily adapted to the forward direction when both premises are strong:

$$\frac{\Gamma ; \varDelta \longrightarrow^0 A \quad \Gamma' ; \varDelta' \longrightarrow^0 B \quad (\varDelta = \varDelta')}{\Gamma, \Gamma' ; \varDelta \longrightarrow^0 A \& B}$$

If one premiss is weak and the other strong, the weak zone must be a subset of the strong zone:

$$\frac{\Gamma ; \varDelta \longrightarrow^0 A \quad \Gamma' ; \varDelta' \longrightarrow^1 B \quad (\varDelta' \subseteq \varDelta)}{\Gamma, \Gamma' ; \varDelta \longrightarrow^0 A \& B}$$

If both premisses are weak, then the conclusion is also weak, but what resources are present in the conclusion? In the ground case, we can simply take the maximal multiplicity for each proposition on the two premisses. To see that this is sound, simply apply weakening to add the missing copies, equalizing the linear contexts in the premisses. It is also complete because the maximum represents the least upper bound. In the free variable calculus this analysis breaks down, because the two propositions in the linear contexts in weak premisses may also be equalized by substitution. In preparation, we therefore introduce a non-deterministic *additive contraction* judgment which is used in the &R rule to generate multiple valid merges of the linear contexts of the premisses.

Definition 4. *The* additive contraction judgement *is of the form* $\varDelta/w + \varDelta'/w' \rightsquigarrow \varDelta''$ *where* $\varDelta$, $\varDelta'$ *and* $\varDelta''$ *are linear contexts, and* w *and* w' *are weak-flags.* $\varDelta$, $\varDelta'$, w, *and* w' *are inputs, and* $\varDelta''$ *is the output. The rules are as follows:*

$$\overline{\cdot/0 + \cdot/0 \rightsquigarrow \cdot} \quad \overline{\cdot/1 + \varDelta/0 \rightsquigarrow \varDelta} \quad \overline{\varDelta/0 + \cdot/1 \rightsquigarrow \varDelta} \quad \overline{\varDelta/1 + \varDelta'/1 \rightsquigarrow \varDelta, \varDelta'}$$

$$\frac{\varDelta/w + \varDelta'/w' \rightsquigarrow \varDelta''}{\varDelta, A/w + \varDelta', A/w' \rightsquigarrow \varDelta'', A}$$

Note that $\rightsquigarrow$ *is non-deterministic because the fourth and fifth rule overlap. Note further that* $\varDelta/1 + \varDelta'/0 \rightsquigarrow \varDelta'$ *iff* $\varDelta \subseteq \varDelta'$, *and for any* $\varDelta''$ *with* $\varDelta \cup \varDelta' \subseteq \varDelta'' \subseteq \varDelta, \varDelta'$, *the judgment* $\varDelta/1 + \varDelta'/1 \rightsquigarrow \varDelta''$ *is derivable.*

The conclusion of a binary multiplicative rule is weak if either of the premisses is weak; thus, the weak-flag of the conclusion is a Boolean-or of those of the premisses. Most unary rules are oblivious to the weakening decoration, which simply survives from the premiss to the conclusion. The exception is ! R, for which it is unsound to have a weak conclusion; there is no derivation of $\cdot ; \top \implies ! \top$, for example.

Left rules with weak premisses require some attention. It is tempting to write the "weak" $\otimes L$ rules as:

$$\frac{\Gamma ; \varDelta, A \longrightarrow^1 C}{\Gamma ; \varDelta, A \otimes B \longrightarrow^1 C} \otimes L_1 \qquad \frac{\Gamma ; \varDelta, B \longrightarrow^1 C}{\Gamma ; \varDelta, A \otimes B \longrightarrow^1 C} \otimes L_2.$$

However, these rules would admit redundant inferences such as the following:

$$\frac{\Gamma ; \varDelta, A, B \longrightarrow^1 C}{\Gamma ; \varDelta, A, A \otimes B \longrightarrow^1 C} \otimes L_2.$$

We might as well have consumed both A and B to form the conclusion, and obtained a stronger result. The sensible strategy is: when A and B are both present, they must *both* be consumed. Otherwise, only apply the rule when one operand is present in a weak sequent. A similar observation can be made about all such rules: there is one weakness-agnostic form, and some possible refined forms to account for weak sequents.

Theorem 5 (soundness).

1. *If $\Gamma ; \Delta \longrightarrow^0 C$, then $\Gamma ; \Delta \Longrightarrow C$.*
2. *If $\Gamma ; \Delta \longrightarrow^1 C$, then $\Gamma ; \Delta' \Longrightarrow C$ for any $\Delta' \supseteq \Delta$.*

Proof (sketch). Structural induction on the forward derivation $\mathcal{F} :: \Gamma ; \Delta \longrightarrow^w C$. The induction hypothesis is applicable for smaller derivations modulo α-renamings of parameters.   □

For the completeness theorem we note that the forward calculus infers a possibly stronger form of the goal sequent.

Theorem 6 (completeness). *If $\Gamma ; \Delta \Longrightarrow C$, then for some $\Gamma' \subseteq \Gamma$:*

1. *either $\Gamma' ; \Delta \longrightarrow^0 C$;*
2. *or $\Gamma' ; \Delta' \longrightarrow^1 C$ for some $\Delta' \subseteq \Delta$*

Proof (sketch). Structural induction on the backward derivation $\mathcal{D} :: \Gamma ; \Delta \Longrightarrow C$.   □

4 Lifting to Free Variable Sequents

The calculus of the previous section uses only ground initial sequents, which is impossible for an implementation of the forward calculus. Continuing with the "recipe" from [16], in this section we present a lifted version of the calculus with explicit unification. We begin, as usual, by fixing a goal sequent $\Gamma_g ; \Delta_g \longrightarrow^w C_g$ and considering only the free subformulas of this goal. In the presentation, the quantified propositions are silently α-renamed as necessary. In this calculus, every proposition on the left and right is accompanied by a substitution for some of its parameters or free term variables. These substitutions are built according the following grammar:

(substitutions)	$\sigma ::= \epsilon$	(identity)
	$\mid \sigma, a_1/a_2$	(param-subst)
	$\mid \sigma, t/x$	(term-subst)

A minor novel aspect of our formulation is that we distinguish parameters (which can be substituted only for each other) and variables (for which we can substitute arbitrary terms, including parameters). The distinction arises from the notion of subformula, since positive universal and negative existential formulas can only ever be instantiated with parameters in a cut-free backward sequent derivation. This sharpening sometimes removes unreachable initial sequents from consideration. Fortunately, the standard notion of most general unifier (written $\mathrm{mgu}(\sigma, \sigma')$) carries over in a straightforward way to this slightly more general setting. We make the customary assumption that substitutions are idempotent. We write $A[\sigma]$ (resp. $t[\sigma]$) for the application of the substitution σ to the free subformula A (resp. term t). Sequents in the free calculus contain formula/substitution pairs, written $A \cdot \sigma$. The composition of σ and ξ, written $\sigma\xi$, has the property $A[\sigma\xi] = (A[\sigma])[\xi]$. The composition of θ with every substitution in a zone Γ or Δ (now containing formula/substitution pairs) is written $\Gamma\theta$ or $\Delta\theta$.

Figure 2 contains the rules of this calculus; we use a double-headed arrow ($\longrightarrow\!\!\!\!\rightarrow$) to distinguish it from the ground forward calculus. The definition of additive contraction needs to be lifted to free subformulas also.

judgmental rules

$$\frac{\theta = \mathrm{mgu}(P[\rho], P')}{\cdot\,;\,P \cdot \rho\theta \longrightarrow^0 P' \cdot \theta}\ \mathrm{init} \qquad \frac{\Gamma\,;\,\Delta, A \cdot \sigma \longrightarrow^w C \cdot \xi}{\Gamma, A \cdot \sigma\,;\,\Delta \longrightarrow^w C \cdot \xi}\ \mathrm{copy}$$

$$\frac{\Gamma, A \cdot \sigma, A \cdot \sigma'\,;\,\Delta \longrightarrow^w C \cdot \xi}{\Gamma\theta, A \cdot \sigma\theta\,;\,\Delta\theta \longrightarrow^w C \cdot \xi\theta}\ \mathrm{contr} \qquad \frac{\Gamma\,;\,\Delta \longrightarrow^w C \cdot \xi}{\Gamma\rho\,;\,\Delta\rho \longrightarrow^w C \cdot \xi\rho}\ \mathrm{ren}$$

multiplicative connectives

$$\frac{\Gamma\,;\,\Delta \longrightarrow^w A \cdot \sigma \quad \Gamma'\,;\,\Delta' \longrightarrow^{w'} B \cdot \sigma'}{\Gamma\theta, \Gamma'\theta\,;\,\Delta\theta, \Delta'\theta \longrightarrow^{w \vee w'} (A \otimes B) \cdot \sigma\theta}\ \otimes R$$

$$\frac{\Gamma\,;\,\Delta, A \cdot \sigma, B \cdot \sigma' \longrightarrow^w C \cdot \xi}{\Gamma\theta\,;\,\Delta\theta, (A \otimes B) \cdot \sigma\theta \longrightarrow^w C \cdot \xi\theta}\ \otimes L \qquad \frac{\Gamma\,;\,\Delta, A_i \cdot \sigma \longrightarrow^1 C \cdot \xi \quad (\forall \rho.\, A_j \cdot \sigma\rho \notin \Delta)}{\Gamma\,;\,\Delta, (A_1 \otimes A_2) \cdot \sigma \longrightarrow^1 C \cdot \xi}\ \otimes L_i$$
$$(i, j) \in \{(1,2), (2,1)\}$$

$$\frac{}{\cdot\,;\,\cdot \longrightarrow^0 1 \cdot \epsilon}\ 1R \qquad \frac{\Gamma\,;\,\Delta \longrightarrow^0 C \cdot \xi}{\Gamma\,;\,\Delta, 1 \cdot \epsilon \longrightarrow^0 C \cdot \xi}\ 1L$$

$$\frac{\Gamma\,;\,\Delta, A \cdot \sigma \longrightarrow^w B \cdot \sigma'}{\Gamma\theta\,;\,\Delta\theta \longrightarrow^w (A \multimap B) \cdot \sigma\theta}\ \multimap R \qquad \frac{\Gamma\,;\,\Delta \longrightarrow^1 B \cdot \sigma \quad (\forall \rho.\, A \cdot \sigma\rho \notin \Delta)}{\Gamma\,;\,\Delta \longrightarrow^1 (A \multimap B) \cdot \sigma}\ \multimap R'$$

$$\frac{\begin{array}{c}\Gamma\,;\,\Delta, B \cdot \sigma \longrightarrow^w C \cdot \xi \\ \Gamma'\,;\,\Delta' \longrightarrow^{w'} A \cdot \sigma' \quad (w = 0 \vee \forall \rho.\, B \cdot \sigma\theta\rho \notin \Delta')\end{array}}{\Gamma\theta, \Gamma'\theta\,;\,\Delta\theta, \Delta'\theta, (A \multimap B) \cdot \sigma\theta \longrightarrow^{w \vee w'} C \cdot \xi\theta}\ \multimap L$$

additive connectives

$$\frac{\begin{array}{c}\Gamma\,;\,\Delta_1 \longrightarrow^{w_1} A \cdot \sigma \\ \Gamma\,;\,\Delta_2 \longrightarrow^{w_2} B \cdot \sigma' \quad (\Delta_1\theta/w_1 + \Delta_2\theta/w_2 \rightsquigarrow \langle \Delta\,;\,\xi \rangle)\end{array}}{\Gamma\theta\xi, \Gamma'\theta\xi\,;\,\Delta \longrightarrow^{w_1 \wedge w_2} (A \& B) \cdot \sigma\theta\xi}\ \& R \qquad \frac{}{\cdot\,;\,\cdot \longrightarrow^1 \top \cdot \epsilon}\ \top R$$

$$\frac{\Gamma\,;\,\Delta, A \cdot \sigma \longrightarrow^w C \cdot \xi}{\Gamma\,;\,\Delta, (A_1 \& A_2) \cdot \sigma \longrightarrow^w C \cdot \xi}\ \& L_i \qquad\qquad i \in \{1, 2\}$$

exponentials

$$\frac{\Gamma\,;\,\cdot \longrightarrow^w A \cdot \sigma}{\Gamma\,;\,\cdot \longrightarrow^0 \,!A \cdot \sigma}\ !R \quad \frac{\Gamma, A \cdot \sigma\,;\,\Delta \longrightarrow^w C \cdot \xi}{\Gamma\,;\,\Delta, !A \cdot \sigma \longrightarrow^w C \cdot \xi}\ !L \quad \frac{\Gamma\,;\,\Delta \longrightarrow^0 C \cdot \xi \quad (\forall \rho.\, A \cdot \rho \notin \Gamma)}{\Gamma\,;\,\Delta, !A \cdot \epsilon \longrightarrow^0 C \cdot \xi}\ !L'$$

quantifiers

$$\frac{\Gamma\,;\,\Delta \longrightarrow^w [a/x]A \cdot (\sigma, b/a)}{\Gamma\,;\,\Delta \longrightarrow^w \forall x.\, A \cdot \sigma}\ \forall R^{\mathrm{b}} \qquad \frac{\Gamma\,;\,\Delta, A \cdot (\sigma, t/x) \longrightarrow^w C \cdot \xi}{\Gamma\,;\,\Delta, \forall x.\, A \cdot \sigma \longrightarrow^w C \cdot \xi}\ \forall L$$

$$\frac{\Gamma\,;\,\Delta \longrightarrow^w A \cdot (\sigma, t/x)}{\Gamma\,;\,\Delta \longrightarrow^w \exists x.\, A \cdot \sigma}\ \exists R \qquad \frac{\Gamma\,;\,\Delta, [a/x]A \cdot (\sigma, b/a) \longrightarrow^w C \cdot \xi}{\Gamma\,;\,\Delta, \exists x.\, A \cdot \sigma \longrightarrow^w C \cdot \xi}\ \exists L^{\mathrm{b}}$$

Note: $\theta = \mathrm{mgu}(\sigma, \sigma')$; ρ is a (fresh) renaming substitution; and premises are variable-disjoint.

Fig. 2. Forward sequent calculus with free subformulas

Definition 7 (lifted additive contraction). *The* lifted additive contraction judgment, *written $\Delta_1/w_1 + \Delta_2/w_2 \rightsquigarrow \langle \Delta ; \xi \rangle$, takes as input the zones Δ_1 and Δ_2, together with their weak flags w_1 and w_2, and produces a contracted zone Δ and its corresponding substitution ξ. The rules for this judgment are as follows.*

$$\overline{\cdot/0 + \cdot/0 \rightsquigarrow \langle \cdot ; \epsilon \rangle} \quad \overline{\cdot/1 + \Delta/0 \rightsquigarrow \langle \Delta ; \epsilon \rangle} \quad \overline{\Delta/0 + \cdot/1 \rightsquigarrow \langle \Delta ; \epsilon \rangle} \quad \overline{\Delta/1 + \Delta'/1 \rightsquigarrow \langle \Delta, \Delta' ; \epsilon \rangle}$$

$$\frac{\theta = \mathrm{mgu}(\sigma, \sigma') \quad \Delta\theta/w + \Delta'\theta/w' \rightsquigarrow \langle \Delta'' ; \xi \rangle}{\Delta, A \cdot \sigma/w + \Delta', A \cdot \sigma'/w' \rightsquigarrow \langle \Delta'', A \cdot \sigma\theta\xi ; \theta\xi \rangle}$$

Lemma 8 (lifted additive contraction).

1. If $\Delta/0 + \Delta'/0 \rightsquigarrow \langle \Delta'' ; \xi \rangle$, then $\Delta\xi = \Delta'\xi = \Delta''$.
2. If $\Delta/0 + \Delta'/1 \rightsquigarrow \langle \Delta'' ; \xi \rangle$, then $\Delta'\xi \subseteq \Delta\xi = \Delta''$.
3. If $\Delta/1 + \Delta'/0 \rightsquigarrow \langle \Delta'' ; \xi \rangle$, then $\Delta\xi \subseteq \Delta'\xi = \Delta''$.
4. If $\Delta/1 + \Delta'/1 \rightsquigarrow \langle \Delta'' ; \xi \rangle$, then $\Delta\xi \subseteq \Delta''$ and $\Delta'\xi \subseteq \Delta''$.   □

Definition 9. *A substitution σ is a* grounding substitution *if for every term-variable $x \in \mathrm{dom}(\sigma)$, the term $x[\sigma]$ contains no term variables.*

Theorem 10 (soundness). *If $\Gamma ; \Delta \longrightarrow^w C \cdot \xi$ and λ is a grounding substitution for Γ, Δ and $C[\xi]$, then $\Gamma[\lambda] ; \Delta[\lambda] \longrightarrow^w C[\xi\lambda]$.*

Proof (sketch). Straightforward induction on the structure of $\mathcal{F} :: \Gamma ; \Delta \longrightarrow^w C \cdot \xi$, using lem. 8 and noting that any grounding unifier must be less general than the mgu.   □

Theorem 11 (completeness).
Suppose $A_1[\sigma_1], A_2[\sigma_2], \ldots ; B_1[\tau_1], B_2[\tau_2] \ldots \longrightarrow^w C[\xi]$ where the A_i, B_j and C are free subformulas of the goal. Then there exist substitutions $\sigma'_1, \sigma'_2, \ldots, \tau'_1, \tau'_2, \ldots, \xi'$ and λ such that:

1. $A_1 \cdot \sigma'_1, A_2 \cdot \sigma'_2, \ldots ; B_1 \cdot \tau'_1, B_2 \cdot \tau'_2, \ldots \longrightarrow^w C \cdot \xi'$; and
2. $\sigma'_i\lambda = \sigma_i; \tau'_j\lambda = \tau_j;$ and $\xi'\lambda = \xi$.

Proof (sketch). Structural induction on the given ground derivation.   □

5   Design of the Implementation

Representation of Sequents and Contraction Linear hypotheses can occur more than once in the linear zone, so for each substitution we also store the *multiplicity* of that substitution; we write this as $A \cdot \sigma^n$ where n is the multiplicity of $A \cdot \sigma$. In the common case of a variable-free proposition, this not only makes the representation of sequents more efficient, but also greatly reduces the non-determinism involved in matching a hypothesis in a premiss. Contraction in the presence of multiplicities is not much different from before; the only change is that we can perform a number of contractions together as a unit.

$$\frac{k = \min(m, n) \quad \theta = \mathrm{mgu}(\sigma, \tau) \quad \Delta_1\theta, A \cdot \sigma\theta^{m-k}/w_1 + \Delta_2\theta, A \cdot \tau\theta^{n-k}/w_2 \rightsquigarrow \langle \Delta ; \xi \rangle}{\Delta_1, A \cdot \sigma^m/w_1 + \Delta_2, A \cdot \tau^n/w_2 \rightsquigarrow \langle \Delta, A \cdot \sigma\theta\xi^k ; \theta\xi \rangle}$$

(Note that $\Delta, A \cdot \sigma^0$ is understood as Δ.)

In the implementation we perform contractions eagerly, that is, after every rule application we calculate the possible contractions in the conclusion of the rule. This allows us to limit contractions to binary rules, and furthermore, consider only the contractions between propositions that originate in different premisses. This is complete because if two hypotheses were to be contractible in the same premiss, then we would already have generated the sequent corresponding to that contraction earlier.

The special case of contracting two weak zones, i.e., $\Delta_1/1 + \Delta_2/1$, can be greatly improved by *first* eagerly contracting propositions that have an invertible unifier. This is complete because a weak $\Delta, A \cdot \sigma, A \cdot \sigma\rho$ is subsumed by a weak $\Delta, A \cdot \sigma$.

Rule Generation. The subformula property gives us the core of the inverse method procedure. We start with all initial sequents of the form $\cdot\,; P \cdot \rho\theta \longrightarrow^0 P' \cdot \theta$, where P is a negative, and P' a positive free atomic subformula of the goal sequent, ρ renames them apart, and $\theta = \text{mgu}(P[\rho], P')$. Next, we name all free subformulas of the goal sequent with unique propositional labels. Then, we specialize all inference rules to these labels as principal formulas before starting the main search procedure.

Subsumption and Indexing. Our prover performs forward, but currently not backward subsumption. Subsumption has to account for linearity and the notion of weak sequent.

Definition 12 (free subsumption). *The* free subsumption relation $\leq$ *between free forward sequents is the smallest relation satisfying:*

$$\left.\begin{array}{l}\left(\Gamma; \Delta \longrightarrow^0 C \cdot \xi\right) \leq \left(\Gamma'; \Delta \longrightarrow^0 C \cdot \xi'\right) \\[2mm] \left(\Gamma; \Delta \longrightarrow^1 C \cdot \xi\right) \leq \left(\Gamma'; \Delta' \longrightarrow^w C \cdot \xi'\right)\end{array}\right\} \text{ for some } \theta \text{ such that } \Gamma\theta\subseteq\Gamma',\, \Delta\theta\subseteq\Delta',\, \text{and } \xi\theta=\xi'$$

The full subsumption check is far too expensive to perform always. Subsumption is usually implemented as a sequence of phases of increasing complexity; Tammet called them *hierarchical tests* [17]. These hierarchical tests are designed to fail as early as possible, as the overwhelming majority of subsumption queries are negative.

Definition 13 (hierarchical tests).
To check if $s = \Gamma; \Delta \longrightarrow^w C \cdot \xi$ subsumes $s' = \Gamma'; \Delta' \longrightarrow^{w'} C' \cdot \xi'$, the following tests are performed in order:

1. *if $w = 0$ and $w' = 1$ then FAIL;*
2. *if $\#\Delta > \#\Delta'$ or $\#\Gamma > \#\Gamma'$ then FAIL (where # count the number of elements);*
3. *respecting multiplicities, if a free subformula L occurs n times in Δ and m times in Δ' and $n > m$ then FAIL; similarly for Γ and Γ';*
4. *if there is no θ for which $C[\xi\theta] = C'[\xi']$, then FAIL;*
5. *if for some $A \cdot \sigma \in \Gamma$ there is no $A \cdot \sigma' \in \Gamma'$ for which $A[\sigma\theta] = A[\sigma']$ (for some θ), then FAIL; similarly for Δ and Δ';*
6. *otherwise attempt the full subsumption test $s \leq s'$.*

Tammet gives examples of other possible tests in [17], particularly tests that consider the depth of terms and statistics such as the number of constants, but we have not so far (for pragmatic reasons) considered them in the linear setting.

For the index we use a global forest of substitution trees [18]. Each inserted sequent is indexed into the substitution tree corresponding to the label of the principal literal, indexed by its corresponding substitution. The leaves of the substitution tree contain the sequents where the indexed formula was the principal formula. To check if a given sequent is subsumed, we look up every formula in the sequent in the index to obtain a collection of subsumption candidates, which are then tested for subsumption using the hierarchical tests above.

Focusing and Lazy Rule Application. Efficient indexing and subsumption algorithms, though important, are not as critical to the design of the prover as the use of derived big-step rules. The inference rules of Fig.2 take tiny steps and thereby produce too many sequents. In our implementation we use a version of focusing [19,2] tailored for forward reasoning to construct derived inference rules with many premises. The essential insight of focusing is that every proof can be converted to one that alternates between two phases–*active* and *focused*. Thinking in the backward direction, during the active phase we break down all connectives whose left or right rules are invertible. This phase has no essential non-determinism. This leads to a so-called *neutral sequent* where we have to choose a formula to focus on, which is then successively decomposed by chaining together non-invertible rules on this particular focus formula. It turns out that in the forward direction we only need to keep neutral sequents if we construct big-step forward rules by analyzing those *frontier propositions* which can occur in neutral sequents and which are also subformulas of the goal. Essentially we simulate a backward focusing phase followed by a backward active phase by inferences in the forward direction. This construction is detailed in [12] for the propositional fragment and can easily be extended to the first-order setting.

We implement a derived rule as a curried function from sequents (premises) to the conclusion of the rule. Each application of a rule to a sequent first tests if the sequent can match the corresponding premiss of the rule; if the match is successful, then the application produces a new partially instantiated rule, or if there are no remaining premisses then it produces a new sequent. The order of arguments of this curried function fixes a particular ordering of the premises of the rule; the search procedure is set up so that any ordering guarantees completeness.

We use a lazy variant of the OTTER loop as the main loop of the search procedure. We maintain two global sets of derived sequents:

- the *active* set containing sequents to be considered as premises of rules; and
- the *inactive* set (sometimes referred to as the *set of support*) that contains all facts that have not yet been transferred to the active set.

The inner loop of the search procedure repeats the following lazy activation step until either the goal sequent is subsumed (in which case the search is successful), or no further rules are applicable to the sequents in the active set and the inactive set is exhausted (in which case the search saturates).

Definition 14 (lazy activation). *To activate the sequent s, i.e., to transfer it from the inactive to the active set, the following steps are performed:*

1. *After renaming, s is inserted into the active set.*
2. *All available rules are applied to s. If these applications produce new rules, R, then the following two steps are performed in a loop until there are no additions to R.*
   (a) *For every sequent s' in the active set, every rule in R is applied to s', and*
   (b) *any new rules generated are added to R.*
3. *The collection of rules R is added to the set of rules.*
4. *All sequents generated during the above applications are tested for subsumption, and the un-subsumed sequents and all their associated contractions are added to the inactive set.*

A sequent is added to the inactive set if it is not globally subsumed by some other sequent derived earlier. In fact, if it is subsumed, then none of its contracts need to be computed. We use the following heuristic for the order of insertion of the contracts of a given sequent: if s is the result of a sequence of contractions from s', then s is considered for insertion in the inactive set before s'.

The initial inactive set and rules are produced uniformly by focusing on the frontier literals of the goal sequent. The collection of (partially applied) rules grows at run-time; this is different from usual implementations of the OTTER loop where the rules are fixed before-hand. On the other hand, rule application is much simpler when each rule is treated as a single-premiss rule (producing sequents or other rules, possibly nothing). Furthermore, the same rule is never applied more than once to any sequent, because a previously derived rule is applied only to newly activated sequents (which were not in the active set before). Thus, the lazy activation strategy implicitly memoizes earlier matches.

Globalization. The final unrestricted zone Γ_g is shared in all branches in a proof of $\Gamma_g ; \Delta_g \Longrightarrow C_g$. One thus thinks of Γ_g as part of the ambient state of the prover, instead of representing it explicitly as part of the current goal. Hence, there is never any need to explicitly record Γ_g or portions of it in the sequents themselves. This gives us the following global and local versions of the copy rule:

$$\frac{\Gamma ; \Delta, A \cdot \sigma \longrightarrow^w C \cdot \xi \quad (\exists \rho. \, A \cdot \sigma\rho \in \Gamma_g)}{\Gamma ; \Delta \longrightarrow^w C \cdot \xi} \; \text{delete} \quad \frac{\Gamma ; \Delta, A \cdot \sigma \longrightarrow^w C \cdot \xi \quad (\forall \rho. \, A \cdot \sigma\rho \notin \Gamma_g)}{\Gamma, A \cdot \sigma ; \Delta \longrightarrow^w C \cdot \xi} \; \text{copy}$$

6 Some Experimental Results

For our experiments we compared a few internal versions of the prover and two provers in the Gandalf family. For comparison purposes, we implemented a purely propositional version of our prover, which performs some additional optimizations that are not possible in the first-order case. The main differences are: contraction in the propositional case always produces exactly one sequent, as opposed to (potentially) exponentially many sequents in the first-order case; furthermore, subsumption is simply a matter of comparing the multiplicities, which is linear instead of quadratic. These properties greatly simplify rule generation and application.

The internal versions of the prover are named **L** followed by a selection of **P** (propositional), **F** (big-step rules using focusing) and **G** (globalization) as suffixes. The default prover is named **L** (first-order, small-step rules, no globalization); **LPF**, for example,

Table 1. Some experimental results

	Gr	Gt	LP	LPF	L	LF
basic	0.06 s	0.08 s	0.024 s	0.018 s	0.058 s	0.037 s
bw-prop	×	×	×	0.001 s	×	0.007 s
coins	0.63 s	×	3.196 s	0.001 s	8.452 s	0.001 s
affine1	$\times^3$	0.01 s	0.003 s	0.001 s	1.645 s	3.934 s
affine2	$\times^3$	×	≈ 12 m	1.205 s	≈ 34 m	4.992 s
qbf1	2.40 s	×	0.013 s	0.001 s	0.038 s	0.002 s
qbf2	×	×	0.037 s	0.001 s	0.512 s	0.060 s
qbf3	×	×	0.147 s	0.003 s	2.121 s	0.820 s

(a) propositional

	L	LF	LFG
bw-fo	×	0.460 s	0.036 s
urn	×	0.413 s	0.261 s
int-gir	×	1.414 s	1.410 s
int-foc	b11m	0.051 s	0.058 s

(b) first-order

is the purely propositional prover with focused big-step rules, but no globalization. In total there are six internal versions, though for space reasons we do not list the running times for every version. These provers are written in Standard ML and are available from the first author's website.[1]

For our experiments the internal versions were compiled using MLTon version 20041119 with the default optimization flags. All time measurements are wall-clock times measured on an unloaded computer with a 2.80GHz Pentium 4 processor with a 512KB L1 cache and 1GB of main memory. Time measurements of less than 0.01 seconds should be taken as unreliable; "×" means no solution was found within around 20 minutes.[2]

For external provers we concentrate on Tammet's Gandalf "nonclassical" distribution (version 0.2), compiled using a packaged version of the Hobbit Scheme compiler. This prover is limited to the propositional fragment of classical linear logic, but comes in two flavors: resolution (**Gr**) and tableau (**Gt**). Neither version incorporates focusing or globalization, and we did not attempt to bound the search for either prover. Another tableau prover for classical propositional multiplicative-exponential linear logic is Lin-TAP [20]; it cannot handle the majority of our examples using & and ⊤, and on the examples that are in the MELL fragment (bw-prop and coins), it does not terminate. For comparatively simpler problems [20, p.14], our prover **LPF** finishes in cumulative time of 0.003 seconds, while LinTAP's total time is 3.30 seconds. Finally, llprover [21] is a demonstration prover for classical first-order linear logic based on the sequent calculus, but it cannot solve any of the examples in table 1.

Purely Propositional Problems The comparisons of this section are restricted to the propositional fragment. They include: simple theorems in linear logic (basic), blocks-world problems for a fixed collection of blocks (bw-prop), a change machine encoding (coins), several affine logic problems encoded in linear logic (affine1 and affine2), and a few hard examples of quantified Boolean formulas compiled to multiplicative-additive linear logic (qbf1, qbf2 and qbf3, in order of increasing complexity) that implement the algorithm of [22]. The results are shown in table 1 (a).

---

[1] http://www.cs.cmu.edu/~kaustuv/

[2] As a check, we have run some of these failing cases for tens of hours, but they eventually exhaust the system memory or our patience.

[3] **Gr** appears to saturate incorrectly in these cases (fails to prove a true proposition), so we have omitted the running time.

It is evident that focusing greatly speeds up both the propositional and first-order cases. For the propositional case, the speedup from the focusing prover to the non-focusing one is between 1.33 (basic) and 597.5 (affine2); for the first-order case, the speedups range from 1.57 (basic) to 408.7 (affine2). Except for bw-prop, these examples were all within the realm of possibility for the small-step provers, though some of them like affine2 severely strain the provers. In the affine1 case the focusing prover **LF** appears to take longer than the small-step prover **L**; this is because this example contains an unprovable proposition for which the inverse method procedure fails to saturate. The test is run for 1000 iterations of the lazy OTTER loop. The focusing prover gets much further than the small-step prover in 1000 iterations, and the delay is due entirely to the fact that the sequents it works with, after even a few dozen iterations, are far more complex than the small-step prover generates in 1000 iterations.

Comparing to Gandalf, the small-step prover **L** is generally competitive with **Gr**: it is slower on some examples (coins), but succeeds on a wider range of problems. **Gt** was uniformly the slowest, taking a long time on even simple problems.

First-Order Problems   Our first-order problems include the following: a first-order blocks world planning example (bw-fo), Dijkstra's urn game (urn), simple intuitionistic first-order propositions encoded as linear propositions, using either Girard's translation (int-gir), or a focusing-aware translation (int-foc) outlined in [12].

The results are shown in table 1 (b). Again, it is fairly obvious that focusing is the dramatic winner, making some problems tractable, and being several orders of magnitude faster for the rest. Adding globalization also seems to have a significant effect here for the examples that are not constructed in an ad-hoc fashion (bw-fo and urn).

7   Conclusion

We have presented a theorem prover for first-order intuitionistic linear logic based on the inverse method which is already practical for a range of examples and significantly improves on prior, more restricted provers. The design is based on general principles that are applicable to both classical linear logic (which is simpler because it admits a one-sided sequent formulation with more symmetries) and affine logic (via weak sequents). Both of these can also be treated by uniform translations to intuitionistic linear logic [15], as can (ordinary) intuitionistic logic [12]. A generalization from a first-order logic to a type theory such as CLF [11] would seem to require mostly a proper treatment of linear higher-order unification constraints, but otherwise be relatively straightforward.

Our prover also leaves room for further high-level and low-level optimizations. In particular, we plan to investigate how to limit the multiplicities of linear hypotheses, either a priori or as a complete heuristic. Some manual experiments seem to indicate that this could have a significant impact on a certain class of problems. The implementation of the prover is *certifying*—it produces independently verifiable proof terms in a type-theory—but the algorithm used to extract these proofs currently lacks a formal presentation and a correctness proof.

References

1. Girard, J.Y.: Linear logic. Theoretical Computer Science **50** (1987) 1–102
2. Andreoli, J.M.: Logic programming with focusing proofs in linear logic. Journal of Logic and Computation **2** (1992) 297–347
3. Galmiche, D., Perrier, G.: Foundations of proof search strategies design in linear logic. In: Symposium on Logical Foundations of Computer Science, St. Petersburg, Russia, Springer-Verlag LNCS 813 (1994) 101–113
4. Galmiche, D.: Connection methods in linear logic and proof nets constructions. Theoretical Computer Science **232** (2000) 213–272
5. Harland, J., Pym, D.J.: Resource-distribution via boolean constraints. In McCune, W., ed.: Proceedings of CADE-14, Springer-Verlag LNAI 1249 (1997) 222–236
6. Cervesato, I., Hodas, J.S., Pfenning, F.: Efficient resource management for linear logic proof search. Theoretical Computer Science **232** (2000) 133–163
7. Pym, D.J., Harland, J.A.: The uniform proof-theoretic foundation of linear logic programming. Journal of Logic and Computation **4** (1994) 175–207
8. Hodas, J.S., Miller, D.: Logic programming in a fragment of intuitionistic linear logic. Information and Computation **110** (1994) 327–365
9. Mints, G.: Resolution calculus for the first order linear logic. Journal of Logic, Language and Information **2** (1993) 59–83
10. Tammet, T.: Proof strategies in linear logic. Journal of Automated Reasoning **12** (1994) 273–304
11. Cervesato, I., Pfenning, F., Walker, D., Watkins, K.: A concurrent logical framework I & II. Technical Report CMU-CS-02-101 and 102, Department of Computer Science, Carnegie Mellon University (2002) Revised May 2003.
12. Chaudhuri, K., Pfenning, F.: Focusing the inverse method for linear logic. In: Proceedings of the 14th Annual Conference on Computer Science Logic (CSL'05). (2005) To appear; an extended version available as Technical Report CMU-CS-05-106, Department of Computer Science, Carnegie Mellon University (2005).
13. Méry, D.: Preuves et Sémantiques dans des Logiques de Ressources. PhD thesis, Université Henri Poincaré, Nancy, France (2004)
14. Donnelly, K., Gibson, T., Krishnaswami, N., Magill, S., Park, S.: The inverse method for the logic of bunched implications. In Baader, F., Voronkov, A., eds.: Proceedings of the 11th International Conference on Logic for Programming, Artificial Intelligence, and Reasoning, Montevideo, Uruguay, Springer LNCS 3452 (2005) 466–480
15. Chang, B.Y.E., Chaudhuri, K., Pfenning, F.: A judgmental analysis of linear logic. Technical Report CMU-CS-03-131R, Carnegie Mellon University (2003)
16. Degtyarev, A., Voronkov, A.: The inverse method. In Robinson, J.A., Voronkov, A., eds.: Handbook of Automated Reasoning. MIT Press (2001) 179–272
17. Tammet, T.: Towards efficient subsumption. In: Proceedings of CADE-15. (1998) 427–441
18. Graf, P.: Term Indexing. Springer LNAI 1053 (1996)
19. Andreoli, J.M.: Focussing and proof construction. Annals of Pure and Applied Logic **107** (2001) 131–163
20. Mantel, H., Otten, J.: LinTAP: A tableau prover for linear logic. In Murray, A., ed.: International Conference TABLEAUX'99, New York, Springer-Verlag LNAI 1617 (1999) 217–231
21. Tamura, N.: Llprover (2005 (last checked)) At: http://bach.istc.kobe-u.ac.jp/llprover.
22. Lincoln, P., Mitchell, J.C., Scedrov, A., Shankar, N.: Decision problems for propositional linear logic. Annals of Pure and Applied Logic **56** (1992) 239–311

The CoRe Calculus

Serge Autexier

Saarland University & German Research Centre for Artificial Intelligence,
(DFKI GmbH), Saarbrücken, Germany
autexier@ags.uni-sb.de

Abstract.
We present the CoRe calculus for contextual reasoning which supports reasoning directly at the assertion level, where proof steps are justified in terms of applications of definitions, lemmas, theorems, or hypotheses (collectively called "assertions") and which is an established basis to generate proof presentations in natural language. The calculus comprises a uniform notion of a logical context of subformulas as well as replacement rules available in a logical context. Replacement rules operationalize assertion level proof steps and technically are generalized resolution and paramodulation rules, which in turn should suit the implementation of automatic reasoning procedures.

1 Introduction

The main application domains of computer-based theorem proving systems are mathematical assistants, mathematical teaching assistants, and hardware and software verification. In these domains, a human guidance of the proof procedures is indispensable, even for theorems that are simple by human standards. For instance the user must provide guidance information about how to explore the search space or specify intermediate lemmas. Therefore, communication between the user and the theorem proving system is crucial. The information provided by the theorem proving system about the proof must be intelligible to the user and the user must convey his/her intentions about how to continue the proof in a manner that is intelligible to the theorem proving system.

Exchanging the information in an intelligible manner is the bottleneck for the communication. A user like a mathematician or software engineer usually has a semantic representation of the problem domain and exploits it to approach and solve proof obligations. They usually have little or no knowledge about formal logic. State of the art automated theorem provers, however, only incorporate deep knowledge about the search space structure based on the syntax and calculus rules. Interactive theorem provers use tactics [11] to incorporate more high-level proof procedures, but these still stick to the syntax and the basic calculus rules. Proof planning [7] has been designed to overcome these limitations. However, in practice it also requires an understanding of the underlying calculus from the user and does not completely overcome the limitations imposed by the lack of abstraction imposed by the underlying calculus.

In this paper we present a calculus for *contextual reasoning* (CoRe) which aims at narrowing the gap between the user and the proof procedures. The

R. Nieuwenhuis (Ed.): CADE 2005, LNAI 3632, pp. 84–98, 2005.

key idea of the calculus is that proof construction proceeds by transformation of (parts of) a formula by applying definitions, lemmas, theorems as well as information contained in the formula without enforcing its decomposition. We have made a point of this idea in the CoRe calculus, where the logical contexts can be statically determined for any part of a formula. The information contained in a logical context is conditioned into *replacement rules*, which formalize the notion of *assertion level* rules.

The assertion level has been introduced by Xiarong Huang in [13] as an abstraction from the pure natural deduction calculus and it is the basis for the generation of proof presentations in natural language close to the style of proofs in mathematical textbooks. The idea is to subsume axioms, definitions, lemmas, and theorems as *assertions*, and the use of a single assertion in the proof search corresponds to a whole proof segment in the underlying calculus. Consider the example assertion taken from [13]: $\forall S_1, S_2 : Set. S_1 \subseteq S_2 \Leftrightarrow \forall x. x \in S_1 \Rightarrow x \in S_2$. This assertion allows us to derive (1) $a \in S_2'$ from $a \in S_1'$ and $S_1' \subseteq S_2'$; (2) $S_1' \not\subseteq S_2'$ from $a \in S_1'$ and $a \notin S_2'$; and (3) $\forall x. x \in S_1' \Rightarrow x \in S_2'$ from $S_1' \subseteq S_2'$.

This paper is organized as follows: In Sec. 2 we recapitulate the basic definitions of higher-order logic, uniform notation, and extensional expansion proofs for higher-order logic [15]. Sec. 3 describes the CoRe calculus and how uniform notation provides a uniform basis to define the logical context of subformulas and to determine replacement rules from the assertions available from the logical context. In Sec. 4 we present an example proof with the CoRe calculus, before addressing related work and concluding in Sec. 5.

2 Preliminaries

2.1 Higher-Order Logic

For the definition of higher-order logic, we use a simple higher-order type system $\mathcal{T}$ [4], composed of a base type ι for individuals, a type o for formulas, and where $\tau \rightarrow \tau'$ denotes the type of functions from τ to τ'. As usual, we assume that the functional type constructor $\rightarrow$ associates to the right.

We annotate constants f_τ and variables x_τ with types τ from $\mathcal{T}$ to indicate their type. A higher-order signature $\Sigma = (\mathcal{T}, \mathcal{F}, \mathcal{V})$ consists of types $\mathcal{T}$, constants $\mathcal{F}$ and variables $\mathcal{V}$, both typed over $\mathcal{T}$. The typed λ-calculus is standard and is defined over a given higher-order signature $\Sigma := (\mathcal{T}, \mathcal{F}, \mathcal{V})$.

Definition 1 (λ-Terms). *Let $\Sigma = (\mathcal{T}, \mathcal{F}, \mathcal{V})$ be a higher-order signature. Then the typed λ-terms $\mathcal{T}_{\Sigma,\mathcal{V}}$ over Σ and $\mathcal{V}$ are: (Var) for all $x_\tau \in \mathcal{V}$, $x \in \mathcal{T}_{\mathcal{F},\mathcal{V}}$ is a variable term of type τ; (Const) for all $c_\tau \in \mathcal{F}$, $c \in \mathcal{T}_{\mathcal{F},\mathcal{V}}$ is a constant term of type τ; (App) if $t : \tau, t' : \tau \rightarrow \tau' \in \mathcal{T}_{\mathcal{F},\mathcal{V}}$ are typed terms, then $(t' \, t) \in \mathcal{T}_{\mathcal{F},\mathcal{V}}$ is an application term of type τ'; (Abs) if $x_\tau \in \mathcal{V}$ and $t : \tau' \in \mathcal{T}_{\mathcal{F},\mathcal{V}}$, then $\lambda x_\tau . t \in \mathcal{T}_{\mathcal{F},\mathcal{V}}$ is an abstraction term of type $\tau \rightarrow \tau'$.*

Definition 2 (Substitutions). *Let $\Sigma = (\mathcal{T}, \mathcal{F}, \mathcal{V})$ be a higher-order signature. A substitution is a type preserving function[1] $\sigma : \mathcal{V} \rightarrow \mathcal{T}_{\mathcal{F},\mathcal{V}}$ that is the identity*

---

[1] i.e. for all variables x_τ, $\sigma(x)$ also has type τ.

function on $\mathcal{V}$ except for finitely many elements from $\mathcal{V}$. This allows for a finite representation of a substitution σ as $\{\sigma(x_1)/x_1, \ldots, \sigma(x_n)/x_n\}$ where $\sigma(y) = y$ if $\forall 1 \le i \le n, y \ne x_i$.

As usual we do not distinguish between a substitution and its homomorphic extension to terms. Given a substitution σ we denote the *domain of σ* by $dom(\sigma) := \{x \in \mathcal{V} \mid \sigma(x) \ne x\}$. Given two substitutions σ and ν, we say that ν is a *σ-refinement* if, and only if, there exists a substitution ρ, such that $\nu = \rho \circ \sigma$. We say that a substitution σ is *ground*, if, and only if, for all $x \in dom(\sigma)$, $\sigma(x)$ contains no free variables. Higher-order λ-terms usually come with certain reduction and expansion rules. We use the β reduction rule and the η expansion rule (see [4]), which give rise to the $\beta\eta$ long normal form, which is unique up to renaming of bound variables (α-equal). Throughout the rest of this paper we assume substitutions are idempotent[2] and all terms are in $\beta\eta$ long normal form.

For the semantics of higher-order logic we use the extensional general models from [12] by taking into account the corrections from [2]. It is based on the notion of *frames* that is a τ-indexed family $\{\overline{D}_\tau\}_{\tau \in \mathcal{T}}$ of nonempty domains, such that $\overline{D}_o = \{\top, \bot\}$ and $\overline{D}_{\tau_1 \rightarrow \tau_2}$ is a collection of functions mapping $\overline{D}_{\tau_1}$ into $\overline{D}_{\tau_2}$. Given a variable assignment ρ, a variable x_τ and an element $e \in \overline{D}_\tau$ we denote by $\rho[e/x]$ that assignment ρ' such that $\rho'(x_\tau) = e$ and $\rho'(y_{\tau'}) = \rho(y_{\tau'})$, if $y_{\tau'} \ne x_\tau$.

Definition 3 (Satisfiability & Validity). *A formula φ is satisfiable if, and only if, there is a model M and a variable assignment ρ such that $M^\rho(\varphi) = \top$. φ is true in a model M if, and only if, for all variable assignments ρ holds $M^\rho(\varphi) = \top$. φ is valid if, and only if, it is true in all models.*

2.2   Uniform Notation

The CORE calculus relies on an extension of extensional expansion proofs and makes use of the concept of polarities and uniform notation (cf. [18,9,17]). Polarities are assigned to formulas and subformulas and are either positive $(+)$ or negative $(-)$. Intuitively, positive polarity of a subformula indicates that it occurs in the succedent of a sequent in a sequent calculus proof and negative polarity is for formulas occurring in the antecedent of a sequent.

Formulas annotated with polarities are called *signed formulas*. Uniform notation assigns *uniform types* to signed formulas which encode their "behavior" in a sequent calculus proof: there are two propositional uniform types α and β, and two types γ and δ for quantification over object variables. A signed formula is of type α if the subformulas obtained by application of the respective sequent calculus decomposition rule on the formula both occur in the same sequent. Signed formulas are of type β, if the decomposition of the signed formula gives rise to a split of the sequent calculus proof and the obtained subformulas occur in different sequents. γ-type signed formulas indicate that the bound variable is freely instantiable, while δ-type signed formulas are those for which the *Eigenvariable*

---

[2] i.e. for all terms t holds $\sigma(\sigma(t)) = \sigma(t)$.

α	α_0	α_1
$(\varphi \vee \psi)^+$	φ^+	ψ^+
$(\varphi \Rightarrow \psi)^+$	φ^-	ψ^+
$(\varphi \wedge \psi)^-$	φ^-	ψ^-
$(\neg\varphi)^+$	φ^-	—
$(\neg\varphi)^-$	φ^+	—

β	β_0	β_1
$(\varphi \wedge \psi)^+$	φ^+	ψ^+
$(\varphi \vee \psi)^-$	φ^-	ψ^-
$(\varphi \Rightarrow \psi)^-$	φ^+	ψ^-

γ	$\gamma_0(c)$
$(\forall x.\varphi)^-$	$(\varphi[x/t])^-$
$(\exists x.\varphi)^+$	$(\varphi[x/t])^+$

δ	$\delta_0(c)$
$(\forall x.\varphi)^+$	$(\varphi[x/c])^+$
$(\exists x.\varphi)^-$	$(\varphi[x/c])^-$

ϵ	ϵ_0	ϵ_1
$(s \Leftrightarrow t)^-$	s	t
$(s = t)^-$	s	t

ζ	ζ_0	ζ_1
$(s \Leftrightarrow t)^+$	s	t
$(s = t)^+$	s	t

Fig. 1. Extended Uniform Notation

condition must hold. We call γ-*variable* (resp. δ-*variable*) variables bound on some γ-type signed formula (resp. δ-type). In Fig. 1 we give the list of signed formulas for each uniform type.

An important intuitive concept is equality and equivalence and we want to treat those as first-class citizens by supporting their use as rewrite rules. For instance, given an equation $s = t$ and a formula $\varphi(s)$ it is natural to allow the rewrite of $\varphi(s)$ to $\varphi(t)$. Similarly we want to support the rewriting with equivalence, i.e. to apply $P \Leftrightarrow Q$ on $\varphi(P)$ to obtain $\varphi(Q)$. Note that we cannot assign polarities to P and Q in $P \Leftrightarrow Q$, while P in $\varphi(P)$ may well have a polarity. Furthermore, the uniform notion of rules obtained from uniform notation is restricted to logical refinement rules and does not capture equivalence rules. In order to capture equations and equivalences we introduce two new uniform types ϵ and ζ respectively for negative and positive equations and equivalences (see the rightmost tables in Fig. 1).

In the following we agree to denote by $\alpha^p(F^q, G^r)$ signed formulas of polarity p, uniform type α, and subformulas F and G with respective polarities q and r according the tables in Fig. 1, including the unary version $\alpha^p(F^q)$. Note that the subformulas are not necessarily direct subformulas, as for instance $\alpha^+(F^-, G^+)$ denotes $(F \Rightarrow G)^+$ but also $(\neg F \vee G)^+$. By abuse of notation we also allow the replacement of F^q and G^r by new formulas. Example: if $\alpha^+(F^q, G^r)$ is $(A^- \Rightarrow B^+)^+$, then $\alpha^p(C, G^r)$ denotes $(C^- \Rightarrow B^+)^+$. We use an analogous notation for formulas of the other uniform types. Furthermore we define $\overline{\alpha}^p(F^q) := \alpha^p(F^q)$ and for $n > 1$, $\overline{\alpha}^p(F_1^{p_1}, \ldots, F_n^{p_n}) := \alpha^p(F_1^{p_1}, \overline{\alpha}^{p_2}(F_2^{p_2}, \ldots, F_n^{p_n}))$. Analogously we define $\overline{\beta}$, but also add the case $n = 0$ by defining $\overline{\beta}^+() := \top^+$ and $\overline{\beta}^-() := \bot^-$.

In the rest of this article we are mainly concerned with signed formulas. To ease the presentation we extend the notion of satisfiability to signed formulas. In order to motivate this definition consider a sequent $\psi_1, \ldots, \psi_n \vdash \varphi$. It represents the proof status that we have to prove φ from the ψ_i. In terms of polarities, all the ψ_i have negative polarity while φ has positive polarity. The ψ_i are the assumptions and thus we consider the models that satisfy those formulas and prove that those models also satisfy φ. Hence, we define that a model M satisfies a *negative* formula ψ_i^- if, and only if, M satisfies ψ_i. From there we derive the dual definition for positive formulas, namely that a model M satisfies a positive formula F^+, if, and only if, M *does not* satisfy F.

Definition 4 (Satisfiability of Signed Formulas). *Let F^p be a signed formula of polarity p, M a model and ρ an assignment. Then we define:* $M^\rho \models F^+$ *holds if, and only if,* $M^\rho \not\models F$. *Conversely, we define* $M^\rho \models F^-$ *holds if, and only* if, $M^\rho \models F$. We extend that notion to sets of signed formulas $\mathcal{F}$ by: $M^\rho \models \mathcal{F}$, if, and only if, for all $F^p \in \mathcal{F}$, $M^\rho \models F^p$.

Lemma 1. *Let M be a model, ρ a variable assignment, F^q, G^r signed formulas of polarities q, r. Then $M^\rho \models \alpha^p(F^q, G^r)$ holds if, and only if, both $M^\rho \models F^q$ and $M^\rho \models G^r$ hold; $M^\rho \models \beta^p(F^q, G^r)$ holds, if, and only if, $M^\rho \models F^q$ or $M^\rho \models G^r$ holds.*

Remark 1 (Notational Conventions). We denote formulas by capital Latin letters $A, B, \ldots, F, G, \ldots$, and formulas with holes by Greek letters $\varphi(.)$, which denotes λ-abstractions $\lambda x.\varphi(x)$ where x occurs *exactly* once in φ. Similarly, we define $\psi(.,.)$ to denote $\lambda x.\lambda y.\psi(x,y)$ and x and y occur exactly once in ψ. [3]

Definition 5 (Literals in Signed Formulas). *Let $\varphi(F)^p$ be a signed formula of polarity p. We say that F is a literal in $\varphi(F)^p$ if using the rules from Fig. 1 we can assign a polarity to F, but not to its subterms.*

2.3 Extensional Expansion Proofs

For the CORE calculus we build upon an extension of extensional expansion proofs from [3]. They rely on extensional expansion trees which are defined in [15] for formulas without equivalences, equations and the atoms $\top$ and $\bot$ and where all negations are around the literals. They are constructed for a formula F along the tree structure of the formula. It follows closely that tree structure, except for existential quantifiers, where it allows to have subtrees for arbitrary many instances t of the quantifier. Each t is called an *expansion term* and the number of instances is the multiplicity of that quantifier. For a universal quantifier the subtree is created for the main formula instantiated with a *selected parameter*. The non-instantiated formula represented by a subtree is called the *shallow formula*. To each subtree we can associate a *deep formula* which represents the formula where all quantifiers have been instantiated with expansion terms. Equations are handled via Leibniz' definition of equality in [15]. It also adds a rule to support reasoning with respect to functional and Boolean extensionality, which gives rise to *extensional expansion trees*. We denote the extensional expansion trees with shallow formula F and multiplicity μ by $\triangle_F^\mu$. The expansion terms in an expansion tree give rise to a substitution, which is $\triangle_F^\mu$-*admissible*, if the transitive closure of the relation induced by the hierarchy of quantifiers from the tree structure together with the relation induced by the substitution among introduction nodes of expansion terms and selected parameters is irreflexive. An extensional expansion tree $\triangle_F^\mu$ is an extensional expansion proof for F, if, and only if, (1) its deep formula is a tautology (i.e., all paths through the deep formula are unsatisfiable) and (2) the associated substitution σ is $\triangle_F^\mu$-admissible.

In [3] we extend that calculus (1) by allowing for equivalences, equations and the atoms $\top$ and $\bot$; (2) by using polarities to overcome the restrictions on the position of negations[4]; (3) by including a rule to dynamically increase the

---

[3] This is similar to the notation used by Schütte in [16].

[4] Here we follow the *indexed formula trees* of Wallen [18] for first-order modal logics and fragments thereof.

multiplicities on the fly in order to avoid guessing the initial multiplicities and restart. Increasing the multiplicities thereby copies relevant existing connections and results in a renaming of existential (γ) variables and universal (δ) variables.

Theorem 1 (Soundness & Completeness with Dynamic Increase of Multiplicities). *F is valid if, and only if, from an extensional expansion tree for the signed formula F^+ with singular multiplicities we can derive an extensional expansion tree and a $\triangle_F^\mu$-admissible substitution σ such that all paths in $\triangle_F^\mu$ are unsatisfiable.*

3   CoRe Calculus

The CoRe calculus relies on the idea to have a single signed formula representing the state of the proof and to use polarities and uniform types to manipulate the subformulas. In order to check the admissibility of substitutions it uses the extensional expansion trees from [3] (see previous section). For a closed signed formula F^p it constructs an extensional expansion tree with singular multiplicities and with γ- and δ-variables. From such an initial extensional expansion tree, we take its deep formula with free γ- and δ-variables.

As an example consider the closed formula representing the structural induction axiom $\forall p_{\iota \to o}. (\forall n_\iota . n = 0 \Rightarrow p(n)) \Rightarrow ((\forall n, n'_\iota . (n = n'+1 \land p(n')) \Rightarrow p(n)) \Rightarrow \forall n_\iota . p(n))$. Writing free γ-variables in capital letters, δ-variables in lower-case letters and performing the necessary renamings to avoid name clashes, the free variable representation of the extensional expansion tree for the negative version of the formula and with singular multiplicities is $((n_1 = 0 \Rightarrow P(n_1)) \Rightarrow (((n_2 = n_3 + 1 \land P(n_3)) \Rightarrow P(n_2)) \Rightarrow P(N)))^-$. Note that the free γ- and δ-variables are bound in the corresponding extensional expansion tree. The extensional expansion tree this formula stems from can be used to check the admissibility of any substitution for this formula, even if the formula is further transformed.

The obtained signed formulas have the property that they do not contain any quantifier with defined polarity, but may well contain quantifiers inside literals. We denote this kind of signed formulas as *quantifier-free formulas*.

Definition 6 (Quantifier-Free & Ground Signed Formulas). *Let F^p be a signed formula of polarity P. We say that F^p is quantifier free (QF), if F^p does not contain any signed subformula of uniform type δ or γ. If F^p additionally contains no free γ-variables, then F^p is ground.*

An example for a ground signed formula which contains quantifiers inside literals is the following definition of the predicate of even natural numbers ($\mathsf{Even}_{\iota \to o} = \lambda x_\iota . \exists n_\iota . x = 2 \times n)^p$. Given a closed formula F, we denote by $\triangle_F^1$ the initial extensional expansion tree of singular multiplicity for F and by F_X^+ the quantifier free formula obtained from $\triangle_F^1$. Using polarities and uniform types we can define relationships between subformulas in signed QF-formulas as well as the *set of set of paths* through signed quantifier-free formulas.

Definition 7 (α- and β-Related Signed Formulas). *Let $\psi(F,G)^p$ be a signed formula of polarity p. We say that F and G are α-related (resp. β-related) in $\psi(F,G)^p$ if, and only if, the smallest signed subformula of $\psi(F,G)^p$ which contains both F and G is of uniform type α (resp. β).*

Definition 8 (Paths in Signed QF-Formula). *Let F^p be a signed quantifier-free formula of polarity p. A path in F^p is a sequence $\ll F_1^{p_1}, \ldots, F_n^{p_n} \gg$ of α-related signed subformulas of F^p. The sets $\mathcal{P}(F^p)$ of paths through F^p is the smallest set containing $\{\ll F^p \gg\}$ and which is closed under the two operations:*

(α) *If $P \cup \{\ll \Gamma, \alpha^p(G^q, H^r) \gg\} \in \mathcal{P}(F^p)$, then $P \cup \{\ll \Gamma, G^q, H^r \gg\} \in \mathcal{P}(F^p)$.*
(β) *If $P \cup \{\ll \Gamma, \beta^p(G^q, H^r) \gg\} \in \mathcal{P}(F^p)$, then $P \cup \{\ll \Gamma, G^q \gg, \ll \Gamma, H^r \gg\} \in \mathcal{P}(F^p)$.*

The following lemma establishes the basic relationship between the paths in the extensional expansion tree for a closed formula and the paths through the corresponding quantifier-free signed formula.

Lemma 2. *Let F be a closed formula, $\triangle_F^1$ the extensional expansion tree with singular multiplicity for F^+ and F_X^+ the QF-formula for $\triangle_F^1$. Then the set of all literal paths through $\triangle_F^1$ are in $\mathcal{P}(F_X^+)$.*

This allows us to derive from the existence of satisfiable (ground) literal paths in the extensional expansion tree the existence of (ground) literal paths in the QF-formula. The preservation of the existence of satisfiable ground paths is the property ensuring the soundness of any further transformation performed on the QF-formula. Since it is cumbersome to always have to reason about the existence of ground paths *only* at the level of literals, we lift the level to paths at an arbitrary, less granular level of paths.

Definition 9 (Satisfiable & Unsatisfiable Ground Paths). *A path p is ground if all contained signed formulas are ground. A ground path p is satisfiable if there exists a model M such that $M \models p$. Otherwise p is unsatisfiable.*

Corollary 1. *Let p be a path, which contains either $\top^+$, or $\perp^-$ or signed formulas F^- and F^+. Then any ground path for p is unsatisfiable.*

The following lemma establishes then that we can choose the level of granularity on which we want to reason about the satisfiability of a complete set of paths.

Lemma 3. *Let F^p be a signed formula, and $P, P' \in \mathcal{P}(F^p)$ complete sets of ground paths through F^p. Then it holds: P contains a satisfiable ground path if, and only if, P' contains a satisfiable ground path.*

3.1 Admissible Replacement Rules

The logical context of some subformula in a signed formula is determined by collecting all α-related subformulas. Now we are concerned with determining the possible rules which can be generated from the available subformulas. To motivate this, consider the goal sequent $A \Rightarrow (B \Rightarrow C) \vdash C$. Applying $A \Rightarrow (B \Rightarrow C)$ to C means that the goal to prove C is replaced by the goal to prove A and B. In this case both occurrences of C have opposite polarities and are α-related via $\vdash$. Furthermore, the new subgoals, i.e. the positive occurrences of A and B, can be determined statically from the formula by collecting all the formulas that are β-related to the negative occurrence of C. This enables generating rules from a formula by fixing the left-hand side, e.g. the negative C. The right-hand side of the rule is then the list of all formulas that are β-related to the left-hand side, and we write this as a rule $C^- \rightarrow \langle A^+, B^+ \rangle$. Analogously, if there is a negative equation or a negative equivalence in the context, i.e. an ϵ-type formula $\epsilon(s,t)$, we obtain the rules $s \rightarrow \langle t, F_1^{p_1}, \ldots, F_n^{p_n} \rangle$ and $t \rightarrow \langle s, F_1^{p_1}, \ldots, F_n^{p_n} \rangle$ where the $F_i^{p_i}$ are the formulas β-related to $\epsilon(s,t)$. This rule contains the information, that some goal formula $\varphi(s)$ where s has an arbitrary polarity – even no polarity – can be refined to the subgoals $\varphi(t), F_1^{p_1}, \ldots, F_n^{p_n}$.

Before formalizing the notion of replacement rules, we introduce a mechanism which allows to weaken parts of α-type signed subformulas.

Definition 10 (Weakening of Signed Formulas). *Let F^p be a signed formula. The set $\mathcal{W}(F^p)$ of weakened signed formulas for F^p is defined recursively over the structure of F: (Atom) $\mathcal{W}(F^p) := \{F^p\}$ if F^p is a literal; (α) $\mathcal{W}(\alpha^p(G^q, H^r)) := \{\alpha^p(G_w^q, H_w^r) \mid G_w^q \in \mathcal{W}(G^q), H_w^r \in \mathcal{W}(H^r)\} \cup \mathcal{W}(G^q) \cup \mathcal{W}(H^r)$; ($\beta$) $\mathcal{W}(\beta^p(G^q, H^r)) := \{\beta^p(G_w^q, H_w^r) \mid G_w^q \in \mathcal{W}(G^q), H_w^r \in \mathcal{W}(H^r)\}$.*

Lemma 4. *Let M be a model, F^p a signed formula and σ a ground substitution such that $\sigma(F)^p$ is ground. Then for any $G^p \in \mathcal{W}(F^p)$ it holds: If $M \models \sigma(F^p)$ then $M \models \sigma(G^p)$.*

Proof. By structural induction over F and using Lemma 1. □

Corollary 2 (Connectable Signed Formulas). *Let F^p and G^{-p} be two signed formulas. If there exists an $F_w^p \in \mathcal{W}(F^p)$ which is α-equal to some $F_w^{-p} \in \mathcal{W}(G^{-p})$, then for any ground substitution σ such that $\sigma(F^p)$ and $\sigma(G^p)$ are ground there exists no model M which satisfies both $\sigma(F^p)$ and $\sigma(G^{-p})$. We say that F^p and G^{-p} are connectable.*

Example 1. As an example consider the negative formula $A \vee (B \wedge C)^-$ and the positive formula $A \vee (C \vee D)^+$. The respective sets of weakened signed formulas are $\mathcal{W}(A \vee (B \wedge C)^-) = \{A \vee (B \wedge C)^-, A \vee B^-, A \vee C^-\}$ and $\mathcal{W}(A \vee (C \vee D)^+) = \{A \vee (C \vee D)^+, A \vee C^+, A \vee D^+, A^+, C^+, D^+\}$. Thus, the formulas are connectable, since $A \vee C^- \in \mathcal{W}(A \vee (B \wedge C)^-)$ and $A \vee C^+ \in \mathcal{W}(A \vee (C \vee D)^+)$.

Definition 11 (Subformula Conditions). *Let $\varphi(F^q)^p$ be a signed formula of polarity p, $G_1^{p_1}, \ldots, G_n^{p_n}$ be all maximal signed formulas that are β-related to F^q in $\varphi(F^q)^p$. Then the conditions of F^q are $\mathcal{C}_{\varphi(.)^p} := \mathcal{W}(G_1^{p_1}) \times \ldots \times \mathcal{W}(G_n^{p_n})$.*

Replacement rules are of two kinds: the first kind are those where the left-hand side is a subformula with a polarity, and the second kind result from ϵ-type formulas. The former are called *resolution replacement rules*, while the latter are called *rewriting replacement rules*.

Definition 12 (Admissible Resolution Replacement Rules). *Given the signed formula $\psi(L^{-p}, G^p)^q$ of polarity q and containing the two signed subformulas L and G of opposite polarities, let $\psi'(L^{-p}, G^p)^r$ be the smallest signed formula containing both L^{-p} and G^p and $(R_1^{p_1}, \ldots, R_n^{p_1}) \in \mathcal{C}_{\psi'(.,G^p)^r}$. Then if L^{-p} and G^p are α-related in $\psi(L^{-p}, G^p)^q$, then $L^{-p} \to \langle R_1^{p_1}, \ldots, R_n^{p_n} \rangle$ is an admissible resolution replacement rule for G^p.*

Definition 13 (Admissible Rewriting Replacement Rules). *Given the signed formula $\psi(\epsilon^-(s,t), G^p)^q$ of polarity q and containing the two signed subformulas $\epsilon^-(s,t)$ and G^p, let $\psi'(\epsilon^-(s,t), G^p)^r$ be the smallest signed formula containing both $\epsilon^-(s,t)$ and G^p and $(R_1^{p_1}, \ldots, R_n^{p_1}) \in \mathcal{C}_{\psi'(.,G^p)^r}$. Then if $\epsilon^-(s,t)$ and G^p are α-related in $\psi(\epsilon^-(s,t), G^p)^q$, then $s \to \langle t, R_1^{p_1}, \ldots, R_n^{p_n} \rangle$ and $t \to \langle s, R_1^{p_1}, \ldots, R_n^{p_n} \rangle$ are admissible rewriting replacement rules for G^p.*

A CORE proof state is denoted by $\triangle; \sigma \triangleright F$ and consists of an extensional expansion tree $\triangle$, a $\triangle$-admissible substitution σ and a QF-formula F. We say that $\triangle; \sigma \triangleright F$ *is satisfiable* if, and only if, for all $\triangle$-admissible ground σ-refinements ν there exists a satisfiable path in $\nu(F)^+$. In order to mimic a textual top-down proof development style, the CORE calculus rules $\frac{\pi}{\pi'}$ are to be read top-down[5], i.e. the problem of proving π is reduced to prove π'. The rules are given in Fig. 2 and consist of three parts separated by dotted lines: The upper part presents the major calculus rules, the middle part consists of the propositional simplification rules, and the lower part gives an extra rule for the application of rewriting replacement rule. The rules from the first two parts are the kernel of the CORE calculus necessary for completeness. The last rule is added for convenience, but is admissible. We only illustrate some of the rules in detail and the meaning of the other rules should follow easily from these descriptions.

The *Weak$_L$* rule allows to reduce the goal to prove a positive formula φ in which occurs a binary α-type subformula $\alpha^p(F^q, G^r)$ to the goal to prove the formula φ, where the binary subformula is replaced by the left conjunct F^q. The surrounding $\overline{\alpha}^p$ is used to adapt the polarities by adding a negation, in case $p \neq q$. Example applications of this rule are: $\triangle; \sigma \triangleright C \wedge (A \Rightarrow B) \vdash_C^{Weak_L} \triangle; \sigma \triangleright C \wedge B$, or $\triangle; \sigma \triangleright C \wedge (A \Rightarrow B) \vdash_C^{Weak_L} \triangle; \sigma \triangleright C \wedge \neg(A)$. This rule changes neither the extensional expansion tree nor the substitution.

The *Leibniz*-rule is used to expand an equality $s =_\tau t$ into Leibniz definition of equality $\forall P_{\tau \to o}. P(s) \Rightarrow P(t)$, for any type τ. Therefore we determine the extensional expansion subtree $\triangle_{s=t^p}$ for $(s = t)^p$ and apply the respective Leibniz-Introduction-rule of extensional expansion proofs. This adds an initial extensional expansion subtree $\triangle_{\forall P. P(s) \Rightarrow P(t)}$ conjunctively to $\triangle_{s=t^p}$. From that new subtree we take the corresponding QF-formula $P(s) \Rightarrow P(t)$ and α-relate it to $s = t^p$ to obtain the new QF-formula $\varphi(\alpha^p(s = t^p, P(s) \Rightarrow P(t)))$. Depending on whether the polarity p is positive or not, P is a new δ-variable or a new γ-variable. An example for this rule is: $\triangle_{\triangle_{s(X)+Y=s(X+Y)^-}}; \sigma \triangleright (X' + 0 = X' \wedge s(X) + Y = s(X + Y)) \Rightarrow s(s(a)) + 0 = s(s(a)) \vdash_C^{Leibniz} \triangle_{\alpha^-(\triangle_{s(X)+Y=s(X+Y)}, \triangle_{\forall P. P(s(X)+Y) \Rightarrow P(s(X+Y))})}; \sigma \triangleright$

---

[5] Unlike sequent calculus rules.

$$\frac{\triangle; \sigma \triangleright \top}{q.e.d.} \; Axiom \quad \frac{\triangle; \sigma \triangleright \varphi(F^p)}{\triangle; \sigma \triangleright \varphi(\alpha^p(F^p, F^p))} \; Contract \quad \frac{\triangle; \sigma \triangleright \varphi((F \Leftrightarrow G)^+)}{\triangle; \sigma \triangleright \varphi(((F \Rightarrow G) \wedge (G \Rightarrow F))^+)} \; \zeta\text{-}Elim$$

$$\frac{\triangle; \sigma \triangleright \varphi(\alpha^p(F^q, G^r))}{\triangle; \sigma \triangleright \varphi(\overline{\alpha}^p(F^q))} \; Weak_L \quad\quad \frac{\triangle; \sigma \triangleright \varphi(\alpha^p(F^q, G^r))}{\triangle; \sigma \triangleright \varphi(\overline{\alpha}^p(G^r))} \; Weak_R$$

$$\frac{\triangle_{\triangle_{s=t^p}}; \sigma \triangleright \varphi(s = t^p)}{\triangle_{\alpha^p(\triangle_{s=t^p}, \triangle_{\forall P. P(s) \Rightarrow P(t)})}; \sigma \triangleright \varphi(\alpha^p(s = t, P(s) \Rightarrow P(t)))} \; Leibniz$$

$$\frac{\triangle_{s=t^p}; \sigma \triangleright \varphi(s = t^p)}{\triangle_{\alpha^p(s=t^p, \lambda x. s = \lambda x. t)}; \sigma \triangleright \varphi(\alpha^p(s = t, \lambda x. s = \lambda x. t))} \; f\text{-}Ext$$

if x local to $s = t$ in $\varphi(s = t^p)$.

$$\frac{\triangle_{A \Leftrightarrow B^p}; \sigma \triangleright \varphi(A \Leftrightarrow B^p)}{\triangle_{\alpha^p(A \Leftrightarrow B^p, \lambda x. A = \lambda x. B)}; \sigma \triangleright \varphi(\alpha^p(A \Leftrightarrow B, \lambda x. A = \lambda x. B))} \; b\text{-}Ext$$

if x local to $A \Leftrightarrow B$ in $\varphi(A \Leftrightarrow B^p)$.

$$\frac{\triangle; \sigma \triangleright F}{\sigma'(\triangle); \sigma' \circ \sigma \triangleright \sigma'(F)} \; Subst \quad\quad \frac{\triangle; \sigma \triangleright \varphi(F^p)}{\triangle; \sigma \triangleright \varphi(\overline{\beta}^p(V_1^{p_1}, \ldots, V_n^{p_n}))} \; Res$$

if $\sigma' \circ \sigma$ admissible wrt. $\triangle$.   If $U^{-p} \to \langle V_1^{p_1}, \ldots, V_n^{p_n} \rangle$ admissible for F^p and U^{-p} and F^p are connectable.

$$\frac{\triangle_{\triangle_1 \cdots \triangle_n}; \sigma \triangleright \varphi(\psi_1(\overrightarrow{V_1}), \ldots, \psi_n(\overrightarrow{V_k}))}{\triangle_{\triangle_1 \triangle_1' \cdots \triangle_n, \triangle_n'}; \sigma' \circ \sigma \triangleright \varphi(\alpha(\psi_1(\overrightarrow{V_1}), \psi_1(\rho(\overrightarrow{V_1}))), \ldots, \alpha(\psi_n(\overrightarrow{V_k}), \psi_n(\rho(\overrightarrow{V_k}))))} \; \mu\text{-}Inc$$

where ρ is the renaming of γ- and δ-variables declared in the $\triangle_i$, $(\overrightarrow{V_i})_{i=1\ldots k}$ is a disjoint partition of $dom(\rho)$, such that for all i, all the variables in $\overrightarrow{V_i}$ occur only in $\psi_i(\overrightarrow{V_i})$, and $\sigma' := [\rho(\sigma(x))/\rho(x) \mid$ for all γ-variables $x \in dom(\rho)]$.

$$\frac{\triangle_{\triangle_{F^p}}; \sigma \triangleright \varphi(F^p)}{\triangle_{\triangle_C}; \rho \circ \sigma \triangleright \varphi(\beta^p(\alpha^p(A^-, F^p), \alpha^p(A^+, F^p)))} \; Cut \; A$$

where $\triangle_{F^p}$ is the minimal subtree containing all literals in F^p, $\overrightarrow{x}'$ are the free variables of A not bound above $\triangle_{F^p}$, $\rho := [\overrightarrow{x}'/\overrightarrow{x}]$, and $\triangle_C$ is the subtree $\gamma^p \overrightarrow{x}'. \beta^p(\alpha^p(A^-, \triangle_{F^p}), \alpha^p(A^+, \triangle_{F^p}))$.

. . . .

$$\frac{\triangle; \sigma \triangleright \varphi(\top \vee F)}{\triangle; \sigma \triangleright \varphi(\top)} \; \vee_L^\top \quad \frac{\triangle; \sigma \triangleright \varphi(F \vee \top)}{\triangle; \sigma \triangleright \varphi(\top)} \; \vee_R^\top \quad \frac{\triangle; \sigma \triangleright \varphi(\bot \vee F)}{\triangle; \sigma \triangleright \varphi(F)} \; \vee_L^\bot \quad \frac{\triangle; \sigma \triangleright \varphi(F \vee \bot)}{\triangle; \sigma \triangleright \varphi(F)} \; \vee_R^\bot$$

$$\frac{\triangle; \sigma \triangleright \varphi(\top \Rightarrow F)}{\triangle; \sigma \triangleright \varphi(F)} \; \Rightarrow_L^\top \quad \frac{\triangle; \sigma \triangleright \varphi(F \Rightarrow \top)}{\triangle; \sigma \triangleright \varphi(\top)} \; \Rightarrow_R^\top \quad \frac{\triangle; \sigma \triangleright \varphi(\bot \Rightarrow F)}{\triangle; \sigma \triangleright \varphi(\top)} \; \Rightarrow_L^\bot \quad \frac{\triangle; \sigma \triangleright \varphi(F \Rightarrow \bot)}{\triangle; \sigma \triangleright \varphi(\neg(F))} \; \Rightarrow_R^\bot$$

$$\frac{\triangle; \sigma \triangleright \varphi(\top \wedge F)}{\triangle; \sigma \triangleright \varphi(F)} \; \wedge_L^\top \quad \frac{\triangle; \sigma \triangleright \varphi(F \wedge \top)}{\triangle; \sigma \triangleright \varphi(F)} \; \wedge_R^\top \quad \frac{\triangle; \sigma \triangleright \varphi(\bot \wedge F)}{\triangle; \sigma \triangleright \varphi(\bot)} \; \wedge_L^\bot \quad \frac{\triangle; \sigma \triangleright \varphi(F \wedge \bot)}{\triangle; \sigma \triangleright \varphi(\bot)} \; \wedge_R^\bot$$

. . . .

$$\frac{\triangle_{\epsilon(s,t)}; \sigma \triangleright \varphi(L(s)^p)}{\triangle_{\alpha^-(\epsilon(s,t), \forall P. \beta^-(L(s)^{-p}, L(t)^p))}; [\lambda x. L(x)/P] \circ \sigma \triangleright \varphi(\overline{\beta}^p(L(t)^p, V_1^{p_1}, \ldots, V_n^{p_n}))} \; Rew$$

If $L(s)^p$ is a literal, $s \to \langle t, V_1^{p_1}, \ldots, V_n^{p_n} \rangle$ is admissible for $L(s)^p$, and $\epsilon(s,t)$ is the equation or equivalence this rule results from.

Fig. 2. The CoRe Calculus

$$\triangle_{\alpha(\triangle_F,\triangle_{F'})};\sigma' \circ \sigma \rhd$$

$$x \xrightarrow{\rho} \rho(x) = x'$$

$$\sigma \downarrow \quad \downarrow \sigma'$$

$$\sigma(x) \xrightarrow{\rho} \sigma'(x') := \rho(\sigma(x))$$

$$\begin{pmatrix} X' + 0 = X' \wedge (\mathsf{s}(X) + Y = \mathsf{s}(X+Y)) \\ \wedge (\mathsf{s}(\mathsf{s}(X) + Y) + 0 = \mathsf{s}(\mathsf{s}(X) + Y) \\ \Rightarrow \mathsf{s}(\mathsf{s}(X+Y)) + 0 = \mathsf{s}(\mathsf{s}(X+Y)))) \\ \wedge (\mathsf{s}(U) + V = \mathsf{s}(U+V) \\ \wedge (\mathsf{s}(\mathsf{s}(U) + V) + 0 = \mathsf{s}(\mathsf{s}(U) + V) \\ \Rightarrow \mathsf{s}(\mathsf{s}(U+V)) + 0 = \mathsf{s}(\mathsf{s}(U+V)))) \end{pmatrix}$$
$$\Rightarrow \mathsf{s}(\mathsf{s}(a)) + 0 = \mathsf{s}(\mathsf{s}(a))$$

Fig. 3. (a) Construction of σ' (b) Proof state after increase of multiplicity

$(X'+0 = X' \wedge (\mathsf{s}(X)+Y = \mathsf{s}(X+Y) \wedge (P(\mathsf{s}(X)+Y) \Rightarrow P(\mathsf{s}(X+Y)))))) \Rightarrow \mathsf{s}(\mathsf{s}(a))+0 = \mathsf{s}(\mathsf{s}(a))$, where P is a new γ-variable.

The functional and Boolean extensionality rules *f-Ext* and *b-Ext* require the variable x to be local with respect to the equation (resp. equivalence). This intuitively means that if x is a γ-variable, then it does not occur in a part which is β-related to the equation (resp. equivalence). If x is a δ-variable, then it does not occur in a part which is α-related. For a formalization we refer to [3].

The rule *Res* is the central rule operationalizing assertion level reasoning. Assertions G can be applied to some subformula F^p if it is possible to compile from G an admissible resolution replacement rule $U^{-p} \rightarrow \langle V_1^{p_1}, \ldots, V_n^{p_n} \rangle$. Such a rule is applicable, if F^p and U^{-p} are connectable, and the application consists of replacing F^p by the disjunction of the subgoals $V_1^{p_1}, \ldots, V_n^{p_n}$, which is represented by $\overline{\beta}^p(V_1^{p_1}, \ldots, V_n^{p_n})$. Examples for this rule are: $\triangle; \sigma \rhd (A \Rightarrow B) \Rightarrow B \vdash_C^{Res} \triangle; \sigma \rhd (A \Rightarrow B) \Rightarrow A$ by the rule $B^- \rightarrow \langle A^+ \rangle$, $\triangle; \sigma \rhd (A \vee (B \wedge C)) \Rightarrow B \vdash_C^{Res} \triangle; \sigma \rhd (A \vee (B \wedge C)) \Rightarrow \neg(A)$ by the rule $B^- \rightarrow \langle A^- \rangle$, or $\triangle; \sigma \rhd (A \Rightarrow B) \Rightarrow B \vdash_C^{Res} \triangle; \sigma \rhd (A \Rightarrow \bot) \Rightarrow B$ by the rule $B^+ \rightarrow \langle . \rangle$ where by definition $\overline{\beta}^-(.) = \bot^-$.

Finally, the μ-*Inc*-rule allows to increase the multiplicities of γ-quantifiers in the extensional expansion tree. This rule is essentially necessary to be combined with the substitution rule in order to preserve formulas containing those variables that will be substituted. Assume $\{x_1, \ldots, x_n\}$ are those γ-variables that will be substituted and whose associated subtrees $\triangle_i$ are maximal with respect to any other substituted γ-variable. The $\triangle_i$ are the immediate subtrees in $\triangle$ of the γ-quantifier that introduced x_i. The increase of the multiplicity of these γ-quantifiers with respect to the x_i results in copies $\triangle_i'$ of the $\triangle_i$, where all γ- and δ-variables declared in $\triangle_i$ have been renamed. This renaming ρ is injective and maps γ-variables to γ-variables and δ-variables to δ-variables. Furthermore, it holds $\rho(x_i) = x_i'$ for all $1 \leq i \leq n$. The extension σ' of the substitution σ is obtained by making the diagram on the left hand side of Fig. 3 commute for any γ-variable x in the domain of ρ. Finally, the renaming is propagated to the QF-formula by considering a partitioning $(\overrightarrow{V_i})_{i=1\ldots k}$ of the renamed γ- and δ-variables $(dom(\rho))$, such that for all $1 \leq i \leq k$ there is a subformula $\psi_i(\overrightarrow{V_i})$ containing *all* occurrences of the variables in $\overrightarrow{V_i}$. The renaming ρ is applied to these $\psi_i(\overrightarrow{V_i})$ to obtain $\psi_i(\rho(\overrightarrow{V_i}))$, which in turn are added conjunctively to $\psi_i(\overrightarrow{V_i})$. Consider the proof state obtained before by the *Leibniz*-rule: We first apply the substitution $\sigma(P) = \lambda x.\mathsf{s}(x + 0) = \mathsf{s}(x)$ and then increase the multiplicity of the γ-quantifier for X. This results in the proof state shown on the right hand side of Fig. 3. Thereby $\rho := [U/X, V/Y, Q/P]$, and $\sigma' := [\rho(\sigma(P))/\rho(P)] = [\lambda x.s(x+0) = s(x)/Q]$.

Definition 14 (CoRe Calculus). *Let π, π' be CoRe proof states. $\pi \vdash_C \pi'$ is a CoRe proof step if, and only if, there is a CoRe calculus rule $\frac{\pi}{\pi'}$ (cf. Fig. 2). A CoRe derivation is a sequence $\pi_1 \vdash_C \pi_2 \vdash_C \ldots \vdash \pi_n$ of CoRe proof steps (as usual we define $\vdash_C^*$ as the reflexive transitive closure of $\vdash_C$).*

Let F be a closed formula, $\triangle_I$ the initial extensional expansion tree for F^+, F_X the corresponding QF-formula and id the empty substitution. Then F is provable in CoRe, if, and only if, there is a derivation $\triangle_I; id \triangleright F_X \vdash_C^* \triangle; \sigma \triangleright \top$.

Theorem 2 (Soundness & Completeness). *The CoRe calculus is sound and complete.*

Proof (Sketch). (Soundness) In order to prove the soundness of the CoRe calculus, we prove (1) for any closed formula F, if F is not valid, then the initial proof state $\triangle_I; id \triangleright F_X$ for F is satisfiable (by Theorem 1 and Lemma 2); (2) each CoRe calculus rule (except the *Axiom*-rule) preserves the satisfiability of the proof state; (3) proof states on which the CoRe *Axiom*-rule is applicable are unsatisfiable (a single path with $\top^+$ is unsatisfiable by Corollary 1).

(Completeness) By completeness of extensional expansion proofs [15] we can assume for any valid closed formula F that we have guessed the right multiplicities for γ-type quantifiers, the right substitution σ, and the necessary applications of *Leibniz*, *f-Ext*, *b-Ext*, ζ-*Elim*, and *Cut*.[6] All paths in the resulting QF-formula F_P are (propositionally) unsatisfiable. That is from $\triangle_I; Id \triangleright F_X$ we can derive a proof state $\triangle_P; \sigma \triangleright F_P$. In a second phase we show that from $\triangle_P; \sigma \triangleright F_P$ we can derive $\triangle; \sigma \triangleright \top$ by proving that *path resolution* [14] is admissible using *Res*, *Contract*, *Weakening* and the simplification rules. The completeness then finally follows from the completeness of path resolution for propositional logic. $\square$

4   Example

We illustrate the CoRe calculus by proving the following theorem over the natural numbers: $\forall n. \sum_{i=1}^{n} i^3 = (\sum_{i=1}^{n} i)^2$. The function symbols have the standard meaning. Besides the necessary definitions and the structural induction axiom for the natural numbers, we assume specific lemmas, in order to keep the proof short. The axioms and lemmas, i.e. the assertions we assume, are given in Fig. 4.

The initial proof state consists of an extensional expansion tree $\triangle_0$ for the conjunction of the assertions implying the conjecture, an empty substitution, and the quantifier free formula resulting from $\triangle_0$. The initial proof state then is

---

[6] *Cut* is used to simulate the extensionality rule from [15]. *Cut* is not admissible in the CoRe-calculus, but would probably be if we add the rules ξ and b (see [5]).

$$\forall p.\forall x.\exists y.\,(p(0) \wedge (p(y) \Rightarrow p(s(y)))) \Rightarrow p(x) \quad (1)$$
$$\forall n.\,\textstyle\sum_{i=1}^{0} i^n = 0 \quad (2)$$
$$\forall n,m.\,\textstyle\sum_{i=1}^{s(m)} i^n = s(m)^n + \textstyle\sum_{i=1}^{m} i^n \quad (3)$$
$$\forall a,b.\,(a+b)^2 = (a^2 + (2 \times b) \times a) + b^2 \quad (4)$$
$$\forall a,b,c.\,a = b \Rightarrow a + c = b + c \quad (5)$$

$$\forall n.\,\textstyle\sum_{i=1}^{n} i^1 = \tfrac{n \times s(n)}{2} \quad (6)$$
$$\forall m.\,2 \times \tfrac{m}{2} = m \quad (7)$$
$$\forall q.\,s(q)^3 = s(q)^2 + (q \times s(q)) \times s(q) \quad (8)$$
$$0 = (0)^2 \quad (9)$$

Fig. 4. Axioms and lemmas assumed in the proof of $\forall n.\,\sum_{i=1}^{n} i^3 = (\sum_{i=1}^{n} i^1)^2$

$$\triangle_0;\mathrm{id}\,\triangleright\;\left(\begin{array}{ll}
(P(0) \wedge (P(y) \Rightarrow P(s(y)))) \Rightarrow P(X) & (1)\quad A' = B' \Rightarrow A' + C' = B' + C' \quad (5)\\
\sum_{i=1}^{0} i^N = 0 & (2)\quad \sum_{i=1}^{N'} i^1 = \tfrac{N' \times s(N')}{2} \quad (6)\\
\sum_{i=1}^{s(M)} i^N = s(M)^N + \sum_{i=1}^{M} i^N & (3)\quad 2 \times \tfrac{M'}{2} = M' \quad (7)\\
(A+B)^2 = (A^2 + (2 \times B) \times A) + B^2 & (4)\quad s(Q)^3 = s(Q)^2 + (Q \times s(Q)) \times s(Q) \quad (8)\\
& \qquad 0 = (0)^2 \quad (9)
\end{array}\right.$$
$$\Rightarrow \textstyle\sum_{i=1}^{n} i^3 = (\sum_{i=1}^{n} i^1)^2$$

Due to lack of space, we introduce a macro step *RuleApplication from H* for the application of some assertion H. Such a macro step consists of applying a replacement rule that can be obtained from H by (1) increasing the multiplicities with respect to the variables that need to be instantiated to apply the rule, (2) apply the necessary substitution, (3) apply the replacement rule, and finally (4) "remove" the instantiated assertion by weakening. The proof is then as follows:

1. By *RuleApplication from (1)*
$$\triangle_1;\sigma_1 \triangleright\; \left((1) \wedge \textstyle\sum_{i=1}^{0} i^N = 0 \wedge (3) \wedge (4) \wedge (5) \wedge (6) \wedge (7) \wedge (8) \wedge (9)\right)$$
$$\Rightarrow \textstyle\sum_{i=1}^{0} i^3 = (\sum_{i=1}^{0} i^1)^2 \wedge (\sum_{i=1}^{y'} i^3 = (\sum_{i=1}^{y'} i^1)^2) \Rightarrow (\sum_{i=1}^{s(y')} i^3 = (\sum_{i=1}^{s(y')} i^1)^2)$$

2. By *RuleApplication from (2)(2×)*
$$\triangle_2;\sigma_2 \triangleright\; \left((1) \wedge (2) \wedge (3) \wedge (4) \wedge (5) \wedge (6) \wedge (7) \wedge (8) \wedge 0 = (0)^2\right)$$
$$\Rightarrow 0 = (0)^2 \wedge (\sum_{i=1}^{y'} i^3 = (\sum_{i=1}^{y'} i^1)^2) \Rightarrow (\sum_{i=1}^{s(y')} i^3 = (\sum_{i=1}^{s(y')} i^1)^2)$$

3. By *RuleApplication from (9)*
$$\triangle_3;\sigma_3 \triangleright\; \left((1) \wedge (2) \wedge (3) \wedge (4) \wedge (5) \wedge (6) \wedge (7) \wedge (8) \wedge (9)\right)$$
$$\Rightarrow \top \wedge (\sum_{i=1}^{y'} i^3 = (\sum_{i=1}^{y'} i^1)^2) \Rightarrow (\sum_{i=1}^{s(y')} i^3 = (\sum_{i=1}^{s(y')} i^1)^2)$$

4. By *Simplify by $\wedge_L^{\top}$*
$$\triangle_4;\sigma_4 \triangleright\; \left((1) \wedge (2) \wedge \textstyle\sum_{i=1}^{s(M)} i^N = s(M)^N + \sum_{i=1}^{M} i^N \wedge (4) \wedge (5) \wedge (6) \wedge (7) \wedge (8) \wedge (9)\right)$$
$$\Rightarrow (\sum_{i=1}^{y'} i^3 = (\sum_{i=1}^{y'} i^1)^2) \Rightarrow (\sum_{i=1}^{s(y')} i^3 = (\sum_{i=1}^{s(y')} i^1)^2)$$

5. By *RuleApplication from (3)*
$$\triangle_5;\sigma_5 \triangleright\; \left((1) \wedge (2) \wedge \textstyle\sum_{i=1}^{s(M)} i^N = s(M)^N + \sum_{i=1}^{M} i^N \wedge (4) \wedge (5) \wedge (6) \wedge (7) \wedge (8) \wedge (9)\right)$$
$$\Rightarrow (\sum_{i=1}^{y'} i^3 = (\sum_{i=1}^{y'} i^1)^2) \Rightarrow (s(y')^3 + \sum_{i=1}^{y'} i^3 = (\sum_{i=1}^{s(y')} i^1)^2)$$

6. By *RuleApplication from (3)*
$$\triangle_6;\sigma_6 \triangleright\; \left((1) \wedge (2) \wedge (3) \wedge (A+B)^2 = (A^2 + (2 \times B) \times A) + B^2 \wedge (5) \wedge (6) \wedge (7) \wedge (8) \wedge (9)\right)$$
$$\Rightarrow (\sum_{i=1}^{y'} i^3 = (\sum_{i=1}^{y'} i^1)^2) \Rightarrow (s(y')^3 + \sum_{i=1}^{y'} i^3 = (s(y') + \sum_{i=1}^{y'} i^1)^2)$$

7. By *RuleApplication from (4)*
$$\triangle_7;\sigma_7 \triangleright\; \left((1) \wedge (2) \wedge (3) \wedge (4) \wedge (5) \wedge (6) \wedge (7) \wedge (8) \wedge (9)\right)$$
$$\Rightarrow (\sum_{i=1}^{y'} i^3 = (\sum_{i=1}^{y'} i^1)^2) \Rightarrow s(y')^3 + \sum_{i=1}^{y'} i^3 =$$
$$(s(y')^2 + (2 \times \sum_{i=1}^{y'} i^1) \times s(y')) + (\sum_{i=1}^{y'} i^1)^2$$

8. By *RuleApplication from $\sum_{i=1}^{y'} i^3 = (\sum_{i=1}^{y'} i^1)^2$*
$$\triangle_8;\sigma_8 \triangleright\; \left((1) \wedge (2) \wedge (3) \wedge (4) \wedge A' = B' \Rightarrow A' + C' = B' + C' \wedge (6) \wedge (7) \wedge (8) \wedge (9)\right)$$
$$\Rightarrow (\sum_{i=1}^{y'} i^3 = (\sum_{i=1}^{y'} i^1)^2) \Rightarrow s(y')^3 + (\sum_{i=1}^{y'} i^1)^2 =$$
$$(s(y')^2 + (2 \times \sum_{i=1}^{y'} i^1) \times s(y')) + (\sum_{i=1}^{y'} i^1)^2$$

9. By *RuleApplication from (5)*
$$\triangle_9;\sigma_9 \triangleright\; \left((1) \wedge (2) \wedge (3) \wedge (4) \wedge (5) \wedge \textstyle\sum_{i=1}^{N'} i^1 = \tfrac{N' \times s(N')}{2} \wedge (7) \wedge (8) \wedge (9)\right)$$
$$\Rightarrow (\sum_{i=1}^{y'} i^3 = (\sum_{i=1}^{y'} i^1)^2) \Rightarrow s(y')^3 = s(y')^2 + (2 \times \sum_{i=1}^{y'} i^1) \times s(y')$$

10. By *RuleApplication from (6)*

$\triangle_{10};\sigma_{10} \triangleright \Big((1) \wedge (2) \wedge (3) \wedge (4) \wedge (5) \wedge (6) \wedge 2 \times \frac{M'}{2} = M' \wedge (8) \wedge (9) \Big)$
$\Rightarrow (\sum_{i=1}^{y'} i^3 = (\sum_{i=1}^{y'} i^1)^2) \Rightarrow s(y')^3 = s(y')^2 + (2 \times \frac{y' \times s(y')}{2}) \times s(y')$

11. By *RuleApplication from (7)*

$\triangle_{11};\sigma_{11} \triangleright \Big((1) \wedge (2) \wedge (3) \wedge (4) \wedge (5) \wedge (6) \wedge (7) \wedge s(Q)^3 = s(Q)^2 + (Q \times s(Q)) \times s(Q) \wedge (9) \Big)$
$\Rightarrow (\sum_{i=1}^{y'} i^3 = (\sum_{i=1}^{y'} i^1)^2) \Rightarrow s(y')^3 = s(y')^2 + (y' \times s(y')) \times s(y')$

12. By *RuleApplication from (8)*

$\triangle_{12};\sigma_{12} \triangleright \big((1) \wedge (2) \wedge (3) \wedge (4) \wedge (5) \wedge (6) \wedge (7) \wedge (8) \wedge (9) \big) \Rightarrow (\sum_{i=1}^{y'} i^3 = (\sum_{i=1}^{y'} i^1)^2) \Rightarrow \top$

13. By *Simplify by* $2\times \Rightarrow_R^\top$: $\triangle_{13};\sigma_{13} \triangleright \top$

14. By *Axiom: q.e.d*

Discussion. The proof illustrates how the CORE calculus supports direct reasoning at the assertion level. The main focus here is less on the fact that we can represent a proof at the assertion level, but rather on the support offered by the calculus to perform a proof at the assertion level. Indeed, representing such a proof is also possible, for instance, in a natural deduction calculus, provided that it comes with a strong equality substitution rule and that the syntactical structure of the assertions fits the decomposition structure of the calculus rules. However, formulas must typically be decomposed in order to apply the contained knowledge, as for instance in the application of the induction hypothesis in step 8 in the example proof. Moreover, there is no support which actively suggests *how* the assertions can be applied. The CORE calculus provides this information as replacement rules, which are directly read out from the assertions.

5   Related Work and Conclusion

We presented the CORE calculus for contextual reasoning. It uses proof theoretic information to statically determine the available assertions for arbitrary subformulas. The proof theoretic information is further used to operationalize the application of the assertions via replacement rules. Therefore the CORE calculus actively supports direct reasoning on the assertion level.

The technical details of the calculus, for instance, that replacement rules are in principle non-normalform resolution and paramodulation rules, can mostly be hidden from the user. However, they should ease the implementation of automated proof procedures, and any proof constructed on their basis would immediately be available as an assertion level proof, which should ease the understanding of the proofs for a human user. Although in this paper we focused on higher-order logic, in [3] we define CORE calculi on top of the matrix calculi for most of the first-order modal logics considered in [18].

The CORE calculus should also provide a suitable basis to accommodate a deduction modulo [8] approach, since it allows to implement a *contextual* congruence on propositions which takes the logical context of subformulas into account through the uniform mechanism to synthesize all available rules from it. It also is in the Deep Inference paradigm for calculi, which is studied for instance in the Calculus of Structures [6]. However, rather than studying proof theoretic properties using deep structural manipulation rules, the CORE calculus aims at supporting a reasoning based on assertions through replacement rules. Finally, it is also related to Focusing Proof Construction [1] which derives sequent calculus macro steps from available assertions. Unlike replacement rules, the focusing proof steps can only be determined for top-level formulas, only be applied to top-level formulas, and do not include conditional equivalences and equations.

Future work consists of treating equality as a primitive concept, define CoRE calculi for further logics (especially intuitionistic logics), provide transformations of CoRE proofs into sequent calculus proofs to ease proof checking, and investigate automated reasoning procedures for CoRE by extending ordering-based automated reasoning techniques like [10] to non-normalform resolution and paramodulation.

References

1. J.-M. Andreoli. Focusing and proof construction. *Annals of Pure and Applied Logic*, 107(1):131–163, 2000.
2. P. B. Andrews. General models, descriptions, and choice in type theory. *The Journal of Symbolic Logic*, 37(2):385–397, June 1972.
3. S. Autexier. *Hierarchical Contextual Reasoning*. PhD thesis, Computer Science Department, Saarland University, Saarbrücken, Germany, 2003.
4. H.Barendregt. *The λ-Calculus - Its Syntax and Semantics*. North Holland, 1984.
5. C. Benzmüller, C. E. Brown, and M. Kohlhase, Semantic Techniques for Higher-Order Cut-Elimination. SEKI Tech. Report SR-2004-07, Saarland University, 2004.
6. K. Brünnler. *Deep Inference and Symmetry in Classical Proofs*. Logos, 2004.
7. A. Bundy. The use of explicit plans to guide inductive proofs. DAI Research Report 349, Department of Artificial Intelligence, University of Edinburgh, 1987.
8. G. Dowek, Th. Hardin, and C. Kirchner. Theorem proving modulo. Rapport de Recherche 3400, INRIA, April 1998.
9. M. Fitting. Tableau methods of proof for modal logics. *Notre Dame Journal of Formal Logic*, XIII:237–247, 1972.
10. H. Ganzinger and J. Stuber. Superposition with equivalence reasoning and delayed clause normal form transformation. In F. Baader, editor, *Automated Deduction — CADE-19*, volume 2741 of *LNCS*, pages 335–349. Springer-Verlag, 2003.
11. M. J. Gordon, A. J. Milner, and C. P. Wadsworth. *Edinburgh LCF – A mechanised logic of computation*. Springer Verlag, 1979. LNCS 78.
12. L. Henkin. Completeness in the theory of types. *The Journal of Symbolic Logic*, 15:81–91, 1950.
13. X. Huang. *Human Oriented Proof Presentation: A Reconstructive Approach*. Number 112 in DISKI. Infix, Sankt Augustin, Germany, 1996.
14. N. V. Murray and E. Rosenthal. Inference with path resolution and semantic graphs. *Journal of the Association of Computing Machinery*, 34(2):225–254, 1987.
15. F. Pfenning. *Proof Transformation in Higher-Order Logic*. Phd thesis, Carnegie Mellon University, 1987.
16. K. Schütte. *Proof Theory. (Originaltitel: Beweistheorie)*, volume 255 of *Die Grundlehren der mathematischen Wissenschaften*. Springer, 1977.
17. R. M. Smullyan. *First-Order Logic*, volume 43 of *Ergebnisse der Mathematik*. Springer-Verlag, Berlin, 1968.
18. L. Wallen. *Automated proof search in non-classical logics: efficient matrix proof methods for modal and intuitionistic logics*. MIT Press series in AI, 1990.

Simulating Reachability Using First-Order Logic with Applications to Verification of Linked Data Structures

T. Lev-Ami[1], N. Immerman[2,*], T. Reps[3,**], M. Sagiv[1],
S. Srivastava[2,*], and G. Yorsh[1,***]

[1] School of Comp. Sci., Tel Aviv Univ.
{tla,msagiv,gretay}@post.tau.ac.il
[2] Dept. of Comp. Sci. Univ. of Massachusetts, Amherst
{immerman,siddharth}@cs.umass.edu
[3] Comp. Sci. Dept., Univ. of Wisconsin
reps@cs.wisc.edu

Abstract. This paper shows how to harness existing theorem provers for first-order logic to automatically verify safety properties of imperative programs that perform dynamic storage allocation and destructive updating of pointer-valued structure fields. One of the main obstacles is specifying and proving the (absence) of reachability properties among dynamically allocated cells.

The main technical contributions are methods for simulating reachability in a conservative way using first-order formulas—the formulas describe a superset of the set of program states that can actually arise. These methods are employed for semi-automatic program verification (i.e., using programmer-supplied loop invariants) on programs such as mark-and-sweep garbage collection and destructive reversal of a singly linked list. (The mark-and-sweep example has been previously reported as being beyond the capabilities of ESC/Java.)

1 Introduction

This paper explores how to harness existing theorem provers for first-order logic to prove reachability properties of programs that manipulate dynamically allocated data structures. The approach that we use involves simulating reachability in a conservative way using first-order formulas—i.e., the formulas describe a superset of the set of program states that can actually arise.

Automatically establishing safety and liveness properties of sequential and concurrent programs that permit dynamic storage allocation and low-level pointer manipulations is challenging. Dynamic allocation causes the state space to be infinite; moreover, a program is permitted to mutate a data structure by destructively updating pointer-valued fields of nodes. These features remain even if a programming language has good capabilities for data abstraction. Abstract-datatype operations are implemented using

---

* Supported by NSF grant CCR-0207373 and a Guggenheim fellowship.
** Supported by ONR under contracts N00014-01-1-{0796,0708}.
*** Partially supported by the Israeli Academy of Science.

R. Nieuwenhuis (Ed.): CADE 2005, LNAI 3632, pp. 99–115, 2005.
© Springer-Verlag Berlin Heidelberg 2005

loops, procedure calls, and sequences of low-level pointer manipulations; consequently, it is hard to prove that a data-structure invariant is reestablished once a sequence of operations is finished [1]. In languages such as Java, concurrency poses yet another challenge: establishing the absence of deadlock requires establishing the absence of any cycle of threads that are waiting for locks held by other threads.

Reachability is crucial for reasoning about linked data structures. For instance, to establish that a memory configuration contains no garbage elements, we must show that every element is reachable from some program variable. Other cases where reachability is a useful notion include

- Specifying acyclicity of data-structure fragments, i.e., every element reachable from node n cannot reach n
- Specifying the effect of procedure calls when references are passed as arguments: only elements that are reachable from a formal parameter can be modified
- Specifying the absence of deadlocks
- Specifying safety conditions that allow establishing that a data-structure traversal terminates, e.g., there is a path from a node to a sink-node of the data structure.

The verification of such properties presents a challenge. Even simple decidable fragments of first-order logic become undecidable when reachability is added [2,3]. Moreover, the utility of monadic second-order logic on trees is rather limited because (i) many programs allow non-tree data structures, (ii) expressing postconditions of procedures (which is essential for modular reasoning) requires referring to the pre-state that holds before the procedure executes, and thus cannot, in general, be expressed in monadic second-order logic on trees—even for procedures that manipulate only singly-linked lists, such as the in-situ list-reversal program shown in Fig. 1 , and (iii) the complexity is prohibitive.

While our work was actually motivated by our experience using abstract interpretation – and, in particular, the TVLA system [4,5,6] – to establish properties of programs that manipulate heap-allocated data structures, in this paper, we consider the problem of verifying data-structure operations, assuming that we have user-supplied loop invariants. This is similar to the approach taken in systems like ESC/Java [7], and Pale [8].

The contributions of the paper can be summarized as follows:

Handling FO(TC) formulas using FO theorem provers. We want to use first-order theorem provers and we need to discuss the transitive closure of certain binary predicates, f. However, first-order theorem provers cannot handle transitive closure. We solve this conundrum by adding a new relation symbol f_{tc} for each such f, together with first-order axioms that assure that f_{tc} is interpreted correctly. The theoretical details of how this is done are presented in Sections 3 and 4. The fact that we are able to handle transitive closure effectively and reasonably automatically is a major contribution and quite surprising.

As explained in Section 3, the axioms that we add to control the behavior of the added predicates, f_{tc}, must be sound but not necessarily complete. One way to think about this is that we are simulating a formula, χ, in which transitive closure occurs, with a pure first-order formula χ'. If our axioms are not complete then we are allowing χ' to denote more stores than χ does. This is motivated by the fact that abstraction can be an aid in the verification of many properties; that is, a definite answer can sometimes be obtained even when information has been lost (in a conservative manner). This means that our methods are sound but potentially incomplete.

If χ' is proven valid in FO then χ is also valid in FO(TC); however, if we fail to prove that χ' is valid, it is still possible that χ is valid: the failure would be due to the incompleteness of the axioms, or the lack of time or space for the theorem prover to complete the proof.

It is easy to write a sound axiom, $T_1[f]$, that is "complete" in the very limited sense that every finite, acyclic model satisfying $T_1[f]$ must interpret f_{tc} as the reflexive, transitive closure of its interpretation of f. However, in practice this is not worth much because, as is well-known, finiteness is not expressible in first-order logic. Thus, the properties that we want to prove do not follow from $T_1[f]$. We do prove that $T_1[f]$ is complete for positive transitive-closure properties. The real difficulties lie in proving properties involving the negation of $TC[f]$.

Induction axiom scheme. To solve the above problem, we add an induction axiom scheme. Although in general, there is no complete, recursively-enumerable axiomatization of transitive closure, we have found that on the examples we have tried, T_1 plus induction allows us to automatically prove all of our desired properties. We think of the axioms that we use as aides for the first-order theorem prover that we employ (Spass [9]) to prove the properties in question. Rather than giving Spass many instances of the induction scheme, our experience is that it finds the proof faster if we give it several axioms that are simpler to use than induction. As already mentioned, the hard part is to show that certain paths do not exist.

Coloring axiom schemes. In particular, we use three axiom schemes, having to do with partitioning memory into a small set of colors. We call instances of these schemes "coloring axioms". Our coloring axioms are simple, and are *easily proved using* Spass *(in under ten seconds) from the induction axioms*. For example, the first coloring axiom scheme, **NoExit**$[A, f]$, says that if no f-edges leave color class, A, then no f-paths leave A. It turns out that the **NoExit** axiom scheme implies – and thus is equivalent to – the induction scheme. However, we have found in practice that explicitly adding other coloring axioms (which are consequences of **NoExit**) enables Spass to prove properties that it otherwise fails at.

We first assume that the programmer provides the colors by means of first-order formulas with transitive closure. Our initial experience indicates that the generated coloring axioms are useful to Spass. In particular, it provides the ability to verify programs like the mark phase of a mark-and-sweep garbage collector. This example has been previously reported as being beyond the capabilities of ESC/Java. TVLA also succeeds on this example; however our new approach provides verification methods that can in some instances be more precise than TVLA.

Prototype implementation. Perhaps most exciting, we have implemented the heuristics for selecting colors and their corresponding axioms in a prototype using Spass. We have used this to automatically choose useful color axioms and then verify several small heap-manipulating programs. More work needs to be done here, but the initial results are very encouraging.

Strengthening Nelson's results. Greg Nelson considered a set of axiom schemes for reasoning about reachability in function graphs, i.e., graphs in which there is at most one f-edge leaving any node [10]. He left open the question of whether his axiom schemes were complete for function graphs. We show that Nelson's axioms are provable from T_1 plus our induction axioms. We also show that Nelson's axioms are not complete: in fact, they do not imply **NoExit**.

Outline. The remainder of the paper is organized as follows: Section 2 explains our notation and the setting; Section 3 introduces the induction axiom scheme and fills in our formal framework; Section 4 states the coloring axiom schemes; Section 5 explains the details of our heuristics; Section 6 describes some related work; Section 7 describes some future directions.

2   Preliminaries

This section defines the basic notations used in this paper and the setting.

2.1   Notation

Syntax: A relational **vocabulary** $\tau = \{p_1, p_2, \ldots, p_k\}$ is a set of relation symbols, each of fixed arity. We write first-order formulas over τ with quantifiers $\forall$ and $\exists$, logical connectives $\wedge$, $\vee$, $\rightarrow$, $\leftrightarrow$, and $\neg$, where atomic formulas include: equality, $p_i(v_1, v_2, \ldots v_{a_i})$, and $\mathrm{TC}[f](v_1, v_2)$, where $p_i \in \tau$ is of arity a_i and $f \in \tau$ is binary. Here $\mathrm{TC}[f](v_1, v_2)$ denotes the existence of a finite path of 0 or more f edges from v_1 to v_2. A formula without TC is called a **first-order** formula.

We use the following precedence of logical operators: $\neg$ has highest precedence, followed by $\wedge$ and $\vee$, followed by $\rightarrow$ and $\leftrightarrow$, and $\forall$ and $\exists$ have lowest precedence.

Semantics: A **model**, $\mathcal{A}$, of vocabulary τ, consists of a non-empty universe, $|\mathcal{A}|$, and a relation $p^{\mathcal{A}}$ over the universe interpreting each relation symbol $p \in \tau$. We write $\mathcal{A} \models \varphi$ to mean that the formula φ is true in the model $\mathcal{A}$.

2.2   Setting

We are primarily interested in formulas that arise while proving the correctness of programs. We assume that the programmer specifies pre and post-conditions for procedures and loop invariants using first-order formulas with transitive closure on binary relations. The transformer for a loop body can be produced automatically from the program code.

For instance, to establish the partial correctness with respect to a user-supplied specification of a program that contains a single loop, we need to establish three properties: First, the loop invariant must hold at the beginning of the first iteration; i.e., we must show that the loop invariant follows from the precondition and the code leading to the loop. Second, the loop invariant provided by the user must be maintained; i.e., we must show that if the loop invariant holds at the beginning of an iteration and the loop condition also holds, the transformer causes the loop invariant to hold at the end of the iteration. Finally, the postcondition must follow from the loop invariant and the condition for exiting the loop.

In general, these formulas are of the form

$$\psi_1[\tau] \wedge T[\tau, \tau'] \rightarrow \psi_2[\tau']$$

where τ is the vocabulary of the before state, τ' is the vocabulary of the after state, and T is the transformer, which may use both the before and after predicates to describe the meaning of the module to be executed. If symbol f denotes the value of a predicate before the operation then f' denotes the value of the same predicate after the operation.

An interesting special case is the proof of the maintenance formula of a loop invariant. This has the form:

$$LC[\tau] \wedge LI[\tau] \wedge T[\tau, \tau'] \rightarrow LI[\tau']$$

Here LC is the condition for entering the loop and LI is the loop invariant. $LI[\tau']$ indicates that the loop invariant remains true after the body of the loop is executed.

The challenge is that the formulas of interest contain transitive closure; thus, the validity of these formulas cannot be directly proven using a theorem prover for first-order logic.

3 Axiomatization of Transitive Closure

The original formula that we want to prove, χ, contains transitive closure, which first-order theorem provers cannot handle. To address this problem, we replace χ by the new formula, χ', where all appearances of $TC[f]$ have been replaced by the new binary relation symbol, f_{tc}.

We show in this paper that from χ', we can often automatically generate an appropriate first-order axiom, σ, with the following two properties:

1. if $\sigma \rightarrow \chi'$ is valid in FO then χ is valid in FO(TC).
2. A theorem prover successfully proves that $\sigma \rightarrow \chi'$ is valid in FO.

We now explain the theory behind this process. A **TC model**, $\mathcal{A}$, is a model such that if f and f_{tc} are in the vocabulary of $\mathcal{A}$, then $(f_{tc})^{\mathcal{A}} = (f^{\mathcal{A}})^{\star}$; i.e., $\mathcal{A}$ interprets f_{tc} as the reflexive, transitive closure of its interpretation of f.

A first-order formula φ is **TC valid** iff it is true in all TC models. We say that an axiomatization, Σ, is **TC sound** if every formula that follows from Σ is TC valid. Since first-order reasoning is sound, Σ is TC sound iff every $\sigma \in \Sigma$ is TC valid.

We say that Σ is **TC complete** if for every TC-valid φ, $\Sigma \models \varphi$. If Σ is TC complete and TC sound, then for all first-order φ,

$$\Sigma \models \varphi \quad \Leftrightarrow \quad \varphi \text{ is TC valid}$$

Thus a TC-complete set of axioms proves exactly the first-order formulas, χ', such that the corresponding FO(TC) formula, χ, is valid.

All the axiomatizations that we consider are TC sound. There is no recursively enumerable TC-complete axiom system (see [11,12]).

3.1 Some TC-Sound Axioms

We begin with our first TC axiom scheme. For any binary relation symbol, f, let,

$$T_1[f] \quad \equiv \quad \forall u, v . f_{tc}(u, v) \ \leftrightarrow \ (u = v) \vee \exists w . f(u, w) \wedge f_{tc}(w, v)$$

We first observe that $T_1[f]$ is "complete" in a very limited way for finite, acyclic graphs, i.e., $T_1[f]$ exactly characterizes the meaning of f_{tc} for all finite, acyclic graphs. The reason this is limited, is that it does not give us a complete set of first-order axioms because, as is well known, there is no first-order axiomatization of "finite".

Proposition 1. *Any finite and acyclic model of* $T_1[f]$ *is a TC model.*

Proof: Let $\mathcal{A} \models T_1[f]$ where $\mathcal{A}$ is finite and acyclic. Let $a_0, b \in |\mathcal{A}|$. Assume there is an f-path from a_0 to b. Since $\mathcal{A} \models T_1[f]$, it is easy to see that $\mathcal{A} \models f_{tc}(a_0, b)$.

Conversely, suppose that $\mathcal{A} \models f_{tc}(a_0, b)$. If $a_0 = b$, then there is a path of length 0 from a_0 to b. Otherwise, by $T_1[f]$, there exists an $a_1 \in |\mathcal{A}|$ such that $\mathcal{A} \models f(a_0, a_1) \wedge f_{tc}(a_1, b)$. Note that $a_1 \neq a_0$ since $\mathcal{A}$ is acyclic. If $a_1 = b$ then there is an f-path of length 1 from a to b. Otherwise there must exist an $a_2 \in |\mathcal{A}|$ such that $\mathcal{A} \models f(a_1, a_2) \wedge f_{tc}(a_2, b)$ and so on, generating a set $\{a_1, a_2, \ldots\}$. None of the a_i can be equal to a_j, for $j < i$, by acyclicity. Thus, by finiteness, some $a_i = b$. Hence $\mathcal{A}$ is a TC model. □

Let $T_1'[f]$ be the $\leftarrow$ direction of $T_1[f]$:

$$T_1'[f] \quad \equiv \quad \forall u, v . f_{tc}(u, v) \ \leftarrow \ (u = v) \vee \exists w . f(u, w) \wedge f_{tc}(w, v)$$

Proposition 2. *Let* f_{tc} *occur only positively in* φ. *If* φ *is TC valid, then* $T_1'[f] \models \varphi$.

Proof: Suppose that $T_1'[f] \not\models \varphi$. Let $\mathcal{A} \models T_1'[f] \wedge \neg \varphi$. Note that f_{tc} occurs only negatively in $\neg \varphi$. Furthermore, since $\mathcal{A} \models T_1'[f]$, it is easy to show by induction on the length of the path, that if there is an f-path from a to b in $\mathcal{A}$, then $\mathcal{A} \models f_{tc}(a, b)$. Define $\mathcal{A}'$ to be the model formed from $\mathcal{A}$ by interpreting f_{tc} in $\mathcal{A}'$ as $(f^{\mathcal{A}})^\star$. Thus $\mathcal{A}'$ is a TC model and it only differs from $\mathcal{A}$ by the fact that we have removed zero or more pairs from $(f_{tc})^{\mathcal{A}}$ to form $(f_{tc})^{\mathcal{A}'}$. Because $\mathcal{A} \models \neg \varphi$ and f_{tc} occurs only negatively in $\neg \varphi$, it follows that $\mathcal{A}' \models \neg \varphi$, which contradicts the assumption that φ is TC valid. □

Proposition 2 shows that proving positive facts of the form $f_{tc}(u, v)$ is easy; it is the task of proving that paths do not exist that is more subtle.

Proposition 1 shows that what we are missing, at least in the acyclic case, is that there is no first-order axiomatization of finiteness. Traditionally, when reasoning about the natural numbers, this problem is mitigated by adding induction axioms. We next introduce an induction scheme that, together with T_1, seems to be sufficient to prove any property we need concerning TC.

Notation: In general, we will use F to denote the set of all binary relation symbols, f, such that $TC[f]$ occurs in a formula we are considering. If $\varphi[f]$ is a formula in which f occurs, let $\varphi[F] = \bigwedge_{f \in F} \varphi(f)$. Thus, for example, $T_1[F]$ is the conjunction of the axiom $T_1[f]$ for all binary relation symbols, f, under consideration.

Definition 1. *For any first-order formulas $Z(u)$, $P(u)$, and binary relation symbol, f, let the **induction principle**, $\mathbf{IND}[Z, P, f]$, be the following first-order formula:*

$$(\forall z \,.\, Z(z) \rightarrow P(z)) \;\wedge\; (\forall u, v \,.\, P(u) \wedge f(u, v) \rightarrow P(v))$$
$$\rightarrow \forall u, z \,.\, Z(z) \wedge f_{\mathrm{tc}}(z, u) \rightarrow P(u)$$

The induction principle says that if every zero point satisfies P, and P is preserved when following edges, then every point reachable from a zero point satisfies P. Obviously this principle is sound.

As an easy application of the induction principle, consider the following cousin of $T_1[f]$,

$$T_2[f] \quad \equiv \quad \forall u, v \,.\, f_{\mathrm{tc}}(u, v) \;\leftrightarrow\; (u = v) \vee \exists w \,.\, f_{\mathrm{tc}}(u, w) \wedge f(w, v)$$

It is easy to see that neither of $T_1[f]$, $T_2[f]$ implies the other. However, in the presence of the induction principle they do imply each other. For example, it is easy to prove $T_2[f]$ from $T_1[f]$ using $\mathbf{IND}[Z, P, f]$ where $Z(v) \equiv v = u$ and $P(v) \equiv u = v \vee \exists w \,.\, f_{\mathrm{tc}}(u, w) \wedge f(w, v)$. Here, for each u we use $\mathbf{IND}[Z, P, f]$ to prove by induction that every v reachable from u satisfies the right-hand side of $T_2[f]$.

A related axiom scheme that we have found useful is the transitivity of reachability:

$$\mathbf{Trans}[f] \quad \equiv \quad \forall u, v, w \,.\, f_{\mathrm{tc}}(u, w) \wedge f_{\mathrm{tc}}(w, v) \rightarrow f_{\mathrm{tc}}(u, v)$$

4   Coloring Axioms

We next describe three TC-sound axioms schemes that are not implied by $T_1[F] \wedge T_2[F]$, and are provable from the induction principle of Section 3. We will see in the sequel that these coloring axioms are very useful in proving that paths do not exist, permitting us to verify a variety of algorithms. In Section 5, we will present some heuristics for automatically choosing particular instances of the coloring axiom schemes that enable us to prove our goal formulas.

The first coloring axiom scheme is the NoExit axiom scheme:

$$(\forall u, v \,.\, A(u) \wedge \neg A(v) \rightarrow \neg f(u, v)) \quad \rightarrow \quad \forall u, v \,.\, A(u) \wedge \neg A(v) \rightarrow \neg f_{\mathrm{tc}}(u, v) \quad (1)$$

for any first-order formula $A(u)$, and binary relation symbol, f, $\mathbf{NoExit}[A, f]$ says that if no f-edge leaves color class A, then no f-path leaves color class A.

Observe that although it is very simple, $\mathbf{NoExit}[A, f]$ does not follow from $T_1[f] \wedge T_2[f]$. Let $G_1 = (V, f, f_{\mathrm{tc}}, A)$ consist of two disjoint cycles: $V = \{1, 2, 3, 4\}$, $f = \{\langle 1, 2 \rangle, \langle 2, 1 \rangle, \langle 3, 4 \rangle, \langle 4, 3 \rangle\}$, and $A = \{1, 2\}$. Let f_{tc} have all 16 possible edges. Thus G_1 satisfies $T_1[f] \wedge T_2[f]$ but violates $\mathbf{NoExit}[A, f]$. Even for acyclic models, $\mathbf{NoExit}[A, f]$ does not follow from $T_1[f] \wedge T_2[f]$ because there are infinite models in which the implication does not hold (see [12]).

$\mathbf{NoExit}[A, f]$ follows easily from the induction principle: if no edges leave A, then induction tells us that everything reachable from a point in A satisfies A. Similarly, $\mathbf{NoExit}[A, f]$ implies the induction axiom, $\mathbf{IND}[Z, A, f]$, for any formula Z.

The second coloring axiom scheme is the GoOut axiom: for any first-order formulas $A(u)$, $B(u)$, and binary relation symbol, f, **GoOut**$[A, B, f]$ says that if the only edges leaving color class A are to B, then any path from a point in A to a point not in A must pass through B.

$$(\forall u, v \,.\, A(u) \wedge \neg A(v) \wedge f(u, v) \rightarrow B(v)) \rightarrow$$
$$\forall u, v \,.\, A(u) \wedge \neg A(v) \wedge f_{tc}(u, v) \rightarrow \exists b \,.\, B(b) \wedge f_{tc}(u, b) \wedge f_{tc}(b, v) \qquad (2)$$

To see that **GoOut**$[A, B, f]$ follows from the induction principle, assume that the only edges out of A enter B. For any fixed u in A, we prove by induction that any point v reachable from u is either in A or has a predecessor, b in B, that is reachable from u.

The third coloring axiom scheme is the **NewStart** axiom, which is useful in the context of dynamically changing graphs: for any first-order formula $A(u)$, and binary relation symbols f and g, think of f as the previous edge relation and g as the current edge relation. **NewStart**$[A, f, g]$ says that if there are no new edges between A nodes, then any new path from A must leave A to make its change:

$$(\forall u, v \,.\, A(u) \wedge A(v) \wedge g(u, v) \rightarrow f(u, v)) \rightarrow$$
$$\forall u, v \,.\, g_{tc}(u, v) \wedge \neg f_{tc}(u, v) \rightarrow \exists b \,.\, \neg A(b) \wedge g_{tc}(u, b) \wedge g_{tc}(b, v) \qquad (3)$$

NewStart$[A, f, g]$ follows from the induction principle by a proof that is similar to the proof of **GoOut**$[A, B, f]$

We remark that the spirit behind our consideration of the coloring axioms is similar to that found in a paper of Greg Nelson's in which he introduced a set of reachability axioms for a functional predicate, f, i.e., there is at most one f edge leaving any point [10]. Nelson asked whether his axiom schemes are complete for the functional setting. We remark that Nelson's axiom schemes are provable from T_1 plus our induction principle. However, Nelson's axiom schemes are not complete: we constructed a functional graph satisfying Nelson's axioms but violating **NoExit**$[A, f]$, (see [12]).

At least one of Nelson's axiom schemes does seem orthogonal to our coloring axioms and may be useful in certain proofs. Nelson's fifth axiom scheme states that the points reachable from a given point are linearly ordered. The soundness of the axiom scheme is due to the fact that f is functional. We make use of a simplified version of Nelson's ordering axiom scheme: Let **Func**$[f] \equiv \forall u, v, w \,.\, f(u, v) \wedge f(u, w) \rightarrow v = w$; then,

Order$[f] \equiv$ **Func**$[f] \rightarrow \forall u, v, w \,.\, f_{tc}(u, v) \wedge f_{tc}(u, w) \rightarrow f_{tc}(v, w) \vee f_{tc}(w, v)$

5 Heuristics for Using the Coloring Axioms

This section presents heuristics for using the coloring axioms. Toward that end, it answers the following questions:

- How can the coloring axioms be used by a theorem prover to prove χ?
- When should a specific instance of a coloring axiom be given to the theorem prover while trying to prove χ?
- What part of the process can be automated?

We first present a running example that will be used in later sections to illustrate the heuristics. We then explain how the coloring axioms are useful, describe the search space for useful axioms, give an algorithm for exploring this space, and conclude by discussing a prototype implementation we have developed that proves the example presented and others.

5.1   Reverse Specification

The heuristics described in Sections 5.2–5.4 are illustrated on problems that arise in the verification of partial correctness of a list reversal procedure. Other examples proven using this technique can be found in the full version of this paper [12].

The procedure reverse, shown in Fig. 1, performs in-place reversal of a singly linked list, destructively updating the list. The precondition requires that the input list be acyclic and unshared. For simplicity, we assume that there is no garbage. The post-condition ensures that the resulting list is acyclic and unshared. Also, it ensures that the nodes reachable from the formal parameter on entry to reverse are exactly the nodes reachable from the return value of reverse at the exit. Most importantly, it ensures that each edge in the original list is reversed in the returned list.

The specification for reverse is shown in Fig. 2. We use unary predicates to represent program variables and binary predicates to represent data-structure fields. Fig. 2(a) defines some shorthands. To specify that a unary predicate z can point to a single node at a time and that a binary predicate f of a node can point to at most one node (a partial function), we use $unique[z]$ and $func[f]$. To specify that there are no cycles of f-fields in the graph, we use $acyclic[f]$. To specify that the graph does not contain nodes shared by f-fields, (i.e., nodes with 2 or more incoming f-fields), we use $unshared[f]$. To specify that all nodes in the graph are reachable from z_1 or z_2 by following f-fields, we use $total[z_1, z_2, f]$. Another helpful shorthand is $r_{x,f}(v)$ which specifies that v is reachable from the node pointed to by x using f-edges.

```
Node reverse(Node x){
 [0]  Node y = null;
 [1]  while (x != null){
 [2]    Node t = x.next;
 [3]    x.next = y;
 [4]    y = x;
 [5]    x = t;
 [6]  }
 [7]  return y;
}
```

Fig. 1. A simple Java-like implementation of the in-place reversal of a singly-linked list

The precondition of the reverse procedure is shown in Fig. 2(b). We use the predicates xe and ne to record the values of the variable x and the next field at the beginning of the procedure. The precondition requires that the list pointed to by x be acyclic and unshared. It also requires that $unique[z]$ and $func[f]$ hold for all unary predicates z that represent program variables and all binary predicates f that represent fields, respectively. For simplicity, we assume that there is no garbage, i.e., all nodes are reachable from x.

The post-condition is shown in Fig. 2(c). It ensures that the resulting list is acyclic and unshared. Also, it ensures that the nodes reachable from the formal parameter x on entry to the procedure are exactly the nodes reachable from the return value y at the

exit. Most importantly, we wish to show that each edge in the original list is reversed in
the returned list (see Eq. (11)).

A loop invariant is given in Fig. 2(d). It describes the state of the program at the
beginning of each loop iteration. Every node is in one of two disjoint lists pointed by x
and y (Eq. (12)). The lists are acyclic and unshared. Every edge in the list pointed to by
x is exactly an edge in the original list (Eq. (14)). Every edge in the list pointed to by y
is the reverse of an edge in the original list (Eq. (15)). The only original edge going out
of y is to x (Eq. (16)).

The transformer is given in Fig. 2(e), using the primed predicates n', x', and y' to
describe the values of predicates n, x, and y, respectively, at the end of the iteration.

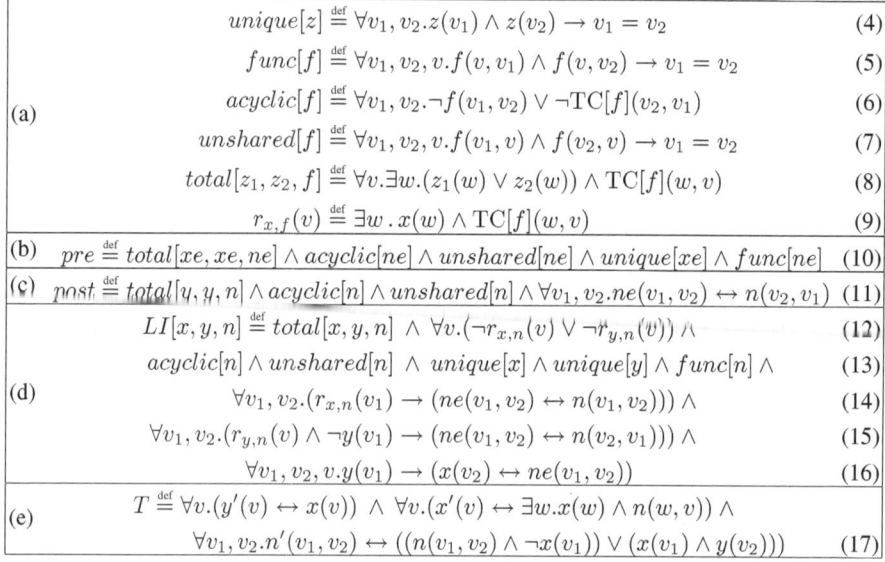

Fig. 2. Example specification of reverse procedure: (a) shorthands, (b) precondition pre, (c) postcondition $post$, (d) loop invariant $LI[x, y, n]$, (e) transformer T (effect of the loop body)

5.2 Proving Formulas Using the Coloring Axioms

All the coloring axioms have the form $A \equiv P_A \rightarrow C_A$, where P_A and C_A are closed
formulas. We call P_A the axiom's premise and C_A the axiom's conclusion. For an axiom
to be useful, the theorem prover will have to prove the premise (as a subgoal) and then
use the conclusion in the proof of the goal formula χ. For each of the coloring axioms,
we now explain when the premise can be proved, how its conclusion can help, and give
an example.

NoExit. The premise $P_{\mathbf{NoExit}}[C, f]$ states that there are no f-edges exiting color class
C. When C is a unary predicate appearing in the program, the premise is sometimes a
direct result of the loop invariant. Another color that will be used heavily throughout
this section is reachability from a unary predicate, i.e., unary reachability, formally

defined in Eq. (9). Let's examine two cases. $P_{\mathbf{NoExit}}[r_{x,f}, f]$ is immediate from the definition of $r_{x,f}$ and the transitivity of f_{tc}. $P_{\mathbf{NoExit}}[r_{x,f}, f']$ actually states that there is no f path from x to an edge for which f' holds but f doesn't, i.e., a change in f' with respect to f. Thus, we use the absence of f-paths to prove the absence of f'-paths. In many cases, the change is an important part of the loop invariant, and paths from and to it are part of the specification.

A sketch of the proof by refutation of $P_{\mathbf{NoExit}}[r_{x',n}, n']$ that arises in the reverse example is given in Fig. 3. The numbers in brackets are the stages of the proof.

1. The negation of the premise expands to:

$$\exists u_1, u_2, u_3 . x'(u_1) \wedge n_{\mathrm{tc}}(u_1, u_2) \wedge \neg n_{\mathrm{tc}}(u_1, u_3) \wedge n'(u_2, u_3)$$

2. Since u_2 is reachable from u_1 and u_3 is not, by transitivity of n_{tc}, we have $\neg n(u_2, u_3)$.
3. By the definition of n' in the transformer, the only edge in which n differs from n' is out of x (one of the clauses generated from Eq. (17) is $\forall v_1, v_2 . \neg n'(v_1, v_2) \vee \neg n(v_1, v_2) \vee x(v_1))$. Thus, $x(u_2)$ holds.
4. By the definition of x' it has an incoming n edge from x. Thus, $n(u_2, u_1)$ holds.

The list pointed to by x must be acyclic, whereas we have a cycle between u_1 and u_2; i.e., we have a contradiction. Thus, $P_{\mathbf{NoExit}}[r_{x',n}, n']$ must hold.

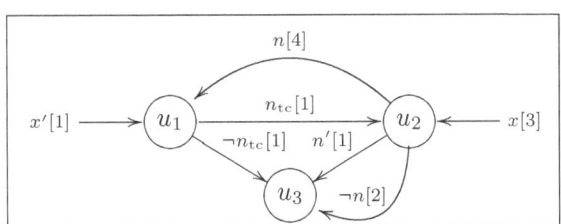

Fig. 3. Proving $P_{\mathbf{NoExit}}[r_{x,n}, n']$

$C_{\mathbf{NoExit}}[C, f]$ states there are no f paths (f_{tc} edges) exiting C. This is useful because proving the absence of paths is the difficult part of proving formulas with TC.

GoOut. The premise $P_{\mathbf{GoOut}}[A, B, f]$ states that all f edges going out of color class A, go to B. When A and B are unary predicates that appear in the program, again the premise sometimes holds as a direct result of the loop invariant. An interesting special case is when B is defined as $\exists w . A(w) \wedge f(w, v)$. In this case the premise is immediate. Note that in this case the conclusion is provable also from T_1. However, from experience, the axiom is very useful for improving performance (2 orders of magnitude when proving the acyclic part of reverse's post condition).

$C_{\mathbf{GoOut}}[A, B, f]$ states that all paths out of A must pass through B. Thus, under the premise $P_{\mathbf{GoOut}}[A, B, f]$, if we know that there is a path from A to somewhere outside of A, we know that there is a path to there from B. In case all nodes in B are reachable from all nodes in A, together with the transitivity of f_{tc} this means that the nodes reachable from B are exactly the nodes outside of A that are reachable from A.

For example, $C_{\text{GoOut}}[y', y, n']$ allows us to prove that only the original list pointed to by y is reachable from y' (in addition to y' itself).

NewStart. The premise $P_{\text{NewStart}}[C, g, h]$ states that all g edges between nodes in C are also h edges. This can mean the iteration has not added edges or has not removed edges according to the selection of h and g. In some cases, the premise holds as a direct result of the definition of C and the loop invariant.

$C_{\text{NewStart}}[C, g, h]$ means that every g path that is not an h path must pass outside of C. Together with $C_{\text{NoExit}}[C, g]$, it proves there are no new paths within C.

For example, in reverse the **NewStart** scheme can be used as follows. No outgoing edges were added to nodes reachable from y. There are no n or n' edges from nodes reachable from y to nodes not reachable from y. Thus, no paths were added between nodes reachable from y. Since the list pointed to by y is acyclic before the loop body, we can prove that it is acyclic at the end of the loop body.

We can see that **NewStart** allows the theorem prover to reason about paths within a color, and the other axioms allow the theorem prover to reason about paths between colors. Together, given enough colors, the theorem prover can often prove all the facts that it needs about paths and thus prove the formula of interest.

5.3   The Search Space of Possible Axioms

To answer the question of when we should use a specific instance of a coloring axiom when attempting to prove the target formula, we first define the search space in which we are looking for such instances. The axioms can be instantiated with the colors defined by an arbitrary unary formula (one free variable) and one or two binary predicates. First, we limit ourselves to binary predicates for which TC was used in the target formula. Now, since it is infeasible to consider all arbitrary unary formulas, we start limiting the set of colors we consider.

The initial set of colors to consider are unary predicates that occur in the formula we want to prove. Interestingly enough, these colors are enough to prove that the postcondition of mark and sweep is implied by the loop invariant, because the only axiom we need is **NoExit**$[marked, f]$.

An immediate extension that is very effective is reachability from unary predicates, as defined in Eq. (9). Instantiating all possible axioms from the unary predicates appearing in the formula and their unary reachability predicates, allows us to prove reverse. For a list of the axioms needed to prove reverse, see Fig. 4. Other example are presented in [12]. Finally, we consider Boolean combinations of the above colors. Though not used in the examples shown in this paper, this is needed, for example, in the presence of sharing or when splicing two lists together.

All the colors above are based on the unary predicates that appear in the original formula. To prove the reverse example, we needed x' as part of the initial colors. Table 1 gives a heuristic for finding the initial colors we need in cases when they cannot be deduced from the formula, and how it applies to reverse

An interesting observation is that the initial colors we need can, in many cases, be deduced from the program code. As in the previous section, we have a good way for deducing paths between colors and within colors in which the edges have not changed.

NoExit$[r_{x',n}, n']$	**GoOut**$[x, x', n]$	**NewStart**$[r_{x',n},\ n, n']$	**NewStart**$[r_{x',n}, n', n]$
NoExit$[r_{x',n'}, n]$	**GoOut**$[x, y, n']$	**NewStart**$[r_{x',n'}, n, n']$	**NewStart**$[r_{x',n'}, n', n]$
NoExit$[r_{y,n}, n']$		**NewStart**$[r_{y,n},\ n, n']$	**NewStart**$[r_{y,n}, n', n]$
NoExit$[r_{y,n'}, n]$		**NewStart**$[r_{y,n'}, n, n']$	**NewStart**$[r_{y,n'}, n', n]$

Fig. 4. The instances of coloring axioms used in proving reverse

Table 1. (a) Heuristic for choosing initial colors. (b) Results of applying the heuristic on reverse

Group	Criteria
Roots[f]	All changes are reachable from one of the colors using f_{tc}
StartChange[f,g]	All edges for which f and g differ start from a node in these colors
EndChange[f,g]	All edges for which f and g differ end at a node in these colors

(a)

Group	Colors	Group	Colors
$Roots[n]$	$x(v),\ y(v)$	$StartChange[n, n']$	$x(v)$
$Roots[n']$	$x'(v),\ y'(v)$	$EndChange[n, n']$	$y(v),\ x'(v)$

(b)

The program usually manipulates fields using pointers, and can traverse an edge only in one direction. Thus, the unary predicates that represent the program variables (including the temporary variables) are in many cases what we need as initial colors.

5.4 Exploring the Search Space

When trying to automate the process of choosing colors, the problem is that the set of possible colors to choose from is doubly-exponential in the number of initial colors; giving all the axioms directly to the theorem prover is infeasible. In this section, we define a heuristic algorithm for exploring a limited number of axioms in a directed way. Pseudocode for this algorithm is shown in Fig. 5. The operator $\vdash$ is implemented as a call to a theorem prover.

Because the coloring axioms have the form $A \equiv P_A \rightarrow C_A$, the theorem prover must prove P_X or the axiom is of no use. Therefore, the pseudocode works iteratively, trying to prove P_A from the current $\psi \wedge \Sigma$, and if successful it adds C_A to Σ.

The algorithm tries colors in increasing levels of complexity. $BC(i, C)$ gives all the Boolean combinations of the predicates in C up to size i. After each iteration we try to prove the goal formula. Sometimes we need the conclusion of one axiom to prove the premise of another. The **NoExit** axioms are particularly useful for proving $P_{\textbf{NewStart}}$. Therefore, we need a way to order instantiations so that axioms useful for proving the premises of other axioms are acquired first. The ordering we chose is based on phases: First, try to instantiate axioms from the axiom scheme **GoOut**. Second, try to instantiate axioms from the axiom scheme **NoExit**. Finally, try to instantiate axioms from the axiom scheme **NewStart**. For **NewStart**$[c, f, g]$ to be useful, we need to be able to show that there are either no incoming f paths or no outgoing f paths from c. Thus,

```
explore(Init, χ) {
    Let χ = ψ → φ
    Σ := {Trans[f], Order[f] | f ∈ F}
    Σ := Σ ∪ {T₁[f], T₂[f] | f ∈ F}
    C := {r_{c,f}(v) | c ∈ Init, f ∈ F}
    C := C ∪ Init
    i := 1
    forever {
        C' := BC(i, C)
        phase1(C', Σ, ψ)
        phase2(C', Σ, ψ)
        phase3(Σ, ψ)
        if Σ ∧ ψ ⊢ φ
            return SUCCESS
        i := i + 1
    }
}
```

```
phase1(C, Σ, ψ) {
    foreach f ∈ F, c_s ≠ c_e ∈ C
        if Σ ∧ ψ ⊢ P_GoOut[c_s, c_e, f]
            Σ := Σ ∪ {C_GoOut[c_s, c_e, f]}
}
phase2(C, Σ, ψ) {
    foreach f ∈ F, c ∈ C
        if Σ ∧ ψ ⊢ P_NoExit[c, f]
            Σ := Σ ∪ {C_NoExit[c, f]}
}
phase3(Σ, ψ) {
    foreach C_NoExit[c, f] ∈ Σ, g ≠ f ∈ F
        if Σ ∧ ψ ⊢ P_NewStart[c, f, g]
            Σ := Σ ∪ {C_NewStart[c, f, g]}
}
```

Fig. 5. An iterative algorithm for instantiating the axiom schemes. Each iteration consists of three phases that augment the axiom set Σ

we only try to instantiate such an axiom when either $P_{\mathbf{NoExit}}[c, f]$ or $P_{\mathbf{NoExit}}[\neg c, f]$ were proven.

5.5 Implementation

The algorithm presented here was implemented using a `Perl` script and the `Spass` theorem prover [9] and used successfully to verify the example programs of Section 5.1.

The method described above can be optimized. For instance, if C_A has already been added to the axioms, we do not try to prove P_A again. These details are important in practice, but have been omitted for brevity.

When trying to prove the different premises, `Spass` may fail to terminate if the formula that it is trying to prove is invalid. Thus, we limit the time that `Spass` can spend proving each formula. It is possible that we will fail to acquire useful axioms this way.

6 Related Work

Shape Analysis. This work was motivated by our experience with TVLA [4,5], which is a generic system for abstract interpretation [13]. The TVLA system is more auto-matic than the methods described in this paper since it does not rely on user-supplied loop invariants. However, the techniques presented in the present paper are potentially more precise due to the use of full first-order reasoning. It can be shown that the **NoExit** scheme allows to infer reachability at least as precisely as evaluation rules for 3-valued logic with Kleene semantics. In the future, we hope to develop an efficient non-interactive theorem prover that enjoys the benefits of both approaches. An interesting observation is that the colors needed in our examples to prove the formula are the same unary predicates used by TVLA to define its abstraction. This similarity may, in the future, help us find better ways to automatically instantiate the required axioms. In particular, inductive logic programming has recently been used to learn formulas to use in TVLA abstractions [14], which holds out the possibility of applying similar methods to further automate the approach of the present paper.

Decidable Logics. Decidable logics can be employed to define properties of linked data structures: Weak monadic second-order logic has been used in [15,8] to define properties of heap-allocated data structures, and to conduct Hoare-style verification using programmer-supplied loop invariants in the PALE system [8]. A decidable logic called L_r (for "logic of reachability expressions") was defined in [16]. L_r is rich enough to express the shape descriptors studied in [17] and the path matrices introduced in [18].

The present paper does not develop decision procedures, but instead suggests methods that can be used in conjunction with existing theorem provers. Thus, the techniques are incomplete and the theorem provers need not terminate. However, our initial experience is that the extra flexibility gained by the use of first-order logic with transitive closure is promising. For example, we can prove the correctness of imperative destructive list-reversal specified in a natural way and the correctness of mark and sweep garbage collectors, which are beyond the scope of Mona and L_r.

Indeed, in [19], we have tried to simulate existing data structures using decidable logics and realized that this can be tricky because the programmer may need to prove a specific simulation invariant for a given program. Giving an inaccurate simulation invariant causes the simulation to be unsound. One of the advantages of the technique described in the present paper is that soundness is guaranteed no matter which axioms are instantiated. Moreover, the simulation requirements are not necessarily expressible in the decidable logic.

Other First-Order Axiomatizations of Linked Data Structures. The closest approach to ours that we are aware of was taken by Nelson as we describe in the full version of the paper [12]. This also has some follow-up work by Leino and Joshi [20]. Our impression from their write-up is that Leino and Joshi's work can be pushed forward by using our coloring axioms.

Dynamic Maintenance of Transitive Closure. Another orthogonal but promising approach to transitive closure is to maintain reachability relations incrementally as we make unit changes in the data structure. It is known that in many cases, reachability can be maintained by first-order formulas [21,22] and even sometimes by quantifier-free formulas [23]. Furthermore, in these cases, it is often possible to automatically derive the first-order update predicates using finite differencing [24].

7   Conclusion

This paper reports on our initial attempts at applying the methodology that has been described; hence, only preliminary conclusions can be drawn.

As mentioned earlier, proving the absence of paths is the difficult part of proving formulas with TC. The promise of the approach is that it is able to handle such formulas effectively and reasonably automatically, as shown by the fact that it can successfully handle the programs described in Section 5 and the full version of the paper [12]. Many issues remain for further work, such as,

- Establishing whether $T_1[F]$ plus the induction scheme is complete for interesting subclasses of formulas (e.g. functional graphs).
- Exploring other heuristics for identifying color classes.
- Exploring variations of the algorithm given in Fig. 5 for instantiating coloring axioms.
- Exploring the use of additional axiom schemes, such as two of the schemes from [10], which are likely to be useful when dealing with predicates that are partial functions. Such predicates arise in programs that manipulate singly-linked or doubly-linked lists—or, more generally, data structures that are acyclic in one or more "dimensions" [25] (i.e., in which the iterated application of a given field selector can never return to a previously visited node).

Thanks to Aharon Abadi and Roman Manevich for interesting suggestions.

References

1. Hoare, C.: Recursive data structures. Int. J. of Comp. and Inf. Sci. **4** (1975) 105–132
2. Grädel, E., M.Otto, E.Rosen: Undecidability results on two-variable logics. Archive of Math. Logic **38** (1999) 313–354
3. Immerman, N., Rabinovich, A., Reps, T., Sagiv, M., Yorsh, G.: The boundery between decidability and undecidability of transitive closure logics. In: CSL'04. (2004)
4. Lev-Ami, T., Sagiv, M.: TVLA: A system for implementing static analyses. In: Static Analysis Symp. (2000) 280–301
5. Sagiv, M., Reps, T., Wilhelm, R.: Parametric shape analysis via 3-valued logic. Trans. on Prog. Lang. and Syst. (2002)
6. Reps, T., Sagiv, M., Wilhelm, R.: Static program analysis via 3-valued logic. In: CAV. (2004) 15–30
7. Flanagan, C., Leino, K., Lillibridge, M., Nelson, G., Saxe, J., Stata, R.: Extended static checking for java. In: SIGPLAN Conf. on Prog. Lang. Design and Impl. (2002)
8. Møller, A., Schwartzbach, M.: The pointer assertion logic engine. In: SIGPLAN Conf. on Prog. Lang. Design and Impl. (2001) 221–231
9. Weidenbach, C., Gaede, B., Rock, G.: Spass & flotter version 0.42. In: CADE-13: Proceedings of the 13th International Conference on Automated Deduction, Springer-Verlag (1996) 141–145
10. Nelson, G.: Verifying reachability invariants of linked structures. In: Symp. on Princ. of Prog. Lang. (1983) 38–47
11. Avron, A.: Transitive closure and the mechanization of mathematics. In: Thirty Five Years of Automating Mathematics, Kluwer Academic Publishers (2003) 149–171
12. Lev-Ami, T., Immerman, N., Reps, T., Sagiv, M., Srivastava, S., Yorsh, G.: Simulating reachability using first-order logic with applications to verification of linked data structures. Available at "http://www.cs.tau.ac.il/~tla/2005/papers/cade05full.pdf" (2005)

13. Cousot, P., Cousot, R.: Abstract interpretation: a unified lattice model for static analysis of programs by construction or approximation of fixpoints. In: POPL '77: Proceedings of the 4th ACM SIGACT-SIGPLAN symposium on Principles of programming languages, ACM Press (1977) 238–252

14. Loginov, A., Reps, T., Sagiv, M.: Abstraction refinement via inductive learning. In: Proc. Computer-Aided Verif. (2005)

15. Elgaard, J., Møller, A., Schwartzbach, M.: Compile-time debugging of C programs working on trees. In: European Symp. On Programming. (2000) 119–134

16. Benedikt, M., Reps, T., Sagiv, M.: A decidable logic for describing linked data structures. In: European Symp. On Programming. (1999) 2–19

17. Sagiv, M., Reps, T., Wilhelm, R.: Solving shape-analysis problems in languages with destructive updating. Trans. on Prog. Lang. and Syst. **20** (1998) 1–50

18. Hendren, L.: Parallelizing Programs with Recursive Data Structures. PhD thesis, Cornell Univ., Ithaca, NY (1990)

19. Immerman, N., Rabinovich, A., Reps, T., Sagiv, M., Yorsh, G.: Verification via structure simulation. In: Proc. Computer-Aided Verif. (2004) 281–294

20. Leino, R.: Recursive object types in a logic of object-oriented programs. Nordic J. of Computing **5** (1998) 330–360

21. Dong, G., Su, J.: Incremental and decremental evaluation of transitive closure by first-order queries. Inf. & Comput. **120** (1995) 101–106

22. Patnaik, S., Immerman, N.: Dyn-FO: A parallel, dynamic complexity class. Journal of Computer and System Sciences **55** (1997) 199–209

23. Hesse, W.: Dynamic Computational Complexity. PhD thesis, Department of Computer Science, UMass, Amherst (2003)

24. Reps, T., Sagiv, M., Loginov, A.: Finite differencing of logical formulas for static analysis. In: European Symp. On Programming. (2003) 380–398

25. Hendren, L., Hummel, J., Nicolau, A.: Abstractions for recursive pointer data structures: Improving the analysis and the transformation of imperative programs. In: SIGPLAN Conf. on Prog. Lang. Design and Impl., New York, NY, ACM Press (1992) 249–260

Privacy-Sensitive Information Flow with JML

Guillaume Dufay[1], Amy Felty[1], and Stan Matwin[1,2]

SITE, University of Ottawa. Ottawa, Ontario K1N 6N5, Canada
Institute of Computer Science, Polish Academy of Sciences, Warsaw, Poland
2   {gdufay, afelty, stan}@site.uottawa.ca

Abstract. In today's society, people have very little control over what kinds of personal data are collected and stored by various agencies in both the private and public sectors. We describe an approach to addressing this problem that allows individuals to specify constraints on the way their own data is used. Our solution uses formal methods to allow developers of software that processes personal data to provide assurances that the software meets the specified privacy constraints. In the domain of privacy, it is often not sufficient to express properties of interest as a relation between the input and output of a program as is done for general program correctness. Here we consider a stronger class of properties that allows us to express constraints on information flow. In particular, we can express that an algorithm does not leak any information from particular "sensitive" values. We describe a general methodology for expressing this kind of information flow property as Hoare-style program verification judgments. We begin with the Java Modelling Language (JML), which is a behavioral interface specification language designed for Java, and we extend the language to include new concepts and keywords for expressing such properties. We use the Krakatoa tool which starts from JML-annotated Java programs, generates proof obligations in the Coq Proof Assistant, and helps to automate their proofs. We extend the Krakatoa tool to understand our extensions to JML and to generate the new form of required proof obligations. We illustrate our method on several data mining algorithms implemented in Java.

1   Introduction

Privacy is one of the main concerns expressed about modern computing, especially in the Internet context. People and groups are concerned by the practice of gathering information without explicitly informing the individuals that data about them is being collected. Oftentimes, even when people are aware that their information is being collected, it is used for purposes other than the ones stated at collection time. The last concern is further aggravated by the power of modern database and data mining operations which allow inferring, from combined data sets, knowledge of which the person is not aware, and would have never consented to generating and disseminating. People have no ownership of their own data: it is not easy for someone to exclude themselves from, e.g. direct marketing campaigns, where the targeted individuals are selected by data mining models.

R. Nieuwenhuis (Ed.): CADE 2005, LNAI 3632, pp. 116–130, 2005.
© Springer-Verlag Berlin Heidelberg 2005

One of the main concepts that has emerged from research on societal and legal aspects of privacy is the idea of Use Limitation Principle (ULP). That principle states that the data should be used only for the explicit purpose for which it has been collected [15]. Our work addresses this question from the technical standpoint. We provide tool support for verifying that this principle is indeed upheld by organizations that perform data mining operations on personal data.

In our setting, users can express individual preferences about what can and cannot be done with their data. We have not yet addressed the question of how users express preferences, though our approach allows any data properties that can be expressed syntactically in formal logic. Certainly, a user-friendly language, easy to handle by an average person, needs to be designed. It could initially have the form of a set of options from which an individual would make choices. We assume that an organization that writes data mining software must provide guarantees that individuals' constraints are met, and that these guarantees come in the form of formal proofs about the source code. Organizations who use the data mining software are given an executable binary. An independent agency, whose purpose is to verify that privacy constraints are met, obtains the binary from the software user, obtains the source code and proof from the software developer, checks the proof, and verifies that the binary is a compiled version of the source code. The details of the architecture just described can be found in another paper [13]. The scenario just described involves using our techniques to guard against malicious code. Our approach can also apply to a setting in which trust is not an issue, for example, within a company that wants to insure that its software release is free of privacy flaws. In this paper, we concentrate on extending the class of privacy constraints which can be handled, and providing tool support for proving these properties.

In our previous work [3,13], we considered privacy properties that could be expressed as requirements on the input-output relation. Additionally, we showed how to incorporate constraints on operations that could potentially violate privacy by overloading the output so that a trace of such operations was evident in the result. Here we consider a stronger class of properties that allows us to express constraints on information flow. In particular, we are interested in properties that express that an algorithm does not leak any information from particular sensitive values, or that a program never writes such sensitive data to a file. Many such properties can be handled through the framework of *non-interference*. Non-interference [7] is a high-level property of programs that guarantees the absence of illegal information flow at runtime. More precisely, non-interference requires distinguishing between public input/output and sensitive input/output. A program that satisfies non-interference will be such that sensitive inputs of that program have no influence on public outputs. Thus, with the non-interference framework, it is straightforward to express the expected properties that an algorithm does not leak any sensitive information or that a value is never written into a file (considering this value as a sensitive input and the file as a public output). The examples we present in this paper include properties constraining information flow both with and without the use of non-interference.

In addition to increasing the class of privacy constraints we can express, we present a new approach to proving such properties. We start from the Weka repository of Java code which implements a variety of data mining algorithms [22]. We simplify the code somewhat, both for illustration purposes and so that we can work within the limitations of current tools that support our approach. Also, we must modify the code to include checks that the privacy constraints that we allow users to specify are met. We then annotate the code with JML (Java Modeling Language) [11] assertions, which express Hoare-style [9] preconditions, postconditions, loop invariants, etc. We use the Krakatoa [12] tool to generate proof obligations in Coq [14] and to partially automate their proofs. Successful completion of these proof obligations ensures that the Java code satisfies its JML specifications. The new approach has two advantages over our old approach [3,13]. First, we work directly with Java programs. Previously, we started with Java code, and translated it to the ML-like language used in Coq. Second, we hope that using tools engineered for program verification will improve the ability to automate proofs in the privacy domain.

To handle the kinds of information-flow properties we are interested in, we extend the expressive power of JML. We also extend Krakatoa to generate the proof obligations required for the extra expressive power, and to help automate their proofs.

Contents of the Paper. In Section 2, we present the tools used in our approach to building proofs of privacy constraints of data mining algorithms. In Section 3, we present our first example; it is a simple one which serves to illustrate our approach using JML and Krakatoa. In Section 4, we discuss a nearest neighbor classification algorithm, and present JML annotations which guarantee that the result value of a data-mining program does not reveal any sensitive information by constraining it to a set of public values. In Section 5, we discuss how non-interference can be used to handle a larger class of privacy-sensitive properties, and how we extend JML and Krakatoa to provide support for non-interference. In Section 6, we apply these results to a Naive Bayes classification algorithm. Finally, in Section 7, we conclude and discuss future work.

2   Tools

First, we will introduce Weka, a Java library of data-mining algorithms, and the notion of classifiers; then, the JML assertion language in which privacy properties are expressed; and finally, the Krakatoa tool, that we use to verify JML-annotated Java programs.

2.1   Weka and Classifiers

Since we target data mining software, we have decided to apply our approach to selected classification modules of Weka (Waikato Environment for Knowledge Analysis) [22]. Weka is an open-source library of data mining algorithms, written in Java, providing a rich set of mining functions, data preprocessing operations, evaluation procedures, and GUIs. Weka has become a tool of choice, commonly used in the data mining community. Classification is one of the basic data mining tasks, and Weka provides Java implementations of all the main classification algorithms. In this paper, we illustrate our approach to privacy-sensitive information flow with two commonly used classification tools, the so called Nearest Neighbor and Naive Bayes classifiers.

The classification task can be defined as follows: given a finite training set $T = \{\langle x_i, y_i \rangle | x_i \in D, y_i \in C\}$, where D denotes the data (set of instances), and C denotes the set of classes, find a function $c : D \mapsto C$ such that for each pair $\langle x', y' \rangle \in T, y' = c(x')$, and furthermore, c will correctly classify unseen examples (i.e. examples that will only arrive in the future, and as such cannot be included in the training set T.) c is referred to as a classifier, and the task of finding c is known as learning c from T, or —alternatively— as the classifier induction task. Usually, $D = A_1 \times \ldots \times A_n$, where $A_i, i = 1, \ldots, n$, is an attribute domain. A_i's are either sets of discrete (nominal) values, or subsets of $\mathbb{R}$. Each $x_i \in D$ can therefore be seen as $x_i = a_{i,1}, \ldots, a_{i,n}$, where $a_{i,j} \in A_j$.

2.2   The Java Modeling Language

The Java Modeling Language [11] (JML) is a behavioral interface specification language designed for Java. It relies on the design by contract approach [16] to guarantee that a program satisfies its specification during runtime. These specifications are given as annotations of the Java source file. More precisely, they are included as special Java comments, either after the symbols //@ or enclosed between /*@ and @*/. For example, the general schema for the annotation of a method is the following:

```
/*@ behavior
  @    requires <precondition>;
  @    modifiable <modified fields and variables>;
  @    ensures <postcondition if no exception raised>;
  @    signals(E) <postcondition when exception E raised>; @*/
```

The underlying model is a an extension of Hoare-Floyd logic [9]: if the precondition holds at the beginning of the method call, then postconditions (with and without exceptions) will hold after the call.

Preconditions and postconditions express first-order logic statements, with a syntax following the Java syntax. Thus, they can easily be written by a programmer. The Java syntax is enriched with special keywords: \result and \old(<expr>) to denote respectively the return value of a method, and the value of a given expression before the execution of a method; and \forall, \exists, ==> to denote respectively universal quantification, existential quantification and logical implication. If the modifiable clause is omitted, it means by convention that the method is side-effect free.

Apart from methods specification, it is also possible to annotate a program with class invariants (predicates on the fields of a class that hold at any time in the class) using the keyword invariant, loop invariants (inside the code of a method with loops) using the keyword loop_invariant, and assertions (that must hold at the given point of the program) using the keyword assert.

Finally, when annotating a program, it might be useful to introduce new variables to keep track of certain aspects or computations. Instead of adding them to the program itself, thus adding new code, it is possible to define variables that will only be used for specification. These variables, called *ghost* variables, are defined in a JML annotation with the keyword ghost and assigned to a Java expression with the keyword set.

2.3   Krakatoa

Once a program has been annotated with JML, these annotations can be verified either during runtime (an exception will be raised if they do not hold) or statically, with a static checker or a theorem prover, given the semantics of the program. A wide range of tools can be used to achieve this goal. Among these, the LOOP Tool [21] will work with the PVS theorem prover, Jack [10] with Atelier B, or Krakatoa [12], that we chose, with Coq [14].

The Krakatoa tool provides a generic model in Coq of the Java runtime environment. Annotated Java programs are not translated directly to Coq, but to the Why [5] input language (an annotated ML-like language), without any loss of precision. Then, Krakatoa relies on static analysis and weakest preconditions calculus of the stand-alone Why tool to generate Coq proof obligations, corresponding to requirements the Java program must respect to meet its specifications. Some of these proof obligations can be discharged automatically through Coq built-in or Krakatoa provided tactics. The remaining proof obligations have to be completed manually, through the interactive proof mechanism of Coq. In some cases, preconditions or loop invariants of the annotated Java program might not be strong enough to prove the postcondition of a method and need to be modified. Proof obligations are then regenerated, but completed proofs not affected by these modifications are kept.

The successful completion of all proof obligations is sufficient to ensure that a program satisfies its specifications. However, the Why tool can also perform a final step, called validation, to embed each functional translation of the methods of Java program with its specification into a Coq term whose type corresponds to the JML specification of that method. This term can be given as a certificate of the soundness of the whole process.

3   A First Example: Joining Two Database Tables

This sections presents an example of a JML-annotated Java program. We redo the example described in [3], where the program was written and proved within Coq. This example serves to illustrate our new approach as well as compare it to the old one. The program performs a database join operation. The data from two sets (Payroll and Employee) is joined into a single set (Combined), ignoring the data from individuals that do not want their data to be used in a join operation. For example, individuals with an exceptionally high salary may not want their `Payroll` information in the same record as their address and phone number. Such detailed records may contain enough information to identify them or to make them the target of certain kinds of direct-marketing campaigns. In this example, such individuals can express that they want to opt out of this operation.

The data structures for this example are standard Java classes. For instance, the payroll notion is captured by the following class that contains the employee ID `PID` it refers to, the salary, and a boolean `JoinInd` which indicates if the person who owns the data has given the permission to use the data in a join operation.

```
class Payroll {
    public int PID;
    public int Salary;
    public boolean JoinInd;
};
```

The result of the join is stored in a `Combined` class, that gathers the data from the classes `Payroll` and `Employee` (which contains, among other fields, name and `EID` which records the employee ID). We can notice at this point that the constructor for the class `Combined` is annotated with a JML specification. It prevents the creation of a `Combined` class for the users that do not allow it (the field `JoinInd` has to be true), and ensures that the field `JoinInd` is unchanged in the created class.

```
class Combined {
  public Payroll m_payroll;
  public Employee m_employee;

  /*@ public normal_behavior
    @ requires p != null && p.JoinInd == true && e.EID == p.PID;
    @ ensures m_payroll.JoinInd == p.JoinInd; @*/
  public Combined(Employee e, Payroll p) {
    m_employee = e;
    m_payroll = p;
} };
```

Note that the assertion above includes a statement that the employee ID fields of e and p are the same. We did not need this in the version in [3] since only one copy of the ID was kept in the new `Combined` record. This is a minor difference which has little effect on the proofs.

The algorithm that iterates though a set of `Payroll` records to perform a join operation is given below:

```
/*@ public normal_behavior
  @ requires Ps != null && Es != null;
  @ ensures (\forall int i; 0 <= i && i < \result.length; \result != null &&
  @    (\result[i] != null ==> \result[i].m_payroll.JoinInd == true)); @*/
public Combined[] join(Payroll[] Ps, Employee[] Es) {
  Combined tab[] = new Combined[Ps.length];

  /*@ loop_invariant
    @ 0 <= i && i <= Ps.length &&
    @ (\forall int j; 0 <= j && j < i && j < tab.length;
    @    (tab[j] != null ==> tab[j].m_payroll.JoinInd == true));
    @ decreases Ps.length-i; @*/
  for (int i=0; i < Ps.length; i++)
```

```
if (Ps[i] != null)
   tab[i] = checkJoinIndAndfindEmployee(Ps[i], Es);
else
   tab[i] = null;

return tab;
}
```

The specification of this algorithm expresses the same property as in [3], but here it is expressed in JML, which uses Java-like syntax and refers directly to variables occurring in the program. The particular property that is expressed is that all data that took part in the join was in fact permitted to do so by the owners of the data. In [3], this property was expressed directly as a formula in Coq. Here, the requirements on individual methods taken together express this property. The method checkJoinIndAndfindEmployee, whose code is not given here, takes a single Payroll record Ps[i] and the entire list of Employee records Es as arguments. If (1) a record is found such that Ps[i].EID matches the employee ID value in one of the records in Es, and (2) Ps[i].JoinInd has value **true**, then a new Combined record is created and returned. Otherwise **null** is returned.

Note that the loop inside the join method had to be annotated, like any loop in the Hoare-logic formalism. Also, the method checkJoinIndAndfindEmployee had to be annotated with precondition, postcondition, and loop invariant since it is called by join and extends the results of the call to the constructor of the class Combined.

After going through Krakatoa, most of the generated proof obligations are automatically solved by Coq. In the JML annotations from the code above, we have omitted some dynamic type information (such as Ps is an instance of Payroll[]) that was needed to complete the proof. The fact that we had to manually insert this information is due to current limitations of Krakatoa that will be fixed in the near future. Around 100 lines of proof were entered to discharge the remaining proof obligations, which is slightly less than the length of the proofs in [3]. Although the difference is not really significant due to the limited size of the example, we believe that this approach leads to smaller proofs and to an increased confidence in the whole engineering process.

4   A Simple Data Privacy Preserving Classifier

In this section, we will describe how to enforce the value of a data-mining algorithm not to reveal any sensitive information, by constraining the output to a set of public values.

The following algorithm, the nearest neighbor classifier algorithm, has been extracted from the Weka library but, to keep proofs simpler, unwanted features for the purpose of this example have been removed (such as method calls related to the Weka graphic interface, or checks for an incorrect or incomplete data set) and data are accessed directly, not through objects. For this particular classifier, the returned value of the *class attribute* (in a data-mining context, the attribute that the classifier aims at determining) for the given instance instance, is the value of the class attribute for the instance from the training set m_Train determined as the nearest of the given instance. The corresponding distance is calculated from the non-class attributes of both.

The specifications for this algorithm constrain the result value to be one of the class attributes (in the row classIndex of instances), thus preventing leaks of any other value of the dataset. This kind of specification can be used for other classifiers on a finite set of class attributes and that return an element of this set. It would be possible to also prevent a particular instance to be used in the algorithm based on the owner requirements, as done in Section 3, but it is supposed in this example that sensitive information resides in non-class attributes and that the class attribute can be public.

```
/*@ public normal_behavior
  @ requires instance != null && m_Train != null && numInstances > 0 &&
  @   instance.length == numAttributes && m_Train.length == numInstances &&
  @   (\forall int i; 0 <= i && i < numInstances;
  @     m_Train[i] != null && m_Train[i].length == numAttributes);
  @ ensures (\exists int i; 0 <= i && i < m_Train.length;
  @   \result == m_Train[i][classIndex]); @*/
public double classifyInstance(double[] instance) throws Exception {
    double dist, minDistance, classValue = 0.0;
    boolean first = true;

    buildClassifier();
    updateMinMax(instance);

    /*@ loop_invariant
      @ 0 <= i && i <= numInstances &&
      @ ((first == true && i == 0) ||
      @ (\exists int j; 0 <= j && j < i; classValue == m_Train[j][classIndex]));
      @ decreases numInstances - i; @*/
    for (int i = 0; i < numInstances; i++)  {
        dist = distance(instance, m_Train[i]);
        if (first || dist < minDistance) {
          minDistance = dist;
          classValue = m_Train[i][classIndex];
          first = false;
        }
    }

    return classValue;
}
```

Proof obligations for this algorithm do not lead to particular problems, they just follow the structure of the code and the annotations. A total of 180 lines of manually entered proof scripts is needed for buildClassifier, classifyInstance and the auxiliary functions involved such as updateMinMax and distance.

This approach leads to simple specifications and proofs for which the result is constrained to a known set. However, in cases where the set result is infinite, a stronger framework is needed, such as the one provided by non-interference.

5   Privacy Through Non-interference

As explained in the introduction, non-interference distinguishes public inputs (resp. outputs) and sensitive inputs (resp. outputs) and prevents leaks from sensitive inputs to public outputs. For example, if we consider the input/output variables x as public and y as sensitive, the program x = y*2 is interferent (direct flow from y to x), whereas the program x = y; x = 0 is not (it is impossible for an attacker to guess the value of y by observing x at the end of the execution). It is also possible to get interference through indirect information flow, for instance the following program is interferent (it is possible to guess the nullity of y):

if (y != 0) **then** x = 1; **else** x = 2;

Finally, interference can be observable through termination of programs (termination-sensitive) or timing leaks. For the sake of this paper, we will not consider these possibilities.

5.1   The General Framework

Non-interference can be enforced through type systems [20,1]. However, in practice these type systems turn out to be laborious to use and they can reject obvious non-interferent programs. Instead, we prefer to follow the approach described in [2] that proceeds by self-composition of the program and that can be described using the Hoare-style logic of JML (and then integrated in the tools we have used so far to study privacy).

Self-composition proceeds by duplicating the code of a program, with two sets of inputs. Thus imperative pointer-free program $P(x, y)$ with public input/output variables x and sensitive input/output y will be non-interferent if forall x_1, x_2, y_1, y_2 we have the following Hoare formula:

$$\{x_1 = x_2\}\ P(x_1, y_1); P(x_2, y_2)\ \{x_1 = x_2\}$$

where ; is usual sequential composition. This formula expresses the fact that the output values of the public variables are independent from the values of the sensitive variables.

More generally, dependencies between the parameters of the program, before and after the execution, are characterized by a relation called *L-equivalence*. In the above Hoare formula, this relation is simple equality. By allowing more general L-equivalence relations between public and sensitive variables, it becomes possible to capture the notion of *declassification* [4,18] within the same framework. Declassification allows leaks, in a controlled way, of sensitive global information to public variables. Indeed declassification would allow a data-mining algorithm to compute over sensitive variables and yield public results that do not give any specific information about any of these sensitive variables. An example of such a use is given in the example of Section 6 in the context of data-mining.

5.2   Extension of Krakatoa

In order to be able to handle non-interference with Krakatoa following the previous framework, some modifications have to be made to both JML and Krakatoa. First, we wish to distinguish pre and post-conditions related to the normal execution of the program and those related to non-interference. For this purpose, we have introduced two optional keywords for method specifications: `requires_ni` and `ensures_ni`. Then, to define L-equivalence relations in the pre and post-conditions of the self-composed program, we need to distinguish variables for each of the two runs of the program. Therefore, we have introduced two keywords `\ni1(<var>)` and `\ni2(<var>)`. Finally, annotations inside the code are also required to exist in three variants, one for normal execution and one for each run of the program for non-interference (these annotations are not necessarily the same for each run). For example, the keywords `loop_invariant_ni1` and `loop_invariant_ni2` are available to distinguish loop invariants to be used for each run of the program.

Krakatoa is modified to recognize these new keywords. When a method is annotated with non-interference specifications, it generates the code for the self-composed method with the corresponding specifications and the appropriate variable names. A particular case appears for method invocations inside a method body. Indeed, the non-interference results from the invoked method will only be available in the second copy of the self-composed code (the two runs must have occurred). In addition, it is necessary to modify the program, with ghost variables, to keep track of the invoked method parameters and result values of the first run (they can be modified later on by assignment and thus would not be available anymore). These values will be used as values of the variables of the first run, in the non-interference results of the invoked method of the second run.

6   Non-interference for the Naive Bayes Classifier

In this section, we will illustrate how the idea of non-interference can be applied on a data-mining algorithm, the naive Bayes classifier, to express a privacy property.

The Naive Bayes classifier predicts the class of an instance $x = a_1, \ldots, a_n$ (`mTrain[i]` in the code below) as

$$c(x) = argmax_{c_j \in C} P(c_j) \prod P(a_i|c_j)$$

i.e. the class of x is obtained by estimating the probabilities of all classes for given attribute values of x. These estimates, known as *priors*, are known from the training set. Probability estimates are approximated by counting frequencies of different classes for given attribute values. The training data needs therefore to be summarized in a table (the variable `probs` in the code below), which keeps the count of the number of instances with specific attribute values for each class.

From a data privacy point of view, we will assume that some people do not want their data (more precisely the corresponding instances in the training set T) to be used by the classifier for this particular class. Having one's data used by a classifier means that this particular individual stands out in T, and can be targeted (by the use of so called data drilling operations) in marketing campaigns, sampling routines, etc. It might be reasonable to object to this. Thus, the output class value of the algorithm, that is a public result, should not depend on these sensitive data. The entire set of training instances can not be considered as fully sensitive data since some information has to be gained from it to classify the instance; it can not be considered as public data either due to the restrictions given above. Rather, the training set should be considered as sensitive data with part of these data (instances that are allowed by their owner to be used in a classifier algorithm) being declassified for non-interference. Then, the L-equivalence relation for this algorithm will expect:

- from the input m_Train, the training set of instances, an array of m_NumInstances instances, to be such that the corresponding duplicated variables for self-composition \nil(m_Train) and \ni2(m_Train) agree on the values that can be used in the classifier algorithm;
- from the public inputs instance, the instance to classify, to be such that the duplicated variables are equal;
- from the public input m_inst_Allow, an array of m_NumInstances booleans that express for a given index whether the instance at the corresponding index in m_Train can be used in the classifier, to be such that the duplicated variables are equal;
- from the public output \result, an array of m_NumClassValues probabilities, to be such that the duplicated variables are equal.

More formally, the JML specification for the naive Bayes algorithm is:

```
/*@ public normal_behavior
  @ requires_ni (\forall int i; i <= 0 && i < m_NumInstances;
  @   ((\nil(m_inst_Allow)[i] == true) ==>
  @     (\forall int j; j <= 0 && j < m_numAttributes;
  @       \nil(m_Train)[i][j] == \ni2(m_Train)[i][j]))) &&
  @   (\nil(instance)[i] == \ni2(instance)[i]) &&
  @   (\nil(m_inst_Allow)[i] == \ni2(m_inst_Allow)[i]);
  @ ensures_ni (\forall int i; 0 <= i && i < m_NumClassValues;
  @   \nil(\result[i]) == \ni2(\result[i])); @*/
```

The structure of the code for this classifier relies on two main methods: buildClassifier that initializes the classifier with the training set data, and distributionForInstance that uses the previously built classifier to classify a given instance. The buildClassifier method must ensure that two training sets that verify the conditions given above will generate equal probability estimators:

```
/*@ public normal_behavior
  @ requires_ni (\forall int i; i <= 0 && i < m_NumInstances;
  @   (\nil(m_inst_Allow)[i] == \ni2(m_inst_Allow)[i]) &&
  @   ((\nil(m_inst_Allow)[i] == true) ==>
  @     (\forall int j; j <= 0 && j < m_numAttributes;
  @       \nil(m_Train)[i][j] == \ni2(m_Train)[i][j])));
  @ ensures_ni instance != null && instance.length == m_NumAttributes &&
  @   (\forall int i; 0 <= i && i < m_NumClassValues;
  @   \nil(m_ClassDistribution).getProbability(i) ==
  @   \ni2(m_ClassDistribution).getProbability(i)) &&
  @   (\forall int attIndex; 0 <= attIndex && attIndex < m_NumAttributes;
  @     (\forall int i; 0 <= i && i < m_NumClassValues;
  @       (\forall int j; 0 <= j && j < m_NumValues[attIndex];
  @         \nil(m_Distributions)[attIndex][i].getProbability(j) ==
  @         \ni2(m_Distributions)[attIndex][i].getProbability(j)))); @*/
public void buildClassifier() {
  m_ClassDistribution = new DiscreteEstimator(m_NumClassValues, true);
```

```
for (int attIndex = 0; attIndex < m_NumAttributes; attIndex++)
  for (int j = 0; j < m_NumClassValues; j++) {
    m_Distributions[attIndex][j] =
      new DiscreteEstimator(m_NumValues[attIndex], true);
  }
for (int i = 0; i < m_NumInstances; i++) {
  if (m_inst_Allow[i] == true) {
    for (int attIndex = 0; attIndex < m_NumAttributes; attIndex++) {
      int distr_idx = m_Train[i][m_ClassIndex];
      m_Distributions[attIndex][distr_idx].
              addValue(m_Train[i][attIndex]);
    }
    m_ClassDistribution.addValue(inst[m_ClassIndex]);
  }
}
```

The `distributionForInstance` method will now compute the probability
distribution `probs` (an array of `m_NumClassValues` values), such that `probs[i]`
is equal to the probability for the instance `instance` to be classified as the ith
class value. Based on the post-conditions of the previous method, the following
specifications will ensure for non-interference that two equal probability distri-
butions will be generated at the end of self-composition.

```
/*@ public normal_behavior
  @ requires_ni (\forall int i; i <= 0 && i < m_NumInstances;
  @    (\ni1(instance)[i] == \ni2(instance)[i])) &&
  @    <ensures_ni of buildClassifier >;
  @ ensures_ni (\forall int i; 0 <= i && i < m_NumClassValues;
  @    \ni1(\result[i]) == \ni2(\result[i])); @*/
public void distributionForInstance(double[] instance) {

  /*@ loop_invariant_ni1
    @    0 <= \ni1(j) && \ni1(j) <= \ni1(m_NumClassValues) &&
    @    (\forall int i; 0 <= i && i < \ni1(j);
    @    \ni1(probs[i]) == \ni1(m_ClassDistribution).getProbability(i) &&
    @    \ni1(probs_save[0][i]) == \ni1(probs[i]));
    @ decreases \ni1(m_NumClassValues) - \ni1(j);
    @ loop_invariant_ni2
    @    0 <= \ni2(j) && \ni2(j) <= \ni2(m_NumClassValues) &&
    @    (\forall int i; 0 <= i && i < \ni2(j);
    @    \ni2(probs[i]) == \ni1(probs_save[0][i]) &&
    @    \ni2(probs_save[0][i]) == \ni1(probs_save[0][i]));
    @ decreases \ni2(m_NumClassValues) - \ni2(j); @*/
  for (int j = 0; j < m_NumClassValues; j++) {
    probs[j] = m_ClassDistribution.getProbability(j);
    //@ set prob_save[0][j] = probs[j];
  }

  /*@ loop_invariant_ni1 <...> ;
    @ loop_invariant_ni2 <...> ; @*/
  for (int attIndex = 0; attIndex < m_NumAttributes; attIndex++) {

    /*@ loop_invariant_ni1
      @    0 <= \ni1(j) && \ni1(j) <= \ni1(m_NumClassValues) &&
      @    (\forall int i; 0 <= i && i < \ni1(j);
      @    \ni1(probs[i]) == \ni1(probs_save[attIndex][i]) *
      @    \ni1(m_Distributions)[0][i].getProbability(\ni1(instance[attIndex])) &&
      @    \ni1(probs_save[attIndex+1][i]) == \ni1(probs[i])) &&
      @    (\forall int i; \ni1(j) <= i && i < \ni1(m_NumClassValues) ;
      @    \ni1(probs[i]) == \ni1(probs_save[attIndex][i]));
      @ loop_invariant_ni2
      @    0 <= \ni2(j) && \ni2(j) <= \ni2(m_NumClassValues) &&
      @    (\forall int i; 0 <= i && i < \ni2(j);
      @    \ni2(probs[i]) == \ni1(probs[i]) &&
      @    \ni2(probs_save[attIndex+1][i]) == \ni1(probs_save[attIndex+1][i])) &&
      @    (\forall int i; \ni2(j) <= i && i < \ni2(m_NumClassValues) ;
```

```
      @    \ni2(probs[i]) == \nil(probs_save[attIndex][i])); @*/
   for (int j = 0; j < m_NumClassValues; j++) {
      probs[j] *= m_Distributions[attIndex][j].getProbability(instance[attIndex]);
      //@ set prob_save[attIndex+1][j] = probs[j];
   }
 }
 return(probs);
}
```

Non-interference specifications have been given for all methods involved in this example (including the three methods from the class DiscreteEstimator devoted to representing probability estimators). Note however that the specifications given above are not complete due to the lack space in the sense that some loop invariants are not shown (they are similar to the ones given) and that Krakatoa requires some additional information, also not shown, about bound limits for arrays, types and non-nullity of objects, loop variants (the index used in the loop) and modified objects.

The resolution of generated proof obligations proceeds by matching values of the second run to the corresponding values of the first run. To do so, it is necessary to keep track of the successive values assigned, which is the role in the specifications of extra variables (for example the array probs_save to keep track of values of the variable probs inside the loop), that can be declared as JML ghost variables. Proofs are not yet completed due to the presence of method invocations involving arrays. Although we are able to use non-interference results in those cases, we are currently working on automatically generating assertions related to the extra ghost variables (arrays) needed to store parameters and result values of the invoked method. The **if** condition inside one loop does not cause any particular problems. The specifications just require the use of logical implication to reason about the value of the test, as was done in Section 3.

Although statements of generated proof obligations can be very long due to the various loops involved (one statement of a proof obligation is over 700 lines long), individual subgoal statements are very concise and the required total length of proof script that had to be given manually for the distributionForInstance method and methods from the class DiscreteEstimator is about 200 lines.

7   Conclusion

We have presented several ways to enforce privacy-sensitive information flow with JML, which we have illustrated on data-mining algorithms. We first extended results from a previous paper to integrate them into the JML framework. We then proposed a way to prevent leaks from sensitive variables when the set of possible results is finite. Finally, we applied the framework of non-interference to provide a stronger means to express and enforce privacy properties. To do so, we extended JML specifications with new specific keywords, but we kept the underlying Hoare-Floyd style verification mechanism. We have completed all proofs in Coq of the generated proof obligations for the first two examples. For the more complex example which uses non-interference, we provided specifications for all methods, but proof obligations for one method could not be completed due to current limitations of the tools. However, we completed proofs for all others methods, thus providing a proof of concept of our methodology.

Related Work. One of the most comprehensive tools related to information flow for Java is JFlow [17]. This tool acts as a compiler to statically and dynamically check programs. It relies on a concept of security levels for variables, which is not sufficient for our purpose, i.e. to catch the kind of declassification we are dealing with. Concerning non-interference, although research in this domain is very active (see [19] for a survey), most of the work done remains theoretical. On the practical side, applications of non-interferent programs are currently limited to security issues in smart cards. [6] is one of the more advanced contributions in this area, using JFlow and Esc/Java. [8] is another example of work exploiting JFlow information-flow policy to address privacy. Although this work does not address data mining in particular, it may be possible to integrate this kind of approach with ours, when dealing with simpler declassification properties, to improve the scope of privacy concerns that can be enforced. Our use of non-interference for JML to express privacy of data-mining algorithms is a novel, promising application area.

Further Work. Future development of our work will aim at first to address the limitations explained in Section 6 concerning the inclusion of non-interference results from called methods inside the proofs, and scaling the approach to more complex Java features and algorithms of the Weka library. Another interesting development would be to automatically generate loop invariants, which can be tedious to write, but are needed for proofs of non-interference. Indeed, the invariant for the two copies of the code of a self-composed program follow the same pattern, and it can be determined statically which variable of the second run corresponds to which variable of the first.

Acknowledgments

The authors acknowledge the support of the Natural Sciences and Engineering Research Council of Canada, and of the Ontario Centres of Excellence, Inc (CITO Division).

References

1. Anindya Banerjee and David A. Naumann. Secure information flow and pointer confinement in a Java-like language. In *Proceedings of the 15th IEEE Computer Security Foundations Workshop*, page 253. IEEE Computer Society, 2002.
2. Gilles Barthe, Pedro D'Argenio, and Tamara Rezk. Secure information flow by self-composition. In *Proceedings of the 17th IEEE Computer Security Foundations Workshop*, pages 100–114. IEEE Computer Society, 2004.

3. Amy Felty and Stan Matwin. Privacy-oriented data mining by proof checking. In *Proceedings of the Sixth European Conference on Principles of Data Mining and Knowledge Discovery*, volume 2431 of *LNCS*. Springer-Verlag, August 2002.
4. E. Ferrari, P. Samarati, E. Bertino, and S. Jajodia. Providing flexibility in information flow control for object-oriented systems. In *Proceedings of the 1997 IEEE Symposium on Security and Privacy*, May 1997.
5. Jean-Christophe Filliâtre. Why: A multi-language multi-prover verification tool. Research Report 1366, LRI, Université Paris Sud, March 2003.
6. Pablo Giambiagi and Mads Dam. Verification of confidentiality properties for Java Card applets. Manuscript, 2003.
7. Joseph A. Goguen and José Meseguer. Security policies and security models. In *Proceedings of 1982 IEEE Symposium on Security and Privacy*, pages 11–20, 1982.
8. Katia Hayati and Martín Abadi. Language-based enforcement of privacy policies. In *Proceedings of Privacy Enhancing Technologies Workshop (PET 2004)*, LNCS. Springer-Verlag, 2004.
9. C. A. R. Hoare. An axiomatic basis for computer programming. *Commununications of ACM*, 12(10):576–580, 1969.
10. JACK: Java Applet Correctness Kit. http://www-sop.inria.fr/everest/soft/Jack/jack.html.
11. Gary T. Leavens, Albert L. Baker, and Clyde Ruby. JML: A notation for detailed design. In *Behavioral Specifications of Businesses and Systems*, pages 175–188. Kluwer Academic Publishers, 1999. http://www.jmlspecs.org.
12. Claude Marché, Christine Paulin-Mohring, and Xavier Urbain. The Krakatoa Tool for Certification of Java/JavaCard Programs annotated in JML. *Journal of Logic and Algebraic Programming*, 58(1–2):86–106, 2004.
13. Stan Matwin, Amy Felty, István Hernádvölgyi, and Venanzio Capretta. Privacy in data mining using formal methods. In *Proceedings of the Seventh International Conference on Typed Lambda Calculi and Applications*, volume 2841 of *LNCS*. Springer-Verlag, 2005.
14. Coq Development Team. The Coq Proof Assistant Reference Manual – Version 8.0. http://coq.inria.fr/doc/, 2004.
15. Information and Privacy Commissioner/Ontario. Data mining: Staking a claim on your privacy. http://www.ipc.on.ca/scripts/index.asp?action=31&P_ID=11387, January 1998.
16. Bertrand Meyer. *Object-Oriented Software Construction*. Prentice Hall, 2 edition, 1997.
17. Andrew C. Myers. JFlow: Practical mostly-static information flow control. In *Symposium on Principles of Programming Languages (POPL)*, pages 228–241, January 1999.
18. Andrew C. Myers, Andrei Sabelfeld, and Steve Zdancewic. Enforcing robust declassification. In *Proceedings of the 17th IEEE Computer Security Foundations Workshop*, pages 172–186. IEEE Computer Society, 2004.
19. Andrei Sabelfeld and Andrew C. Myers. Language-based information-flow security. *IEEE J. Selected Areas in Communications*, 21(1):5–19, January 2003.
20. Geoffrey Smith. A new type system for secure information flow. In *Proceedings of the 14th IEEE Computer Security Foundations Workshop*, page 115. IEEE Computer Society, 2001.
21. The LOOP Project. http://www.cs.kun.nl/sos/research/loop/.
22. I.H. Witten and E. Frank. *Data Mining*. Morgan Kaufmann, 2000.

The Decidability of the First-Order Theory of Knuth-Bendix Order

Ting Zhang, Henny B. Sipma, and Zohar Manna*

Computer Science Department,
Stanford University, Stanford, CA 94305-9045
{tingz, sipma, zm}@theory.stanford.edu

Abstract. Two kinds of orderings are widely used in term rewriting and theorem proving, namely *recursive path ordering* (RPO) and *Knuth-Bendix ordering* (KBO). They provide powerful tools to prove the termination of rewriting systems. They are also applied in ordered resolution to prune the search space without compromising refutational completeness. Solving ordering constraints is therefore essential to the successful application of ordered rewriting and ordered resolution. Besides the needs for decision procedures for quantifier-free theories, situations arise in constrained deduction where the truth value of quantified formulas must be decided. Unfortunately, the full first-order theory of recursive path orderings is undecidable. This leaves an open question whether the first-order theory of KBO is decidable. In this paper, we give a positive answer to this question using quantifier elimination. In fact, we shall show the decidability of a theory that is more expressive than the theory of KBO.

1 Introduction

Two kinds of orderings are widely used in term rewriting and theorem proving. One is *recursive path ordering* (RPO) which is based on syntactic precedence [9]. The other is *Knuth-Bendix ordering* (KBO) which is of hybrid nature; it relies on numerical values assigned to symbols as well as syntactic precedence [13]. In ordered term rewriting, a strategy built on ordering constraints can dynamically orient an equation, at the time of instantiation, even if the equation is not uniformly orientable. This provides a powerful tool to prove the termination of rewriting systems [6]. In ordered resolution and paramodulation, ordering constraints are used to select maximal literals to perform resolution. It also serves as enabling conditions for inference rules and such conditions can be inherited from previous inferences at each deduction step. This helps to prune redundancy of the search space without compromising refutational completeness [25].

Solving ordering constraints is therefore essential to the successful application of ordered rewriting and ordered resolution. The decision procedures for

---

* This research was supported in part by NSF grants CCR-01-21403, CCR-02-20134, CCR-02-09237, CNS-0411363, and CCF-0430102, by ARO grant DAAD19-01-1-0723, and by NAVY/ONR contract N00014-03-1-0939.

R. Nieuwenhuis (Ed.): CADE 2005, LNAI 3632, pp. 131–148, 2005.

quantifier-free constraints of both types of orderings have been well-studied [3,12,23,22,24,14,15]. However, situations arise where we need to decide the truth values of quantified formulas on those orderings, especially in the $\exists^*\forall^*$ fragment. Examples include checking the soundness of simplification rules in constrained deduction [7]. Unfortunately, the full first-order theory of recursive path orderings is undecidable [28,7] except for the special case where the language only has unary functions and the precedence order is total [21]. Until now it has been an open question whether the first-order theory of Knuth-Bendix order is decidable (RTA open problem ♯99). Here we answer this question affirmatively by showing that an extended theory of term algebras with Knuth-Bendix order admits quantifier elimination.

The basic framework is the combination of term algebras with Presburger arithmetic. The extended language has two sorts; the integer sort $\mathbb{Z}$ and the term sort TA. Intuitively, the language is the set-theoretic union of the language of term algebras and the language of Presburger arithmetic. Formulas are formed from *term literals* and *integer literals* using logical connectives and quantifications. The combination is tightly coupled in the following sense. We have a *weight function* mapping terms to integers as well as various *boundary functions* mapping integers to terms. In addition, the Knuth-Bendix order is expanded in two directions. First, the order is decomposed into three disjoint suborders depending on which of three conditions is used in the definition. Secondly, all orders (including the suborders) are extended to gap orders, which assert the least number of distinct objects between two terms. Moreover, as Knuth-Bendix order is recursively defined on a lexicographic extension of itself, gap orders are extended to tuples of terms. Thus we actually establish the decidability of a richer theory.

Related Work and Comparison. Presburger arithmetic (PA) was first shown to be decidable in 1929 by the quantifier elimination method [10]. Efficient algorithms were later discovered by Cooper [8] and further improved in [26].

The decidability of the first-order theory of term algebras was first shown by Mal'cev using quantifier elimination [20]. This result was proved again later in different settings [19,5,11,4,2,27,17,18,29,30].

Quantifier elimination has been used to obtain decidability results for various extensions of term algebras. [19] shows the decidability of the theory of infinite and rational trees. [4] presents an elimination procedure for term algebras with membership predicate in the regular tree language. [2] presents an elimination procedure for structures of feature trees with arity constraints. [27] shows the decidability of term algebras with queues. [18] shows the decidability of term powers, which are term algebras augmented with coordinatewise-defined predicates. [29] extends the quantifier elimination procedure in [11] for term algebras with constant weight function.

The decidability of the theory of RPO has been well-studied. [3] proves the decidability of the quantifier-free theory of total lexicographic path ordering (LPO, a variant of RPO). A similar result holds for RPO [12]. [23] (resp. [22]) establishes the NP-completeness for the quantifier-free theory of LPO (resp.

RPO). A more efficient algorithm for the quantifier-free theory of RPO is given in [24]. [28,7] show the undecidability of the first-order theory of LPO and the undecidability of the first-order theory of RPO in case of partial precedence. The decidability of the first-order theory of RPO (LPO) in case of unary signature and total precedence is due to [21]. The decidability of the first-order theory of RPO in case of total precedence remains open.

Recently some partial decidability results for the theory of KBO have been obtained. [14] shows the decidability of the quantifier-free theory of term algebras with KBO. [15] improves the algorithm and shows that the quantifier-free theory of KBO is NP-complete. Analogous to [21], [16] shows the decidability of the first-order theory of KBO in the case where all functions are unary.

In this paper, we show the general decidability result for an extended theory of KBO with arbitrary function symbols and weight functions. The method combines the extraction of integer constraints from term constraints with a reduction of quantifiers on term variables to quantifiers on integer variables.

Paper Organization. Section 2 defines term algebras. Section 3 introduces the theory of term algebras with Knuth-Bendix ordering and presents the technical machinery for eliminating quantifiers. Section 4 presents the main contribution of this paper: it expands the elimination procedure in [29] for the extended theory of KBO and proves its correctness. Section 5 briefly explains how to adapt the elimination procedure to the special case where the language contains a unary function of weight 0. Section 6 concludes with some ideas for future work. Due to space limitation all proofs have been omitted from this paper. An extended version of this paper, which includes a detailed description of notation and terminology, and all proofs, is available from the first author's webpage.

2   Term Algebras

We present a general language and structure of term algebras. *In this paper we assume that the signature of our language is finite.* For notation convenience, we do not distinguish syntactic terms in the language from semantic terms in the corresponding structure. The meaning should be clear from the context.

Definition 1. *A term algebra* $\mathfrak{A}_{\mathsf{TA}} : \langle \mathsf{TA}; C, \mathcal{A}, \mathcal{S}, \mathcal{T} \rangle$ *consists of*

1. TA: *The term domain, which exclusively consists of terms recursively built up from constants by applying constructors. The type of a term t, denoted by* $\mathsf{type}(t)$, *is the outmost constructor of t. We say that t is α-typed (or is an α-term) if $\alpha = \mathsf{type}(t)$.*
2. C: *A finite set of constructors:* $\alpha, \beta, \gamma, \ldots$ *The arity of α is denoted by* $\mathsf{ar}(\alpha)$.
3. $\mathcal{A}$: *A finite set of constants:* $a, b, c, \ldots$ *We require $\mathcal{A} \neq \emptyset$ and $\mathcal{A} \subseteq C$. For $a \in \mathcal{A}$,* $\mathsf{ar}(a) = 0$ *and* $\mathsf{type}(a) = a$.
4. $\mathcal{S}$: *A finite set of selectors. For a constructor α with arity $k > 0$, there are k selectors* $\mathsf{s}_1^\alpha, \ldots, \mathsf{s}_k^\alpha$ *in $\mathcal{S}$. We call s_i^α $(1 \leq i \leq k)$ the ith α-selector. For a term x, $\mathsf{s}_i^\alpha(x)$ returns the ith component of x if x is an α-term and x itself otherwise.*

5. $\mathcal{T}$: *A finite set of testers. For each constructor α there is a corresponding tester Is_α.*
   For a term x, $\mathsf{Is}_\alpha(x)$ is true if and only if x is an α-term. Note that for a constant a,
   $\mathsf{Is}_a(x)$ is just $x = a$. In addition there is a special tester Is_A such that $\mathsf{Is}_A(x)$ is true
   if and only if x is a constant.

We use $\mathcal{L}_{\mathsf{TA}}$ to denote the language of term algebras.

Proposition 1 (Axiomatization of Term Algebras). *Let $\bar{z}_\alpha$ abbreviate $z_1, \ldots, z_{\mathsf{ar}(\alpha)}$.*
The following formula schemes, in which variables are implicitly universally quantified
over TA, axiomatize $\mathrm{Th}(\mathfrak{A}_{\mathsf{TA}})$.

A1. $t(x) \neq x$, *if t is built solely by constructors and t properly contains x.*
A2. $\alpha(x_1 \ldots, x_{\mathsf{ar}(\alpha)}) \neq \beta(y_1, \ldots, y_{\mathsf{ar}(\beta)})$, *if $\alpha, \beta \in C$ and $\alpha \not\equiv \beta$.*
A3. $\alpha(x_1, \ldots, x_{\mathsf{ar}(\alpha)}) = \alpha(y_1, \ldots, y_{\mathsf{ar}(\alpha)}) \rightarrow \bigwedge_{1 \leq i \leq \mathsf{ar}(\alpha)} x_i = y_i$.
A4. $\mathsf{Is}_\alpha(x) \leftrightarrow \exists \bar{z}_\alpha \alpha(\bar{z}_\alpha) = x$, *if $\alpha \in C \setminus \mathcal{A}$; $\mathsf{Is}_a(x) \leftrightarrow x = a$, if $a \in \mathcal{A}$.*
A5. $\mathsf{Is}_A(x) \leftrightarrow \bigvee_{a \in \mathcal{A}} \mathsf{Is}_a(x)$.
A6. $\mathsf{s}_i^\alpha(x) = y \leftrightarrow \exists \bar{z}_\alpha \big(\alpha(\bar{z}_\alpha) = x \wedge y = z_i \big) \vee \big(\forall \bar{z}_\alpha (\alpha(\bar{z}_\alpha) \neq x) \wedge x = y \big)$.

This set of axioms is a variant of the axiomatization given in [11].

Selectors and testers can be defined by constructors and vice versa. One
direction has been shown by (A4-A6), which are pure definitional axioms. The
other direction follows from the equivalence of $\bigwedge_{i=1}^k \mathsf{s}_i^\alpha(x) = x_i \wedge \mathsf{Is}_\alpha(x)$ and
$x = \alpha(x_1, \ldots, x_k)$. For simplicity, from now on we assume $\mathcal{L}_{\mathsf{TA}}$ only has selector
functions, and we use $x = \alpha(x_1, \ldots, x_k)$ only in discussions at the semantic level.

We write $\alpha = (\mathsf{s}_1^\alpha, \ldots, \mathsf{s}_k^\alpha)$ $(k > 0)$ to mean that α is a constructor of arity k,
and $\mathsf{s}_1^\alpha, \ldots, \mathsf{s}_k^\alpha$ are the corresponding selectors of α. We use L to denote selector
sequences. If $L = \mathsf{s}_1, \ldots, \mathsf{s}_n$, Lx stands for $\mathsf{s}_1(\ldots(\mathsf{s}_n(x)\ldots))$, and we say that the
depth of x in Lx is n. The depth of x in a formula φ is the maximum depth of x in
the selector terms in φ, denoted by $\mathsf{depth}_\varphi(x)$.

3   Term Algebras with Knuth-Bendix Order

In this section we introduce the theory of term algebras with KBO and present
the technical machinery needed in the quantifier elimination procedure.

Let Σ be a finite signature in the constructor language (i.e., $\Sigma = C$ in Def.
1) and $W : \Sigma \to \mathbb{N}$ a weight function. We expand $\mathsf{dom}(W)$ to TA by recursively
defining $W(\alpha(t_1, \ldots, t_k)) = W(\alpha) + \sum_{i=1}^k W(t_i)$. Let $<^\Sigma$ be a linear precedence order
on symbols in Σ. We enumerate all symbols in the decreasing $<^\Sigma$-order such that
$\alpha_1 >^\Sigma \alpha_2 >^\Sigma \ldots >^\Sigma \alpha_{|\Sigma|}$.

Definition 2 (Knuth-Bendix Order [13]). *A Knuth-Bendix order (KBO) $<^{\mathsf{kb}}$ (pa-*
rameterized with a weight function W and a precedence order $<^\Sigma$) is defined recursively
such that for $u, v \in \mathsf{TA}$, $u <^{\mathsf{kb}} v$ if and only if one of the following conditions holds:
(i) $W(u) < W(v)$, (ii) $W(u) = W(v)$ and $\mathsf{type}(u) <^\Sigma \mathsf{type}(v)$, (iii) $W(u) = W(v)$,
$u \equiv \alpha(u_1, \ldots, u_k)$, $v \equiv \alpha(v_1, \ldots, v_k)$ and

$$(\exists i) \Big[1 \leq i \leq k \wedge u_i <^{\mathsf{kb}} v_i \wedge \forall j (1 \leq j < i \rightarrow u_j = v_j) \Big]. \tag{1}$$

The KBO $<^{kb}$ is a well-founded total order on TA [13,1]. To guarantee well-foundedness, two *compatibility conditions* for W and $<^\Sigma$ are required: (i) $W(a) > 0$ for any constant a, and (ii) a unary function of weight 0, if present, should be the maximum in $<^\Sigma$. Let us denote by $\bot$ the smallest term with respect to $<^{kb}$. It follows from (i) and (ii) that $\bot$ must be an atom and so it can be determined when W and $<^\Sigma$ are given. By (ii) if a unary function of weight 0 exists, it must be unique. For presentation simplicity, we assume that $W(\alpha_1) > 0$. However, the existence of such function actually simplifies our decision procedure. We defer the discussion to Sec. 5.

Definition 3. *The structure of term algebras with KBO is* $\mathfrak{A}_{kb} = \langle \mathfrak{A}_{TA}; <^{kb} \rangle$. *Let* $\mathscr{L}_{kb}$ *denote the language of* $\mathfrak{A}_{kb}$.

3.1 Proof Plan

We shall show the decidability of $\text{Th}(\mathfrak{A}_{kb})$ by quantifier elimination. The procedure relies on the following two ideas: *solved form* and *depth reduction*.

1. *Solved Form.* A quantifier-free formula $\varphi(x, \bar{y})$ is *solved* in x if it is in the form

$$\bigwedge_{i \leq m} u_i <^{kb} x \ \wedge \ \bigwedge_{j \leq n} x <^{kb} v_j \ \wedge \ \varphi'(\bar{y}), \tag{2}$$

   where x does not appear in u_i, v_i and φ'. It is not hard to argue that $(\exists x)\,\varphi(x, \bar{y})$ simplifies to

$$\bigwedge_{i \leq m, j \leq n} u_i <^{kb}_2 v_j \ \wedge \ \varphi'(\bar{y}) \tag{3}$$

   where $<^{kb}_n$, called *gap order*, is an extension of $<^{kb}$ such that $x <^{kb}_n y$ states there is an increasing chain from x to y with at least $n-1$ elements in between [10, page 196]. It is clear that the elimination of $\exists x$, the transformation from (2) to (3), becomes straightforward once the matrix $\varphi(x, \bar{y})$ is solved in x, or equivalently, $\text{depth}_\varphi(x) = 0$. That leads us to the notion of *depth reduction*.

2. *Depth Reduction.* Let us first consider the simple case where x is α-typed for a proper constructor α and all occurrences of x have depth greater than 0. By introducing new variables $x_1, \ldots, x_{ar(\alpha)}$ (called the descendants of x) to represent x, we can rewrite $\exists x \varphi(x, \bar{y})$ to

$$\exists x_1, \ldots, \exists x_{ar(\alpha)} \varphi'(x_1, \ldots, x_{ar(\alpha)}, \bar{y}), \tag{4}$$

   where $\varphi'(x_1, \ldots, x_{ar(\alpha)}, \bar{y})$ is obtained from $\varphi(x, \bar{y})$ by substituting x_i for $\mathsf{s}^\alpha_i x$ ($1 \leq i \leq ar(\alpha)$). It is clear that $\text{depth}_{\varphi'}(x_i) < \text{depth}_\varphi(x)$. If all occurrences of x have the same depth, then by repeating the process we can generate a formula solved in $\bar{x}^*$ where $\bar{x}^*$ are descendants of x. A difficulty arises when not all occurrences of x have equal depth. So eventually we meet the situation where some occurrences of x have depth 0 and some do not. Here we have to represent all occurrences of x of depth 0 in terms of $\mathsf{s}^\alpha_1(x), \ldots, \mathsf{s}^\alpha_{ar(\alpha)}(x)$. This amounts to reducing literals of the form $x <^{kb}_n t$ and literals of the form

$t <_n^{kb} x$ to quantifier-free formulas using $s_1^\alpha(x), \ldots, s_{ar(\alpha)}^\alpha(x)$. After that we can introduce new variables and do quantifier manipulation just as in the simple case to bring $\exists x \varphi(x, \bar{y})$ into the form of (4). Therefore depth reduction essentially depends on the reduction of $x <_n^{kb} t$ and reduction of $t <_n^{kb} x$. In order to carry out the reduction we need to extend the language as follows:

(a) We decompose $<^{kb}$ into three disjoint suborders $<^w$, $<^p$ and $<^l$, each of which is also extended to gap orders.

(b) We introduce Presburger arithmetic explicitly in order to define *counting constraints* to count how many distinct terms there are at certain weight, and define *boundary functions* to *delineate* gap orders.

(c) The reduction of literals like $x <_n^{kb} t$ or $t <_n^{kb} x$ eventually comes down to resolving relations between two terms of the same weight and of the same type. So we need to extend all aforementioned notions to tuples of terms of the same total weight.

In the rest of this section we define these extensions.

3.2   Decomposition of Knuth-Bendix Order

Definition 4. *A Knuth-Bendix order $<^{kb}$ can be decomposed into three disjoint orders, a weight order $<^w$, a precedence order $<^p$, and a lexicographical order $<^l$, as follows:*

$$u <^w v \Leftrightarrow W(u) < W(v),$$
$$u <^p v \Leftrightarrow W(u) = W(s) \ \& \ \text{type}(u) <^\Sigma \text{type}(v),$$
$$u <^l v \Leftrightarrow W(u) = W(v) \ \& \ \text{type}(u) = \text{type}(v) \ \& \ u <^{kb} v,$$

such that $u <^{kb} v$ is equivalent to $u <^w v \lor u <^p v \lor u <^l v$. We write $u <^{pl} v$ as an abbreviation for $u <^p v \lor u <^l v$.

3.3   Gap Orders

To express formulas of the form $\exists x (u <^\# x <^\# v)$ in a quantifier-free language we need to extend all aforementioned orders to "gap" orders.

Definition 5 (Gap Orders). *Define $<_n^{kb}$ $(n \geq 0)$ such that*

$$u <_n^{kb} v \Leftrightarrow (\exists u_1, \ldots, \exists u_n)\left[u <^{kb} u_1 <^{kb} \ldots <^{kb} u_n \leq^{kb} v \right].$$

For $\# \in \{w, p, l, pl\}$, define $<_n^\#$ such that $u <_n^\# v \Leftrightarrow u <_n^{kb} v \wedge u <^\# v$, and $u \leq_n^\# v$ such that $(u <_n^\# v) \wedge \neg(u <_{n+1}^\# v)$.

A gap order $u <_n^\# v$ $(n \geq 1)$ states that "u is less than v w.r.t. $<^\#$, and there are *at least* $n - 1$ elements in between." Similarly, $u \leq_n^\# v$ $(n \geq 1)$ states that "u is less than v w.r.t. $<^\#$, and there are *exactly* $n - 1$ elements in between". Note that $<_1^\#$ is just $<^\#$, $<_0^\#$ is $\leq^\#$, $\leq_0^\#$ is $=$, and we have $u <_n^\# v \Leftrightarrow u <_{n+1}^\# v \lor u \leq_n^\# v$.

Example 1. The formula $\exists x (u <^l x <^l v)$ reduces to $u <_2^l v$ if u, v do not contain x.

3.4   Boundary Functions

Consider the formula $u \leq_1^w v$. Intuitively it states "$W(u) < W(v)$ and there are no terms z such that $u <^{kb} z <^{kb} v$, that is, u is the largest term of weight $W(u)$ and v is the smallest term of weight $W(v)$". To express this we introduce *boundary functions*.

Definition 6 (Boundary Functions). *Let $n, p > 0$. The following functions are called boundary functions:*

1. $0^w : \mathbb{N} \to \mathsf{TA}$ *such that* $0^w(n)$ *is the smallest term (w.r.t. $<^{kb}$) of weight n,*
2. $0^p : \mathbb{N}^2 \to \mathsf{TA}$ *such that* $0^p(n, p)$ *is the smallest term (w.r.t. $<^{kb}$) of weight n and type α_p,*

where, for all of the above, $f(n) = \perp$ and $f(n, p) = \perp$, if no such term exists.

Similarly we define $1^w : \mathbb{N} \to \mathsf{TA}$ and $1^p : \mathbb{N}^2 \to \mathsf{TA}$ as the largest terms with the corresponding properties. We write $0^\sharp_{(...)}$ for $0^\sharp(...)$ and $1^\sharp_{(...)}$ for $1^\sharp(...)$. Terms having one of these functions as root symbol are called *boundary terms*. A literal of the form $u \star v$, where $\star$ is either equality or a gap order, is *open* if both u and v are ordinary terms in TA, *closed* if both u and v are boundary terms, and *half-open* otherwise.

3.5   Integer Extension of Term Algebras

To be able to express the boundary terms in the formal language, we extend term algebras with Presburger arithmetic (PA).

Definition 7. *The structure of term algebras with integers is* $\mathfrak{A}_{\mathsf{TA}}^{\mathbb{Z}} = \langle \mathfrak{A}_{\mathsf{TA}}; \mathfrak{A}_{\mathbb{Z}}; (.)^w \rangle$, *where $\mathfrak{A}_{\mathbb{Z}}$ is Presburger arithmetic and $(.)^w$ denotes the weight function.*

We call terms of sort TA (resp. $\mathbb{Z}$) TA-*terms* (resp. *integer terms*), similarly for variables and quantifiers. We also use "term" for "TA" when there is no confusion. A TA-term can occur inside the weight function. Such occurrence is called *integer occurrence* to be distinguished from the normal *term occurrence*. From now on, we freely use integer terms t^w to form Presburger formulas, and we use $\mathsf{depth}_\varphi(x)$ to denote the maximum depth of term occurrences of x in φ.

Example 2. The formula $(\exists x : \mathsf{TA}) \left[0^w_{(x^w)} <^{pl} x <^{pl} 1^w_{(x^w)} \right]$ states that there exists a term $t \in \mathsf{TA}$ such that there are at least three elements with the same weight as t (including t itself). Note that the first and the third occurrences of x are integral while the second one is an ordinary term.

The truth value of the formula in Ex. 2 relies on the number of distinct TA-terms of a certain weight. This is the essential use of Presburger arithmetic.

Definition 8 (Counting Constraint). *A counting constraint is a predicate $\mathsf{CNT}_n^\alpha(z)$ that states there are at least $n+1$ different α-terms of weight z. $\mathsf{CNT}_n(z)$ is similarly defined with α-terms replaced by TA-terms. We write Tree^α (resp. Tree) for CNT_0^α (resp. CNT_0).*

Counting constraints play a central role in our elimination procedure; it helps reduce term quantifiers to integer quantifiers.

Example 3. The formula from Ex. 2 is reduced to $(\exists z : \mathbb{Z})\, \mathsf{CNT}_2(z)$.

It was proved in [14,29] that counting constraints can be expressed in PA.

Example 4. Consider $\mathfrak{A}_{\mathsf{list}}^{\mathbb{Z}} = \langle \mathfrak{A}_{\mathsf{list}}; \mathfrak{A}_{\mathbb{Z}}; (.)^{\mathsf{w}} \rangle$ where $\mathfrak{A}_{\mathsf{list}} = \langle \mathsf{list}, \mathsf{cons}, \mathsf{car}, \mathsf{cdr}, a \rangle$ is the LISP list structure with the only atom a, and $(.)^{\mathsf{w}}$ is a constant weight function equal to 1. It has been shown in [30] that $\mathsf{CNT}_n^{\mathsf{cons}}(x)$ is $x \geq 2m - 1 \wedge 2 \nmid m$ where m is the least number such that the m-th *Catalan number* $C_m = \frac{1}{m}\binom{2m-2}{m-1}$ is greater than n. This is not surprising as C_m gives the number of binary trees with m leaves (that tree has $2m - 1$ nodes).

3.6   Extension of Knuth-Bendix Order

Definition 9. *The structure of term algebras with KBO, extended with gap orders, boundary functions and Presburger arithmetic, is*

$$\mathfrak{A}_{\mathsf{kb}^+}^{\mathbb{Z}} = \langle \mathfrak{A}_{\mathsf{kb}}; \mathfrak{A}_{\mathbb{Z}}; <_n^{\sharp}, \leq_n^{\sharp}, \sharp \in \{\mathsf{kb}, \mathsf{w}, \mathsf{p}, \mathsf{l}, \mathsf{pl}\}, n \geq 0; 0_{(...)}^*, 1_{(...)}^*, * \in \{\mathsf{w}, \mathsf{p}\}\rangle.$$

We denote by $\mathscr{L}_{\mathsf{kb}^+}$ the language extending $\mathscr{L}_{\mathsf{kb}}$ with gap orders and boundary terms and by $\mathscr{C}_{\mathbb{Z}}$ the language of Presburger arithmetic (including weight functions on terms). The complete language is denoted by $\mathscr{L}_{\mathsf{kb}^+}^{\mathbb{Z}}$.

3.7   Tuples of Terms

The extensions for tuples of terms are defined as follows:

Definition 10 (Orders on Tuples). *Let* $\bar{u} = \langle u_1, \ldots, u_k \rangle, \bar{v} = \langle v_1, \ldots, v_k \rangle$ *such that* $\Sigma_{i=1}^k \mathsf{W}(u_i) = \Sigma_{i=1}^k \mathsf{W}(v_i)$. *The lexicographical extension* $<^{k;\mathsf{kb}}$ $(k \geq 1)$ *of* $<^{\mathsf{kb}}$ *on k-tuples of the same weight is defined such that* $\bar{u} <^{k;\mathsf{kb}} \bar{v}$ *if and only if (1) holds.*

Definition 11 (Suborders on Tuples). *Let* $\bar{u} = \langle u_1, \ldots, u_k \rangle, \bar{v} = \langle v_1, \ldots, v_k \rangle \in \mathsf{TA}^k$, $\sharp \in \{\mathsf{w}, \mathsf{p}, \mathsf{l}, \mathsf{pl}\}$. *We define those composite orders on tuples as follows.*

$$\bar{u} <^{k;\sharp} \bar{v} \leftrightarrow u_1 <^{\sharp} v_1 \vee (u_1 = v_1 \wedge \langle u_2, \ldots, u_k \rangle <^{k-1;\mathsf{kb}} \langle v_2, \ldots, v_k \rangle)$$

We say that $\bar{u} <^{k;\sharp} \bar{v}$ *is proper if* $u_1 <^{\sharp} v_1$ *and we have* $\bar{u} <^{k;\mathsf{kb}} \bar{v} \leftrightarrow \bar{u} <^{k;\mathsf{w}} \bar{v} \vee \bar{u} <^{k;\mathsf{p}} \bar{v} \vee \bar{u} <^{k;\mathsf{l}} \bar{v}$.

Definition 12 (Gap Orders between Tuples). *We define* $<_n^{k;\mathsf{kb}}$ $(k \geq 1; n \geq 0)$ *such that*

$$\bar{u} <_n^{k;\mathsf{kb}} \bar{v} \leftrightarrow (\exists \bar{u}_1, \ldots, \exists \bar{u}_n : \mathsf{TA}^k)\left[\bar{u} <^{k;\mathsf{kb}} \bar{u}_1 <^{k;\mathsf{kb}} \ldots <^{k;\mathsf{kb}} \bar{u}_n \leq^{k;\mathsf{kb}} \bar{v}\right].$$

For $\sharp \in \{\mathsf{w}, \mathsf{p}, \mathsf{l}, \mathsf{pl}\}$, *define* $<_n^{k;\sharp}$ *such that* $\bar{u} <_n^{k;\sharp} \bar{v} \leftrightarrow \bar{u} <_n^{k;\sharp} \bar{v} \wedge \bar{u} <^{k;\sharp} \bar{v}$, *and* $\bar{u} \leq_n^{k;\sharp} \bar{v}$ *such that* $(\bar{u} <_n^{k;\sharp} \bar{v}) \wedge \neg(\bar{u} <_{n+1}^{k;\sharp} \bar{v})$. *Again note that* $<_1^{k;\sharp}$ *is just* $<^{k;\sharp}$, $<_0^{k;\sharp}$ *is* $\leq^{k;\sharp}$, $\leq_0^{k;\sharp}$ *is* $=$, *and* $\bar{u} <_n^{k;\sharp} \bar{v} \leftrightarrow \bar{u} <_{n+1}^{k;\sharp} \bar{v} \vee \bar{u} \leq_n^{k;\sharp} \bar{v}$.

Definition 13 (Tuple Boundary Functions). *Let $k, n, m, p > 0$. Define partial functions:*

1. $\bar{0}^{k;kb} : \mathbb{N} \to \mathsf{TA}^k$ *($k \geq 1$) such that $\bar{0}^{k;kb}(n)$ is the smallest k-tuple (w.r.t. $<^{k;kb}$) of weight n.*
2. $\bar{0}^{k;w} : \mathbb{N}^2 \to \mathsf{TA}^k$ *($k \geq 1$) such that $\bar{0}^{k;w}(n, m)$ is the smallest k-tuple (w.r.t. $<^{k;kb}$) of weight n and the first component has weight m.*
3. $\bar{0}^{k;p} : \mathbb{N}^3 \to \mathsf{TA}^k$ *($k \geq 1$) such that $\bar{0}^{k;p}(n, m, p)$ is the smallest k-tuple (w.r.t. $<^{k;kb}$) of weight n and the first component has weight m and type α_p.*

Similarly we define $\bar{1}^{k;kb} : \mathbb{N} \to \mathsf{TA}^k$, $\bar{1}^{k;w} : \mathbb{N}^2 \to \mathsf{TA}^k$ and $\bar{1}^{k;p} : \mathbb{N}^3 \to \mathsf{TA}^k$ to be the largest k-tuples with the corresponding properties. As before these functions are made total by assigning $\langle \perp, \ldots \perp \rangle$ to undefined values. We write $\bar{0}^{k;\#}_{(\ldots)}$ for $\bar{0}^{k;\#}(\ldots)$ and $\bar{1}^{k;\#}_{(\ldots)}$ for $\bar{1}^{k;\#}(\ldots)$. Terms having one of these functions as root symbol are called *boundary tuples*. As before we call a literal $\bar{u} \star \bar{v}$ *open* if both $\bar{u}$ and $\bar{v}$ are ordinary tuples, *closed* if both $\bar{u}$ and $\bar{v}$ are boundary tuples, and *half-open* otherwise.

To avoid unnecessary complications, we choose to treat tuples (including boundary tuples) as "syntactic sugar"; they are only used in the intermediate steps of the reduction. Lemma 5 shows that literals containing tuples can be reduced to formulas in $\mathscr{L}^{\mathbb{Z}}_{kb^+}$.

3.8 Delineated Gap Order Completion

Revisiting the transformation from (2) to (3), we see that the number of gap orders in (3) is quadratic in the number of gap orders in (2). This complicates the termination proof for the elimination procedure. Nevertheless, we can avoid this difficulty by postulating the relative positions of parameters. This leads to the notion of *order completion*.

Definition 14 (Gap Order Completion). *A gap order completion (GOC) φ' of a conjunction of literals $\varphi(t_1, \ldots, t_n)$ is chain $t_{f(1)} \trianglerighteq \ldots \trianglerighteq t_{f(n)}$, where f is a permutation function on $\{1, \ldots, n\}$ and $\trianglerighteq$ stands for $=, \leq^{\#}_n$ or $<^{\#}_n$ ($\# \in \{w, p, l, pl\}$, $n \geq 1$).*

Example 5. A possible GOC of $\varphi(x, y, z) : x <^w_9 y \wedge x <^{pl} z \wedge z <^w y$ is $x <^{pl}_5 z <^w_4 y$.

However, gap order completions are not sufficient. It is quite clear to see $(\exists x : \mathsf{TA})[u <^w x <^p v]$ implies $u <^w_2 v$. But for the converse to hold, $v \neq 0^w_{(vw)}$ is required. As another example, $(\exists x : \mathsf{TA})[u <^p x <^p v]$ implies $u <^p_2 v$, but not vice versa. In order to preserve equivalence, intuitively, we need to "delineate" a GOC to make sure ordinary terms in different *intervals* (a notion to be define precisely soon) are not related in any gap orders. For example, consider the linear order $x_1 <^w_{n_1} x_2 <^p_{n_2} x_3 <^l_{n_3} x_4$ The order imposed may be viewed as follows

The weight of x_1 is strictly lower than that of x_2, x_3, and x_4. The weight of x_2, x_3, and x_4 is the same, but the precedence of x_2 is lower than that of x_3 and x_4. Finally, x_3 is smaller than x_4 in the lexicographic order. We call a maximal list of elements with the same weight a w-interval, and similarly a maximal list of elements with the same weight and precedence order a p-interval. Thus, the second w-interval above has two inner p-intervals.

We want to avoid relating ordinary elements at different levels in different intervals. Therefore we augment the gap order completion with boundary terms, called a delineated gap order completion.

Definition 15 (Delineated Gap Order Completion). *A delineated gap order completion (DGOC) is a GOC in which if there occurs the following pattern* $v_1 \unrhd_{n_1}^\sharp u \unrhd_{n_2}^\natural v_2$, *where* $n_1, n_2 > 0$, $\unrhd$ *stands for either* $<$ *or* $\leq$, $\sharp, \natural \in \{$w, p, l, pl$\}$, *and* u *is an ordinary term in* $\mathscr{L}_{kb}$, *then either* $\sharp \equiv \natural \equiv$ pl *or* $\sharp \equiv \natural \equiv$ l. *I.e., ordinary terms do not delineate two intervals unless they are asserted equal to boundary terms.*

Example 6. Revisit Ex. 5. A possible DGOC of $\varphi(x, y, z)$ is

$$\varphi'(x, y, z) : \underbrace{0_{(x^w)}^w <_1^{pl} x <_5^{pl} z <_2^{pl} 1_{(x^w)}^w}_{\text{w-interval}} <_1^w \underbrace{0_{(y^w)}^w <_1^{pl} y <_1^{pl} 1_{(y^w)}^w}_{\text{w-interval}}$$

Now we have $(\exists z : \mathsf{TA})\varphi'(x, y, z) \leftrightarrow 0_{(x^w)}^w <_1^{pl} x <_7^{pl} 1_{(x^w)}^w <_1^w 0_{(y^w)}^w <_1^{pl} y <_1^{pl} 1_{(y^w)}^w$.

Lemma 1 (Delineated Gap Order Completion). *Any conjunction of positive literals in* $\mathscr{L}_{kb^+}$ *is equivalent to a finite disjunction of delineated gap order completions.*

Now we state a sequence of lemmas which will justify the elimination procedure given in the next section. These lemmas share the following common features: (i) they state the soundness of symbolic transformations for formulas in *primitive form*, a special prenex form where the prefix only consists of existential quantifiers and the matrix is a conjunction of literals; (ii) a formula φ is transformed to a finite disjunction $\bigvee_i \varphi_i$ where for any i, φ_i is in primitive form and the matrix of φ_i contains no more open gap order literals than that of φ does. To save space, we omit these conditions in the description of each lemma.

In principle, boundary terms can appear in the weight function or in selectors, selector terms can occur in the weight function, and the weight function can be used to construct boundary terms. Repeating this process we can build more and more complex terms. The following lemma eliminates this superficial complication. From now on, we assume that boundary terms are not properly embedded in other terms.

Lemma 2 (Depth Reduction of Boundary Terms). *Any formula in* $\mathscr{L}_{kb^+}^Z$ *can be effectively reduced to an equivalent formula in which no boundary terms appear inside selectors or the weight function.*

The following lemma states that we can always assume that all *term occurrences* of a TA-variable have the same depth, and hence we are able to reduce them all together to depth 0.

Lemma 3 (Depth Reduction). *Let $\star \in \{<_n^{\mathsf{kb}}, <_n^{\mathsf{w}}, <_n^{\mathsf{p}}, <_n^{\mathsf{l}}, <_n^{\mathsf{pl}}, \leq_n^{\mathsf{kb}}, \leq_n^{\mathsf{w}}, \leq_n^{\mathsf{p}}, \leq_n^{\mathsf{l}}, \leq_n^{\mathsf{pl}}\}$. If x is of type α_p with $\alpha_p = (\mathsf{s}_1^{\alpha_p}, \ldots, \mathsf{s}_k^{\alpha_p})$ and t is an arbitrary term, then $x \star t$ $(t \star x)$ can be effectively reduced to an equivalent quantifier-free formula $\varphi(\mathsf{s}_1^{\alpha_p} x, \ldots, \mathsf{s}_k^{\alpha_p} x)$ (in $\mathscr{L}_{\mathsf{kb}^+}^{\mathbb{Z}}$) in which x does not appear and $\mathsf{s}_i^{\alpha_p} x$ $(1 \leq i \leq k)$ is not inside selectors.*

As we mentioned before, this is the main battlefield of quantifier elimination. To streamline the proof, we introduce the following two lemmas.

Lemma 4 (Term Reduction). *Let $\star \in \{<_n^{\mathsf{kb}}, <_n^{\mathsf{w}}, <_n^{\mathsf{p}}, <_n^{\mathsf{l}}, <_n^{\mathsf{pl}}, \leq_n^{\mathsf{kb}}, \leq_n^{\mathsf{w}}, \leq_n^{\mathsf{p}}, \leq_n^{\mathsf{l}}, \leq_n^{\mathsf{pl}}\}$.*

1. *If x is an ordinary term of type α_p with $\alpha_p = (\mathsf{s}_1^{\alpha_p}, \ldots, \mathsf{s}_k^{\alpha_p})$ and t is either a boundary term or an ordinary term not containing x, then $x \star t$ $(t \star x)$ can be effectively reduced to an equivalent quantifier-free formula $\varphi(\mathsf{s}_1^{\alpha_p} x, \ldots, \mathsf{s}_k^{\alpha_p} x)$ in which x does not occur and $\mathsf{s}_i^{\alpha_p} x$ $(1 \leq i \leq k)$ is not inside selectors.*
2. *If $x \star t$ $(t \star x)$ is closed, i.e., both t and x are boundary terms, then it can be effectively reduced to an equivalent Presburger formula.*

Lemma 4 states that literals containing non-atom terms can be expressed only using the components of those terms. The reduction eventually comes down to the success of decomposing relations between tuples of the same weight, as is stated by the following lemma.

Lemma 5 (Tuple Reduction). *Let $\star \in \{<_n^{k;\mathsf{kb}}, <_n^{k;\mathsf{w}}, <_n^{k;\mathsf{p}}, <_n^{k;\mathsf{l}}, <_n^{k;\mathsf{pl}}, \leq_n^{k;\mathsf{kb}}, \leq_n^{k;\mathsf{w}}, \leq_n^{k;\mathsf{p}}, \leq_n^{k;\mathsf{l}}, \leq_n^{k;\mathsf{pl}}\}$, and U, V be k-tuples of the same weight.*

1. *If $U = \langle u_1, \ldots, u_k \rangle$ is an ordinary tuple, then $U \star V$ $(V \star U)$ can be effectively reduced to an equivalent quantifier-free formula $\varphi(u_1, \ldots, u_k)$ (in $\mathscr{L}_{\mathsf{kb}^+}^{\mathbb{Z}}$) in which u_i $(1 \leq i \leq k)$ does not occur inside selectors.*
2. *If $U \star V$ $(V \star U)$ is a closed tuple, i.e., both U and V are boundary tuples, then it can be effectively reduced to an equivalent Presburger formula.*

Lemma 6 (Elimination of Term Variables). *Let x be a term variable, $\varphi_{\mathsf{kb}^+}(x)$ a conjunction of literals in $\mathscr{L}_{\mathsf{kb}^+}$ with $\mathsf{depth}_{\varphi_{\mathsf{kb}^+}}(x) = 0$, and $\varphi_{\mathbb{Z}}(x)$ a Presburger formula in which x occurs inside the weight function. Then $(\exists x : \mathsf{TA})[\varphi_{\mathsf{kb}^+}(x) \wedge \varphi_{\mathbb{Z}}(x)]$ can be effectively reduced to $\varphi'_{\mathsf{kb}^+} \wedge \varphi'_{\mathbb{Z}}$ in which x does not occur and φ'_{kb^+} is quantifier-free.*

Lemma 6 states that we can remove term quantifiers by reducing them to integer quantifiers. The next lemma guarantees the elimination of integer quantifiers.

Lemma 7 (Elimination of Integer Variables). *Let z be an integer variable, $\varphi_{\mathsf{kb}^+}(z)$ a conjunction of literals in $\mathscr{L}_{\mathsf{kb}^+}$ where z occurs inside boundary terms, and $\varphi_{\mathbb{Z}}(z)$ a Presburger formula. Then $(\exists z : \mathbb{Z})[\varphi_{\mathsf{kb}^+}(z) \wedge \varphi_{\mathbb{Z}}(z)]$ can be effectively reduced to $\varphi'_{\mathsf{kb}^+} \wedge \varphi'_{\mathbb{Z}}$ where no z occurs and φ'_{kb^+} is quantifier-free.*

4   Quantifier Elimination for $\mathrm{Th}(\mathfrak{A}_{\mathsf{kb}^+}^{\mathbb{Z}})$

In this section we extend the quantifier elimination procedure for $\mathrm{Th}(\mathfrak{A}_{\mathsf{TA}}^{\mathbb{Z}})$ [29] to an elimination procedure for $\mathrm{Th}(\mathfrak{A}_{\mathsf{kb}^+}^{\mathbb{Z}})$. First we introduce some notations to simplify the algorithm description.

4.1 Primitive Form

It is well-known that eliminating arbitrary quantifiers reduces to eliminating existential quantifiers from *primitive formulas* of the form

$$(\exists x)\, \varphi(x, \bar{y}) \equiv (\exists x) \big[A_i(x, \bar{y}) \wedge \cdots \wedge A_n(x, \bar{y}) \big], \tag{5}$$

where $A_i(x, \bar{y})$ are $(1 \leq i \leq n)$ literals [11]. We also assume that $A_i(x, \bar{y})$ are not of the form $x = t$ in case t does not contain x, as $(\exists x)[x = t \wedge \varphi'(x, \bar{y})]$ simplifies to $\varphi'(t, \bar{y})$. In addition we can assume A_i are positive literals. The details of elimination of negation are given in the extended version of this paper.

4.2 Nondeterminism

All transformations are carried out on formulas of the form (5). Each step of the transformations manipulates (5) to produce a version of the same form (or multiple versions of the same form in case disjunctions are introduced), and thus in each step $(\exists x)\varphi(x, \bar{y})$ refers to the updated version rather than to the original input formula. Whenever we say "guess ψ", we mean to add a finite disjunction $\bigvee_i \varphi_i$, which is valid in the context and contains ψ as a disjunct, to $\varphi(\bar{x}, \bar{y})$. It should be understood that an implicit disjunctive splitting is carried out and we work on each resultant "simultaneously".

4.3 Type Completion

We say a selector term $s_i^\alpha(t)$ is *proper* if $\mathsf{Is}_\alpha(t)$ holds. We can make selector terms proper with type information.

Definition 16 (Type Completion). *φ' is a type completion of φ if φ' is obtained from φ by conjoining tester predicates such that for any term t in φ, exactly one type of tester predicate $\mathsf{Is}_\alpha(t)$ ($\alpha \in C$) is in φ'.*

Example 7. Let $\alpha, \beta \in C$, $\alpha \not\equiv \beta$ and $\alpha = (s_1^\alpha)$. A possible type completion for $y = s_1^\alpha(x)$ is $y = s_1^\alpha(x) \wedge \mathsf{Is}_\beta(x) \wedge \mathsf{Is}_\beta(s_1^\alpha(x)) \wedge \mathsf{Is}_\beta(y)$, which simplifies to $y = x \wedge \mathsf{Is}_\beta(x) \wedge \mathsf{Is}_\beta(y)$ by Axioms (A4) and (A6). Another type completion is $y = s_1^\alpha(x) \wedge \mathsf{Is}_\alpha(x) \wedge \mathsf{Is}_\beta(s_1^\alpha(x)) \wedge \mathsf{Is}_\beta(y)$ in which the selector term is proper. As the third example, a type completion could be $y = s_1^\alpha(x) \wedge \mathsf{Is}_\alpha(x) \wedge \mathsf{Is}_\alpha(s_1^\alpha(x)) \wedge \mathsf{Is}_\beta(y)$ which simplifies to false.

We assume that all formulas are type-complete. In particular, all selector terms are (simplified to) proper ones. The reason behind this assumption is that a symbolic transformation can always be carried out to replace a non-type-complete formula φ by an equivalent finite disjunction of type completions of φ. In terms of efficiency, however, one would prefer doing the on-the-fly disjunctive splitting when the type information of a specific term is needed. We also assume that every type completion is sound with respect to types. Certain type completion of φ may be contradictory due to type conflicts. For example, $\mathsf{Is}_A(x) \wedge \mathsf{Is}_\alpha(s(x))$ ($\alpha \in C \setminus \mathcal{A}$) is unsatisfiable. Nevertheless, unsatisfiable disjuncts will not affect soundness of the transformation and they can be easily detected and removed. At last, note that we omit listing tester literals unless they are needed for correctness proof.

4.4   Elimination Procedure

The elimination procedure consists of the following two algorithms:

Algorithm 1 (Elimination of Integer Variables).
We assume that formulas with quantifiers on integer variables are in the form

$$(\exists \bar{z} : \mathbb{Z}) \left[\varphi_{\mathbb{Z}}(\bar{x}, \bar{y}, \bar{z}) \wedge \varphi_{\mathsf{kb}^+}(\bar{x}, \bar{y}, \bar{z}) \right], \tag{6}$$

where $\bar{y}, \bar{z}$ are integer variables, $\bar{x}$ are term variables. Note that $\bar{x}$ may occur inside the weight function in $\varphi_{\mathbb{Z}}(\bar{x}, \bar{y}, \bar{z})$ and $\bar{y}, \bar{z}$ may appear inside boundary terms in $\varphi_{\mathsf{kb}^+}(\bar{x}, \bar{y}, \bar{z})$.
Repeatedly apply the following subprocedures (A') and (B') to (6) until $\bar{z} = \emptyset$.

(A') *If none of $\bar{z}$ appears inside any boundary terms, then $\varphi_{\mathsf{kb}^+}(\bar{x}, \bar{y}, \bar{z})$ is just $\varphi_{\mathsf{kb}^+}(\bar{x}, \bar{y})$, which can be moved out of $\exists \bar{z}$. We then obtain*

$$(\exists \bar{z} : \mathbb{Z}) \left[\varphi_{\mathbb{Z}}(\bar{x}, \bar{y}, \bar{z}) \right] \wedge \varphi_{\mathsf{kb}^+}(\bar{x}, \bar{y}).$$

Since $(\exists \bar{z} : \mathbb{Z})[\varphi_{\mathbb{Z}}(\bar{x}, \bar{y}, \bar{z})]$ is in $\mathscr{L}_{\mathbb{Z}}$, we can proceed to remove the block of existential quantifiers using Cooper's method ([8,26]). In fact, we can defer the elimination until all term quantifiers are gone.
(B') *If for some $z \in \bar{z}$, z occurs inside some boundary terms, we eliminate z by Lemma 7.*

Algorithm 2 (Elimination of Term Variables).
We assume that formulas with quantifiers on term variables are in the form

$$(\exists \bar{x} : \mathsf{TA}) \left[\varphi_{\mathsf{kb}^+}(\bar{x}, \bar{y}, \bar{z}) \wedge \varphi_{\mathbb{Z}}(\bar{x}, \bar{y}, \bar{z}) \right], \tag{7}$$

where $\bar{x}, \bar{y}$ are term variables, $\bar{z}$ are integer variables. Note that $\bar{z}$ may occur inside boundary terms in $\varphi_{\mathsf{kb}^+}(\bar{x}, \bar{y}, \bar{z})$, and $\bar{x}, \bar{y}$ may occur inside the weight function in $\varphi_{\mathbb{Z}}(\bar{x}, \bar{y}, \bar{z})$.
Repeatedly apply the following subprocedures (A) and (B) to (7) until $\bar{x} = \emptyset$.

(A) **Depth Reduction. Repeat** *(a),(b),(c) in the order* **while** *$(\forall \mathbf{x} \in \bar{x})$* depth$_{\varphi_{\mathsf{kb}^+}}(\mathbf{x}) > 0$.
  (a) SELECTION. *Select a α-typed variable $x \in \bar{x}$ for some $\alpha = (\mathsf{s}_1^\alpha, \ldots, \mathsf{s}_{\mathsf{ar}(\alpha)}^\alpha)$. This selection is always possible as* depth$_{\varphi_{\mathsf{kb}^+}}(x) > 0$. *We require that in the next run of (a), we choose one of the variables generated by this run of (b). I.e., the variable selection is done in depth-first manner. This is crucial to guarantee that a run eventually leaves (A). Let $\bar{x}' \equiv \bar{x} \setminus x$.*
  (b) DECOMPOSITION. *We rewrite (7) to:*

$$\left(\exists \bar{x}', x_1, \ldots, x_{\mathsf{ar}(\alpha)}, x : \mathsf{TA} \right) \left[\mathsf{ls}_\alpha(x) \wedge \bigwedge_{1 \le i \le \mathsf{ar}(\alpha)} \mathsf{s}_i^\alpha(x) = x_i \right.$$
$$\left. \wedge \varphi_{\mathsf{kb}^+}(\bar{x}, \bar{y}, \bar{z}) \wedge \varphi_{\mathbb{Z}}(\bar{x}, \bar{y}, \bar{z}) \right]. \tag{8}$$

  (c) SIMPLIFICATION. *Exhaustively apply the following simplification rules to φ_{kb^+} and $\varphi_{\mathbb{Z}}$ in (8):*
   (1) *replace $\mathsf{s}_i^\alpha(x)$ by x_i $(1 \le i \le \mathsf{ar}(\alpha))$;*
   (2) *replace x^{w} by $\sum_{i=1}^{\mathsf{ar}(\alpha)} x_i^{\mathsf{w}} + \mathsf{W}(\alpha)$;*

(3) replace $x <_n^{\#} t$ by DEPTH-REDUCTION ($x <_n^{\#} t$);

(4) similar for $t <_n^{\#} x$, $x \leq_n^{\#} t$ and $t \leq_n^{\#} x$.

The existence of DEPTH-REDUCTION follows from Lemma 3. Let the resulting formula be

$$\left(\exists \bar{x}', x_1, \ldots, x_{\mathrm{ar}(\alpha)}, x : \mathsf{TA}\right) \left[\mathsf{Is}_\alpha(x) \wedge \bigwedge_{1 \leq i \leq \mathrm{ar}(\alpha)} \mathsf{s}_i^\alpha(x) = x_i \right.$$

$$\left. \varphi'_{\mathsf{kb}^+}(\bar{x}', \mathsf{s}_1^\alpha(x), \ldots, \mathsf{s}_{\mathrm{ar}(\alpha)}^\alpha(x), \bar{y}, \bar{z}) \wedge \varphi'_{\mathbb{Z}}(\bar{x}', \mathsf{s}_1^\alpha(x), \ldots, \mathsf{s}_{\mathrm{ar}(\alpha)}^\alpha(x), \bar{y}, \bar{z}) \right]. \quad (9)$$

It is now clear that if x occurs in φ'_{kb^+} and $\varphi'_{\mathbb{Z}}$ it occurs inside some of $\mathsf{s}_1^\alpha(x), \ldots, \mathsf{s}_{\mathrm{ar}(\alpha)}^\alpha(x)$. Since

$$\left(\forall x_1, \ldots, x_{\mathrm{ar}(\alpha)} : \mathsf{TA}\right)\left(\exists x : \mathsf{TA}\right)\left[\mathsf{Is}_\alpha(x) \wedge \bigwedge_{1 \leq i \leq \mathrm{ar}(\alpha)} \mathsf{s}_i^\alpha(x) = x_i \right]$$

is valid in $\mathfrak{A}_{\mathsf{TA}}$, we can replace in (9), $\mathsf{s}_1^\alpha(x), \ldots, \mathsf{s}_{\mathrm{ar}(\alpha)}^\alpha(x)$, respectively, by $x_1, \ldots, x_{\mathrm{ar}(\alpha)}$, and hence remove $\bigwedge_{1 \leq i \leq \mathrm{ar}(\alpha)} \mathsf{s}_i^\alpha(x) = x_i$, $\mathsf{Is}_\alpha(x)$ together with $\exists x$, obtaining

$$\left(\exists \bar{x}', x_1, \ldots, x_{\mathrm{ar}(\alpha)} : \mathsf{TA}\right) \left[\varphi'_{\mathsf{kb}^+}(\bar{x}', x_1, \ldots, x_{\mathrm{ar}(\alpha)}, \bar{y}, \bar{z}) \right.$$

$$\left. \wedge \varphi'_{\mathbb{Z}}(\bar{x}', x_1, \ldots, x_{\mathrm{ar}(\alpha)}, \bar{y}, \bar{z}) \right]. \quad (10)$$

(B) Elimination. Repeat (B) while $(\exists x \in x)$ depth$_{\varphi_{\mathsf{kb}^+}}(x) = 0$.

Take the x as in the guard condition, guess a DGOC for all terms related with x in gap order literals (by Lemma 1) and then eliminate x by Lemma 6.

Theorem 1. Th($\mathfrak{A}_{\mathsf{kb}^+}^{\mathbb{Z}}$) is decidable, and hence so is Th($\mathfrak{A}_{\mathsf{kb}}$).

Example 8. Let us go through an example with emphasis on the depth reduction. Due to space limitation, we only show *one* simple trace of the reduction. Consider in the LISP list structure the following formula

$$(\exists x) \left[\mathsf{car}(x) <_2^! \mathsf{cdr}(\mathsf{cdr}(x)) \wedge \mathsf{cdr}(\mathsf{cdr}(\mathsf{car}(x))) <_3^! y \right], \quad (11)$$

where depth$_{(11)}(x) = 3$. At the first run of (A), we introduce fresh variables x_1 and x_2 to replace $\mathsf{car}(x)$ and $\mathsf{cdr}(x)$, respectively. By a standard quantifier manipulation we obtain

$$(\exists x_1 \exists x_2) \left[x_1 <_2^! \mathsf{cdr}(x_2) \wedge \mathsf{cdr}(\mathsf{cdr}(x_1)) <_3^! y \right], \quad (12)$$

where depth$_{(12)}(x_1) = 2$ and depth$_{(12)}(x_2) = 1$, both less than depth$_{(11)}(x)$. In the second run of (A), we pick x_1 and replace $x_1 <_2^! \mathsf{cdr}(x_2)$ by $\mathsf{car}(x_1) = \mathsf{car}(\mathsf{cdr}(x_2)) \wedge \mathsf{cdr}(x_1) <_2^! \mathsf{cdr}(\mathsf{cdr}(x_2))$ (which is one of several choices). We obtain

$$(\exists x_2 \exists x_{11} \exists x_{12}) \left[x_{11} = \mathsf{car}(\mathsf{cdr}(x_2)) \wedge x_{12} <_2^! \mathsf{cdr}(\mathsf{cdr}(x_2)) \wedge \mathsf{cdr}(x_{12}) <_3^! y \right]. \quad (13)$$

At this point we have depth$_{(13)}(x_{11}) = 0$ and the run enters (B). In this case we can immediately remove $\exists x_{11}$, obtaining

$$(\exists x_2 \exists x_{12}) \left[x_{12} <_2^! \mathsf{cdr}(\mathsf{cdr}(x_2)) \wedge \mathsf{cdr}(x_{12}) <_3^! y \right], \quad (14)$$

where $\mathsf{depth}_{(14)}(x_{12}) = 1$ and $\mathsf{depth}_{(14)}(x_2) = 2$. At the third run of (A), we select x_{12}. The run could give us

$$(\exists x_2 \exists x_{121} \exists x_{122}) \left[x_{121} = \mathsf{car}(\mathsf{cdr}(\mathsf{cdr}(x_2))) \wedge x_{122} <^!_2 \mathsf{cdr}(\mathsf{cdr}(x_2)) \wedge x_{122} <^!_3 y \right], \quad (15)$$

which as before by (B) simplifies to

$$(\exists x_2 \exists x_{122}) \left[x_{122} <^!_2 \mathsf{cdr}(\mathsf{cdr}(x_2)) \wedge x_{122} <^!_3 y \right]. \quad (16)$$

Still we have $\mathsf{depth}_{(16)}(x_{122}) = 0$ which justifies another run of (B). Let us take a gap order completion $x_{122} <^!_2 \mathsf{cdr}(\mathsf{cdr}(x_2)) <^!_1 y$ (which again is just one of many choices) and rewrite (16) to

$$(\exists x_2 \exists x_{122}) \left[x_{122} <^!_2 \mathsf{cdr}(\mathsf{cdr}(x_2)) <^!_1 y \right]. \quad (17)$$

With the help of boundary functions, (17) reduces to

$$(\exists x_2) \left[0^{\mathsf{w}}_{((\mathsf{cdr}(\mathsf{cdr}(x_2)))^{\mathsf{w}})} <^!_2 \mathsf{cdr}(\mathsf{cdr}(x_2)) <^!_1 y \right]. \quad (18)$$

The fourth and the fifth runs of (A) (with the same trick of quantifier manipulation) give us

$$(\exists x_{222}) \left[0^{\mathsf{w}}_{(x^{\mathsf{w}}_{222})} <^!_2 x_{222} <^!_1 y \right]. \quad (19)$$

After that the run comes back again to (B) as $\mathsf{depth}_{(19)}(x_{222}) = 0$. Here we have to reduce term quantifiers to integer quantifiers in that x_{222} also appears in boundary terms. By Lemma 6, (19) is equivalent to

$$(\exists z) \left[0^{\mathsf{w}}_{(z)} <^!_3 y \wedge \mathsf{Tree}^{\mathsf{cons}}(z) \right], \quad (20)$$

which simplifies to $0^{\mathsf{w}}_{(y^{\mathsf{w}})} <^!_3 y \wedge \mathsf{Tree}^{\mathsf{cons}}(y^{\mathsf{w}})$, and in turn to

$$0^{\mathsf{w}}_{(y^{\mathsf{w}})} <^!_3 y, \quad (21)$$

as $0^{\mathsf{w}}_{(y^{\mathsf{w}})} <^!_3 y$ implies $\mathsf{Tree}^{\mathsf{cons}}(y^{\mathsf{w}})$. It is not hard to verify that (21) implies (11) as desired. (We do not have equivalence because this is just one trace of reduction.)

We note that the depth reduction of a variable is at the expense of increasing the depth of a term on the other side of a relation. This happens when φ contains $x \star t$ (or $t \star x$) and $\mathsf{depth}_\varphi(x) > 0$. For example, from (12) to (13), the depth of x_2 increases by 1. Moreover, the depth reduction in general introduces more existential quantifiers and more equalities in the matrix (e.g., also in the reduction from (12) to (13)). In each transformation, however, the number of open gap order literals in each resulting primitive formula is no more than that in the original (primitive) formula. Moreover, the final elimination procedure removes at least one open gap order literal if the eliminated variable occurs in such literals (e.g., from (17) to (18) and from (19) to (20)). When all open gap order literals are gone, the depths of terms will be strictly decreasing. This forces the run to eventually leave (A) and from then on to stay in (B) until all existential quantifiers are removed.

5 Presence of a 0-weight unary function

As mentioned earlier, the presence of a unary function α_0 of weight 0 in Σ simplifies the elimination procedure. Intuitively, the existence of α_0 makes $<^w$ and $<^p$ dense almost everywhere except around atoms. This follows from the fact that $1^w_{(m)}$ and $1^p_{(m,p)}$ are undefined (i.e., no maximum) except when α_p is an atom and $m = W(\alpha_p)$. Accordingly, if u is not an atom, then for any $n \geq 1$, $u <^w_n v$ (resp. $u <^p_n v$) is equivalent to $u <^w v$ (resp. $u <^p v$). Also, it suffices for $\mathcal{L}^{\mathbb{Z}}_{kb^+}$ to only have lower boundary functions in order to decompose gap orders. More details are given in the extended version of this paper.

6 Conclusion

We showed the decidability of the first-order theory of term algebras with Knuth-Bendix order by quantifier elimination. Our method combines the extraction of integer constraints from term constraints with the reduction of quantifiers on term variables to quantifiers on integer variables. In fact, we established the decidability of a much more expressive theory.

Two problems related to practical complexity need further investigation. First, as a rule of thumb, more expressive power means higher complexity. Even if the theoretical complexity bound is the same, in practice the efficiency will be compromised. It is worthwhile to search for the smallest extension of KBO which admits quantifier elimination. Second, the elimination is intrinsically limited to processing quantified variables one at a time. We plan to extend the method in [30] to eliminate a block of quantifiers of the same kind in one step. We believe this will be a significant improvement in pragmatic terms, since in most applications the quantifier alternation depth is small.

We also plan to investigate the decidability issue of the first-order theory of KBO in the term domain with variables [13,1].

Acknowledgments

We thank Aaron Bradley for his comments on an earlier version of this paper. We thank the anonymous referees for their careful reading and suggestions.

References

1. Franz Baader and Tobias Nipkow. *Term Rewriting and All That*. Cambridge University Press, Cambridge, UK, 1999.
2. Rolf Backofen. A complete axiomatization of a theory with feature and arity constraints. *Journal of Logical Programming*, 24(1&2):37–71, 1995.
3. Hubert Comon. Solving symbolic ordering constraints. *International Journal of Foundations of Computer Science*, 1(4):387–411, 1990.
4. Hubert Comon and Catherine Delor. Equational formulae with membership constraints. *Information and Computation*, 112(2):167–216, 1994.

5. Hubert Comon and Pierre Lescanne. Equational problems and disunification. *Journal of Symbolic Computation*, 7:371–425, 1989.
6. Hubert Comon and Ralf Treinen. Ordering constraints on trees. In *Proceedings of the 19th International Colloquium on Trees in Algebra and Programming (CAAP'94)*, volume 787 of *Lecture Notes in Computer Science*, pages 1–14, Edinburgh, U.K., Apr 1994. Springer-Verlag.
7. Hubert Comon and Ralf Treinen. The first-order theory of lexicographic path orderings is undecidable. *Theoretical Computer Science*, 176(1-2):67–87, 1997.
8. D. C. Cooper. Theorem proving in arithmetic without multiplication. In *Machine Intelligence*, volume 7, pages 91–99. American Elsevier, 1972.
9. Nachum Dershowitz. Orderings for term-rewriting systems. *Theoretical Computer Science*, 7:279–301, 1982.
10. H. B. Enderton. *A Mathematical Introduction to Logic*. Academic Press, 2001.
11. Wilfrid Hodges. *Model Theory*. Cambridge University Press, Cambridge, UK, 1993.
12. Jean-Pierre Jouannaud and Mitsuhiro Okada. Satisfiability of systems of ordinal notation with the subterm property is decidable. In *Proceedings of the 18th International Colloquium on Automata, Languages and Programming (ICALP'91)*, volume 510 of *Lecture Notes in Computer Science*, pages 455–468. Springer-Verlag, 1991.
13. Donald E. Knuth and Peter Bendix. Simple word problems in universal algebras. In *Computational Problems in Abstract Algebra*, pages 263–297. Pergamon Press, 1970. Reprinted in *Automation of Reasoning, Vol. 2* Jürgen Siekmann and G. Wrightson, editors, pp. 342-376, Springer-Verlag, 1983.
14. Konstantin Korovin and Andrei Voronkov. A decision procedure for the existential theory of term algebras with the Knuth-Bendix ordering. In *Proceedings of the 15th IEEE Symposium on Logic in Computer Science (LICS'00)*, pages 291 – 302, IEEE Computer Society Press, 2000.
15. Konstantin Korovin and Andrei Voronkov. Knuth-Bendix constraint solving is NP-complete. In *Proceedings of 28th International Colloquium on Automata, Languages and Programming (ICALP'01)*, volume 2076 of *Lecture Notes in Computer Science*, pages 979–992. Springer-Verlag, 2001.
16. Konstantin Korovin and Andrei Voronkov. The decidability of the first-order theory of the Knuth-Bendix order in the case of unary signatures. In *Proceedings of the 22th Conference on Foundations of Software Technology and Theoretical Computer Science, (FSTTCS'02)*, volume 2556 of *Lecture Notes in Computer Science*, pages 230–240. Springer-Verlag, 2002.
17. Viktor Kuncak and Martin Rinard. On the theory of structural subtyping. Technical Report MIT-LCS-TR-879, Massachusetts Institute of Technology, January 2003.
18. Viktor Kuncak and Martin Rinard. The structural subtyping of non-recursive types is decidable. In *Proceedings of the 18th IEEE Symposium on Logic in Computer Science (LICS'03)*, pages 96–107. IEEE Computer Society Press, 2003.
19. M. J. Maher. Complete axiomatizations of the algebras of finite, rational and infinite tree. In *Proceedings of the 3th IEEE Symposium on Logic in Computer Science (LICS'88)*, pages 348–357. IEEE Computer Society Press, 1988.
20. A. I. Mal'cev. Axiomatizable classes of locally free algebras of various types. In *The Metamathematics of Algebraic Systems, Collected Papers*, chapter 23, pages 262–281. North Holland, 1971.
21. Paliath Narendran and Michael Rusinowitch. The theory of total unary RPO is decidable. In *Proceedings of the 1st International Conference on Computational Logic (CL 2000)*, volume 1861 of *Lecture Notes in Artificial Intelligence*, pages 660–672. Springer-Verlag, 2000.

22. Paliath Narendran, Michael Rusinowitch, and Rakesh M. Verma. RPO constraint solving is in NP. In *Proceedings of the 12th International Workshop on Computer Science Logic (CSL'98)*, volume 1584 of *Lecture Notes in Computer Science*, pages 385 – 398. Springer-Verlag, 1999.
23. Robert Nieuwenhuis. Simple LPO constraint solving methods. *Information Processing Letters*, 47(2):65–69, 1993.
24. Robert Nieuwenhuis and J. Rivero. Solved forms for path ordering constraints. In *Proceedings of 10th International Conference on Rewriting Techniques and Applications (RTA'99)*, volume 1631 of *Lecture Notes in Computer Science*, pages 1–15. Springer-Verlag, 1999.
25. Robert Nieuwenhuis and Albert Rubio. Theorem proving with ordering and equality constrained clauses. *Journal of Symbolic Computation*, 19(4):321–351, 1995.
26. C. R. Reddy and D. W. Loveland. Presburger arithmetic with bounded quantifier alternation. In *Proceedings of the 10th Annual Symposium on Theory of Computing*, pages 320–325. ACM Press, 1978.
27. Tatiana Rybina and Andrei Voronkov. A decision procedure for term algebras with queues. *ACM Transactions on Computational Logic*, 2(2):155–181, 2001.
28. Ralf Treinen. A new method for undecidability proofs of first order theories. *Journal of Symbolic Computation*, 14:437–457, 1992.
29. Ting Zhang, Henny Sipma, and Zohar Manna. Decision procedures for recursive data structures with integer constraints. In *Proceedings of the 2nd International Joint Conference on Automated Reasoning (IJCAR'04)*, volume 3097 of *Lecture Notes in Computer Science*, pages 152–167. Springer-Verlag, 2004.
30. Ting Zhang, Henny Sipma, and Zohar Manna. Term algebras with length function and bounded quantifier alternation. In *Proceedings of the 17th International Conference on Theorem Proving in Higher Order Logics (TPHOLs'04)*, volume 3223 of *Lecture Notes in Computer Science*, pages 321–336. Springer-Verlag, 2004.

Well-Nested Context Unification[*]

Jordi Levy[1], Joachim Niehren[2], and Mateu Villaret[3]

[1] IIIA, CSIC, Barcelona, Spain
[2] INRIA Futurs, LIFL, Lille, France
[3] IMA, Universitat de Girona, Spain

Abstract. Context unification (CU) is the open problem of solving context equations for trees. We distinguish a new decidable variant of CU– *well-nested CU* – and present a new unification algorithm that solves well-nested context equations in non-deterministic polynomial time. We show that minimal well-nested solutions of context equations can be composed from the material present in the equation (see Theorem 1). This property is wishful when modeling natural language ellipsis in CU.

1 Introduction

Context unification (CU) is the problem of solving equations with context variables over the structure of finite trees [Com92, Lev96, NPR97, SSS99, SS02]. CU is the natural generalisation of *word unification* (WU) [Mak77]. It can equally be understood as a variant of *linear second-order unification* [Lev96, LV00].

Whether CU is decidable is a long-standing open problem. So far, only a number of fragments of CU could be shown decidable. The most prominent fragment is WU proved decidable in [Mak77] and PSPACE in [Pla99]. Two further decidability results were shown for stratified CU [SS02] and for the 2 variable fragment [SSS99]. All these decidable fragments are defined over syntactic properties of the equations considered. In contrast to these results, the present paper introduces a new decidable variant, *well-nested context unification*, whose definition relies on semantic properties of solutions.

The well-nestedness restriction is motivated by an applications of CU [NPR97] in the field of compositional semantics of natural language, whose original goal was to improve on previous approaches to ellipsis resolution based on higher-order unification [DSP91, GK96]. The equational language of CU here serves as a uniform modelling language for ellipsis and scope underspecification phenomena. This approach then led to the development of the *constraint language for lambda structures* (CLLS) [EKN01]. The *parallelism constraints* feature of the CLLS, has been shown equally expressive than CU [NK01], hence its decidability is still unknown.

[*] This research has been partially supported by the CYCIT projects iDEAS (TIN2004-04343), Mulog (TIN2004-07933-C03-01) and Asilan (TIN2004-07672-C03-01), the Mostrare project of INRIA Futurs, the Universities of Lille 1 and 3, and the projects Tralala and ACIMDD of the ACI masses de données.

R. Nieuwenhuis (Ed.): CADE 2005, LNAI 3632, pp. 149–163, 2005.

The first decidable fragment of CLLS that is sufficiently expressive for the envisaged application to computational semantics (even though not for all aspects of ellipsis modelling) is the language of *well-nested parallelism constraints* [EN03], whose satisfiability problem is NP-complete. Modelling languages with lower algorithmic complexity are not known for these applications.

Unfortunately, however, the NP-algorithm for well-nested parallelism constraints remains questionable in practice. It relies on repeatedly solving *full dominance constraints*, which is exponentially less efficient in theory and practice than solving *normal dominance constraints* [BDMN04]. Normal dominance constraints in turn are sufficient for modelling pure scope underspecification.

Well-nested CU is properly less expressive than well-nested parallelism constraints because the later subsume (full) dominance constraints, whereas well-nested CU does not. We show that minimal well-nested solutions of CU equations can always be composed from the material present in the equations (see Theorem 1). This surprising property is wanted for ellipsis modelling, where it means that elided parts of ellipsis can always be reconstructed from what was uttered elsewhere. CLLS does not satisfy this condition. In particular, some well-nested parallelism constraints that we can not express in well-nested CU fail to have this property.

We contribute a new unification algorithm that solves well-nested CU equations in non-deterministic polynomial time. Our algorithm guesses how to construct minimal well-nested solutions from the given equation. We show NP-hardness of well-nested CU satisfiability by encoding string matching [Ang80].

The paper proceeds as follows, in Section 2 we illustrate how well-nested CU can solve ellipses, in Sections 3 and 4 we provide basic definitions to introduce in Section 5 well-nested CU and prove that size-minimal solutions have a polynomial representation. In Section 6 we show its NP-completeness proving that guessed solutions can be checked in polynomial time.

2   Ellipsis Resolution

To illustrate the usage of CU in ellipsis resolution, we consider an example from [Sag76] which contains two VP-ellipsis with a nice nesting structure, but without scope underspecification:

(1) Mary can't go to Princeton in the fall, (2) but she can in the spring,
(3) although if she does then ...

Sentence *(2)* means that Mary can go to Princeton in the spring, so the phrase *go to Princeton* is elided. Sentence *(3)* states that something will happen if Mary goes to Princeton in the spring; *go to Princeton in the spring* is elided here.

Following [Mon73], we can represent the semantics of sentences by lambda terms, i.e., in higher-order logics:

$$x_1 = mary@\lambda m(in\_the\_fall@(can't@m@go\_to\_princeton))$$
$$x_2 = mary@\lambda m(in\_the\_spring@(can@m@go\_to\_princeton))$$
$$x_3 = (mary@\lambda m(in\_the\_spring@(do@m@go\_to\_princeton))) \rightarrow x_4$$

The variables x_1, x_2, x_3 denote the respective meanings of the three sentences, while x_4 stands for the meaning of the consequence that Mary fears.

A non-trivial question is how to resolve the nested ellipses automatically. This problem is typically split into two part [DSP91, GK96, NPR97, EKN01, EN03]. First one has to infer the nesting structure of the ellipsis, second, one has to reconstruct the elided parts from the nesting structure. In this paper, we only treat the second problem, resolving ellipsis given their nesting structure. The nesting structure of the example is indicated in Fig. 1.

The dashed box represents the first elided part, the larger dotted box the second one. Note that one occurrence of the dashed box is nested inside of the dotted box, while the second one is completely outside of it. These boxes are well-nested because they do not properly overlap.

When looking at the trees formed by the abstract syntax of the lambda-terms, the boxes become contexts, i.e., trees where a subtree has been substi-

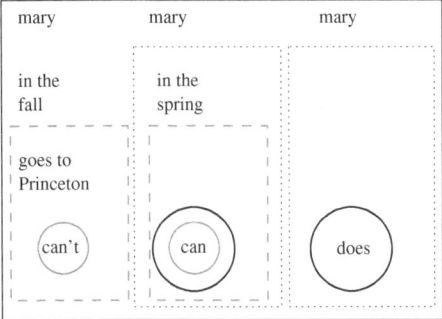

Fig. 1. Nesting sketches for 2

tuted by a hole marker •. The lambda terms can then be described by context equations of CU. Let X be a context variable for the dashed box and Y for the dotted box. We can then describe the meaning of the ellipsis by the following equations:

$$x_1 \overset{?}{=} mary@\lambda m(in\_the\_fall@X(can't)) \qquad X \overset{?}{=} (\bullet@m)@go\_to\_princeton$$
$$x_2 \overset{?}{=} mary@\lambda m(Y(can)) \qquad\qquad Y \overset{?}{=} in\_the\_spring@X$$
$$x_3 \overset{?}{=} (mary@\lambda m(Y(do))) \to x_4$$

Similar semantical descriptions can be inferred compositionally from the syntax of the sentence [EKN01, NPR97]. Resolving the ellipsis amounts to find minimal size unifiers of the equations. Such solutions are well-nested, i.e., correspond to instantiating context variables with contexts that do not properly overlap, moreover, they are made of the material already present in the equations.

3   Well-Nested Segments

Given a signature Γ of symbols, we write $\mathcal{T}_\Gamma$ for the set of terms over Γ. The *size* $|t|$ of a term t is the number of its symbols in Γ. We identify positions in terms by their relative address from the root, using the Dewey's decimal notation. We denote the set of positions of a term t by $\mathsf{pos}(t)$. The word's prefix ordering $p \preceq q$ is also known as the *dominance ordering* on positions, which holds if p is an ancestor or equal to q. For all terms t and positions $p \in \mathsf{pos}(t)$, we let $\mathsf{lab}_t(p)$ denote the symbol of Γ at position p of t.

A *segment* $\langle p, q \rangle$ is a pair of positions such that $p \preceq q$. A segment $\langle p, q \rangle$ belongs to a term t if $p, q \in \mathsf{pos}(t)$. Every segment of a term t distinguishes a set of positions of t: $\mathsf{pos}_t(\langle p, q \rangle) = \{r \in \mathsf{pos}(t) \mid p \preceq r, q \npreceq r\}$

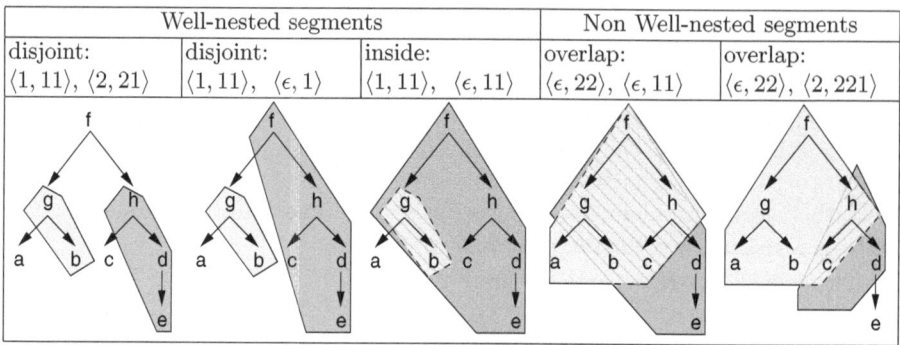

Well-nested segments			Non Well-nested segments	
disjoint: $\langle 1, 11\rangle, \langle 2, 21\rangle$	disjoint: $\langle 1, 11\rangle,\ \langle \epsilon, 1\rangle$	inside: $\langle 1, 11\rangle,\ \langle \epsilon, 11\rangle$	overlap: $\langle \epsilon, 22\rangle, \langle \epsilon, 11\rangle$	overlap: $\langle \epsilon, 22\rangle, \langle 2, 221\rangle$

Fig. 2. Relationships between segments

Definition 1. *Let S_1 and S_2 be two segments of t. We say that S_1 is inside of S_2 in t, if $\mathsf{pos}_t(S_1) \subseteq \mathsf{pos}_t(S_2)$. The segments S_1 and S_2 are disjoint in t, if $\mathsf{pos}_t(S_1) \cap \mathsf{pos}_t(S_2) = \emptyset$. The segments S_1 and S_2 are well-nested, if one of them lies inside the other in t, or if they are disjoint.*

Examples for the possible segment relationship are given in Fig. 2. A non-empty segment $\langle p, q\rangle$ of t is inside of a segment $\langle p', q'\rangle$ of t if, and only if, $p' \preceq p$ and $q \preceq q'$. Notice that $\mathsf{pos}_t(\langle p, p\rangle) = \emptyset$ for any position $p \in \mathsf{pos}(t)$.

4   Context Unification

We introduce the variant of context unification (CU) that we will use. Let Σ be a signature of constants: $f, g, \ldots$ with non-zero arity, and $a, b, \ldots$ with arity zero. Let Vars be a countable infinite set of *context variables* $X, Y, Z, \ldots$ with arity one[1], and let $\bullet$ be a symbol of arity 0 that we call the *hole marker*. The signature of CU-terms $s, t, \ldots$ over Σ is $\Gamma(\Sigma) = \Sigma \uplus \mathsf{Vars} \uplus \{\bullet\}$, i.e.:

$$s, t ::= a \mid \bullet \mid X(t) \mid f(t_1, \ldots, t_n)$$

The set of variables occurring in a term t is $\mathsf{Var}(t)$.

A *tree* over Σ is a term over $\Gamma(\Sigma)$ that does not contain the hole marker (but possibly variables). A *context* over Σ is a term over $\Gamma(\Sigma)$ that contains a *unique* occurrence of the hole marker, at a position denoted by $\mathsf{hole}(t)$. We write Tree_Σ for the set of trees over Σ and Cont_Σ for the set of contexts over Σ.

Given a tree t and a segment $\langle p, q\rangle$ of t, we write $t\langle p, q\rangle$ for the context in t that starts at p and has its hole at q. For instance $g(f(a, b), c))\langle 1, 1 \cdot 2\rangle = f(a, \bullet)$. Note that $\mathsf{pos}_t(\langle p, q\rangle) = \{p \cdot r \mid r \in \mathsf{pos}(t\langle p, q\rangle)\} \setminus \{q\}$.

The hole marker is like a λ-bound variable, for instance the context $f(g(a, \bullet))$ corresponds to the linear higher-order term $\lambda x. f(f(a, x))$. This view is formally adapted in linear second-order unification [Lev96, LV00].

---

[1] Note that we could add variables of arbitrary arities, but, although the adaptation of this results to this more general framework is not straight forward, we do not do so for sake of simplicity.

While avoiding more general higher-order syntax here, we will nevertheless use contexts $t \in \mathsf{Cont}_\Sigma$ as functions $t : \mathcal{T}_{\Gamma(\Sigma)} \to \mathcal{T}_{\Gamma(\Sigma)}$ which map terms s to terms $t(s)$ by substituting the hole marker in t by s. For instance,

$$f(g(a, \bullet))(b) = f(g(a, b)) \qquad \text{or} \qquad f(g(a, \bullet))(f(\bullet)) = f(g(a, f(\bullet)))$$

Note that $t(\mathsf{Tree}_\Sigma) \subseteq \mathsf{Tree}_\Sigma$ and $t(\mathsf{Cont}_\Sigma) \subseteq \mathsf{Cont}_\Sigma$ since the unique hole of t will be filled by either a tree or a context.

A *context substitution* is a function $\sigma : \mathsf{Vars} \to \mathsf{Cont}_\Sigma$ such that $\sigma(X) \neq X$ only for a finite set of variables $X \in \mathsf{Vars}$. We lift substitutions from variables to functions on terms $\sigma : \mathcal{T}_{\Gamma(\Sigma)} \to \mathcal{T}_{\Gamma(\Sigma)}$ while using contexts as functions:

$$\begin{aligned} \sigma(X(s)) &= \sigma(X)(\sigma(s)), & \sigma(a) &= a, \\ \sigma(f(t_1, \ldots, t_n)) &= f(\sigma(t_1), \ldots, \sigma(t_n)), & \sigma(\bullet) &= \bullet. \end{aligned}$$

Contexts $\sigma(X)$, substituted for variables X, are immediately applied so that their holes are filled. Thus, $\sigma(\mathsf{Tree}_\Sigma) \subseteq \mathsf{Tree}_\Sigma$ and $\sigma(\mathsf{Cont}_\Sigma) \subseteq \mathsf{Cont}_\Sigma$. The composition of two substitutions σ_1 and σ_2, written as $\sigma_1 \circ \sigma_2$, is defined by $(\sigma_1 \circ \sigma_2)(t) = \sigma_1(\sigma_2(t))$. Substitutions are usually represented as $[X_1 \mapsto \sigma(X_1), \ldots, X_n \mapsto \sigma(X_n)]$, where the X_i's are the variables for which $\sigma(X_i) \neq X_i$. As we will see in the next section, we can get more compact representations by means of composition.

A *context equation* e is a term $s \overset{?}{=} t$ with $s, t \in \mathsf{Tree}_\Sigma$. Equations are terms over the signature $\Sigma \cup \mathsf{Var} \cup \{\overset{?}{=}\}$, therefore the notions of positions and segments apply to equation, as to all other types of terms. We apply substitutions to equations such that the result of $\sigma(s \overset{?}{=} t)$ is the equation $\sigma(s) \overset{?}{=} \sigma(t)$.

A *unifier* of an equation $s \overset{?}{=} t$ is a substitution σ satisfying $\sigma(s) = \sigma(t)$. A unifier σ of e is said to be *ground* if $\sigma(e)$ does not contain any variable, and *size-minimal* if it minimizes $\sum_{X \in \mathsf{Var}(e)} |\sigma(X)|$ while satisfying $\sigma(X) = X$ for all $X \notin \mathsf{Var}(e)$.

CU is often seen as the satisfiability problem of context equations, i.e.: given a fixed signature Σ, the problem of deciding whether a given system of context equations over Σ has a unifier. For the sake of simplicity, we restrict ourselves to a single context equation as input. This does not affect expressiveness.

CU can alternatively be considered as a constraint solving problem, i.e., the problem to enumerate all unifiers of a given equation. Rather than enumerating all unifiers, one usually prefers to enumerate only the most general unifiers (from which all others can be obtained by instantiation). In this paper, we will consider a similar variant. We will be able to enumerate all size-minimal unifiers in a compact representation by using compositions of substitutions.

Definition 2 (Correspondence function). *Let e be an equation with unifier σ. Positions in e and $\sigma(e)$ correspond through the function $\mathsf{c}_\sigma^e : \mathsf{pos}(e) \to \mathsf{pos}(\sigma(e))$ that satisfies for all positions $p \cdot i \in \mathsf{pos}(e)$ and variables $X \in \mathsf{Var}(e)$:*

$$\mathsf{c}_\sigma^e(\epsilon) = \epsilon \qquad \text{and} \qquad \mathsf{c}_\sigma^e(p \cdot i) = \begin{cases} \mathsf{c}_\sigma^e(p) \cdot i & \text{if } \mathsf{lab}_e(p) \in \Sigma \\ \mathsf{c}_\sigma^e(p) \cdot \mathsf{hole}(\sigma(X)) & \text{if } \mathsf{lab}_e(p) \in \mathsf{Var} \end{cases}$$

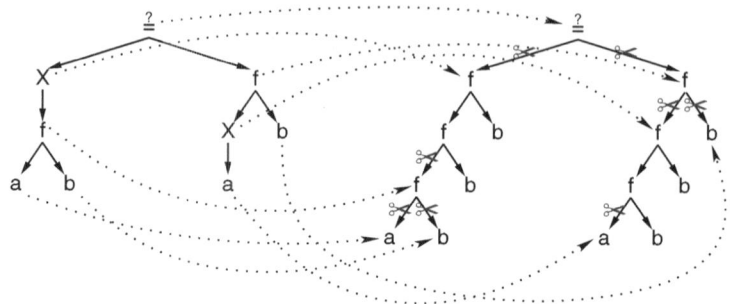

Fig. 3. The equation $e = X(f(a, b)) \stackrel{?}{=} f(X(a), b)$ and $\sigma(e)$, where σ is the non-well nested unifier $[X \mapsto f(f(\bullet, b), b)]$, with the correspondence function c_σ^e and the "cuts"

Figure 3 represents the correspondence function of a given equation and unifier. Notice that if $\langle p, q \rangle$ is a segment of e then $\langle c_\sigma^e(p), c_\sigma^e(q) \rangle$ is a segment of the equation $\sigma(e)$. Furthermore, if $\mathsf{lab}_e(p) = X$ then $\sigma(X) = \sigma(e) \langle c_\sigma^e(p), c_\sigma^e(p \cdot 1) \rangle$.

We next define an equivalence relation on positions of $\sigma(e)$ that must carry the same labels. An occurrence of variable X in an equation e is a position p that satisfies $\mathsf{lab}_e(p) = X$.

Definition 3 (Equivalent positions). *Let e be an equation with unifier σ. Let $\approx_\sigma^e \subseteq \mathsf{pos}(\sigma(e)) \times \mathsf{pos}(\sigma(e))$ be the least equivalence relation satisfying:*

1. *corresponding positions in both sides of the instantiated equation are equivalent: all position p such that $1 \cdot p, 2 \cdot p \in \mathsf{pos}(\sigma(e))$ satisfy $\quad 1 \cdot p \approx_\sigma^e 2 \cdot p$*
2. *corresponding positions in instances of different occurrences of the same variable are equivalent: all $X \in \mathsf{Var}(e)$, $q \in \mathsf{pos}(\sigma(X)) \setminus \mathsf{hole}(\sigma(X))$, and occurrences p_1 and p_2 of X in e satisfy $\quad c_\sigma^e(p_1) \cdot q \approx_\sigma^e c_\sigma^e(p_2) \cdot q$*

Lemma 1. *If σ is a solution of equation e then every pair of equivalent positions $p_1 \approx_\sigma^e p_2$ has the same label, i.e. $\mathsf{lab}_{\sigma(e)}(p_1) = \mathsf{lab}_{\sigma(e)}(p_2)$.*

5   Well-Nested Unifiers

In this section we define well-nested unifiers, and show how to represent well-nested unifiers of equations in polynomial space depending on the size of the equation.

It is well known that most general first-order unifiers can be exponentially sized on the size of the problem. Take as example the problem $g(X_1, \ldots, X_{n-1}) \stackrel{?}{=} g(f(X_2, X_2), \ldots, f(X_n, X_n))$. Fortunately, these unifiers can be represented in polynomial space as a composition of substitutions. In our example: $[X_{n-1} \mapsto f(X_n, X_n)] \circ \cdots \circ [X_1 \mapsto f(X_2, X_2)]$. Moreover, the terms on the right of the arrows, $f(X_i, X_i)$ in our example, are subterms of the original unification problem. These two properties are used by practical implementations

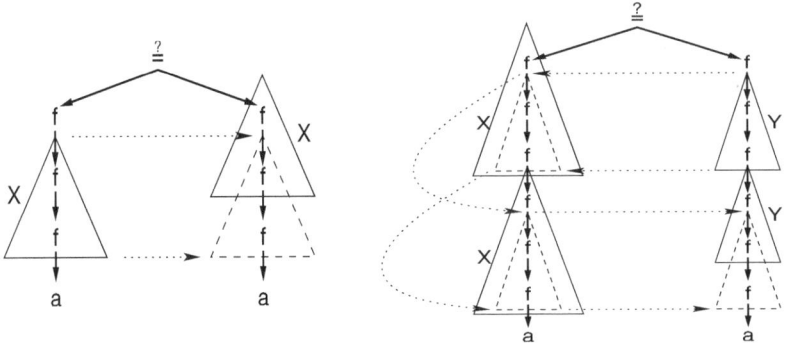

Fig. 4. Forbidden overlaps

of first-order unification. Theorem 1 states a similar property for well-nested CU, that does not hold neither for general CU nor for word unification.

Size-minimal and most general well-nested unifiers do not introduce new constants not occurring in the original equation. In the application to ellipsis resolution, this is expected to hold for all size-minimal unifiers. In general CU, however, this property does not hold. For instance, the equation $X(a)\overset{?}{=}Y(b)$ has as many size-minimal unifiers of the form $[X \mapsto f(\bullet, b), Y \mapsto f(a, \bullet)]$ as binary symbols $f \in \Sigma$. This makes CU dependent of the signature.

If we allow the use of n-ary context variables, like in linear second-order unification, then we get this property back, but only for most general unifiers (not necessarily size-minimal ones). For instance, we get a unique most general unifier $[X \mapsto Z(\bullet, b), Y \mapsto Z(a, \bullet)]$ for our example, that uses a fresh binary context variable Z, but that does not introduce new constants.

Roughly speaking, a unifier σ of an equation e is well-nested if all segments of $\sigma(e)$ that encompass variable occurrences in e are pairwise well-nested. Consider for instance the equation $f(X(a))\overset{?}{=}X(f(a))$. Well-nestedness forbids unifiers like $[X \mapsto f(f(\bullet))]$ since the segments encompassing the two occurrences of X overlap when overlaying the two sides of the instantiated equation. The situation is illustrated in Fig. 4 (left)

More indirect overlaps are raised by "reflection" in the instances of other variable occurrences. Consider for instance the equation $X(X(a))\overset{?}{=}f(Y(Y(f(a))))$ and the unifier $[X \mapsto f(f(f(\bullet))), Y \mapsto f(f(\bullet))]$ in Fig. 4 (right). Here the instances of the occurrences of Y are nested into those of X. However, if we overlay the both instances of X, then we see that the two instances of Y overlap. Thus, the unifier is not well-nested.

In order to formally define well-nested unifiers we need to extend the equivalence relation $\approx_\sigma^e$ on positions in $\sigma(e)$ to an equivalence relation $\equiv_\sigma^e$ on segments of $\sigma(e)$.

Definition 4 (Equivalent segments). *Let $\equiv_\sigma^e$ be the least equivalence relation on segments of $\sigma(e)$ that satisfies:*

1. corresponding segments on both sides of the instantiated equation are equivalent: *for all segments* $\langle 1 \cdot p, 1 \cdot q \rangle$ *of* $\sigma(e)$, $\langle 1 \cdot p, 1 \cdot q \rangle \equiv_\sigma^e \langle 2 \cdot p, 2 \cdot q \rangle$
2. segments encompassing different occurrences of the same variable are equivalent: *all occurrence* p_1, p_2 *of the same variables* $X \in \mathsf{Var}(e)$ *in* e *satisfy* $\langle c_\sigma^e(p_1), c_\sigma^e(p_1 \cdot 1) \rangle \equiv_\sigma^e \langle c_\sigma^e(p_2), c_\sigma^e(p_2 \cdot 1) \rangle$
3. corresponding subsegments in equivalent segments are equivalent: *for all equivalent segments* $\langle p_1, q_1 \rangle \equiv_\sigma^e \langle p_2, q_2 \rangle$ *of* $\sigma(e)$, *and all segments* $\langle p, q \rangle$ *satisfying that* $\langle p_1 \cdot p, p_1 \cdot q \rangle$ *is a segment of* $\sigma(e)$ *inside* $\langle p_1, q_1 \rangle$, $\langle p_1 \cdot p, p_1 \cdot q \rangle \equiv_\sigma^e \langle p_2 \cdot p, p_2 \cdot q \rangle$

If $\langle p, q \rangle$ is a segment of $\sigma(e)$ just containing one position (this must be p), then $\langle p, q \rangle \equiv_\sigma^e \langle p', q' \rangle$ if, and only if, $p \approx_\sigma^e p'$. This shows that the equivalence on segments indeed extends on the equivalence on positions.

Definition 5 (Well-nested CU). *Let* σ *be a unifier of an equation* e. *The* $\langle \sigma, e \rangle$-*images of variables in* e *are the segments* $\langle q, r \rangle$ *of* $\sigma(e)$ *such that there exists a variable-labeled position* p *in the equation* e *whose corresponding segment in* $\sigma(e)$ *is equivalent:*

$$\langle q, r \rangle \;\equiv_\sigma^e\; \langle c_\sigma^e(p), c_\sigma^e(p \cdot 1) \rangle$$

We call a unifier σ *of* e well-nested *if the set of all* $\langle \sigma, e \rangle$-*images of variables in* e *are pairwise well-nested.* Well nested CU *is the problem of deciding whether a given equation has a well-nested unifier or not.*

In what follows, we will show how to solve this problem by enumerating all well-nested size-minimal ground unifier of a given input equation.

Definition 6 (Normal unifer). *We say that a well-nested unifier is* normal *if it is size-minimal among all well-nested unifiers.*

It is important that normal unifiers are required to be size-minimal *among* well-nested unifiers, since size-minimal ground unifiers may not always need to be well-nested. As a consequence of the fact that non-ground well-nested unifiers can be made smaller by instantiating variables to the empty context, we have:

Lemma 2. *Normal unifiers are ground.*

We next recall a nice property of WU, that we will extend to CU, in order to prove the existence of compact representations of normal CU unifiers. We do not actually know who was the first in stating this property for WU, but to our knowledge, it was first proved in [PR98]. The set of positions in the equation e define "cuts" in $\sigma(e)$ throughout the function c_σ^e: formally, a position $p \in \sigma(e)$ is a *cut* if it is in the range of c_σ^e. These "cuts" limit the possible subwords that a size-minimal unifier may contain:

Proposition 1 (Lemma 6 of [PR98]). *If* σ *is a minimal unifier of a word equation* $t \overset{?}{=} u$, *then each subword of* $\sigma(t)$ *has an occurrence "over a cut".*

In [PR98] cuts are located *between* two consecutive positions, *cutting* the word into two pieces (see the scissors in Fig. 3). We *locate* the cuts in the positions on the range of c_σ^e (in Fig. 3, just bellow the scissors, where the dotted arrows representing c_σ^e points). In WU, "having an occurrence over a cut" means that there is an(other) occurrence of *the same* subword containing a cut in an inner point or in one of their extremes.

The analogous result fails for general CU but carries over in the well-nested case (see Lemma 5). Moreover, we generalise it by restricting the relation "has an(other) occurrence" to "there is an(other) $\equiv_\sigma^e$ equivalent segment", and, if the segment contains two or more positions, by restricting cuts to properly cut the context into two *non-empty* terms, i.e., discarding cuts on the extremes. To prove this result, we need two previous lemmas. Lemma 4 plays the same role as the Lemma 4 of [PR98].

Lemma 3. *Given a term t, and a marking function $m : \mathrm{pos}(t) \to \{\mathsf{rem}, \mathsf{pres}\}$, let $\mathsf{pres}(t, m)$ be a term for which there exists an embedding from the set of preserved nodes of t to $\mathrm{pos}(\mathsf{pres}(t, m))$, i.e. a bijective morphism*

$$f_{t,m} : \{p \in \mathrm{pos}(t) \mid m(p) = \mathsf{pres}\} \to \mathrm{pos}(\mathsf{pres}(t, m))$$

preserving the tree structure (ancestor, brother and label relations). Then, if such term $\mathsf{pres}(t, m)$ exists, then it is unique.

Given an equation e and a well-nested unifier σ, let $m : \mathrm{pos}(\sigma(e)) \to \{\mathsf{rem}, \mathsf{pres}\}$ be a marking function satisfying:

1. *For any pair of positions $p, q \in \mathrm{pos}(\sigma(e))$, if $p \approx_\sigma^e q$ then $m(p) = m(q)$, i.e., equivalent positions are both removed or both preserved.*
2. *For any $p \in \mathrm{pos}(e)$, we have $m(c_\sigma^e(p)) = \mathsf{pres}$, i.e. cuts are preserved.*
3. *The term $\mathsf{pres}(\sigma(e), m)$ exists, i.e. after the removing process you get a term.*

Then, there exists a well-nested unifier σ' of e such that $\sigma'(e) = \mathsf{pres}(\sigma(e), m)$. Moreover, if there exists some removed node, then σ' is size-smaller than σ.

Proof. First, notice that if $\mathsf{pres}(t, m)$ exists there must be a unique outermost preserved position in t, and the function $f_{t,m}^{-1}$ has to map ϵ to it. Moreover, for any $p \in \mathrm{pos}(\mathsf{pres}(t, m))$ there must be $\mathrm{arity}(\mathrm{lab}_t(f_{t,m}^{-1}(p)))$ many preserved outermost bellow $f_{t,m}^{-1}(p)$ positions in t, and $f_{t,m}^{-1}$ has to map $p \cdot i$ to the i-th of them. Therefore, $f_{t,m}^{-1}$ is unique, so its inverse, and $\mathsf{pres}(t, m)$.

Now, notice that a unifier σ of e can be characterised by e, the term $\sigma(e)$ and the correspondence function c_σ^e. In our case, we characterise σ' by the correspondence function $c_{\sigma'}^e = f_{\sigma(e),m} \circ c_\sigma^e$.

Since cuts are not removed, this function maps any position of e to a position in $\mathsf{pres}(\sigma(e), m)$, which on the other side, is a well-formed term.

For any $p, q \in \mathrm{pos}(e)$ with $\mathrm{lab}_e(p) = \mathrm{lab}_e(q) \in \mathrm{Var}(e)$ we have $\sigma(e)\langle c_\sigma^e(p), c_\sigma^e(p \cdot 1)\rangle = \sigma(e)\langle c_\sigma^e(q), c_\sigma^e(q \cdot 1)\rangle$. Now, since the positions of these two contexts are $\approx_\sigma^e$-equivalent one to one, and equivalent positions are both removed or preserved, we have $\mathsf{pres}(\sigma(e), m)\langle c_{\sigma'}^e(p), c_{\sigma'}^e(p \cdot 1)\rangle = \mathsf{pres}(\sigma(e), m)\langle c_{\sigma'}^e(q), c_{\sigma'}^e(q \cdot 1)\rangle$. Therefore, we can define a substitution σ' mapping X to $\mathsf{pres}(\sigma(e), m)\langle \mathsf{c}_{\sigma'}^e(p), \mathsf{c}_{\sigma'}^e(p \cdot 1)\rangle$, for any of the occurrences p of X.

It can also be proved that $p \approx_{\sigma'}^e q \Leftrightarrow f_{\sigma(e),m}^{-1}(p) \approx_\sigma^e f_{\sigma(e),m}^{-1}(q)$. This makes $\langle \sigma, e \rangle$-images to correspond to $\langle \sigma', e \rangle$-images throughout $f_{\sigma(e),m}$. Therefore, if σ is well-nested, then σ' is well-nested, too. $\square$

Lemma 4. *If σ is a normal unifier of an equation e, then every equivalence class C of $\approx_\sigma^e$ is cut by some constant in e, i.e., there is a position $p \in \mathsf{pos}(e)$ such that $\mathsf{c}_\sigma^e(p) \in C$ and $\mathsf{lab}_e(p) \in \Sigma$.*

Proof. All positions of C are labeled alike in $\sigma(e)$ according to Lemma 1. Let $f = \mathsf{lab}_{\sigma(e)}(p)$ for all positions $p \in C$. Now, assume that the statement does not hold, so all positions in C belong to segments that are denotations of some variables. There are three cases:

If f is unary, the proof is quite similar to the proof given in [PR98]: if C does not contain cuts of a unary constant of e, then we can remove these occurrences of the constant f in C, which would result into a smaller unifier. As it is stated in Lemma 3, this replacement only involves changes in the instantiations of the variables, by deleting the nodes of the equivalence class. Moreover, this removing process preserves the well-nestedness property, because the relative position between segments do not change with the process.

If arity(f) ≥ 2, the proof is more subtle, and requires the unifier to be well-nested. For every $p \in C$, consider the longest path q_p in e such that $\mathsf{c}_\sigma^e(q_p) \preceq p$. First, notice that, since we have assumed that C does not contain cuts corresponding to constants, all q_p correspond to variables $X_p = \mathsf{lab}_e(q_p)$ and, second, notice that p is inside the segment $\langle \mathsf{c}_\sigma^e(q_p), \mathsf{c}_\sigma^e(q_p \cdot 1)\rangle$.

Now, we can prove the following property: for any $p, p' \in C$, and integer numbers i, i', if $p \cdot i \preceq \mathsf{c}_\sigma^e(q_p \cdot 1)$ and $p' \cdot i' \preceq \mathsf{c}_\sigma^e(q_{p'} \cdot 1)$, then $i = i'$. To prove it, assume $i \neq i'$. Since p and p' are related by the $\approx_\sigma^e$ equivalence relation, and this is the transitive closure of a more restrictive relation, we can find two positions $p'', p''' \in C$, related by this more restrictive relation, and such that $p'' \cdot i \preceq \mathsf{c}_\sigma^e(q_{p''} \cdot 1)$ and $p''' \cdot i' \preceq \mathsf{c}_\sigma^e(q_{p'''} \cdot 1)$. According to the $\approx_\sigma^e$ definition, for these new positions, either $X_{p''} = X_{p'''}$, or $p'' = 1 \cdot r$ and $p''' = 2 \cdot r$, for some r. In the first case, we have $i = i'$. In the second case, if $i \neq i'$, we would have two bad-nested variables $X_{p''}$ and $X_{p'''}$. Notice that, in this point of the proof, we make use of the well-nestedness property.

We can conclude that there exists a number i such that, for every $p \in C$, every integer $j \neq i$, and every sequence r, there are not cuts of the form $p \cdot j \cdot r$. (Notice that this does not imply that there exist cuts of the form $p \cdot i \cdot r$). Now, using again Lemma 3, we can replace all the contexts $\sigma(e)\langle p, p \cdot i\rangle$ by the empty context, because they do not contain cuts. Again this removing process preserves the well-nestedness property, and results in a smaller well-nested unifier, which contradicts the assumption.

If arity(f) $= 0$ we can not apply the previous reasoning because $\sigma(e)\langle p, p \cdot i\rangle$ is not a context. However, if C does not contains cuts corresponding to constants of e, then C does not contain cuts at all (it could only contain cuts corresponding to first-order variables, but we do not consider them). Then, the set of parents of $p \in C$, i.e., the set of positions $\{q \mid \exists p \in C . \exists j \in \mathbb{N} . p = q \cdot j\}$ is an equivalence class of positions C'. This class of positions C' can contain cuts corresponding to context variables, but not to constants (this would imply that C contains cuts). Therefore, since the nodes of C' are labelled with a nonzero arity constant, we can apply some of the two previous cases to C', and reach a contradiction. $\square$

We want to remark that Lemma 4 does not hold for general CU. For instance, the context unification equation $e = X(a) \overset{?}{=} Y(b)$ has a size-minimal unifier $\sigma = [X \mapsto f(\bullet, b), Y \mapsto f(a, \bullet)]$. The set of positions $\{1, 2\}$ is an $\approx_\sigma^e$-equivalence class (all of them corresponding to the same constant f), but there exists no f-labeled node in e that could generate a f-labelled cut in $\sigma(e)$. This size-minimal unifier, however, is not well-nested.

Lemma 5. *If σ is a normal unifier of an equation e, then, for any non-empty segment $\langle p, q \rangle$ of $\sigma(e)$, there exists an $\equiv_\sigma^c$ equivalent segment $\langle p', q' \rangle$ over a cut $c_\sigma^e(r)$, i.e. there exist $r \in \mathrm{pos}(e)$ and $\langle p', q' \rangle \equiv_\sigma^e \langle p, q \rangle$ such that $c_\sigma^e(r)$ is inside $\langle p', q' \rangle$.*
Moreover, if $\langle p, q \rangle$ contains just one position, then we can restrict $\mathrm{lab}_e(p')$ to be a unary constant, and if $\langle p, q \rangle$ contains two or more positions, then we can restrict r to satisfy $c_\sigma^e(r) \neq p'$.

Proof. Consider a segment $\langle p, q \rangle$ containing just one position, this must be p, and $q = p \cdot 1$. Consider the $\approx_\sigma^e$-equivalence class defined by p. By Lemma 4, there exists a cut $p' \approx_\sigma^e p$ corresponding to a constant $f = \mathrm{lab}_e(c_\sigma^{e\,-1}(p'))$ in the same equivalence class. Then, the segment $\langle p', p' \cdot 1 \rangle$ fulfils our requirements, because $p \approx_\sigma^e p'$ implies $\langle p, p \cdot 1 \rangle \equiv_\sigma^e \langle p', p' \cdot 1 \rangle$.

Consider a segment $\langle p, q \rangle$ containing more than one position. Suppose that the lemma does not hold, then any segment $\langle p', q' \rangle$, in the same $\equiv_\sigma^e$-equivalence class as $\langle p, q \rangle$, does not contain any cut, except for possibly p'. Comparing the definitions of $\approx_\sigma^e$ and $\equiv_\sigma^e$, in such conditions, we have $\langle p, p \cdot s \rangle \equiv_\sigma^e \langle p', p' \cdot s \rangle$ implies $p \approx_\sigma^e p'$. Now, by Lemma 4, there exists a position $p' \approx_\sigma^e p$ over a cut corresponding to a constant. Therefore, for any integer i, the position $p' \cdot i$ is also a cut. Now, since $\langle p', q' \rangle$ contains more than one position, it must contain $p' \cdot i$, for some i, hence it contains another cut, apart from p', which contradicts the initial supposition. $\square$

Lemma 6. *If σ is a normal unifier of an equation e, then, for every variable $X \in \mathrm{Var}(e)$, there exists a segment $\langle p, q \rangle$ in e such that $\sigma(X) = \sigma(e\langle p, q \rangle)$, where the segment $e\langle p, q \rangle$ is not of the form $Y(\bullet)$.*

Proof. There are several cases. If $\sigma(X)$ is the empty context $\bullet$ then, for any empty segment $\langle p, p \rangle$ of e, we have $\sigma(X) = \sigma(e\langle p, p \rangle) = \sigma(\bullet) = \bullet$.

If $\sigma(X) = f(\bullet)$, we can assume that X occurs in e, say at position p, and $\sigma(X)$ occurs in $\sigma(e)$, at segment $\langle c_\sigma^e(p), c_\sigma^e(p \cdot 1) \rangle = \langle c_\sigma^e(p), c_\sigma^e(p) \cdot 1 \rangle$, otherwise σ would not be size-minimal. Now, using Lemma 5, we have an occurrence r in e such that $\langle c_\sigma^e(r), c_\sigma^e(r) \cdot 1 \rangle \equiv_\sigma^e \langle p, p \cdot 1 \rangle$ and $\mathrm{lab}_e(r) = f$. Therefore, $\sigma(X) = \sigma(e\langle r, r \cdot 1 \rangle) = f(\bullet)$.

If $\sigma(X)$ contains two or more positions, let p be an occurrence of X in e, and $\langle c_\sigma^e(p), c_\sigma^e(p \cdot 1) \rangle$ be the corresponding occurrence of $\sigma(X)$ in $\sigma(e)$. By Lemma 5 there exists an occurrence $\langle p', q' \rangle$ of $\sigma(X)$ over a cut: $\langle p', q' \rangle \equiv_\sigma^e \langle c_\sigma^e(p), c_\sigma^e(p \cdot 1) \rangle$, and there exists a position r in e such that $c_\sigma^e(r)$ is inside $\langle p', q' \rangle$ and $c_\sigma^e(r) \neq p'$. Now, we will prove that p' and q' are also cuts.

Suppose that p' is not a cut. Let $s \in \mathsf{pos}(e)$ be the longest position such that $c_\sigma^e(s) \preceq p'$. Then, $c_\sigma^e(s) \prec p' \prec c_\sigma^e(s \cdot 1)$, hence s corresponds to a variable occurrence. Since $\langle c_\sigma^e(s), c_\sigma^e(s \cdot 1) \rangle$ does not contain other cuts than $c_\sigma^e(s)$, we have $c_\sigma^e(s \cdot 1) \preceq c_\sigma^e(r)$. This gives two overlapping segments $\langle p', q' \rangle$ and $\langle c_\sigma^e(s), c_\sigma^e(s \cdot 1) \rangle$ that violate the well-nestedness condition. A similar argument allows us to prove that q' is a cut.

We can conclude that $\langle c_\sigma^{e\,-1}(p'), c_\sigma^{e\,-1}(q') \rangle$ is a segment of e containing a position r with $r \neq c_\sigma^{e\,-1}(p')$ and $r \neq c_\sigma^{e\,-1}(q')$, therefore fulfilling the conditions of the lemma.   □

Theorem 1. *Normal unifiers σ of a context equation e have a representation of the form:*

$$\sigma = [X_1 \mapsto e\langle p_1, q_1 \rangle] \circ \cdots \circ [X_n \mapsto e\langle p_n, q_n \rangle]$$

where $\langle p_i, q_i \rangle$ are segments of e and X_i variables of $\mathsf{Var}(e)$ that do not occur in $e\langle p_j, q_j \rangle$ for $j \leq i$.
This representation is polynomial in $|e|$, and does not use constants not occurring in e.

Proof. By Lemma 6, for any variable X_i of the equation, we can find a segment $\langle p_i, q_i \rangle$ of the equation such that $\sigma(X_i) = \sigma(e\langle p_i, q_i \rangle)$, where $\langle p_i, q_i \rangle$ is something else than just one variable. For any variable X_j occurring in $e\langle p_i, q_i \rangle$ we say that $X_i > X_j$. The transitive closure of this relation results into an irreflexive relation. Now, Let $X_1 > X_2 > \cdots > X_n$ be a total ordering of the variables of the equation compatible with this transitive and irreflexive ordering. This ordering is used to express the σ as it is stated in the theorem.   □

6   Well-Nested Context Unification is in NP

In this section we prove that given an equation e, and a substitution σ that can be represented in polynomial space on $|e|$ in the form $[X_1 \mapsto t_1] \circ \cdots \circ [X_n \mapsto t_n]$, we can check if σ is a unifier of e in polynomial time on $|e|$. This result is not trivial, since $|\sigma(e)|$ can be exponential in $|e|$. For instance, the equation:

$$X_n(X_{n-1}(\cdots X_1(a) \cdots)) \overset{?}{=} X_{n-1}(X_n(X_{n-2}(X_{n-2}(\cdots f(a) \cdots))))$$

has as unique unifier that we can represent as follows:

$$\sigma = [X_1 \mapsto f(\bullet)] \circ [X_2 \mapsto X_1(X_1(\bullet))] \circ \cdots \circ [X_n \mapsto X_{n-1}(X_{n-1}(\bullet))]$$

This representation has linear size on n, like the equation. However $\sigma(e) = f^{2^n - 1}(a) \overset{?}{=} f^{2^n - 1}(a)$ has exponential size on n.

To overcome this problem we construct a context free grammar generating a preorder traversal of the term $\sigma(t)$, and another for $\sigma(u)$. Both grammars will be of polynomial size (not so the word generated by them). Then, we can use a result (see Lemma 7) due to Plandowski that allows us to check the equality of the words generated by the two grammars in polynomial time on the size of the grammars.

A *context-free grammar (CFG)* is a 4-tuple (Σ, N, P, S), where Σ is an alphabet of *terminal* symbols, N is an alphabet of *non-terminal* symbols, P is a finite set of rules, and $S \in N$ is the *start symbol*. We will not distinguish a particular start symbol, and we will represent a context free grammars as a 3-tuple (Σ, N, P).

Definition 7. *A context free grammar $G = (\Sigma, N, P)$ generates a word $w \in \Sigma^*$ if there exists a non-terminal symbol $A \in N$ such that w belongs to the language defined by (Σ, N, P, A). In such case, we also say that A generates w.*

We say that a context free grammar is a singleton CFG *if it is not recursive and every non-terminal symbol occurs in the left-hand side of exactly one rule. Then, every non-terminal symbol $A \in N$ generates just one word, noted w_A.*

Plandowski [Pla94, Pla95] defines singleton grammars, but he calls them *grammars defining set of words*. He proves the following result.

Lemma 7 ([Pla95], Theorem 33). *The word equivalence problem for singleton context-free grammars is defined as follows: given a grammar and two non-terminal symbols A and B, to decide whether $w_A = w_B$. This problem can be solved in polynomial worst-case time on the size of the grammar.*

In order to translate trees and contexts into sequences, we will use a pair of nonterminal symbols X^L and X^R for each context variable X. Then, the translation function is defined by:

$$\mathsf{trans}(a) = a \qquad \mathsf{trans}(f(t_1, \ldots, t_n)) = f\,\mathsf{trans}(t_1) \cdots \mathsf{trans}(t_n)$$
$$\mathsf{trans}(\bullet) = \bullet \qquad \mathsf{trans}(X(t)) = X^L\,\mathsf{trans}(t)\,X^R$$

Given a unification equation $t \overset{?}{=} u$, and a guessed substitution $[X_1 \mapsto v_1] \circ \cdots \circ [X_n \mapsto v_n]$, where $v_i \in \mathsf{Cont}_\Sigma$, we generate the following grammar:

$$\left. \begin{array}{l} A \to \mathsf{trans}(t) \\ B \to \mathsf{trans}(u) \end{array} \right| \left. \begin{array}{l} X_i^L \to \mathsf{left}(\mathsf{trans}(v_i)) \\ X_i^R \to \mathsf{right}(\mathsf{trans}(v_i)) \end{array} \right\} \text{ for every } i = 1, \ldots, n$$

where $\mathsf{left}(\alpha \bullet \beta) = \alpha$ and $\mathsf{right}(\alpha \bullet \beta) = \beta$.

In the case of our example, the grammar would be:

$$A \to X_n^L\,X_{n-1}^L \cdots X_1^L\,a\,X_1^R \cdots X_{n-1}^R\,X_n^R$$
$$B \to X_{n-1}^L\,X_{n-1}^L\,X_{n-2}^L\,X_{n-2}^L \cdots f\,a \cdots X_{n-2}^R\,X_{n-2}^R\,X_{n-1}^R\,X_{n-1}^R$$
$$\begin{array}{ll} X_n^L \to X_{n-1}^L\,X_{n-1}^L & \bigg| \quad X_n^R \to X_{n-1}^R\,X_{n-1}^R \\ X_{n-1}^L \to X_{n-2}^L\,X_{n-2}^L & \bigg| \quad X_{n-1}^R \to X_{n-2}^R\,X_{n-2}^R \\ \qquad \cdots & \qquad\quad \cdots \\ X_1^L \to f & \bigg| \quad X_1^R \to \epsilon \end{array}$$

Lemma 8. *Given a context equation e, and a substitution of the form:*

$$\sigma = [X_1 \mapsto v_1] \circ \cdots \circ [X_n \mapsto v_n]$$

where X_i are distinct context variables, v_i are contexts, and X_i does not occurs in $v_1 \ldots v_i$, we can check if it is a unifier in polynomial time on $|e| + |\sigma|$.

Proof. Since every constant has a unique arity, then for every pair of terms $t, u \in \mathcal{T}_{\Gamma(\Sigma)}$, $t = u$ if, and only if, $\mathsf{trans}(t) = \mathsf{trans}(u)$.

Now, we can prove, by induction on k, that the grammar

$$\left. \begin{array}{l} A \to \mathsf{trans}(t) \\ B \to \mathsf{trans}(u) \end{array} \right| \quad \left. \begin{array}{l} X_i^L \to \mathsf{left}(\mathsf{trans}(v_i)) \\ X_i^R \to \mathsf{right}(\mathsf{trans}(v_i)) \end{array} \right\} \text{ for every } i = n - k, \ldots, n$$

using A as the start symbol, can generate $\mathsf{trans}([X_{n-k} \mapsto v_{n-k}] \circ \cdots \circ [X_n \mapsto v_n]t)$, and similarly for B and u. Finally, using Lemma 7, we simply have to check if A and B generate the same sequence, in polynomial time on the size of the grammar, that is polynomial on $|e| + |\sigma|$. □

Theorem 2. *Deciding if a context equation has a well-nested unifier is NP-complete.*

Proof. Given an equation e, we can guess a representation for normal unifiers in polynomial time (Theorem 1), and then check in polynomial time whether it is represents indeed a unifier (Lemma 8). Finally, it can be proved that checking if the unifier represented in the form of Theorem 1 is well-nested can also be done in polynomial time using similar techniques.

For NP-hardness, it is sufficient to note that well-nested CU subsumes string matching [Ang80]. This is since the standard encoding of string matching will produce context equations that only have well-nested unifiers. □

7   Conclusions

Well-nested context unifiers are a kind of context unifiers with interest in computational linguistics. Here we prove that decidability of the existence of a well-nested context unifier is NP-complete. Additionally, we prove that well-nested size-minimal unifiers can be represented in polynomial space, as a composition of substitutions where each one of them instantiate a context variable by a *segment* of the original equation (a similar result holds for most general unifiers in first-order unification). As a direct consequence, well-nested size-minimal unifiers do not use constants not occurring in the original equation, a wishful property for computational linguistic applications. All these results are extensible to word unification.

In the future, we plan to study the relationship between well-nested context unification and well-nested parallelism constraints, and to look for a more efficient algorithm to compute well-nested unifiers than the brute force guessing.

Acknowledgments. We acknowledge the suggestions of the anonymous referees and the fruitful discussions with Katrin Erk.

References

[Ang80]   D. Angluin. Finding patterns common to a set of strings. *Journal of Computer and Systems Sciences*, 21:46–62, 1980.

[BDMN04]   M. Bodirsky, D. Duchier, S. Miele, and J. Niehren. A new algorithm for normal dominance constraints. In *ACM-SIAM Symposium on Discrete Algorithms*, pages 54–78, 2004.

[Com92]   H. Comon. Completion of rewrite systems with membership constraints. In *Int .Coll. on Automata Languages and Progr.*, vol. 623 of *LNCS*, 1992.

[DSP91]   M. Dalrymple, S. Shieber, and F. Pereira. Ellipsis and higher-order unification. *Linguistics & Philosophy*, 14:399–452, 1991.

[EKN01]   M. Egg, A. Koller, and J. Niehren. The constraint language for lambda structures. *Journal of Logic, Language, and Information*, 10:457–485, 2001.

[EN03]   K. E. Erk and J. Niehren. Well-nested parallelism constraints for ellipsis resolution. In *11th Conf. of the European Chapter of the Ass. of Comp. Ling.*, pages 115–122, 2003.

[GK96]   C. Gardent and M. Kohlhase. Higher–order coloured unification and natural language semantics. In *34th Meet. of the Ass. for Comput. Ling.*, 1996.

[Lev96]   J. Levy. Linear second-order unification. In *7th Int. Conf. on Rewriting Techniques and Applications*, vol. 1103 of *LNCS*, pages 332–346, 1996.

[LV00]   J. Levy and M. Villaret. Linear second-order unification and context unification with tree-regular constraints. In *11th Int. Conf. on Rewriting Techniques and Applications*, vol. 1833 of *LNCS*, pages 156–171, 2000.

[Mak77]   G. S. Makanin. The problem of solvability of equations in a free semigroup. *Math. USSR Sbornik*, 32(2):129–198, 1977.

[Mon73]   R. Montague. The proper treatment of quantification in ordinary English. In *Approaches to Natural Language*. Dordrecht, 1973.

[NK01]   J. Niehren and A. Koller. Dominance constraints in context unification. In *Logical Aspects of Computational Linguistics*, vol. 2014 of *LNAI*, pages 199–218, 2001.

[NPR97]   J. Niehren, M. Pinkal, and P. Ruhrberg. A uniform approach to underspecification and parallelism. In *35th Meeting of the Association of Computational Linguistics*, pages 410–417, 1997.

[Pla94]   W. Plandowski. Testing equivalence of morphisms in context-free languages. In J. van Leeuwen, editor, *Proc. of the 2nd Annual European Symposium on Algorithms*, vol. 855 of *LNCS*, pages 460–470, 1994.

[Pla95]   W. Plandowski. *The Complexity of the Morphism Equivalence Problem for Context-Free Languages*. PhD thesis, Department of Mathematics, Informatics and Mechanics, Warsaw University, 1995.

[Pla99]   W. Plandowski. Satisfiability of word equations with constants is in PSPACE. In *40th IEEE Found. of Comp. Science*, pages 495–500, 1999.

[PR98]   W. Plandowski and W. Rytter. Application of lempel-ziv encodings to the solution of words equations. In *25th ICALP*, vol. 1443 of *LNCS*, pages 731–742, 1998.

[Sag76]   I. Sag. *Deletion and logical form*. PhD thesis, MIT, Cambridge, 1976.

[SS02]   M. Schmidt-Schauß. A decision algorithm for stratified context unification. *Journal of Logic and Computation*, 12:929–953, 2002.

[SSS99]   M. Schmidt-Schauß and K. U. Schulz. Solvability of context equations with two context variables is decidable. In *16th Int. Conf. on Automated Deduction*, LNAI, pages 67–81, 1999.

Termination of Rewrite Systems
with Shallow Right-Linear, Collapsing,
and Right-Ground Rules[*]

Guillem Godoy[1] and Ashish Tiwari[2]

[1] Technical University of Catalonia,
Jordi Girona 1, Barcelona, Spain
ggodoy@lsi.upc.es
[2] SRI International, Menlo Park, CA 94025
tiwari@csl.sri.com

Abstract. We show that termination is decidable for rewrite systems that contain shallow and right-linear rules, collapsing rules, and right-ground rules. This class of rewrite systems is expressive enough to include interesting rules. Our proof uses the fact that this class of rewrite systems is known to be regularity-preserving and hence the reachability and join-ability problems are decidable. Decidability of termination is obtained by analyzing the nonterminating derivations.

1 Introduction

Term rewriting systems are Turing-complete models of computation that specify rules for replacing certain patterns in terms by equivalent, in some cases simpler, other terms. Simpler models of computation result by imposing additional constraints on the form of terms in a rewrite system. For instance, if variables are not allowed, we get *ground* term rewrite system, which have been extensively studied, mainly via mapping them to tree automata [2]. More complex models of computation arise by allowing restricted variable occurrences in the term rewrite system (or the tree automata transitions).

Termination is one of the central properties of rewrite systems. Termination guarantees that any expression cannot be infinitely rewritten, and hence, the existence of a normal form for it. As we go from simple to more general classes of rewrite systems, the complexity of deciding termination increases until it becomes undecidable. For example, while termination is decidable in polynomial time for ground term rewriting systems [13, 16], it is undecidable for general rewrite systems and string rewrite systems [13]. It is, therefore, fruitful to identify the decidability barrier and study decidability issues for some intermediate classes, especially if these classes are expressive enough to capture interesting rules.

---
[*] The first author was partially supported by Spanish Min. of Educ. and Science by the LogicTools project (TIN2004-03382). The second author was supported in part by the National Science Foundation under grants ITR-CCR-0326540 and CCR-0311348.

R. Nieuwenhuis (Ed.): CADE 2005, LNAI 3632, pp. 164–176, 2005.
© Springer-Verlag Berlin Heidelberg 2005

There are several negative, and few positive, results on decidability of termination for classes of rewrite system. Termination is undecidable for even one (non-linear) rule [4]. Termination is usually established using well-founded orderings [6]. It is undecidable whether a single term rewriting rule can be proved terminating using a simplification ordering [15] or a monotonic ordering total on ground terms [7]. On the other hand, several powerful techniques and implementations exist that can automatically prove termination of many rewriting systems [8, 3, 12, 1]. These systems are based on combinations of techniques such as the use of well-founded term orderings, transformations, semantic interpretations, and dependency-pairs. The success of these tools suggests the natural question: is there an interesting and large class of rewrite systems for which termination is indeed decidable?

In this context, we consider term rewriting systems that contain shallow right-linear rules, collapsing rules, and right-ground rules. In a shallow right-linear rule $l \rightarrow r$, every variable occurs at most once in r, and all variables in l, r occur at depth 0 or 1. Some examples of shallow right-linear rules are $0 \wedge x \rightarrow 0$, $x \wedge x \rightarrow x$, $1 \wedge x \rightarrow x$, $1 \vee x \rightarrow 1$, $x \vee x \rightarrow x$, $0 \wedge x \rightarrow x$, $x \wedge y \rightarrow y \wedge x$ and $x \vee y \rightarrow y \vee x$. A rule of the form $l \rightarrow x$, where x is a variable, is called collapsing. For example, $\neg(\neg x) \rightarrow x$ is a collapsing rule. A rule $l \rightarrow r$, where r is a ground term, is a right-ground rule. For example, $x \wedge (\neg x) \rightarrow 0$, $x \vee (\neg x) \rightarrow 1$ are right-ground rules.

Our proof of decidability of termination relies on the decidability of reachability and joinability. Takai, Kaji, and Seki [17] showed that right-linear finite-path-overlapping systems effectively preserve recognizability. The class of rewrite systems defined by shallow right-linear, collapsing, and right-ground rules, is right-linear and finite-path-overlapping, and hence it follows that the reachability and joinability problems for this class is decidable. We point out here that reachability is known to be undecidable for linear TRS's, and also for shallow TRS's [14].

In this paper, we prove the decidability of termination for TRS's that contain shallow right-linear, collapsing, or right-ground rules. We use the decidability of reachability and joinability for this class as a black box. For termination, we give a checkable characterization based on some reachability conditions and the termination of a restricted rewrite system related with the original one.

In Section 2 we introduce some basic notions and notations. In Section 3 we present termination-preserving transformations that replace the shallow right-linear rules by flat right-linear rules and that replace the right-ground rules by right-constant rules. This section is quite easy and similar to parts of other previous works, but not identical, and allows us to simplify the arguments in the rest of the paper. In Section 4 we characterize the termination property for flat right-linear systems that contain additional collapsing and right-constant rules, and prove its decidability.

2 Preliminaries

We use standard notation from the term rewriting literature. A signature Σ is a (finite) set of function symbols, which is partitioned as $\cup_i \Sigma_i$ such that

$f \in \Sigma_n$ if the arity of f is n. Symbols in Σ_0, called *constants*, are denoted by a, b, c, d, with possible subscripts. The elements of a set $\mathcal{V}$ of variable symbols are denoted by x, y with possible subscripts. The set $\mathcal{T}(\Sigma, \mathcal{V})$ of *terms* over Σ and $\mathcal{V}$, *position p* in a term, *subterm $t|_p$* of term t at position p, and the term $t[s]_p$ obtained by replacing $t|_p$ by s are defined in the standard way. For example, if t is $f(a, g(b, h(c)), d)$, then $t|_{2.2.1} = c$, and $t[d]_{2.2} = f(a, g(b, d), d)$. We write $p_1 \succ p_2$ (or, $p_2 \prec p_1$) if p_2 is a proper prefix of p_1. By *Vars(t)* we denote the set of all variables occurring in t. The *height* of a term s is 0 if s is a variable or a constant, and $1 + max_i height(s_i)$ if $s = f(s_1, \ldots, s_m)$. Usually we will denote a term $f(t_1, \ldots, t_n)$ by the simplified form $ft_1 \ldots t_n$, and $t[s]_p$ by $t[s]$ when p is clear by the context or not important.

A substitution σ is sometimes presented explicitly as $\{x_1 \mapsto t_1, \ldots, x_n \mapsto t_n\}$. We assume standard definitions for a *rewrite rule $l \to r$*, a *rewrite system R*, the *one step rewrite relation at position p induced by R* $\to_{R,p}$, and the *one step rewrite relation induced by R* (at any position) $\to_R$. If $p = \lambda$, then the rewrite step $\to_{R,p}$ is said to be applied *at the topmost position* (at the root) and is denoted by $s \to_R^r t$; it is denoted by $s \to_R^{nr} t$ otherwise.

The notations $\leftrightarrow$, $\to^+$, and $\to^*$, are standard. R is terminating if no infinite derivation $s_1 \to_R s_2 \to \cdots$ exists. A term t is *reachable* from s by R (or, R-reachable) if $s \to_R^* t$. A term s is *R-irreducible* (or, in R-normal form) if there is no term t such that $s \to_R t$. We denote by $s \to^! t$ the fact that an irreducible term t is reachable from s by the $\to$ relation. If s is a term and S is a set of terms, then we define $Reach(s) = \{t : s \to_R^* t\}$ and $Reach(S) = \bigcap_{s \in S} Reach(s)$. A set S of two or more terms is *R-joinable* if $Reach(S) \neq \emptyset$. A *(rewrite) derivation* or *proof* (from s) is a sequence of rewrite steps (starting from s), that is, a sequence $s \to_R s_1 \to_R s_2 \to_R \ldots$.

A term t is called *ground* if t contains no variables. It is called *shallow* if all variable positions in t are at depth 0 or 1. It is called *linear* if every variable occurs at most once in t. A rule $l \to r \in R$ is called *right-ground* if r is ground, and *collapsing* if r is a variable. It is called *shallow right-linear* if the term r is linear, and both l, r are shallow, and *flat* if both l, r are height 0 or 1 terms.

3 Termination-Preserving Transformations

Let R be such that for every rule $l \to r \in R$, either r is ground, or r is a variable, or $l \to r$ is shallow and right-linear. Henceforth, we also assume that in every rule $l \to r$ we have $Vars(r) \subseteq Vars(l)$ (this is usual for rewrite rules, and without this property the corresponding rewrite system is trivially non-terminating).

By replacing non-constant ground terms by new constants, as described formally by the following two transformation rules, we can transform the rewrite system R into a rewrite system R' such that for every rule $l \to r \in R'$, either r is a constant, or r is a variable, or $l \to r$ is flat and right-linear.

$$\frac{l \to r[s]}{l \to r[c], c \to s} \qquad \frac{l[s] \to r}{l[c] \to r, s \to c} \qquad \text{if } s \text{ is non-constant and} \atop \text{ground; } c \text{ a new constant}$$

We show in Lemma 3 that application of these two transformation rules is terminating, and hence we can exhaustively apply them. As an optimization we can replace multiple instances of s on LHS (equivalently, RHS) by the *same* constant c, but we use the unoptimized version here to keep proofs simple. We prove below that applying these two transformation rules preserves termination. We remark here that this transformation is very similar (though not identical) to the one that preserves confluence, see [11].

Lemma 1. *Let R' be obtained from R using one of the two transformation rules described above. For every derivation $t_1 \to_R t_2 \to_R t_3 \to_R \cdots$, there exists a derivation $t_1 \to_{R'}^+ t_2 \to_{R'}^+ t_3 \to_{R'}^+ \cdots$.*

Proof. Suppose $t_i \to_{l \to r, \sigma, p} t_{i+1}$, where $l \to r \in R$. If $l \to r$ is also present in R', then clearly $t_i \to_{R'} t_{i+1}$. If not, then suppose $r|_q = s$ and $l \to r \in R$ is replaced by $l \to r[c]_q$ and $c \to s$ in R'. In this case, $t_i \to_{l \to r[c], \sigma, p} t_{i+1}[c]_{p.q} \to_{c \to s, id, p.q} t_{i+1}[s]_{p.q} = t_{i+1}$. In the other case, suppose $l|_q = s$ and $l \to r \in R$ is replaced by $l[c]_q \to r$ and $s \to c$ in R'. Now we have $t_i \to_{s \to c, id, p.q} t_i[c]_{p.q} \to_{l[c] \to r, \sigma, p} t_{i+1}$. This completes the proof. □

Lemma 2. *Let R' be obtained from R using one of the two transformation rules described above. Let c be the new constant that names some non-constant ground term s. For every infinite derivation $t_1 \to_{R'} t_2 \to_{R'} \cdots$ over $\mathcal{T}(\Sigma \cup \{c\}, \mathcal{V})$, there exists an infinite derivation $t_1\sigma \to_R^* t_2\sigma \to_R^* \cdots$ over $\mathcal{T}(\Sigma, \mathcal{V})$, where σ is $\{c \mapsto s\}$ and is applied as a substitution interpreting c as a variable.*

Proof. Suppose $R' = (R - \{l[s] \to r\}) \cup \{l[c] \to r, s \to c\}$. Consider the step $t_i \to_{l' \to r', \rho, p} t_{i+1}$, where $l' \to r' \in R'$. There are three cases. (1) If $l' \to r' \in R$, then since the constant c is new and not present in R, it follows that $t_i\sigma \to_{l' \to r', \rho\sigma, p} t_{i+1}\sigma$. (2) If $l' = l[c]$ and $r' = r$, then we can use the rewrite rule $l[s] \to r$ from R to get $t_i\sigma \to_{l[s] \to r, \rho\sigma, p} t_{i+1}\sigma$. (3) If $l' = s$ and $r' = c$, then $t_i\sigma = t_{i+1}\sigma$ and there is no corresponding step in the R-derivation. Since $\{s \to c\}$ is terminating, case (1) and (2) happen infinitely often, and hence the derivation $t_1\sigma \to_R^* t_2\sigma \to_R^* \cdots$ is an infinite derivation. Finally, we complete the proof by saying that the argument for the case when $R' = (R - \{l \to r[s]\}) \cup \{l \to r[c], c \to s\}$ can be done similarly. □

Let $R \vdash R'$ denote that R' is obtained from R using an application of the either of the two transformation rules.

Lemma 3. *Every derivation $R_1 \vdash R_2 \vdash R_3 \vdash \cdots$ is necessarily finite. If R_1 is such that for every $l \to r \in R$, either r is ground, or r is a variable, or $l \to r$ is shallow and r is linear, then the final rewrite system R obtained as $R_1 \vdash^! R$ is such that for every $l \to r \in R$, either r is a constant, or r is a variable, or $l \to r$ is flat and r is linear.*

Proof. Let $measure(R)$ be the multiset consisting of the depths of l and r for every $l \to r \in R$. If $R \vdash R'$, then $measure(R) >^m measure(R')$, where $>^m$ is the multiset extension of the regular greater-than $>$ ordering on the naturals. Hence the relation $\vdash$ is well-founded. If $R_1 \vdash^! R$, and R violates the second claim, then one of the two transformation rules will be applicable on R, thus contradicting that R is the normal form of R w.r.t $\vdash$.   □

The following theorem is now an easy consequence of Lemma 1, Lemma 2, and Lemma 3.

Theorem 1. *If R is any collection of right-ground rules, right-variable rules, and shallow and right-linear rules, then R can be transformed into R' such that R' is a collection of right-constant rules, right-variable rules, and flat and right-linear rules. Furthermore, R is terminating if and only if R' is terminating.*

Additionally, with a transformation identical to the one presented in [11], we can encode function symbols with nonzero arity using just one function symbol, say f, with sufficiently large arity m. Hence, we can assume that $\Sigma = \Sigma_0 \cup \{f\}$, where f is of arity m. This encoding preserves termination and is done just to simplify the proofs.

4   Termination

As a consequence of Theorem 1, we can without loss of generality assume that all rules in R are right-constant, right-variable, or flat and right-linear; and that Σ contains only one non-constant function symbol f of arity m.

Termination is decidable for right-ground term rewriting systems [5] and also for the more general class that also has right-variable rules [9]. An important idea used in [9] for handling nonlinear left-hand side terms is that of treating sets of constants (generalized to *terms* in this paper) as first-class objects (terms). Intuitively, a set $S \subseteq \mathcal{T}(\Sigma, \mathcal{V})$ of terms *represents* any (all) terms that are R-reachable from every term $s \in S$. For example, under this interpretation, the rewrite rule $fxx \to fax$ can rewrite fS_1S_2 to $fa(S_1 \cup S_2)$, *if* there is some term R-reachable from every term in $S_1 \cup S_2$. This is the basis for Definition 1.

A second observation we make in this paper is that the termination of R can be decomposed into termination of right-constant or right-variable rules and the termination of flat and right-linear rules. In particular, this means that the rules $l \to r$ where $depth(l) = depth(r) = 1$, called *permutation rules*, play an important role in characterizing the termination of the rewrite system R.

Definition 1. *[UJoin, UPerm] Let R be a flat right-linear TRS over Σ. Define an infinite set $K = \{S : \emptyset \neq S \subseteq \mathcal{T}(\Sigma, \mathcal{V})$ is R-joinable$\}$ of new constants (every set in K represents a constant) called set-constants. The rewrite systems $UJoin(R)$, $Perm(R)$, and $UPerm(R)$ (over $\Sigma \cup K$) are defined as follows:*

$UJoin(R) = \{S \to \{c\} : S \in K, \ c \in \Sigma_0, \ c \in Reach(S), \ and \ S \neq \{c\}\}$

$Perm(R) = \{l \to r \in R : depth(l) = depth(r) = 1\}$

$UPerm(R) = \{fS_1 \ldots S_m \to fT_1 \ldots T_m : \exists fs_1 \ldots s_m \to ft_1 \ldots t_m \in Perm(R),$

$\qquad S_p \in K, T_p \in K \ for \ all \ p \in \{1, \ldots, m\} \ AND$

$\qquad S_p = \{s_p\} \ whenever \ s_p \in \Sigma_0 \ AND$

$\qquad T_p = \{t_p\} \ whenever \ t_p \in \Sigma_0 \ AND$

$\qquad \forall x \in Vars(fs_1 \ldots s_m), \ \bigcup_{s_i = x} S_i \ is \ R\text{-}joinable, \ and \ if \ some \ t_j \ is \ x,$

$\qquad then \ T_j \ is \ \bigcup_{s_i = x} S_i\}$

Note that the set K is infinite. The rewrite system $UPerm(R)$ is ground (though constants of the form $\{x\} \in K$ may appear in it). The sets $UPerm(R)$ and $UJoin(R)$ are theoretical constructions possibly containing infinitely many rules. The termination characterization of Lemma 6 applies the rewrite system $UPerm(R) \cup UJoin(R)$ only on terms $fS_1 \ldots S_m$ where each $S_i \subseteq \Sigma_0$. Hence, the relevant rules of $UPerm(R) \cup UJoin(R)$ are those that only contain set-constants S s.t. $S \subseteq \Sigma_0$.

Example 1. If $R = \{fxxx \to fxc_1c_2, \ c_2 \to c_1\}$, then the set $UJoin(R)$ restricted to set-constants over Σ_0 is $\{\{c_2\} \to \{c_1\}, \{c_1, c_2\} \to \{c_1\}\}$. The set $UPerm(R)$ restricted similarly contains the 27 rules: $\{fS_1S_2S_3 \to fS\{c_1\}\{c_2\} :$ $\emptyset \neq S_1, S_2, S_3 \subseteq \{c_1, c_2\}, S = S_1 \cup S_2 \cup S_3\}$. The full set $UPerm(R)$ contains several other rules, for example, rules of the form $fSSS \to fSc_1c_2$, where S is any subset of terms that are R-joinable.

The notion of set-constants and the new definition of rewriting (using $Unions$) induced by R, and captured by $UPerm(R)$, allows us to
(a) project certain infinite R-derivations onto infinite $(UPerm(R) \cup UJoin(R))$-derivations over ground terms (Lemma 4 and Example 2); and
(b) lift infinite $(UPerm(R) \cup UJoin(R))$-derivations starting from a flat ground term to an infinite R-derivation (Lemma 5 and Example 3).
The consequence of these two lemmas is that termination of R restricted to derivations with only permutation steps at root positions is equivalent to termination with $(UPerm(R) \cup UJoin(R))$-derivations starting from flat ground terms.

4.1   Projecting Rewrite Derivations

Consider an infinite derivation $fcc(gb) \to_R fc(gb)b \to_R \cdots$ using some rewrite system R containing the permutation rule $fxxy \to fxyb$. We will first project this derivation onto the $(UPerm(R) \cup UJoin(R))$-derivation $f\{c\}\{c\}\{gb\} \to$ $f\{c\}\{gb\}\{b\} \to \cdots$, and thereafter remove the non-constant subterm gb from it by carefully analyzing the role of gb in the subsequent derivation.

Lemma 4. *Let* $s_1 = fs_{11} \ldots s_{1m} \to_R s_2 \to \cdots$ *be an infinite R-derivation that contains infinitely many top steps, where all of them are with rules in $Perm(R)$.*

Then there is an infinite ($UPerm(R) \cup UJoin(R)$)-derivation starting from a term of the form $fS_1 \ldots S_m$, where every S_i is included in Σ_0 and is R-joinable.

Proof. We project the derivation $s_1 \rightarrow s_2 \cdots$ onto an infinite ($UPerm(R) \cup UJoin(R)$)-derivation $t_1 \rightarrow^* t_2 \cdots$.

First, if s_1 is of the form $fs_{11} \ldots s_{1m}$, then let t_1 be $f\{s_{11}\} \ldots \{s_{1m}\}$. We inductively define t_{i+1} as follows. If $s_i \rightarrow s_{i+1}$ is a non-root rewrite step, then $t_{i+1} = t_i$. If $s_i \rightarrow s_{i+1}$ is a root rewrite step using a permutation rule $l \rightarrow r$, then, we define $t_{i+1}|_p$ as $\{s_{i+1}|_p\}$ if $r|_p$ is a constant (and hence equal to $s_{i+1}|_p$), and as $\bigcup_{l|_{p'}=r|_p} t_i|_{p'}$ if $r|_p$ is a variable.

By construction, all t_i's are flat terms where the depth 1 constants are sets S that contain either constants of the original signature or the initial subterms $s_{11}, \ldots, s_{1m}$, i.e., if $t_i|_p = S$, then $S \subseteq \{s_{11}, \ldots, s_{1m}\} \cup \Sigma_0$.

We prove by induction on i that every $s_i|_p$ is reachable from all terms in $t_i|_p$, that is, $s_i|_p \in Reach(t_i|_p)$ for all p. This is trivially true for s_1 and t_1. If s_{i+1} is obtained from s_i using a non-root rewrite step, then it follows that $t_i = t_{i+1}$ (by definition), $s_i|_p \in Reach(t_i|_p)$ (by induction hypothesis), and $s_i|_p \rightarrow s_{i+1}|_p$, which together implies that $s_{i+1}|_p \in Reach(t_{i+1}|_p)$. If s_{i+1} is obtained from s_i with a root rewrite step using a permutation rule $l \rightarrow r$, then, (i) for positions p s.t. $r|_p \in \Sigma_0$ the result directly follows since for such p we have $t_{i+1}|_p$ is $\{s_{i+1}|_p\}$, and (ii) for positions p s.t. $r|_p \in V$ we have $t_{i+1}|_p = \bigcup_{l|_{p'}=r|_p} t_i|_{p'}$ (by definition), each $s_i|_{p'} \in Reach(t_i|_{p'})$ (by induction hypothesis), and $s_{i+1}|_p$ coincides with all $s_i|_{p'}$ such that $l|_{p'} = r|_p$, which together implies that $s_{i+1}|_p \in Reach(t_{i+1}|_p)$.

Now we show that if $s_i \rightarrow s_{i+1}$ with a permutation rule $l \rightarrow r$, then $t_i \rightarrow^*_{UJoin(R)} \rightarrow_{UPerm(R)} t_{i+1}$, where the last step is done with the $UPerm(R)$ rule $fS_1 \ldots S_m \rightarrow t_{i+1}$ constructed from $l \rightarrow r$ by setting $S_j = \{l|_j\}$ if $l|_j$ is a constant, and $S_j = t_i|_j$ is $l|_j$ is a variable. The term t_i may differ from $fS_1 \ldots S_m$ at positions p such that $l|_p$ is a constant. For each such position p, S_p coincides with $\{s_i|_p\}$, and by the previous fact, this $s_i|_j$ is reachable from all terms in $t_i|_p$, and hence, a $UJoin(R)$ rule $t_i|_p \rightarrow S_p$ exists. Hence we conclude that $t_i \rightarrow^*_{UJoin(R)} fS_1 \ldots S_m \rightarrow_{UPerm(R)} t_{i+1}$.

Since the derivation $s_1 \rightarrow \ldots$ contains infinite root rewrite steps with permutation rules, the derivation $t_1 \rightarrow \ldots$ is also infinite. By right-linearity of R and the definition of $UPerm(R)$ and $UJoin(R)$, the number of occurrences of the non-constant terms $\{s_{11}, \ldots, s_{1m}\}$ in the set-constants S cannot increase in the infinite derivation $t_1 \rightarrow \ldots$. If some non-constant s_{1j}'s are persisting, then the sets in which they occur can only become larger. Choose i large enough so that the sets containing non-constant terms do not change any more in the derivation $t_i \rightarrow t_{i+1} \rightarrow \cdots$. We can map this infinite derivation into a new one over flat terms, in which all sets contain only constants, by eliminating the non-constant s_{1j} occurrences. Before doing it, we pick a fixed constant $c \in \Sigma_0$. Now, if S contains some non-constant s_{1j} and also some constants, then we just remove such s_{1j} from the set S. If S contains no constants, but only terms such as s_{1j}, then we replace S by $\{c\}$. With these replacements, it is easily verified that the derivation $t_i \rightarrow t_{i+1} \rightarrow \cdots$ is transformed into a new infinite rewriting derivation $t'_i \rightarrow t'_{i+1} \rightarrow \cdots$ with all terms flat and all set-constants only containing constants from Σ_0. This completes the proof. □

Example 2. Let $\Sigma = \{f, g, a, b, c\}$, where $arity(f) = 3$, $arity(g) = 1$, and arity of all other symbols is 0. Consider the rewrite system:

$$R = \{fxxy \rightarrow fxyb, \ fxyd \rightarrow fcxa, \ a \rightarrow gb, \ b \rightarrow d, \ gb \rightarrow c\}.$$

Note that a normalizes to c and b normalizes to d. Consider the following infinite R-derivation obtained by successively normalizing the depth 1 subterms (denoted by superscript $*, nr$) and applying the appropriate permutation rule from R (denoted by superscript r_1, r_2 for the two rules respectively).

$$fccgb \rightarrow^{*,nr} fccc \rightarrow^{r_1} fccb \rightarrow^{*,nr} fccd \rightarrow^{r_1} fcdb$$
$$\rightarrow^{*,nr} fcdd \rightarrow^{r_2} fcca \rightarrow^{*,nr} fccc \rightarrow^{r_1} \cdots$$

We project this derivation in two steps. In the first step, we ignore the nr-steps and use the derived $UPerm(R)$ rule to perform the r-steps. Note that we need to use the $UJoin(R)$ rule $\{b\} \rightarrow \{d\}$ below. We get the following derivation:

$$f\{c\}\{c\}\{gb\} \rightarrow^{r_1} f\{c\}\{gb\}\{b\} \quad \rightarrow^{r_1} f\{c, gb\}\{b\}\{b\} \quad \rightarrow^{nr} f\{c, gb\}\{b\}\{d\}$$
$$\rightarrow^{r_2} f\{c\}\{c, gb\}\{a\} \quad \rightarrow^{r_1} f\{c, gb\}\{a\}\{b\} \quad \rightarrow^{r_1} f\{a, c, gb\}\{b\}\{b\}$$
$$\rightarrow^{nr} f\{a, c, gb\}\{b\}\{d\} \rightarrow^{r_2} f\{c\}\{a, c, gb\}\{a\} \rightarrow^{r_1} f\{a, c, gb\}\{a\}\{b\}$$
$$\rightarrow^{r_1} f\{a, c, gb\}\{b\}\{b\} \rightarrow^{nr} \cdots$$

In the second step, we notice that the set-constants in the derivation starting from $f\{a, c, gb\}\{b\}\{b\}$ do not change, and hence, we forget the nonconstants in the sets and get the following $(UPerm(R) \cup UJoin(R))$-derivation starting from $f\{a, c\}\{b\}\{b\}$.

$$f\{a, c\}\{b\}\{b\} \rightarrow^{nr} f\{a, c\}\{b\}\{d\} \rightarrow^{r_2} f\{c\}\{a, c\}\{a\} \rightarrow^{r_1} f\{a, c\}\{a\}\{b\}$$
$$\rightarrow^{r_1} f\{a, c\}\{b\}\{b\} \rightarrow^{nr} \cdots$$

This is the required nonterminating derivation.

4.2 Lifting Rewrite Derivations

We next prove the converse of Lemma 4 under the assumption that $UJoin(R)$ is terminating. First we need the notions of a position being *related to* and *going to* other positions in a permutation rule. For example, in the rule $fxxy \rightarrow fxyb$, position 1 is related to position 2 in the left-hand side term and positions 1 and 2 both go to position 1 on the right-hand side term. The goes-to relation is well-defined since rules are right-linear. We naturally generalize this to $UPerm(R)$-rules below.

Definition 2. *Let $fS_1 \ldots S_m \rightarrow fT_1 \ldots T_m$ be a rule in $UPerm(R)$, and let $fs_1 \ldots s_m \rightarrow ft_1 \ldots t_m$ be the rule of R from which it is constructed. (Since a rule in $UPerm(R)$ can be constructed from different rules in R, we are assuming an implicit arbitrary selection).*

We say that $i_1, \ldots, i_k$ are positions related to *i in $fS_1 \ldots S_m \to fT_1 \ldots T_m$ if s_i is a variable and $s_{i_1} = \ldots = s_{i_k} = s_i$. We say that position i goes to* position *j in $fS_1 \ldots S_m \to fT_1 \ldots T_m$ if s_i is a variable and $t_j = s_i$. We say that i is an* original constant position *in $fS_1 \ldots S_m \to fT_1 \ldots T_m$ if s_i is a constant.*

Lemma 5. *Suppose $UJoin(R)$ is terminating. Let $s_1 \to s_2 \to \cdots$ be an infinite ($UPerm(R) \cup UJoin(R)$)-derivation, where $s_1 = fS_1 \ldots S_m$ and every $S_i \subseteq \Sigma_0$ is a set of R-joinable constants. Then, R is nonterminating.*

Proof. We associate a sequence of positions $i_1, i_2, \ldots$ with every depth 1 position i in s_1 as follows: $i_1 = i$ and for every $j \geq 1$, (a) $i_{j+1} = i_j$ if the rewrite rule used in $s_j \to s_{j+1}$ is from $UJoin(R)$, (b) i_j goes to i_{j+1} if the rewrite rule $s_j \to s_{j+1}$ is in $UPerm(R)$, and (c) i_{j+1} is undefined (and the sequence terminates) if the position i_j does not go to any position in the rule $s_j \to s_{j+1} \in UPerm(R)$. Note that this sequence is uniquely defined for every i since R is right-linear. Thus, the sequence associated with i can be either finite or infinite. It is easy to prove inductively that, if $i_1 \ldots$ is the sequence associated with $i = i_1$ in s_1, then, for all i_k in this sequence, the set $s_k|_{i_k}$ is R-joinable and $Reach(s_1|_{i_1}) \supseteq Reach(s_k|_{i_k}) \neq \emptyset$ (on terms over the original signature). We can similarly associate a sequence of positions with any depth 1 position i in any term s_j (by considering the infinite derivation $s_j \to s_{j+1} \to \cdots$).

We now define the *use* of a depth 1 position i in s_1.

$$
\begin{aligned}
use(s_1, i) &= \{c\} && \text{if } i_1 \ldots i_k \text{ is the sequence associated with } i, \ s_k|_{i_k} = \{c\}, \\
&&& \text{and } i_k \text{ is an original constant position in } s_k \to s_{k+1} \\
&= \bigcup_{j \in J} s_k|_j && \text{if } i_1 \ldots i_k \text{ is the sequence associated with } i, \text{ and } J \text{ is the} \\
&&& \text{set of all positions related to } i_k \text{ in the rule } s_k \to s_{k+1}. \\
&= \bigcup_{j \geq 1} s_j|_{i_j} && \text{if the sequence } i_1 \ldots \text{ associated with } i \text{ is infinite}
\end{aligned}
$$

From the definition of *use*, it is easy to see that $use(s_1, i)$ is R-joinable and $Reach(use(s_1, i)) \subseteq Reach(s_k|_{i_k})$ for all k. An important property of the *use* function is that if $i_1, i_2, \ldots, i_k, \ldots$ is the sequence associated with position $i = i_1$ in s_1, then $use(s_k, i_k) = use(s_1, i = i_1)$ for all such k.

We wish to map terms s_i over the extended signature to terms t_i over the original signature, and hence we need to find a concrete representation term for each $s_k|_{i_k}$. Therefore, define $Choice(\{c\}) = c$ if $c \in \Sigma_0$, and $Choice(S) = t$ if $S \neq \{c\}$ for any $c \in \Sigma_0$ and t is *any* (selected) term in $Reach(S)$. We map every term s_i of the original infinite derivation into a new term

$$s_i' = f(Choice(use(s_i, 1)), \ldots, Choice(use(s_i, m)))$$

over the original signature. Our intention is to show that we have an infinite derivation $s_1' \to_R^* s_2' \to_R^* s_3' \ldots$, proving then that R is nonterminating.

If a rewrite step $s_i \to s_{i+1}$ is done with a rule of $UJoin(R)$, then $s_i' = s_{i+1}'$ by the definition of *use*, and hence $s_i' \to_R^* s_{i+1}'$ trivially. For finishing the proof it will be enough to show that if a rewrite step $s_i \to s_{i+1}$ is done with a rule of $UPerm$, then $s_i' \to_R^+ s_{i+1}'$ (note that there are infinitely many steps of this kind in the derivation $s_1 \rightarrow s_2 \rightarrow \ldots$ since by assumption $UJoin(R)$ is terminating). The rule used in this step is precisely $s_i \rightarrow s_{i+1}$ since $UPerm$ is ground. Let $l \rightarrow r$ be the rule in R from which it is constructed. For every variable position j in l, let $j_1, \ldots, j_k$ be the positions related to j in the rule $l \rightarrow r$. By the definition of use, the sets $use(s_i, j_1), \ldots, use(s_i, j_k)$ are identical, and hence, the terms $Choice(use(s_i, j_1)), \ldots, Choice(use(s_i, j_k))$ are identical. Moreover, if the variable appears in r at position p, then all $j_1 \ldots j_k$ go to p in the rewrite rule, and hence $Choice(use(s_{i+1}, p))$ is also the same term. Therefore, rewriting s_i' with $l \rightarrow r$ produces a term, say s', that coincides with s_{i+1}' in all the depth 1 positions that are variable positions in r. For the rest of positions, s' contains constants that coincide with the corresponding singleton sets at the same positions in s_{i+1}. That is, for any other position p, $s'|_p = c$ for some $c \in \Sigma_0$. In this case, $s_{i+1}|_p = \{c\}$. But, it is the case that $c \rightarrow_R^* Choice(use(s_{i+1}, p))$, and hence, $s' \rightarrow_R^* s_{i+1}'$, which proves that $s_i' \rightarrow_R^+ s_{i+1}'$. □

Example 3. Consider the rewrite system R and the following infinite ($UPerm(R)$ $\cup UJoin(R)$)-derivation from Example 2:

$$f\{a,c\}\{b\}\{b\} \rightarrow^{nr} f\{a,c\}\{b\}\{d\} \rightarrow^{r_2} f\{c\}\{a,c\}\{a\} \rightarrow^{r_1} f\{a,c\}\{a\}\{b\}$$
$$\rightarrow^{r_1} f\{a,c\}\{b\}\{b\} \rightarrow^{nr} \cdots$$

The sequence associated with term $f\{a,c\}\{b\}\{b\}$, call it s_1, and position 1 is the infinite sequence $1, 1, 2, 1, 1, 1, 2, 1, 1, \ldots$; whereas the sequence associated with s_1 and position 2 is the finite sequence $2, 2$ and the sequence associated with s_1 and position 3 is the finite sequence $3, 3$. Therefore, $use(s_1, 1) = \{a, c\}$, $use(s_1, 2) = \{b\}$, and $use(s_1, 3) = \{d\}$. We can set the *Choice* function so that $Choice(\{a, c\}) = c$, $Choice(\{b\}) = b$, and $Choice(\{d\}) = d$. Lifting the terms using the $Choice(use(\_, \_))$ function, we get the following infinite R-derivation:

$$fcbd \rightarrow^{r_2} fcca \rightarrow^{*,nr} fccc \rightarrow^{r_1} fccb \rightarrow^{r_1} fcbb$$
$$\rightarrow^{*,nr} fcbd \rightarrow^{r_2} \cdots$$

Note that we have to apply $\rightarrow^{*,nr}$ steps (here $a \rightarrow^* c$ steps since we chose c as $Choice(\{a, c\})$) to go from an intermediate term (for example, $fcca$) to the lifting of the next term ($fccc$).

4.3   Deciding Termination

The following lemma characterizes termination of a rewrite system R that contains only right-constant, right-variable, or flat and right-linear rules.

Lemma 6. *R is terminating iff the following three conditions are satisfied:*

1. *There is no insertion rule $x \rightarrow r \in R$.*
2. *The rewrite system $UPerm(R) \cup UJoin(R)$ terminates starting from any flat term of the form $f S_1 \ldots S_n$ where every S_i contains only R-joinable constants from Σ.*
3. *It is not the case that $c \rightarrow_R^+ C[c]$ for any constant c and context $C[\_]$.*

Proof. $\Rightarrow$: Suppose R is terminating. If either of conditions (1) or (3) are violated, then the rewrite system R is clearly nonterminating. Now suppose conditions (1) and (3) are satisfied but condition (2) is violated and there is an infinite rewriting derivation $s_1 \rightarrow s_2 \rightarrow s_3 \rightarrow \cdots$ with $UPerm \cup UJoin(R)$ starting from a term of the form $s_1 = fS_1 \ldots S_m$, where every S_i is joinable and $S_i \subseteq \Sigma$. Condition (3) implies that $UJoin(R)$ is terminating. This fact together with Lemma 5 implies that R is nonterminating, a contradiction.

$\Leftarrow$: We prove by contradiction. Suppose the three conditions are satisfied but R is nonterminating. We compare nonterminating derivations by the size of their initial terms. For the case of two derivations starting from constants, we compare them by comparing the constants with the following ordering: d is smaller than c if $c \rightarrow_R^+ C[d]$ for some context $C[\_]$ (by condition (3) this is a well founded ordering). We consider a minimal nonterminating derivation:

$$s = s_1 \rightarrow s_2 \rightarrow s_3 \rightarrow \cdots$$

Consider the top rewrite steps in this derivation. If any of these top steps are collapsing, then they can be commuted with the next rewrite step and moved to the right. Repeatedly doing this would result in an infinite derivation without top collapsing steps and with the same initial term. We can, therefore, assume that there are no top collapsing steps in the above derivation. Moreover, there are no applications of rewrite rules of the form $l \rightarrow c$ at the top, since otherwise, from that point on we obtain a smaller derivation (either by the size or the constant ordering). This observation, along with condition (1), means that we can assume that all top steps in the above infinite derivation have to be applications of the permutation rules $f \ldots \rightarrow f \ldots$, with the exception that the first top step can be the application of a rule of the form $c \rightarrow f \ldots$.

There are two cases:

(a) there are finitely many top rewrite steps, or
(b) there are infinitely many top rewrite steps.

Case (a): Suppose there are no top rewrite steps applied after reaching term s_i. Clearly, there is an infinite derivation starting from some subterm $s_i|_p$ of s_i where $p \in \{1, \ldots, m\}$. Since all root rewrites are done using left-constant or permutation rules, it follows that the term $s_i|_p$ is reachable from some subterm $s_1|_{p'}$ with $p' \in \{1, \ldots, m\}$, or it is reachable from some constant c such that $s_1 \rightarrow_R^+ C[c]$. In the former case, there is an infinite derivation starting from a strictly smaller term $s_1|_{p'}$. In the latter case, there is a smaller infinite derivation starting from c (either in the size or the constant ordering).

Case (b): In this case we assume that s_1 is not a constant. Otherwise, we consider the same derivation but starting from s_2. Lemma 4 shows that there is an infinite $(UPerm(R) \cup UJoin(R))$-derivation starting from a ground term $fS_1 \ldots S_m$, where $S_i \subseteq \Sigma_0$ are R-joinable. This contradicts Condition (2). $\qquad \square$

Example 4. The rewrite system $R = \{fxxx \rightarrow fxc_1c_2, c_2 \rightarrow c_1\}$ is nonterminating because the rewrite system $\{f\{c_1\}\{c_1\}\{c_2\} \rightarrow f\{c_1, c_2\}\{c_1\}\{c_2\}$, $\{c_1, c_2\} \rightarrow$

$\{c_1\}\}$, which is contained in the union of $UJoin(R)$ and $UPerm(R)$, does not terminate starting from $f\{c_1\}\{c_1\}\{c_2\}$.

Finally, we show that the three conditions characterizing termination of R can be decided using the decidability of R-reachability and R-joinability and the decidability of termination of ground TRSs. This result subsumes our previous termination decidability result [9].

Theorem 2. *The termination property for TRS's containing only shallow right-linear rules, arbitrary collapse rules, and right-ground rules is decidable.*

Proof. Using Theorem 1, any such TRS can be transformed to a TRS R that contains only flat right-linear rules, arbitrary collapse rules, and right-constant rules, while preserving termination. Hence, decidability reduces to checking the three conditions of Lemma 6.

Condition (1) is trivially checkable. For the decidability of condition (3), we consider any constant c and distinguish two cases: checking if $c \to_R^+ c$ and checking if $c \to_R^* C[c]$ for some non-empty context $C[\_]$. For the first case, note that, since $Vars(r) \subseteq Vars(l)$ is satisfied, the number of different terms reachable from c in one rewrite step is finite; and hence, we can check if c is reachable from every one of them by the decidability of R-reachability. For the second case, note that, since R is regularity-preserving, the set of terms reachable from c is recognizable. We can now check condition (3) by checking emptiness of the intersection of this set with the set of terms in which c occurs at non-root position, which is recognizable, too.

For the decidability of condition (2), note that the rewrite systems $UJoin(R)$ and $UPerm(R)$ restricted to set-constants S s.t. $S \subseteq \Sigma_0$ can be constructed, due to the fact that R-reachability and R-joinability are decidable, and that the number of different S s.t. $S \subseteq \Sigma_0$ is finite. Now, this case reduces to checking termination of a ground TRS, which is decidable. $\square$

5   Conclusion

In this paper, we showed that termination is decidable for rewrite systems that contain right-ground, collapsing, or shallow right-linear rewrite rules. The proof is especially elegant since it is modular over the decidability results for reachability and joinability [17]. We also prove some properties about rewriting using shallow right-linear TRSs, which are used to prove the main results of this paper. Using these intuitions, we have shown elsewhere [10] that confluence is decidable for shallow and right-linear rewrite systems.

It will be interesting to explore the possibility of extending the class of rewrite systems without compromising decidability of termination. Another direction for future work would be investigating the termination of rewriting modulo certain axioms such as associativity and commutativity.

Acknowledgments. We thank the reviewers for their helpful comments.

References

[1] C. Borralleras, M. Ferreira, and A. Rubio. Complete monotonic semantic path orderings. In *Conf. on Automated Deduction, CADE*, volume 1831 of *LNAI*, pages 346–364, 2000.

[2] H. Comon, M. Dauchet, R. Gilleron, F. Jacquemard, D. Lugiez, S. Tison, and M. Tommasi. Tree automata techniques and applications. Available on: http://www.grappa.univ-lille3.fr/tata, 1997.

[3] E. Contejean, C. Marche, B. Monate, and X. Urbain. CiME. http://cime.lri.fr/.

[4] M. Dauchet. Simulation of Turing machines by a regular rewrite rule. *Theoretical Computer Science*, 103(2):409–420, 1992.

[5] N. Dershowitz. Termination of linear rewriting systems. In *Automata, Languages and Programming, 8th Colloquium, ICALP*, volume 115 of *LNCS*, pages 448–458, 1981.

[6] N. Dershowitz. Orderings for term rewriting systems. *Theoretical computer science TCS*, 17(3):279–301, 1982.

[7] A. Geser, A. Middeldorp, E. Ohlebusch, and H. Zantema. Relative undecidability in term rewriting, Part 1: The termination hierarchy. *Information and Computation*, 178(1):101–131, 2002.

[8] J. Giesl, R. Thiemann, P. Schneider-Kamp, and S. Falke. Automatic termination proofs with AProVE. In *Rewriting Techniques and Applications, RTA*, volume 3091 of *LNCS*, pages 210–220, 2004.

[9] G. Godoy and A. Tiwari. Deciding fundamental properties of right-(ground or variable) rewrite systems by rewrite closure. In *Intl. Joint Conf. on Automated Deduction, IJCAR*, volume 3097 of *LNAI*, pages 91–106, 2004.

[10] G. Godoy and A. Tiwari. Confluence of shallow right-linear rewrite systems. In *Computer Science Logic, CSL*, 2005. To appear.

[11] G. Godoy, A. Tiwari, and R. Verma. Deciding confluence of certain term rewriting systems in polynomial time. *Annals of Pure and Applied Logic*, 130(1-3):33–59, 2004.

[12] N. Hirokawa and A. Middeldorp. Tsukuba termination tool. In *Conf. on Rewriting Techniques and Applications, RTA*, volume 2706 of *LNCS*, pages 311–320, 2003.

[13] G. Huet and D. S. Lankford. *On the uniform halting problem for term rewriting systems*. INRIA, Le Chesnay, France, 1978. Technical Report 283.

[14] F. Jacquemard. Reachability and confluence are undecidable for flat term rewriting systems. *Inf. Process. Lett.*, 87(5):265–270, 2003.

[15] A. Middeldorp and B. Gramlich. Simple termination is difficult. *Applicable Algebra in Engineering, Communication, and Computing AAECC*, 6(2):115–128, 1995.

[16] D. A. Plaisted. Polynomial time termination and constraint satisfaction tests. In *Rewriting Techniques and Applications, RTA*, volume 690 of *LNCS*, pages 405–420, 1993.

[17] T. Takai, Y. Kaji, and H. Seki. Right-linear finite path overlapping term rewriting systems effectively preserve recognizability. In *Rewriting Techniques and Applications, RTA*, volume 1833 of *LNCS*, pages 246–260, 2000.

The OWL Instance Store: System Description

Sean Bechhofer, Ian Horrocks, and Daniele Turi

Information Management Group, School of Computer Science,
The University of Manchester, Manchester, UK
{lastname}@cs.manchester.ac.uk

Abstract. We describe the instance store, a system for reasoning about individuals (i.e., instances of classes) in OWL ontologies. By using a hybrid reasoner/database architecture, our system is able to perform efficient reasoning over large volumes of instance data, as required by many real world applications.

1 Introduction

Ontologies, with their intuitive taxonomic structure and class based semantics, are widely used in domains like bio- and medical-informatics, where there is a tradition in establishing taxonomies of terms. The recent W3C recommendation of *OWL* [8] as the language of choice for web ontologies also underlines the long term vision that ontologies will play a central role in the semantic web. Most importantly, as shown in [9], most of the available OWL ontologies can be captured in *OWL-DL*—a subset of OWL for which highly optimised Description Logic [5] reasoners can be used to support ontology design and deployment.

Unfortunately, existing reasoners (and tools), while successful in dealing with the (relatively small and static) *class* level information in ontologies, fail when presented with the large volumes of *instance* level data often required by realistic applications, hampering the use of reasoning over ontologies beyond the class level.

The system we present—the *instance Store* (*iS*)—addresses this problem using a hybrid database/reasoner architecture: a relational database is used to persist instances, while a class level (i.e. 'TBox') reasoner is used to infer ontological information about the classes they belong to; moreover, part of this ontological information is also persisted in the database. The *iS* only supports a very limited form of reasoning about individuals, i.e., answering instance retrieval queries w.r.t. an ontology and a set of axioms asserting class-instance relationships, and it is clear that from a theoretical point of view this could be reduced to pure TBox reasoning. The *iS* is, however, able to process *much* larger numbers of individuals than it is currently possible using standard Description Logic reasoners. Moreover, this kind of reasoning turns out to be useful in a wide range of applications, in particular those where domain models are used to structure and investigate large data sets.

2 Architecture and Interface

There is a long tradition of coupling databases to knowledge representation systems in order to perform reasoning, most notably the work in [10]. However, in our architecture

R. Nieuwenhuis (Ed.): CADE 2005, LNAI 3632, pp. 177–181, 2005.

initialise(database: Database, reasoner: OWLReasoner, ontology: OWLOntology)
addAssertion(instance: URI, class: OWLDescription)
retrieve(query: OWLDescription): Set⟨URI⟩

Fig. 1. The *iS* API

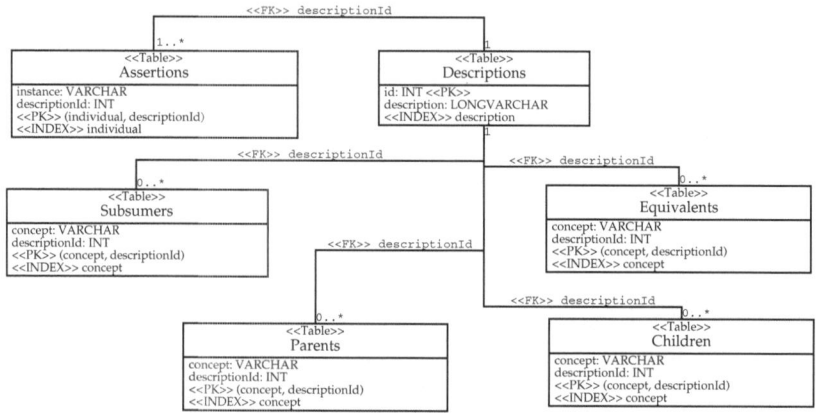

Fig. 2. Database Schema for *iS*

we do not use the standard approach of associating a table (or view) with each class and property. Instead, we have a fixed and relatively simple schema that is independent of the structure of the ontology and of the instance data. The *iS* is, therefore, agnostic about the provenance of data, and uses a new, dedicated database for each ontology (although the schema is always the same).

The basic functionality and the database schema of the *iS* system are illustrated in Figure 1 and Figure 2 respectively. At start-up, the initialise method is called w.r.t. a relational database, an OWL class reasoner, and a class level (i.e., not containing instances) OWL ontology. The method creates the schema for database if needed (ie if the *iS* is new), parses ontology and loads it into the reasoner.

To populate the *iS*, one calls the addAssertion method repeatedly. Each assertion states that instance (a URI) belongs to class, which is an arbitrary OWL description (not involving other instances). Once one has populated the *iS* with some—possibly millions of—instances, one can query it using the retrieve method. A query again consists of an arbitrary OWL class description, and the result is the set of all instances belonging to the query class.

3 Implementation

We have implemented our system in Java[1]. The communication with the reasoner is implemented using the DIG interface [7]. This allows the *iS* system to be, again, fully

---

[1] Source code, binaries, GUI, and test suites are publicly available from SourceForge [3].

agnostic of the actual reasoner used; indeed, we have used FaCT [14], Racer [12], and FaCT++ [1] in our applications, sometimes using all of them at different times for the same *iS*. As for the database system, we have used *MySQL*, *Oracle* and *Hypersonic*, accessed either through *JDBC* or through *Hibernate*

The key algorithms in the Java code itself are those for addAssertion and retrieve. Our starting point is the 'semantic indexing' of [15], taking the atomic classes in the ontology as indexing concepts. In order to improve performance we also cache additional information about descriptions: for every description D used in a class-instance assertion or query, we store D in the Descriptions table, compute (using a TBox reasoner) the location of D in the class hierarchy, and cache all named (atomic) concepts that

- subsume D (storing them in the Subsumers table);
- are equivalent to D (storing them in the Equivalents table);
- are parents (direct subsumers) of D (storing them in the Parents table); or
- are children (direct subsumees) of D (storing them in the Children table).

Caching this information avoids the need to traverse the class hierarchy (and issue many DB queries) when answering instance retrieval queries. With this data in place, the speed of retrieval for a query Q depends on whether:

1. Q is referenced in Equivalents ($\Rightarrow$ virtually immediate answer);
2. Q subsumes the conjunction of its parents[2] ($\Rightarrow$ fast answer);
3. there is a set I of individuals, each of which is an instance of *all* of the parents of Q and not an instance of *any* child of Q ($\Rightarrow$ speed of answer depends on size of I).

Note that almost all the reasoning needed in retrieval is performed by means of (single) SQL queries, with the exception of the last case where the reasoner is needed for as many subsumption tests as the size of I. For more details visit the *iS* website [4].

Clearly, the performance gains obtained by caching classified descriptions come at the expenses of maintenance: changes at the class level of the ontology require costly updates to the *iS*. The whole system, however, is geared towards scalability and fast retrieval times, and the applications below demonstrate that this is useful in realistic scenarios.

4   Applications and Performance

The first application we describe illustrates the performance of the *iS* w.r.t. a real world problem with more that half a million instances. The Gene Ontology (*GO*) consortium publishes every month a database [2] of gene products referring to terms in a large (tens of thousands of classes) ontology. The structural simplicity of the ontology (little more than a taxonomy of classes) means that its transitive closure can be precomputed and stored in the database so that, when a client searches for the gene products whose descriptions are subsumed by a set of terms, the answer can be returned without using

---

[2] Every concept is always subsumed by the conjunction of its parents, hence this effectively checks whether Q is equivalent to the conjunction of its parents.

any reasoner. Together with other functionality provided by the database, this provides biologists with a service which is highly valued and widely used.

To test the *iS*, we mined (the SWISS-PROT fragment of) the Gene Ontology database extracting 653,762 gene product descriptions which we loaded in the *iS* using the addAssertion method (in 23,750 seconds using FaCT++). In our mining we exploited the fact that gene terms form three more or less separate taxonomies of 'processes', 'components' and 'functions'. We therefore added three corresponding new properties (also known as *roles*) to the gene ontology and described gene products using them. For instance, we asserted that *1433 CANAL* is an instance of the class of gene products that **take part in** *intracellular signalling cascade*, are **part of** *chloroplast*, and have the **function** of *protein domain specific binding activity*. (We denote roles in bold and classes in italics.)

This does not take into account annotations and other information present in the GO database, but our aim was simply to test a large set of realistic and interesting data. Extensions in the structure of the ontology (as envisaged in GONG [17]) would allow more complex assertions to be made and more complex queries to be asked .

Our results are very encouraging. We have tested our GO *iS* against various queries formulated by domain experts. Their descriptions are similar in structure to the description of the assertion for the above gene product *1433 CANAL*, i.e. a conjunction of processes, components and functions (each conjunct possibly empty), and the retrieval times range between less than a second and few seconds depending on the factors discussed in Section 3. The queries cover all three cases mentioned in the previous section, thus including run-time calls to the reasoner for subsumption checks.

More bioinformatics applications of the *iS* include its use to guide gene annotation [6] and, more recently, to investigate the structure of data mined from the InterPro database of protein families [16].

We also built another example [11] of *iS* within the proof-of-concept project *MONET*, where mathematical web-services are envisaged to register to a broker using the *iS* to perform service matching. A typical service description specifies the 'GAMS' classification of the service, the problem it solves, input and output formats, the directives it accepts, the software used to implement it, and the algorithm it implements. All this involves several classes and roles in nested conjunctions from an ontology containing thousands of classes interconnected by means of tens of roles. The structural richness of the ontology means that services can then be matched using, e.g., a bibliographic reference to their implemented algorithm. The MONET *iS* contains too few instances for its performance to be significant, however it illustrates the expressivity of our approach.

5   Conclusions and Future Work

The architectural choices made in the implementation of *iS* ensure that we use appropriate technologies for appropriate tasks. It is clear that at some point the reasoner must be used in order to retrieve individuals, but in our approach it is only used when necessary. Databases are well suited to handling large volumes of data and are optimised for the performance of operations such as joins and intersections.

The functionality of the *iS* is limited, but is sufficient to support several interesting applications, and allows us to deal with volumes of instance data that cannot, to the best of our knowledge, be handled by any other reasoner.

In the present *iS*, roles are allowed to appear at class level as in the GO role **take part in**, but no role assertion between instances is allowed, i.e., we cannot assert that instance *x* is related via role **r** to instance *y*. We are currently working on an extension of the *iS* that uses the *precompletion* technique [13] to overcome this limitation (although at the cost of some restrictions on the structure of the ontology).

References

1. FaCT++. http://owl.man.ac.uk/factplusplus.
2. Gene Ontology Database. http://www.godatabase.org/dev/database.
3. Instance Store API. http://sourceforge.net/projects/instancestore.
4. Instance Store website. http://instancestore.man.ac.uk.
5. Franz Baader, Diego Calvanese, Deborah McGuinness, Daniele Nardi, and Peter Patel-Schneider, editors. *The Description Logic Handbook — Theory, Implementation and Applications*. Cambridge University Press, 2003.
6. M. Bada, D. Turi, R. McEntire, and R. Stevens. Using Reasoning to Guide Annotation with Gene Ontology Terms in GOAT. *SIGMOD Record (special issue on data engineering for the life sciences)*, June 2004.
7. Sean Bechhofer. The DIG description logic interface: DIG/1.1. In *Proceedings of the 2003 Description Logic Workshop (DL 2003)*, 2003.
8. Sean Bechhofer, Frank van Harmelen, Jim Hendler, Ian Horrocks, Deborah L. McGuinness, Peter F. Patel-Schneider, and Lynn Andrea Stein. OWL Web Ontology Language Reference. Technical Report REC-owl-ref-20040210, The Worldwide Web Consortium, February 2004.
9. Sean Bechhofer and Raphael Volz. Patching Syntax in OWL Ontologies. In *Proceedings of 3rd International Semantic Web Conference (ISWC'04)*, Hiroshima, Japan, 2004.
10. Alexander Borgida and Ronald J. Brachman. Loading data into description reasoners. In *Procs ACM SIGMOD Int'l Conf. on Management of Data*, pages 217–226, 1993.
11. Olga Caprotti, Mike Dewar, and Daniele Turi. Mathematical service matching using Description Logic and OWL. In *Proceedings of 3rd International Conference on Mathematical Knowledge Management (MKM'04)*, volume 3119 of *LNCS*. Springer-Verlag, 2004.
12. V. Haarslev and R. Moller. Description of the RACER system and its applications. In R. Gore, A. Leitsch, and T. Nipkow, editors, *Procs of IJCAR 2001*, volume 2083 of *Lecture Notes in Artificial Intelligence*. Springer-Verlag Inc., 2001.
13. Bernhard Hollunder. Consistency checking reduced to satisfiability of concepts in terminological systems. 18(2–4):133–157, 1996.
14. I. Horrocks. Using an expressive description logic: FaCT or fiction? In *Procs 6th Int'l Conf. on Principles of Knowledge Representation and Reasoning (KR'98)*, pages 636–647, 1998.
15. A. Schmiedel. Semantic indexing based on description logics. In F. Baader, M. Buchheit, M.A. Jeusfeld, and W. Nutt, editors, *Reasoning about structured objects: knowledge representation meets databases. Proceedings of the KI'94 Workshop KRDB'94*, September 1994.
16. K. Wolstencroft, P. Lord, L. Tabernero, A. Brass, and R. Stevens. Intelligent classification of proteins using an ontology. Submitted for publication, 2005.
17. Chris J. Wroe, Robert D. Stevens, Carole A. Goble, and Michael Ashburner. A methodology to migrate the gene ontology to a description logic environment using daml+oil. In *Proceedings of the 8th Pacific Symposium on Biocomputing (PSB)*, Hawaii, January 2003.

Temporal Logics over Transitive States

Boris Konev[1], Frank Wolter[1], and Michael Zakharyaschev[2]

[1] Department of Computer Science,
University of Liverpool,
Liverpool L69 7ZF, U.K.
{b.konev, frank}@csc.liv.ac.uk
[2] Department of Computer Science,
King's College London,
Strand, London WC2R 2LS, U.K.
mz@dcs.kcl.ac.uk

Abstract. We investigate the computational behaviour of 'two-dimensional' propositional temporal logics over $(\mathbb{N}, <)$ (with and without the next-time operator $\bigcirc$) that are capable of reasoning about states with transitive relations. Such logics are known to be undecidable (even Π_1^1-complete) if the domains of states with those relations are assumed to be constant. Motivated by applications in the areas of temporal description logic and specification & verification of hybrid systems, in this paper we analyse the computational impact of allowing the domains of states to expand. We show that over finite expanding domains (with an arbitrary, tree-like, quasi-order, or linear transitive relation) the logics are recursively enumerable, but undecidable. If these finite domains eventually become constant then the resulting $\bigcirc$-free logics are decidable (but not in primitive recursive time); on the other hand, when equipped with $\bigcirc$ they are not even recursively enumerable. Finally, we show that temporal logics over infinite expanding domains as above are undecidable even for the language with the sole temporal operator 'eventually.' The proofs are based on Kruskal's tree theorem and reductions of reachability problems for lossy channel systems.

1 Introduction

Temporal logics are used in computer science and artificial intelligence to model states (of soft or hardware, data or knowledge bases, spatial regions, multi-agent systems, etc.) changing over time. (For uniformity, we can think of such states as first- or higher-order structures of some fixed signature.) Perhaps the best known example is LTL, the propositional linear temporal logic of infinite sequences $\sigma_0 \sigma_1 \ldots$ of states, equipped with temporal operators like $\square_F$ 'always in the future' or $\bigcirc$ 'at the next state.' LTL is decidable in PSPACE [28], reasoning with this logic can be mechanised using tableaux [30] or resolution [6], with the existing provers performing reasonably well [15,27].

However, being *propositional*, LTL is only capable of reasoning about states of some *fixed finite size* which must be known in advance. This restriction seriously

R. Nieuwenhuis (Ed.): CADE 2005, LNAI 3632, pp. 182–203, 2005.

limits the scope of applications of LTL in areas where *infinite* or *arbitrarily finite* states are required. Typical examples of such applications are:

- verification and specification of 'infinite state systems' such as real-time systems, hybrid (dynamical) systems, broadcast protocols, and channel systems;
- spatio-temporal representation and reasoning in artificial intelligence (where the states modelling space are usually either unbounded or infinite);
- temporal data or knowledge bases, e.g., 'dynamic' ontologies or temporal entity relationship models (where states are finite, but one cannot impose *a priory* any upper bound on their size);
- distributed multi-agent systems.

The obvious idea to cope with unbounded states by means of 'upgrading' propositional temporal logic to first-order one is extremely expensive: first-order LTL is not recursively enumerable (in fact, it is Π_1^1-hard; see, e.g., [7,8]), and so we cannot even have a semi-decision procedure.

The attempts to 'tame' first-order temporal logic in the fields of temporal data and knowledge bases, multi-agent systems and spatio-temporal representation and reasoning have led to *semi-decidable* and *decidable* fragments that can be obtained by imposing certain independence and locality restrictions.

The monodic fragment of first-order LTL allows applications of temporal operators to formulas with at most one free variable [14]. Thus, in the framework of this fragment we can only control the temporal change of properties—i.e., *unary predicates*—of states, while binary, ternary, etc. relations can change *arbitrarily*. The full monodic fragment turns out to be semi-decidable [32], and if we restrict the first-order part to a decidable fragment (for example, to the two-variable or guarded fragments), then the resulting monodic fragment is usually decidable as well. The simplest interesting fragment of this sort is the one-variable first-order LTL (Sistla and German [29] considered it in the context of verification). Various spatio-temporal logics based on spatial formalisms like $\mathcal{RCC}$-8, $\mathcal{BRCC}$-8, etc. can be encoded in the one-variable first-order temporal logic [8,9] and therefore inherit its good computational properties. Monodic fragments of this kind are usually decidable in *elementary* time [8], and both tableau- and resolution-based provers have been developed and implemented for monodic temporal logics [19,17,16].

The idea of monodicity is based on two conditions: the 'positive'

Mono$^+$: temporal constraints can be imposed on unary predicates

and the 'negative'

Dya$^-$: no temporal constraint can be imposed on n-nary predicates for $n \geq 2$.

Having in mind possible applications of temporal logic mentioned above, condition **Dya$^-$** appears too restrictive. In fact, in temporal knowledge bases (say, temporal description logics), more sophisticated spatio-temporal formalisms, in particular, dynamic topological logics (designed for reasoning about safety and liveness properties of hybrid systems), or infinite state systems we *do need* to control *binary relations*, for instance, to ensure that some of them do not change in time or can only expand.

Thus, we are facing the problem of weakening **Dya**⁻ without compromising too much the good computational behaviour of the monodic fragment.

Some limits for such a weakening are well-known. For example, one cannot simply replace **Dya**⁻ with **Dya**⁺, for $n = 2$, because even the monadic two-variable fragment of first-order temporal logic with *one constant binary relation* is not recursively enumerable. So it seems that without imposing extra constraints on the language no weakening of **Dya**⁻ can result in new and interesting decidable temporal logics.

Mono⁺ **and locality.** The strongest existing *decidable* temporal logics that are capable of controlling binary relations of unbounded states replace **Dya**⁻ with some *locality conditions* which can be characterised as follows:

(1) over time, binary relations can be constant or expanding, or can change arbitrarily,

(2) within states, these binary relations can satisfy some *local* constraints like reflexivity, symmetry, the triangle inequality (for metric), but not transitivity,

(3) the language referring to binary relations is *local* in the sense that we are only allowed to quantify along these relations like in modal or description logic.

Basically, conditions (2)–(3) mean that every satisfiable formula φ of our language can be satisfied in a model where the length of any strict path of the form $x_0 R x_1 \ldots R x_n$ is bounded by an elementary function depending on φ.

The resulting formalisms can be regarded as extensions of propositional LTL with propositional modal-like (or description logic) operators over states. Typical examples are temporal description logics [25,31] (where the states are described by the standard $\mathcal{ALC}$ and its extensions), temporal epistemic logics [12,8] (with state languages like $\mathbf{S5}_m$), and a number of temporal metric logics [18]. The satisfiability problem for such logics is often *non-elementary* (but primitive recursive).

It is worth noting that by adding to the language a *non-local* state operator (such as the universal modality) we immediately obtain an undecidable logic.

Transitive states. The most important example of a non-local constraint on binary relations is *transitivity* which occurs naturally in almost all the examples mentioned above: words in the channel of a channel system are linearly ordered (and therefore based on a transitive structure), expressive description logics allow transitive relations to model, e.g., the part-of relation, quasi-ordered structures representing topological spaces are transitive, common knowledge operators in epistemic logics are interpreted by transitive relations.

Unfortunately, *even a single transitive relation* which does not change over time (and interprets the 'modal' operator from (3) in the standard way) leads to an undecidable (even Π_1^1-complete) temporal logic [8,11]. This also holds true if we impose on the transitive relation some extra conditions like linearity, reflexivity, being a tree, etc. Undecidability strikes even for the language with sole temporal operator $\Diamond_F$ (without next-time or until) and even if we are only interested in safety properties (that is, interpret the language not over $\mathbb{N}$ but over arbitrary finite initial segments of $\mathbb{N}$).

The proofs of these 'negative' results heavily use the *constant domain assumption* according to which the domains of all states coincide—and, therefore, if uRv holds in some state (where R is the transitive relation and u, v are some state elements) then uRv must hold in *all* states. Being quite natural in many cases, this constant domain assumption may become inadequate for some other applications. For example, in temporal data and knowledge bases new objects may be created, which gives us states with *expanding domains*: that uRv holds in some state σ only means that it also holds in all *subsequent* states, while before σ elements u and v may not exist. Similarly, topological dynamic systems with continuous functions give rise to states with expanding domains [21,18]. And constructive logics like first-order intuitionistic logic can only be interpreted in models with expanding domains.

It is known that logics with expanding domains are reducible to logics with constant domains; see, e.g., [8]. A major open problem was whether the former can actually have better computational properties than the latter. A partial affirmative answer (for logics with finite flows of time) was obtained in [10]. Here we investigate this problem in full generality.

We show that over finite expanding domains (with an arbitrary, tree-like, quasi-order, or linear transitive relation) the logics are recursively enumerable, but undecidable. If these finite domains eventually become constant then the resulting $\bigcirc$-free logics are decidable (but not in primitive recursive time); on the other hand, when equipped with $\bigcirc$ they are not even recursively enumerable. (Decidability can also be recovered for full **LTL** if we consider only safety properties, that is, models with finite flows of time [10].) Finally, we show that temporal logics over infinite expanding domains as above are undecidable even for the language with the sole temporal operator 'eventually.' The proofs are based on Kruskal's tree theorem and reductions of reachability problems for lossy channel systems.

2   Temporal Models with Expanding Domains

We begin by introducing the intended semantics for our temporal language discussed above. The only *flow of time* we deal with in this paper is $(\mathbb{N}, <)$. *States* are first-order structures with one transitive binary relation and countably many unary predicates. More precisely, let $\mathfrak{S}$ be a function which associates with every $x \in \mathbb{N}$ a structure

$$\mathfrak{S}(x) = (W_x, R_x, P_x^1, P_x^2, \dots) \tag{1}$$

where W_x is a nonempty set, $R_x \subseteq W_x \times W_x$, and $P_x^i \subseteq W_x$ for all i. We will call $\mathfrak{S}$ a *temporal model with expanding domains*, or an *e-model*, for short, if it satisfies the following conditions: whenever $x < y$ then

- $W_x \subseteq W_y$ and
- for all $u, v \in W_x$, we have uR_xv iff uR_yv.

(see Fig. 1).

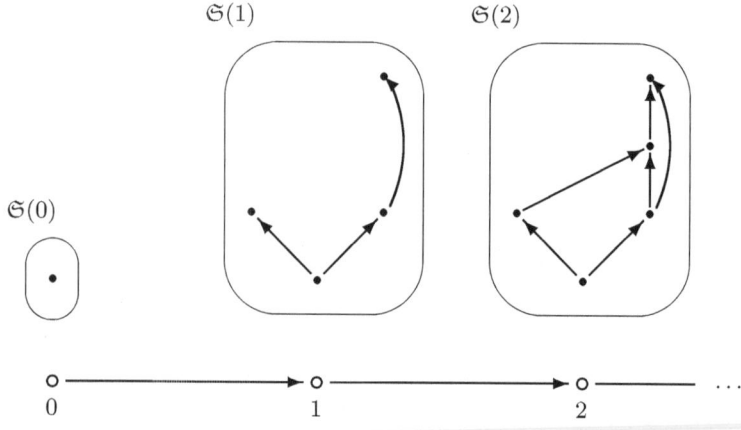

Fig. 1. An e-model $\mathfrak{S}$

We consider two propositional languages $\mathcal{TL}$ and $\mathcal{TL}_\bigcirc$ for speaking about e-models. The former contains the *temporal operator* $\Diamond_F$ (and its dual $\Box_F$), the modal diamond $\Diamond$ (and its dual box $\Box$) interpreted over the binary relations in the states, as well as *state variables* (unary predicates) $p_1, p_2, \ldots$ and the Booleans. $\mathcal{TL}_\bigcirc$ extends this language with the *next-time* operator $\bigcirc$. Thus, the formulas φ of $\mathcal{TL}_\bigcirc$ can be defined by taking

$$\varphi ::= p_i \mid \neg\varphi \mid \varphi_1 \wedge \varphi_2 \mid \Diamond\varphi \mid \Diamond_F\varphi \mid \bigcirc\varphi$$

To simplify inductive proofs, we do not include in the language $\mathcal{TL}_\bigcirc$ the 'until' operator $\mathcal{U}$. As there is a satisfiability preserving reduction of formulas with $\mathcal{U}$ to $\mathcal{TL}_\bigcirc$-formulas (see, e.g., [5] or Section 7 below), all our results for $\mathcal{TL}_\bigcirc$ hold for the language with $\mathcal{U}$ as well.

Given an e-model $\mathfrak{S}$ of the form (1), $x \in \mathbb{N}$ and $u \in W_x$, we define the *truth relation* $\mathfrak{S}, (x, u) \models \varphi$ (or simply $(x, u) \models \varphi$, if understood) inductively as follows:

- $(x, u) \models p_i$ iff $u \in P_x^i$,
- $(x, u) \models \Diamond\psi$ iff there exists $v \in W_x$ such that uR_xv and $(x, v) \models \psi$,
- $(x, u) \models \Diamond_F\psi$ iff there exists $y \in \mathbb{N}$ such that $x < y$ and $(y, u) \models \psi$,
- $(x, u) \models \bigcirc\psi$ iff $(x + 1, u) \models \psi$,

plus the standard clauses for the Boolean connectives.

We say that φ is *satisfied* in $\mathfrak{S}$ if $(x, u) \models \varphi$ for some $x \in \mathbb{N}$ and $u \in W_x$; φ is *valid* in $\mathfrak{S}$ ($\mathfrak{S} \models \varphi$, in symbols) if $(x, u) \models \varphi$ holds for every pair (x, u) with $u \in W_x$. If all formulas from a set $\Sigma \subseteq \mathcal{L}_\circ$ are valid in $\mathfrak{S}$ then we write $\mathfrak{S} \models \Sigma$.

In this paper, we consider the following classes of e-models $\mathfrak{S}$ of the form (1):

- $\mathcal{A}$, the class of all e-models,
- $\mathcal{QO}$, the class of e-models with quasi-ordered states, that is, each R_x is transitive and reflexive,
- $\mathcal{T}$, the class of e-models $\mathfrak{S}$ where each (W_x, R_x) is a tree,
- $\mathcal{L}$, the class of e-models $\mathfrak{S}$ where each (W_x, R_x) is a strict linear order.
- the subclasses $\mathcal{A}_{\text{fin}}, \mathcal{QO}_{\text{fin}}, \mathcal{T}_{\text{fin}}, \mathcal{L}_{\text{fin}}$ of the classes above that only have *finite* states,
- the subclasses $\mathcal{C}^c$ of the above classes $\mathcal{C}$ containing only models with *eventually constant* domains in the sense that there exists $n \in \mathbb{N}$ such that $(W_x, R_x) = (W_n, R_n)$ for all $x \geq n$.

Let $\mathcal{C}$ be one of the classes of e-models defined above. Our goal is to investigate the computational properties of the logics

$$\mathsf{Log}\mathcal{C} \quad = \quad \{\varphi \in \mathcal{TL} \mid \forall \mathfrak{S} \in \mathcal{C} \ \mathfrak{S} \models \varphi\}$$

and

$$\mathsf{Log}_\circ\mathcal{C} \quad = \quad \{\varphi \in \mathcal{TL}_\circ \mid \forall \mathfrak{S} \in \mathcal{C} \ \mathfrak{S} \models \varphi\}.$$

Our starting point is the results from [11,8] according to which the corresponding logics under the *constant domain assumption* are not recursively enumerable, with some of them being actually Π_1^1-complete. By allowing models with expanding domains, we hope to obtain more positive results.

Our hopes are not groundless. We will use Kruskal's tree theorem to prove the following:

Theorem 1. *Let* $\mathcal{C} \in \{\mathcal{A}_{\text{fin}}, \mathcal{QO}_{\text{fin}}, \mathcal{T}_{\text{fin}}, \mathcal{L}_{\text{fin}}\}$. *Then* $\mathsf{Log}_\circ\mathcal{C}$ *(and therefore* $\mathsf{Log}\mathcal{C}$*) is recursively enumerable.*

It remains an open problem whether the same can be proved for the corresponding classes of models with not necessarily finite states. However, none of these logics is decidable:

Theorem 2. *Let* $\mathcal{C} \in \{\mathcal{A}, \mathcal{QO}, \mathcal{T}, \mathcal{L}, \mathcal{A}_{\text{fin}}, \mathcal{QO}_{\text{fin}}, \mathcal{T}_{\text{fin}}, \mathcal{L}_{\text{fin}}\}$. *Then* $\mathsf{Log}\mathcal{C}$ *(and therefore* $\mathsf{Log}_\circ\mathcal{C}$*) is undecidable.*

This result is proved by encoding the undecidable ω-reachability problem for lossy channel systems (see below for definitions).

Consider now the impact of the assumption that eventually the states are constant. In this case we reveal a crucial difference between full **LTL** and **LTL** with sole temporal operator $\Diamond_F$:

Theorem 3. *Let* $\mathcal{C} \in \{\mathcal{A}^{\mathsf{c}}_{\mathsf{fin}}, \mathcal{QO}^{\mathsf{c}}_{\mathsf{fin}}, \mathcal{T}^{\mathsf{c}}_{\mathsf{fin}}, \mathcal{L}^{\mathsf{c}}_{\mathsf{fin}}\}$. *Then*

(i) $\mathsf{Log}\mathcal{C}$ *is decidable (but not in primitive recursive time), while*

(ii) $\mathsf{Log}_{\circ}\mathcal{C}$ *is not recursively enumerable.*

The proofs are based on Kruskal's tree theorem, a reduction of the non-primitive recursive reachability problem for lossy channel systems, and a reduction of the undecidable Post correspondence problem (PCP).

Below we only present the proofs of these theorems for the class $\mathcal{L}_{\mathsf{fin}}$ of finite strict linear orders. It is not completely trivial to extend these proofs to arbitrary transitive structures or quasi-orders. For instance, to deal with branching and/or reflexive states, constructions from [10] should be combined with the techniques introduced in the present paper. Also, the applications of Higman's lemma [13] we use here have to be replaced by the corresponding applications of Kruskal's tree theorem [22]. To prove Theorem 3 (ii) in full generality, the undecidable 'master problem' used in this paper (reachability for non-lossy channel systems) should be replaced by PCP as in [8,18].

3   Recursive Enumerability

In this section we prove the following:

Theorem 4. $\mathsf{Log}_{\circ}\mathcal{L}_{\mathsf{fin}}$ *is recursively enumerable.*

Given a $\mathcal{IL}_{\circ}$-formula φ, let $sub\,\varphi$ be the set of all subformulas of φ and their negations. Denote by $\boldsymbol{T}_{\varphi}$ the set of *Boolean types* $\boldsymbol{t}$ over $sub\,\varphi$, where

- $\neg\psi \in \boldsymbol{t}$ iff $\psi \notin \boldsymbol{t}$, for every $\neg\psi \in sub\,\varphi$, and
- $\chi \wedge \psi \in \boldsymbol{t}$ iff $\chi \in \boldsymbol{t}$ and $\psi \in \boldsymbol{t}$, for every $\chi \wedge \psi \in sub\,\varphi$.

A $\boldsymbol{T}_{\varphi}$-*word* $\mathfrak{T} = \langle T, <, l \rangle$ is a finite strict linear order $\langle T, < \rangle$ with a *labelling function* l which assigns to every $u \in T$ a type $l(u) \in \boldsymbol{T}_{\varphi}$. A $\boldsymbol{T}_{\varphi}$-word $\mathfrak{T} = \langle T, <, l \rangle$ is said to be *coherent* if, for every $\Diamond\psi \in sub\,\varphi$ and every $u \in T$, we have $\Diamond\psi \in l(u)$ iff there exists a $v \in T$ such that $u < v$ and $\psi \in l(v)$.

Consider a function $\mathfrak{f}$ associating with every natural number x a coherent $\boldsymbol{T}_{\varphi}$-word $\mathfrak{f}(x) = \langle T_x, <_x, l_x \rangle$. A *run* r through $\mathfrak{f}$ is a function with the domain

$$\mathrm{dom}\, r = \{k \in \mathbb{N} \mid k \geq m\},$$

for some $m \in \mathbb{N}$, such that $r(x) \in T_x$ for all $x \in \mathrm{dom}\, r$ and

- for every $x \in \mathrm{dom}\, r$ and every $\Diamond_F\psi \in sub\,\varphi$, we have $\Diamond_F\psi \in r(x)$ iff there exists $y > x$ such that $\psi \in r(y)$;
- for all $x \in \mathrm{dom}\, r$ and all $\bigcirc\psi \in sub\,\varphi$, we have $\bigcirc\psi \in r(x)$ iff $\psi \in r(x+1)$.

If n is the minimal number of $\mathrm{dom}\, r$ then we say that r *starts at* n.

For a set $\mathfrak{R}$ of runs through $\mathfrak{f}$, we say that the pair $(\mathfrak{f}, \mathfrak{R})$ is a *quasimodel for* φ if the following conditions are satisfied:

(q0) $\varphi \in l_0(w)$ for the minimal w in T_0,

(q1) for all $x \in \mathbb{N}$ and $w \in T_x$, there is a unique run $r \in \mathfrak{R}$ such that $r(x) = w$.

(q2) for all $r, r' \in \mathfrak{R}$ and all $x, y \in \mathrm{dom}\, r \cap \mathrm{dom}\, r'$, $r(x) <_x r'(x)$ iff $r(y) <_y r'(y)$.

Lemma 1. *A $\mathcal{TL}_{\bigcirc}$-formula φ is satisfiable in an e-model from $\mathcal{L}_{\mathsf{fin}}$ iff there exists a quasimodel for φ.*

Proof. We only show the implication ($\Leftarrow$) and leave the (basically trivial) other direction to the reader.

Given a quasimodel $(\mathfrak{f}, \mathfrak{R})$ for φ, define

$$\mathfrak{S}(x) = (W_x, R_x, P_x^1, P_x^2, \dots)$$

by taking, for $x \in \mathbb{N}$,

- $W_x = \{r \in \mathfrak{R} \mid x \in \operatorname{dom} r\}$,
- $rR_x r'$ iff $r(x) <_x r'(x)$, whenever $x \in \operatorname{dom} r \cap \operatorname{dom} r'$,
- $P_x^i = \{r \in \mathfrak{R} \mid x \in \operatorname{dom} r, p_i \in l_x(r(x))\}$.

Clearly, $\mathfrak{S}$ is an e-model from $\mathcal{L}_{\mathsf{fin}}$. By a straightforward induction on the construction of $\psi \in sub\,\varphi$ one can show that $(x, r) \models \psi$ iff $\psi \in r(x)$. The claim of the lemma follows now from **(q0)**.

Of course, the unsurprising Lemma 1 simply reformulates the notion of satisfiability in $\mathcal{L}_{\mathsf{fin}}$ into the language of quasimodels. However, this language will be convenient for showing that actually we can effectively enumerate those formulas that do not have quasimodels.

Suppose we are given a quasimodel $(\mathfrak{f}, \mathfrak{R})$ for φ as above. Formulas of the form $\Diamond_F \psi$ that occur in some $l_x(w)$, $x \in \mathbb{N}$, will be called *eventualities in* $\mathfrak{f}(x)$. We say that an eventuality $\Diamond_F \psi \in l_x(w)$ is *realised at* $y > x$ if y is the minimal number such that $\psi \in l_y(r(y))$, where r is that unique run in $\mathfrak{R}$ for which $r(x) = w$. An eventuality is *realised until* z (or *in the interval* (n, m)) if it is realised at some $y < z$ (at some $y \in (n, m)$, respectively).

We say that $\mathfrak{f}(y) = \langle T_y, <_y, l_y \rangle$ is *embeddable into* $\mathfrak{f}(z) = \langle T_z, <_z, l_z \rangle$, where $y < z$, if there exists an injective map $g : T_y \to T_z$ such that, for all $u, v \in T_y$,

- $u <_y v$ iff $g(u) <_z g(v)$,
- $l_z(g(u)) = l_y(u)$.

If $x < y < z$ and $\mathfrak{f}(y)$ is embeddable into $\mathfrak{f}(z)$ by a map g respecting the runs through $\mathfrak{f}(x)$ in the sense that $g(r(y)) = r(z)$ whenever $x \in \operatorname{dom} r$ then we say that $\mathfrak{f}(y)$ is x-*embeddable into* $\mathfrak{f}(z)$

Let $\ell(\varphi)$ be the *length* of φ, say, $\ell(\varphi) = |sub\,\varphi|$ and let $s(n, \varphi) = (\ell(\varphi) + 1)^{n+1}$.

Lemma 2. *A $\mathcal{TL}_{\bigcirc}$-formula φ is satisfiable in an e-model from $\mathcal{L}_{\mathsf{fin}}$ iff there is a quasimodel $(\mathfrak{f}, \mathfrak{R})$ for φ such that*

(A) *$|T_n| \leq s(n, \varphi)$, where $\mathfrak{f}(n) = \langle T_n, <_n, l_n \rangle$, $n \in \mathbb{N}$, and*

(B) *for the sequence $0 = k_0 < k_1 < k_2 < \dots$ of minimal numbers such that all eventualities in $\mathfrak{f}(k_i)$ are realised until k_{i+1}, if $k_i < n < m < k_{i+1}$ and $\mathfrak{f}(n)$ is k_i-embeddable into $\mathfrak{f}(m)$, then some eventuality from $\mathfrak{f}(k_i)$ is realised in the interval (n, m).*

Proof. Suppose that a quasimodel $(\mathfrak{h}, \mathfrak{Q})$ for φ is given, with $\mathfrak{h}(n) = \langle T_n, <_n, l_n \rangle$, $n \in \mathbb{N}$. Define two operations shrink and delete on $(\mathfrak{h}, \mathfrak{Q})$.

Shrink makes T_n of size $\leq s(n, \varphi)$ provided that $|T_{n-1}| \leq s(n-1, \varphi)$ or $n = 0$. If $n = 0$, then set $T = \{w\}$, where w is minimal in $\langle T_0, <_0 \rangle$. If $n > 0$, then let $T \subseteq T_n$ be the set of points $w \in T_n$ such that there exists a run $r \in \mathfrak{Q}$ with $r(n) = w$ and $n - 1 \in \operatorname{dom} r$.

Define $T'_n \subseteq T_n$ by adding to T the set of all $<_n$-maximal points $u \in T_n$ such that there is some $w \in T$, $w <_n u$, with $\Diamond \psi \in l_n(w)$ and $\psi \in l(u)$. It should be clear that the size of T'_n is as required. Denote by $<'_n$ and l'_n the restrictions of $<_n$ and l_n to T'_n, respectively. Clearly, $\langle T'_n, <'_n, l'_n \rangle$ is coherent. Now define $\mathfrak{h}'$ by taking, for $m \in \mathbb{N}$,

$$\mathfrak{h}'(m) = \begin{cases} \mathfrak{h}(m) & \text{if } m \neq n, \\ \langle T'_n, <'_n, l'_n \rangle & \text{if } m = n. \end{cases}$$

Finally, define a set $\mathfrak{Q}'$ of runs as follows: we put r to $\mathfrak{Q}'$ if $r \in \mathfrak{R}$ and $n \notin \operatorname{dom} r$, or if $n \in \operatorname{dom} r$ and $r(n) \in T'_n$; and if $n \in \operatorname{dom} r$ but $r(n) \notin T'_n$ then we put to $\mathfrak{Q}'$ the restriction of r to $\{n+1, \dots\}$. It is easy to see that $(\mathfrak{h}', \mathfrak{Q}')$ is a quasimodel for φ.

Delete removes a part of the quasimodel between $\mathfrak{h}(n)$ and $\mathfrak{h}(m)$, $n < m$, if the former is embeddable in the latter. More precisely, let $x < n < m$ and $\mathfrak{h}(n)$ is x-embeddable in $\mathfrak{h}(m)$ by some injection g. Construct a new quasimodel $(\mathfrak{h}', \mathfrak{Q}')$ as follows. First we set

$$\mathfrak{h}'(k) = \begin{cases} \mathfrak{h}(k) & \text{if } k < n, \\ \mathfrak{h}(k + m - n) & \text{if } k \geq n \end{cases}$$

(that is we 'cut off' the words $\mathfrak{h}(n), \dots, \mathfrak{h}(m-1)$ from the original quasimodel). And then we construct $\mathfrak{Q}'$ by putting into it runs r' defined by taking

- if $r \in \mathfrak{R}$ starts at $k \in [n, m]$ then r' starts at n and $r'(n + y) = r(m + y)$, $y \geq 0$;
- if $r \in \mathfrak{R}$ starts at $k > m$ then r' starts at $n + k - m$ and $r'(n + k - m + y) = r(k + y)$, $y \geq 0$;
- if $r \in \mathfrak{R}$ starts at $k < n$ then there is $r_1 \in \mathfrak{R}$ such that $g(r(n)) = r_1(m)$, and we set

$$r'(k) = \begin{cases} r(k) & \text{if } k < n, \\ r_1(k + m - n) & \text{if } k \geq n. \end{cases}$$

It is not hard to check that $(\mathfrak{h}', \mathfrak{Q}')$ is still a quasimodel for φ.

Using these two operations we can transform any given quasimodel $(\mathfrak{h}, \mathfrak{Q})$ for φ into a quasimodel $(\mathfrak{f}, \mathfrak{R})$ for φ satisfying the conditions of the lemma. We begin by shrinking $\mathfrak{h}(0)$ and finding the minimal k_1 such that all eventualities in the resulting $\mathfrak{h}(0)$ are realised until k_1. Then we shrink the $\mathfrak{h}(i)$, for $0 < i \leq k_1$, and delete a part of the quasimodel (if such a part exists) between $\mathfrak{h}(n)$ and $\mathfrak{h}(m)$, $0 < n < m < k_1$, such that $\mathfrak{h}(m)$ is 0-embeddable into $\mathfrak{h}(n)$ and no eventuality from $\mathfrak{h}(0)$ is realised in the interval (n, m). Note that, due to 0-embeddability of

$\mathfrak{h}(n)$ into $\mathfrak{h}(m)$, in the resulting quasimodel every eventuality from $\mathfrak{h}(0)$ is realised until some $k'_1 \leq k_1$. Then, we repeat the procedure. After finitely many iterations we end up with a quasimodel for φ with the first segment $[0, k_1]$ satisfying the conditions of the lemma. We then proceed with considering the word k_1, etc.

Now, to conclude the proof of Theorem 4, it is enough to show that there is an algorithm which, when applied to a $\mathcal{TL}_{\bigcirc}$-formula φ, eventually stops iff φ is not satisfiable. The existence of such an algorithm can be proved using Lemma 2, Higman's lemma [13] and König's lemma.

The algorithm explores all possible ways of constructing a quasimodel for a given φ *satisfying* the conditions of Lemma 2. By condition (A), the choice of T_φ-words for the nth position in such a quasimodel is bounded by some recursive function $s'(n, \varphi)$. We claim that all possible ways of constructing a first segment $[0, k_1]$ satisfying the conditions of Lemma 2 must come to an end (exhaust all possible choices) after some step N_1. Indeed, suppose otherwise, i.e., for every $n \in \mathbb{N}$, we can have a sequence of T_φ-words $\mathfrak{f}(0), \ldots, \mathfrak{f}(n)$ satisfying (A), (B) and such that not all eventualities in $\mathfrak{f}(0)$ are realised until n. Then, by (A) and König's lemma, there exists an infinite sequence such that condition (A) holds, at least one of the eventualities from $\mathfrak{f}(0)$ is not satisfied, and if $n < m$ and $\mathfrak{f}(n)$ is 0-embeddable into $\mathfrak{f}(m)$ then some eventuality from $\mathfrak{f}(0)$ is realised in the interval (n, m). Let m be the smallest number such that all eventualities in $\mathfrak{f}(0)$ realised in this sequence are actually realised until m (such a number exists because there are only finitely many such eventualities). But then, by Higman's lemma, we must have some i, j, for $m < i < j$, such that $\mathfrak{f}(i)$ is 0-embeddable in $\mathfrak{f}(j)$, contrary to condition (B).

If we fail to construct at least one first segment satisfying Lemma 2, then φ is not satisfiable. Otherwise we try to extend successful first segments to realise the eventualities of their last word, again complying with conditions (A) and (B), and so forth. Clearly, φ is not satisfiable iff this algorithm eventually stops.

4 Decidability

We now show that if we consider satisfiability in models with eventually constant finite domains then we can obtain a decidable logic, provided that its language does not contain the next-time operator.

Theorem 5. $\mathsf{Log}\mathcal{L}^{\mathsf{c}}_{\mathsf{fin}}$ *is decidable, but not in primitive recursive time.*

The crucial difference between $\mathsf{Log}\mathcal{L}^{\mathsf{c}}_{\mathsf{fin}}$ and $\mathsf{Log}\mathcal{L}_{\mathsf{fin}}$ is revealed by the following:

Lemma 3. *A $\mathcal{TL}$-formula φ is satisfiable in an e-model from $\mathcal{L}^{\mathsf{c}}_{\mathsf{fin}}$ iff there is a quasimodel $(\mathfrak{f}, \mathfrak{R})$ for φ such that, for some $N \in \mathbb{N}$,*

(a) $|T_n| \leq s(n, \varphi)$, *where $\mathfrak{f}(n) = \langle T_n, <_n, l_n \rangle$ and $n < N$,*
(b) *there are no $n < m < N$ such that $\mathfrak{f}(n)$ is embeddable into $\mathfrak{f}(m)$,*
(c) *for all $n \geq N$, $|T_n| = |T_N|$ and there are some $N = n_1 < \cdots < n_k$ such that the set $A_i = \{m \geq N \mid \mathfrak{f}(n_i) = \mathfrak{f}(m)\}$ is infinite for each n_i, and every $\mathfrak{f}(n)$, for $n \geq N$, belongs to some A_i.*

Proof. Since e-models in $\mathcal{L}_{\mathrm{fin}}^{\mathrm{c}}$ have finite states with eventually constant domains, we may assume that φ is satisfied in a quasimodel $(\mathfrak{h}, \mathfrak{Q})$ satisfying condition (c) for some $N \in \mathbb{N}$. By applying operations **shrink** and **delete** from the proof of Lemma 2 (with plain 'embeddable' instead of 'x-embeddable') to the T_φ-words from the segment $\mathfrak{h}(0), \ldots, \mathfrak{h}(N-1)$ as many times as possible (the number N will become smaller after each application of **delete**), we will eventually construct a quasimodel as required.

Now, using the same argument as in the previous section (involving Higman's and König's lemmas), we can effectively construct finitely many initial segments $\mathfrak{f}(0), \ldots, \mathfrak{f}(n)$, satisfying (a) and (b) above, of possible quasimodels for φ. For each such segment, take the final state $\mathfrak{f}(n) = (T_n, <_n, l_n)$ and suppose that $w_0 <_n \cdots <_n w_m$ are all elements of T_n. Consider the formula

$$\chi_{\mathfrak{f}} = \bar{l}_n(w_0) \wedge \diamondsuit\left(\bar{l}_n(w_1) \wedge \diamondsuit\left(\bar{l}_n(w_2) \wedge \diamondsuit(\cdots \diamondsuit \bar{l}_n(w_m)\cdots)\right)\right),$$

where $\bar{l}_n(w) = \bigwedge\{\psi \mid \psi \in l_n(w)\}$. It should be clear that φ is satisfiable iff, for at least one of the constructed segments $\mathfrak{f}(0), \ldots, \mathfrak{f}(n)$, the formula $\chi_{\mathfrak{f}}$ is satisfiable in a quasimodel $\mathfrak{f}(n+1), \mathfrak{f}(n+2), \ldots$ (with some set $\mathfrak{R}$ of runs) satisfying condition (c) of Lemma 3.

Observe now that the temporal operators $\Box_F$ and $\diamondsuit_F$ in such quasimodels behave like **S5** modalities: for all $m > n$ and all $w \in T_m$, we have $\diamondsuit_F \psi \in l_m(w)$ iff there is $k > n$ such that $\psi \in l_k(r(k))$, where $r(m) = w$. Thus, we can complete the decidability part of the proof of Theorem 5 if we can prove the following.

Let $\mathcal{C}$ be the class of bimodal models of the form $(W, R, (\mathfrak{V}_x \mid x \in V))$, where $V \neq \emptyset$, (W, R) is a finite strict linear order, and $\mathfrak{V}_x$, for $x \in V$, is a valuation in W (i.e., a map from the set of propositional variables into the set of subsets of W). In other words, we have $|V|$ (not necessarily distinct) models based on (W, R). Define the truth relation $(x, u) \models \varphi$ for $\mathcal{TL}$-formulas in such a model by taking for $x \in V$ and $u \in W$:

- $(x, u) \models p_i$ iff $u \in \mathfrak{V}_x(p_i)$,
- $(x, u) \models \diamondsuit_F \psi$ iff there exists $y \in V$ such that $(y, u) \models \psi$,
- $(x, u) \models \diamondsuit \psi$ iff there exists $v \in W$ such that uRv and $(x, v) \models \psi$,

plus the standard clauses for the Booleans. (In fact, we have defined bimodal models based on product frames of the form $(V, V \times V) \times (W, R)$; see [8] for details.)

Proposition 1. *The satisfiability problem for $\mathcal{TL}$-formulas in models from $\mathcal{C}$ is decidable. (Moreover, if a formula is satisfiable, then it satisfiable in a finite model from $\mathcal{C}$).*

This proposition can be proved using the quasimodel technique from [8]. The second half of Theorem 5 can be proved in the same way as in [10] using a reduction of the non-primitive recursive reachability problem for lossy channel systems from [26].

In the next section we will show that the addition of the 'next-time' operator $\bigcirc$ results in a logic for $\mathcal{L}_{\mathsf{fin}}^{\mathsf{c}}$ that is not even recursively enumerable. Notice that the decidability proof given above breaks down for $\bigcirc$ when we observe that 'on the tail' the temporal operators behave like **S5** modalities: this is not the case for $\bigcirc$. Lemma 3, however, still holds for the language with $\bigcirc$.

5   Undecidable Problems for Channel Systems

Our proofs of undecidability and non-recursive enumerability (Theorems 2 and 3) proceed by reduction of suitable reachability problems for channel systems. We briefly discuss the required problems in this section; for further information on channel systems the reader is referred to [2,4,26].

A *single channel system* is a triple $S = \langle Q, \Sigma, \Delta \rangle$, where $Q = \{q_1, \ldots, q_n\}$ is a finite set of *control states*, $\Sigma = \{a, b, \ldots\}$ is a finite alphabet of *messages*, and $\Delta \subseteq Q \times \{?, !\} \times \Sigma \times Q$ is a finite set of *transitions*.

A *configuration* of S is a pair $\gamma = \langle q, \boldsymbol{w} \rangle$, where $q \in Q$ and $\boldsymbol{w} \in \Sigma^*$. Say that a configuration $\gamma' = \langle q', \boldsymbol{w}' \rangle$ is the result of a *perfect transition* of S from $\gamma = \langle q, \boldsymbol{w} \rangle$ and write $\gamma \to_p \gamma'$ if

- there is $(q, !, u, q') \in \Delta$ such that $\boldsymbol{w}' = u\boldsymbol{w}$, or
- there is $(q, ?, u, q') \in \Delta$ such that $\boldsymbol{w} = \boldsymbol{w}'u$.

The *reachability problem* for channel systems is formulated as follows: given a channel system S and two states q_0 and q_f, decide whether there is a computation starting from $\langle q_0, \epsilon \rangle$ and reaching q_f, where ϵ is the empty word. This reachability problem is obviously *recursively enumerable*. However, similarly to the halting problem for Turing machines we have the following result that was proved in [2]:

Theorem 6. *The reachability problem for channel systems is undecidable.*

We say that γ' is a result of a *lossy transition* from γ and write $\gamma \to_\ell \gamma'$ if

$$\gamma \ \sqsupseteq \ \gamma_1 \to_p \gamma_2 \ \sqsupseteq \ \gamma'$$

for some γ_1 and γ_2, where $\langle q, \boldsymbol{w} \rangle \sqsupseteq \langle q', \boldsymbol{w}' \rangle$ iff $\boldsymbol{w}'$ is a subword of $\boldsymbol{w}$ and $q = q'$.

The ω-*reachability problem* for (lossy) channel systems is formulated as follows: given a channel system S and two states q_0 and q_f, decide whether for every $n \in \mathbb{N}$ there exists a lossy computation of S starting with $\langle q_0, \epsilon \rangle$ and reaching q_f at least n times. The proof of the next theorem was kindly suggested by Ph. Schnoebelen.

Lemma 4. *The ω-reachability problem for lossy channel systems is undecidable.*

Proof. We prove this lemma by reduction of the undecidable *boundedness problem* [24]: given a channel system S, determine whether the set of configurations of S that are reachable from $\langle q_0, \epsilon \rangle$ is *finite*.

Given a channel system S, we construct a system S' in such a way that S is bounded, that is, has only finitely many configurations reachable from $\langle q_0, \epsilon \rangle$, iff S' has the ω-reachability property. The set of states of S' extends that of S with one new additional state q_{rec}, and the set of transitions of S' is that of S plus non-deterministic transitions from every state of S into q_{rec}. Being in q_{rec}, the system reads one symbol from the channel and stays in q_{rec}. It should be clear now that there exists a (lossy) computation of S' starting with $\langle q_0, \epsilon \rangle$ and reaching q_{rec} arbitrary many times iff S is unbounded.

6   Non-recursive Enumerability

Here we show that the addition of the next-time operator to $\mathcal{TL}$ immediately destroys the decidability result of Theorem 5 for $\mathsf{Log}\mathcal{L}^c_{fin}$.

Theorem 7. $\mathsf{Log}_{\bigcirc}\mathcal{L}^c_{fin}$ *is not recursively enumerable.*

Proof. Given a channel system S, control states q_0 and q_f, we construct a $\mathcal{TL}_{\bigcirc}$-formula φ_{S,q_0,q_f} which is satisfiable in a model from $\mathcal{L}^c_{fin}$ iff a computation started from $\langle q_0, \epsilon \rangle$ reaches q_f. Since the reachability problem for channel systems is undecidable, but recursively enumerable, this will show that the set $\mathsf{Log}_{\bigcirc}\mathcal{L}^c_{fin}$ cannot be recursively enumerable.

With a slight abuse of notation, we use the propositional variables

- δ, for every instruction $\delta \in \Delta$,
- a, for every $a \in \Sigma$,
- q, for every $q \in Q$,
- m, a marker,
- end, a marker for 'end of word' or 'empty word.'

Let w stand for $\bigvee_{a\in\Sigma} a$, and let $\Box^+_F \psi = \psi \wedge \Box_F \psi$, $\Box^+ \psi = \psi \wedge \Box \psi$, and $\Diamond^+_F \psi = \psi \vee \Diamond_F \psi$, $\Diamond^+ \psi = \psi \vee \Diamond \psi$.

Intuitively, our encoding of the reachability problem works as follows. First we 'mark' infinitely many states by making marker m true everywhere in these states (and false in all others).

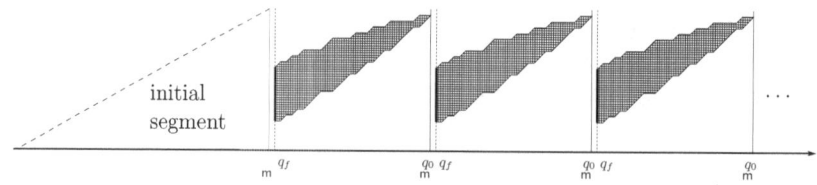

Fig. 2. Encoding of the perfect channel reachability problem

This can be achieved by using the formulas:

$$\Box^+_F \Diamond^+_F m \tag{2}$$

$$\Box^+_F ((m \rightarrow \Box m) \wedge (\neg m \rightarrow \Box \neg m)) \tag{3}$$

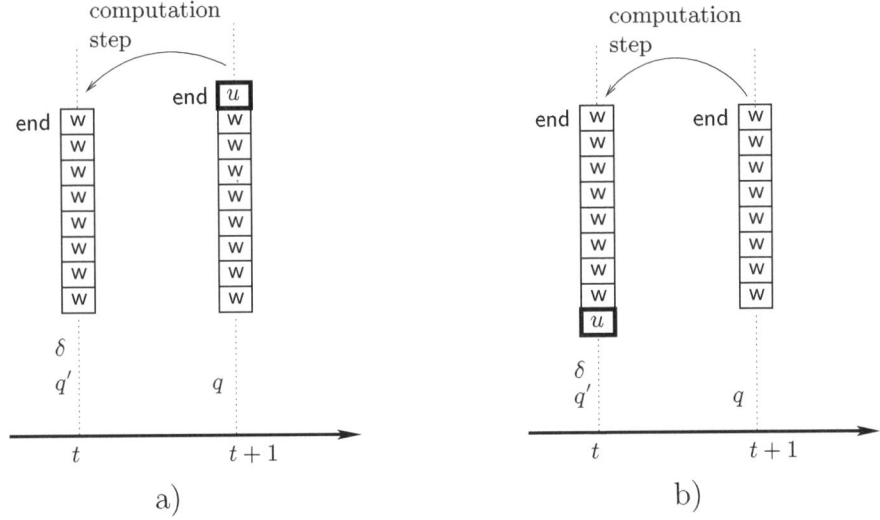

Fig. 3. Encoding one transition of a channel system. a) $\delta = (q, ?, u, q')$: symbol u is read from the end of the channel; b) $\delta = (q, !, u, q')$: symbol u is written at the beginning of the channel

Between any two markers, we simulate from right to left (that is, from future to past) a computation of the channel system S starting with $\langle q_0, \epsilon \rangle$ and reaching control state q_f; see Fig. 2. At every moment x we write the contents of the channel on the linear order $(W_x, <_x)$ as a word without 'gaps.' We mark its end with end, and if the word is empty then end will hold somewhere:

$$\Box_F^+ \left(\Diamond^+ \mathsf{end} \wedge \Box^+ (\mathsf{end} \rightarrow \Box \neg \mathsf{end}) \right) \tag{4}$$

$$\Box_F^+ \Box^+ \left((\mathsf{w} \wedge \Box \neg \mathsf{w}) \rightarrow \mathsf{end} \right) \tag{5}$$

$$\Box_F^+ \Box^+ \neg \left(\mathsf{w} \wedge \Diamond (\neg \mathsf{w} \wedge \Diamond \mathsf{w}) \right) \tag{6}$$

At every marked state, the system is in control state q_0 and the channel is empty. Moreover, this initial configuration is not obtained from any previous state by any instruction δ:

$$\Box_F^+ \Box^+ \left(\mathsf{m} \rightarrow (q_0 \wedge \neg \mathsf{w} \wedge \bigwedge_{\delta \in \Delta} \neg \delta) \right) \tag{7}$$

At every non-marked state the system is in a certain control state q which results from the previous state by means of an application of some instruction δ:

$$\Box_F^+ \Box^+ \left(\bigvee_{q \in Q} q \wedge \bigwedge_{q \neq q'} (q \rightarrow \neg q') \wedge \bigwedge_{q \in Q} (q \rightarrow \Box q) \right) \tag{8}$$

$$\Box_F^+ \Box^+ \left(\neg \mathsf{m} \rightarrow (\bigvee_{\delta \in \Delta} \delta \wedge \bigwedge_{\delta \neq \delta'} (\delta \rightarrow \neg \delta') \wedge \bigwedge_{\delta \in \Delta} (\delta \rightarrow \Box \delta)) \right) \tag{9}$$

The following formula ensures that words are encoded properly and that the contents of channels does not change arbitrarily:

$$\Box_F^+\Box^+\left(\bigwedge_{a\in\Sigma} (a \to \bigcirc(w \to a)) \land \bigwedge_{a\neq a'} (a \to \neg a')\right) \tag{10}$$

Finally, we encode the effect of instructions δ; see Fig. 3. For every instruction $\delta = (q, !, u, q')$, take

$$\Box_F^+(\delta \to \bigcirc q) \tag{11}$$

$$\Box_F^+\left(\delta \to q' \land \Diamond^+(u \land \neg\bigcirc w) \land \Box^+\left(w \to \Box(w \leftrightarrow \bigcirc w)\right)\right) \tag{12}$$

This formula says that we add u to the beginning of the word encoded at the next moment of time and that nothing else changes. Similarly, for every instruction $\delta = (q, ?, u, q')$, take

$$\Box_F^+\left(\delta \to \bigcirc\left(q \land \Diamond(u \land \mathsf{end})\right)\right) \tag{13}$$

$$\Box_F^+\left(\delta \to \left(q' \land \Diamond^+(\mathsf{end} \land \bigcirc(\Diamond\mathsf{end} \land \Box\Box\neg\mathsf{end})) \land \right.\right.$$
$$\left.\left. \Box^+(w \to \bigcirc w) \land \Box^+(\bigcirc(w \land \neg\mathsf{end}) \to w)\right)\right) \tag{14}$$

This formula says that we delete u from the end of the word encoded at the next moment of time and that nothing else changes. To make sure that the final state of the computations is q_f, we need one more formula

$$\Box_F^+(\mathsf{m} \to \bigcirc q_f) \tag{15}$$

Note that (15) together with (7) and (8) also ensure that there cannot be two marked adjacent states.

Let φ_{S,q_0,q_f} be the conjunction of formulas (2)–(15). It is not difficult to show that if there exists a computation of S starting from $\langle q_0, \epsilon\rangle$ and reaching q_f, then φ_{S,q_0,q_f} is satisfied in a model with constant domains such that between any two markers the computation of S is simulated. Conversely, suppose that φ_{S,q_0,q_f} is satisfied in a model from $\mathcal{L}_{\mathsf{fin}}^c$. Take two successive marked states n_1 and n_2 such that $W_{n_1} = W_{n_2}$ (i.e., the domain does not change between n_1 and n_2). Then a computation of S starting with q_0 and reaching q_f is simulated between n_2 and n_1.

This completes the proof of Theorem 7.

7   Undecidability

The encoding of *perfect* channel systems in the previous section was only possible because we were considering models with eventually constant domains. In models with expanding domains we can only simulate *lossy* computations of channel systems. Actually, a very simple modification of the formula φ_{S,q_0,q_f} above is enough to prove that $\mathsf{Log}_\bigcirc\mathcal{L}_{\mathsf{fin}}$ is undecidable. We begin by showing how to do this, and after that explain how to remove $\bigcirc$ in order to prove undecidability of $\mathsf{Log}\mathcal{L}_{\mathsf{fin}}$.

Proposition 2. *For any channel system S and states q_0 and q_f, one can construct a $\mathcal{TL}_\bigcirc$-formula φ_{S,q_0,q_f} which is satisfiable in a model from $\mathcal{L}_{\mathrm{fin}}$ iff, for every $n \in \mathbb{N}$, there exists a lossy computation of S starting with $\langle q_0, \epsilon \rangle$ and reaching q_f at least n times.*

Proof. As above we use markers m that are true in infinitely many states and simulate a *lossy* computation between any two marked states. However, instead of forcing these computations to reach q_f, now we ensure that, for every n, there exist two marked states such that a computation between them reaches q_f at least n times. This will be enforced by the formula $\psi_{\omega\text{-rec}}$ which replaces the conjunct (15) in φ_{S,q_0,q_f}. The formula $\psi_{\omega\text{-rec}}$ is defined as the conjunction of (16)–(19) below.

First we introduce an auxiliary variable s that cannot be true on two different elements of W_x

$$\square_F^+ \square^+ (\mathsf{s} \to \square\neg\mathsf{s}), \tag{16}$$

and if s is true on some $u \in W_x$, then q_f is also true there

$$\square_F^+ \square^+ (\mathsf{s} \to q_f) \tag{17}$$

The variable s is used for 'counting.' Whenever marker m is true, we can guarantee that at the next moment of time there exists a *new* domain point where s is true:

$$\square_F^+ \square^+ \big(\mathsf{m} \to \square(\square \perp \to \bigcirc\Diamond\mathsf{s})\big) \tag{18}$$

(Here we use the fact that the domains can expand.) The next formula together with (18) ensure that if s is true n times in some interval between two markers, then in the next interval it must be true at least $n+1$ times:

$$\square_F^+ \square^+ \big(\mathsf{s} \to \square_F(\mathsf{m} \to \neg\mathsf{m}\mathcal{U}\mathsf{s})\big) \tag{19}$$

Using the standard technique (see, e.g., [5]) formula (19), containing the 'until' operator $\mathcal{U}$, can be replaced with the following $\mathcal{TL}_\bigcirc$-formula which is satisfiable iff (19) is satisfiable:

$$\square_F^+\square^+\big(\mathsf{s} \to \square_F(\mathsf{m} \to \bigcirc p \wedge \Diamond_F \mathsf{s})\big) \wedge \square_F^+\square^+\big((p \to \neg\mathsf{m} \vee \mathsf{s}) \wedge (p \to \bigcirc(p \vee \mathsf{s}))\big)$$

where p is a fresh variable.

We are now in a position to prove the following:

Theorem 8. $\mathrm{Log}\mathcal{L}_{\mathrm{fin}}$ *is undecidable.*

Proof. Given a channel system S and states q_0, q_f, we construct, by modifying the formula φ_{S,q_0,q_f} above, a $\mathcal{TL}$-formula ψ_{S,q_0,q_f} which is satisfiable in a model from $\mathcal{L}_{\mathrm{fin}}$ iff, for every $n \in \mathbb{N}$, there exists a lossy computation of S starting with $\langle q_0, \epsilon \rangle$ and reaching q_f at least n times.

Although the language $\mathcal{TL}$ does not contain the next-time operator, we can simulate 'locally' some of its properties. Let v_i be a fresh propositional variable. Then we have $(x, u) \models \Diamond_F \mathsf{v}_i \wedge \square_F \square_F \neg \mathsf{v}_i$ iff

- $(x+1, u) \models \mathsf{v}_i$,
- $(y, u) \not\models \mathsf{v}_i$ for all $y > x + 1$.

Thus, at point (x, u) we can refer to the next point $(x+1, u)$ along the time axis. However, this can be done only *once* for given u and v_i. We denote the resulting 'one-off' next-time operator by O_i and use it as an extra temporal operator of our language $\mathcal{TL}$ bearing in mind that every occurrence of $\mathsf{O}_i \varphi$ should be replaced with $\diamondsuit_F \mathsf{v}_i \wedge \square_F \square_F \neg \mathsf{v}_i$, and that $\square_F^+ \square^+ (\mathsf{v}_i \to \varphi)$ should be added as a conjunct to the whole formula.

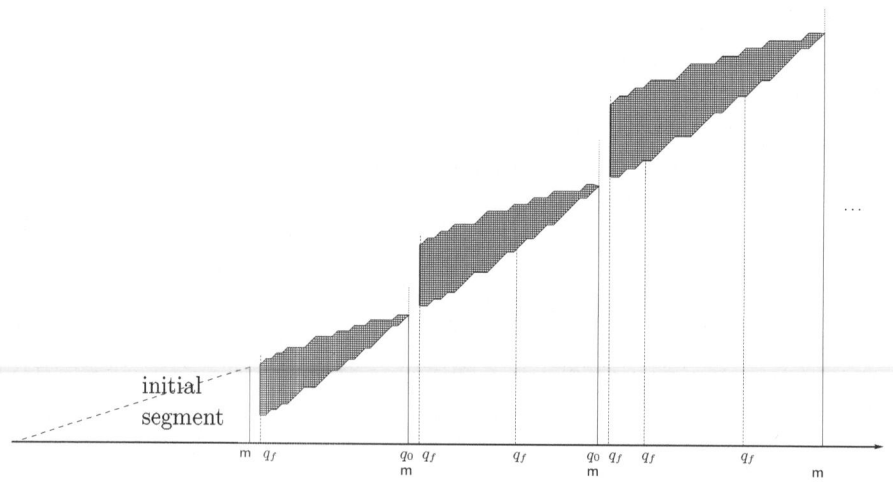

Fig. 4. Encoding of the lossy channel ω-reachability problem

The $\mathcal{TL}$-encoding of the ω-reachability problem for lossy channels is done in almost the same way as in the proofs of Theorem 7 and Proposition 2. In every next interval between two occurrences of the marker m, we model a computation of the channel system S visiting the state q_f at least one time more than in the previous interval, and the contents of the channel is written on the linear order as a word without gaps. Note, however, that if some point $u \in W_x$ is used for writing a word at time point x, it will never be used again for encoding words in other intervals—simply because our 'surrogate' next-time operators cannot be reused. Fortunately, this is not a real problem: by expanding the domain we can always find the required 'fresh' points; see Fig. 4.

The modification ψ_{S,q_0,q_f} of φ_{S,q_0,q_f} we need keeps conjuncts (2)–(9) intact. We add the conjunct

$$\square_F^+ \square^+ \neg \big(\mathsf{w} \wedge \diamondsuit_F (\neg \mathsf{w} \wedge \diamondsuit_F \mathsf{w}) \big) \tag{20}$$

saying that for any given domain point the set of time points with a symbol from Σ written on it is a (possibly empty) interval. In particular, symbols from

Σ cannot be written on the same domain points in different intervals between markers. Further, replace (10)–(13) with the following formulas (21)–(25):

$$\Box_F^+\Box^+\left(\bigwedge_{a\in\Sigma}(a\to\Box_F(w\to a))\wedge\bigwedge_{a\neq a'}(a\to\neg a')\right)\tag{21}$$

For every instruction $\delta=(q,!,u,q')$,

$$\Box_F^+(\delta\to\Diamond^+\bigcirc_1 q)\tag{22}$$

$$\Box_F^+\left(\delta\to q'\wedge\Diamond^+(u\wedge\Box_F\neg w)\wedge\Box^+\left(w\to\Box(w\leftrightarrow\Diamond_F w)\right)\right)\tag{23}$$

and for every instruction $\delta=(q,?,u,q')$,

$$\Box_F^+\left(\delta\to\Diamond^+\bigcirc_2\left(q\wedge\Diamond(u\wedge end)\right)\right)\tag{24}$$

$$\Box_F^+\left(\delta\to q'\wedge\Diamond^+(end\wedge\Box_F\Diamond end)\wedge\Box^+(w\to\Diamond_F w)\right)\tag{25}$$

Note that formulas (22) and (24) may force introduction of new domain points.

We also have to replace formulas (16)–(19) with some other formulas expressing the same property of m and q_f: the number of occurrences of q_f between adjacent markers m is growing in time. In formulas (26)–(27) below, $p\wedge\Box_F\neg p$ plays the same role as the variable s in (16)–(17):

$$\Box_F^+\Box^+\left((p\wedge\Box_F\neg p)\to\Box\neg(p\wedge\Box_F\neg p)\right)\tag{26}$$

$$\Box_F^+\Box^+\left((p\wedge\Box_F\neg p)\to q_f\right)\tag{27}$$

The following formulas (28)–(32) guarantee that for every $N\in\mathbb{N}$, there are adjacent marked $t_1<t_2$ such that the number of time points $t\in(t_1,t_2)$ for which $(t,u)\models p\wedge\Box_F\neg p$, for some $u\in W_t$, is $\geq N$:

$$\Box_F^+\Box^+\left(m\wedge p\to\Box_F(m\to\neg p)\right)\tag{28}$$

$$\Box_F^+\Box^+(\neg p\to\Box_F\neg p)\tag{29}$$

$$\Box_F^+\Box^+\left(p\wedge\Box_F\neg p\to\Box^+(\Box\perp\to\bigcirc_3\Diamond p)\right)\tag{30}$$

$$\Box_F^+\Box^+\left(m\to\Box^+(\Box\perp\to\bigcirc_4\Diamond p)\right)\tag{31}$$

$$\Box_F^+\Box^+(p\wedge\Box_F\neg p\to\neg m)\tag{32}$$

This claim is proved by induction on N. We only show the basis of induction $N=1$ and indicate how to extend it to the inductive step.

Let t_0 be the first marked time point. By (31), there is $u\in W_{t_0+1}$ such that $(t_0+1,u)\models p$. Two cases are possible now. First, if $(t,u)\models p\wedge\Box_F\neg p$ holds for some $t>t_0$ before the next marked point, then we are done. Otherwise, $(t_1,u)\models p$ for the next marked time point t_1. Let t_2 be the first marked point after t_1. Then, by (28) and (29), $(t,u)\models\neg p$ for all $t\geq t_2$. It follows that for some t with $t_1\leq t<t_2$ we must have $(t,u)\models p\wedge\Box_F\neg p$. In view of (32), $t\neq t_1$.

For the inductive step we use (30) to ensure that the number of points with $p \wedge \Box_F \neg p$ in the next interval between two marked points is at least the same as in the previous one, while (31) adds one more point of this kind.

8   An Application to Dynamic Topological Logic

Dynamic topological logic was introduced in 1997 (see, e.g., [20,1,21]) as a logical formalism for describing the behaviour of dynamical systems, e.g., in order to specify liveness and safety properties of hybrid systems [3]. Roughly, (some aspects of) the behaviour of such systems are modelled by means of a *topology* $\mathfrak{T}$ on a *space* Δ and a *continuous function* f acting on Δ. What we are interested in is the asymptotic behaviour of iterations of f, in particular, the orbits $w, f(w), f^2(w), \ldots$ of states $w \in \Delta$. The language $\mathcal{TL}_\bigcirc$ provides a natural formalism for speaking about such iterations, with propositional variables interpreted as subsets of Δ, the modal operator $\Diamond$ interpreted as the topological *closure* operator $\mathbb{C}$ on $\mathfrak{T}$, and the temporal operators $\Diamond_F$ and $\bigcirc$ as iterations of the function f.

More formally, by a *dynamic topological model* we understand a structure

$$\mathfrak{M} = (\Delta, \mathfrak{T}, f, P^1, P^2, \ldots),$$

where Δ is a space with topology $\mathfrak{T}$, $f : \Delta \to \Delta$ is a continuous function with respect to this topology, and $P^i \subseteq \Delta$ for all i. For a $\mathcal{TL}_\bigcirc$-formula φ and $w \in \Delta$, the *truth relation* $\mathfrak{M}, w \models \varphi$ is defined as follows:

$$\begin{aligned}
\mathfrak{M}, w \models p_i \quad &\text{iff} \quad w \in P^i, \\
\mathfrak{M}, w \models \Diamond\varphi \quad &\text{iff} \quad w \in \mathbb{C}\{v \in \Delta \mid \mathfrak{M}, v \models \varphi\}, \\
\mathfrak{M}, w \models \bigcirc\varphi \quad &\text{iff} \quad \mathfrak{M}, f(w) \models \varphi, \\
\mathfrak{M}, w \models \Diamond_F\varphi \quad &\text{iff} \quad \mathfrak{M}, f^n(w) \models \varphi \text{ for some } n \in \mathbb{N}.
\end{aligned}$$

A formula φ is *valid in* $\mathfrak{M}$ if $\mathfrak{M}, w \models \varphi$ for every $w \in \Delta$.

Every quasi-order (Δ, R) gives rise to a topological space with the interior operator $\mathbb{I}$ defined by $\mathbb{I}(X) = \{x \in X \mid \forall y \in \Delta \ (xRy \to y \in X)\}$ (as usual, $\mathbb{C}(X) = X \setminus \mathbb{I}(\Delta \setminus X)$). Such spaces are known as *Aleksandrov spaces*. For Aleksandrov spaces, the truth relation for $\Diamond$ can be defined in a more familiar Kripke-style way:

$$\mathfrak{M}, w \models \Diamond\varphi \quad \text{iff} \quad \mathfrak{M}, v \models \varphi \text{ for some } v \in \Delta \text{ such that } wRv.$$

Moreover, it is easy to see that a function f is continuous with respect to this topology iff $\forall w, v \in \Delta \ (wRv \to f(w)Rf(v))$.

By the *dynamic topological logic of Aleksandrov spaces* we understand the set of $\mathcal{TL}_\bigcirc$-formulas that are valid in all dynamic topological model based on Aleksandrov spaces.

Theorem 9. *The dynamic topological logic of Aleksandrov spaces is undecidable.*

Proof. Using the techniques developed in [10], one can show that every $\mathcal{TL}_{\bigcirc}$-formula φ is satisfiable in an e-model from $\mathcal{QO}$ iff φ has a dynamic topological model based on an Aleksandrov space.

A lot of problems related to dynamic topological logics remain open. For example, is the dynamic topological logic of Aleksandrov spaces recursively enumerable? Is it finitely axiomatisable? Is the dynamic topological logic of arbitrary topological spaces decidable and/or axiomatisable?

9   Conclusion

Being a very attractive and powerful formalism for representation of and reasoning about systems with changing states, first-order temporal logic is notorious for its bad computational behaviour. This applies, in particular, to first-order temporal logics which can represent non-local constraints on binary relations such as transitivity. The present paper makes one more step in the search for fundamental reasons that could explain this phenomenon and thereby help in finding maximal 'well-behaved' fragments. Here we investigate the potential computational impact of relaxing the standard constant domain assumption by allowing states to expand over time. We consider the standard propositional temporal logic **LTL** equipped with an additional 'modal' operator for speaking about transitive relations over states. This fragment of first-order temporal logic comes from temporal description logic, specification & verification of hybrid systems, and some other areas. The main results of our research, given by Theorems 1–3 above, show that by allowing expanding domains we can indeed end up with logics having better computational properties. The logics still remain extremely complex, but sometimes they become recursively enumerable or even decidable, which makes them a subject for various theorem proving techniques.

It is worth noting that the same results can be proved for the language containing additionally a modal operator interpreted by the converse R_x^{-1} of R_x in each state $\mathfrak{S}(x)$. Also, as this language interpreted over strict linear orders is expressively complete for the two-variable fragment of first-order logic [23], we can reformulate our results as decidability/undecidability results for the monodic fragment of the two-variable first-order temporal logic over e-models based on finite or arbitrary strict linear orders.

Acknowledgements. We are grateful to A. Bovykin, Ph. Schnoebelen, and the members of the London Logic Forum for stimulating discussions, comments and suggestions.

The work on this paper was partially supported by U.K. EPSRC grants no. GR/R42474/01, GR/S63175/01, GR/S61973/01/01, and GR/S63182/01.

References

1. S. Artemov, J. Davoren, and A. Nerode. Modal logics and topological semantics for hybrid systems. Technical Report MSI 97-05, Cornell University, 1997.
2. D. Brand and P. Zafiropulo. On communicating finite-state machines. *Journal of the ACM*, 30:323–342, 1983.
3. J. Davoren and A. Nerode. Logics for hybrid systems. *Proceedings of the IEEE*, 88:985–1010, 2000.
4. A. Finkel. Decidability of the termination problem for completely specified protocols. *Distributed Computing*, 7:129–135, 1994.
5. M. Fisher. A resolution method for temporal logic. In J. Myopoulos and R. Reiter, editors, *Proceedings of IJCAI'91*, pages 99–104. Morgan Kaufman, 1991.
6. M. Fisher, C. Dixon, and M. Peim. Clausal temporal resolution. *ACM Transactions on Computational Logic (TOCL)*, 2(1):12–56, 2001.
7. D. Gabbay, I. Hodkinson, and M. Reynolds. *Temporal Logic*, volume 1. Oxford University Press, 1994.
8. D. Gabbay, A. Kurucz, F. Wolter, and M. Zakharyaschev. *Many-Dimensional Modal Logics: Theory and Applications*, volume 148 of *Studies in Logic*. Elsevier, 2003.
9. D. Gabelaia, R. Kontchakov, A. Kurucz, F. Wolter, and M. Zakharyaschev. Combining spatial and temporal logics: expressiveness vs. complexity. *Journal of Artificial Intelligence Research*, 23:167–243, 2005.
10. D. Gabelaia, A. Kurucz, F. Wolter, and M. Zakharyaschev. Non-primitive recursive decidability of products of modal logics with expanding domains. Manuscript. Available at http://www.dcs.kcl.ac.uk/staff/mz, 2004.
11. D. Gabelaia, A. Kurucz, F. Wolter, and M. Zakharyaschev. Products of 'transitive' modal logics. *Journal of Symbolic Logic*, 2005. In print. Draft available at http://www.dcs.kcl.ac.uk/staff/mz.
12. J. Halpern and M. Vardi. The complexity of reasoning about knowledge and time I: lower bounds. *Journal of Computer and System Sciences*, 38:195–237, 1989.
13. G. Higman. Ordering by divisibility in abstract algebras. *Proceedings of the London Mathematical Society*, 2:326–336, 1952.
14. I. Hodkinson, F. Wolter, and M. Zakharyaschev. Decidable fragments of first-order temporal logics. *Annals of Pure and Applied Logic*, 106:85–134, 2000.
15. U. Hustadt and B. Konev. TRP++ 2.0: A temporal resolution prover. In F. Baader, editor, *Automated Deduction. Proceedings of the 19th International Conference on Automated Deduction (CADE-19)*, volume 2741 of *Lecture Notes in Computer Science*, pages 274–278. Springer, 2003.
16. U. Hustadt, B. Konev, A. Riazanov, and A. Voronkov. **TeMP**: A temporal monodic prover. In *Proceedings IJCAR 2004*, volume 3097 of *LNAI*, pages 326–330. Springer, 2004.
17. B. Konev, A. Degtyarev, C. Dixon, M. Fisher, and U. Hustadt. Towards the implementation of first-order temporal resolution: the expanding domain case. *Information and Computation*, 2005. In print. Available as Technical Report ULCS-03-005, The University of Liverpool, Department of Computer Science.
18. B. Konev, R.. Kontchakov, F. Wolter, and M. Zakharyaschev. On dynamic topological and metric logics. Manuscript. Available at http://www.dcs.kcl.ac.uk/staff/mz, 2005.
19. R. Kontchakov, C. Lutz, F. Wolter, and M. Zakharyaschev. Temporalising tableaux. *Studia Logica*, 76:91–134, 2004.

20. P. Kremer and Mints. Dynamic topological logic. *Bulletin of Symbolic Logic*, 3:371–372, 1997.
21. P. Kremer and G Mints. Dynamic topological logic. *Annals of Pure and Applied Logic*, 131:133–158, 2005.
22. J.B. Kruskal. Well-quasi-orderings, the tree theorem, and Vázsonyi's conjecture. *Transactions of the American Mathematical Society*, 95:210–225, 1960.
23. C. Lutz, U. Sattler, and F. Wolter. Modal logic and the two-variable fragment. In *Proceedings of Computer Science Logic (CSL 2001)*, pages 262–276. Lecture Notes in Computer Science 2141, Springer, 2001.
24. R. Mayr. Undecidable problems in unreliable computations. *Theoretical Computer Science*, 297:337–354, 2003.
25. K. Schild. Combining terminological logics with tense logic. In *Proceedings of the 6th Portuguese Conference on Artificial Intelligence*, pages 105–120, Porto, 1993.
26. Ph. Schnoebelen. Verifying lossy channel systems has nonprimitive recursive complexity. *Information Processing Letters*, 83:251–261, 2002.
27. S. Schwendimann. *Aspects of Computational Logic*. PhD thesis, Universität Bern, Switzerland, 1998.
28. A. Sistla and E. Clarke. The complexity of propositional linear temporal logics. *Journal of the Association for Computing Machinery*, 32:733–749, 1985.
29. A. Sistla and S. German. Reasoning with many processes. In *Proceedings of the Second IEEE Symposium on Logic in Computer Science*, pages 138–153, 1987.
30. P. Wolper. The tableau method for temporal logic: An overview. *Logique et Analyse*, 28:119–152, 1985.
31. F. Wolter and M. Zakharyaschev. Temporalizing description logics. In D. Gabbay and M. de Rijke, editors, *Frontiers of Combining Systems II*, pages 379–401. Studies Press/Wiley, 2000.
32. F. Wolter and M. Zakharyaschev. Axiomatizing the monodic fragment of first-order temporal logic. *Annals of Pure and Applied Logic*, 118:133–145, 2002.

Deciding Monodic Fragments
by Temporal Resolution[*]

Ullrich Hustadt[1], Boris Konev[1], and Renate A. Schmidt[2]

[1] Department of Computer Science, University of Liverpool, UK
{U.Hustadt, B.Konev}@csc.liv.ac.uk
[2] School of Computer Science, University of Manchester, UK
Renate.Schmidt@manchester.ac.uk

Abstract. In this paper we study the decidability of various fragments of monodic first-order temporal logic by temporal resolution. We focus on two resolution calculi, namely, monodic temporal resolution and fine-grained temporal resolution. For the first, we state a very general decidability result, which is independent of the particular decision procedure used to decide the first-order part of the logic. For the second, we introduce refinements using orderings and selection functions. This allows us to transfer existing results on decidability by resolution for first-order fragments to monodic first-order temporal logic and obtain new decision procedures. The latter is of immediate practical value, due to the availability of **TeMP**, an implementation of fine-grained temporal resolution.

1 Introduction

Temporal logics have long been recognised as introducing appropriate languages for specifying a wide range of important computational properties in computer science and artificial intelligence [6]. However, until recently, the practical use of temporal logics has largely been restricted to propositional temporal logics. First-order temporal logic has generally been avoided as no complete proof system can exist for this logic. However, recent work by Hodkinson, Wolter, and Zakharyaschev [11] shows that a specific fragment of first-order temporal logic, called the *monodic* fragment, or *monodic* first-order temporal logic, has the completeness property. This initial result was followed by an examination of the monodic fragment in terms of decidable subclasses, automated deduction, and applications.

Hodkinson et al. [10,11,23] show the decidability of the monadic, two-variable, fluted, and loosely guarded fragments of monodic first-order temporal logic without equality as well as the decidability of the monodic packed fragment of first-order temporal logic with equality. Kontchakov et al. [17] have developed a framework for devising tableau decision procedures for such decidable monodic first-order temporal logics. Using this framework they present tableau decision

---

[*] Supported by EPSRC (grant GR/L87491) and the Nuffield foundation (grant NAL/00841/G30).

R. Nieuwenhuis (Ed.): CADE 2005, LNAI 3632, pp. 204–218, 2005.

procedures for the one-variable fragment and for the fragment corresponding to the modal logic $S4_u$ of monodic first-order temporal logic.

In parallel, Degtyarev, Dixon, Fisher, Hustadt, and Konev have investigated monodic first-order temporal logic in the context of resolution. Degtyarev et al. [5] present a temporal resolution calculus, called *monodic temporal resolution*, for this logic, which is then used in [4] to establish a general decidability result for temporal resolution. Decidability of all the classes from Hodkinson et al., as well as, the Gödel class and the dual Maslov class $\overline{K}$ fragments of monodic first-order temporal logic are shown to be immediate consequences of this general result. Konev et al. [14,15] devise the *fine-grained resolution calculus* as an alternative resolution calculus for monodic first-order temporal logic which is more amenable to mechanisation. This calculus forms the basis of the temporal monodic theorem prover **TeMP** presented in [12].

In this paper we focus on decidability results in the context of the fine-grained resolution calculus. To motivate why the general decidability result obtained in the context of monodic temporal resolution does not easily carry over to the fine-grained resolution calculus, we first provide a brief presentation of both calculi and some basic results about them, including the mentioned decidability result. A main contribution of this paper is the introduction of refinements of fined-grained resolution which incorporate advanced techniques developed in the context of first-order resolution (e.g. [1,19]) into temporal resolution, namely orderings and selection functions. We prove completeness of this refined calculus by simulating derivations in the monodic temporal resolution calculus. The refined calculus, called *ordered fine-grained resolution with selection*, allows us to transfer decidability results and decision procedures obtained for fragments of first-order logic to the corresponding fragments of monodic first-order temporal logic.

2   First-Order Temporal Logic

The language of First-Order Temporal Logic, **FOTL**, is an extension of classical first-order logic by temporal operators for a discrete linear model of time (isomorphic to $\mathbb{N}$, that is, the most commonly used model of time). The signature of **FOTL** (without equality and function symbols) consists of a countably infinite set of *variables* x_0, x_1, ..., a countably infinite set of *constants* c_0, c_1, ..., a non-empty set of *predicate symbols* P, P_0, ..., each with a fixed arity ≥ 0, the *propositional operators* $\top$, $\neg$, $\vee$, the *quantifiers* $\exists x_i$ and $\forall x_i$, and the *temporal operators* $\square$ ('always in the future'), $\lozenge$ ('eventually in the future'), $\bigcirc$ ('at the next moment'), and U ('until'). The set of formulae of **FOTL** is defined as follows: $\top$ is a **FOTL** formula; if P is an n-ary predicate symbol and t_1, ..., t_n are variables or constants, then $P(t_1,\ldots,t_n)$ is an *atomic* **FOTL** formula; if φ and ψ are **FOTL** formulae, then so are $\neg\varphi$, $\varphi \vee \psi$, $\exists x\varphi$, $\forall x\varphi$, $\square\varphi$, $\lozenge\varphi$, $\bigcirc\varphi$, and $\varphi \, U \, \psi$. We also use $\bot$, $\wedge$, and $\Rightarrow$ as additional operators, defined using $\top$, $\neg$, and $\vee$. Free and bound variables of a formula are defined in the standard way, as well as the notions of open and closed formulae. Given a formula φ, we write

$\varphi(x_1, \ldots, x_n)$ to indicate that all the free variables of φ are among $x_1, \ldots, x_n$. As usual, a *literal* is either an atomic formula or its negation.

Formulae of this logic are interpreted over structures $\mathfrak{M} = (D_n, I_n)_{n \in \mathbb{N}}$ that associate with each element n of $\mathbb{N}$, representing a moment in time, a first-order structure $\mathfrak{M}_n = (D_n, I_n)$ with its own non-empty domain D_n and interpretation I_n. An *assignment* $\mathfrak{a}$ is a function from the set of variables to $\bigcup_{n \in \mathbb{N}} D_n$. The application of an assignment to terms is defined in the standard way, in particular, $\mathfrak{a}(c) = c$ for every constant c. The *truth relation* $\mathfrak{M}_n \models^{\mathfrak{a}} \varphi$ is defined (only for those $\mathfrak{a}$ such that $\mathfrak{a}(x) \in D_n$ for every variable x) as follows:

$\mathfrak{M}_n \models^{\mathfrak{a}} \top$

$\mathfrak{M}_n \models^{\mathfrak{a}} P(t_1, \ldots, t_n)$ iff $(I_n(\mathfrak{a}(t_1)), \ldots, I_n(\mathfrak{a}(t_n))) \in I_n(P)$

$\mathfrak{M}_n \models^{\mathfrak{a}} \neg\varphi$ iff not $\mathfrak{M}_n \models^{\mathfrak{a}} \varphi$

$\mathfrak{M}_n \models^{\mathfrak{a}} \varphi \vee \psi$ iff $\mathfrak{M}_n \models^{\mathfrak{a}} \varphi$ or $\mathfrak{M}_n \models^{\mathfrak{a}} \psi$

$\mathfrak{M}_n \models^{\mathfrak{a}} \exists x \varphi$ iff $\mathfrak{M}_n \models^{\mathfrak{b}} \varphi$ for some assignment $\mathfrak{b}$ that may differ from $\mathfrak{a}$ only in x and such that $\mathfrak{b}(x) \in D_n$

$\mathfrak{M}_n \models^{\mathfrak{a}} \forall x \varphi$ iff $\mathfrak{M}_n \models^{\mathfrak{b}} \varphi$ for every assignment $\mathfrak{b}$ that may differ from $\mathfrak{a}$ only in x and such that $\mathfrak{b}(x) \in D_n$

$\mathfrak{M}_n \models^{\mathfrak{a}} \bigcirc\varphi$ iff $\mathfrak{M}_{n+1} \models^{\mathfrak{a}} \varphi$

$\mathfrak{M}_n \models^{\mathfrak{a}} \Diamond\varphi$ iff there exists $m \geq n$ such that $\mathfrak{M}_m \models^{\mathfrak{a}} \varphi$

$\mathfrak{M}_n \models^{\mathfrak{a}} \Box\varphi$ iff for all $m \geq n$, $\mathfrak{M}_m \models^{\mathfrak{a}} \varphi$

$\mathfrak{M}_n \models^{\mathfrak{a}} \varphi \, \mathsf{U} \, \psi$ iff there exists $m \geq n$ such that $\mathfrak{M}_m \models^{\mathfrak{a}} \psi$ and $\mathfrak{M}_i \models^{\mathfrak{a}} \varphi$ for every $i, n \leq i < m$

In this paper we make the *expanding domain assumption*, that is, $D_n \subseteq D_m$ if $n < m$, and we assume that the interpretation of constants is *rigid*, that is, $I_n(c) = I_m(c)$ for all $n, m \in \mathbb{N}$.

The set of valid formulae of this logic is not recursively enumerable. However, the set of valid *monodic* formulae is known to be finitely axiomatisable [23]. A formula φ of **FOTL** is called *monodic* if any subformula of φ of the form $\bigcirc\psi$, $\Box\psi$, $\Diamond\psi$, or $\psi_1 \, \mathsf{U} \, \psi_2$ contains at most one free variable. For example, the formulae $\forall x \Box \exists y P(x, y)$ and $\forall x \Box P(x, c)$ are monodic, while $\forall x \forall y (P(x, y) \Rightarrow \Box P(x, y))$ is not monodic.

Every monodic temporal formula can be transformed into an equi-satisfiable normal form, called *divided separated normal form (DSNF)* [14].

Definition 1. *A monodic temporal problem* P *in divided separated normal form* (DSNF) *is a quadruple* $\langle \mathcal{U}, \mathcal{I}, \mathcal{S}, \mathcal{E} \rangle$, *where*

1. *the universal part* $\mathcal{U}$ *and the initial part* $\mathcal{I}$ *are finite sets of first-order formulae;*
2. *the step part* $\mathcal{S}$ *is a finite set of clauses of the form* $p \Rightarrow \bigcirc q$, *where* p *and* q *are propositions, and* $P(x) \Rightarrow \bigcirc Q(x)$, *where* P *and* Q *are unary predicate symbols and* x *is a variable; and*
3. *the eventuality part* $\mathcal{E}$ *is a finite set of formulae of the form* $\Diamond L(x)$ *(a non-ground eventuality clause) and* $\Diamond l$ *(a ground eventuality clause), where* l *is a propositional literal and* $L(x)$ *is a unary non-ground literal with variable* x *as its only argument.*

With each monodic temporal problem $\langle \mathcal{U}, \mathcal{I}, \mathcal{S}, \mathcal{E} \rangle$ we associate the FOTL formula $\mathcal{I} \wedge \Box \mathcal{U} \wedge \Box \forall x \mathcal{S} \wedge \Box \forall x \mathcal{E}$. When we talk about particular properties of a temporal problem (e.g., satisfiability, validity, logical consequences, etc) we refer to properties of this associated formula.

The transformation to DSNF is based on using a renaming and unwinding technique which substitutes non-atomic subformulae and replaces temporal operators by their fixed point definitions as described, for example, in [8]. A step in this transformation which is of relevance for the results presented here is the following: We recursively rename each innermost open subformula $\xi(x)$, whose main connective is a temporal operator, by $P_\xi(x)$, where $P_{\xi(x)}$ is a new unary predicate, and rename each innermost closed subformula ζ, whose main connective is a temporal operator, by p_ζ, where p_ζ is a new propositional variable. In the terminology of [11] $P_\xi(x)$ and p_ζ are called the *surrogates* of $\xi(x)$ and ζ, respectively. Renaming introduces formulae defining $P_\xi(x)$ and p_ζ of the following form (since we are only interested in satisfiability, we use implications instead of equivalences for renaming positive occurrences of subformulae, see also [20]):

$$(a) \ \Box \forall x (P_\xi(x) \Rightarrow \xi(x)) \qquad \text{and} \qquad (b) \ \Box (p_\zeta \Rightarrow \zeta).$$

If the main connective of $\xi(x)$ or ζ is either $\Box$ or $\mathcal{U}$, then the formula will be replaced by its fixed point definition. If the main connective of $\xi(x)$ or ζ is either the $\bigcirc$ or $\Diamond$ operator, the defining formula will be simplified further to obtain step or eventuality clauses.

Theorem 1 (see [4], Theorem 1). *Any monodic first-order temporal formula can be transformed into an equi-satisfiable monodic temporal problem in DSNF with at most a linear increase in the size of the problem.*

In the next section we briefly recall the temporal resolution calculus first developed in [5] and we present a general decidability result for this calculus.

3   Monodic Temporal Resolution

The monodic temporal resolution calculus does not directly operate on the formulae and clauses of a monodic temporal problem P, but, as described next, operates on merged derived step clauses and full merged step clauses computed from the constant flooded form of P. Let P $= \langle \mathcal{U}, \mathcal{I}, \mathcal{S}, \mathcal{E} \rangle$ be a monodic temporal problem, then the temporal problem $\mathsf{P}^c = \langle \mathcal{U}, \mathcal{I}, \mathcal{S}, \mathcal{E}^c \rangle$ where $\mathcal{E}^c = \mathcal{E} \cup \{ \Diamond L(c) \mid \Diamond L(x) \in \mathcal{E}, c \text{ is a constant in P} \}$ is the *constant flooded form* of P. (Strictly speaking, P^c is not in DSNF: We have to rename ground eventualities by propositions.) Evidently, P^c is satisfiability equivalent to P. Let

$$P_{i_1}(x) \Rightarrow \bigcirc M_{i_1}(x), \ldots, P_{i_k}(x) \Rightarrow \bigcirc M_{i_k}(x) \tag{1}$$

be a subset of the set of step clauses of P^c. Then formulae of the form

$$P_{i_j}(c) \Rightarrow \bigcirc M_{i_j}(c) \qquad \text{and} \qquad \exists x \bigwedge_{j=1}^{k} P_{i_j}(x) \Rightarrow \bigcirc \exists x \bigwedge_{j=1}^{k} M_{i_j}(x), \tag{2}$$

where c is a constant in P^c and $j = 1, \ldots, k$, are called *derived* step clauses.[1] Note that formulae of the form (2) are logical consequences of (1). Let $\{ \Phi_1 \Rightarrow$

---

[1] In [4] derived step clauses, are termed e-derived step clauses.

$\bigcirc\Psi_1,\ldots,\Phi_n \Rightarrow \bigcirc\Psi_n\}$ be a set of derived step clauses or *ground* step clauses in P^c. Then $(\bigwedge_{i=1}^n \Phi_i) \Rightarrow \bigcirc(\bigwedge_{i=1}^n \Psi_i)$ is called a *merged derived step clause*.

Let $\mathcal{A} \Rightarrow \bigcirc\mathcal{B}$ be a merged derived step clause, let $P_1(x) \Rightarrow \bigcirc M_1(x), \ldots,$ $P_k(x) \Rightarrow \bigcirc M_k(x)$ be a subset of the original step clauses in P^c, and let $\mathcal{A}(x) \overset{\mathrm{def}}{=} \mathcal{A} \wedge \bigwedge_{i=1}^k P_i(x)$, $\mathcal{B}(x) \overset{\mathrm{def}}{=} \mathcal{B} \wedge \bigwedge_{i=1}^k M_i(x)$. Then $\forall x(\mathcal{A}(x) \Rightarrow \bigcirc\mathcal{B}(x))$ is called a *full merged step clause*.

In what follows, $\mathcal{A} \Rightarrow \bigcirc\mathcal{B}$ and $\mathcal{A}_i \Rightarrow \bigcirc\mathcal{B}_i$ denote merged derived step clauses, $\forall x(\mathcal{A}(x) \Rightarrow \bigcirc\mathcal{B}(x))$ and $\forall x(\mathcal{A}_i(x) \Rightarrow \bigcirc\mathcal{B}_i(x))$ denote full merged step clauses, and $\mathcal{U}$ denotes the (current) universal part of a monodic temporal problem P. We now define the temporal resolution calculus, $\mathfrak{I}_e$, for the expanding domain case. The inference rules of $\mathfrak{I}_e$ are the following.

- *Step resolution rule w.r.t. $\mathcal{U}$:*

$$\frac{\mathcal{A} \Rightarrow \bigcirc\mathcal{B}}{\neg\mathcal{A}} \; (\bigcirc_{res}^{\mathcal{U}}), \quad \text{if } \mathcal{U} \cup \{\mathcal{B}\} \models \bot.$$

- *Termination rule w.r.t. $\mathcal{U}$ and $\mathcal{I}$:*

$$\frac{}{\bot} \; (\bot_{res}^{\mathcal{U}}), \quad \text{if } \mathcal{U} \cup \mathcal{I} \models \bot.$$

- *Eventuality resolution rule w.r.t. $\mathcal{U}$:*

$$\frac{\forall x(\mathcal{A}_1(x) \Rightarrow \bigcirc\mathcal{B}_1(x)) \quad \ldots \quad \forall x(\mathcal{A}_n(x) \Rightarrow \bigcirc\mathcal{B}_n(x)) \quad \Diamond L(x)}{\forall x \bigwedge_{i=1}^n \neg\mathcal{A}_i(x)} \; (\Diamond_{res}^{\mathcal{U}}),$$

where $\forall x(\mathcal{A}_i(x) \Rightarrow \bigcirc\mathcal{B}_i(x))$ are full merged step clauses such that for every i, $1 \le i \le n$, the *loop* side conditions $\forall x(\mathcal{U} \wedge \mathcal{B}_i(x) \Rightarrow \neg L(x))$ and $\forall x(\mathcal{U} \wedge \mathcal{B}_i(x) \Rightarrow \bigvee_{j=1}^n(\mathcal{A}_j(x)))$ are valid.[2]
The set of full merged step clauses, satisfying the loop side conditions, is called a *loop* in $\Diamond L(x)$ and the formula $\bigvee_{j=1}^n \mathcal{A}_j(x)$ is called a *loop formula*.

- *Ground eventuality resolution rule w.r.t. $\mathcal{U}$:*

$$\frac{\mathcal{A}_1 \Rightarrow \bigcirc\mathcal{B}_1 \quad \ldots \quad \mathcal{A}_n \Rightarrow \bigcirc\mathcal{B}_n \quad \Diamond l}{\bigwedge_{i=1}^n \neg\mathcal{A}_i} \; (\Diamond_{res}^{\mathcal{U}}),$$

where $\mathcal{A}_i \Rightarrow \bigcirc\mathcal{B}_i$ are merged derived step clauses such that for every i, $1 \le i \le n$, the *loop* side conditions $\mathcal{U} \wedge \mathcal{B}_i \models \neg l$ and $\mathcal{U} \wedge \mathcal{B}_i \models \bigvee_{j=1}^n \mathcal{A}_j$ are valid. The notions of *ground loop* and *ground loop formula* are defined similarly to the case above.

Let P be a temporal problem. By TRes(P) we denote the set of all possible conclusions of the inference rules above applied to P^c.

Definition 2 (Derivation). *Let* $P = \langle \mathcal{U}, \mathcal{I}, \mathcal{S}, \mathcal{E} \rangle$ *be a monodic temporal problem. A derivation from* P *is a sequence of universal parts,* $\mathcal{U} = \mathcal{U}_0 \subset \mathcal{U}_1 \subset \mathcal{U}_2 \subset \cdots$, *such that* $\mathcal{U}_{i+1}$ *is obtained from* $\mathcal{U}_i$ *by applying an inference rule to* $\langle \mathcal{U}_i, \mathcal{I}, \mathcal{S}, \mathcal{E}^c \rangle$ *and adding its conclusion to* $\mathcal{U}_i$. *The* $\mathcal{I}$, $\mathcal{S}$ *and* $\mathcal{E}^c$ *parts of the temporal problem are not changed during a derivation.*

---

[2] In the case $\mathcal{U} \models \forall x \neg L(x)$, the *degenerate clause*, $\top \Rightarrow \bigcirc\top$, can be considered as a premise of this rule; the conclusion of the rule is then $\neg\top$ and the derivation successfully terminates.

A derivation *terminates* if, and only if, either a contradiction is derived, in which case we say that the derivation *terminates successfully*, or if no new formulae can be derived by further inference steps. Any derivation can be continued yielding a terminating derivation. Note that since there exist only finitely many different full merged step clauses, the number of different conclusions of the inference rules of monodic temporal resolution is finite. Therefore, every derivation is finite.

A derivation $\mathcal{U} = \mathcal{U}_0 \subset \mathcal{U}_1 \subset \mathcal{U}_2 \subset \cdots \subset \mathcal{U}_n$ from $\langle \mathcal{U}, \mathcal{I}, \mathcal{S}, \mathcal{E} \rangle$ is called *fair* (we adopt terminology from [1]) if for any $i \geq 0$ and formula $\varphi \in \mathrm{TRes}(\langle \mathcal{U}_i, \mathcal{I}, \mathcal{S}, \mathcal{E}^c \rangle)$, there exists $j \geq i$ such that $\varphi \in \mathcal{U}_j$.

It is important to note that all the inference rules have side conditions which are first-order problems. For example, consider a temporal problem $\mathsf{P} = \langle \mathcal{U}, \mathcal{I}, \mathcal{S}, \mathcal{E} \rangle$, where only $\mathcal{I}$ is non-empty, that is, P is simply a first-order problem. Then the only inference rule applicable is the termination rule. If the rule can be applied, then a single application of the rule would derive a contradiction indicating that P is unsatisfiable. If the rule cannot be applied, because $\mathcal{I}$ is not contradictory, then the derivation terminates without a single inference step being performed, indicating that P is satisfiable. This also illustrates why all derivations can be finite although the satisfiability problem of monodic FOTL is only semi-decidable.

So, in general, the side conditions of our inference rules are only semi-decidable and in the case a side condition is false, it may happen that the test of this side condition does not terminate. To ensure fairness we must make sure that each such test cannot indefinitely block the investigation of alternative applications of inference rules in a derivation.

Theorem 2 (see [4, Theorem 10]). *The rules of $\mathfrak{I}_e$ preserve satisfiability over expanding domains. A monodic temporal problem P is unsatisfiable over expanding domains iff any fair derivation in $\mathfrak{I}_e$ from P^c terminates successfully.*

4   Decidability by Monodic Temporal Resolution

Monodic temporal resolution provides a decision procedure for a class of monodic FOTL formulae provided that there exists a first-order decision procedure for the side conditions of all inference rules. Examination of the side conditions shows that we are interested in the satisfiability of (i) the conjunction of the (current) universal part and the initial part, and (ii) the conjunction of the (current) universal part and sets of monadic formulae built from predicate symbols which occur in the step and eventuality part of a temporal problem. At the same time, in each step of the derivation the universal part is extended by monadic formulae from the conclusion of the inference rule applied in the inference step. So, after imposing restrictions on the form of the universal and initial parts of a class of temporal problems, we can guarantee decidability of this class.

However, formalising which fragments of monodic FOTL are decidable by monodic temporal resolution is slightly more complex, since we have to take our "rename and unwind" transformation to divided separated normal form into account, as the following example illustrates.

Example 1. Let $\varphi(x, y, z, u)$ be the first-order formula $Q_1(x, y, z) \vee Q_2(y, z) \vee Q(x, y, z, u)$. Then the formula $\exists x \forall y \forall z \exists u \varphi(x, y, z, u)$ belongs to the dual of Maslov's class K which is decidable. In contrast, consider the temporal formula $\exists x \Box \Diamond \forall y \forall z \exists u \varphi(x, y, z, u)$ with the same φ. Once transformed to an equisatisfiable temporal problem $\mathsf{P} = \langle \mathcal{U}, \mathcal{I}, \mathcal{S}, \mathcal{E} \rangle$ in DSNF, the universal part $\mathcal{U}$ contains the formula $\forall x (P_\varphi(x) \Rightarrow \forall y \forall z \exists u \varphi(x, y, z, u))$ which does not belong to Maslov's class K. (It belongs to the undecidable Surányi class $\forall^3 \exists$ [2].)

To solve this problem, we define decidable fragments in terms of surrogates.

Definition 3 (Temporalisation by Renaming). *Let $\mathfrak{C}$ be a class of first-order formulae. Let φ be a monodic temporal formula in negation normal form (that is, the only Boolean connectives are conjunction, disjunction and negation, and negations are only applied to atoms). Let $\overline{\varphi}$ denote the formula that results from φ by replacing all of its subformulae whose main connective is a temporal operator and which is not within a scope of another temporal operator with their surrogates.*

We say that φ belongs to the class $\mathcal{T}_{ren}\mathfrak{C}$ if

1. *$\overline{\varphi}$ belongs to $\mathfrak{C}$ and*
2. *for every subformula of the form $\mathcal{T}\psi$, where $\mathcal{T}$ is a temporal operator (or of the form $\psi_1 \mathcal{T} \psi_2$ if $\mathcal{T}$ is binary), either $\overline{\psi}$ is a closed formula belonging to $\mathfrak{C}$ or the formula $\forall x (P(x) \Rightarrow \overline{\psi})$, where P is a new unary predicate symbol, belongs to $\mathfrak{C}$ (analogous conditions for ψ_1, ψ_2).*

Note that the formulae indicated in the first and second items of the definition exactly match the shape of the formulae contributing to $\mathcal{U}$ when we reduce a temporal formula to the normal form by renaming the complex expressions and replacing temporal operators by their fixed point definitions.

Theorem 3 (Decidability by Temporal Resolution). *Let $\mathfrak{C}$ be a decidable class of first-order formulae which does not contain equality and functional symbols, but possibly contains constants, such that*

- *$\mathfrak{C}$ is closed under conjunction;*
- *$\mathfrak{C}$ contains universal monadic formulae.*

Then $\mathcal{T}_{ren}\mathfrak{C}$ is decidable.

Proof. [See also [4, Theorem 8.3]] After reduction to DSNF, all formulae from $\mathcal{U}$ belong to $\mathfrak{C}$. The (monadic) formulae from side conditions and the (monadic) formulae generated by temporal resolution rules belong to $\mathfrak{C}$. Therefore, testing the applicability of one of the temporal resolution rules becomes decidable. Given that all derivations are finite, due to the finiteness of the set of merged derived step clauses and full merged step clauses, decidability follows.   ❑

A consequence of Theorem 3 is the decidability of a wide range of temporal monodic classes. These include the monadic, two-variable, fluted, guarded, and loosely guarded fragments of monodic first-order temporal logic which have also been shown to be decidable in [11,23]. In addition, decidability also follows for other classes, for example, the class $\mathcal{T}_{ren}\exists^*\forall^2\exists^*$ and the class $\mathcal{T}_{ren}\overline{K}$ where $\overline{K}$ is the dual of Maslov's class K [18]. Moreover, combining the constructions from [10] and the saturation-based decision procedure for the guarded fragment with equality [9], it is possible to build a temporal resolution decision procedure for the monadic guarded and loosely guarded fragments with equality [16].

5   Monadic Fine-Grained Temporal Resolution

The main drawback of monadic temporal resolution is that the notion of merged derived step clauses and full merged step clauses is quite involved and that the search for merged step clauses to which one of the deduction rules can successfully be applied is computationally hard, in general it is only semi-decidable. The idea underlying the *monadic fine-grained temporal resolution calculus*, fine-grained resolution for short, is to refine the deduction rules of $\mathfrak{I}_e$ in such a way that they perform much smaller steps, but with decidable side conditions for their applicability. Of course, the price that has to be paid is that derivations are no longer guaranteed to be finite.

In more detail, fine-grained resolution differs from the calculus $\mathfrak{I}_e$ in two aspects. First, instead of the step resolution and the termination rule of $\mathfrak{I}_e$, we use a set of deduction rules operating on clausified problems. Second, we use a particular algorithm, called FG-BFS, to determine the loops to which we apply the ground and non-ground eventuality resolution rule of $\mathfrak{I}_e$.

Definition 4. *Let* $P = \langle \mathcal{U}, \mathcal{I}, \mathcal{S}, \mathcal{E} \rangle$ *be a monadic temporal problem. The clausification* $\mathrm{Cls}(P)$ *of* P *is a quadruple* $\langle \mathcal{U}', \mathcal{I}', \mathcal{S}', \mathcal{E} \rangle$ *such that (i)* $\mathcal{U}'$ *is a set of clauses, called* universal clauses, *obtained by clausification of* $\mathcal{U}$; *(ii)* $\mathcal{I}'$ *is a set of clauses, called* initial clauses, *obtained by clausification of* $\mathcal{I}$; *(iii)* $\mathcal{S}'$ *is the smallest set of step clauses such that all step clauses from* $\mathcal{S}$ *are in* $\mathcal{S}'$ *and for every non-ground step clause* $P(x) \Rightarrow \bigcirc L(x)$ *in* $\mathcal{S}$ *and every constant* c *occurring* P, *the clause* $P(c) \Rightarrow \bigcirc L(c)$ *is in* $\mathcal{S}'$.

Example 2. Let $P = \langle \mathcal{U}, \mathcal{I}, \mathcal{S}, \mathcal{E} \rangle$ where $\mathcal{U} = \{\exists x Q(x)\}$, $\mathcal{I} = \{P(c)\}$, $\mathcal{S} = \{P(x) \Rightarrow \bigcirc Q(x)\}$, and $\mathcal{E} = \emptyset$. Then $\mathrm{Cls}(P) = \langle \mathcal{U}', \mathcal{I}', \mathcal{S}', \mathcal{E} \rangle$ where $\mathcal{U}' = \{Q(d)\}$ with d a Skolem constant, $\mathcal{I}' = \{P(c)\}$, and $\mathcal{S}' = \{P(x) \Rightarrow \bigcirc Q(x), P(c) \Rightarrow \bigcirc Q(c)\}$.

During a derivation more general *step* clauses can be derived, which are of the form $C \Rightarrow \bigcirc D$, where C is a *conjunction* of propositions, atoms of the form $P(x)$ and ground formulae of the form $P(c)$, where P is a unary predicate symbol and c is a constant such that c occurs in the input formula, and D is a *disjunction* of arbitrary literals.

Let us first define the deduction rules of fine-grained step resolution which replace the step resolution and the termination rule of $\mathfrak{I}_e$. In the following, we assume that different premises and conclusions of the deduction rules have no variables in common; variables may be renamed if necessary.

(1) *First-order resolution between two universal clauses.* Defined as standard first-order resolution between two clauses. The result is a universal clause.

(2) *First-order factoring on a universal clause.* Again, defined as standard first-order factoring on a clause. The result is a universal clause.

(3) *First-order resolution between an initial and a universal clause, between two initial clauses, and factoring on an initial clause.* Defined in analogy to the two deduction rules above only that the result is an initial clause.

(4) *Fine-grained step resolution.*

$$\frac{C_1 \Rightarrow \bigcirc(D_1 \vee L) \quad C_2 \Rightarrow \bigcirc(D_2 \vee \neg M)}{(C_1 \wedge C_2)\sigma \Rightarrow \bigcirc(D_1 \vee D_2)\sigma} \ ,$$

where $C_1 \Rightarrow \bigcirc(D_1 \vee L)$ and $C_2 \Rightarrow \bigcirc(D_2 \vee \neg M)$ are step clauses and σ is a most general unifier of the literals L and M *such that σ does not map variables from C_1 or C_2 into a constant or a functional term.*[3]

$$\frac{C_1 \Rightarrow \bigcirc(D_1 \vee L) \quad D_2 \vee \neg M}{C_1\sigma \Rightarrow \bigcirc(D_1 \vee D_2)\sigma} \ ,$$

where $C_1 \Rightarrow \bigcirc(D_1 \vee L)$ is a step clause, $D_2 \vee \neg M$ is a universal clause, and σ is a most general unifier of the literals L and M *such that σ does not map variables from C_1 into a constant or a functional term.*

(5) *Fine-grained step factoring.*

$$\frac{C \Rightarrow \bigcirc(D \vee L \vee M)}{C\sigma \Rightarrow \bigcirc(D \vee L)\sigma} \ ,$$

where σ is a most general unifier of the literals L and M *such that σ does not map variables from C into a constant or a functional term.*

(6) *Clause conversion.* A step clause of the form $C \Rightarrow \bigcirc\bot$ is rewritten to the universal clause[4] $\neg C$.

Besides the rules above we still need the eventuality resolution rule and the ground eventuality resolution rule of $\mathfrak{I}_e$. However, we use a particular algorithm, called FG-BFS (for fine-grained breadth-first search), to find loop formulae, that is, to find a disjunction of the left-hand sides of full merged step clauses that together with an eventuality literal forms the premises for the ground and non-ground eventuality resolution rules. This algorithm internally uses the deduction rules above with the exception of the clause conversion rule.

Let *fine-grained resolution* be the calculus consisting of the rules (1) to (6) above, together with the ground and non-ground eventuality resolution rules, restricted to loops found by the FG-BFS algorithm. We denote this calculus by $\mathfrak{I}_{FG}$. The calculus can be extended by first-order redundancy elimination rules, e.g. tautology and subsumption deletion, as well as analogous rules for step clauses.

---

[3] This restriction justifies skolemisation of the universal part: Skolem constants from one moment of time do not propagate to the previous moment.

[4] Here, and further, $\neg(L_1(x) \wedge \cdots \wedge L_k(x))$ abbreviates $(\neg L_1(x) \vee \cdots \vee \neg L_k(x))$.

A (linear) *derivation* in $\mathfrak{I}_{FG}$ from the clausification $\text{Cls}(\mathsf{P}^c)$ of a constant flooded monodic temporal problem P^c is a sequence of clauses $C_1, \ldots$ such that each clause C_i is either an element of $\text{Cls}(\mathsf{P}^c)$ or else the conclusion by a deduction rule applied to clauses from premises $C_1, \ldots, C_{i-1}$. A derivation $C_1, \ldots, C_m$ is also called a *proof of C_m*. A proof of the empty clause is called a *refutation*. A derivation $C_1, \ldots, C_m$ *terminates* iff for any derivation $C_1, \ldots, C_m, C_{m+1}$, the clause C_{m+1} is a variant of a clause in $\text{Cls}(\mathsf{P}^c) \cup \{C_1, \ldots, C_m\}$.

Theorem 4 ([15] Theorems 5 and 9). *Fine-grained resolution is sound and complete for constant flooded monodic temporal problems over expanding domains.*

For the class of problems where all the literals in a problem are propositional or ground, fine-grained resolution is a decision procedure, as the inference steps performed by it are exactly those performed by the clausal temporal calculus [8] for propositional linear-time temporal logic, which is an exponential time decision procedure for the satisfiability problem of that logic. However, for all the classes mentioned at the end of Section 4, termination of fine-grained resolution cannot be guaranteed. So, in analogy to the approach taken to obtain resolution decision procedure for decidable fragments of first-order logic, we develop sound and complete *refinements* of fine-grained resolution to ensure termination of derivations. We assume that we are given an *atom ordering* $\succ$, that is, a total and well-founded ordering on ground first-order atoms which is stable under substitution, and a *selection function* S which maps any first-order clause C to a (possibly empty) subset of its negative literals. An atom ordering $\succ$ is extended to literals by $(\neg)A \succ (\neg)B$ if $A \succ B$ and $\neg A \succ A$. A literal L is called (strictly) maximal w.r.t. a clause C iff there exists a ground substitution σ such that for all $L' \in C$: $L\sigma \succeq L'\sigma$ ($L\sigma \succ L'\sigma$). A literal L is *eligible* in a clause $L \vee C$ if either it is selected in $L \vee C$, or no literal is selected in C and L is maximal w.r.t. C.

The atom ordering $\succ$ and the selection function S are used to restrict the applicability of the deduction rules of fine-grained resolution as follows.

(1) *First-order ordered resolution with selection between two universal clauses*

$$\frac{C_1 \vee A \quad \neg B \vee C_2}{(C_1 \vee C_2)\sigma} \ ,$$

if σ is the most general unifier of A and B, $A\sigma$ is eligible in $(C_1 \vee A)\sigma$, and $\neg B\sigma$ is eligible in $(\neg B \vee C_2)\sigma$.

(2) *First-order ordered positive factoring with selection*

$$\frac{C_1 \vee A \vee B}{(C_1 \vee A)\sigma} \ ,$$

if σ is the most general unifier of A and B, and $A\sigma$ is eligible in $(C_1 \vee A \vee B)\sigma$.

(3) *First-order ordered resolution with selection between an initial and a universal clause, between two initial clauses, and ordered positive factoring with selection on an initial clause.* These are defined in analogy to the two deduction rules above with the only difference that the result is an initial clause.

(4) *Ordered fine-grained step resolution with selection.*

$$\frac{C_1 \Rightarrow \bigcirc(D_1 \vee A) \quad C_2 \Rightarrow \bigcirc(D_2 \vee \neg B)}{(C_1 \wedge C_2)\sigma \Rightarrow \bigcirc(D_1 \vee D_2)\sigma} ,$$

where $C_1 \Rightarrow \bigcirc(D_1 \vee L)$ and $C_2 \Rightarrow \bigcirc(D_2 \vee \neg M)$ are step clauses, σ is a most general unifier of the literals L and M such that σ does not map variables from C_1 or C_2 into a constant or a functional term, $A\sigma$ *is eligible in* $(D_1 \vee A)\sigma$, *and* $\neg B\sigma$ *is eligible in* $(D_2 \vee \neg B)\sigma$.

$$\frac{C_1 \Rightarrow \bigcirc(D_1 \vee L) \quad D_2 \vee \neg M}{C_1\sigma \Rightarrow \bigcirc(D_1 \vee D_2)\sigma} ,$$

where $C_1 \Rightarrow \bigcirc(D_1 \vee L)$ is a step clause, $D_2 \vee \neg M$ is a universal clause, and σ is a most general unifier of the literals L and M such that σ does not map variables from C_1 into a constant or a functional term, $N\sigma$ *is eligible in* $(D_2 \vee \neg N)\sigma$, *and* $L\sigma$ *is eligible in* $(D_1 \vee L)\sigma$.

(5) *Ordered fine-grained positive step factoring with selection.*

$$\frac{C \Rightarrow \bigcirc(D \vee A \vee B)}{C\sigma \Rightarrow \bigcirc(D \vee A)\sigma} ,$$

where σ is a most general unifier of the atoms A and B such that σ does not map variables from C into a constant or a functional term, and $A\sigma$ *is eligible in* $(D \vee A \vee B)\sigma$.

(6) *Clause conversion.* A step clause of the form $C \Rightarrow \bigcirc\bot$ is rewritten to the universal clause $\neg C$.

Let *ordered fine-grained resolution with selection* be the calculus consisting of the rules (1) to (6) above, together with the ground and non-ground eventuality resolution rules, restricted to loops found by the FG-BFS algorithm which now uses the rules (1) to (5) above instead of their unrefined variants. We denote this calculus by $\mathfrak{I}_{FG}^{S,\succ}$. Again, the calculus can be extended by first-order redundancy elimination rules as well as analogous rules for step clauses.

Note that for ordered fine-grained step resolution with selection, the ordering and selection function only influence which literals on the right-hand side of a step clause are eligible, literals on the left-hand side are not taken into account.

Theorem 5. *Ordered fine-grained resolution with selection is sound and complete for constant flooded monodic temporal problems over expanding domains.*

Proof. [Sketch] Soundness of $\mathfrak{I}_{FG}^{S,\succ}$ is straightforward as any derivation in $\mathfrak{I}_{FG}^{S,\succ}$ is also a derivation in $\mathfrak{I}_{FG}$, which is sound according to Theorem 4.

The proof of completeness proceeds along the lines of the completeness proof of $\mathfrak{I}_{FG}$ presented in [14]. Assume that $\mathsf{P}^c = \langle \mathcal{U}_0, \mathcal{I}, \mathcal{S}, \mathcal{E} \rangle$ is a constant flooded monodic temporal problem and $\Delta = \mathcal{U}_0, \dots, \mathcal{U}_n$ is a derivation from P^c in $\mathfrak{I}_e$ such that $\mathcal{U}_n$ contains $\bot$, that is, $\mathfrak{I}_e$ is able to derive a contradiction from P^c. By induction on the length of the derivation we show that this derivation can be simulated by $\mathfrak{I}_{FG}^{S,\succ}$. We construct a refutation $\Delta' = C_0^1, \dots, C_0^{n_0}, \dots, C_n^1, \dots, C_n^{n_k}$

of the clausification $\mathrm{Cls}(\mathsf{P}^c)$ of P^c where each step in Δ will correspond to one or more steps in Δ'. At the start Δ just consists of $\mathcal{U}_0$ and the corresponding derivation Δ' consists of all the clauses $C_0^1, \ldots, C_0^{n_0}$ in $\mathrm{Cls}(\mathsf{P}^c)$. Let $U(\Delta')$ and $I(\Delta')$ denote the set of all universal and initial clauses in Δ', respectively. By the fact that clausification preserves satisfiability, $\mathcal{U}_0$ is satisfiable iff $U(\Delta')$ is satisfiable and $\mathcal{U}_0 \cup \mathcal{I}$ is satisfiable iff $U(\Delta') \cup I(\Delta')$ is satisfiable. Furthermore, if $\mathcal{U}_0$ would contain $\bot$, then Δ' would contain the empty clause.

Now, in each step of Δ a first-order formula u_i, $1 \leq i \leq n$, is added to $\mathcal{U}_{i-1}$ to obtain $\mathcal{U}_i$, where u_i is the conclusion of one the deduction rules of $\mathfrak{I}_e$ applied to $\langle \mathcal{U}_{i-1}, \mathcal{I}, \mathcal{S}, \mathcal{E} \rangle$. We show that using $\mathfrak{I}_{FG}^{s,\succ}$ we can derive a clause $C_i^{n_i}$ from the clauses in the derivation Δ' constructed so far such that the universal closure of $C_i^{n_i}$ implies u_i. This also implies that $\mathcal{U}_i$ is satisfiable iff $U(\Delta') \cup \{C_i^{n_i}\}$ is satisfiable and $\mathcal{U}_i \cup \mathcal{I}$ is satisfiable iff $U(\Delta') \cup I(\Delta') \cup \{C_i^{n_i}\}$ is satisfiable. We then add $C_i^{n_i}$ and all intermediate clauses $C_i^1, \ldots, C_i^{n_i-1}$ used in its derivation to Δ'. To show the existence and derivability of $C_i^{n_i}$ we consider which deduction rule of $\mathfrak{I}_e$ has been used to derive u_i.

Suppose u_i has been derived by an application of the termination rule (which implies that u_i is $\bot$). Then the set $\mathcal{U}_{i-1} \cup \mathcal{I}$ of first-order formulae is unsatisfiable, which, by induction hypothesis, implies that $U(\Delta') \cup I(\Delta')$ is unsatisfiable. By completeness of first-order ordered resolution with selection (see, e.g. [1]), we will be able to derive the empty clause from the clauses in $U(\Delta') \cup I(\Delta')$ using the resolution and factoring rules of $\mathfrak{I}_{FG}^{s,\succ}$ for universal and initial clauses, that is, rules (1) to (3), and extend Δ' accordingly.

Suppose u_i has been derived by an application of the step resolution rule. Then there is a merged derived step clause $\mathcal{A} \Rightarrow \bigcirc \mathcal{B}$ such that the formula $\mathcal{U}_{i-1} \cup \{\mathcal{B}\}$ is unsatisfiable. The merged derived step clause $\mathcal{A} \Rightarrow \bigcirc \mathcal{B}$ is constructed from some step clauses $p_j \Rightarrow \bigcirc q_j$, $1 \leq j \leq m_1$, $P_k(c_k) \Rightarrow \bigcirc Q_k(c_k)$, $1 \leq k \leq m_2$, and $P_l(x_l) \Rightarrow \bigcirc Q_l(x_l)$, $1 \leq l \leq m_3$, in $\mathcal{S}$ which are also present in Δ'. Define a set $L(\mathcal{B})$ of literals as $\{q_j \mid 1 \leq j \leq m_1\} \cup \{Q_k(c_k) \mid 1 \leq k \leq m_2\} \cup \{Q_l(x_l) \mid 1 \leq l \leq m_3\}$. Again, due to the completeness of first-order ordered resolution with selection, a derivation of the empty clause from $L(\mathcal{B}) \cup U(\Delta')$ exists. Then inspection of the rules (4) and (5) for ordered fine-grained step resolution with selection and for ordered right positive factoring with selection, respectively, shows that we can also construct a derivation from $\{p_j \Rightarrow \bigcirc q_j \mid 1 \leq j \leq m_1\} \cup \{P_k(c_k) \Rightarrow \bigcirc Q_k(c_k) \mid 1 \leq k \leq m_2\} \cup \{P_l(x_l) \Rightarrow \bigcirc Q_l(x_l) \mid 1 \leq l \leq m_3\} \cup U(\Delta')$. This will not be a derivation of the empty clause, but of a step clause $\mathcal{P} \Rightarrow \bigcirc \bot$ where $\mathcal{P}$ is a conjunction of literals in $\{P_j \mid 1 \leq j \leq m_1\} \cup \{P_k(c_k) \mid 1 \leq k \leq m_2\} \cup \{P_l(x_l) \mid 1 \leq l \leq m_3\}$, though not necessarily all of them. An application of rule (6) for clause conversion allows us to derive the universal clause $\neg \mathcal{P}$. We can show that the universal closure of $\neg \mathcal{P}$ implies u_i. We add all the clauses in the derivation of $\mathcal{P} \Rightarrow \bigcirc \bot$ to Δ' as $C_i^1, \ldots, C_i^{n_i-1}$ for some n_i, and also add $\neg \mathcal{P}$ as $C_i^{n_i}$.

Finally, concerning the ground and non-ground eventuality resolution rules and the use of the FG-BFS algorithm to compute loops, we simply observe that using rules (1) to (5) in the algorithm will not change the loops the algorithm will compute. This follows from the considerations above. $\square$

6   Decidability by Ordered Fine-Grained Resolution with Selection

Ordered fine-grained resolution with selection allows us to transfer decidability results and decision procedures obtained for fragments of first-order logic to the corresponding fragments of monodic first-order temporal logic.

We present two examples. First, we consider the temporalisation $\mathcal{T}_{ren}\,GF$ of the guarded fragment by renaming according to Definition 3 and we show how a decision procedure can be constructed from the procedure for the guarded fragment developed in [9]. Second, using the same approach we derive a decision procedure for $\mathcal{T}_{ren}\overline{K}$ and $\mathcal{T}_{ren}\overline{DK}$ based on the procedure for $\overline{K}$ and $\overline{DK}$ developed in [13] ($\overline{DK}$ is the class containing all conjunctions of formulae of the class $\overline{K}$).

Ganzinger and de Nivelle [9] use the following ordering $\succ_{GF}$ and selection function S_{GF} to decide the guarded fragment: $\succ_{GF}$ is an arbitrary lexicographic path ordering on terms and atoms based on a precedence $\succ$ on function and predicate symbols such that $f \succ c \succ p$ for any non-constant function symbol f, constant c, and predicate symbol p. The selection function S_{GF} selects one of the guards in any clause that is non-functional[5] and contains at least one guard; it selects one of the functional negative literals in a clause containing such literals; and it does not select any literal in a clause containing a positive functional literal but no negative functional literal. On guarded clauses, that is, the class of clauses which contains the clause normal form of any guarded formula, the selection function S_{GF} is well-defined. In addition, the decision procedure in [9] requires that in the computation of the clausification of guarded formulae *structural transformation* [7,20] is used to introduce surrogates for universally quantified subformulae. Let ST_{GF} denote this transformation.

We can use exactly the same ordering and selection function to obtain a decision procedure for $\mathcal{T}_{ren}\,GF$.

Theorem 6. *Let $\succ_{GF}$ and S_{GF} be the ordering and selection function defined above. Then $\mathfrak{I}_{FG}^{S_{GF},\succ_{GF}}$ decides the satisfiability problem of $\mathcal{T}_{ren}\,GF$.*

Proof. By Theorem 5, $\mathfrak{I}_{FG}^{S_{GF},\succ_{GF}}$ is sound and complete. It remains to show termination. Let φ be a formula in $\mathcal{T}_{ren}\,GF$ and P^c be the corresponding constant flooded temporal problem. In analogy to [9], we use the structural transformation ST_{GF} in the computation of the clausification of P^c. Let P^c_{Cls} denote $\mathrm{Cls}(\mathrm{ST}_{GF}(\mathsf{P}^c))$. First, we give a syntactical characterisation of the clauses in P^c_{Cls} and of the clauses we might have derived from it. To do so, we extend the notion of a guarded clause to step clauses as follows. A step clause $C \Rightarrow \bigcirc D$ is guarded iff the first-order clause $\neg C \vee D$ is guarded and C is monadic. Then all the universal, initial, and step clauses in P^c_{Cls} are guarded. We can also show that all inference steps possible by $\mathfrak{I}_{FG}^{S_{GF},\succ_{GF}}$ on guarded (step) clauses will result in a guarded (step) clause. Second, the number of guarded (step) clauses (up to variable renaming) over the signature Σ of P^c_{Cls} is finitely bounded, more

---

[5] An expression is *functional* if it contains a constant or a function symbol, and *non-functional* otherwise.

precisely, there is a double exponential upper bound in the size of Σ on their number. Consequently, any derivation from P^c_{Cls} will either eventually produce the empty clause or no new clauses can be added to the derivation.   ❑

Our second example is a decision procedure for $\mathcal{T}_{ren}\overline{\mathrm{K}}$ and $\mathcal{T}_{ren}\overline{\mathrm{DK}}$ based on the resolution decision procedure for $\overline{\mathrm{K}}$ and $\overline{\mathrm{DK}}$ by Hustadt and Schmidt [13]. The procedure uses an atom ordering $\succ_K$ which is a recursive path ordering based on a total precedence $\succ$ on function and predicate symbols which basically gives precedence to symbols of greater arity. The selection function S_K maps any clause to the empty set. The decision procedure also uses an additional inference rule, namely *splitting*, to perform case analysis on clauses consisting of variable-disjoint subclauses. While it is possible to extend the calculus $\mathfrak{I}^{S,\succ}_{FG}$ by a splitting inference rule, it is easier to use splitting through new predicate symbols instead [3,21]. Here, whenever we have a clause $C \vee D$ such that C and D are variable-disjoint, we replace it by two clauses $C \vee p$ and $\neg p \vee D$, where p is a new predicate symbol of arity 0 smaller than any other predicate symbol. Finally, the procedure requires the use of structural transformation in the computation of the clausification of formulae in $\overline{\mathrm{K}}$ and $\overline{\mathrm{DK}}$. Here, certain occurrences of one-variable literals with constant or duplicate variable arguments have to replaced by surrogates (see [13] for details). Let $\mathrm{ST}_{\overline{K}}$ denote this transformation.

Theorem 7. *Let $\succ_K$ and S_K be the ordering and selection function defined above. Then $\mathfrak{I}^{S_K,\succ_K}_K$ decides the satisfiability problem of $\mathcal{T}_{ren}\overline{\mathrm{K}}$ and $\mathcal{T}_{ren}\overline{\mathrm{DK}}$.*

Proof.   Along the lines of the proof of Theorem 6. Let φ be a formula in $\mathcal{T}_{ren}\overline{\mathrm{K}}$ or $\mathcal{T}_{ren}\overline{\mathrm{DK}}$, let P^c be the corresponding constant flooded temporal problem, and let P^c_{Cls} be $\mathrm{Cls}(\mathrm{ST}_{\overline{K}}(\mathsf{P}^c))$. The characterisation of clauses in P^c_{Cls} and of the clauses we derive from it is based on the notions of (strongly) k-regular and (strongly) CDV-free clauses introduced in [13]. Again, we need to extend these notions to step clauses. We can then show that all universal, initial, and step clauses in P^c_{Cls} are strongly CDV-free and k-regular, or strongly k-regular if φ belongs to $\mathcal{T}_{ren}\overline{\mathrm{DK}}$. Inference steps restricted by $\succ_K$ and S_K will also only derive clauses with these properties. There is a double exponential upper bound on the number of (strongly) k-regular, (strongly) CDV-free clauses in the size of the signature of P^c_{Cls}. This shows termination of any derivation in $\mathfrak{I}^{S_K,\succ_K}_K$.   ❑

7   Future Work

One motivation for our interest in classes decidable by ordered fine-grained resolution with selection is that with the theorem prover **TeMP** [12] we have an implementation of fine-grained resolution. **TeMP** takes advantage of the arithmetic translation of temporal problems which allows us to use a first-order theorem prover, in our case Vampire, to implement the inference rules of fine-grained resolution. Consequently, we will have to transfer the restrictions imposed by ordered fine-grained resolution with selection to the level of the first-order theorem prover employed by **TeMP** to realise the decision procedures presented in this paper.

References

1. L. Bachmair and H. Ganzinger. Resolution theorem proving. In Robinson and Voronkov [22], chapter 2, pp. 19–99.
2. E. Börger, E Grädel, and Yu. Gurevich. *The Classical Decision Problem.* Springer, 1997.
3. H. de Nivelle. Splitting through new proposition symbols. In *Proc. LPAR 2001*, vol. 2250 of *LNAI*, pp. 172–185. Springer, 2001.
4. A. Degtyarev, M. Fisher, and B. Konev. Monodic temporal resolution. *ACM Transactions on Computational Logic.* To appear.
5. A. Degtyarev, M. Fisher, and B. Konev. Monodic temporal resolution. In *Proc. CADE-19*, vol. 2741 of *LNAI*, pp. 397–411. Springer, 2003.
6. E. A. Emerson. Temporal and modal logic. In J. van Leeuwen, editor, *Handbook of Theoretical Computer Science*, chapter 16, pp. 997–1072. Elsevier, 1990.
7. C. Fermüller, A. Leitsch, U. Hustadt, and T. Tammet. Resolution decision procedures. In Robinson and Voronkov [22], chapter 25, pp. 1791–1850.
8. M. Fisher, C. Dixon, and M. Peim. Clausal temporal resolution. *ACM Transactions on Computational Logic*, 2(1):12–56, 2001.
9. H. Ganzinger and H. de Nivelle. A superposition decision procedure for the guarded fragment with equality. In *Proc. LICS'99*, pp. 295–304. IEEE, 1999.
10. I. Hodkinson. Monodic packed fragment with equality is decidable. *Studia Logica*, 72(2):185–197, 2002.
11. I. Hodkinson, F. Wolter, and M. Zakharyaschev. Decidable fragments of first-order temporal logics. *Annals of Pure and Applied Logic*, 106:85–134, 2000.
12. U. Hustadt, B. Konev, A. Riazanov, and A. Voronkov. **TeMP**: A temporal monodic prover. In *Proc. IJCAR 2004*, vol. 3097 of *LNAI*, pp. 326–330. Springer, 2004.
13. U. Hustadt and R. A. Schmidt. Maslov's class K revisited. In *Proc. CADE-16*, vol. 1632 of *LNAI*, pp. 172–186. Springer, 1999.
14. B. Konev, A. Degtyarev, C. Dixon, M. Fisher, and U. Hustadt. Mechanising first-order temporal resolution. *Information and Computation.* To appear. Also available as Technical Report ULCS-03-023, Dep. Comp. Sci., Univ. Liverpool, 2003.
15. B. Konev, A. Degtyarev, C. Dixon, M. Fisher, and U. Hustadt. Towards the implementation of first-order temporal resolution: the expanding domain case. In *Proc. TIME-ICTL 2003*, pp. 72–82. IEEE, 2003.
16. B. Konev, A. Degtyarev, and M. Fisher. Handling equality in monodic temporal resolution. In *Proc. LPAR 2003*, vol. 2850 of *LNAI*, pp. 214–228. Springer, 2003.
17. R. Kontchakov, C. Lutz, F. Wolter, and M. Zakharyaschev. Temporalising tableaux. *Studia Logica*, 76(1):91–134, 2004.
18. S. Ju. Maslov. The inverse method for establishing deducibility for logical calculi. In V. P. Orevkov, editor, *The Calculi of Symbolic Logic I: Proceedings of the Steklov Institute of Mathematics, number 98 (1968)*, pp. 25–96. American Math. Soc., 1971.
19. R. Nieuwenhuis and A. Rubio. Paramodulation-based theorem proving. In Robinson and Voronkov [22], chapter 7, pp. 371–443.
20. A. Nonnengart and Ch. Weidenbach. Computing small clause normal forms. In Robinson and Voronkov [22], chapter 6, pp. 335–370.
21. A. Riazanov and A. Voronkov. Splitting without backtracking. In *Proc. IJCAI 2001*, pp. 611–617. Morgan Kaufmann, 2001.
22. A. Robinson and A. Voronkov, editors. *Handbook of Automated Reasoning.* Elsevier, 2001.
23. F. Wolter and M. Zakharyaschev. Axiomatizing the monodic fragment of first-order temporal logic. *Annals of Pure and Applied logic*, 118:133–145, 2002.

Hierarchic Reasoning in Local Theory Extensions[*]

Viorica Sofronie-Stokkermans[**]

Max-Planck-Institut für Informatik, Stuhlsatzenhausweg 85, Saarbrücken, Germany
sofronie@mpi-sb.mpg.de

Abstract. We show that for special types of extensions of a base theory, which we call *local*, efficient hierarchic reasoning is possible. We identify situations in which it is possible, for an extension T_1 of a theory T_0, to express the decidability and complexity of the universal theory of T_1 in terms of the decidability resp. complexity of suitable fragments of the theory T_0 (universal or $\forall\exists$). These results apply to theories related to data types, but also to certain theories of functions from mathematics.

1 Introduction

Many problems in mathematics and computer science and, in particular, problems involving reasoning in and about complex systems, can be reduced to proving the satisfiability of conjunctions of literals modulo some background theory. This theory may be quite complex: it can for instance be the extension of a base theory with additional functions (free, monotone, or recursively defined) or a combination of theories. It is therefore extremely important to find methods for efficient reasoning in extensions and combinations of theories.

In this paper we address the problem of reasoning in extensions of theories. We show that for special types of theory extensions, which we call *local*, hierarchic reasoning in which a theorem prover for the base theory is used as a "black box" is possible. Many theories important for computer science or mathematics are local extensions of a base theory. Examples are theories of data structures, e.g. theories of lists (or arrays cf. [6]); but also theories of monotone functions or of functions satisfying the Lipschitz conditions at a given point. We identify situations where the decidability of the universal theory of an extension of a theory is a consequence of the decidability of a certain fragment of the base theory.

The notion of local extension of a theory which we introduce in this paper generalizes the notion of *locality of a theory* introduced by Givan and McAllester [7,8], and of locality of equational theories studied by Ganzinger [4]. For local theories, validity of ground Horn clauses can be checked in polynomial time.

---

[*] This work was partly supported by the German Research Council (DFG) as part of the Transregional Collaborative Research Center "Automatic Verification and Analysis of Complex Systems" (SFB/TR 14 AVACS). See www.avacs.org for more information.
[**] Many of the results presented here are the result of joint work with Harald Ganzinger[†].

R. Nieuwenhuis (Ed.): CADE 2005, LNAI 3632, pp. 219–234, 2005.

Similar ideas also occurred in algebra. To prove that the uniform word problem for lattices is decidable in polynomial time, Skolem (1920) used the following idea: replace the lattice operations $\vee$ and $\wedge$ by ternary relations $r_\vee$ and $r_\wedge$, required to be functional, but not necessarily total. The lattice axioms were translated to a relational form, by flattening them and then replacing every atom of the form $x \vee y \approx z$ with $r_\vee(x, y, z)$ (similarly for $\wedge$-terms). Additional axioms were added, stating that equality is an equivalence and that the relations are compatible with equality and functional. This new presentation, consisting only of Horn, function-free clauses, can be used for deciding in polynomial time the uniform word problem for lattices. The correctness and completeness of the method relies on the fact that every partially-ordered set (where $\vee$ and $\wedge$ are partially defined) embeds into a lattice. The idea described above was extended by Burris [2] to quasi-varieties of algebras. He proved that if a quasi-variety axiomatized by a set $\mathcal{K}$ of Horn clauses has the property that *every finite partial algebra which is a partial model of the axioms in $\mathcal{K}$ can be extended to a total algebra model of $\mathcal{K}$* then the uniform word problem for $\mathcal{K}$ is decidable in polynomial time.

In [4], Ganzinger established a link between proof theoretic and semantic concepts for polynomial time decidability of uniform word problems. He defined two notions of locality for equational Horn theories, and established relationships between these notions of locality and corresponding semantic conditions, referring to embeddability of partial algebras into total algebras.

Our paper continues this line of research. Its main contributions are the following. First, we generalize in several ways the notion of locality of an equational theory:

- We consider local extensions $\mathcal{T}_0 \subseteq \mathcal{T}_1$, where the base theory $\mathcal{T}_0$ can be arbitrary. If $\mathcal{T}_0$ is the empty theory the original notion of locality is recovered.
- In defining locality of a theory extension $\mathcal{T}_0 \subseteq \mathcal{T}_1$ by a set $\mathcal{K}$ of formulae we allow $\mathcal{K}$ to be an arbitrary set of clauses (not necessarily Horn).

Second, we relate the extended notions of locality we consider with semantic properties, involving embeddability of partial algebras into total algebras.

Third, we use these results for hierarchic reasoning in local theory extensions, and identify situations in which this allows us to express the complexity of the universal theory of the extension as a function of the complexity of appropriate fragments of the base theory. We also sketch a possibility of extending the results beyond universally quantified formulae.

Structure of the Paper: Section 2 contains basic notions and notations. In Section 3, embeddability conditions are introduced and illustrated by examples; in Section 4 notions of locality of an extension are defined. The main contributions of the paper are contained in Sections 5 and 6: In Section 5 we establish links between various notions of locality of a theory extension and semantic properties, involving embeddability of partial algebras into total algebras. This helps to identify cases in which suitable locality conditions for theory extensions hold.

In Section 6 we establish parameterized complexity results of the universal theory of the extension in terms of the complexity of fragments of the base theory. Section 7 sketches a possibility of going beyond the universal fragment.

1.1   Idea

We illustrate the idea of our approach. Let $\mathbb{R} \cup \mathsf{L}_f^{\lambda_1} \cup \mathsf{L}_g^{\lambda_2}$ be the extension of the theory $\mathbb{R}$ of reals with function symbols f, g satisfying the following axioms:

$$(\mathsf{L}_f^{\lambda_1}) \quad |f(x) - f(c_0)| \leq \lambda_1 \cdot |x - c_0| \qquad (\mathsf{L}_g^{\lambda_2}) \quad |g(x) - g(c_0)| \leq \lambda_2 \cdot |x - c_0|$$

where c_0, λ_1, and λ_2 are constants and the free variable x is, in both cases, implicitly universally quantified. We want to prove:

$$\mathbb{R} \cup (\mathsf{L}_f^{\lambda_1}) \cup (\mathsf{L}_g^{\lambda_2}) \models \forall x(|f(x) + g(x) - (f(c_0) + g(c_0))| \leq (\lambda_1 + \lambda_2) \cdot |x - c_0|).$$

Standard theorem provers for first order logic cannot always be used in such situations, as these can usually handle only approximations of the theory of real numbers. Provers for reals do not know about additional functions. The Nelson-Oppen method for reasoning in combinations of theories cannot be used either. The method we propose reduces the task of proving the formula above to the problem of checking the satisfiability of a set of constraints over $\mathbb{R}$ as follows:

Negate. Let $\mathcal{K} = (\mathsf{L}_f^{\lambda_1}) \cup (\mathsf{L}_g^{\lambda_2})$. Note that $\mathbb{R} \cup \mathcal{K} \models \forall x C(x)$ (where $C(x)$ is $(|f(x) + g(x) - (f(c_0) + g(c_0))| \leq (\lambda_1 + \lambda_2) \cdot |x - c_0|)$) if and only if $\mathbb{R} \cup \mathcal{K} \cup G$ is unsatisfiable, where $G = |f(c) + g(c) - (f(c_0) + g(c_0))| \not\leq (\lambda_1 + \lambda_2) \cdot |c - c_0|$ is the set of ground clauses obtained from $\neg \forall x C(x)$ by Skolemization.

Take Ground Instances of Extension Axioms. We will show that $\mathbb{R} \cup \mathcal{K} \cup G$ is satisfiable if and only if $\mathbb{R} \cup \mathcal{K}[G] \cup G$ has a partial model in which all terms in the set $\mathsf{st}(\mathcal{K}, G)$ consisting of all ground subterms in $\mathcal{K}$ or in G are defined (and hence $f(c_0), g(c_0), f(c), g(c)$ are defined). ($\mathcal{K}[G]$ denotes the set of all instances of $\mathcal{K}$ in which the terms starting with f or g are in $\mathsf{st}(\mathcal{K}, G)$.) We compute $\mathcal{K}[G] \cup G$ and flatten replacing the ground terms starting with f or g with new constants:

$$\begin{aligned}
(\mathcal{K}[G] \cup G)_{\mathsf{flat}} := \; & f(c) \approx d \; \wedge \; f(c_0) \approx d_0 \; \wedge \; g(c) \approx e \; \wedge \; g(c_0) \approx e_0 \; \wedge \\
& |d - d_0| \leq \lambda_1 \cdot |c - c_0| \; \wedge \; |d_0 - d_0| \leq \lambda_1 \cdot |c_0 - c_0| \; \wedge \\
& |e - e_0| \leq \lambda_2 \cdot |c - c_0| \; \wedge \; |e_0 - e_0| \leq \lambda_2 \cdot |c_0 - c_0| \; \wedge \\
& |(d + e) - (d_0 + e_0)| \leq (\lambda_1 + \lambda_2) \cdot |c - c_0|
\end{aligned}$$

Relational Translation. We compute the relational translation of the clauses above, using instead of f and g the functional binary predicates r^f and r^g:

$$\begin{aligned}
(\mathcal{K}[G] \cup G)^* := \; & r^f(c, d) \; \wedge \; r^f(c_0, d_0) \; \wedge \; r^g(c, e) \; \wedge \; r^g(c_0, e_0) \; \wedge \\
& |d - d_0| \leq \lambda_1 \cdot |c - c_0| \; \wedge \; |d_0 - d_0| \leq \lambda_1 \cdot |c_0 - c_0| \; \wedge \\
& |e - e_0| \leq \lambda_2 \cdot |c - c_0| \; \wedge \; |e_0 - e_0| \leq \lambda_2 \cdot |c_0 - c_0| \; \wedge \\
& |(d + e) - (d_0 + e_0)| \leq (\lambda_1 + \lambda_2) \cdot |c - c_0|
\end{aligned}$$

$$\mathsf{Fun} := x_1 \approx x_2 \wedge R(x_1, y_1) \wedge R(x_2, y_2) \rightarrow y_1 \approx y_2 \quad \text{for } R = r^f \text{ or } R = r^g$$

We will show that we only need to consider those instances Fun^* of Fun in which the r^f resp. r^g literals are the ground literals already occurring in $(\mathcal{K}[G] \cup G)^*$, and that $\mathbb{R} \cup (\mathcal{K}[G] \cup G)^* \cup \mathsf{Fun}^*$ has a (relational) model if and only if the following set of constraints in $\mathbb{R}$ is satisfiable:

$$\{\lambda_1{>}0, \ \lambda_2{>}0, \ (c{\approx}c_0 \rightarrow d{\approx}d_0), \ (c{\approx}c_0 \rightarrow e{\approx}e_0), \ (|d - d_0| \leq \lambda_1 \cdot |c - c_0|),$$
$$(|e - e_0| \leq \lambda_2 \cdot |d - c_0|), \ |(d + e) - (d_0 + e_0)| \not\leq (\lambda_1 + \lambda_2) \cdot |c - c_0|\}.$$

We proved the unsatisfiability of this set of non-linear constraints using the REDLOG demo [3] (we considered the disjunction over the cases $c \leq c_0$ and $c > c_0$ and used quantifier elimination).

2   Basic Notions and Notations

Local Theories. The notion of *local theory* was introduced in [7,8] by Givan and McAllester. A local theory is a set of Horn clauses $\mathcal{K}$ such that, for any ground Horn clause C, $\mathcal{K} \models C$ only if already $\mathcal{K}[C] \models C$ (where $\mathcal{K}[C]$ is the set of instances of $\mathcal{K}$ in which all terms are subterms of ground terms in either $\mathcal{K}$ or C). In [4], Ganzinger defined *locality* and *stable locality* of equational Horn theories, and established relationships between these notions of locality and embeddability of partial algebras into total algebras.

Total and Partial Algebras. We now present some generalities on partial algebras. Further details on partial algebras can be found in [1].

A *partial Σ-algebra* is a structure $(A, \{f_A\}_{f \in \Sigma})$, where A is a non-empty set and for every $f \in \Sigma$ with arity n, f_A is a partial function from A^n to A. A *(total) Σ-algebra* is a partial Σ-algebra where all functions f_A are total. In what follows we usually denote with the same symbol both an algebra and its support.

The notion of evaluating a term t with respect to a variable assignment $\beta : X \rightarrow A$ for its variables in a partial algebra A is the same as for total algebras, except that this evaluation is undefined if $t = f(t_1, \ldots, t_n)$ and either one of $\beta(t_i)$ is undefined, or else $(\beta(t_1), \ldots, \beta(t_n))$ is not in the domain of f_A.

A total map $h : A \rightarrow B$ between partial Σ-algebras A and B is called a *weak Σ-homomorphism* if whenever $f_A(a_1, \ldots, a_n)$ is defined in A, then also $f_B(h(a_1), \ldots, h(a_n))$ is defined in B and $h(f_A(a_1, \ldots, a_n)) = f_B(h(a_1), \ldots, h(a_n))$. A partial algebra A weakly embeds into a (total) algebra B if there exists an injective weak Σ-homomorphism from A to B.

In what follows we will consider structures $(A, \{f_A\}_{f \in \Sigma}, \{P_A\}_{P \in \mathsf{Pred}})$, where Pred is a set of predicate symbols and $(A, \{f_A\}_{f \in \Sigma})$ is a partial Σ-algebra. We will refer to this type of structures as Π-algebras (or Π-models), where $\Pi = (\Sigma, \mathsf{Pred})$. We say that a partial Π-algebra A weakly embeds into a (total) Π-algebra B if there exists $i : A \rightarrow B$ which is an injective weak Σ-homomorphism from A to B, and an embedding with respect to Pred.

We define *Evans validity* in structures $(A, \{f_A\}_{f \in \Sigma}, \{P_A\}_{P \in \mathsf{Pred}})$, where Pred is a set of predicate symbols and $(A, \{f_A\}_{f \in \Sigma})$ is a partial Σ-algebra. In what follows the symbol $\approx$ standing for formal equality will be considered to be symmetric also syntactically, so $s \approx t$ denotes at the same time also $t \approx s$. Let $\beta : X \rightarrow A$.

(1) $(A, \beta) \models t \approx s$ if and only if (a) $\beta(t)$ and $\beta(s)$ are both defined and equal; or
  (b) $\beta(s)$ is defined, $t = f(t_1, \ldots, t_n)$ and $\beta(t_i)$ is undefined for at least one of the direct subterms of t; or (c) both $\beta(s)$ and $\beta(t)$ are undefined.

(2) $(A, \beta) \models t \not\approx s$ if and only if (a) $\beta(t)$ and $\beta(s)$ are both defined and different; or (b) at least one of $\beta(s)$ and $\beta(t)$ is undefined.

(3) $(A, \beta) \models P(t_1, \ldots, t_n)$ if and only if (a) $\beta(t_1), \ldots, \beta(t_n)$ are all defined and $(\beta(t_1), \ldots, \beta(t_n)) \in P_A$; or (b) at least one of $\beta(t_1), \ldots, \beta(t_n)$ is undefined.

(4) $(A, \beta) \models \neg P(t_1, \ldots, t_n)$ if and only if (a) $\beta(t_1), \ldots, \beta(t_n)$ are all defined and $(\beta(t_1), \ldots, \beta(t_n)) \notin P_A$; or (b) at least one of $\beta(t_1), \ldots, \beta(t_n)$ is undefined.

(A, β) *satisfies a clause* C (notation: $(A, \beta) \models C$) if $(A, \beta) \models L$ for at least one literal L in C. A *satisfies* C (notation: $A \models C$) if $(A, \beta) \models C$ for all assignments β. A *satisfies a set of clauses* $\mathcal{K}$ (notation: $A \models \mathcal{K}$) if $A \models C$ for all $C \in \mathcal{K}$.

The notion of *weak validity* is obtained from Evans validity by replacing conditions (1)(b) and (c) in the definition of truth of equality atoms with condition

(b')   at least one of $\beta(s)$, $\beta(t)$ is undefined.

Validity of non-equality literals is the same. The notion of weak validity extends to clauses and sets of clauses in the usual way. We use the notation: $(A, \beta) \models_w L$ for a literal L; $(A, \beta) \models_w C$; $A \models_w C$ for a clause C, etc.

Example 1. Let A be a partial Σ-algebra, where $\Sigma = \{\mathsf{car}/1, \mathsf{nil}/0\}$. Assume that nil_A is defined and $\mathsf{car}_A(\mathsf{nil}_A)$ is not defined. Then $A \not\models \mathsf{car}(\mathsf{nil}) \approx \mathsf{nil}$ (since $\mathsf{car}_A(\mathsf{nil})$ is undefined in A, but nil is defined in A); and $A \models \mathsf{car}(\mathsf{nil}) \not\approx \mathsf{nil}$, $A \models_w \mathsf{car}(\mathsf{nil}) \approx \mathsf{nil}$, $A \models_w \mathsf{car}(\mathsf{nil}) \not\approx \mathsf{nil}$ (because one term is not defined in A).

Theory Extensions. In this paper we consider extensions of theories, in which the signature is extended by new *function symbols*. For the sake of simplicity we assume that the set of predicate symbols remains unchanged in the extension. A theory can be regarded as a set of formulae. Then extension with a set of formulae is set union. In what follows we regard theories as sets of formulae. [1]

Let $\mathcal{T}_0$ be an arbitrary theory with signature $\Pi_0 = (\Sigma_0, \mathsf{Pred})$, where the set of function symbols is Σ_0. We consider extensions $\mathcal{T}_1$ of $\mathcal{T}_0$ with signature $\Pi = (\Sigma, \mathsf{Pred})$, where the set of function symbols is $\Sigma = \Sigma_0 \cup \Sigma_1$. We assume that $\mathcal{T}_1$ is obtained from $\mathcal{T}_0$ by adding a set $\mathcal{K}$ of (universally quantified) clauses.

A *partial model* of $\mathcal{T}_1$ with totally defined Σ_0 function symbols is a partial Π-algebra A where (i) the reduct $A_{|\Pi_0}$ of A to the signature Π_0 is a model of $\mathcal{T}_0$ (in the classical sense, i.e. all operations in Σ_0 are total); (ii) A satisfies (in the Evans sense) all clauses in $\mathcal{K}$.

A partial Π-algebra A is a *weak partial model* of $\mathcal{T}_1$ with totally defined Σ_0 function symbols if (i) $A_{|\Pi_0}$ is a (classical) model of $\mathcal{T}_0$ and (ii') A weakly satisfies all clauses in $\mathcal{K}$.

---

[1] If a theory $\mathcal{T}_0$ is regarded as a collection of models, then its extension with a set $\mathcal{K}$ of formulae consists of all structures (in the extended signature) which are models of $\mathcal{K}$ and whose reduct to the signature of $\mathcal{T}_0$ is in $\mathcal{T}_0$. All the notions defined in this paper can easily be reformulated to a setting in which $\mathcal{T}_0$ is a collection of models.

In what follows, if the base theory T_0 and its signature are clear from the context, we will refer to *partial models* of T_1, resp. *weak partial models* of T_1. We will denote by $\mathsf{PMod}(\Sigma_1, T_1)$ the class of all partial models of T_1 in which the functions in Σ_1 are partial, and all other function symbols are total; and by $\mathsf{PMod_w}(\Sigma_1, T_1)$ the class of all weak partial models of T_1 in which the Σ_1 functions are partial and all the other function symbols are total. We denote by $\mathsf{PMod^f}(\Sigma_1, T_1)$, resp. $\mathsf{PMod_w^f}(\Sigma_1, T_1)$ the class of all finite partial models (resp. weak partial models) of T_1, with total Σ_0 functions, and partial Σ_1 functions. $\mathsf{Mod}(T_1)$ denotes the class of all models of T_1 in which all functions in $\Sigma_0 \cup \Sigma_1$ are totally defined. Note that $\mathsf{Mod}(T_1) \subseteq \mathsf{PMod}(\Sigma_1, T_1) \subseteq \mathsf{PMod_w}(\Sigma_1, T_1)$.

3   Embeddability

For theory extensions $T_0 \subseteq T_1 = T_0 \cup \mathcal{K}$, where $\mathcal{K}$ is a set of clauses, and for the classes of partial algebras mentioned above we consider the following conditions:

(Emb)   Every $A \in \mathsf{PMod}(\Sigma_1, T_1)$ weakly embeds into a total model of T_1.
(Emb$_w$)   Every $A \in \mathsf{PMod_w}(\Sigma_1, T_1)$ weakly embeds into a total model of T_1.

Weaker conditions, which only refer to embeddability of *finite* partial models can also be defined. These will be denoted by (Emb^f), resp. (Emb$_w^f$). We also define stronger notions of embeddability, which we call *completability*.

(Comp)   Every $A \in \mathsf{PMod}(\Sigma_1, T_1)$ weakly embeds into a total model B of T_1
                 such that $A_{|\Pi_0}$ and $B_{|\Pi_0}$ are isomorphic
                 (or, more generally: elementarily equivalent).

(Comp^f), (Comp$_w$) and (Comp$_w^f$) are defined analogously.

Example 1. *We present several examples of theory extensions for which embedding conditions among those mentioned above hold.*

(1) **Shallow extensions:** *Suppose that $T_0 \subseteq T_1$ is a shallow theory extension, i.e. $T_1 = T_0 \cup \mathcal{K}$, where $\mathcal{K}$ is a set consisting only of clauses in which partial function symbols occur only in equality atoms, only positively and only at the root of terms. Assume that all extension functions are declared partial. Then the extension $T_0 \subseteq T_1$ satisfies the embeddability condition (Comp) [6]. Extensions with functions defined by tail recursions are shallow [6].*
(2) **Extensions with free functions:** *Any extension of a theory T_0 with a set of free function symbols satisfies condition (Comp$_w$).*
(3) **Extensions with selector functions:** *Let T_0 be a theory with signature $\Pi_0 = (\Sigma_0, \mathsf{Pred})$, let $c \in \Sigma_0$ with arity n, and let $\Sigma_1 = \{s_1, \ldots, s_n\}$ consist of n unary function symbols. Let $T_1 = T_0 \cup \mathsf{Sel}$ (a theory with signature $\Pi = (\Sigma_0 \cup \Sigma_1, \mathsf{Pred})$) be the extension of T_0 with the set Sel of clauses below. Then the extension $T_0 \subseteq T_1$ satisfies condition (Comp).*

If in addition T_0 satisfies the (universally quantified) formula $\mathsf{Inj}(c)$ (i.e. c is injective in T_0) then the extension $T_0 \subseteq T_1$ satisfies condition $(\mathsf{Comp_w})$.

$$(\mathsf{Sel}) \qquad s_1(c(x_1, \ldots, x_n)) \approx x_1$$

$$\ldots$$

$$s_n(c(x_1, \ldots, x_n)) \approx x_n$$

$$x \approx c(x_1, \ldots, x_n) \to c(s_1(x), \ldots, s_n(x)) \approx x$$

$$(\mathsf{Inj}(c)) \qquad c(x_1, \ldots, x_n) \approx c(y_1, \ldots, y_n) \to (\bigwedge_{i=1}^{n} x_i \approx y_i)$$

(4) **Extensions with monotone functions:** *Let T_0 be one of the following theories: (1) $\mathcal{P}$ (posets), (2) $\mathcal{T}$ (totally-ordered sets), (3) $\mathcal{DO}$ (dense totally-ordered sets), (4) $\mathcal{S}$ (semilattices), (5) $\mathcal{L}$ (lattices), (6) $\mathcal{DL}$ (distributive lattices), (7) $\mathcal{B}$ (Boolean algebras), (8) $\mathbb{R}$ (theory of reals).*

Let Mon_f be the monotonicity axiom:

$$(\mathsf{Mon}_f) \qquad \bigwedge_{i=1}^{n} x_i \leq y_i \to f(x_1, \ldots, x_n) \leq f(y_1, \ldots, y_n).$$

The extension $T_0 \subseteq T_0 \cup \mathsf{Mon}_f$ satisfies condition $(\mathsf{Emb_w})$ in the cases (1)–(5); satisfies condition $(\mathsf{Comp_w^f})$ in the cases (6) and (7); and satisfies condition $(\mathsf{Comp_w})$ in case (8).

(5) **Lipschitz functions:** *The extension $\mathbb{R} \subseteq \mathbb{R} \cup (\mathsf{L}_f^{\lambda})$ of $\mathbb{R}$ with a unary function which is λ-Lipschitz in a point x_0 (for $\lambda > 0$) satisfies condition $(\mathsf{Comp_w})$.*

$$(\mathsf{L}_f^{\lambda}) \qquad \forall x \ |f(x) - f(x_0)| \leq \lambda \cdot |x - x_0|$$

4   Local Theory Extensions

We now define two notions of locality of a theory extension which generalize the notion of local equational theory studied by Ganzinger in [4] and of locality of a theory in general [7,8].

Let $\mathcal{K}$ be a set of clauses in the signature $\Pi = (\Sigma_0 \cup \Sigma_1, \mathsf{Pred})$. In what follows, when we refer to sets G of ground clauses we assume that they are in the signature $\Pi^c = (\Sigma \cup \Sigma_c, \mathsf{Pred})$, where $\Sigma = \Sigma_0 \cup \Sigma_1$, and Σ_c is a set of new constants.

If Ψ is a set of ground $\Sigma_0 \cup \Sigma_1 \cup \Sigma_c$-terms, where Σ_c is a set of (new) constants, we denote by $\mathcal{K}_\Psi$ the set of all instances of $\mathcal{K}$ in which all terms starting with a Σ_1-function symbol are ground terms in Ψ. We denote by $\mathcal{K}^\Psi$ the set of all instances of $\mathcal{K}$ in which all variables occurring below a Σ_1-function symbol are instantiated with ground terms in the set $T_{\Sigma_0}(\Psi)$ of Σ_0-terms generated by Ψ.

If G is a set of ground clauses and $\Psi = \mathsf{st}(\mathcal{K}, G)$ is the set of ground subterms occurring in either $\mathcal{K}$ or G then we write $\mathcal{K}[G] := \mathcal{K}_\Psi$, and $\mathcal{K}^{[G]} := \mathcal{K}^\Psi$.

We identify the following types of locality of a theory extension $T_0 \subseteq T_1$, where $T_1 = T_0 \cup \mathcal{K}$ with $\mathcal{K}$ a set of (universally quantified) clauses:

(Loc)   For every set G of ground clauses $\mathcal{T}_1 \cup G \models \perp$ iff $\mathcal{T}_0 \cup \mathcal{K}[G] \cup G$ has no weak partial model in which all terms in $\mathsf{st}(\mathcal{K}, G)$ are defined.

(SLoc)   For every set G of ground clauses $\mathcal{T}_1 \cup G \models \perp$ iff $\mathcal{T}_0 \cup \mathcal{K}^{[G]} \cup G$ has no partial model in which all terms in $\mathsf{st}(\mathcal{K}, G)$ are defined.

Weaker notions (Loc^f), resp. (SLoc^f) can be defined if we require that the respective conditions hold only for *finite* sets G of ground clauses. An extension $\mathcal{T}_0 \subseteq \mathcal{T}_1$ is *local (stably local)* if it satisfies condition (Loc^f) (resp. (SLoc^f)). A local (stably local) theory [4] is a local (stably local) extension of the empty theory.

5   Locality and Embeddability

We establish links between the notions of locality and embeddability. This extends the results established for local equational theories in [4].

Let $\mathcal{T}_0$ be an arbitrary theory with signature $\Pi_0 = (\Sigma_0, \mathsf{Pred})$. Let $\mathcal{T}_1$ be an extension of $\mathcal{T}_0$ by a set $\mathcal{K}$ of clauses in signature $\Pi = (\Sigma_0 \cup \Sigma_1, \mathsf{Pred})$. Under appropriate assumptions, locality implies embeddability. The converse, which is proved in this section, will be used to provide examples of local theory extensions.

5.1   Flattening of Goals

We first show that in the locality condition we can assume, w.l.o.g., that G consists only of flat and linear (resp. purified) clauses.

We say that a ground clause is Σ_1-*flat* if only constants appear as arguments of function symbols in Σ_1. A Σ_1-flat ground clause is Σ_1-*linear* if whenever a constant occurs in two terms in the clause starting with function symbols in Σ_1, the two terms are identical, and if no term starting with a function symbol in Σ_1 contains two occurrences of the same constant.

Any set G of ground clauses in a signature Σ containing Σ_1 can be transformed into a set $G_{\mathsf{flin}(\Sigma_1)}$ of ground clauses in which subterms starting with function symbols in Σ_1 are flat and linear. This can be done by introducing, in a bottom-up manner, new constants for subterms occurring below functions in Σ_1, and adding the corresponding definitions to the set of clauses. A set G of ground clauses can be transformed into a *purified* set of clauses $G_{\mathsf{sep}(\Sigma_1)}$ (i.e. the function symbols in Σ_1 are separated from the other symbols) by introducing, in a bottom-up manner, new constants c_t for subterms $t = f(g_1, \ldots, g_n)$ with $f \in \Sigma_1$, g_i ground $\Sigma_0 \cup \Sigma_c$-terms (where Σ_c is a set of constants which contains the constants introduced by flattening), together with corresponding definitions $c_t \approx t$. These transformations preserves satisfiability and unsatisfiability with respect to total algebras, and also with respect to partial algebras in which all ground subterms which are flattened are defined.

Lemma 1. *Let $\mathcal{K}$ be a set of clauses containing only Σ_1-flat ground subterms. Assume that for any set G of Σ_1-flat and Σ_1-linear (resp. purified, resp. flat,* *linear and purified) ground clauses, if $\mathcal{T}_0 \cup \mathcal{K} \cup G \models \perp$ then $\mathcal{T}_0 \cup \mathcal{K}[G] \cup G$ has no partial algebra model in which all terms in $\mathsf{st}(\mathcal{K}, G)$ are defined. Then the extension $\mathcal{T}_0 \subseteq \mathcal{T}_0 \cup \mathcal{K}$ satisfies condition* (Loc).

5.2 Embeddability of Weak Partial Models Implies Locality

We prove that for extensions which are Σ_1-flat and Σ_1-linear embeddability of weak partial models into total models implies locality.

A non-ground clause is Σ_1-*flat* if function symbols (including constants) do not occur as arguments of function symbols in Σ_1. A Σ_1-flat non-ground clause is called Σ_1-*linear* if whenever a variable occurs in two terms in the clause which start with function symbols in Σ_1, the two terms are identical, and if no term which starts with a function in Σ_1 contains two occurrences of the same variable.

Theorem 2. *Let $\mathcal{K}$ be a set of clauses which are Σ_1-flat and Σ_1-linear, and let $\mathcal{T}_1 = \mathcal{T}_0 \cup \mathcal{K}$. Then the following hold:*

(1) If the extension $\mathcal{T}_0 \subseteq \mathcal{T}_1$ satisfies (Emb$_w$) *then it satisfies* (Loc).
(2) Assume that $\mathcal{T}_0$ is a locally finite universal theory, and that $\mathcal{K}$ contains only finitely many ground subterms. If the extension $\mathcal{T}_0 \subseteq \mathcal{T}_1$ satisfies (Emb$_w^f$), *then $\mathcal{T}_0 \subseteq \mathcal{T}_1$ satisfies* (Loc^f).

Proof: (1) Assume that $\mathcal{T}_0 \cup \mathcal{K}$ is not a local extension of $\mathcal{T}_0$. Then there exists a set G of ground clauses (with additional constants) such that $\mathcal{T}_0 \cup \mathcal{K} \cup G \models \perp$ but $\mathcal{T}_0 \cup \mathcal{K}[G] \cup G$ has a weak partial model P in which all terms in $\mathsf{st}(\mathcal{K}, G)$ are defined. By Lemma 1 we can assume w.l.o.g. that $G = G_0 \cup G_1$, where G_0 contains no function symbols in Σ_1 and G_1 consists of ground unit clauses of the form $f(c_1, \dots, c_n) \approx c$, where $c_1, \dots, c_n, c$ are constants in $\Sigma_0 \cup \Sigma_c$ and $f \in \Sigma_1$.[2]

We construct another structure, A, having the same support as P, which inherits all relations in **Pred** and all maps in $\Sigma_0 \cup \Sigma_c$ from P, but on which the domains of definition of the Σ_1-functions are restricted as follows: for every $f \in \Sigma_1$, $f_A(a_1, \dots, a_n)$ is defined if and only if there exist constants $c^1, \dots, c^n$ such that $f(c^1, \dots, c^n)$ is in $\mathsf{st}(\mathcal{K}, G)$ and $a^i = c_P^i$ for all $i \in \{1, \dots, n\}$. In this case we define $f_A(a_1, \dots, a_n) := f_P(c_P^1, \dots, c_P^n)$. The reduct of A to $(\Sigma_0 \cup \Sigma_c, \mathsf{Pred})$ coincides with that of P. Thus, A is a model of $\mathcal{T}_0 \cup G_0$. By the way the operations in Σ_1 are defined in A it is clear that A satisfies G_1, so A satisfies G.

To show that $A \models_w \mathcal{K}$ we use the fact that if D is a clause in $\mathcal{K}$ and $\beta : X \to A$ is an assignment in which $\beta(t)$ is defined for every term t occurring in D, then (by the way Σ_1-functions are defined in A) we can construct a substitution σ with $\sigma(D) \in \mathcal{K}[G]$ and $\beta \circ \sigma = \beta$. As $(P, \beta) \models_w \sigma(D)$ we can infer $(A, \beta) \models_w D$.

As $A \models_w \mathcal{K}$, A weakly embeds into a total algebra B satisfying $\mathcal{T}_0 \cup \mathcal{K}$. But then $B \models G$, so $B \models \mathcal{T}_0 \cup \mathcal{K} \cup G$, which is a contradiction.

---

[2] All results below hold if only purified goals are considered; flattening and linearity of goals is not absolutely necessary.

(2) Proof similar to (1), with the difference that we start with a finite set G of ground clauses, and as support for A we take $\{t_P \mid t \in T_{\Sigma_0}(\mathsf{st}(\mathcal{K}, G))\}$; all operations and relations are defined as above. $\mathcal{T}_0$ is a universal theory, so $A_{|\Pi_0}$ (a Π_0-substructure of $P_{|\Pi_0}$) is also a model of $\mathcal{T}_0$. As $\mathsf{st}(\mathcal{K}, G)$ is finite and $\mathcal{T}_0$ is locally finite, A is finite, so (Emb_w^f) is sufficient to find a total model B of $\mathcal{T}_0 \cup \mathcal{K} \cup G$. □

Example 2. *The following theory extensions $\mathcal{T}_0 \subseteq \mathcal{T}_1$ are local:*

(1) Any extension $\mathcal{T}_1$ of a theory $\mathcal{T}_0$ with a set of free function symbols.

(2) Any extension $\mathcal{T}_1$ of a theory $\mathcal{T}_0$ with a set of selectors $\{s_1, \ldots, s_n\}$ for an n-ary function c which is injective in $\mathcal{T}_0$.

(3) The extension of any of the theories $\mathcal{T}_0$ in Example 1(4) with monotone functions: $\mathcal{T}_1 = \mathcal{T}_0 \cup \mathsf{Mon}_f$, where Mon_f is the monotonicity axiom of the n-ary function f.

(4) The extension of the theory of reals with a λ-Lipschitz function at x_0.

Proof: (Sketch) Locality follows from the embeddability properties in Example 1. Ad (3): To prove condition (Loc^f) when $\mathcal{T}_0$ is $\mathcal{DL}$ or $\mathcal{B}$ it suffices to show that (Emb_w^f) holds, because these theories are universal and locally finite. □

5.3   Embeddability of Evans Partial Models Implies Stable Locality

We now show that, for an extension $\mathcal{T}_1 = \mathcal{T}_0 \cup \mathcal{K}$ of a universal theory $\mathcal{T}_0$, embeddability of Evans partial models into total models implies stable locality.

Theorem 3. *Let $\mathcal{T}_0$ be a universal theory and $\mathcal{K}$ be a set of clauses. Then:*

(1) If the extension $\mathcal{T}_0 \subseteq \mathcal{T}_1$ satisfies (Emb) then it satisfies (SLoc).

(2) Assume that $\mathcal{T}_0$ is a locally finite universal theory, and that $\mathcal{K}$ contains only finitely many ground subterms. If the extension $\mathcal{T}_0 \subseteq \mathcal{T}_1$ satisfies (Emb^f), then $\mathcal{T}_0 \subseteq \mathcal{T}_1$ satisfies (SLoc^f).

Proof: The proof is similar to that of Theorem 2. The first difference is in the construction of the partial model A of $\mathcal{T}_0 \cup \mathcal{K} \cup G$. Let $A = \{t_P \mid t \in T_{\Sigma_0}(\mathsf{st}(\mathcal{K}, G))\}$. Define the functions and relations in Π_0 as in P. If $f \in \Sigma_1$ is an n-ary function and $t_P^1, \ldots, t_P^n \in A$, then $f_A(t_P^1, \ldots, t_P^n)$ is defined and equal to t_P if and only if $t_P = f(t^1, \ldots, t^n)_P \in A$. $A_{|\Pi_0}$ is a Π_0-substructure of $P_{|\Pi_0}$. As $\mathcal{T}_0$ is a universal theory, $A_{|\Pi_0}$ is a total model of $\mathcal{T}_0$. As all terms in $\mathsf{st}(\mathcal{K}, G)$ are defined both in P and in A, and $P \models G$, A satisfies all clauses in G. To show that A satisfies $\mathcal{K}$ note that every assignment $\beta : X \to A$ defines at least one substitution $\sigma : X \to T_{\Sigma_0}(\mathsf{st}(\mathcal{K}, G))$ such that $(\sigma(t))_P = \beta(t)$. Then $\sigma(D)$ is an instance of D in $\mathcal{K}^{[G]}$, so $P \models \sigma(D)$, hence $(A, \beta) \models D$. It follows that A satisfies $\mathcal{T}_0 \cup \mathcal{K} \cup G$. The existence of a total model of $\mathcal{T}_0 \cup \mathcal{K} \cup G$ follows from (Emb).

(2) If G is a finite set of clauses then the additional conditions guarantee that A is finite, so only embeddability of finite partial models is necessary in the proof. □

Example 3. *(1) A shallow extension of a universal theory is stably local.*
(2) Let T_0 be a universal theory with signature $\Pi_0 = (\Sigma_0, \mathsf{Pred})$, and let $c \in \Sigma_0$ be a function symbol with arity n. Let $\Pi = (\Sigma_0 \cup \Sigma_1, \mathsf{Pred})$, where $\Sigma_1 = \{s_1, \ldots, s_n\}$. Let $T_1 = T_0 \cup \mathcal{K}_{\mathsf{sel}}$ be the extension of T_0 with the selector axioms for the unary functions $s_1, \ldots, s_n$. Then $T_0 \subseteq T_1$ satisfies condition (SLoc).

6   Relational Encodings, Decidability and Complexity

The locality conditions we consider relate satisfiability in total models to satisfiability of certain ground instances with respect to partial models. We can replace reasoning about partially defined functions with reasoning about relations.

For the signature $\Pi = (\Sigma_0 \cup \Sigma_1, \mathsf{Pred})$ let Π^* denote the signature $(\Sigma_0, \Sigma_1^* \cup \mathsf{Pred})$, where every n-ary function symbol f in Σ_1 is replaced by an $(n{+}1)$-ary relation symbol r^f. If A is a Π-algebra, its relational variant is the Π^*-structure A^* for which $r_A^f(a_1, \ldots, a_n, a)$ if and only if $f_A(a_1, \ldots, a_n)$ is defined and equal to a.

The idea of the relational translation is to replace each atom $f(c_1, \ldots, c_n) \approx c$ with the $r^f(c_1, \ldots, c_n, c)$.

We use the relational translation to establish relationships between the decidability resp. complexity of the universal clause theory of the extension and the decidability resp. complexity of a suitable fragment of the base theory.

6.1   Flattening and Relational Encoding

The locality conditions defined in Section 4 require that $T_1 \cup G$ is satisfiable (where G is a set of ground clauses) if and only if $T_0 \cup \mathcal{K}*[G] \cup G$ has a (Evans, weak, finite) partial model with additional properties, where, depending on the notion of locality, $\mathcal{K} * [G]$ is $\mathcal{K}[G]$ or $\mathcal{K}^{[G]}$. In these sets of clauses the function symbols in Σ_1 only occur at the root of ground terms. Therefore, they can be flattened as explained in Section 5.1. They can also be purified (i.e. the function symbols in Σ_1 are separated from the other symbols) by introducing, in a bottom-up manner, new constants c_t for subterms $t = f(g_1, \ldots, g_n)$ with $f \in \Sigma_1$, g_i ground $\Sigma_0 \cup \Sigma_c$-terms (where Σ_c is a set of constants which contains the constants introduced by flattening, resp. purification), together with corresponding definitions $c_t \approx t$. The set of clauses thus obtained has the form $\mathcal{K}_0 \cup G_0 \cup D$, where D is a set of ground unit clauses of the form $f(g_1, \ldots, g_n) \approx c$, where $f \in \Sigma_1$, c is a constant, $g_1, \ldots, g_n$ are ground terms without function symbols in Σ_1; and $\mathcal{K}_0$ and G_0 are clauses without function symbols in Σ_1. (If we flatten and then purify $\mathcal{K} * [G] \cup G$ we ensure that D consists of ground unit clauses of the form $f(c_1, \ldots, c_n) \approx c$, where $f \in \Sigma_1$, and $c_1, \ldots, c_n, c$ are constants.) These flattening and purification transformations preserve both satisfiability and unsatisfiability with respect to total algebras, and also with respect to partial algebras in which all ground subterms which are flattened are defined.

For the sake of simplicity in what follows we will always flatten and then purify G and $\mathcal{K} * [G]$. All results also hold if only purification is applied.

Lemma 4. *Let $\mathcal{K}$ be a set of clauses and G a ground clause, and let $\mathcal{K}_0 \cup G_0 \cup D$ be obtained from $\mathcal{K}*[G] \cup G$ by flattening and purification, as explained above. Then the following are equivalent:*

(1) $\mathcal{T}_0 \cup \mathcal{K}*[G] \cup G$ has a partial model in which all terms in $\mathsf{st}(\mathcal{K}, G)$ are defined.

(2) $\mathcal{T}_0 \cup \mathcal{K}_0 \cup G_0 \cup D$ has a partial model with all terms in $\mathsf{st}(\mathcal{K}_0, G_0, D)$ defined.

(3) $\mathcal{T}_0 \cup \mathcal{K}_0 \cup G_0 \cup \mathsf{Fun}(D^*) \cup D^*$ has a relational model, where $D^* = \{ r^f(c_1, \ldots, c_n, c) \mid (f(c_1, \ldots, c_n) \approx c) \in D \}$ and $\mathsf{Fun}(D^*) = \{ \bigwedge_{i=1}^{n} c_i \approx d_i \wedge r^f(c_1, \ldots, c_n, c) \wedge r^f(d_1, \ldots, d_n, d) \rightarrow c \approx d \mid f \in \Sigma_1, r^f(c_1, \ldots, c_n, c), r^f(d_1, \ldots, d_n, d) \in D^* \}$.

(4) $\mathcal{T}_0 \cup \mathcal{K}_0 \cup G_0 \cup N_0$ has a (total) model, where $N_0 = \{ \bigwedge_{i=1}^{n} c_i \approx d_i \rightarrow c = d \mid r^f(c_1, \ldots, c_n, c), r^f(d_1, \ldots, d_n, d) \in D^* \}$.

6.2 Decidability and Complexity

Let $\mathcal{T}_0$ be an arbitrary Π_0-theory, where $\Pi_0 = (\Sigma_0, \mathsf{Pred})$ and let $\mathcal{T}_1 = \mathcal{T}_0 \cup \mathcal{K}$, where $\mathcal{K}$ is a finite set of clauses in a signature $\Pi = (\Sigma_0 \cup \Sigma_1, \mathsf{Pred})$.

Theorem 5. *Assume that the theory extension $\mathcal{T}_0 \subseteq \mathcal{T}_1$ either (1) satisfies condition (Loc^f), or else (2) satisfies condition (SLoc^f) and $\mathcal{T}_0$ is locally finite. Then:*

(a) If all variables in the clauses in $\mathcal{K}$ occur below some function symbol from Σ_1 and if the universal theory of $\mathcal{T}_0$ is decidable, then the universal theory of $\mathcal{T}_1$ is decidable.

(b) If the $\forall\exists$ theory of $\mathcal{T}_0$ is decidable then the universal theory of $\mathcal{T}_1$ is decidable.

Proof: It is sufficient to show that the universal clause theory of $\mathcal{T}_1$ is decidable. We present the proofs under hypotheses (1) and (2) in parallel.

Let C be a clause in the signature Π with variables $x_1, \ldots, x_n$. Obviously, $\mathcal{T}_0 \cup \mathcal{K} \models \forall x_1 \ldots x_n C$ if and only if $\mathcal{T}_0 \cup \mathcal{K} \cup G$ is unsatisfiable, where G is the set of ground unit clauses obtained from $\exists x_1 \ldots x_n \neg C$ by Skolemization. By the locality assumption, the last statement is equivalent to saying that $\mathcal{T}_0 \cup \mathcal{K}*[G] \cup G$ has no (weak) partial model in which all terms in $\mathsf{st}(\mathcal{K}, G)$ are defined (where $\mathcal{K}*[G]$ is $\mathcal{K}[G]$ in the case of local extensions and $\mathcal{K}^{[G]}$ for stably local extensions). Let $\mathcal{K}_0 \cup G_0 \cup D$ be the flattened form of $\mathcal{K}*[G] \cup G$. By Lemma 4 we know that $\mathcal{T}_0 \cup \mathcal{K}*[G] \cup G$ has no (weak) partial model in which all terms in $\mathsf{st}(\mathcal{K}, G)$ are defined if and only if $\mathcal{T}_0 \cup \mathcal{K}_0 \cup G_0 \cup N_0$ has a total model, where $N_0 = \{ \bigwedge_{i=1}^{n} c_i \approx d_i \rightarrow c \approx d \mid f(c_1, \ldots, c_n) \approx c, f(d_1, \ldots, d_n) \approx d \in D \}$.

Flattening and purification increase the size (i.e. the total number of symbols) of clauses only by a linear factor. So the size of $\mathcal{K}_0$ is linear in the size of $\mathcal{K}*[G]$ and the size of $G_0 \cup D$ is linear in the size of $G \cup \mathcal{K}$. N_0 contains at most $|D|^2$ clauses, so the number of clauses in N_0 is quadratic in the number of Σ_1 ground terms occurring in $\mathcal{K}$ and G. The maximal length of the clauses in N_0 is $m + 1$, where m is the maximal arity of a function symbol in Σ_1. The only difference between (1) and (2) is the number of clauses in $\mathcal{K}*[G]$.

(1) For a local extension, $\mathcal{K}*[G] = \mathcal{K}[G]$. If $\mathcal{K}$ is finite then $\mathcal{K}[G]$ has at most $n_c \cdot n_t \cdot |\mathsf{st}(\mathcal{K}, G)|$ clauses, where n_c is the number of clauses in $\mathcal{K}$ and n_t the maximal number of distinct Σ_1 terms in a clause in $\mathcal{K}$. Then $\mathcal{K}_0 \cup G_0 \cup N_0$ is finite, of size polynomial in the size of $\mathcal{K} \cup G$.

(2) For a stably local extension, $\mathcal{K} * [G] = \mathcal{K}^{[G]}$. If $\mathcal{K}$ is finite and $\mathcal{T}_0$ is locally finite then there are only finitely many equivalence classes in $\mathcal{T}_{\Sigma_0}(\text{st}(\mathcal{K}, G))$ with respect to equality modulo $\mathcal{T}_0$ (say $n_{\mathcal{K},\mathcal{G}}$). If we only choose the representatives for instantiation in $\mathcal{K}^{[G]}$, the resulting set of clauses is finite, of size polynomial in $n_{\mathcal{K},\mathcal{G}}$ and in the size of $\mathcal{K} \cup \mathcal{G}$. Then $\mathcal{K}_0 \cup G_0 \cup N_0$ is finite.

The proof now continues for both local and stably local extensions:

(a) Assume that for every clause in $\mathcal{K}$, every variable occurs below a Σ_1 function. Then $\mathcal{K}[G]$ (and $\mathcal{K}_0$) consists only of ground clauses. If checking the satisfiability of (existentially quantified) conjunctions of clauses w.r.t. $\mathcal{T}_0$ is decidable, then the universal clause theory of $\mathcal{T}_1$ is decidable, and its complexity is determined by the complexity of satisfiability checking for sets of clauses in $\mathcal{T}_0$ and the size of $\mathcal{K}_0 \cup G_0 \cup N_0$. The problem of checking the satisfiability of conjunctions of clauses is decidable iff the universal theory of $\mathcal{T}_0$ is decidable: if $\{k_1, \ldots, k_m\}$ is the set of all constants that occur in $\mathcal{K}_0 \cup G_0 \cup N_0$ the following are equivalent:

(i) $\mathcal{T}_0 \cup \bigwedge_{C \subseteq \mathcal{K}_0 \cup G_0 \cup N_0} C(k_1, \ldots, k_m) \models \perp$.
(ii) $\mathcal{T}_0 \cup \exists x_1, \cdots x_m (\bigwedge_{C \in \mathcal{K}_0 \cup G_0 \cup N_0} C(x_1, \ldots, x_m)) \models \perp$.
(iii) $\mathcal{T}_0 \models \forall x_1, \cdots x_m (\bigvee_{C \in \mathcal{K}_0 \cup G_0 \cup N_0} \neg C(x_1, \ldots, x_m))$.

(b) If some variables in clauses in $\mathcal{K}$ do not occur below Σ_1-function symbols then the clauses in $\mathcal{K}_0$ are not necessarily ground: they contain variables $\{y_1, \ldots, y_k\}$, and constants in $\{c_1, \ldots, c_n\}$. The following statements are equivalent:

(i) $\mathcal{T}_0 \cup \left(\bigwedge_{C \in \mathcal{K}_0} \forall y_1 \ldots \forall y_k \, C(c_1, \ldots, c_n, y_1, \ldots, y_k) \wedge \bigwedge_{C \in G_0 \cup N_0} C(c_1, \ldots, c_n) \right) \models \perp$.

(ii) $\mathcal{T}_0 \cup \exists x_1 \ldots x_n \left(\bigwedge_{C \in \mathcal{K}_0} \forall y_1 \ldots y_k C(x_1 \ldots x_n, y_1 \ldots y_k) \wedge \bigwedge_{C \in G_0 \cup N_0} C(x_1 \ldots x_n) \right) \models \perp$

(iii) $\mathcal{T}_0 \models \forall x_1 \ldots x_n \left(\bigvee_{C \in \mathcal{K}_0} \exists y_1 \ldots y_k \neg C(x_1 \ldots x_n, y_1 \ldots y_k) \vee \bigvee_{C \in G_0 \cup N_0} \neg C(x_1 \ldots x_n) \right)$.

If the $\forall \exists$ fragment of $\mathcal{T}_0$ is decidable then we can use this and the equivalence of (i) and (iii) to check whether $\mathcal{T}_0 \cup \mathcal{K}_0 \cup G_0 \cup N_0$ is satisfiable. The size of $\mathcal{K}_0 \cup G_0 \cup N_0$ and the complexity of the $\forall \exists$ fragment of $\mathcal{T}_0$ then determine the complexity of the universal fragment of $\mathcal{T}_1$.   □

Corollary 6. *Let $\mathcal{T}_0$ be a theory for which the satisfiability of a set of ground clauses of size n can be checked in time at most $g(n)$, and let $\mathcal{T}_0 \subseteq \mathcal{T}_0 \cup \mathcal{K}$ be a local theory extension where in every clause in $\mathcal{K}$ each variable occurs below some extension function. The validity of a set of clauses in the extension can be checked in time $g(c \cdot n^2)$, where c is a constant. This holds for:*

(1) Extensions with free function symbols (alternative proof of results in [5,10]).
(2) Extensions with monotone functions (see also Example 4)
  – If $\mathcal{T}_0$ is the theory $\mathcal{DL}$ (of distributive lattices) or $\mathcal{B}$ (of Boolean algebras) the complexity of the universal clause theory of an extension of $\mathcal{T}_0$ with monotone functions is in co-NP.
  – If $\mathcal{T}_0$ is the theory $\mathcal{L}$ (of lattices) or $\mathcal{SL}$ (of semilattices) then the complexity of the universal clause theory of an extension of $\mathcal{T}_0$ with monotone functions is in co-NP, and that of the universal Horn theory of $\mathcal{T}_1$ is in PTIME.

(3) Extensions of theories of injective constructors with selectors.

(4) Extensions of $\mathbb{R}$ with Lipschitz functions: *the universal clause theory is in* EXPTIME *(an example was already presented in Section 1.1).*

Example 4. *Let T_0 be a theory (with a binary predicate $\leq$), and T_1 a local extension of T_0 with two monotone functions f and g. Consider the following problem:*

$$T_0 \cup \mathsf{Mon}_f \cup \mathsf{Mon}_g \models \forall x, y, z, u, v (x \leq y \wedge f(y \vee z) \leq g(u \wedge v) \rightarrow f(x) \leq g(v))$$

The problem reduces to the problem of checking whether $T_0 \cup \mathsf{Mon}_f \cup \mathsf{Mon}_g \cup G \models \bot$, where $G = c_0 \leq c_1 \wedge f(c_1 \vee c_2) \leq g(c_3 \wedge c_4) \wedge f(c_0) \not\leq g(c_4)$.

After flattening, using the locality of the extension $T_0 \subseteq T_1$, making the relational translation, and computing N_0, we obtain the following set of clauses:

$c_0 \leq c_1$	$r_f(d_1, e_1)$	$d_1 = c_0 \rightarrow e_1 = e_3$	$d_1 \leq c_0 \rightarrow e_1 \leq e_3$
$d_1 = c_1 \vee c_2$	$r_f(c_0, e_3)$	$d_2 = c_4 \rightarrow e_2 = e_4$	$c_0 \leq d1 \rightarrow e_3 \leq e_1$
$d_2 = c_3 \wedge c_4$	$r_g(c_4, e_4)$		$d_2 \leq c_4 \rightarrow e_2 \leq e_4$
$e_1 \leq e_2$	$r_g(d_2, e_2)$		$d_4 \leq d_2 \rightarrow e_4 \leq e_2$
$e_3 \not\leq e_4$			

(1) Assume T_0 is $\mathcal{DL}$ or $\mathcal{B}$. The universal clause theory of $\mathcal{DL}$ (resp. $\mathcal{B}$) is the theory of the two element lattice (resp. two element Boolean algebra), so testing Boolean satisfiability is sufficient. (This is in NP.) We proved unsatisfiability using SPASS *[11], but SAT solvers such as, e.g.* CHAFF *[9], can be used as well.*

(2) If $T_0 = \mathcal{L}$ we can reduce the problem above to the problem of checking the satisfiability of a set of ground Horn clauses (via the relational translation of Skolem described in the introduction). This can be checked in PTIME.

(3) If $T_0 = \mathbb{R}$ we first need to explain what $\vee$ and $\wedge$ are. For this, we replace $d_1 = c_1 \vee c_2$ with $(c_1 \leq c_2 \rightarrow d_1 = c_2) \wedge (c_2 < c_1 \rightarrow d_1 = c_2)$ and similarly for $d_2 = c_3 \wedge c_4$. We proved unsatisfiability using the REDLOG *demo [3].*

We can therefore conclude that in all cases above:
$$T_1 \models \forall x, y, z, u, v (x \leq y \wedge f(y \vee z) \leq g(u \wedge v) \rightarrow f(x) \leq g(v)).$$

7   Beyond the Universal Fragment

Analyzing the proof of Theorems 2 and 3 we notice that the embeddability conditions (Comp) and (Comp$_w$) imply, in fact, stronger locality conditions. Consider a theory extension $T_0 \subseteq T_0 \cup \mathcal{K}$ with a set $\mathcal{K}$ of formulae of the form $\forall x_1 \ldots x_n (\Phi(x_1, \ldots, x_n) \vee C(x_1, \ldots, x_n))$, where $\Phi(x_1, \ldots, x_n)$ is an *arbitrary first-order formula* in the base signature Π_0 with free variables $x_1, \ldots, x_n$, and $C(x_1, \ldots, x_n)$ is a *clause* in the signature Π.

We can extend the notion of locality of an extension accordingly:

(ELoc)   For every formula $\Gamma = \Gamma_0 \cup G$, where Γ_0 is a Π_0-sentence and G is a set of ground clauses, $T_1 \cup \Gamma \models \bot$ iff $T_0 \cup \mathcal{K}[\Gamma] \cup \Gamma$ has no weak partial model in which all terms in $\mathsf{st}(\mathcal{K}, G)$ are defined.

A stable locality condition (ESLoc) can be defined similarly. The proofs of Theorems 2 and 3 can be adapted with minimal changes to prove a stronger result:

Theorem 7. *(1) Assume all terms of $\mathcal{K}$ starting with a Σ_1 function are flat and linear. If the extension $\mathcal{T}_0 \subseteq \mathcal{T}_1$ satisfies (Comp$_w$) then it satisfies (ELoc).*
(2) Assume that $\mathcal{T}_0$ is a universal theory. If the extension $\mathcal{T}_0 \subseteq \mathcal{T}_1$ satisfies (Comp) then it satisfies (ESLoc).

Proof: (Idea) By (Comp), the partial model and its total completion have supports whose Π_0-reducts are isomorphic, hence elementarily equivalent. Therefore the (weak) embedding guaranteed by (Comp$_w$) resp. (Comp) preserves and reflects the truth of all first-order formulae in the base signature.   □

Further generalizations are possible (concerning both the form of the set of extension formulae, and the form of the goals). This is work in progress.

8   Conclusions

We introduced notions of locality for theory extensions and showed that for local theory extensions we may regard w.l.o.g. the extension functions as functional relations. Using a relational translation we identified situations where it is possible to express the decidability (complexity) of an extension $\mathcal{T}_1$ in terms of the decidability (complexity) of a fragment of the base theory $\mathcal{T}_0$ (universal or $\forall\exists$). These results apply to theories of data types and to some theories of functions from algebra or mathematical analysis.

There seem to exist relationships with results on combinations of non stably infinite theories [10]. The result on extensions of an arbitrary theory with free functions which we obtain as an example was discovered independently in a different context by Ganzinger [5] and Tinelli and Zarba [10]. However, here we go beyond analyzing mere combinations of theories: we look at proper extensions, in which the extension axioms contain functions from the base theory. In this paper we restrict ourselves to one-sorted theories. Similar results can be obtained in a many-sorted framework.

Acknowledgements. Many of the results presented here are the direct or indirect result of discussions and joint work with Harald Ganzinger during the last years. Some of the ideas on local extensions sketched in [6] are now presented in full detail and generalized. I thank Uwe Waldmann for discussions on partial algebras, and Carsten Ihlemann for commenting on an early draft of this paper. Many thanks to the referees for their helpful comments.

References

1. P. Burmeister. *A Model Theoretic Oriented Approach to Partial Algebras: Introduction to Theory and Application of Partial Algebras, Part I*, volume 31 of *Mathematical Research*. Akademie-Verlag, Berlin, 1986.
2. S. Burris. Polynomial time uniform word problems. *Mathematical Logic Quarterly*, 41:173–182, 1995.
3. A. Dolzmann and T. Sturm. Redlog: Computer algebra meets computer logic. *ACM SIGSAM Bulletin*, 31(2):2–9, 1997.

4. H. Ganzinger. Relating semantic and proof-theoretic concepts for polynomial time decidability of uniform word problems. In *Proc. 16th IEEE Symposium on Logic in Computer Science (LICS'01)*, pages 81–92. IEEE Computer Society Press, 2001.

5. H. Ganzinger. Shostak light. In A. Voronkov, editor, *Automated Deduction – CADE-18*, LNCS 2392, pages 332–346, Copenhagen, Denmark, 2002. Springer.

6. H. Ganzinger, V. Sofronie-Stokkermans, and U. Waldmann. Modular proof systems for partial functions with weak equality. In *Proc. International Joint Conference on Automated Reasoning (IJCAR'04), LNCS 3097*, pages 168–182. Springer, 2004.

7. R. Givan and D. McAllester. New results on local inference relations. In *Principles of Knowledge Representation and reasoning: Proceedings of the Third International Conference (KR'92)*, pages 403–412. Morgan Kaufmann Press, 1992.

8. D. McAllester. Automatic recognition of tractability in inference relations. *Journal of the Association for Computing Machinery*, 40(2):284–303, 1993.

9. M. Moskewicz, C. Madigan, Y. Zhao, L. Zhang, and S. Malik. Chaff: Engineering an efficient SAT solver. In *Proceedings of the 39th Design Automation Conference (DAC 2001)*, pages 530–535. ACM, 2001.

10. C. Tinelli and C.G. Zarba. Combining non-stably infinite theories. *Journal of Automated Reasoning*, 2005. to appear.

11. Ch. Weidenbach, B. Gaede, and G. Rock. SPASS & FLOTTER Version 0.42. In M.A. McRobie and J.K. Slaney, editors, *Proceedings of CADE-13*, LNCS 1104, pages 141–145. Springer Verlag, 1996.

Proof Planning for First-Order Temporal Logic

Claudio Castellini[1,2] and Alan Smaill[2]

[1] LIRA-Lab, University of Genova, Italy
[2] School of Informatics, University of Edinburgh, Scotland

Abstract. Proof planning is an automated reasoning technique which improves proof search by raising it to a meta-level. In this paper we apply proof planning to First-Order Linear Temporal Logic (**FOLTL**), which can be seen as a quantified version of Linear Temporal Logic, overcoming its finitary limitation. Automated reasoning in **FOLTL** is hard, since it is non-recursively enumerable; but its expressiveness can be exploited to precisely model the behaviour of complex, infinite-state systems. In order to demonstrate the potentiality of our technique, we introduce a case-study inspired by the Feature Interactions problem and we model it in **FOLTL**; we then describe a set of methods which tackle and solve the validation problem for a number of properties of the model; and lastly we present a set of experimental results showing that the methods we propose capture the common patterns in the proofs presented, guide the search at the object level and let the overall system build large and highly structured proofs. This paper to some extent improves over previous work that showed how proof planning can be used to detect such interactions.

1 Introduction

Conceived in the early 80s by Bundy, *Proof Planning* [2] has proved along the years to be a sophisticated, effective technique for doing automated reasoning in complex frameworks, where standard theorem proving can do little; especially, e.g., in mathematical reasoning, proof by induction [4] and non-standard analysis [23]. (For more details about proof planning, see also [3,22,24]).

In proof planning search is raised to a meta-level. Rather than exploring a space of inference rules applied backwards to a goal formula, as is standard in theorem proving, in proof planning first a *proof plan* is generated, roughly comparable to a proof tree, but in which nodes are labelled by possibly unsound macro-steps of reasoning (*methods*) rather than by inference rules. Standard examples of methods are case-splits and induction schemas. If a proof plan is found, its soundness is verified by extracting a proof from it; this is accomplished by having *tactics* attached to methods, which (partially) specify how a single method is translated to a set of inference rules. The key idea is that the meta-search space is typically orders of magnitude smaller than the original one, and little or no backtracking is likely to occur. This enables proof planning to tackle logics (and the associated problems) normally beyond the capacity of standard theorem provers. Of course, no claim of completeness can be made about the

R. Nieuwenhuis (Ed.): CADE 2005, LNAI 3632, pp. 235–249, 2005.

process — proof planning can fail either at the planning level (no plan could be found) or at the proving level (no proof of the input formula could be extracted from the plan); therefore this technique is best applied to very complex logics for which an incomplete approach is better than no approach at all.

In this paper we apply proof planning to such a complex logic, First-Order Linear Temporal Logic (**FOLTL**), which can be seen as the first-order-style quantified counterpart of *Linear Temporal Logic* (LTL). **FOLTL** is not only undecidable but indeed non-recursively enumerable [19]; but it is also very expressive, so that it can be used, as is the case here, to precisely model complex systems beyond the reach of finitary methods like model-checking, automata- or LTL-based methods. As far as we know, so far there are no effective, general-purpose automated reasoning approaches to **FOLTL**[1]; the aim of this paper is to show that proof planning can actually make the situation better.

We choose to study a well-known problem in Formal Methods, Feature Interactions in Telecommunication Systems (FIs, see [7]). We build an abstract **FOLTL** model of part of the problem and devise a set of proof planning methods which let us validate a number of interesting properties of the model. Although the case-study is not yet ready to be presented as a general solution to the problem of FIs, we believe it is an interesting application of proof planning, and can be extended and refined to eventually become a tool for Formal Methods, possibly and likely in combination with push-button techniques such as model-checking. In view of this, we recall that previous work, e.g., [17,16,11] already showed how proof planning can actually be used to *detect* FIs, although that usually requires reasoning by refutation; whereas, so far, our approach only works by proving statements.

Proof plans can be compared to sketches of human proofs, reflecting the intuitions of the mathematician, while the details are left to a subsequent phase. The experimental results we show indicate that this actually is the case, at least for the case-study considered: several similar properties are planned using the same sets of methods, and the proofs then generated share a common structure. Moreover, the approach requires a reasonable amount of computer resources, and an amount of human intervention which, though initially high, decreases as more and more problems are tackled.

The paper is structured as follows: Section 2 gives some preliminaries about **FOLTL** and proof planning; Section 3 introduces our case-study and the way we build the model of the problem; Section 4 describes the methods devised to solve it; Section 5 shows the experimental results and comments on them; lastly Section 6 contains comparison with related work, conclusions and future work.

2   Preliminaries

In this Section we first sketch our presentation of **FOLTL** and the sequent calculus we will be using. For more details, the reader can refer to [10].

---

[1] Interesting results have been obtained, though, in applying clausal resolution to the *monodic* fragment of **FOLTL**, see, e.g., [13].

FOLTL. Our presentation of **FOLTL** extends first-order logic with the unary temporal operators $\square$ ("always"), $\lozenge$ ("eventually") and $\bigcirc$ ("next") and the binary temporal operators $\mathcal{U}$ ("until") and $\mathcal{W}$ ("weak until"). The semantics is standard [1], assuming constant domains and rigid designators; this means that no objects in the domain of quantification, $\mathcal{D}$, are created or destroyed along time, and that the only "dynamic" objects are predicates. An example of **FOLTL** formula, akin to those we are about to see in our case-study, is this:

$$\forall x.\square\,[state_1(x) \supset (state_1(x)\,\mathcal{W}\,(trans_{12}(x) \vee \exists t.trans_{13}(x, t)))]$$

Assuming the domain of quantification represents individuals which can be, at any given time, in a certain state, the informal (but accurate) reading of the above example is: any individual, at any time, being in state 1, will remain in that state forever, or will eventually take a transition to state 2, or to state 3. In general, $p\,\mathcal{W}\,q$ stands for "either p will hold forever, or eventually q will hold, with p holding meanwhile". The standard semantics of $\mathcal{W}$ is exploited here to make sure that the individual will stay in state 1 while it is waiting for something to happen, or that no transitions will ever be taken. Notice that the way states and transitions are represented is a free mixture of first-order predicates and possibly quantifiers.

The proof system we use is a sequent calculus based upon those presented in [12] for quantified modal logics; it is an extension of **QS4.3**, sound and complete for the quantified logic of reflexive, transitive, weakly connected frames, with rules for $\bigcirc, \mathcal{U}$ and $\mathcal{W}$. The resulting calculus is called $\mathcal{C}_{\mathbf{FOLTL}}$ and its soundness follows straightforwardly from the semantics of the operators modelled.

3   Case-Study

By using **FOLTL** complex systems can be modelled and verified with no finitary limitation; but also, we use its expressivity to keep the model small and intuitive. Typically, in a large telephone network, if a customer subscribes to one or more features (such as, e.g., ring-back when free, reject anonymous calls etc.) it can be the case that *interactions* arise among different features, if not between a feature and the basic service, where an interaction is an unexpected, unwanted behaviour arising from insufficient, inconsistent and/or wrong specifications. Hence the need of detecting interactions as soon as possible, e.g., at specification time, or as well, to validate the model with respect to some property stating that no interactions arise. The problem has received great attention both from the academical and the industrial world, see, e.g., [18]; it is complex: any user, from an unbounded pool, can subscribe to any feature(s), and each feature must behave correctly for each user, in any possible scenario. Traditionally, it has been solved by approximating the scenario in a finitary way and then using well-known techniques such as model-checking [17,8] or Boolean satisfiability (see [7] for an exhaustive survey); but in this case a positive answer is not definitive; if, e.g., the approximation assumes there are 3 users, an interaction involving more users would go undetected.

The case-study we present is an abstraction and simplification of FIs as stated in [9], to our knowledge one of the most effective approaches so far to the problem; in particular we show how *compatibility* of properties of a basic call service plus a feature can be verified. The phone network is modelled as a set of *users*, each of which enjoys the abilities of answering the phone, dialling a number etc., the so-called *Basic Call Service* (BCS). The environment must take care of establishing connections among users. We model the generic user as a set of **FOLTL** formulae defining the behaviour of the automaton given in Figure 1.

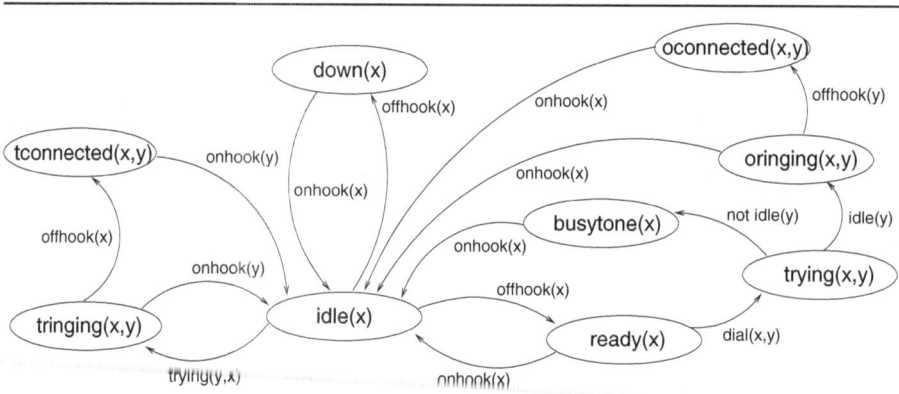

Fig. 1. A graphical representation of the BCS automaton

In the Figure, the ovals and arrows are labelled respectively by states in which a user can be (e.g., *idle*) and transitions that take a user to another state (e.g., *offhook*). The behaviour of the generic user with BCS is then enforced via a set of **FOLTL** formulae:

1. *(Initial state)* Every user is initially idle:
   $\forall x.idle(x)$
2. *(Progress)* For each state, either the user remains in the state forever, or a transition happens; e.g.,
   $\forall x.\Box\ idle(x) \supset idle(x)\ \mathcal{W}\ (offhook(x) \vee \exists t.trying(t, x))$
3. *(Trigger)* For each state and transition, if they happen simultaneously then the user will be in a new state at the next instant; e.g.,
   $\forall x.\Box\ idle(x) \wedge offhook(x) \supset (\bigcirc ready(x) \vee \bigcirc down(x))$

Additionally, a few first-order invariants are needed:

1. *(Persistence of states)* Each user is always at least in a state, e.g.,
   $\forall \overline{x}.\Box\ \bigvee state(\overline{x})$
   where $state(\overline{x})$ denotes the generic state predicate such as *idle*, *ready* and so on.

2. *(System axioms)* relating states to one another, e.g.,
$\forall x, y.\Box\ oconnected(x, y) \leftrightarrow tconnected(y, x)$

The above model concisely and intuitively models the behaviour of a user overcoming a number of standard pitfalls; for instance, as already noted in [9], the use of $\mathcal{W}$ in progress formulae is much better than, e.g., $\Box(p \supset \Diamond q)$. Indeed this formulation is too weak, since (a) it would be true in a scenario in which the user hears the busy tone *later on*, not necessarily as a result of this very call; (b) it would be false in a scenario in which the user failed to progress infinitely often, that is, for some reason the network took an infinite time to process her call. In fact $\mathcal{W}$ specifies *what must hold* while we are waiting for an event to happen, and we can also be satisfied if the event *never happens*. It seems reasonable, in this case, not to force any fairness constraint on the system — it seems legal to have a user waiting for something to happen forever; fairness constraints could be anyway imposed on any transition(s) by using $\mathcal{U}$, which forces the releasing event to eventually happen.

Notice also that this model (i) enforces some subtle properties of a real phone network, such as, e.g., that a user that has been called (the *terminator*) cannot terminate a call, whereas the user who has called (the *originator*) can — this is the customary behaviour of a standard phone network, modelled via two predicates *tconnected* and *oconnected*; (ii) enjoys a high degree of non-determinism: a state can have more than one successor state even if the action is the same, and as well a user can permanently remain in the current state; (iii) there is no restriction on the number of users.

We also introduce a simple feature called *Originating Call Screening* (OCS), according to which a user subscribing to OCS has a predefined list of users, calling whom is prohibited. A new predicate, $ocs(x, y)$, declares that user x has user y on his screening list; an axiom is added, stating that nobody can be on his own's screening list: $\forall x.\Box\ \neg ocs(x, x)$; and, in order to prevent calling a screened user, the trigger formula determining the transition from *ready* to *trying* is modified to $\forall x, y.\Box\ ready(x) \wedge dial(x, y) \wedge \neg ocs(x, y) \supset \bigcirc trying(x, y)$. The revised version of BCS is called BCS'.

3.1   Properties

In this Subsection we list the properties we are interested in proving. They resemble the properties stated in [9] and are expressed, as the model is, in **FOLTL**; the goal is to prove that the model enjoys them, which is achieved through standard logical implication.

Reachability. There are capabilities the system must enjoy at least under suitable (good) conditions; for example, it must be eventually possible to connect any two users, if the originator dials, if the line is available, if the terminator hangs up and so on. These properties correspond to looking for a path in the graph of Figure 1 from the initial state to the required state. Assume we can somehow collect all good conditions in a formula $\phi(x)$; then we want to prove that:

Reach1. Under suitable conditions, any user can get ready:

$\forall x.\phi(x) \supset \Diamond \, ready(x)$

Reach2. Under suitable conditions, any user can connect any other user:

$\forall x.\phi(x) \supset \Diamond \, \exists t.oconnected(x,t)$

Reach3. Under suitable conditions, any user can be connected to any other user:

$\forall x.\phi(x) \supset \Diamond \, \exists t.tconnected(t,x)$

First-Order Properties. By a first-order property we denote a $\Box$-formula not containing temporal operators other than $\bigcirc$, and we are interested in checking whether the property always holds. Thus we look at:

FO1. The user we are *trying* to dial is the same as the user we have just dialled:

$\forall x,y.\Box \, (ready(x) \wedge dial(x,y)) \supset \bigcirc trying(x,y)$

FO2. If I am ringing y and she hangs up, I will be next connected to her:

$\forall x,y.\Box \, (oringing(x,y) \wedge offhook(y)) \supset \bigcirc oconnected(x,y)$

FO3. If I am ringing y and she hangs up, she will be next connected to me:

$\forall x,y.\Box \, (oringing(x,y) \wedge offhook(y)) \supset \bigcirc tconnected(y,x)$

These properties are similar to trigger formulae, but in general they can have a quite more complex first-order structure.

Weak-Until Properties. An interesting class of properties employ the $\mathcal{W}$ operator. We look at:

WU1. If I dial myself, I will hear the busy tone before getting back to idle:

$\forall x.\Box \, (ready(x) \wedge dial(x,x)) \supset (\neg idle(x) \, \mathcal{W} \, busytone(x))$

The semantics of the operator here helps establishing what *must not hold* between two events. A slightly different kind of weak-until properties are:

WU2. If I am trying to connect y, I will keep on trying until I will hear the busy tone or I will be ringing her:

$\forall x,y.\Box \, trying(x,y) \supset trying(x,y) \, \mathcal{W} \, (busytone(x) \vee oringing(x,y))$

WU3. If I am ready, I will stay ready until I will get back to idle or I will be trying to connect to someone:

$\forall x,y.\Box \, ready(x) \supset ready(x) \, \mathcal{W} \, (idle(x) \vee \exists t.trying(x,t))$

Notice that, although **WU2** and **WU3** may look similar to progress properties, they are indeed different, since in general the event on the right-hand side of $\mathcal{W}$ *cannot* immediately be found in a progress formula.

OCS. Once we add OCS to the system we have to state and prove the characteristic property of OCS itself:

OCS. Assuming user x has a user *alice* on his screening list, x can never be connected to t as originator:

$\forall x.\Box \, (ocs(x, alice) \supset \neg oconnected(x, alice))$

When the user enjoys OCS, we expect some of the above properties to be still provable, while some others are not. In particular, we are interested in proving that **WU1** is still valid (no user may have himself on his screening list), whereas **WU3** is no longer (a *ready* user trying to dial a screened user will never be *trying* to connect to her).

4   Proof Planning for the Case-Study

We employ the proof planner λCℓAM, written in λProlog (see [25]). A proof plan is a tree whose nodes are labelled by pairs (method, sequent); every method is associated to a tactic. In the style of [15], a (basic) tactic is a rule of inference plus some operational content (for instance, which formula in the sequent the rule must be applied to); more complex (compound) tactics enforce, e.g., repeated, conditional and exhaustive application of tactics. The object-level theorem prover we have devised, FTL, written in λProlog too, is actually tactic-based, in order to be seamlessly integrated with λCℓAM. First the input sequent is translated into λCℓAM's internal syntax and λCℓAM's planning engine is called; if a plan is found, it is translated to a tactic tree, which FTL runs on the input sequent. If the result is a proof of the sequent, soundness of $\mathcal{C}_{\mathbf{FOLTL}}$ rules ensures the sequent is valid. Due to space limitations, we explain in detail the first sets of methods only, tailored for reachability properties, and informally sketch the others. The interested reader is referred, once again, to [10].

Reachability. This method mimics *backward-reachability*:

```
method exists_path
  repeat:
    1 (init) if we are in the initial state, stop; otherwise,
    2 (trig) find a trigger formula leading to the current state;
             then for each associated transition,
    3 (prog) find a progress formula leading to the current transition;
             for each associated state, make it the current state and
             go back to 1.
```

Methods such as this are called *compound* since they apply other methods, indicated in parentheses. The loop at steps 1-3 takes care of finding all possible trigger and progress formulae that may lead to the current state. The hope is that, eventually, a path from *idle* to it will be found. Notice that the above scheme only shows the *operational content* of the method, without committing to the *structure* of the proof plan; in general, every method may open several branches, leading to the construction of a proof plan as a tree. Then, if the proof plan is found, in order to build a proof of the goal formula, tactics attached to each method are glued together, building a tactic tree; and the tactic tree is finally executed, possibly leading to a proof of the goal formula. Tactics take care of filling the gaps left by the proof plan.

An example will clarify. Consider **Reach1** (and Figure 1). We start from *ready*; since it is not the initial state (step 1), we find all trigger formulae that can lead to it. There is just one: $\forall x. \Box \; idle(x) \land offhook(x) \supset (\bigcirc ready(x) \lor \bigcirc down(x))$. So, in order to get to *ready*, *idle* and *offhook* must have happened in the past (step 2). The only progress formula related to this is $\forall x.\Box\ idle(x) \supset idle(x)\ \mathcal{W}\ (offhook(x) \vee \exists t.trying(t,x))$ (step 3). So we now know that, if the user is *idle*, under suitable conditions, it will get to *ready*. Since *idle* is the initial state, we are done.

Proof planning here also takes care of dynamically building the assumptions $\phi(x)$ needed to prove the statement. Initially, $\phi(x)$ is instantiated with a logical variable[2]. While looking for the plan, methods (trig) and (prog) neglect all transitions not leading toward the state we want to reach, and collect them into $\phi(x)$. Consider the application of method (trig) in the above example: the method simply "forgets" that *idle* and *offhook* may lead to *down*, as well as to *ready*, and records it in $\phi(x)$. When it comes to building the tactic tree, $\phi(x)$ is used in the hypotheses, and the method forces a tactic called close_tac to close the proof branch related to the neglected transition. At the object level, one can view this tactic as a very carefully controlled application of the cut rule: assuming that the right transitions are taken, the statement can be proved. In general, for each trigger formula found by (trig) having n transitions (for instance, there are three possible transitions out of *idle*) the resulting proof employs close_tac $n - 1$ times. Something analogous happens with method (prog), since in general one can nondeterministically progress to more than one transition from each state.

In this case, proof planning literally "directs" the search at the object level and builds the correct assumptions on-the-fly.

First-order properties. In order to plan and prove these properties we use the "persistence of states" invariant (see Section 3) with the following compound method:

```
method all_paths
  repeat until closed:
    1 (inv)   introduce an invariant in the hypotheses
    2 (mlor)  open a branch for each disjunction
    3 (mutex) either close by mutual exclusion, or
    4 (pl)    try and close the branch by first-order logic
```

The method works like this: open up n branches using the invariant in the hypotheses; then, in $n - 1$ branches, close thanks to the detection of mutual exclusion, while the remaining branch is closed by first-order reasoning. The latter task is devoted to the object-level theorem prover and therefore delayed to the proving phase. Mutual exclusion between states (i.e., that no user can be simultaneously in two different states) is realized via a further method which detects the presence in the hypotheses of two state predicates; the associated tactic selects the appropriate system axioms and works by propositional reasoning.

Notice how this method somehow resembles a well-known inductive reasoning technique which consists of showing that a certain property is preserved over all possible states of the system, in case there is a finite number of (classes of)

---

[2] Recall that the system is written in a higher-order language, so that logical variables may well stand for predicates and formulae.

them. Here the invariant consists of an n-ary disjunction appearing among the hypotheses, and that is how n search branches are opened.

Weak-until properties. Once again, the proving strategy is inspired by model checking, this time by *forward reachability*. Consider Property **WU1**: we start from the state specified in the antecedent of the goal (in this case, $ready(x)$) and find a trigger formula telling us what happens if we take the transition specified in the same place (in this case, $dial(x, x)$). Open one branch for each transition found; then for each branch, that is, following each possible path forward, check whether we have reached the state on the right hand side of the $\mathcal{W}$ in the goal (in this case, $busytone(x)$). If it is the case, stop. Otherwise, find a progress formula and identify what transitions can be taken from this state; again, open a branch for each possible transition and, for each one, close the branch by mutual exclusion detection. Then go back to the beginning.

Slightly simpler than the previous one, Properties **WU2** and **WU3** are proved in a similar way, but trying to identify a trigger formula corresponding to the required goal. If this is not the case, the same method seen above is employed to let the system progress.

OCS is proved using a slight variation of the previously mentioned invariant. The associated method opens a search branch for each possible state user x is in; all of them but one are closed by detection of mutual exclusion, while the remaining one goes through by propositional reasoning — this is reasonable, since the use of the OCS axiom $\forall x.\Box \neg ocs(x, x)$ should match with the condition $\neg ocs(x, x)$ when dealing with the transition from $ready$ to $trying$.

As far as **WU1** is concerned, we expect the very same methods employed to prove its validity with BCS to carry the proof on in this case; on the other hand, proving that **WU3** interacts with OCS is slightly more complex; we prove that user *alice*, *not* being on anyone's screening list, validates the property. This is due to the fact that our approach cannot work, so far, by refutation; therefore, if a property *fails* to be provable, there is no way to tell whether that is due to incompleteness of the system or to an interaction actually arising. The formula stating this should be provable via the same set of methods used for Property **WU2**, and in fact it is.

Discussion. Although applied to this particular case-study, it is worth noting that the methods outlined above are in principle applicable to any model formalised in **FOLTL** along the guidelines given in Section 3. In fact, the main points in building the model and devising a proof planning strategy for it are those of (i) exploiting the expressivity and complexity of **FOLTL** in order to accurately model the behaviour of the system, and (ii) taking advantage of the shape of the goal formulae in order to try and plan / prove them.

5 Experimental Results

Since our approach is not push-button, it seems fair to give an overview of the time spent by the *user* in devising the approach, beside showing CPU times.

We adopt Cantu et al.'s three-fold classification of the time required by the user [14]: human time is divided into *User Time*, spent in formalising a problem, *Proof Time*, spent in tuning proof techniques without modifying the tool, and *Tool Time*, used for debugging. The properties outlined in Section 4 have been verified by the system λCLAM/FTL; Table 1 shows the results. Columns contain, for each property, data about the proof plan and the proof (depth d, number of nodes #N, CPU Time in seconds), total CPU Time in seconds, and human time required to devise the solution (User, Proof, Tool time and total, in man-hours). The last two rows show averages and totals. All experiments were run on a PC equipped with an AMD K6 200MHz processor, 256 MB on board memory and Linux 2.4.7. We employed a patched version of the λProlog environment Teyjus v1.0-b33 and λCLAM v4.0.0 (2002). The heap space of the λProlog compiler / simulator was raised to 512 MB in order to avoid heap overflow.

Table 1. Experimental results

Property	Proof plan			Proof				Human time			
	d	#N	Time	d	#N	Time		U	P	T	
Reach1	13	15	11	23	31	2	**13**	2	100	200	**302**
Reach2	19	21	24	66	92	7	**31**	1	10	1	**12**
Reach3	15	17	15	38	52	3	**18**	1	1	1	**3**
FO1	28	44	49	39	322	17	**66**	4	10	20	**34**
FO2	28	44	58	39	321	20	**78**	1	1	2	**4**
FO3	28	44	58	39	327	20	**78**	1	1	5	**7**
WU1	17	19	20	48	97	10	**30**	10	70	100	**180**
WU2	14	16	11	41	112	14	**25**	4	10	10	**24**
WU3	14	16	11	43	111	14	**25**	1	1	1	**3**
BCS'+WU1	17	19	21	57	112	13	**34**	1	1	1	**3**
BCS'+OCS	32	80	76	41	341	96	**172**	8	5	20	**33**
BCS'+WU3	14	16	11	47	110	20	**31**	20	10	10	**40**
Averages	20	29.6	30.4	43.4	169	19.7		3.5	18.3	30.9	
Totals			365			236	**601**	54	220	371	**645**

We now comment on each single experiment, analysing the structure of the plans and proofs obtained and the CPU and human time needed.

Reachability. The planner finds a path from *idle* to the required state, and the depth of each plan is related to the distance on the graph. Consider Figure 1: to prove **Reach1** the proof planner needs discover that any user can get to *ready* from *idle* in "one step"; analogously for **Reach2** (four steps) and **Reach3** (two steps). Timings, depths and numbers of nodes roughly reflect this proportion; the structures of plans and proofs also look very similar from a qualitative point of view. This is a clear indication that the employed methods capture the common structure in the three proofs. Unsurprisingly, planning time dominates proof checking time, as expected, and **Reach2** is the hardest.

The ratio between the depth and number of nodes of both the plans and the proofs are low, meaning that the associated trees are quite narrow; the planner is actually guiding the search in an efficient way, i.e., cutting away useless search branches. As far as human time is concerned, consider **Reach1**: it was no great problem to invent the proof plans (User time 2 man-hours) but it was quite hard to build the correct machinery, both in terms of methods (Proof time 100 man-hours) and in terms of adjusting the system (Tool time 200 man-hours). In fact this was the very first attempt, and, as expected, took a long time to set up. The times scale down radically, however, if we proceed on to the other properties, especially because the very same set of methods, with slight modifications, work fine for all three of them.

First-order properties. Proof plans and proofs of these Properties present remarkable similarities in structure. The high number of nodes of the proofs comes from 8 similar search branches opened up by the use of an invariant. Notice also that these proofs do *not* include the proof of the invariant itself. On a smaller scale, there is a pattern in human times which is similar to that one can see for the previous set of Properties. Tool time appears a little larger (5 man-hours) for Property **FO3** since it was necessary to code and use a system axiom in that case, in order to have the proof go through.

WU1. This problem required a big effort in human terms, as shown in the Table, since it was necessary for the first time to devise a way of proving an invariant with a $\mathcal{W}$ operator in it. In particular, a number of different methods were required, and it was not clear in the beginning how to translate the intuitive ideas to tactics.

WU2, WU3. Another two similar plans and proofs, proved by the same set of methods. The first one required some effort on the human side, while the second was proved quite easily. Actually, the experience gathered for **WU1** helped.

BCS'. That **WU1** still holds with OCS on could be proved with little or no modification to the methods explained above. As one can see, the human time required was small. If compared with the figures above, for BCS+**WU1**, the proof is somehow deeper and larger because of the added complexity of the OCS feature. Validating **OCS** requires the largest effort of the whole benchmark set, due to the use of the OCS invariant, which introduces complexity in each branch of both the proof plan and the proof. Lastly, **WU3** proved to be quite complex, as is witnessed by the human time. Actually, there is still no systematic way of determining how to detect an interaction when there is one, and this single problem needed some 20 man-hours to find out how to discover it.

Statistics. Consider the "Averages" row. One can see that the average proof plan is 20 nodes deep and contains about 30 nodes in total (ratio: 0.66): proof plans are narrow and deep and not very large overall. This indicates that the proof planner chooses the right methods quite easily, also taking into account that basically no backtracking happens. In short, the abstract search space is tractable. On the other hand, the average *proof* is about 43 nodes deep and has

169 nodes in total (ratio: 0.25), which seems to suggest that there is a lot more "decoration" in a proof than in a proof plan — this agrees with the idea that the proof plan abstracts away much more than is allowed in a proof. Considering that here the search space is infinite and the object logic is non recursively enumerable, such a depth is remarkable. The plan is directing the search, which is the idea behind proof planning.

Proof planning time dominates over proof checking time by a factor of 3 to 2. This is sensible as well, since most of the "intelligence" of the system lies in the plan rather than in the proof, although the tactics in the methods can be rather involved, let alone requiring some degree of automation themselves. For instance, in some places the planner closes a branch assuming it can be closed at the object level too via propositional reasoning — the mutual exclusion detection method works exactly like this — but then the object level theorem prover must exhaustively apply propositional reasoning in order to carry the proof to the end. Most of the time spent by the planner is required for reasoning on the shape of the formulae present in the sequent; in this case, higher order unification plays a leading role.

The whole set of benchmarks can be solved on a rather slow machine in something more than 10 minutes of CPU time, and the total human time required to set the machinery up was some 4 man-months full-time, assuming one man-month full-time is 160 man-hours; since this is a novel approach, such an effort appears reasonable, since it also takes into account the time spent to debug a system which is still prototypical. To this extent, it is worth noting that there is definite dominance of Tool time over Proof time, and of Proof time over User time: detecting and fixing bugs is harder than designing methods and tactics, at least in the initial phase of the development of a novel approach.

Notice, lastly, that the human time reported in Table 1 is *not* time spent in human interaction with the system; once the methods and tactics have been devised, the process is automatic, if it runs to the end. Rather, one can think of User and Proof Time as time spent by the user in programming the planning and proof search of the system, and of Tool Time as time spent in debugging the system. This approach is quite different from standard interactive theorem proving.

6   Conclusions, Related and Future Work

This paper outlines a new application field and methodology for Proof Planning, by showing that **FOLTL** can be used to model complex systems, and that proof planning applied to this logic can be used to prove interesting properties of such models. In particular, (i) we have built an abstract model of part of a well-known problem of Formal Methods, that of Feature Interactions, using the complexity and expressivity of **FOLTL** to keep the model both accurate and small; (ii) we have devised a set of general-purpose proof planning methods which, applied to such a model, lead to the verification of a number of interesting properties of the model itself; (iii) a set of experimental results shows that the approach is vi-able: solving the problems required a reasonable amount of computer resources; human time, though quite high in absolute terms, is shown to decrease steadily once the initial attempts are made. In particular, it is worth noting that the proofs obtained under the guidance of proof planning are remarkably structured and that most of the useless search is cut away in the planning phase, thanks to the fact that the structure of proofs is captured by the methods employed.

Calder and Miller's work (see e.g., [5,8,9]) is the main source of inspiration to the experimental test-set presented in this paper. It is difficult to quantitatively compare the results obtained by Calder and Miller and ours, since (1) the machines used are rather different, and (2) there is no indication on the human time required by Calder and Miller's approach in their papers; from a qualitative point of view our approach has, in general, a precise advantage over Calder and Miller's (and any other model-checking-based approach), since our proofs use no finitary approximation whatsoever and need find no suitable abstraction for that — this characteristic comes "for free" from the use of **FOLTL**. But it must as well be remarked that, in [6], the authors extend their approach to an unbounded number of users, thanks to an abstraction-based technique. Moreover, their model is much more detailed and realistic than ours, also thanks to the use of a well-established modelling language such as ProMeLa [20,21]. As a final remark, notice that in [9] the authors solve the problem for a wider set of properties than ours.

The present paper can be seen as an extension and a generalisation of the preliminary result of [11]; in particular, in that paper, the required User Time was unacceptably long, and concentrated in a single, big tactic, containing something like 150 basic tactics. Some of them had to be applied to a precise formula in the antecedents or consequent of a sequent — that is, the user had to specify not only what sequent rule was to be used, but also on which formula. Moreover, the *order* in which basic tactics appeared in that tactic was absolutely crucial. One wrong position and the execution would not go through any more, preventing the system from proving soundness of the proof plan. In short, the methods devised were non reusable and had very little generality. In this work, we have built a set of methods which can be used, with little or no modification, to tackle analogous problems for any **FOLTL** model resembling the one presented in Section 3. For instance, under such an assumption, method exists_path (see Section 4) will be able to find and prove reachability properties in a number of cases.

Future work, in fact, will mainly focus along two orthogonal directions: on one hand, finding more problems like the case-study presented, in order to prove the extensibility and generality of the approach; on the other hand, extending the model toward the full set of features, and making it more concrete, possibly by extracting it out of a formal specification. Also, a systematic way of *detecting* interactions still has to be devised: the main drawback of our approach seems, right now, that it cannot work by refutation, requiring statements to be proved in order to catch interactions; but the work reported in, e.g., [17,16,11,10] shows how the theorem-proving approach can be used to detect interactions, and that is one of the main lines of future research.

References

1. Martin Abadí and Zohar Manna. Nonclausal deduction in first-order temporal logic. *Journal of the ACM*, 37(2):279–317, April 1990.
2. A. Bundy. The use of explicit plans to guide inductive proofs. In R. Lusk and R. Overbeek, editors, *9th International Conference on Automated Deduction*, pages 111–120. Springer-Verlag, 1988. Longer version available from Edinburgh as DAI Research Paper No. 349.
3. A. Bundy. Proof planning. In B. Drabble, editor, *Proceedings of the 3rd International Conference on AI Planning Systems, (AIPS) 1996*, pages 261–267, 1996. also available as DAI Research Report 886.
4. A. Bundy, A. Stevens, F. van Harmelen, A. Ireland, and A. Smaill. Rippling: A heuristic for guiding inductive proofs. *Artificial Intelligence*, 62:185–253, 1993. Also available from Edinburgh as DAI Research Paper No. 567.
5. M. Calder and E. Magill, editors. *Feature Interactions in Telecommunications and Software Systems VI*. IOS Press, 2000.
6. M. Calder and A. Miller. Automated verification of any number of concurrent, communicating processes. In *Proceedings of the 17th IEEE Automated Software Engineering (ASE 2002)*, pages 227–230, 2002.
7. Muffy Calder, Mario Kolberg, Evan H. Magill, and Stephan Reiff-Marganiec. Feature interaction: A critical review and considered forecast. *Computer Networks*, 2002.
8. Muffy Calder and Alice Miller. Using SPIN for feature interaction analysis — A case study. In M.B. Dwyer, editor, *Model checking software: 8th International SPIN Workshop, Toronto, Canada, May 19-20, 2001: proceedings*, pages 143–162. Springer, 2001. Lecture Notes in Computer Science No. 2057.
9. Muffy Calder and Alice Miller. Feature interaction detection by pairwise analysis of LTL properties. Submitted to FMSD - Formal Methods in Software Development. Still under review at the time of writing (2005)., April 2002.
10. Claudio Castellini. *Automated reasoning in quantified modal and temporal logics.* PhD thesis, School of Informatics, University of Edinburgh (UK), 2005.
11. Claudio Castellini and Alan Smaill. Proof planning for feature interactions: a preliminary report. In Andrei Voronkov and Matthias Baaz, editors, *Proceedings of LPAR, Logic for Programming Artificial intelligence and Reasoning (Tbilisi, Georgia)*, volume 2514 of *Lecture Notes in Computer Science*, pages 102–114. Springer, 2002.
12. Claudio Castellini and Alan Smaill. A systematic presentation of quantified modal logics. *Logic Journal of the IGPL*, 10(6):571–599, November 2002.
13. Anatoli Degtyarev, Michael Fisher, and Boris Konev. Monodic temporal resolution. In Franz Baader, editor, *Proceedings of CADE-19, International Conference on Automated Deduction, Miami, FL (USA)*, 2003.
14. F. J. Cantu, A. Bundy, A. Smaill, and D. Basin. Experiments in automating hardware verification using inductive proof planning. In M. Srivas and A. Camilleri, editors, *First international conference on formal methods in computer-aided design*, volume 1166 of *Lecture Notes in Computer Science*, pages 94–108, Palo Alto, CA, USA, November 1996. Springer Verlag.
15. A. Felty. Implementing tactics and tacticals in a higher-order logic programming language. *Journal of Automated Reasoning*, 11(1):43–81, 1993.

16. Amy Felty. Temporal logic theorem proving and its application to the feature interaction problem. In E. Giunchiglia and F. Massacci, editors, *Issues in the Design and Experimental Evaluation of Systems for Modal and Temporal Logics*, number D11 14/01 in Technical Report, University of Siena, 2001.

17. Amy P. Felty and Kedar S. Namjoshi. Feature specification and automated conflict detection. In *Feature Interactions Workshop*. IOS Press, 2000.

18. Nancy Griffeth, Ralph Blumenthal, Jean-Charles Gregoire, and Tadashi Ohta. A feature interaction benchmark for the first feature interaction detection contest. *Computer Networks (Amsterdam, Netherlands: 1999)*, 32(4):389–418, April 2000.

19. I. Hodkinson, F. Wolter, and M. Zakharyaschev. Decidable fragments of first order temporal logics. *Annals of Pure and Applied Logic*, 106:85–134, 2000.

20. G. J. Holzmann. Design and validation of protocols: a tutorial. *Computer Networks and ISDN Systems*, 25(9):981–1017, April 1993. also in: Proc. 11th PSTV91, INWG/IFIP, Stockholm, Sweden.

21. G.J. Holzmann. The model checker spin. *IEEE Trans. on Software Engineering*, 23(5):279–295, May 1997. Special issue on Formal Methods in Software Practice.

22. Manfred Kerber. Proof planning: A practical approach to mechanized reasoning in mathematics. In Wolfgang Bibel and Peter H. Schmitt, editors, *Automated Deduction, a Basis for Application – Handbook of the German Focus Programme on Automated Deduction*, chapter III.4, pages 77–95. Kluwer Academic Publishers, Dordrecht, The Netherlands, 1998.

23. E. Maclean, J. Fleuriot, and A. Smaill. Proof-planning non-standard analysis. In *Proceedings of the Seventh International Symposium on Artificial Intelligence and Mathematics*, Fort Lauderdale, Florida, 2002.

24. Erica Melis and Jörg Siekmann. Knowledge-based proof planning. *Artificial Intelligence*, 115(1):65–105, 1999.

25. Gopalan Nadathur and Dale Miller. Higher-order logic programming. In D. Gabbay, C. Hogger, and A. Robinson, editors, *Handbook of Logic in AI and Logic Programming, Volume 5: Logic Programming*. Springer Verlag, Oxford, 1986.

System Description: MULTI
A Multi-strategy Proof Planner

Andreas Meier and Erica Melis

German Research Center for Artificial Intelligence (DFKI), Saarbrücken, Germany
{ameier, melis}@dfki.de
http://www.activemath.org/{~ameier,~melis}

1 Introduction

The CASC competitions among automated theorem provers show that there is no single system that outperforms all other systems in all domains. One reaction to this observation is the combination of several systems in a competitive (e.g., the SSCPA system) or cooperative manner (e.g., the CSSCPA and TECHS systems). Thereby, general-purpose heuristics select promising systems to be executed and promising intermediate results to be exchanged. Typically, exchanged results are restricted to clauses or equations that are accepted by all systems. The use of particular domain-specific services for particular subtasks such as the construction of mathematical objects and their flexible cooperation with other services guided by mathematically motivated control knowledge is not possible.

Multi-strategy proof planning [10] provides a framework for the flexible collaboration of independent services, so-called *strategies*, that can realize various operations on proof plans guided by explicitly represented control knowledge. This paper describes the realization of multi-strategy proof planning in the MULTI system and illustrates its application in conducted case studies. Although MULTI focuses to proof planning techniques, its architecture and concepts are generally applicable for a flexible and knowledge-based cooperation of independent services in theorem proving.

2 Proof Planning and Multi-Strategy Proof Planning

Proof planning [1,11] is a theorem proving technique, which constructs a proof at the abstract level of so-called *methods*, i.e., tactics enriched by explicit pre- and postconditions. Methods result from the analysis of common structures or common procedures of a family of proofs. They can encode not only general proof steps but also steps particular to a mathematical domain. Mathematically motivated heuristics are encoded in the control knowledge employed in the search for a solution plan consisting of a sequence or hierarchy of methods.

The idea of multi-strategy proof planning is to combine services instead of performing simple proof planning at the level of methods only. These separate but flexibly collaborating services, so-called *strategies*, realize various proof plan modifications and refinements. For instance, strategies can realize different

R. Nieuwenhuis (Ed.): CADE 2005, LNAI 3632, pp. 250–254, 2005.

kinds of backtracking or they can provide particular services of external systems that can be flexibly integrated with strategies that apply methods. Thereby, the search for applicable strategies introduces an additional hierarchical level into proof planning that allows for the incorporation of further mathematically motivated knowledge.

3   The MULTI System

Algorithms and Strategies. MULTI distinguishes algorithms and strategies. *Algorithms* are independent and parameterized proof plan refinement and modification processes. Currently, MULTI employs 6 different algorithms that all work on partial proof plans: PPlanner for method introduction, InstVar for the instantiation of variables, BackTrack for the deletion of steps from the plan under construction, Exp for the expansion of complex methods, ATP for closing a goal by calling traditional automated theorem provers, and CPlanner for case-based planning. The former three algorithms are decoupled and parameterized functionalities of a simple proof planning process, whereas the latter three algorithms introduce new functionalities. The framework is open to introduce further algorithms that can contribute to proof plan construction.

Strategies specify different services or behaviors of the algorithms. Strategies result from instantiations of the parameters of an algorithm. For instance, the parameters of PPlanner include a set of methods. When such a strategy is executed, then PPlanner introduces only steps that use the methods specified in the strategy. A parameter of InstVar is the function that determines how the instantiation for a variable is computed. A parameter of BackTrack is the function that computes a set of refinement steps that will be deleted from the proof plan.

Technically, strategies are implemented as data structures with three slots: (1) an application condition stating when the strategy is applicable, i.e., the legal applicability knowledge, (2) the algorithm which is employed by the strategy, and (3) the parameter setting with which the algorithm is employed. Examples of strategies can be found in [7,8].

Strategic Control. The heuristic knowledge of the utility of the application of strategies under certain conditions is separated from the strategies. It is declaratively encoded in *strategic control rules*, which provide the basis for meta-reasoning and a global guidance in MULTI. Technically, a control rule is an IF-THEN pair, where the IF-part is a predicate about the proof planning status and the THEN-part is an action that ranks or prunes possible strategy applications.

Architecture. In order to allow for the flexible cooperation of the independent strategies guided by meta-reasoning, MULTI uses a blackboard architecture. In blackboard systems [5], independent components, so-called knowledge sources, collaborate to solve a problem whose solution state is placed on a blackboard that all knowledge sources can access. Figure 1 shows the blackboard architecture of MULTI, which is similar to the *BB1* blackboard system (see [5]). The architecture has two blackboards and two kinds of knowledge sources, one for the proof planning problem and one for the control problem, i.e., the problem of deciding which strategy to apply next.

The proof blackboard contains the current proof plan as well as the history of the proof planning process. The strategies are the knowledge sources that work on this proof blackboard. The control blackboard contains job offers, demands, and a memory. When a strategy's condition part is satisfied, the strategy posts its applicability information, i.e., a *job offer*, onto the control blackboard. The *MetaReasoner* is the knowledge source working on the control blackboard. It ranks

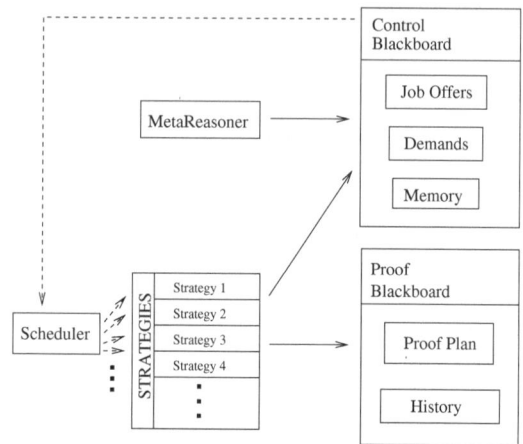

Fig. 1. MULTI's blackboard architecture

the job offers by evaluating the strategic control-rules. The *Scheduler* looks up the control blackboard, takes the highest ranked job offer and executes it.

A strategy that is executed changes the proof plan and records its steps in the history. Executed strategies can also post demands and memory entries onto the control blackboard, which allows to interleave strategy executions. For instance, if the currently executed strategy S can continue only, if another strategy S' is executed first, then S is interrupted, its status is saved in the memory, and a demand for S' is placed onto the control blackboard. After the execution of S' the strategy S can be re-invoked from the memory.

How MULTI Operates. In a nutshell, MULTI operates according to the following cycle, in which no order or sequence of strategies is hard-coded:

- *Job Offers:* Applicable strategies post their applicability (for the current partial plan) as 'job offers' onto the control blackboard.
- *Guidance:* Strategic control rules are evaluated to rank the job offers.
- *Invocation:* The strategy with the highest ranked job offer is invoked.
- *Execution:* The strategy works on the proof blackboard and can place new demands and memory entries onto the control blackboard.

Discussion. To achieve more autonomous strategies the heuristic knowledge could be part of the strategies as well. This would result in a multi-agent approach where the strategies negotiate with each other which strategy to apply next. Our reason for separating heuristic utility and legal feasibility knowledge and for implementing a blackboard architecture is the availability of global control knowledge that favors a central control mechanism over many interacting and locally deciding agents.

4   Experiences and Case Studies

MULTI's strategy level allows for the formalization and incorporation of proof knowledge in strategies and strategic control rules that is beyond the method-level and its control. The knowledge encoded can be diverse. For instance, strategies can describe different techniques to prove a class of problems. Strategies can also describe different ways of backtracking or different ways of constructing mathematical objects to instantiate variables. Strategic control rules can describe, for instance, in which order to try several applicable strategies. They can also guide failure handling. In order to illustrate the advantages of MULTI's strategy level we briefly discuss its application to the residue class domain (see [8] for a detailed description). Other major case studies conducted with MULTI tackle ϵ–δ–proofs [7] and permutation group problems [2].

Residue class conjectures classify given residue class structures wrt. their algebraic category. An example theorem is "the residue class structure $(\mathbb{Z}_5, \bar{+})$ is associative". Other problems from this domain concern the isomorphy of two algebraic structures, e.g., "the residue class structures $(\mathbb{Z}_5, \bar{+})$ and $(\mathbb{Z}_5, \bar{*})$ are not isomorphic". We proved about 19.000 residue class conjectures with MULTI.

Although the problems in this domain are within the range of difficulty a traditional automated theorem prover can handle (see experiments in [8]), it is nevertheless an interesting case study for proof planning, since with MULTI it is possible to generate substantially different and intuitive proofs based on entirely different proof ideas. The parameters of PPlanner can configure strategies taking advantage of different proof settings that mirror proof techniques existing in mathematics. For instance, we realized the three intuitive techniques *naive case-split*, *equational reasoning*, and *reduction to known facts* for this domain in three PPlanner strategies, with different sets of methods and control rules. The availability of three techniques to tackle the problems of the domain extends the robustness of the combined proof planning approach in the sense that if one strategy does not succeed, then another one may continue. A strategic control rule attempts to use the generally most efficient strategy first and the most reliable one last. Moreover, since MULTI supports the switching of strategies, different subproblems in one proof process can be tackled by different strategies.

Reasoning about impasses is a natural ingredient of meta-reasoning at the strategic level: strategic control rules can analyze and exploit failures to guide subsequent strategy applications (see also [7]). In the residue class domain, there exists knowledge of suitable backtrack points for certain failures and proof situations. This knowledge is encoded into strategic control rules, which guide the application of different BackTrack strategies that realize the appropriate back-tracking. This considerably prunes the traversed search space.

MULTI's concept of strategies and algorithms explicitly supports the incorporation of external systems to provide services and to solve subtasks. For instance, to prove that two given residue class structures are not isomorphic, we realized discriminating in MULTI, which requires finding a property P such that P holds for the one structure but not for the other. Discriminating is mainly realized by a main PPlanner strategy, which, however, is interleaved with the call of 3 different

254   A. Meier and E. Melis

external systems in different strategies (see [9] for details). The computation of P is provided by the theory formation system HR [3], which is called in an Inst-Var strategy. In order to show that P holds for the one structure but not for the other, computer algebra systems are employed within another InstVar strategy. Finally, the resolution prover SPASS is used in a strategy of ATP to prove that $\forall X.\forall Y.P(X) \wedge \neg P(Y) \Rightarrow X \not\sim Y$ holds (while this step is fairly obvious for a mathematician, it is crucial for a formal proof).

5 Conclusion and Availability

We presented the multi-strategy proof planner MULTI. The most important features of MULTI are the usage of independent strategies realizing various proof plan services and their flexible cooperation guided by declaratively represented strategic meta-reasoning. MULTI differs from other approaches of proof planning embodied by λCl^AM [12] and ISAPLANNER [4] (1) by its notion of strategies that covers diverse services such as different kinds of backtracking or calls of external systems and (2) by its flexible combination of strategies that is not pre-defined in compound hierarchies of steps as in λCl^AM and ISAPLANNER.

MULTI is implemented in Allegro Common Lisp as a component of the mathematical assistant system ΩMEGA [6]. Its source code is available as part of the ΩMEGA source code, see http://www.ags.uni-sb.de/~omega. Further information can be found at http://www.ags.uni-sb.de/~ameier/multi.html.

References

1. A. Bundy. The Use of Explicit Plans to Guide Inductive Proofs. In *Proc. of CADE-9*, pages 111–120, 1988.
2. A. Cohen, S. Murray, M. Pollet, and V. Sorge. Certifying solutions to permutation group problems. In *Proc. of CADE-19*, pages 258–273, 2003.
3. S. Colton. The HR program for theorem generation. In *Proc. of CADE-18*, pages 280–284, 2002.
4. L. Dixon and J.D. Fleuriot. IsaPlanner: A prototype proof planner in Isabelle. In *Proc. of CADE-19*, pages 279–283, 2003.
5. R. Engelmore and T. Morgan, editors. *Blackboard Systems*. Addison-Wesley, 1988.
6. The OMEGA Group. Proof Development with OMEGA. In *Proc. of CADE-18*, pages 144–149, 2002.
7. A. Meier and E. Melis. Failure reasoning in multiple-strategy proof planning. *Electronic Notes in Theoretical Computer Science*, To appear 2005.
8. A. Meier, M. Pollet, and V. Sorge. Comparing Approaches to Explore the Domain of Residue Classes. *Journal of Symbolic Computation*, 34(4):287–306, 2002.
9. A. Meier, V. Sorge, and S. Colton. Employing Theory Formation to Guide Proof Planning. In *Proc. of Joint AISC and Calculemus 2002*, pages 275–289, 2002.
10. E. Melis and A. Meier. Proof planning with multiple strategies. In *Proc. of the First Intern. Conference on Computational Logic (CL2000)*, pages 644–659, 2000.
11. E. Melis and J. Siekmann. Knowledge-Based Proof Planning. *Artificial Intelligence*, 115(1):65–105, 1999.
12. J.D.C. Richardson, A. Smaill, and I.M. Green. System description: Proof planning in higher-order logic with λClam. In *Proc. of CADE-15*, pages 129–133, 1998.

Decision Procedures Customized for Formal Verification*

Randal E. Bryant and Sanjit A. Seshia

School of Computer Science, Carnegie Mellon University, Pittsburgh, PA
{Randy.Bryant, Sanjit.Seshia}@cs.cmu.edu

Abstract. The UCLID verifier models a hardware or software system as an abstract state machine, where the state variables can be Boolean or integer values, or functions mapping integers to integers or Booleans. The core of the verifier consists of a decision procedure that checks the validity of formulas over the combined theories of uninterpreted functions with equality and linear integer arithmetic. It operates by transforming a formula into an equisatisfiable Boolean formula and then invoking a SAT solver. This approach has worked well for the class of logic and the types of formulas encountered in verification.

1 Introduction

Formal verification of high-level system models requires decision procedures that scale to very large formulas, but over a limited set of theories. Compared to other efforts in developing automated theorem provers, we are not trying to prove the great theorems of mathematics. Rather, our task is to show that a hardware or software designer has successfully implemented a set of interlocks and state consistency constraints to ensure safe operation. Typical verification tasks include proving that a distributed cache memory system maintains a consistent view of the global shared memory, or that a pipelined microprocessor correctly implements a sequential instruction set. By exploiting the limited expressiveness of our modeling language, we have developed a decision procedure that greatly outperforms more general-purpose procedures within its limited domain.

This paper presents our UCLID verifier and its decision procedure. It describes some of the performance-enhancing features of the decision procedure that could have broader applicability.

2 The UCLID Verifier and Decision Procedure

The UCLID verifier represents systems at a more abstract level than existing model checkers [9]. Like a model checker, we view a system as having a collection of state variables that are updated by each step of system operation, but we allow a richer set of state variables than simple Boolean values. UCLID supports the following state variable types:

---

* This research was supported by the Semiconductor Research Corporation, Contract RID 1029.001, and by ARO grant DAAD19-01-1-0485.

R. Nieuwenhuis (Ed.): CADE 2005, LNAI 3632, pp. 255–259, 2005.

- Boolean
- Integer
- Functions mapping integers to integers (referred to as *function* state elements).
- Functions mapping integers to Booleans (referred to as *predicate* state elements).

We use integer state variables to abstract data words and addresses. We also abstract most operations on data using uninterpreted functions. The goal is to capture the control logic of a system in complete detail, but abstract as much of the data processing logic as possible.

The combination of functional state elements and integers (to describe pointers) allows us to model a rich variety of memory structures, including random-access arrays, queues, stacks, and content-addressable memories [4]. That is, we view a memory (or an array) as a function mapping addresses (respectively, indices) to the value stored at the indicated location. A write operation at a location mutates the function, changing the value returned by future read operations from that location. We represent a mutated function as a lambda expression. Functional state elements provide greater modeling generality than the theories of arrays provided by other decision procedures, in that we can model memories in which multiple elements are updated simultaneously.

Whereas a conventional model checker can only verify one particular system configuration, e.g., an eight-node shared memory system, the abstraction capabilities of UCLID enable it to verify an entire class of systems via parametric modeling. By defining the model in terms of a symbolic constant n, we can verify an n-node memory system or a system with an n-entry FIFO buffer, for all possible values of n.

UCLID supports only a limited set of operations on integers, namely addition by a constant, and comparison for equality and ordering. These operations are sufficient to express the pointer manipulations for representing different memory structures. More complex operations must be abstracted as uninterpreted functions. As a result, the UCLID decision procedure need only support a combination of two theories:

- Uninterpreted functions with equality
- Difference logic, with predicates of the form $x \le y + c$, where x and y are integer variables, and c is a numeric constant.

More recently, we have generalized from difference logic to arbitrary linear predicates over integers [13].

The UCLID decision procedure [8] is invoked to prove the validity of a formula (in the above combination of theories) indicating that the desired system property holds. It operates by performing a series of transformations to yield an equi-satisfiable propositional formula, on which a Boolean satisfiability solver is invoked. First, it applies beta reduction to eliminate the lambda expressions describing mutations to the functional state variables. This can potentially cause an exponential blow-up, but we have not experienced this in practice. Second, uninterpreted functions are eliminated using our own technique based on conditional expression expansion [2,3]. Finally, the difference constraints can be encoded into propositional logic using either a *direct* approach, where Boolean variables are used to represent the valuations of different ordering predicates, or with a *small-model* approach in which bounded-range integer variables are encoded using bit-level representations. We automatically combine and select between these two methods using a procedure that was optimized via machine learning [14,12].

3   Useful Ideas

In the course of developing UCLID, we have uncovered several design principles that could prove useful in other contexts.

Develop Custom Decision Procedures. Although it is very appealing to have a decision procedure that supports as many theories as possible, this generality incurs a significant performance penalty. In addition, some theories, such as integer linear arithmetic, are difficult to implement using general-purpose techniques for supporting combinations of theories [11].

We can tune the decision procedure to exploit the characteristics of formulas generated for a specific application. In UCLID, we have devised ways to exploit the sparse structure of linear predicates [13], the sparse structure of equality and ordering predicates [5,6], and the polarity of equations [2,3].

Of course, creating a customized decision procedure for every possible application could incur considerable redundancy and overspecialization. With UCLID, we have sought to support a sufficiently rich set of modeling capabilities to capture a wide variety of systems. The ability to abstract functionality with uninterpreted functions has proved especially important in providing a general modeling capability.

Preserve Conditional Structure. The formulas UCLID derives by symbolic execution of a system model yields conditional expressions of the form $ITE(\phi, T_1, T_2)$, where T_1 and T_2 are integer expressions describing two possible outcomes, depending on whether formula ϕ evaluates to true or false. The standard approach to proving the validity of a formula ψ containing conditional expressions of the form shown above, is to "flatten" out the conditionals. That is, a symbolic integer constant u is introduced, and ψ is rewritten as

$$[(\phi \wedge u = T_1) \vee (\neg\phi \wedge u = T_2)] \Rightarrow \psi'$$

where ψ' is identical to ψ, except that the conditional expression has been replaced by the constant u. Expanding the implication gives a formula of the form:

$$(\phi \wedge u \neq T_1) \vee (\neg\phi \wedge u \neq T_2) \vee \psi'$$

In performing this transformation, we have introduced two equality tests having negative polarity in the formula. This limits the ability to exploit *positive equality* [2,3], cases where we can prove the validity of a formula by considering only a restricted set interpretations of the uninterpreted functions and symbolic constants. By directly supporting conditional expressions with our decision procedure, the polarity structure of the original expressions is more clearly preserved.

Preserve Boolean Structure. Classical ground decision procedures only handle a formula consisting of a conjunction of literals, where each literal is either an atomic formula or its negation [11]. To deal with formulas containing disjunctions, they must flatten the formula into disjunctive normal form (DNF). This often leads to exponential blow-up.

Even recent decision procedures [1,7,15] using SAT solvers at their core can be viewed as dynamically generating conjunctions of literals. Only by "learning" clauses expressing theory-specific constraints do they avoid enumerating a complete DNF representation of the formula.

UCLID, by contrast, maintains the Boolean structure of the original formula while transforming it to propositional logic. Its *eager encoding* approach (described below) avoids any enumeration of a conjunction that does not also satisfies the theory-specific constraints.

Exploit the Clause Management Capabilities of Modern SAT Solvers. The development of the CHAFF Boolean satisfiability solver [10] has led to a new generation of solvers that can deal with very large sets of clauses. They are remarkably effective at determining that a large formula is unsatisfiable for the case where a small subset of the clauses lead to a contradiction. This structure occurs for our decision procedures based on our *eager* encodings of theory properties into propositional logic [14]. For example, in encoding the equality predicates in a formula, we introduce a sufficient set of predicates to imply all of the transitivity constraints implied by these equalities [5,6]. By exploiting the sparse structure of the equations in the formula, we can avoid the cubic blowup that would be required to completely axiomatize the transitivity of equality.

References

1. G. Audemard, P. Bertoli, A. Cimatti, A. Korniowicz, and R. Sebastiani. A SAT based approach for solving formulas over Boolean and linear mathematical propositions. In Andrei Voronkov, editor, *Conference on Automated Deduction (CADE '02)*, LNCS 2392, pages 195–210, 2002.
2. R. E. Bryant, S. German, and M. N. Velev. Exploiting positive equality in a logic of equality with uninterpreted functions. In N. Halbwachs and D. Peled, editors, *Computer-Aided Verification (CAV '99)*, LNCS 1633, pages 470–482, 1999.
3. R. E. Bryant, S. German, and M. N. Velev. Processor verification using efficient reductions of the logic of uninterpreted functions to propositional logic. *ACM Transactions on Computational Logic*, 2(1):1–41, January 2001.
4. R. E. Bryant, S. K. Lahiri, and S. A. Seshia. Modeling and verifying systems using a logic of counter arithmetic with lambda expressions and uninterpreted functions. In E. Brinksma and K. G. Larsen, editors, *Computer-Aided Verification (CAV '02)*, LNCS 2404, pages 78–92, 2002.
5. R. E. Bryant and M. N. Velev. Boolean satisfiability with transitivity constraints. In A. Emerson and P. Sistla, editors, *Computer-Aided Verification (CAV '00)*, LNCS 1855, pages 85–98, 2000.
6. R. E. Bryant and M. N. Velev. Boolean satisfiability with transitivity constraints. *ACM Transactions on Computational Logic*, 3(4), October 2002.
7. J.-C. Filliâtre, S. Owre, H. Rueß, and N. Shankar. ICS: Integrated Canonizer and Solver. In G. Berry, H. Comon, and A. Finkel, editors, *Computer-Aided Verification (CAV '01)*, LNCS 2102, pages 246–249, 2001.
8. S. K. Lahiri and S. A. Seshia. The UCLID decision procedure. In *Computer-Aided Verification (CAV '04)*, LNCS 3114, pages 475–478, 2004.
9. K. McMillan. *Symbolic Model Checking*. Kluwer Academic Publishers, 1992.

10. M. Moskewicz, C. Madigan, Y. Zhao, L. Zhang, and S. Malik. Chaff: Engineering an efficient SAT solver. In *38th Design Automation Conference (DAC '01)*, pages 530–535, 2001.

11. G. Nelson and D. C. Oppen. Simplification by cooperating decision procedures. *ACM Transactions on Programming Languages and Systems*, 1(2):245–257, 1979.

12. S. A. Seshia. *Adaptive Eager Boolean Encoding for Arithmetic Reasoning in Verification*. PhD thesis, Carnegie Mellon University, 2005.

13. S. A. Seshia and R. E. Bryant. Deciding quantifier-free Presburger formulas using parameterized solution bounds. In 19th *IEEE Symposium on Logic in Computer Science (LICS)*, July 2004.

14. S. A. Seshia, S. K. Lahiri, and R. E. Bryant. A hybrid SAT-based decision procedure for separation logic with uninterpreted functions. In *40th Design Automation Conference (DAC '03)*, pages 425–430, June 2003.

15. A. Stump, C. W. Barrett, and D. L. Dill. CVC: A Cooperating Validity Checker. In Ed Brinksma and Kim Guldstrand Larsen, editors, *14th International Conference on Computer Aided Verification (CAV)*, volume 2404 of *LNCS*, pages 500–504. Springer-Verlag, 2002.

An Algorithm for Deciding BAPA:
Boolean Algebra with Presburger Arithmetic[*]

Viktor Kuncak[1], Huu Hai Nguyen[2], and Martin Rinard[1,2]

[1] MIT CSAIL, Cambridge, USA
[2] Singapore-MIT Alliance

Abstract. We describe an algorithm for deciding the first-order multisorted theory BAPA, which combines 1) Boolean algebras of sets of uninterpreted elements (BA) and 2) Presburger arithmetic operations (PA). BAPA can express the relationship between integer variables and cardinalities of a priory unbounded finite sets, and supports arbitrary quantification over sets and integers.

Our motivation for BAPA is deciding verification conditions that arise in the static analysis of data structure consistency properties. Data structures often use an integer variable to keep track of the number of elements they store; an invariant of such a data structure is that the value of the integer variable is equal to the number of elements stored in the data structure. When the data structure content is represented by a set, the resulting constraints can be captured in BAPA. BAPA formulas with quantifier alternations arise when verifying programs with annotations containing quantifiers, or when proving simulation relation conditions for refinement and equivalence of program fragments. Furthermore, BAPA constraints can be used for proving the termination of programs that manipulate data structures, and have applications in constraint databases.

We give a formal description of a decision procedure for BAPA, which implies the decidability of BAPA. We analyze our algorithm and obtain an elementary upper bound on the running time, thereby giving the first complexity bound for BAPA. Because it works by a reduction to PA, our algorithm yields the decidability of a combination of sets of uninterpreted elements with any decidable extension of PA. Our algorithm can also be used to yield an optimal decision procedure for BA through a reduction to PA with bounded quantifiers.

We have implemented our algorithm and used it to discharge verification conditions in the Jahob system for data structure consistency checking of Java programs; our experience with the algorithm is promising.

1 Introduction

Program analysis and verification tools can greatly contribute to software reliability, especially when used throughout the software development process. Such tools are even more valuable if their behavior is predictable, if they can be applied to partial programs, and if they allow the developer to communicate the design information in the form of specifications. Combining the basic idea of [18] with decidable logics leads to analysis tools that have these desirable properties. Such analyses are precise (because formulas

---

[*] CADE-20.

R. Nieuwenhuis (Ed.): CADE 2005, LNAI 3632, pp. 260–277, 2005.
© Springer-Verlag Berlin Heidelberg 2005

represent loop-free code precisely) and predictable (because the checking of verification conditions terminates either with a realizable counterexample or with a sound claim that there are no counterexamples).

A key challenge in this approach to program analysis and verification is to identify a logic that captures an interesting class of program properties, but is nevertheless decidable. In [29] we identify the first-order theory of Boolean algebras (BA) as a useful language for reasoning about dynamically allocated objects: BA allows expressing generalized typestate properties and reasoning about data structures as dynamically changing sets of objects. (We are interested in BA of all subsets of some set; this theory was shown decidable already in [31, 46], see [22] for the discussion of other models of Boolean algebra axioms.)

The motivation for this paper is the fact that we often need to reason not only about the data structure content, but also about the size of the data structure. For example, we may want to express the fact that the number of elements stored in a data structure is equal to the value of an integer variable that is used to cache the data structure size, or we may want to introduce a decreasing integer measure on the data structure to show program termination. These considerations lead to a natural generalization of the first-order theory of BA of sets, a generalization that allows integer variables in addition to set variables, and allows stating relations of the form $|A| = k$ meaning that the cardinality of the set A is equal to the value of the integer variable k. Once we have integer variables, a natural question arises: which relations and operations on integers should we allow? It turns out that, using only the BA operations and the cardinality operator, we can already define all operations of PA. This leads to the structure BAPA, which properly generalizes both BA and PA.

As we explain in Section 2, a version of BAPA was shown decidable already in [14] (which also proves the well-known Feferman-Vaught theorem [19, Section 9.6] about the products of first-order theories). Recently, a decision procedure for a fragment of BAPA without quantification over sets was presented in [55], cast as a multi-sorted theory. Starting from [29] as our motivation, we have observed in [26] the decidability of the full BAPA (which was initially left open in [55]). An algorithm for a single-sorted version of BAPA was presented independently in [42] as a way of evaluating queries in constraint databases; [42] leaves open the complexity of the satisfiability problem.

Our paper gives the first formal description of a decision procedure for the full first-order theory of BAPA. Furthermore, we analyze our decision procedure and show that it yields an elementary upper bound on the complexity of BAPA. Our result is the first upper complexity bound on BAPA; along with a lower bound from PA, we obtain a good estimate of BAPA worst-case complexity. We have also implemented our decision procedure; we report on our initial experience in using the decision procedure in the context of a system for checking data structure consistency.

Contributions. We summarize the contributions of our paper as follows.

1. As a **motivation** for BAPA, we show in Section 3 how BAPA constraints can be used for program analysis and verification by expressing 1) data structure invariants, 2) the correctness of procedures with respect to their specifications, 3) simulation relations between program fragments, and 4) termination conditions for programs that manipulate data structures.

2. We present an **algorithm** α (Section 4) that translates BAPA sentences into PA sentences by translating set quantifiers into integer quantifiers.
3. We analyze our algorithm α and show that it yields an **elementary upper bound** on the worst-case complexity of the validity problem for BAPA sentences that is close to the bound on PA sentences themselves (Section 5). This is the first complexity bound for BAPA, and is the main contribution of this paper.
4. We discuss our initial experience in using our **implementation** of BAPA to discharge verification conditions generated in the Jahob verification system [23].
5. In addition, we note the following related results:
   (a) PA sentences generated by translating BA sentences without cardinalities can be decided in **optimal** alternating time (Section 5.2);
   (b) Our algorithm extends to **countable sets** with a predicate distinguishing finite and infinite sets (Section 7);
   (c) In contrast to the undecidability of MSOL with equicardinality operator, we identify a **decidable** combination of MSOL over trees with BA (Section 7).

A preliminary version of our results, including the algorithm and complexity analysis appear in [26], which also contains proofs and further details of our results.

2   The First-Order Theory BAPA

Figure 3 presents the syntax of Boolean Algebra with Presburger Arithmetic (BAPA), which is the focus of this paper. We next present some justification for the operations in Figure 3. Our initial motivation for BAPA was the use of BA to reason about data structures in terms of sets [28]. Our language for BA (Figure 1) allows cardinality constraints of the form $|A| = K$ where K is a *constant* integer. Such constant cardinality constraints are useful and enable quantifier elimination for the resulting language [31,46]. However, they do not allow stating constraints such as $|A| = |B|$ for two sets A and B, and cannot represent constraints on changing program variables. Consider therefore the equicardinality relation $A \sim B$ that holds iff $|A| = |B|$, and consider BA extended with relation $A \sim B$. Define the ternary relation $\text{plus}(A, B, C) \iff (|A| + |B| = |C|)$ by the formula $\exists x_1. \exists x_2. x_1 \cap x_2 = \emptyset \wedge A \sim x_1 \wedge B \sim x_2 \wedge x_1 \cup x_2 = C$. The relation $\text{plus}(A, B, C)$ allows us to express addition using arbitrary sets as representatives for natural numbers; $\emptyset$ can represent the natural number zero, and any singleton set can represent the natural number one. (The property of A being a singleton is definable using e.g. the first-order formula $A \neq \emptyset \wedge \forall B. A \cap B = B \Rightarrow (B = \emptyset \vee B = A)$.) Moreover, we can represent integers as equivalence classes of pairs of natural numbers under the equivalence relation $(x, y) \approx (u, v) \iff x + v = u + y$; this construction also allows us to express the unary predicate of being non-negative. The quantification over pairs of sets represents quantification over integers, and quantification over integers with the addition operation and the predicate "being non-negative" can express all PA operations, presented in Figure 2. Therefore, a natural closure under definable operations leads to our formulation of the language BAPA in Figure 3, which contains both sets and integers.

$F ::= A \mid F_1 \wedge F_2 \mid F_1 \vee F_2 \mid \neg F \mid$
$\quad \exists x.F \mid \forall x.F$

$A ::= B_1 = B_2 \mid B_1 \subseteq B_2 \mid$
$\quad \mid B \mid = K \mid \mid B \mid \geq K$

$B ::= x \mid \mathbf{0} \mid \mathbf{1} \mid B_1 \cup B_2 \mid B_1 \cap B_2 \mid B^c$

$K ::= 0 \mid 1 \mid 2 \mid \ldots$

Fig. 1. Formulas of Boolean Algebra (BA)

$F ::= A \mid F_1 \wedge F_2 \mid F_1 \vee F_2 \mid \neg F \mid$
$\quad \exists k.F \mid \forall k.F$

$A ::= T_1 = T_2 \mid T_1 < T_2 \mid C \, \mathsf{dvd} \, T$

$T ::= K \mid T_1 + T_2 \mid K \cdot T$

$K ::= \ldots -2 \mid -1 \mid 0 \mid 1 \mid 2 \ldots$

Fig. 2. Formulas of Presburger Arithmetic (PA)

$F ::= A \mid F_1 \wedge F_2 \mid F_1 \vee F_2 \mid \neg F \mid$

$\quad \exists x.F \mid \forall x.F \mid \exists k.F \mid \forall k.F$

$A ::= B_1 = B_2 \mid B_1 \subseteq B_2 \mid$

$\quad T_1 = T_2 \mid T_1 < T_2 \mid C \, \mathsf{dvd} \, T$

$B ::= x \mid \mathbf{0} \mid \mathbf{1} \mid B_1 \cup B_2 \mid B_1 \cap B_2 \mid B^c$

$T ::= k \mid K \mid \mathsf{MAXC} \mid T_1 + T_2 \mid K \cdot T \mid \mid B \mid$

$K ::= \ldots -2 \mid -1 \mid 0 \mid 1 \mid 2 \ldots$

Fig. 3. Formulas of Boolean Algebra with Presburger Arithmetic (BAPA)

The argument above also explains why we attribute the decidability of BAPA to [14, Section 8], which showed the decidability of BA over sets extended with the equicardinality relation $\sim$, using the decidability of the first-order theory of the addition of cardinal numbers.

The language BAPA has two kinds of quantifiers: quantifiers over integers and quantifiers over sets; we distinguish between these two kinds by denoting integer variables with symbols such as k, l and set variables with symbols such as x, y. We use the shorthand $\exists^+ k.F(k)$ to denote $\exists k.k \geq 0 \wedge F(k)$ and, similarly $\forall^+ k.F(k)$ to denote $\forall k.k \geq 0 \Rightarrow F(k)$. In summary, the language of BAPA in Figure 3 subsumes the language of PA in Figure 2, subsumes the language of BA in Figure 3, and contains non-trivial combination of these two languages in the form of using the cardinality of a set expression as an integer value.

The semantics of operations in Figure 3 is the expected one. We interpret integer terms as integers, and interpret set terms as elements of the powerset of a finite set. The MAXC constant denotes the size of the finite universe $\mathcal{U}$, so we require $\mathsf{MAXC} = |\mathcal{U}|$ in all models. Our results generalize to the Boolean algebra of powersets of a countable set, see Section 7.

3   Applications of BAPA

This section illustrates the importance of BAPA constraints. Section 3.1 shows the uses of BAPA constraints to express and verify data structure invariants as well as procedure preconditions and postconditions. Section 3.2 shows how a class of simulation relation conditions can be proved automatically using a decision procedure for BAPA. Section 3.3 shows how BAPA can be used to express and prove termination conditions for a class of programs.

3.1 Verifying Data Structure Consistency

Figure 4 presents a procedure insert in a language that directly manipulates sets. Such languages can either be directly executed [13] or can arise as abstractions of programs in standard languages [29]. The program in Figure 4 manipulates a global set of objects content and an integer field size. The program maintains an invariant I that the size of the set content is equal to the value of the variable size. The insert procedure inserts an element e into the set and correspondingly updates the integer variable. The requires clause (precondition) of the insert procedure is that the parameter e is a non-null reference to an object that is not stored in the set content. The ensures clause (postcondition) of the procedure is that the size variable after the insertion is positive. Note that we represent references to objects (such as the procedure parameter e) as sets with at most one element. An empty set represents a null reference; a singleton set $\{o\}$ represents a reference to object o. The value of a variable after procedure execution is indicated by marking the variable name with a prime.

The insert procedure maintains an invariant, I, which captures the relationship between the size of the set content and the integer variable size. The invariant I is implicitly conjoined with the requires and the ensures clauses of the procedure. The Hoare triple in Figure 5 summarizes the resulting correctness condition for the insert procedure. Figure 6

```
var content : set;
var size : integer;
invariant I ⟺ (size = |content|);

procedure insert(e : element)
maintains I
requires |e| = 1 ∧ |e ∩ content| = 0
ensures size' > 0
{
    content := content ∪ e;
    size := size + 1;
}
```

Fig. 4. An Example Procedure

$$\{|e| = 1 \land |e \cap \text{content}| = 0 \land \text{size} = |\text{content}|\}$$

$$\text{content} := \text{content} \cup e; \quad \text{size} := \text{size} + 1;$$

$$\{\text{size}' > 0 \land \text{size}' = |\text{content}'|\}$$

Fig. 5. Hoare Triple for insert Procedure

$$\forall e. \forall \text{content}. \forall \text{content}'. \forall \text{size}. \forall \text{size}'.$$
$$(|e| = 1 \land |e \cap \text{content}| = 0 \land \text{size} = |\text{content}| \land$$
$$\text{content}' = \text{content} \cup e \land \text{size}' = \text{size} + 1) \Rightarrow$$
$$\text{size}' > 0 \land \text{size}' = |\text{content}'|$$

Fig. 6. Verification Condition for Figure 5

presents a verification condition corresponding to the Hoare triple in Figure 5. Note that the verification condition contains both set and integer variables, contains quantification over these variables, and relates the sizes of sets to the values of integer variables. Our small example leads to a formula without quantifier alternations; in general, formulas that arise in verification may contain alternations of existential and universal variables over both integers and sets. This paper shows the decidability of such formulas and presents the complexity of the decision procedure.

3.2 Proving Simulation Relation Conditions

BAPA constraints are also useful when proving that a given binary relation on states is a simulation relation between two program fragments. Figure 7 shows one such example. The concrete procedure start1 manipulates two sets: a set of running processes and a set of suspended processes in a process scheduler. The procedure start1 inserts a new process x into the set of running processes R, unless there are already too many running processes. The procedure start2 is a version of the procedure that operates in a more abstract state space: it maintains only the union P of all processes and the number k of running processes. Figure 7 shows a forward simulation relation r between the transition relations for start1 and start2. The standard simulation relation diagram condition is $\forall s_1.\forall s_1'.\forall s_2.(t_1(s_1, s_1') \wedge r(s_1, s_2)) \Rightarrow \exists s_2'. (t_2(s_2, s_2') \wedge r(s_1', s_2'))$. In the presence of preconditions, $t_1(s_1, s_1') = (\mathsf{pre}_1(s_1) \Rightarrow \mathsf{post}_1(s_1, s_1'))$ and $t_2(s_2, s_2') = (\mathsf{pre}_2(s_2) \Rightarrow \mathsf{post}_2(s_2, s_2'))$, and sufficient conditions for simulation relation are:

1. $\forall s_1.\forall s_2.r(s_1, s_2) \wedge \mathsf{pre}_2(s_2) \Rightarrow \mathsf{pre}_1(s_1)$
2. $\forall s_1.\forall s_1'.\forall s_2.\exists s_2'. \ r(s_1, s_2) \wedge \mathsf{post}_1(s_1, s_1') \wedge \mathsf{pre}_2(s_2) \Rightarrow \mathsf{post}_2(s_2, s_2') \wedge r(s_1', s_2')$

Figure 7 shows BAPA formulas that correspond to the simulation relation conditions in this example. Note that the second BAPA formula has a quantifier alternation, which illustrates the relevance of quantifiers in BAPA.

3.3 Proving Termination of Programs

We next show that BAPA is useful for proving program termination. A standard technique for proving termination of a loop is to introduce a ranking function f that maps program state into a non-negative integer, then prove that the value of the function decreases at each loop iteration. In other words, if $t(s, s')$ denotes the relationship between the state at the beginning and the state at the end of each loop iteration, then the condition $\forall s.\forall s'.t(s, s') \Rightarrow f(s) > f(s')$ holds. Figure 8 shows an example program that processes each element of the initial value of set iter; this program can be viewed as manipulating an iterator over a data structure that implements a set. Using the the ability to take cardinality of a set allows us to define a natural ranking function for this program. Figure 9 shows the termination proof based on such ranking function. The resulting termination condition can be expressed as a formula that belongs to BAPA, and can be discharged using our decision procedure. In general, we can reduce the termination problem of programs that manipulate both sets and integers to showing a simulation relation with a fragment of a terminating program that manipulates only integers, which

```
var R : set;
var S : set;
```

```
var P : set;
var k : integer;
```

procedure start1(x)
requires $x \not\subseteq R \wedge |x| = 1 \wedge |R| < \text{MAXR}$
ensures $R' = R \cup x \wedge S' = S$
```
{
  R := R ∪ x;
}
```

procedure start2(x)
requires $x \not\subseteq P \wedge |x| = 1 \wedge k < \text{MAXR}$
ensures $P' = P \cup x \wedge k' = k + 1$
```
{
  P := P ∪ x;
  k := k + 1;
}
```

Simulation relation r:
$r((R, S), (P, k)) = (P = R \cup S \wedge k = |R|)$

Simulation relation conditions in BAPA:
1. $\forall x, R, S, P, k.(P = R \cup S \wedge k = |R|) \wedge (x \not\subseteq P \wedge |x| = 1 \wedge k < \text{MAXR}) \Rightarrow$
   $\qquad (x \not\subseteq R \wedge |x| = 1 \wedge |R| < \text{MAXR})$
2. $\forall x, R, S, R', S', P, k. \exists P', k'.((P = R \cup S \wedge k = |R|) \wedge (R' = R \cup x \wedge S' = S) \wedge$
   $\qquad (x \not\subseteq P \wedge |x| = 1 \wedge k < \text{MAXR})) \Rightarrow$
   $\qquad (P' = P \cup x \wedge k' = k + 1) \wedge (P' = R' \cup S' \wedge k' = |R'|)$

Fig. 7. Proving simulation relation in BAPA

```
var iter : set;
```

procedure iterate()
```
{
  while iter ≠ ∅ do
    var e : set;
    e := choose iter;
    iter := iter \ e;
    process(e);
  done
}
```

Ranking function:
$f(s) = |s|$

Transition relation:
$t(\text{iter}, \text{iter}') = (\exists e. |e| = 1 \wedge e \subseteq \text{iter} \wedge \text{iter}' = \text{iter} \setminus e)$

Termination condition in BAPA:
$\forall \text{iter}. \forall \text{iter}'. (\exists e. |e| = 1 \wedge e \subseteq \text{iter} \wedge \text{iter}' = \text{iter} \setminus e)$
$\qquad \Rightarrow |\text{iter}'| < |\text{iter}|$

Fig. 8. Terminating program

Fig. 9. Termination proof for Figure 8

can be proved terminating using techniques [38]. The simulation relation condition can be proved correct using our BAPA decision procedure whenever the simulation relation is expressible with a BAPA formula.

4   Decision Procedure for BAPA

This section presents our algorithm, denoted α, which decides the validity of BAPA sentences. The algorithm reduces a BAPA sentence to an equivalent PA sentence with the same number of quantifier alternations and an exponential increase in the total size of the formula. This algorithm has several desirable properties:

1. Given the space and time bounds for PA sentences [41], the algorithm α yields reasonable space and time bounds for deciding BAPA sentences (Section 5).
2. The algorithm α does not eliminate integer variables, but instead produces an equivalent quantified PA sentence. The resulting PA sentence can therefore be decided using *any* decision procedure for PA, including the decision procedures based on automata [21, 30].
3. The algorithm α can eliminate set quantifiers from any extension of PA. We thus obtain a technique for adding a particular form of set reasoning to every extension of PA, and the technique preserves the decidability of the extension. One example of decidable theory that extends PA is MSOL over strings, see See Section 7.
4. For simplicity we present the algorithm α as a decision procedure for formulas with no free variables, but the algorithm can be used to transform and simplify formulas with free variables as well, because it transforms one quantifier at a time starting from the innermost one. Because of this feature, we can use the algorithm α to project out local state components from formulas that describe invariants and transition relations, and simplify the resulting formulas.

We next describe the algorithm α for transforming a BAPA sentence F_0 into a PA sentence. As the first step of the algorithm, transform F_0 into prenex form

$$Q_p v_p. \ldots . Q_1 v_1. F(v_1, \ldots, v_p)$$

where F is quantifier-free, and each quantifier $Q_i v_i$ is of one the forms $\exists k, \forall k, \exists y, \forall y$ where k denotes an integer variable and y denotes a set variable.

The next step of the algorithm is to separate F into BA part and PA part. To achieve this, replace each formula $x = y$ where x and y are sets, with the conjunction $x \subseteq y \wedge y \subseteq x$, and replace each formula $x \subseteq y$ with the equivalent formula $|x \cap y^c| = 0$. In the resulting formula, each set x occurs in some term $|t(x)|$. Next, use the same reasoning as when generating disjunctive normal form for propositional logic to write each set expression $t(x)$ as a union of cubes (regions in Venn diagram). The cubes have the form $\bigwedge_{i=1}^{n} x_i^{\alpha_i}$ where $x_i^{\alpha_i}$ is either x_i or x_i^c; there are $m = 2^n$ cubes $s_1, \ldots, s_m$. Suppose that $t(x) = s_{j_1} \cup \ldots \cup s_{j_a}$; then replace the term $|t(x)|$ with the term $\sum_{i=1}^{a} |s_{j_i}|$. In the resulting formula, each set x appears in an expression of the form $|s_i|$ where s_i is a cube. For each s_i introduce a new variable l_i. Then the resulting formula is equivalent to

$$Q_p v_p. \ldots . Q_1 v_1. \\ \exists^+ l_1, \ldots, l_m. \bigwedge_{i=1}^{m} |s_i| = l_i \wedge G_1 \tag{1}$$

where G_1 is a PA formula. Formula (1) is the starting point of the main phase of algorithm α. The main phase of the algorithm successively eliminates quantifiers $Q_1 v_1, \ldots,$ $Q_p v_p$ while maintaining a formula of the form

$$Q_p v_p \ldots Q_r v_r. \\ \exists^+ l_1 \ldots l_q. \bigwedge_{i=1}^{q} |s_i| = l_i \wedge G_r \tag{2}$$

where G_r is a PA formula, r grows from 1 to $p + 1$, and $q = 2^e$ where e for $0 \le e \le n$ is the number of set variables among $v_p, \ldots, v_r$. The list $s_1, \ldots, s_q$ is the list of all 2^e partitions formed from the set variables among $v_p, \ldots, v_r$.

We next show how to eliminate the innermost quantifier $Q_r v_r$ from the formula (2). During this process, the algorithm replaces the formula G_r with a formula G_{r+1} which has more integer quantifiers. If v_r is an integer variable then the number of sets q remains the same, and if v_r is a set variable, then q reduces from 2^e to 2^{e-1}. We next consider each of the four possibilities $\exists k, \forall k, \exists y, \forall y$ for the quantifier $Q_r v_r$.

Consider first the case $\exists k$. Because k does not occur in $\bigwedge_{i=1}^q |s_i| = l_i$, simply move the existential quantifier to G_r and let $G_{r+1} = \exists k.G_r$, which completes the step.

For universal quantifiers, it suffices to let $G_{r+1} = \forall k.G_r$, again because k does not occur in $\bigwedge_{i=1}^q |s_i| = l_i$.

We next show how to eliminate an existential set quantifier $\exists y$ from

$$\exists y.\ \exists^+ l_1 \ldots l_q.\ \bigwedge_{i=1}^q |s_i| = l_i \wedge G_r \tag{3}$$

which is equivalent to $\exists^+ l_1 \ldots l_q.\ (\exists y.\ \bigwedge_{i=1}^q |s_i| = l_i) \wedge G_r$. This is the key step of the algorithm and relies on the following lemma (see [26] for proof).

Lemma 1. *Let $b_1, \ldots, b_n$ be finite disjoint sets, and $l_1, \ldots, l_n, k_1, \ldots, k_n$ be natural numbers. Then the following two statements are equivalent:*

1. *There exists a finite set y such that $\bigwedge_{i=1}^n |b_i \cap y| = k_i \wedge |b_i \cap y^c| = l_i$*
2. $\bigwedge_{i=1}^n |b_i| = k_i + l_i$.

In the quantifier elimination step, assume without loss of generality that the set variables $s_1, \ldots, s_q$ are numbered such that $s_{2i-1} \equiv s_i' \cap y^c$ and $s_{2i} \equiv s_i' \cap y$ for some cube s_i'. Then apply Lemma 1 and replace each pair of conjuncts

$$|s_i' \cap y^c| = l_{2i-1} \wedge |s_i' \cap y| = l_{2i}$$

with the conjunct $|s_i'| = l_{2i-1} + l_{2i}$, yielding formula

$$\exists^+ l_1 \ldots l_q.\ \bigwedge_{i=1}^{q'} |s_i'| = l_{2i-1} + l_{2i} \wedge G_r \tag{4}$$

for $q' = 2^{e-1}$. Finally, to obtain a formula of the form (2) for $r + 1$, introduce fresh variables l_i' constrained by $l_i' = l_{2i-1} + l_{2i}$, rewrite (4) as

$$\exists^+ l_1' \ldots l_{q'}'.\ \bigwedge_{i=1}^{q'} |s_i'| = l_i' \wedge (\exists l_1 \ldots l_q.\ \bigwedge_{i=1}^{q'} l_i' = l_{2i-1} + l_{2i} \wedge G_r)$$

and let

$$G_{r+1} \equiv \exists^+ l_1 \ldots l_q.\ \bigwedge_{i=1}^{q'} l_i' = l_{2i-1} + l_{2i} \wedge G_r$$

This completes the description of elimination of an existential set quantifier $\exists y$.

To eliminate a set quantifier $\forall y$, observe that

$$\neg(\exists^+ l_1 \ldots l_q. \bigwedge_{i=1}^{q} |s_i| = l_i \wedge G_r)$$

is equivalent to $\exists^+ l_1 \ldots l_q. \bigwedge_{i=1}^{q} |s_i| = l_i \wedge \neg G_r$, because the existential quantifier is used as a let-binding, so we may first substitute all values l_i into G_r, then perform the negation, and then extract back the definitions of all values l_i. By expressing $\forall y$ as $\neg \exists y \neg$, we can show that the elimination of $\forall y$ is analogous to elimination of $\exists y$: introduce fresh variables $l'_i = l_{2i-1} + l_{2i}$ and let

$$G_{r+1} \equiv \forall^+ l_1 \ldots l_q. (\bigwedge_{i=1}^{q'} l'_i = l_{2i-1} + l_{2i}) \Rightarrow G_r$$

After eliminating all quantifiers as described above, we obtain a formula of the form $\exists^+ l. |\mathcal{U}| = l \wedge G_{p+1}(l)$. We define the result of the algorithm, denoted $\alpha(F_0)$, to be the PA sentence $G_{p+1}(\mathsf{MAXC})$.

This completes the description of the algorithm α. Given that the validity of PA sentences is decidable [39], the algorithm α is a decision procedure for BAPA sentences.

Theorem 2. *The algorithm α described above maps each* BAPA*-sentence F_0 into an equivalent* PA*-sentence $\alpha(F_0)$.*

Formalization of the Algorithm α. To formalize the algorithm α, we wrote a concise implementation in O'Caml, see [26]. As an illustration, when we run the implementation on the BAPA formula in Figure 6 which represents a verification condition, we immediately obtain the PA formula in Figure 10. Note that the structure of the resulting formula mimics the structure of the original formula: every set quantifier is replaced by the corresponding block of quantifiers over non-negative integers constrained to partition the previously introduced integer variables. Figure 11 presents the correspondence between the set variables of the BAPA formula and the integer variables of the translated PA formula. Note that the relationship $\mathsf{content}' = \mathsf{content} \cup e$ translates into the conjunction of the constraints $|\mathsf{content}' \cap (\mathsf{content} \cup e)^c| = 0 \wedge |(\mathsf{content} \cup e) \cap \mathsf{content}'^c| = 0$, which reduces to the conjunction $l_{100} = 0 \wedge l_{011} + l_{001} + l_{010} = 0$ using the translation of set expressions into the disjoint union of partitions, and the correspondence in Figure 11.

5   Complexity

In this section we analyze the algorithm α from Section 4 and obtain space bounds on BAPA from the corresponding space bounds for PA. We then show that the new decision procedure is optimal for BA if applied to BA formulas. Moreover, by construction, our procedure reduces to the procedure for PA formulas if there are no set quantifiers. In summary, our decision procedure is optimal for BA, does not impose any overhead for pure PA formulas, and the complexity of the general BAPA validity has the same height of the tower of exponentials as the complexity of PA itself.

$\forall^+ l_1.\forall^+ l_0.\ \text{MAXC} = l_1 + l_0 \Rightarrow$
$\forall^+ l_{11}.\forall^+ l_{01}.\forall^+ l_{10}.\forall^+ l_{00}.$
$l_1 = l_{11} + l_{01} \wedge l_0 = l_{10} + l_{00} \Rightarrow$
$\forall^+ l_{111}.\forall^+ l_{011}.\forall^+ l_{101}.\forall^+ l_{001}.$
$\forall^+ l_{110}.\forall^+ l_{010}.\forall^+ l_{100}.\forall^+ l_{000}.$
$l_{11} = l_{111} + l_{011} \wedge l_{01} = l_{101} + l_{001} \wedge$
$l_{10} = l_{110} + l_{010} \wedge l_{00} = l_{100} + l_{000} \Rightarrow$
$\quad \forall size.\forall size'.$
$\quad (l_{111} + l_{011} + l_{101} + l_{001} = 1 \wedge$
$\quad l_{111} + l_{011} = 0 \wedge$
$\quad l_{111} + l_{011} + l_{110} + l_{010} = size \wedge$
$\quad l_{100} = 0 \wedge$
$\quad l_{011} + l_{001} + l_{010} = 0 \wedge$
$\quad size' = size + 1) \Rightarrow$
$\quad\quad (0 < size' \wedge$
$\quad\quad\quad l_{111} + l_{101} + l_{110} + l_{100} = size')$

general relationship:

$$l_{i_1,\ldots,i_k} = |\text{set}_q^{i_1} \cap \text{set}_{q+1}^{i_2} \cap \ldots \cap \text{set}_S^{i_k}|$$
$$q = S - (k - 1)$$
(S is number of set variables)

in this example:

$\text{set}_1 = \text{content}'$
$\text{set}_2 = \text{content}$
$\text{set}_3 = e$

$l_{000} = |\text{content}'^c \cap \text{content}^c \cap e^c|$
$l_{001} = |\text{content}'^c \cap \text{content}^c \cap e|$
$l_{010} = |\text{content}'^c \cap \text{content} \cap e^c|$
$l_{011} = |\text{content}'^c \cap \text{content} \cap e|$
$l_{100} = |\text{content}' \cap \text{content}^c \cap e^c|$
$l_{101} = |\text{content}' \cap \text{content}^c \cap e|$
$l_{110} = |\text{content}' \cap \text{content} \cap e^c|$
$l_{111} = |\text{content}' \cap \text{content} \cap e|$

Fig. 10. The translation of the BAPA sentence from Figure 6 into a PA sentence

Fig. 11. The Correspondence between Integer Variables in Figure 10 and Set Variables in Figure 6

5.1 An Elementary Upper Bound

We next show that the algorithm in Section 4 transforms a BAPA sentence F_0 into a PA sentence whose size is at most exponential and which has the same number of quantifier alternations.

If F is a formula in prenex form, let $\text{size}(F)$ denote the size of F, and let $\text{alts}(F)$ denote the number of quantifier alternations in F. Define the iterated exponentiation function $\exp_k(x)$ by $\exp_0(x) = x$ and $\exp_{k+1}(x) = 2^{\exp_k(x)}$.

Lemma 3. *For the algorithm α from Section 4 there is a constant $c > 0$ such that $\text{size}(\alpha(F_0)) \leq 2^{c \cdot \text{size}(F_0)}$ and $\text{alts}(\alpha(F_0)) = \text{alts}(F_0)$. Moreover, the algorithm α runs in $2^{O(\text{size}(F_0))}$ time and space.*

We next consider the worst-case space bound on BAPA. Recall first the following bound on space complexity for PA.

Fact 1. *[15, Chapter 3] The validity of a PA sentence of length n can be decided in space $\exp_2(O(n))$.*

From Lemma 3 and Fact 1 we conclude that the validity of BAPA formulas can be decided in space $\exp_3(O(n))$. It turns out, however, that we obtain better bounds on BAPA validity by analyzing the number of quantifier alternations in BA and BAPA formulas.

Fact 2. *[41] The validity of a PA sentence of length n and the number of quantifier alternations m can be decided in space $2^{n^{O(m)}}$.*

From Lemma 3 and Fact 2 we obtain our space upper bound, which implies the upper bound on deterministic time.

Theorem 4. *The validity of a* BAPA *sentence of length n and the number of quantifier alternations m can be decided in space* $\exp_2(O(mn))$, *and, consequently, in deterministic time* $\exp_3(O(mn))$.

If we approximate quantifier alternations by formula size, we conclude that BAPA validity can be decided in space $\exp_2(O(n^2))$ compared to $\exp_2(O(n))$ bound for PA from Fact 1. Therefore, despite the exponential explosion in the size of the formula in the algorithm α, thanks to the same number of quantifier alternations, our bound has the same number of exponentials as the bound for PA.

5.2 BA as a Special Case

We next analyze the result of applying the algorithm α to a pure BA sentence F_0. By a pure BA sentence we mean a BA sentence without cardinality constraints, containing only the standard operations $\cap, \cup, ^c$ and the relations $\subseteq, =$. At first, it might seem that the algorithm α is not a reasonable approach to deciding BA formulas given that the best upper bounds for PA [15, Chapter 3] are worse than the corresponding bounds for BA [22]. However, we identify a special form of PA sentences $PA_{BA} = \{\alpha(F_0) \mid F_0 \text{ is in BA}\}$ and show that such sentences can be decided in alternating time optimal for BA [22].

Let F_0 be a pure BA formula and let S be the number of set variables in F_0 (the set variables are the only variables in F_0). Let $l_1, \ldots, l_q$ be the free variables of the formula $G_r(l_1, \ldots, l_q)$ in the algorithm α. Then $q = 2^e$ for $e = S + 1 - r$. Let $w_1, \ldots, w_q$ be integers specifying the values of $l_1, \ldots, l_q$. We then have the following lemma.

Lemma 5. *For each r where $1 \leq r \leq S$, formula $G_r(w_1, \ldots, w_q)$ is equivalent to formula $G_r(\bar{w}_1, \ldots, \bar{w}_q)$ where $\bar{w}_i = \min(w_i, 2^{r-1})$.*

Consider a formula F_0 of size n with S variables. Then $\alpha(F_0) = G_{S+1}$. By Lemma 3, size($\alpha(F_0)$) is $O(nS2^S)$. By Lemma 5, it suffices for the outermost quantified variable of $\alpha(F_0)$ to range over the integer interval $[0, 2^S]$, and the range of subsequent variables is even smaller. Therefore, the value of each of the $2^{S+1} - 1$ variables can be represented in $O(S)$ space. Because $\alpha(F_0)$ has S quantifier alternations, $\alpha(F_0)$ the values of all bound variables can be guessed in alternating time $O(S)$. The truth value of a PA formula for given values of variables can be evaluated in time polynomial in the size of the formula, so deciding $\alpha(F_0)$ can be done in alternating time bounded by $n^a 2^{bS}$ for some constants a, b. Because $S \leq n$, we conclude that the algorithm α can be used to decide a pure BA formula by alternating Turing machine running in time 2^{cn} for some $c > 0$ and performing n alternations. The class of all such problems is called Berman complexity class $STA(*, 2^{cn}, n)$. Theorem 5.6 in [22] shows that BA (even if interpreted only over all finite Boolean algebras) is in fact complete for the class $STA(*, 2^{cn}, n)$. Therefore, our algorithm α allows optimal decision procedure for BA, if the PA decision procedure exploits the special structure of the generated formula $\alpha(F_0)$; this special structure is given by Lemma 5. Note that the class $STA(*, 2^{cn}, n)$ is contained in the deterministic exponential space, which is equal to alternating exponential time, the only difference being that the number of alternations in $STA(*, 2^{cn}, n)$ is restricted to be linear.

6   Experience Using Our Decision Procedure for BAPA

We have experimented with BAPA in the context of Jahob system [23] for verifying data structure consistency of Java programs. Jahob parses Java source code annotated with formulas in Isabelle syntax written in comments, generates verification conditions, and uses decision procedures and theorem provers to discharge these verification conditions. Jahob currently contains interfaces to the Isabelle interactive theorem prover [36], the Simplify theorem prover [12] as well as the Omega Calculator [40] and the LASH [30] decision procedures for PA.

Using Jahob, we have generated verification conditions for several Java program fragments that require reasoning about sets and their cardinalities, for example, to prove the equality between the set representing the number of elements in a list and the integer field size after they have been updated. The formulas arising from examples in Section 3 have also been discharged using our current implementation. By comparing different decision procedures, we have found that Simplify is able to deal with some of the formulas involving only sets or only integers, but not with formulas that relate cardinalities of operations on sets to cardinalities of the individual sets. These formulas can be proved in Isabelle, but require user interaction in terms of auxiliary lemmas. On the other hand, our implementation of the decision procedure automatically discharges these formulas.

Our initial experience indicates that the direct implementation of the basic algorithm works fast as long as the number of set variables is small; typical timings are fractions of a second for 4 or less set variables, less than 10 seconds for 5 variables. More than 5 set variables cause the PA decision procedure to run out of memory. (We have used the Omega Calculator to decide PA formulas because we found that it outperforms LASH in the formulas generated from our examples.) On the other hand, the decision procedure is much less sensitive to the number of integer variables in BAPA formulas, because they translate into the same number of integer variables in the generated PA formula.

Our current implementation makes use of certain formula transformations to reduce the size of the generated PA formula. We found that eliminating set variables by substitution of equals for equals is an effective optimization. We also observed that lifting quantifiers to the top level noticeably improves the performance of the Omega Calculator. These transformations extend the range of formulas that the current system can handle. A possible alternative to the current approach is to interleave the elimination of integer variables with the elimination of the set variables and perform formula simplifications during this process [26, Section 5.2]; this alternative approach does not yield good worse-case complexity bounds but could be useful for subclasses of BAPA formulas.

7   Further Observations

We next sketch some further observations about BAPA, see [26] for details.

Countable Sets. A generalization of BAPA where set variables range over subsets of an arbitrary (not necessarily finite) set is decidable, which follows from the decidability of the first-order theory of the addition of cardinals [14]. We here consider the case of all subsets of a countable set, and argue that the complexity results we have developed so far still apply. We first generalize the language of BAPA and the interpretation of BAPA operations, as follows. Introduce function $\inf(b)$ which returns 0 if b is a finite set and 1 if b is a countable set. Define $|b|$ to be some arbitrary integer (for concreteness, zero) if b is infinite, and the cardinality of b if b is finite. A countable or finite cardinal can therefore be represented in PA using a pair (k, i) of an integer k and an infinity flag i. The relation representing the addition of cardinals $(k_1, i_1) + (k_2, i_2) = (k_3, i_3)$ is then definable by formula

$$(i_1 = 0 \wedge i_2 = 0 \wedge i_3 = 0 \wedge k_1 + k_2 = k_3) \vee ((i_1 \neq 0 \vee i_2 \neq 0) \wedge i_3 = 1 \wedge k_3 = 0)$$

Moreover, we have the following generalization of Lemma 1.

Lemma 6. *Let* $b_1, \ldots, b_n$ *be disjoint sets,* $l_1, \ldots, l_n, k_1, \ldots, k_n$ *be natural numbers, and* $p_1, \ldots, p_n, q_1, \ldots, q_n \in \{0, 1\}$. *Then the following two statements are equivalent:*

1. There exists a set y such that

$$\bigwedge_{i=1}^{n} |b_i \cap y| = k_i \wedge \inf(b_i \cap y) = p_i \wedge |b_i \cap y^c| = l_i \wedge \inf(b_i \cap y^c) = q_i$$

2.

$$\bigwedge_{i=1}^{n} (p_i = 0 \wedge q_i = 0 \Rightarrow |b_i| = k_i + l_i) \wedge (\inf(b_i) = 0 \Leftrightarrow (p_i = 0 \wedge q_i = 0))$$

The algorithm for the case of countable set then generalizes using Lemma 6 in the natural way; the resulting PA formulas are at most polynomially larger than for the finite case, so we obtain the same complexity bounds.

Relationship to MSOL. The monadic second-order logic (MSOL) over strings is a decidable logic that can encode Presburger arithmetic by encoding addition using one successor symbol and quantification over sets. There are two important differences between MSOL over strings and BAPA: (1) BAPA can express relationships of the form $|A| = k$ where A is a set variable and k is an integer variable; such relation is not definable in MSOL over strings; (2) when MSOL over strings is used to represent PA operations, the sets contain binary integer digits whereas in BAPA the sets contain uninterpreted elements. Note also that MSOL extended with a construct that takes a set of elements and returns an encoding of the size of that set is undecidabe, because it could express MSOL with equicardinality, which is undecidable by a reduction from Post correspondence problem. Despite this difference, the algorithm α gives a way to combine MSOL over strings with BA yielding a decidable theory. Namely, α does not impose any upper bound on the complexity of the theory for reasoning about integers, so it implies the decidability of the BAPA extension where the constraints on cardinalities of sets are expressed using relations on integers definable in MSOL over strings; these relations go beyond PA [48, Page 400], [7].

8   Related Work

Our paper is the first result that shows a complexity bound for the first-order theory of BAPA. The decidability for BAPA, presented as BA with equicardinality constraints was shown in [14] (see Section 2). A decision procedure for a special case of BAPA was presented in [55], which allows only quantification over *elements* but not over *sets* of elements. [42] shows the decidability of a single-sorted version of BAPA that only contains the set sort. Note that bound integer variables can be simulated using bound set variables, but there are notational and efficiency reasons to allow integer variables.

Presburger Arithmetic. The original result on decidability of PA is [39]. The best known bound on formula size is [15]. An analysis based on the number of quantifier alternations is presented in [41]. Our implementation uses quantifer-elimination based Omega test [40]. Among the decision procedures for full PA, [9] is the only proof-generating version, and is based on [11]. Decidable fragments of arithmetic that go beyond PA include [6, 21].

Boolean Algebras. The first results on decidability of BA are from [31], [1, Chapter 4] and use quantifier elimination, from which one can derive small model property; [22] gives the complexity of the satisfiability problem. [33] studies unification in Boolean rings. The quantifier-free fragment of BA is shown NP-complete in [32]; see [27] for a generalization of this result using parameterized complexity of the Bernays Schönfinkel-Ramsey class of first-order logic [5, Page 258]. [8] gives an overview of several fragments of set theory including theories with quantifiers but no cardinality constraints and theories with cardinality constraints but no quantification over sets. Among the systems for interactively reasoning about richer theories of sets are Isabelle [36], HOL [17], PVS [37], TPS [2]; first-order frameworks such as Athena [3] can use axiomatizations of sets along with calls to resolution-based theorem provers such as Vampire [51] to reason about sets.

Combinations of Decidable Theories. The techniques for combining *quantifier-free* theories [35, 43] and their generalizations such as [49, 50, 53, 54] are of great importance for program verification. Our paper shows a particular combination result for *quantified formulas*, which add additional expressive power in writing specifications. Among the general results for quantified formulas are the Feferman-Vaught theorem for products [14] and term powers [24, 25]. While we have found quantifiers to be useful in several contexts, many problems can be encoded in quantifier-free formulas, so it is interesting to consider a combination of BAPA with solvers for quantifier-free formulas [16, 47], which would likely improve the efficiency on common verification conditions compared to the current direct use of Omega decision procedure. Description logics [4] support sets with cardinalities as well as relations, but do not support quantification over sets.

Analyses of Dynamic Data Structures. In addition to the new technical results, one of the contributions of our paper is to identify the uses of our decision procedure for verifying data structure consistency. We have shown how BAPA enables the verification tools to reason about sets and their sizes. This capability is particularly important for analyses that handle dynamically allocated data structures where the number of objects is statically unbounded [34, 45, 52]. Recently, these approaches were extended to handle the combinations of the constraints representing data structure contents and constraints representing numerical properties of data structures [10, 44]. Our result provides a systematic mechanism for building precise and predictable versions of such analyses. Among other constraints used for data structure analysis, BAPA is unique in being a complete algorithm for an expressive theory that supports arbitrary quantifiers. In addition to applications in Section 3, possible applications of our decision procedure include query evaluation in constraint databases [42] and loop invariant inference [20].

9   Conclusion

Motivated by static analysis and verification of relations between data structure content and size, we have presented an algorithm for deciding the first-order theory of Boolean algebras with Presburger arithmetic (BAPA), showed an elementary upper bound on the worst-case complexity, implemented the algorithm and applied it to discharge verification conditions. Our experience indicates that the algorithm will be useful as a component of a decision procedure of our data structure verification system.

Acknowledgements. We thank Alexis Bes, Chin Wei-Ngan, Calogero Zarba, Peter Revesz, Andreas Podelski, Bruno Courcelle, Cesare Tinelli, Konstantin Korovin, Stanford REACT group, Berkeley CHESS group, and CADE-20 reviewers on useful comments.

References

1. W. Ackermann. *Solvable Cases of the Decision Problem*. North Holland, 1954.
2. P. B. Andrews, S. Issar, D. Nesmith, and F. Pfenning. The TPS theorem proving system. In *10th CADE*, volume 449 of *LNAI*, pages 641–642, 1990.
3. K. Arkoudas, K. Zee, V. Kuncak, and M. Rinard. Verifying a file system implementation. In *Sixth International Conference on Formal Engineering Methods (ICFEM'04)*, volume 3308 of *LNCS*, Seattle, Nov 8-12, 2004 2004.
4. F. Baader, D. Calvanese, D. McGuinness, D. Nardi, and P. Patel-Schneider, editors. *The Description Logic Handbook: Theory, Implementation and Applications*. CUP, 2003.
5. E. Börger, E. Grädel, and Y. Gurevich. *The Classical Decision Problem*. Springer-Verlag, 1997.
6. M. Bozga and R. Iosif. On decidability within the arithmetic of addition and divisibility. In *FOSSACS'05*, 2005.
7. V. Bruyére, G. Hansel, C. Michaux, and R. Villemaire. Logic and p-recognizable sets of integers. *Bull. Belg. Math. Soc. Simon Stevin*, 1:191–238, 1994.
8. D. Cantone, E. Omodeo, and A. Policriti. *Set Theory for Computing*. Springer, 2001.
9. A. Chaieb and T. Nipkow. Generic proof synthesis for Presburger arithmetic. Technical report, Technische Universität München, October 2003.
10. W.-N. Chin, S.-C. Khoo, and D. N. Xu. Extending sized types with with collection analysis. In *ACM PEPM'03*, 2003.
11. D. C. Cooper. Theorem proving in arithmetic without multiplication. In B. Meltzer and D. Michie, editors, *Machine Intelligence*, volume 7, pages 91–100. Edinburgh University Press, 1972.

12. D. Detlefs, G. Nelson, and J. B. Saxe. Simplify: A theorem prover for program checking. Technical Report HPL-2003-148, HP Laboratories Palo Alto, 2003.
13. R. K. Dewar. Programming by refinement, as exemplified by the SETL representation sublanguage. *ACM TOPLAS*, July 1979.
14. S. Feferman and R. L. Vaught. The first order properties of products of algebraic systems. *Fundamenta Mathematicae*, 47:57–103, 1959.
15. J. Ferrante and C. W. Rackoff. *The Computational Complexity of Logical Theories*, volume 718 of *Lecture Notes in Mathematics*. Springer-Verlag, 1979.
16. H. Ganzinger, G. Hagen, R. Nieuwenhuis, A. Oliveras, and C. Tinelli. DPLL(T): Fast decision procedures. In R. Alur and D. Peled, editors, *16th CAV*, volume 3114 of *LNCS*, pages 175–188. Springer, 2004.
17. M. J. C. Gordon and T. F. Melham. *Introduction to HOL, a theorem proving environment for higher-order logic*. Cambridge University Press, Cambridge, England, 1993.
18. C. A. R. Hoare. An axiomatic basis for computer programming. *Communications of the ACM*, 12(10):576–580, 1969.
19. W. Hodges. *Model Theory*, volume 42 of *Encyclopedia of Mathematics and its Applications*. Cambridge University Press, 1993.
20. D. Kapur. Automatically generating loop invariants using quantifier elimination. In *IMACS Intl. Conf. on Applications of Computer Algebra*, 2004.
21. N. Klarlund, A. Møller, and M. I. Schwartzbach. MONA implementation secrets. In *Proc. 5th International Conference on Implementation and Application of Automata*. LNCS, 2000.
22. D. Kozen. Complexity of boolean algebras. *Theoretical Computer Science*, 10:221–247, 1980.
23. V. Kuncak. The Jahob project web page. http://www.mit.edu/~vkuncak/projects/jahob/, 2004.
24. V. Kuncak and M. Rinard. On the theory of structural subtyping. Technical Report 879, Laboratory for Computer Science, Massachusetts Institute of Technology, 2003.
25. V. Kuncak and M. Rinard. Structural subtyping of non-recursive types is decidable. In *Eighteenth Annual IEEE Symposium on Logic in Computer Science*, 2003.
26. V. Kuncak and M. Rinard. The first-order theory of sets with cardinality constraints is decidable. Technical Report 958, MIT CSAIL, July 2004.
27. V. Kuncak and M. Rinard. Decision procedures for set-valued fields. In *1st International Workshop on Abstract Interpretation of Object-Oriented Languages (AIOOL 2005)*, 2005.
28. P. Lam, V. Kuncak, and M. Rinard. Generalized typestate checking using set interfaces and pluggable analyses. *SIGPLAN Notices*, 39:46–55, March 2004.
29. P. Lam, V. Kuncak, and M. Rinard. Generalized typestate checking for data structure consistency. In *6th International Conference on Verification, Model Checking and Abstract Interpretation*, 2005.
30. LASH. The LASH Toolset. http://www.montefiore.ulg.ac.be/~boigelot/research/lash/.
31. L. Loewenheim. Über mögligkeiten im relativkalkül. *Math. Annalen*, 76:228–251, 1915.
32. K. Marriott and M. Odersky. Negative boolean constraints. Technical Report 94/203, Monash University, August 1994.
33. U. Martin and T. Nipkow. Boolean unification: The story so far. *Journal of Symbolic Computation*, 7(3):275–293, 1989.
34. A. Møller and M. I. Schwartzbach. The Pointer Assertion Logic Engine. In *Proc. ACM PLDI*, 2001.
35. G. Nelson and D. C. Oppen. Simplification by cooperating decision procedures. *ACM TOPLAS*, 1(2):245–257, 1979.
36. T. Nipkow, L. C. Paulson, and M. Wenzel. *Isabelle/HOL: A Proof Assistant for Higher-Order Logic*, volume 2283 of *LNCS*. Springer-Verlag, 2002.

37. S. Owre, J. M. Rushby, and N. Shankar. PVS: A prototype verification system. In D. Kapur, editor, *11th CADE*, volume 607 of *LNAI*, pages 748–752, jun 1992.

38. A. Podelski and A. Rybalchenko. Transition predicate abstraction and fair termination. In *ACM POPL*, 2005.

39. M. Presburger. über die vollständigkeit eines gewissen systems der aritmethik ganzer zahlen, in welchem die addition als einzige operation hervortritt. In *Comptes Rendus du premier Congrès des Mathématiciens des Pays slaves, Warsawa*, pages 92–101, 1929.

40. W. Pugh. The Omega test: a fast and practical integer programming algorithm for dependence analysis. In *Supercomputing '91: Proceedings of the 1991 ACM/IEEE conference on Supercomputing*, pages 4–13. ACM Press, 1991.

41. C. R. Reddy and D. W. Loveland. Presburger arithmetic with bounded quantifier alternation. In *ACM STOC*, pages 320–325. ACM Press, 1978.

42. P. Revesz. Quantifier-elimination for the first-order theory of boolean algebras with linear cardinality constraints. In *Proc. Advances in Databases and Information Systems (ADBIS'04)*, volume 3255 of *LNCS*, September 2004.

43. H. Ruess and N. Shankar. Deconstructing Shostak. In *Proc. 16th IEEE LICS*, 2001.

44. R. Rugina. Quantitative shape analysis. In *Static Analysis Symposium (SAS'04)*, 2004.

45. M. Sagiv, T. Reps, and R. Wilhelm. Parametric shape analysis via 3-valued logic. *ACM TOPLAS*, 24(3):217–298, 2002.

46. T. Skolem. Untersuchungen über die Axiome des Klassenkalküls und über "Produktations- und Summationsprobleme", welche gewisse Klassen von Aussagen betreffen. Skrifter utgit av Vidnskapsselskapet i Kristiania, I. klasse, no. 3, Oslo, 1919.

47. A. Stump, C. Barrett, and D. Dill. CVC: a Cooperating Validity Checker. In *14th International Conference on Computer-Aided Verification*, 2002.

48. W. Thomas. Languages, automata, and logic. In *Handbook of Formal Languages Vol.3: Beyond Words*. Springer-Verlag, 1997.

49. C. Tinelli and C. Zarba. Combining non-stably infinite theories. *Journal of Automated Reasoning*, 2004. (Accepted for publication).

50. A. Tiwari. *Decision procedures in automated deduction*. PhD thesis, Department of Computer Science, State University of New York at Stony Brook, 2000.

51. A. Voronkov. The anatomy of Vampire (implementing bottom-up procedures with code trees). *Journal of Automated Reasoning*, 15(2):237–265, 1995.

52. G. Yorsh, T. Reps, and M. Sagiv. Symbolically computing most-precise abstract operations for shape analysis. In *10th TACAS*, 2004.

53. C. G. Zarba. *The Combination Problem in Automated Reasoning*. PhD thesis, Stanford University, 2004.

54. C. G. Zarba. Combining sets with elements. In N. Dershowitz, editor, *Verification: Theory and Practice*, volume 2772 of *Lecture Notes in Computer Science*, pages 762–782. Springer, 2004.

55. C. G. Zarba. A quantifier elimination algorithm for a fragment of set theory involving the cardinality operator. In *18th International Workshop on Unification*, 2004.

Connecting Many-Sorted Theories

Franz Baader[1] and Silvio Ghilardi[2]

[1] Institut für Theoretische Informatik, TU Dresden
[2] Dipartimento di Scienze dell'Informazione, Università degli Studi di Milano

Abstract. Basically, the connection of two many-sorted theories is obtained by taking their disjoint union, and then connecting the two parts through connection functions that must behave like homomorphisms on the shared signature. We determine conditions under which decidability of the validity of universal formulae in the component theories transfers to their connection. In addition, we consider variants of the basic connection scheme.

1 Introduction

The combination of decision procedures for logical theories arises in many areas of logic in computer science, such as constraint solving, automated deduction, term rewriting, modal logics, and description logics. In general, one has two first-order theories T_1 and T_2 over signatures Σ_1 and Σ_2, for which validity of a certain type of formulae (e.g., universal, existential positive, etc.) is decidable. These theories are then combined into a new theory T over a combination Σ of the signatures Σ_1 and Σ_2. The question is whether decidability transfers from T_1, T_2 to their combination T.

One way of combining the theories T_1, T_2 is to build their union $T_1 \cup T_2$. Both the Nelson-Oppen combination procedure [16,15] and combination procedures for the word problem [19,17,5] address this type of combination, but for different types of formulae to be decided. Whereas the original combination procedures were restricted to the case of theories over disjoint signatures, there are now also solutions for the non-disjoint case [8,22,6,9,11,3], but they always require some additional restrictions since it is easy to see that in the unrestricted case decidability does not transfer. Similar combination problems have also been investigated in modal logic, where one asks whether decidability of (relativized) validity transfers from two modal logics to their fusion [12,20,23,4]. The approaches in [11,3] actually generalize these results from equational theories induced by modal logics to more general first-order theories satisfying certain model-theoretic restrictions: the theories T_1, T_2 must be *compatible* with their shared theory T_0, and this shared theory must be *locally finite* (i.e., its finitely generated models are finite). The theory T_i is compatible with the shared theory T_0 iff (i) $T_0 \subseteq T_i$; (ii) T_0 has a model completion T_0^*; and (iii) every model of T_i embeds into a model of $T_i \cup T_0^*$.

In [13], a new combination scheme for modal logics, called $\mathcal{E}$-connection, was introduced, for which decidability transfer is much simpler to show than in the

R. Nieuwenhuis (Ed.): CADE 2005, LNAI 3632, pp. 278–294, 2005.

case of the fusion. Intuitively, the difference between fusion and $\mathcal{E}$-connection can be explained as follows. A model of the fusion is obtained from two models of the component logics by identifying their domains. In contrast, a model of the $\mathcal{E}$-connection consists of two separate models of the component logics together with certain connecting relations between their domains. There are also differences in the syntax of the combined logic. In the case of the fusion, the Boolean operators are shared, and all operators can be applied to each other without restrictions. In the case of the $\mathcal{E}$-connection, there are two copies of the Boolean operators, and operators of the different logics cannot be mixed; the only connection between the two logics are new (diamond) modal operators that are induced by the connecting relations.

If we want to adapt this approach to the more general setting of combining first-order theories, then we must consider many-sorted theories since only the sorts allow us to keep the domains separate and to restrict the way function symbols can be applied to each other. Let T_1, T_2 be two many-sorted theories that may share some sorts as well as function and relation symbols. We first build the disjoint union $T_1 \uplus T_2$ of these two theories (by using disjoint copies of the shared parts), and then connect them by introducing *connection functions* between the shared sorts. These connection functions must behave like homomorphisms for the shared function and predicate symbols, i.e., the axioms stating this are added to $T_1 \uplus T_2$. This corresponds to the fact that the new diamond operators in the $\mathcal{E}$-connection approach distribute over disjunction and do not change the false formula $\perp$. We call the combined theory obtained this way the *connection* of T_1 and T_2.

This kind of connection between theories has already been considered in automated deduction (see, e.g., [1,24]), but only in very restricted cases where both T_1 and T_2 are fixed theories (e.g., the theory of sets and the theory of integers in [24]) and the connection functions have a fixed meaning (like yielding the length of a list). In categorical logic, this type of connection can be seen as an instance of a general co-comma construction in bicategories associated with theories and syntactic interpretations (see, e.g., [25]). However, in this general setting, computational properties of the combined theories have not been considered yet.

This paper is a first step towards providing general results on the transfer of decidability from component theories to their connection. We start by considering the simplest case where there is just one connection function, and show that decidability transfers whenever certain model-theoretic conditions are satisfied. These conditions are weaker than the ones required in [3] for the case of the union of theories.[1] In addition, both the combination procedure and its proof of correctness are much simpler than the ones in [11,3]. The approach easily extends to the case of several connection functions. We will also consider variants of the general combination scheme where the connection function must satisfy additional properties (like being surjective, an embedding, or an isomorphism), or where a theory is connected with itself. The first variant is, for example, in-

---

[1] Our conditions are in general not weaker than the ones in [11], although this is the case for all the theories we have considered until now.

teresting since the combination result for the union of theories shown in [11] can be obtained from the variant where one has an isomorphism as connection function. The second case is interesting since it can be used to reduce the global consequence problem in the modal logic **K** to propositional satisfiability, which is a surprising result.

2   Notation and Definitions

In this section, we fix the notation and give some important definitions, in particular a formal definition of the connection of two theories.

We use standard *many-sorted first-order logic* (see, e.g., [10]), but try to avoid the notational overhead caused by the presence of sorts as much as possible. Thus, a *signature* Ω consists of a non-empty set of sorts S together with a set of function symbols $\mathcal{F}$ and a set of predicate symbols $\mathcal{P}$. The function and predicate symbols are equipped with arities from S^* in the usual way. For example, if the arity of $f \in \mathcal{F}$ is $S_1 S_2 S_3$, then this means that the function f takes tuples consisting of an element of sort S_1 and an element of sort S_2 as input, and produces an element of sort S_3. We consider logic with equality, i.e., the set of predicate symbols contains a symbol $\approx_S$ for equality in every sort S. Usually, we will just use $\approx$ without explicitly specifying the sort. In this paper we usually assume that signatures are at most countable.

Terms and first-order formulae over Ω are defined in the usual way, i.e., they must respect the arities of function and predicate symbols, and the variables occurring in them are also equipped with sorts. An Ω-*atom* is a predicate symbol applied to (sort-conforming) terms, and an Ω-*literal* is an atom or a negated atom. A *ground* literal is a literal that does not contain variables. We use the notation $\phi(\underline{x})$ to express that ϕ is a formula whose free variables are among the ones in the tuple of variables $\underline{x}$. An Ω-*sentence* is a formula over Ω without free variables. An Ω-*theory* T is a set of Ω-sentences (called the axioms of T). If T, T' are Ω-theories, then we write (by a slight abuse of notation) $T \subseteq T'$ to express that all the axioms of T are logical consequences of the axioms of T'.

From the semantic side, we have the standard notion of an Ω-*structure* $\mathcal{A}$, which consists of non-empty and pairwise disjoint domains A_S for every sort S, and interprets function symbols f and predicate symbols P by functions $f^{\mathcal{A}}$ and predicates $P^{\mathcal{A}}$ according to their arities. By A we denote the union of all domains A_S. Validity of a formula ϕ in an Ω-structure $\mathcal{A}$ ($\mathcal{A} \models \phi$), satisfiability, and logical consequence are defined in the usual way. The Ω-structure $\mathcal{A}$ is a *model* of the Ω-theory T iff all axioms of T are valid in $\mathcal{A}$. If $\phi(\underline{x})$ is a formula with free variables $\underline{x} = x_1, \ldots, x_n$ and $\underline{a} = a_1, \ldots, a_n$ is a (sort-conforming) tuple of elements of A, then we write $\mathcal{A} \models \phi(\underline{a})$ to express that $\phi(\underline{x})$ is valid in $\mathcal{A}$ under the assignment $\{x_1 \mapsto a_1, \ldots, x_n \mapsto a_n\}$. Note that $\phi(\underline{x})$ is valid in $\mathcal{A}$ iff it is valid under all assignments iff its universal closure is valid in $\mathcal{A}$.

An Ω-*homomorphism* between two Ω-structures $\mathcal{A}$ and $\mathcal{B}$ is a mapping $\mu : A \to B$ that is sort-conforming (i.e., maps elements of sort S in $\mathcal{A}$ to elements of sort S in $\mathcal{B}$), and satisfies the condition

$(*)$ $\mathcal{A} \models \alpha(a_1, \ldots, a_n)$ implies $\mathcal{B} \models \alpha(\mu(a_1), \ldots, \mu(a_n))$

for all Ω-atoms $\alpha(x_1, \ldots, x_n)$ and (sort-conforming) elements $a_1, \ldots, a_n$ of A. In case the converse of $(*)$ holds too, μ is called an Ω-*embedding*. Note that an embedding is something more than just an injective homomorphism since the stronger condition must hold not only for the equality predicate, but for all predicate symbols. If the embedding μ is the identity on A, then we say that $\mathcal{A}$ is a Ω-*substructure* of $\mathcal{B}$.

We say that Σ is a subsignature of Ω (written $\Sigma \subseteq \Omega$) iff Σ is a signature that can be obtained from Ω by removing some of its sorts and function and predicate symbols. If $\Sigma \subseteq \Omega$ and $\mathcal{A}$ is an Ω-structure, then the Σ-*reduct* of $\mathcal{A}$ is the Σ-structure $\mathcal{A}_{|\Sigma}$ obtained from $\mathcal{A}$ by forgetting the interpretations of sorts, function and predicate symbols from Ω that do not belong to Σ. Conversely, $\mathcal{A}$ is called an *expansion* of the Σ-structure $\mathcal{A}_{|\Sigma}$ to the larger signature Ω. If $\mu : \mathcal{A} \rightarrow \mathcal{B}$ is an Ω-homomorphism, then the Σ-*reduct* of μ is the Σ-homomorphism $\mu_{|\Sigma} : \mathcal{A}_{|\Sigma} \rightarrow \mathcal{B}_{|\Sigma}$ obtained by restricting μ to the sorts that belong to Σ, i.e., by restricting the mapping to the domain of $\mathcal{A}_{|\Sigma}$.

Given a set X of constant symbols not belonging to the signature Ω, but each equipped with a sort from Ω, we denote by Ω^X the extension of Ω by these new constants. If $\mathcal{A}$ is an Ω-structure, then we can view the elements of A as a set of new constants, where $a \in A_S$ has sort S. By interpreting each $a \in A$ by itself, $\mathcal{A}$ can also be viewed as an Ω^A-structure. The *positive diagram* $\Delta_\Omega^+(\mathcal{A})$ of $\mathcal{A}$ is the set of all ground Ω^A-atoms that are true in $\mathcal{A}$, and the *diagram* $\Delta_\Omega(\mathcal{A})$ of $\mathcal{A}$ is the set of all ground Ω^A-literals that are true in $\mathcal{A}$. *Robinson's diagram theorems* [7] say that there is a homomorphism (embedding) between the Ω-structures $\mathcal{A}$ and $\mathcal{B}$ iff it is possible to expand $\mathcal{B}$ to an Ω^A-structure in such a way that it becomes a model of the positive diagram (diagram) of $\mathcal{A}$.

Basic Connections

In the remainder of this section, we introduce our basic scheme for connecting many-sorted theories, and illustrate it with the example of $\mathcal{E}$-connections of modal logics. Let T_1, T_2 be theories over the respective signatures Ω_1, Ω_2, and let Ω_0 be a common subsignature of Ω_1 and Ω_2. We call Ω_0 the *connecting signature*. In addition, let T_0 be an Ω_0-theory[2] that is contained in both T_1 and T_2. We define the new theory $T_1 >_{T_0} T_2$ (called the *connection of T_1 and T_2 over T_0*) as follows.

The *signature* Ω of $T_1 >_{T_0} T_2$ contains the disjoint union $\Omega_1 \uplus \Omega_2$ of the signatures Ω_1 and Ω_2, where the shared sorts and the shared function and predicate symbols are appropriately renamed, e.g., by attaching labels 1 and 2. Thus, if S (f, P) is a sort (function symbol, predicate symbol) contained in both Ω_1 and Ω_2, then S^i (f^i, P^i) for $i = 1, 2$ are its renamed variants in the disjoint

---

[2] When *defining* the connection of T_1, T_2, the theory T_0 is actually irrelevant; all we need is its signature Ω_0. However, for our decidability transfer results to hold, T_0 and the T_i must satisfy certain model-theoretic properties.

union, where the arities are accordingly renamed. In addition, Ω contains a *new function symbol* h_S of arity $S^1 S^2$ for every sort S of Ω_0.

The *axioms* of $T_1 >_{T_0} T_2$ are obtained as follows. Given an Ω_i-formula ϕ, its renamed variant ϕ^i is obtained by replacing all shared symbols by their renamed variants with label i. The axioms of $T_1 >_{T_0} T_2$ consist of

$$\{\phi^1 \mid \phi \in T_1\} \cup \{\phi^2 \mid \phi \in T_2\},$$

together with the universal closures of the formulae

$$h_S(f^1(x_1, \ldots, x_n)) \approx f^2(h_{S_1}(x_1), \ldots, h_{S_n}(x_n)),$$
$$P^1(x_1, \ldots, x_n) \rightarrow P^2(h_{S_1}(x_1), \ldots, h_{S_n}(x_n)),$$

for every function (predicate) symbol f (P) in Ω_0 of arity $S_1 \ldots S_n S$ ($S_1 \ldots S_n$).

Since the signatures Ω_1 and Ω_2 have been made disjoint, and since the additional axioms state that the family of mappings h_S behaves like an Ω_0-homomorphism, it is easy to see that the *models of* $T_1 >_{T_0} T_2$ are formed by triples of the form $(\mathcal{M}^1, \mathcal{M}^2, h^{\mathcal{M}})$, where $\mathcal{M}^1$ is a model of T_1, $\mathcal{M}^2$ is a model of T_2, and $h^{\mathcal{M}}$ is an Ω_0-homomorphism $h^{\mathcal{M}} : \mathcal{M}^1_{|\Omega_0} \rightarrow \mathcal{M}^2_{|\Omega_0}$ between the respective Ω_0-reducts.

Example 1. The most basic variant of the $\mathcal{E}$-connection scheme introduced in [13] is an instance of our approach if one translates it into the algebraic setting. The abstract description systems considered in [13], which cover all the usual modal and description logics, are closely related to to Boolean-based equational theories (see [2] for details). The theory E is called *Boolean-based equational theory* [3] iff its signature Σ has just one sort, equality is the only predicate symbol, the set of function symbols contains the Boolean operators $\sqcap, \sqcup, \neg, \top, \bot$, and its set of axioms consists of identities (i.e., the universal closures of atoms $s \approx t$) and contains the Boolean algebra axioms.

For example, consider the basic modal logic **K**, where we use only the modal operator $\Diamond$ (since $\Box$ can then be defined). The Boolean-based equational theory $E_{\mathbf{K}}$ corresponding to **K** is obtained from the theory of Boolean algebras by adding the identities $\Diamond(x \sqcup y) \approx \Diamond(x) \sqcup \Diamond(y)$ and $\Diamond(\bot) \approx \bot$.

Let us illustrate the notion of an $\mathcal{E}$-connection also on this simple example. To build the $\mathcal{E}$-connection of **K** with itself, one takes two disjoint copies of **K**, obtained by renaming the Boolean operators and the diamonds, e.g., into $\sqcap_i, \sqcup_i, \neg_i, \top_i, \bot_i, \Diamond_i$ for $i = 1, 2$. The signature of the $\mathcal{E}$-connection contains all these renamed symbols together with a new symbol $\Diamond$. However, it is now a two-sorted signature, where symbols with index i are applied to elements of sort S_i and yield as results an element of this sort. The new symbol has arity $S_1 S_2$.[3] The semantics of this $\mathcal{E}$-connection can be given in terms of Kripke structures. A Kripke structure for the $\mathcal{E}$-connection consists of two Kripke structures $\mathcal{K}_1, \mathcal{K}_2$

---

[3] In the $\mathcal{E}$-connection scheme introduced in [13], there is also an inverse diamond operator $\Diamond^-$ with arity $S_2 S_1$, but the algebraic approach introduced in the present paper cannot treat this case (see the conclusion for a discussion).

for $\mathbf{K}$ over disjoint domains W_1 and W_2, together with an additional connecting relation $E \subseteq W_2 \times W_1$. The symbols with index i are interpreted in $\mathcal{K}_i$, and the new symbol $\Diamond$ is interpreted as the diamond operator induced by E, i.e., for every $X \subseteq W_1$ we have

$$\Diamond(X) := \{x \in W_2 \mid \exists y \in W_1.\ (x,y) \in E \wedge y \in X\}.$$

This interpretation of the new operator implies that it satisfies the usual identities of a diamond operator, i.e., $\Diamond(x \sqcup_1 y) \approx \Diamond(x) \sqcup_2 \Diamond(y)$ and $\Diamond(\bot_1) \approx \bot_2$, and that these identities are sufficient to characterize its semantics. Thus, the equational theory corresponding to the $\mathcal{E}$-connection of $\mathbf{K}$ with itself consists of these two axioms, together with the axioms of $E_{\mathbf{K}_1}$ and $E_{\mathbf{K}_2}$.

Obviously, this theory is also obtained as the connection of the theory $E_{\mathbf{K}}$ with itself, if the connecting signature Ω_0 consists of the single sort of $E_{\mathbf{K}}$, the predicate symbol $\approx$, and the function symbols $\sqcup, \bot$. As theory T_0 we can take the theory of semilattices, i.e., the axioms that say that $\sqcup$ is associative, commutative, and idempotent, and that $\bot$ is a unit for $\sqcup$.

Example 2. The previous example can be varied by including $\sqcap$ in the connecting signature, and taking as theory T_0 the theory of distributive lattices with a least element $\bot$. It is easy to see that this corresponds to the case of an $\mathcal{E}$-connection where the connecting relation E is required to be a partial function.

3   Positive Algebraic Completions and Compatibility

In order to transfer decidability results from the component theories T_1, T_2 to their connection $T_1 >_{T_0} T_2$ over T_0, the theories T_0, T_1, T_2 must satisfy certain model-theoretic conditions, which we introduce below. The most important one is that T_0 has a positive algebraic completion. Before we can define this concept, we must introduce some notions from model theory [7].

The formula ϕ is called *open* iff it does not contain quantifiers; it is called *universal* iff it is obtained from an open formula by adding a prefix of universal quantifiers; and it is called *geometric* iff it is built from atoms by using conjunction, disjunction, true, false, and existential quantifiers.[4]

The main property of geometric formulae is that they are preserved under homomorphisms in the following sense: if $\mu : \mathcal{A} \to \mathcal{B}$ is a homomorphism between Ω-structures and $\phi(x_1, \dots, x_n)$ is a geometric formula over Ω, then $\mathcal{A} \models \phi(a_1, \dots, a_n)$ implies $\mathcal{B} \models \phi(\mu(a_1), \dots, \mu(a_n))$ for all (sort-conforming) $a_1, \dots, a_n \in A$. Open formulae are related to embeddings in various ways. First, they are preserved under building sub- and superstructures, i.e., if $\mathcal{A}$ is a substructure of $\mathcal{B}$, $\phi(x_1, \dots, x_n)$ is an open formula, and $a_1, \dots, a_n \in A$ are sort-conforming, then $\mathcal{A} \models \phi(a_1, \dots, a_n)$ iff $\mathcal{B} \models \phi(a_1, \dots, a_n)$. Moreover, two Ω-theories T, T' entail the same set of open formulae iff every model of T can be embedded into a model of T' and vice versa (see [7] for these and related results).

---

[4] The latter formulae are called "geometric" in categorical logic [14] since they are preserved under inverse image geometric morphisms.

The theory T is a *universal theory* iff its axioms are universal sentences; it is a *geometric theory* iff it can be axiomatized by using universal closures of geometric sequents, where a geometric sequent is an implication between two geometric formulae. Note that any universal theory is geometric since open formulae are conjunctions of clauses and clauses can be rewritten as geometric sequents.

Definition 1. *Let T be a universal and T^* a geometric theory over Ω. We say that T^* is a* positive algebraic completion *of T iff the following properties hold:*

1. *$T \subseteq T^*$;*
2. *every model of T embeds into a model of T^*;*[5]
3. *for every geometric formula $\phi(\underline{x})$ there is an* open *geometric formula $\phi^*(\underline{x})$ such that $T^* \models \phi \leftrightarrow \phi^*$.*

It can be shown that the models of T^* are exactly the algebraically closed models of T (see [2]). In particular, this means that the positive algebraic completion of T is unique, provided that it exists.

When trying to show that Property 3 of Definition 1 holds for given theories T, T^*, it is sufficient to consider *simple existential formulae* $\phi(\underline{x})$, i.e., formulae that are obtained from conjunctions of atoms by adding an existential quantifier prefix. In fact, any geometric formula ϕ can be normalized to a disjunction $\phi_1 \vee \ldots \vee \phi_n$ of simple existential formulae ϕ_i by using distributivity of conjunction and existential quantification over disjunction. In addition, if $T^* \models \phi_i \leftrightarrow \phi_i^*$ for geometric open formulae ϕ_i^* ($i = 1, \ldots, n$), then $\phi_1^* \vee \ldots \vee \phi_n^*$ is also a geometric open formula and $T^* \models (\phi_1 \vee \ldots \vee \phi_n) \leftrightarrow (\phi_1^* \vee \ldots \vee \phi_n^*)$.

The following lemma will turn out to be useful later on.

Lemma 1. *Assume that T, T^* satisfy Property 2 of Definition 1. If $\phi(\underline{x})$ is a simple existential formula and $\phi^*(\underline{x})$ is an open formula, then $T^* \models \phi \rightarrow \phi^*$ implies $T \models \phi \rightarrow \phi^*$.*

This is an immediate consequence of the facts that $\phi \rightarrow \phi^*$ is then equivalent to an open formula, and open formulae are preserved under building substructures.

The first ingredient of our combinability condition is the following notion of compatibility, which is a variant of analogous compatibility conditions introduced in [11,3] for the case of the union of theories.

Definition 2. *Let $T_0 \subseteq T$ be theories over the respective signatures $\Omega_0 \subseteq \Omega_1$. We say that T is T_0-algebraically compatible iff T_0 is universal, has a positive algebraic completion T_0^*, and every model of T embeds into a model of $T \cup T_0^*$.*

The second ingredient is that T_0 must be locally finite, i.e., all finitely generated models of T_0 are finite. To be more precise, we need the following effective variant of local finiteness defined in [11,3]. Let T_0 be a universal theory over the finite signature Ω_0. Then T_0 is called *effectively locally finite* iff for every tuple of variables $\underline{x}$, one can effectively determine terms $t_1(\underline{x}), \ldots, t_k(\underline{x})$ such that, for every further term $u(\underline{x})$, we have that $T_0 \models u \approx t_i$ for some $i = 1, \ldots, k$.

---

[5] Equivalently, T and T^* entail the same universal sentences.

4   The Main Combination Result

We are interested in deciding the universal fragments of our theories, i.e., validity of universal formulae (or, equivalently open formulae) in a theory T. This is the decision problem also treated by the Nelson-Oppen combination method (albeit for the union of theories). It is well known that this problem is equivalent to the problem of deciding whether a set of literals is satisfiable in some model of T. We call such a set of literals a *constraint*.

By introducing new free constants (i.e., constants not occurring in the axioms of the theory), we can assume without loss of generality that such constraints contain no variables. In addition, we can transform any ground constraint into an equisatisfiable set of *ground flat literals*, i.e., literals of the form

$$a \approx f(a_1, \ldots, a_n), \quad P(a_1, \ldots, a_n), \quad \text{or} \quad \neg P(a_1, \ldots, a_n),$$

where $a, a_1, \ldots, a_n$ are (sort-conforming) free constants, f is a function symbol, and P is a predicate symbol (possibly also equality).

Theorem 1. *Let T_0, T_1, T_2 be theories over the respective signatures $\Omega_0, \Omega_1, \Omega_2$, where Ω_0 is a common subsignature of Ω_1 and Ω_2. Assume that $T_0 \subseteq T_1$ and $T_0 \subseteq T_2$, that T_0 is universal and effectively locally finite, and that T_2 is T_0-algebraically compatible. Then the decidability of the universal fragments of T_1 and T_2 entails the decidability of the universal fragment of $T_1 >_{T_0} T_2$.*

To prove the theorem, we consider a finite set Γ of ground flat literals over the signature Ω of $T_1 >_{T_0} T_2$ (with additional free constants), and show how it can be tested for satisfiability in $T_1 >_{T_0} T_2$. Since all literals in Γ are flat, we can divide Γ into three disjoint sets $\Gamma = \Gamma_0 \cup \Gamma_1 \cup \Gamma_2$, where Γ_i $(i = 1, 2)$ is a set of literals in the signature Ω_i (expanded with free constants), and Γ_0 is of the form

$$\Gamma_0 = \{h(a_1) \approx b_1, \ldots, h(a_n) \approx b_n\}$$

for free constants $a_1, b_1, \ldots, a_n, b_n$. Here and in the following we omit the sort index when writing the connection functions h_S.

Proposition 1. *The constraint $\Gamma = \Gamma_0 \cup \Gamma_1 \cup \Gamma_2$ is satisfiable in $T_1 >_{T_0} T_2$ iff there exists a triple $(\mathcal{A}, \mathcal{B}, \nu)$ such that*

1. *$\mathcal{A}$ is an Ω_0-model of T_0, which is generated by $\{a_1^{\mathcal{A}}, \ldots, a_n^{\mathcal{A}}\}$;*
2. *$\mathcal{B}$ is an Ω_0-model of T_0, which is generated by $\{b_1^{\mathcal{B}}, \ldots, b_n^{\mathcal{B}}\}$;*
3. *$\nu : \mathcal{A} \to \mathcal{B}$ is an Ω_0-homomorphism such that $\nu(a_j^{\mathcal{A}}) = b_j^{\mathcal{B}}$ for $j = 1, \ldots, n$;*
4. *$\Gamma_1 \cup \Delta_{\Omega_0}(\mathcal{A})$ is satisfiable in T_1;*
5. *$\Gamma_2 \cup \Delta_{\Omega_0}(\mathcal{B})$ is satisfiable in T_2.*

Proof. The only-if direction is simple. In fact, as noted in Section 2, a model $\mathcal{M}$ of $T_1 >_{T_0} T_2$ is given by a triple $(\mathcal{M}^1, \mathcal{M}^2, h^{\mathcal{M}})$, where $\mathcal{M}^1$ is a model of T_1, $\mathcal{M}^2$ is a model of T_2, and $h^{\mathcal{M}} : \mathcal{M}^1_{|\Omega_0} \to \mathcal{M}^2_{|\Omega_0}$ is an Ω_0-homomorphism between the respective Ω_0-reducts. Assume that this model $\mathcal{M}$ satisfies Γ. We can take as $\mathcal{A}$

the substructure of $\mathcal{M}^1_{|\Omega_0}$ generated by (the interpretations of) $a_1, \ldots, a_n$, as $\mathcal{B}$ the substructure of $\mathcal{M}^2_{|\Omega_0}$ generated by (the interpretations of) $b_1, \ldots, b_n$, and as homomorphism ν the restriction of $h^{\mathcal{M}}$ to $\mathcal{A}$. It is easy to see that the triple $(\mathcal{A}, \mathcal{B}, \nu)$ obtained this way satisfies 1.–5. of the proposition.

Conversely, assume that $(\mathcal{A}, \mathcal{B}, \nu)$ is a triple satisfying 1.–5. of the proposition. Because of 4. and 5., there is an Ω_1-model $\mathcal{N}'$ of T_1 satisfying $\Gamma_1 \cup \Delta_{\Omega_0}(\mathcal{A})$ and an Ω_2-model $\mathcal{N}''$ of T_2 satisfying $\Gamma_2 \cup \Delta_{\Omega_0}(\mathcal{B})$. By Robinson's diagram theorem, $\mathcal{N}'$ has $\mathcal{A}$ as an Ω_0-substructure and $\mathcal{N}''$ has $\mathcal{B}$ as an Ω_0-substructure. We assume without loss of generality that $\mathcal{N}'$ is at most countable and that $\mathcal{N}''$ is a model of $T_2 \cup T_0^*$. The latter assumption is by T_0-algebraic compatibility of T_2, and the former assumption is by the Löwenheim-Skolem theorem since our signatures are at most countable. Let us enumerate the elements of $\mathcal{N}'$ as

$$c_1, c_2, \ldots, c_n, c_{n+1}, \ldots,$$

where we assume that $c_i = a_i^{\mathcal{A}}$ $(i = 1, \ldots, n)$, i.e., $c_1, \ldots, c_n$ are generators of $\mathcal{A}$. We define an increasing sequence of sort-conforming functions $\nu_k : \{c_1, \ldots c_k\} \to N''$ (for $k \geq n$) such that, for every ground $\Omega_0^{\{c_1, \ldots, c_k\}}$-atom α we have

$$\mathcal{N}'_{|\Omega_0} \models \alpha(c_1, \ldots, c_k) \quad \text{implies} \quad \mathcal{N}''_{|\Omega_0} \models \alpha(\nu_k(c_1), \ldots, \nu_k(c_k)).$$

We first take ν_n to be ν. To define ν_{k+1} (for $k \geq n$), let us consider the conjunction $\psi(c_1, \ldots, c_k, c_{k+1})$ of the $\Omega_0^{\{c_1, \ldots, c_{k+1}\}}$-atoms that are true in $\mathcal{N}'_{|\Omega_0}$: this conjunction is finite (modulo taking representative terms, thanks to local finiteness of T_0). Let $\phi(x_1, \ldots, x_k)$ be $\exists x_{k+1}.\psi(x_1, \ldots, x_k, x_{k+1})$ and let $\phi^*(x_1, \ldots, x_k)$ be a geometric open formula such that $T_0^* \models \phi \leftrightarrow \phi^*$.

By Lemma 1, $T_0 \models \phi \to \phi^*$, and thus we have $\mathcal{N}'_{|\Omega_0} \models \phi^*(c_1, \ldots, c_k)$ and also $\mathcal{N}''_{|\Omega_0} \models \phi^*(\nu_k(c_1), \ldots, \nu_k(c_k))$ by the induction hypothesis. Since $\mathcal{N}''_{|\Omega_0}$ is a model of T_0^*, there is a b such that $\mathcal{N}''_{|\Omega_0} \models \psi(\nu_k(c_1), \ldots, \nu_k(c_k), b)$ for some b. We now obtain the desired extension ν_{k+1} of ν_k by setting $\nu_{k+1}(c_{k+1}) := b$. Taking $\nu_\infty = \bigcup_{k \geq n} \nu_k$, we finally obtain a homomorphism $\nu_\infty : \mathcal{N}'_{|\Omega_0} \to \mathcal{N}''_{|\Omega_0}$ such that the triple $(\mathcal{N}', \mathcal{N}'', \nu_\infty)$ is a model of $T_1 >_{T_0} T_2$ that satisfies $\Gamma_0 \cup \Gamma_1 \cup \Gamma_2$.   □

The above proof uses the assumption that T_0 is locally finite. By using heavier model-theoretic machinery, one can also prove the proposition without using local finiteness of T_0 (see [2]). However, since the proof of Theorem 1 needs this assumption anyway (see below), we gave the above proof since it is simpler.

To conclude the proof of Theorem 1, we describe a *non-deterministic decision procedure* that effectively guesses an appropriate triple $(\mathcal{A}, \mathcal{B}, \nu)$ and then checks whether it satisfies 1.–5. of Proposition 1. To guess an Ω_0-model of T_0 that is generated by a finite set X, one uses effective local finiteness of T_0 to obtain an effective bound on the size of such a model, and then guesses an Ω_0-structure that satisfies this size bound. Once the structures $\mathcal{A}, \mathcal{B}$ are given, one can build their diagrams, and use the decision procedures for T_1 and T_2 to check whether 4. and 5. of Proposition 1 are satisfied. If the answer is yes, then $\mathcal{A}, \mathcal{B}$ are also models of T_0: in fact, if for instance $\Gamma_1 \cup \Delta_{\Omega_0}(\mathcal{A})$ is satisfiable in the model $\mathcal{M}$

of T_1, then $\mathcal{M}$ has $\mathcal{A}$ as a substructure, and this implies $\mathcal{A} \models T_0$ because T_0 is universal and $T_0 \subseteq T_1$. Finally, one can guess a mapping $\nu : A \to B$ that satisfies $\nu(a_j^{\mathcal{A}}) = b_j^{\mathcal{B}}$, and then use the diagrams of $\mathcal{A}, \mathcal{B}$ to check whether ν satisfies the homomorphism condition $(*)$.

The proof of Proposition 1 shows that our decidability transfer result can easily be extended to the case of *several connection functions*, possibly going in both directions. In fact, one simply considers several Ω_0-homomorphisms between $\mathcal{A}$ and $\mathcal{B}$ in 3. of the proposition, and extends them separately to homomorphisms between $\mathcal{N}'$ and $\mathcal{N}''$ (see [2] for more details). If there are also connection functions in the other direction (and thus homomorphisms from $\mathcal{B}$ to $\mathcal{A}$), then T_1 must also be T_0-algebraically compatible.

Examples

When trying to axiomatize the positive algebraic completion T_0^* of a given universal theory T_0, it is sufficient to produce for every simple existential formula $\phi(\underline{x})$ an appropriate geometric and open formula $\phi^*(\underline{x})$. Take as theory T_0^* the one axiomatized by T_0 together with the formulae $\phi \leftrightarrow \phi^*$ for every simple existential formula ϕ. In order to complete the job, it is sufficient to show that every model of T_0 embeds into a model of T_0^*. It should also be noted that one can without loss of generality restrict the attention to simple existential formulae with just one existential quantifier since more than one quantifier can then be treated by iterated elimination of single quantifiers.

In the next example we encounter a special case where the formulae $\phi \leftrightarrow \phi^*$ are already valid in T_0. In this case, we have $T_0 = T_0^*$, and thus the model-embedding condition is trivially satisfied. In addition, any theory T with $T_0 \subseteq T$ is T_0-algebraically compatible.

Example 3. Recall from [3] the definition of a Gaussian theory. Let us call a conjunction of atoms an *e-formula*. The universal theory T_0 is *Gaussian* iff for every e-formula $\phi(\underline{x}, y)$ it is possible to compute an e-formula $\psi(\underline{x})$ and a term $s(\underline{x}, \underline{z})$ with fresh variables $\underline{z}$ such that

$$T_0 \models \phi(\underline{x}, y) \leftrightarrow (\psi(\underline{x}) \wedge \exists \underline{z}.(y \approx s(\underline{x}, \underline{z}))). \tag{1}$$

Any Gaussian theory T_0 is its own positive algebraic completion. In fact, it is easy to see that (1) implies $T_0 \models (\exists y.\phi(\underline{x}, y)) \leftrightarrow \psi(\underline{x})$, and thus the comment given above this example applies.

As a consequence, our combination result applies to all the examples of effectively locally finite Gaussian theories given in [3] (e.g., Boolean algebras, vector spaces over a finite field, empty theory over a signature whose sets of predicates consists of $\approx$ and whose set of function symbols is empty): if the universal theory T_0 is effectively locally finite and Gaussian, and T_1, T_2 are arbitrary theories containing T_0 and with decidable universal fragment, then the universal fragment of $T_1 >_{T_0} T_2$ is also decidable.

Example 4. Let T_0 be the theory of semilattices (see Example 1). This theory is obviously effectively locally finite. In the following, we use the disequation $s \sqsubseteq t$

as an abbreviation for the equation $s \sqcup t \approx t$. Obviously, any equation $s \approx t$ can be expressed by the disequations $s \sqsubseteq t \wedge t \sqsubseteq s$.

The theory T_0 has a positive algebraic completion, which can be axiomatized as follows. Let $\phi(\underline{x})$ be a simple existential formula with just one existential quantifier. Using the fact that $z_1 \sqcup \ldots \sqcup z_n \sqsubseteq z$ is equivalent to $z_1 \sqsubseteq z \wedge \ldots \wedge z_n \sqsubseteq z$, it is easy to see that $\phi(\underline{x})$ is T_0-equivalent to a formula of the form

$$\psi(\underline{x}) \wedge \exists y.((y \sqsubseteq t_1) \wedge \cdots \wedge (y \sqsubseteq t_n) \wedge (u_1 \sqsubseteq s_1 \sqcup y) \wedge \cdots \wedge (u_m \sqsubseteq s_m \sqcup y)), \quad (2)$$

where $\psi(\underline{x}), t_i, s_j, u_k$ do not contain y. Let $\phi^*(\underline{x})$ be the formula

$$\psi(\underline{x}) \quad \wedge \quad \bigwedge_{i=1}^{n} \bigwedge_{j=1}^{m} (u_j \sqsubseteq s_j \sqcup t_i), \quad (3)$$

and let T_0^* be obtained from T_0 by adding to it the universal closures of all formulae $\phi \leftrightarrow \phi^*$.

We prove that T_0^* is contained in the theory of Boolean algebras. In fact, the system of disequations (2) is equivalent, in the theory of Boolean algebras, to

$$\psi(\underline{x}) \wedge \exists y.((y \sqsubseteq t_1) \wedge \cdots \wedge (y \sqsubseteq t_n) \wedge (u_1 \sqcap \neg s_1 \sqsubseteq y) \wedge \cdots \wedge (u_m \sqcap \neg s_m \sqsubseteq y), \quad (4)$$

and hence to

$$\psi(\underline{x}) \wedge (u_1 \sqcap \neg s_1 \sqsubseteq t_1 \sqcap \ldots \sqcap t_n) \wedge \cdots \wedge (u_m \sqcap \neg s_m \sqsubseteq t_1 \sqcap \ldots \sqcap t_n). \quad (5)$$

Finally, it is easy to see that (5) and (3) are equivalent.

Since every semilattice embeds into a Boolean algebra [2], this shows that T_0^* is the positive algebraic completion of T_0. In addition, this implies that any Boolean-based equational theory T is T_0-algebraically compatible since T_0^* is contained in T. Consequently, Theorem 1 covers the case of a basic $\mathcal{E}$-connection (see Example 1) for arbitrary classical modal logics as components.

In [2] we show a similar result for the case where the theory T_0 is the theory of distributive lattices with $\bot$. Thus, our result also covers the case of connecting relations that are partial functions (see Example 2).

Complexity Considerations

The complexity of the combined decision procedure described in the proof of Theorem 1 is usually higher than the complexity of the decision procedures for the components. There are two main reasons for this complexity increase. First, one must guess the Ω_0-structures $\mathcal{A}, \mathcal{B}$ as the well as the mapping $\nu : A \to B$. This can be done by a *non-deterministic* procedure whose complexity depends on the bound on the size of Ω_0-models of T_0 with n generators given by the effective local finiteness of T_0. Second, the decision procedures for T_1 and T_2 are respectively applied to $\Gamma_1 \cup \Delta_{\Omega_0}(\mathcal{A})$ and $\Gamma_2 \cup \Delta_{\Omega_0}(\mathcal{B})$. The size of the diagrams again depends on the bound on the size of finitely generated Ω_0-models of T_0.

Let us consider the case where T_0 is the theory of semilattices (see Examples 1 and 4) in more detail. Given generators $a_1, \ldots, a_n$, there are 2^n representative terms, namely all terms of the form $a_{i_1} \sqcup \cdots \sqcup a_{i_k}$ for $\{i_1, \ldots, i_k\} \subseteq \{1, \ldots, n\}$ (where the empty disjunction corresponds to $\bot$). Atoms are of the form $t_1 \approx t_2$ where t_1, t_2 are such representative terms, and thus there are $2^n \cdot 2^n = 2^{2n}$ atoms. One can now guess a possible diagram of an Ω_0-structure by guessing (in non-deterministic exponential time) a subset S of the set of atoms. Given such a subset, the potential diagram is $\Delta_S := \{\alpha \mid \alpha \in S\} \cup \{\neg \alpha \mid \alpha \notin S\}$. Of course, not every such set Δ_S is indeed the diagram of an Ω_0-structure, but the ones that are not will lead to unsatisfiability when satisfiability in T_i of $\Gamma_i \cup \Delta_S$ is tested. Since the size of Δ_S is $O(n \cdot 2^{2n})$, the complexity of this satisfiability test is one exponential higher than the complexity of the satisfiability problem in T_i.

Assume that we have guessed sets S_1, S_2 determining the diagrams of semilattices $\mathcal{A}, \mathcal{B}$ generated by $a_1, \ldots, a_n$ and $b_1, \ldots, b_n$, respectively. Guessing an Ω_0-homomorphism $\nu : \mathcal{A} \to \mathcal{B}$ is not really necessary. In fact, if it exists, such a homomorphism ν is uniquely determined by the requirement that $\nu(a_i) = b_i$ ($i = 1, \ldots, n$) since the semilattice $\mathcal{A}$ is generated by the $a_1, \ldots, a_n$. Obviously, an Ω_0-homomorphism $\nu : \mathcal{A} \to \mathcal{B}$ with $\nu(a_i) = b_i$ exists iff $\alpha(a_1, \ldots, a_n) \in S_1$ implies $\alpha(b_1, \ldots, b_n) \in S_2$ for all Ω_0-atoms $\alpha(x_1, \ldots, x_n)$. Thus, if one first guesses S_1, then one can start with $S_1' := \{\alpha(b_1, \ldots, b_n) \mid \alpha(a_1, \ldots, a_n) \in S_1\}$ and add some additional atoms when guessing S_2.

To sum up, in the case of T_0 being the theory of semilattices, our combined decision procedure has the following complexity. Its starts with a non-deterministic exponential step that guesses potential diagrams Δ_{S_1} and Δ_{S_2} such that the homomorphism condition (∗) is satisfied. Then it tests $\Gamma_i \cup \Delta_{S_i}$ ($i = 1, 2$) for satisfiability in T_i. Since the size of Δ_{S_i} is exponential, the complexity of this step is one exponential higher than the complexity of deciding the universal fragment of T_i. This shows that our combination procedure has the same complexity as the one for $\mathcal{E}$-connections described in [13].

Let us consider the complexity increase caused by the combination procedure in more detail for the complexity class EXPTIME, which is often encountered when considering the global satisfiability problem in modal logic. Thus, assume that the decision procedures for the universal fragments of T_1 and T_2 are in EXPTIME, and that T_0 is the theory of semilattices. The combined decision procedure then generates doubly-exponentially many decision problems of exponential size for the component procedures. Each of these component decision problems can be decided in double-exponential time. Thus, the overall complexity of the combined decision procedure is 2EXPTIME.

5   A Variant of the Connection Scheme

Here we consider a slightly different combination scheme where a theory T is connected with itself rather than with a copy of itself. Let $T_0 \subseteq T$ be theories over the respective signatures $\Omega_0 \subseteq \Omega$. We use $T_{>T_0}$ to denote the theory whose models are models $\mathcal{M}$ of T endowed with a homomorphism $h : \mathcal{M}_{|\Omega_0} \to \mathcal{M}_{|\Omega_0}$.

Thus, the *signature* Ω' of $T_{>T_0}$ is obtained from the signature Ω of T by adding a new function symbol h_S of arity SS for every sort S of Ω_0. The axioms of $T_{>T_0}$ are obtained from the axioms of T by adding

$$h_S(f(x_1,\ldots,x_n)) \approx f(h_{S_1}(x_1),\ldots,h_{S_n}(x_n)),$$
$$P(x_1,\ldots,x_n) \to P(h_{S_1}(x_1),\ldots,h_{S_n}(x_n)),$$

for every function (predicate) symbol f (P) in Ω_0 of arity $S_1\ldots S_n S$ ($S_1\ldots S_n$).

Example 5. An interesting example of a theory obtained as such a connection is the theory $E_{\mathbf{K}}$ corresponding to the basic modal logic $\mathbf{K}$ (see Example 1). In fact, let T be the theory of Boolean algebras, and T_0 the theory of semilattices over the signature Ω_0 as defined in Example 1. If we use the symbol $\Diamond$ for the connection function, then $T_{>T_0}$ is exactly the theory $E_{\mathbf{K}}$.

Theorem 2. *Let T_0, T be theories over the respective signatures Ω_0, Ω, where Ω_0 is a subsignature of Ω. Assume that $T_0 \subseteq T$, that T_0 is universal and effectively locally finite, and that T is T_0-algebraically compatible. Then the decidability of the universal fragment of T entails the decidability of the universal fragment of $T_{>T_0}$.*

To prove the theorem, we consider a finite set $\Gamma \cup \Gamma_0$ of ground flat literals over the signature Ω' of $T_{>T_0}$, where Γ is a set of literals in the signature Ω of T (expanded with free constants), and Γ_0 is of the form

$$\Gamma_0 = \{h(a_1) \approx b_1, \ldots, h(a_n) \approx b_n\}.$$

The theorem is an easy consequence of the following proposition, whose proof is similar to the one of Proposition 1.

Proposition 2. *The constraint $\Gamma \cup \Gamma_0$ is satisfiable in $T_{>T_0}$ iff there exists a triple $(\mathcal{A}, \mathcal{B}, \nu)$ such that*

1. *$\mathcal{A}$ is an Ω_0-model of T_0, which is generated by $\{a_1^{\mathcal{A}}, \ldots, a_n^{\mathcal{A}}\}$;*
2. *$\mathcal{B}$ is an Ω_0-model of T_0, which is generated by $\{b_1^{\mathcal{B}}, \ldots, b_n^{\mathcal{B}}\}$;*
3. *$\nu : \mathcal{A} \to \mathcal{B}$ is an Ω_0-homomorphism such that $\nu(a_j^{\mathcal{A}}) = b_j^{\mathcal{B}}$ for $j = 1, \ldots, n$;*
4. *$\Gamma \cup \Delta_{\Omega_0}(\mathcal{A}) \cup \Delta_{\Omega_0}(\mathcal{B})$ is satisfiable in T.*

Applied to the connection of BA with itself w.r.t. the theory of semilattices considered in Example 5, the theorem shows that deciding the universal theory of $E_{\mathbf{K}}$ can be reduced to deciding the universal theory of BA. It is well-known that deciding the universal theory of $E_{\mathbf{K}}$ is equivalent to deciding global consequence in $\mathbf{K}$, and that deciding the universal theory of BA is equivalent to propositional reasoning. Thus, we have shown the (rather surprising) result that the global consequence problem in $\mathbf{K}$ can be reduced to purely propositional reasoning. However, if we directly apply the non-deterministic combination algorithm suggested by Proposition 2, then the complexity of the obtained decision procedure is worse then the known ExpTime-complexity [20] of the problem. The deterministic combination procedure described below overcomes this problem.

A Deterministic Combination Procedure

As pointed out in [18], Nelson-Oppen style combination procedures can be made deterministic in the presence of a certain convexity condition. Let T be a theory over the signature Ω, and let Ω_0 be a subsignature of Ω. Following [21], we say that T is Ω_0-*convex* iff every finite set of ground Ω^X-literals (using additional free constants from X) T-entailing a disjunction of $n > 1$ ground Ω_0^X-atoms, already T-entails one of the disjuncts. Note that universal Horn Ω-theories are always Ω-convex. In particular, this means that equational theories (like BA) are convex w.r.t. any subsignature.

Let $T_0 \subseteq T$ be theories over the respective signatures Ω_0, Ω, where Ω_0 is a subsignature of Ω. If T is Ω_0-convex, then Theorem 2 can be shown with the help of a deterministic combination procedure. (The same is actually also true for Theorem 1, but will not explicitly be shown here.)

Let $\Gamma \cup \Gamma_0$ be a finite set of ground flat literals (with free constants) in the signature of $T_{>T_0}$; suppose also that Γ does not contain the symbol h and that $\Gamma_0 = \{h(a_1) \approx b_1, \ldots, h(a_n) \approx b_n\}$. We say that Γ is Γ_0-*saturated* iff for every Ω_0-atom $\alpha(x_1, \ldots, x_n)$, $T \cup \Gamma \models \alpha(a_1, \ldots, a_n)$ implies $\alpha(b_1, \ldots, b_n) \in \Gamma$.

Theorem 3. *Let T_0, T be theories over the respective signatures Ω_0, Ω, where Ω_0 is a subsignature of Ω. Assume that $T_0 \subseteq T$, that T_0 is universal and effectively locally finite, and that T is Ω_0-convex and T_0-algebraically compatible. Then the following deterministic procedure decides whether $\Gamma \cup \Gamma_0$ is satisfiable in $T_{>T_0}$ (where Γ, Γ_0 are as above):*

1. *Γ_0-saturate Γ;*
2. *check whether the Γ_0-saturated set $\widehat{\Gamma}$ obtained this way is satisfiable in T.*

The saturation process (and thus the procedure) terminates because T_0 is locally finite. In addition, if $\Gamma \cup \Gamma_0$ is satisfied in a model $\mathcal{M}$ of $T_{>T_0}$, then the reduct of $\mathcal{M}$ to the signature Ω obviously satisfies $\widehat{\Gamma}$. Conversely, if the Γ_0-saturated set $\widehat{\Gamma}$ is satisfiable in T, then one can use $\widehat{\Gamma}$ to construct a triple $(\mathcal{A}, \mathcal{B}, \nu)$ satisfying 1.–4. of Proposition 2 (see [2] for details).

Example 5 (continued). Let us come back to the connection of $T := BA$ with itself w.r.t. the theory T_0 of semilattices, which yields as combined theory the equational theory $E_{\mathbf{K}}$ corresponding to the basic modal logic $\mathbf{K}$. In this case, checking during the saturation process whether $T \cup \Gamma \models \alpha(\underline{a})$ amounts to checking whether a propositional formula ϕ_Γ (whose size is linear in the size of Γ) implies a propositional formula of the form $\psi_1 \Leftrightarrow \psi_2$, where ψ_1, ψ_2 are disjunctions of the propositional variables from $\underline{a}$. Since there are only exponentially many different formulae of the form $\psi_1 \Leftrightarrow \psi_2$, the saturation process needs at most exponentially many such propositional tests, and the size of the intermediate sets Γ and of the Γ_0-saturated set $\widehat{\Gamma}$ is at most exponential. However, all these sets contain only the free constants $\underline{a}$. Since propositional reasoning can be done in time exponential in the number of propositional variables, this shows that both the saturation process and the final satisfiability test of $\widehat{\Gamma}$ in T can be done in time exponential in the number of free constants $\underline{a}$.

Consequently, we have shown that Theorem 3 yields an EXPTIME decision procedure for the global consequence relation in **K**, which thus matches the known worst-case complexity of the problem.

6   Conditions on the Connection Functions

Until now, we have considered connection functions that are arbitrary homomorphisms. In this section we impose the additional conditions that the connection functions be surjective, embeddings, or isomorphisms: in this way, we obtain new combined theories, which we denote by $T_1 >_{T_0}^s T_2, T_1 >_{T_0}^{em} T_2, T_1 >_{T_0}^{iso} T_2$, respectively. For these combined theories one can show combination results that are analogous to Theorem 1: one just needs different compatibility conditions.

To treat embeddings and isomorphisms, we use the compatibility condition introduced in [11,3] for the case of unions of theories (see also the introduction of this paper). Following [11,3], we call this condition T_0-*compatibility* in the following.

Theorem 4. *Let T_0, T_1, T_2 be theories over the respective signatures $\Omega_0, \Omega_1, \Omega_2$, where Ω_0 is a common subsignature of Ω_1 and Ω_2. Assume that $T_0 \subseteq T_1$ and $T_0 \subseteq T_2$, and that T_0 is universal and effectively locally finite.*

1. *If T_2 is T_0-compatible, then the decidability of the universal fragments of T_1 and T_2 entails the decidability of the universal fragment of $T_1 >_{T_0}^{em} T_2$.*
2. *If T_1 and T_2 are T_0-compatible, then the decidability of the universal fragments of T_1 and T_2 entails the decidability of the universal fragment of $T_1 >_{T_0}^{iso} T_2$.*

A proof of this theorem, which is similar to the proof of Theorem 1, can be found in [2]. It is easy to see that the problem of deciding the universal fragment of $T_1 >_{T_0}^{iso} T_2$ is interreducible in polynomial time with the problem of deciding the universal fragment of $T_1 \cup T_2$. Consequently, the proof of part 2. of Theorem 4 yields an alternative proof of the combination result in [11].

To treat $T_1 >_{T_0}^s T_2$, we must dualize the notions "algebraic completion" and "algebraic compatibility" (see [2] for the definitions of these dual notions, and the formulation and proof of the corresponding combination result).

7   Conclusion

We have introduced a new scheme for combining many-sorted theories, and have shown under which conditions decidability of the universal fragment transfers from the component theories to their combination. Though this kind of combination has been considered before in restricted cases [13,1,24], it has not been investigated in the general algebraic setting considered here.

In this paper, we mainly concentrated on the simplest case of connecting many-sorted theories where there is just one connection function. The approach was then extended to the case of several independent connection functions, and to variants of the general combination scheme where the connection function must satisfy additional properties or where a theory is connected with itself.

On the one hand, our results are more general than the combination results for $\mathcal{E}$-connections of abstract description systems shown in [13] since they are not restricted to Boolean-based equational theories, which are closely related to abstract description systems (see Example 1). For instance, we have shown in Example 3 that any pair of theories T_1, T_2 extending a universal theory T_0 that is effectively locally finite and Gaussian satisfies the prerequisites of our transfer theorem. Examples of such theories having nothing to do with Boolean-based equational theories can be found in [3].

On the other hand, in the $\mathcal{E}$-connection approach introduced in [13], one usually considers not only the modal operator induced by a connecting relation E (see Example 1), but also the modal operator induced by its inverse E^{-1}. It is not adequate to express these two modal operators by independent connection functions going in different directions since this does not capture the relationships that must hold between them. For example, if $\Diamond$ is the diamond operator induced by the connecting relation E, and $\Box^-$ is the box operator induced by its inverse E^-, then the formulae $x \to \Box^-\Diamond x$ and $\Diamond\Box^-y \to y$ are valid in the $\mathcal{E}$-connection. In order to express these relationships in the algebraic setting without assuming the presence of the Boolean operators in the shared theory, one can replace the logical implication $\to$ by a partial order $\leq$, and require that $x \leq r(\ell(x))$ and $\ell(r(y)) \leq y$ hold for the connection functions r, ℓ generalizing the diamond and the inverse box operator. If ℓ, r are also order preserving, then this mean that ℓ, r is a pair of *adjoint functions* for the partial order $\leq$. This suggests a new way of connecting theories through pairs of adjoint functions. Again, we can show transfer of decidability provided that certain algebraic conditions are satisfied.

References

1. Farid Ajili and Claude Kirchner. A modular framework for the combination of symbolic and built-in constraints. In *Proceedings of Fourteenth International Conference on Logic Programming*, pages 331–345, Leuven, Belgium, 1997. The MIT Press.
2. Franz Baader and Silvio Ghilardi. Connecting many-sorted theories. LTCS-Report LTCS-05-04, TU Dresden, Germany, 2005.
   See http://lat.inf.tu-dresden.de/research/reports.html.
3. Franz Baader, Silvio Ghilardi, and Cesare Tinelli. A new combination procedure for the word problem that generalizes fusion decidability results in modal logics. In *Proceedings of the Second International Joint Conference on Automated Reasoning (IJCAR'04)*, volume 3097 of *Lecture Notes in Artificial Intelligence*, pages 183–197, Cork (Ireland), 2004. Springer-Verlag.
4. Franz Baader, Carsten Lutz, Holger Sturm, and Frank Wolter. Fusions of description logics and abstract description systems. *Journal of Artificial Intelligence Research*, 16:1–58, 2002.

5. Franz Baader and Cesare Tinelli. A new approach for combining decision procedures for the word problem, and its connection to the Nelson-Oppen combination method. In *Proceedings of the 14th International Conference on Automated Deduction*, volume 1249 of *Lecture Notes in Artificial Intelligence*, pages 19–33, Townsville (Australia), 1997. Springer-Verlag.
6. Franz Baader and Cesare Tinelli. Deciding the word problem in the union of equational theories. *Information and Computation*, 178(2):346–390, 2002.
7. Chen-Chung Chang and H. Jerome Keisler. *Model Theory*. North-Holland, Amsterdam-London, IIIrd edition, 1990.
8. Eric Domenjoud, Francis Klay, and Christophe Ringeissen. Combination techniques for non-disjoint equational theories. In *Proceedings of the 12th International Conference on Automated Deduction*, volume 814 of *Lecture Notes in Artificial Intelligence*, pages 267–281, Nancy (France), 1994. Springer-Verlag.
9. Camillo Fiorentini and Silvio Ghilardi. Combining word problems through rewriting in categories with products. *Theoretical Computer Science*, 294:103–149, 2003.
10. Jean H. Gallier. *Logic for Computer Science: Foundations of Automatic Theorem Proving*. Harper & Row, 1986.
11. Silvio Ghilardi. Model-theoretic methods in combined constraint satisfiability. *Journal of Automated Reasoning*, 33(3–4):221–249, 2004.
12. Marcus Kracht and Frank Wolter. Properties of independently axiomatizable bimodal logics. *The Journal of Symbolic Logic*, 56(4):1469–1485, 1991.
13. Oliver Kutz, Carsten Lutz, Frank Wolter, and Michael Zakharyaschev. $\mathcal{E}$-connections of abstract description systems. *Artificial Intelligence*, 156:1–73, 2004.
14. Michael Makkai and Gonzalo E. Reyes. *First-Order Categorical Logic*, volume 611 of *Lecture Notes in Mathematics*. Springer-Verlag, Berlin, 1977.
15. Greg Nelson. Combining satisfiability procedures by equality-sharing. In W. W. Bledsoe and D. W. Loveland, editors, *Automated Theorem Proving: After 25 Years*, volume 29 of *Contemporary Mathematics*, pages 201–211. American Mathematical Society, Providence, RI, 1984.
16. Greg Nelson and Derek C. Oppen. Simplification by cooperating decision procedures. *ACM Trans. on Programming Languages and Systems*, 1(2):245–257, October 1979.
17. Tobias Nipkow. Combining matching algorithms: The regular case. *Journal of Symbolic Computation*, 12:633–653, 1991.
18. Derek C. Oppen. Complexity, convexity and combinations of theories. *Theoretical Computer Science*, 12:291–302, 1980.
19. Don Pigozzi. The join of equational theories. *Colloquium Mathematicum*, 30(1):15–25, 1974.
20. Edith Spaan. *Complexity of Modal Logics*. PhD thesis, Department of Mathematics and Computer Science, University of Amsterdam, The Netherlands, 1993.
21. Cesare Tinelli. Cooperation of background reasoners in theory reasoning by residue sharing. *Journal of Automated Reasoning*, 30(1):1–31, January 2003.
22. Cesare Tinelli and Christophe Ringeissen. Unions of non-disjoint theories and combinations of satisfiability procedures. *Theoretical Computer Science*, 290(1):291–353, January 2003.
23. Frank Wolter. Fusions of modal logics revisited. In *Advances in Modal Logic*. CSLI, Stanford, CA, 1998.
24. Calogero Zarba. Combining multisets with integers. In *Proc. of the 18th International Conference on Automated Deduction (CADE'18)*, volume 2392 of *Lecture Notes in Artificial Intelligence*, pages 363–376, Copenhagen (Denmark), 2002. Springer-Verlag.
25. Marek W. Zawadowski. Descent and duality. *Ann. Pure Appl. Logic*, 71(2):131–188, 1995.

A Proof-Producing Decision Procedure
for Real Arithmetic

Sean McLaughlin[1] and John Harrison[2]

[1] Computer Science Department, Carnegie-Mellon University,
5000 Forbes Avenue, Pittsburgh PA 15213, USA
seanmcl@cmu.edu
[2] Intel Corporation, JF1-13,
Hillsboro OR 97124, USA
johnh@ichips.intel.com

Abstract. We present a fully proof-producing implementation of a quantifier elimination procedure for real closed fields. To our knowledge, this is the first generally useful proof-producing implementation of such an algorithm. While many problems within the domain are intractable, we demonstrate convincing examples of its value in interactive theorem proving.

1 Overview and Related Work

Arguably the first automated theorem prover ever written was for a theory of linear arithmetic [8]. Nowadays many theorem proving systems, even those normally classified as 'interactive' rather than 'automatic', contain procedures to automate routine arithmetical reasoning over some of the supported number systems like $\mathbb{N}$, $\mathbb{Z}$, $\mathbb{Q}$, $\mathbb{R}$ and $\mathbb{C}$. Experience shows that such automated support is invaluable in relieving users of what would otherwise be tedious low-level proofs. We can identify several very common limitations of such procedures:

- Often they are restricted to proving purely universal formulas rather than dealing with arbitrary quantifier structure and performing general quantifier elimination.
- Often they are not complete even for the supported class of formulas; in particular procedures for the integers often fail on problems that depend inherently on divisibility properties (e.g. $\forall x\, y \in \mathbb{Z}.\ 2x + 1 \neq 2y$)
- They seldom handle non-trivial nonlinear reasoning, even in such simple cases as $\forall x\, y \in \mathbb{R}.\ x > 0 \wedge y > 0 \Rightarrow xy > 0$, and those that do [18] tend to use heuristics rather than systematic complete methods.
- Many of the procedures are standalone decision algorithms that produce no certificate of correctness and do not produce a 'proof' in the usual sense. The earliest serious exception is described in [4].

Many of these restrictions are not so important in practice, since subproblems arising in interactive proof can still often be handled effectively. Indeed, sometimes the restrictions are unavoidable: Tarski's theorem on the undefinability of truth implies that there cannot even be a complete semidecision procedure for nonlinear reasoning over

R. Nieuwenhuis (Ed.): CADE 2005, LNAI 3632, pp. 295–314, 2005.

the integers. At the other end of the tower of number systems, one of the few implementations that has none of the above restrictions is described in [16], but that is for the complex numbers where quantifier elimination is particularly easy.

Over the real numbers, there are algorithms that can in principle perform quantifier elimination from arbitrary first-order formulas built up using addition, multiplication and the usual equality and inequality predicates. In this paper we describe the implementation of such a procedure in the HOL Light theorem prover [14], a recent incarnation of HOL [11]. It is in principle complete, and can handle arbitrary quantifier structure and nonlinear reasoning. For example it is able to prove the criterion for a quadratic equation to have a real root automatically:

$$\forall a\ b\ c.\ (\exists x.\ ax^2 + bx + c = 0) \Leftrightarrow a = 0 \land (b = 0 \Rightarrow c = 0) \lor a \neq 0 \land b^2 \geq 4ac$$

Similar — and indeed more powerful — algorithms have been implemented before, the first apparently being by Collins [7]. However, our algorithm has the special feature that it is integrated into the HOL Light prover and rather than merely asserting the answer it *proves* it from logical first principles.

The second author has previously implemented another algorithm for this subset in proof-producing style [15] but the algorithm was so inefficient that it never managed to eliminate two nested quantifiers and has not been useful in practice. The closest previous work is by Mahboubi and Pottier in Coq [21], who implemented precisely the same algorithm as us — in fact we originally learned of the algorithm itself via Pottier. However, while it appeared to reach a reasonable stage of development, this procedure seems to have been abandoned and there is no version of it for the latest Coq release. Therefore, our algorithm promises to be the first generally useful version that produces proofs.

2   Theoretical Background

In this section we describe the theoretical background in more detail. Some of this material will already be familiar to the reader.

2.1   Quantifier Elimination

We say that a theory T in a first-order language L *admits quantifier elimination* if for each formula p of L, there is a quantifier-free formula q such that $T \models p \Leftrightarrow q$. (We assume that the equivalent formula contains no new free variables.) For example, the well-known criterion for a quadratic equation to have a (real) root can be considered as an example of quantifier elimination in a suitable theory T of reals:

$$T \models (\exists x.\ ax^2 + bx + c = 0) \Leftrightarrow a \neq 0 \land b^2 \geq 4ac \lor a = 0 \land (b \neq 0 \lor c = 0)$$

If a theory admits quantifier elimination, then in particular any closed formula (one with no free variables, such as $\forall x.\ \exists y.\ x < y$) has a T-equivalent that is ground, i.e. contains no variables at all. In many cases of interest, we can quite trivially decide whether a ground formula is true or false, since it just amounts to evaluating a Boolean combination of arithmetic operations applied to constants, e.g. $2 < 3 \Rightarrow 4^2 + 5 < 23$. (One interesting exception is the theory of algebraically closed fields of unspecified characteristic, where quantifiers can be eliminated but the ground formulas cannot in general be evaluated without knowledge about the characteristic.) Consequently quantifier elimination in such cases yields a decision procedure, and also shows that such a theory T is complete, i.e. every closed formula can be proved or refuted from T. For a good discussion of quantifier elimination and many explicit examples, see [19].

2.2   Real-Closed Fields

We consider a decision procedure for the theory of real arithmetic with addition and multiplication. While we will mainly be interested in the real numbers $\mathbb{R}$, the same procedure can be exploited for more general algebraic structures, so-called *real closed fields*. The real numbers are characterized up to isomorphism by the axioms for an ordered field together with some suitable second-order completeness axiom (e.g. 'every bounded nonempty set of reals has a least upper bound'). The real-closed field axioms are those for an ordered field together with the assumptions that every nonnegative element has a square root:

$$\forall x.\ x \geq 0 \Rightarrow \exists y.\ x = y^2$$

and second that all polynomials of odd degree have a root, i.e. we have an infinite set of axioms, one like the following for each *odd* n:

$$\forall a_0, \ldots, a_n.\ a_n \neq 0 \Rightarrow \exists x.\ a_n x^n + a_{n-1} x^{n-1} + \ldots + a_1 x + a_0 = 0.$$

We will implement quantifier elimination for the reals using quite a number of analytic properties. All of these have been rigorously proven in HOL Light starting from a definitional construction of the reals [15]. However, these proofs sometimes rely on the completeness property, which is true for the reals but not for all real-closed fields. With more work, we could in fact show that all these analytic facts follow from the real-closed field axioms alone, and hence make the procedure applicable to other real-closed fields (e.g. the algebraic or computable reals). However, since we don't envisage any practical applications, this is not a high priority.

2.3   Quantifier Elimination for the Reals

A decision procedure for the theory of real closed fields, based on quantifier elimination, was first demonstrated by Tarski [30][1]. However, Tarski's procedure, a generalization of the classical technique due to Sturm [29] for finding the number of real roots of a univariate polynomial, was both difficult to understand and highly inefficient in practice. Many alternative decision methods were subsequently proposed; two that are significantly simpler were given by Seidenberg[27] and Cohen[6].

---

[1] Tarski actually discovered the procedure in 1930, but it remained unpublished for many years afterwards.

Perhaps the most efficient general algorithm currently known, and the first actually to be implemented on a computer, is the Cylindrical Algebraic Decomposition (CAD) method introduced by Collins[7]. A relatively simple but rather inefficient, algorithm is also given in [19] (see [9] for a more leisurely description). Another even simpler but generally rather more efficient algorithm is given by Hörmander [17] based on an unpublished manuscript by Paul Cohen[2] (see also [10,3] and a closely related algorithm due to Muchnik [26,22]). It was this algorithm that we chose to implement.

2.4 Fully-Expansive Decision Procedures

Theorem provers like HOL Light belong to the tradition established in Edinburgh LCF [12], where all theorems must be produced by application of simple primitive logical rules, though arbitrary programmability can be used to compose them. Thus, we need a procedure that does not simply assert that a formula is a quantifier-free equivalent of the input, but *proves* it from first principles.

At first sight, implementing decision procedures such that they produce proofs seems a daunting task. Indeed, it is in general significantly harder than simply writing a standalone 'black box' that returns an answer. However, if we want to really be sure about correctness, the only other obvious alternative, often loosely called 'reflection' [13], is to formally prove a standalone implementation correct. This is generally far more difficult again, and has so far only been applied to relatively simple algorithms. Moreover, it is of no help if one wants an independently checkable proof for other reasons, e.g. for use in proof-carrying code [23].

Even discounting the greater implementation difficulty of a proof-producing decision procedure, what about the cost in efficiency of producing a proof? In many cases of practical interest, neither the implementation difficulty nor the inefficiency need be as bad as it might first appear, because it is easy to arrange for a more computationally intensive phase not so different from a standalone implementation to produce some kind of certificate that can then be checked by the theorem prover. Since inference only needs to enter into the second phase, the overall slowdown is not so large. The first convincing example seems to have been [20], where a pretty standard first-order prover is used to search for a proof, which when eventually found, is translated into HOL inferences. Blum [2] generalizes such observations beyond the realm of theorem proving, by observing that in many situations, having an algorithm produce an easily checkable certificate is an effective way of ensuring result correctness, and more tractable than proving the original program correct.

In a more 'arithmetical' vein, the second author has recently been experimenting with a technique based on real Nullstellensatz certificates to deal with the universal subset of the present theory of reals [24]. This involves a computationally expensive search using a separate semidefinite programming package, but this search usually results in a compact certificate which needs only a few straightforward inferences to verify. For example, using this procedure we can verify the 'universal half' of the quadratic example:

---

[2] 'A simple proof of Tarski's theorem on elementary algebra', mimeographed manuscript, Stanford University 1967.

$$\forall a\ b\ c\ x.\ ax^2 + bx + c = 0 \Rightarrow b^2 - 4ac \geq 0$$

by considering the certificate:

$$b^2 - 4ac = (2ax + b)^2 - 4a(ax^2 + bx + c)$$

Since the first term on the right is a square, and the second is zero by hypothesis, it is clear that the LHS is nonnegative. Almost all the computational cost is in coming up with the appropriate square term and multiple of the input equation to make this identity hold; checking it is then easy.

However, Hörmander's algorithm (in common with all others for the full theory that we are familiar with) does not seem to lend itself to this kind of separation of 'search' and 'checking', and so we need to essentially implement all the steps of the procedure in a theorem-producing way. However, we can make this somewhat more efficient, as well as more intellectually manageable, by proving very general lemmas that apply to large families of special cases. By coding up relatively complicated syntactic structures using logical constructs, we avoid re-proving many analytical lemmas for many different cases. This will be seen more clearly when we look at the implementation of the algorithm in detail.

3   The Algorithm

Our procedure was designed by systematically modifying to produce proofs a standalone implementation of Hörmander's algorithm in OCaml.[3] We will sometimes explain the algorithm with reference to this code, since it shows the detailed control flow explicitly; it is hoped that this will still be clarifying even though it sometimes contains other functions that are not explained. In the next section we consider some of the special problems that arise when reimplementing the procedure in proof-producing style. It is instructive to see the parallels and differences: the basic control flow is all but identical, yet we produce theorems at each stage, replacing ad hoc term manipulation by logical inference.

Note first that, since our terms are built up from constants by negation, addition, subtraction and multiplication, we can rewrite all the atomic formulas in the form $p(x_1, \ldots, x_n) \bowtie 0$ where $p(x_1, \ldots, x_n)$ is a polynomial in $x_1, \ldots, x_n$ and $\bowtie$ is an equality or inequality predicate ($=, \leq, <, \neq$ etc.) It greatly helps if we initially rewrite the polynomials into a canonical representation and maintain this throughout the algorithm. In particular, we regard a multivariate polynomial $p(x_1, \ldots, x_n)$ as a polynomial in x_n with parameters polynomials in $p(x_1, \ldots, x_{n-1})$, each of those in turn regarded as a polynomial in x_{n-1} etc., where the sorting of the variables is determined by the nesting of quantifiers, x_1 being the outermost and x_n the innermost.

3.1   The Role of Sign Matrices

The key idea of the algorithm is to obtain a 'sign matrix' for a set of univariate polynomials $p_1(x), \ldots, p_n(x)$. Such a matrix is a division of the real line into a (possibly

---

[3] Available from http://www.cl.cam.ac.uk/users/jrh/atp in real.ml with some support functions from other files.

empty) ordered sequence of m points $x_1 < x_2 < \cdots < x_m$ representing precisely the zeros of the polynomials, with the rows of the matrix representing, in alternating fashion, the points themselves and the intervals between adjacent pairs and the two intervals at the ends:

$$(-\infty, x_1), x_1, (x_1, x_2), x_2, \ldots, x_{m-1}, (x_{m-1}, x_m), x_m, (x_m, +\infty)$$

and columns representing the polynomials $p_1(x), \ldots, p_n(x)$, with the matrix entries giving the signs, either positive $(+)$, negative $(-)$ or zero (0), of each polynomial p_i at the points and on the intervals. For example, for the collection of polynomials:

$$p_1(x) = x^2 - 3x + 2$$
$$p_2(x) = 2x - 3$$

the sign matrix looks like this:

Point/Interval	p_1	p_2
$(-\infty, x_1)$	$+$	$-$
x_1	0	$-$
(x_1, x_2)	$-$	$-$
x_2	$-$	0
(x_2, x_3)	$-$	$+$
x_3	0	$+$
$(x_3, +\infty)$	$+$	$+$

Note that x_1 and x_3 represent the roots 1 and 2 of $p_1(x)$ while x_2 represents 1.5, the root of $p_2(x)$. However the sign matrix contains no numerical information about the location of the points x_i, merely specifying the order of the roots of the various polynomials and what signs they take there and on the intervening intervals. It is easy to see that the sign matrix for a set of univariate polynomials $p_1(x), \ldots, p_n(x)$ is sufficient to answer any question of the form $\exists x. \, P[x]$ where the body $P[x]$ is quantifier-free and all atoms are of the form $p_i(x) \bowtie_i 0$ for any of the relations $=, <, >, \leq, \geq$ or their negations. We simply need to check each row of the matrix (point or interval) and see if one of them makes each atomic subformula true or false; the formula as a whole can then simply be "evaluated" by recursion.

In order to perform general quantifier elimination, we simply apply this basic operation to all the innermost quantified subformulas first (we can consider a universally quantified formula $\forall x. \, P[x]$ as $\neg(\exists x. \, \neg P[x])$ and eliminate from $\exists x. \, \neg P[x]$). This can then be iterated until all quantifiers are eliminated. The only difficulty is that the coefficients of a polynomial may now contain other variables as parameters; we will consider the univariate case first for simplicity, and then consider the fairly straightforward generalization to the parametrized case.

We will explain the key parts of the algorithm both in English and with reference to the OCaml code. We use a simple representation of the sign matrix as a list of lists of sign values, the sign values belonging to a four-member enumerated type

{Positive, Negative, Zero, Nonzero}. The top-level list corresponds to the sequence of points and intervals, and each sublist gives the sign values for the various polynomials there. For example, the sign matrix given above would be represented by

```
[[Positive; Negative];
 [Zero; Negative];
 [Negative; Negative];
 [Negative; Zero];
 [Negative; Positive];
 [Zero; Positive];
 [Positive; Positive]]
```

3.2   Computing the Sign Matrix

The following simple observation is key. To find the sign matrix for

$$p, p_1, \ldots, p_n$$

it suffices to find one for the set of polynomials

$$p', p_1, \ldots, p_n, q_0, q_1, \ldots, q_n$$

where p', which we will sometimes write p_0 for regularity's sake, is the derivative of p, and q_i is the remainder on dividing p by p_i. For suppose we have a sign matrix for the second set of polynomials. We can proceed as follows.

First, we split the sign matrix into two equally-sized parts, one for the $p', p_1, \ldots, p_n$ and one for the $q_0, q_1, \ldots, q_n$, but for now keeping all the points in each matrix, even if the corresponding set of polynomials has no zeros there. We can now infer the sign of $p(x_i)$ for each point x_i that is a zero of one of the polynomials $p', p_1, \ldots, p_n$, as follows. Since q_k is the remainder of p after division by p_k, $p(x) = s_k(x)p_k(x) + q_k(x)$ for some $s_k(x)$. Therefore, since $p_k(x_i) = 0$ we have $p(x_i) = q_k(x_i)$ and so we can derive the sign of p at x_i from that of the corresponding q_k. If the point x_i is not a zero of one of the $p', p_1, \ldots, p_n$, or we are dealing with an interval, we just arbitrarily assign Nonzero;[4] it will be dealt with in the next step. The following code, given sign matrices pd for $p', p_1, \ldots, p_n$ and qd for $q_0, \ldots, q_n$, gives a corresponding sign matrix for $p, p', p_1, \ldots, p_n$, with the correct signs for p at the points, but not in general at intervals. (Here index gets the position index of the first occurrence of an element in a list, and el gets an indexed element.)

```
let infersign pd qd =
  try let i = index Zero pd in el i qd :: pd
  with Failure _ -> Nonzero :: pd;;
```

Now we can throw away the second sign matrix, giving signs for the $q_0, \ldots, q_n$, and retain the (partial) matrix for $p, p', p_1, \ldots, p_n$. We next 'condense' this matrix to remove points that are not zeros of one of the $p', p_1, \ldots, p_n$, but only of one of the q_i. The signs of the $p', p_1, \ldots, p_n$ in an interval from which some other points have been removed can be read off from any of the subintervals in the original subdivision — they cannot change because there are no zeros for the relevant polynomials there.

---

[4] It may so happen that the value of $p(x_i)$ is actually zero. This does not effect the correctness of the algorithm, as that entry in the sign matrix is eliminated in the next step. We must be somewhat more careful in the proof-producing implementation.

```
let rec condense ps =
  match ps with
    int::pt::other -> let rest = condense other in
                      if mem Zero pt then int::pt::rest else rest
  | _ -> ps;;
```

Now we have a sign matrix with correct signs at all the points that are zeros of the the set of polynomials it involves, but with undetermined signs for p on the intervals, and the possibility that there may be additional zeros of p inside these intervals. But note that since there are certainly no zeros of p' inside the intervals, there can be at most one additional root of p in each interval. Whether there is one can be inferred, for an internal interval (x_i, x_{i+1}), by seeing whether the signs of $p(x_i)$ and $p(x_{i+1})$, determined in the previous step, are both nonzero and are different. If not, we can take the sign on the interval from whichever sign of $p(x_i)$ and $p(x_{i+1})$ is nonzero (we cannot have them both zero, since then there would have to be a zero of p' in between). Otherwise we insert a new point y between x_i and x_{i+1} which is a zero (only) of p, and infer the signs on the new subintervals (x_i, y) and (y, x_{i+1}) from the signs at the endpoints. Other polynomials have the same signs on (x_i, y), y and (y, x_{i+1}) that had been inferred for the original interval (x_i, x_{i+1}). For external intervals, we can use the same reasoning if we temporarily introduce new points $-\infty$ and $+\infty$ and infer the sign of $p(-\infty)$ by flipping the sign of p' on the lowest interval $(-\infty, x_1)$ and the sign of $p(+\infty)$ by copying the sign of p' on the highest interval $(x_n, +\infty)$. (Because the extremal behavior of polynomials is determined by the leading term, and those of p and p' are related by a positive multiple of x.) The following function assumes that these 'infinities' have been added first:

```
let rec inferisign ps =
  match ps with
    pt1::int::pt2::other ->
      let res = inferisign(pt2::other)
      and tint = tl int and s1 = hd pt1 and s2 = hd pt2 in
      if s1 = Positive & s2 = Negative then
        pt1::(Positive::tint)::(Zero::tint)::(Negative::tint)::res
      else if s1 = Negative & s2 = Positive then
        pt1::(Negative::tint)::(Zero::tint)::(Positive::tint)::res
      else if (s1 = Positive or s2 = Negative) & s1 = s2 then
        pt1::(s1::tint)::res
      else if s1 = Zero & s2 = Zero then
        failwith "inferisign: inconsistent"
      else if s1 = Zero then
        pt1::(s2 :: tint)::res
      else if s2 = Zero then
        pt1::(s1 :: tint)::res
      else failwith "inferisign: can't infer sign on interval"
  | _ -> ps;;
```

The overall operation is built up following the above lines. We structure it in such a way that it modifies a matrix and rather than returning it, passes it to a continuation function `cont`. As we will see later, this makes the overall implementation of the algorithm smoother. (Here `unzip` separates a list of pairs into two separate lists, `chop_list` splits a list in two at a numbered position, `replicate k a` makes a list containing `k` copies of `a`, `tl` is the tail of a list and `butlast` returns all but the very last element.)

```
let dedmatrix cont mat =
  let n = length (hd mat) / 2 in
  let mat1,mat2 = unzip (map (chop_list n) mat) in
  let mat3 = map2 inferpsign mat1 mat2 in
  let mat4 = condense mat3 in
  let k = length(hd mat4) in
  let mats = (replicate k (swap true (el 1 (hd mat3))))::mat4@
              [replicate k (el 1 (last mat3))] in
  let mat5 = butlast(tl(inferisign mats)) in
  let mat6 = map (fun l -> hd l :: tl(tl l)) mat5 in
  cont(condense mat6);;
```

3.3 Multivariate Polynomials

Note that this reasoning relies only on fairly straightforward observations of real analysis. Essentially the same procedure can be used even for multivariate polynomials,
treating other variables as parameters while eliminating one variable. The only slight
complication is that instead of literally dividing one polynomial s by another one p:

$$s(x) = p(x)q(x) + r(x)$$

we may instead have only a pseudo-division

$$a^k s(x) = p(x)q(x) + r(x)$$

where a is the leading coefficient of p, in general a polynomial in the other variables. In
this case, to infer the sign of $p(x)$ from that of $r(x)$ where $q(x) = 0$ we need to know
that $a \neq 0$ and what its sign is. Determining this may require a number of case-splits
over signs or zero-ness of polynomials in other variables, complicating the formula if
we then eliminate other variables. We will maintain an environment of sign hypotheses
sgns, and when we perform pseudo-division, we will check that $a \neq 0$ and make sure
that the signs of $s(x)$ and $r(x)$ are the same, by negating $r(x)$ or multiplying it by a
when necessary, depending on how much we know about the sign of a and whether k
is odd or even. (Here pdivide is a raw syntactic pseudo-division operation with no
check on the nature of the divisor's head coefficient.)

```
let pdivides vars sgns q p =
  let s = findsign vars sgns (head vars p) in
  if s = Zero then failwith "pdivides: head coefficient is zero" else
  let (k,r) = pdivide vars q p in
  if s = Negative & k mod 2 = 1 then poly_neg r
  else if s = Positive or k mod 2 = 0 then r
  else poly_mul (tl vars) (head vars p) r;;
```

We will also need to case-split over positive/negative status of coefficients, and the
following function fits a case-split into the continuation-passing framework; there is a
similar function split_zero used as well:

```
let split_sign vars sgns pol cont_p cont_n =
  let s = findsign vars sgns pol in
  if s = Positive then cont_p sgns
  else if s = Negative then cont_n sgns
  else if s = Zero then failwith "split_sign: zero polynomial" else
  let ineq = Atom(R(">",[pol; Fn("0",[])])) in
  Or(And(ineq,cont_p (assertsign vars sgns (pol,Positive))),
      And(Not ineq,cont_n (assertsign vars sgns (pol,Negative))));;
```

We have explained a recursive algorithm for determining a sign matrix, but we haven't explained where to stop. If we reach a constant polynomial, then the sign will be determined (perhaps after case splitting) independent of the main variable, so we want to be able to insert a fixed sign at a certain place in a sign matrix. Again, we use a continuation-based interface:

```
let matinsert i s cont mat = cont (map (insertat i s) mat);;
```

The main loop will use the continuation to convert the finally determined sign matrix (of which there may be many variants because of case splitting) to a formula. However, note that because of the rather naive case-splitting, we may reach situations where an inconsistent set of sign assumptions is made — for example $a < 0$ and $a^3 > 0$ or just $a^2 < 0$. This can in fact lead to the 'impossible' situation that the sign matrix has two zeros of some $p(x)$ with no zero of $p'(x)$ in between them — which in inferisign will generate an exception. We do not want to actually fail here, but we are at liberty to return whatever formula we like, such as $\bot$. This is dealt with by the following exception-trapping function:

```
let trapout cont m =
  try cont m with Failure "inferisign: inconsistent" -> False;;
```

The main loop is organized as mutually recursive functions. The main function matrix assumes that the signs of all the leading coefficients of the polynomials are known, i.e. in sgns. If the set of polynomials is empty, we just apply the continuation to the trivial sign matrix, remembering the error trap. If there is a constant among the polynomials, we remove it and set up the continuation so the appropriate sign is re-inserted. Otherwise, we pick the polynomial with the highest degree, which will be the p in our explanation above, and recurse to splitzero, adding logic to rearrange the polynomials so that we can assume p is at the head of the list.

```
let rec matrix vars pols cont sgns =
  if pols = [] then trapout cont [[]] else
  if exists (is_constant vars) pols then
    let p = find (is_constant vars) pols in
    let i = index p pols in
    let pols1,pols2 = chop_list i pols in
    let pols' = pols1 @ tl pols2 in
    matrix vars pols' (matinsert i (findsign vars sgns p) cont) sgns
  else
    let d = itlist (max ** degree vars) pols (-1) in
    let p = find (fun p -> degree vars p = d) pols in
    let p' = poly_diff vars p and i = index p pols in
    let qs = let p1,p2 = chop_list i pols in p'::p1 @ tl p2 in
    let gs = map (pdivides vars sgns p) qs in
    let cont' m = cont(map (fun l -> insertat i (hd l) (tl l)) m) in
    splitzero vars qs gs (dedmatrix cont') sgns
```

The function splitzero simply case-splits over the zero status of the coefficients of the polynomials in the list pols, assuming those in dun are already fixed:

```
and splitzero vars dun pols cont sgns =
  match pols with
    [] -> splitsigns vars [] dun cont sgns
  | p::ops -> if p = Fn("0",[]) then
              let cont' = matinsert (length dun) Zero cont in
```

```
            splitzero vars dun ops cont' sgns
   else split_zero (tl vars) sgns (head vars p)
        (splitzero vars dun (behead vars p :: ops) cont)
        (splitzero vars (dun@[p]) ops cont)
```

When all polynomials are dealt with, we recurse to another level of splitting where we split over positive-negative status of the coefficients that are already determined to be nonzero:

```
and splitsigns vars dun pols cont sgns =
  match pols with
    [] -> dun
  | p::ops -> let cont' = splitsigns vars (dun@[p]) ops cont in
              split_sign (tl vars) sgns (head vars p) cont' cont'
```

That is the main loop of the algorithm; we start with a continuation that will perform an appropriate test on the sign matrix entries for each literal in the formula, and we construct the sign matrix for all polynomials that occur in the original formula.

4 Proof-Producing Implementation

The proof-producing version, by design, follows the same structure. In this section we concentrate on interesting design decisions for this variant.

4.1 Polynomials

Our canonical representation of polynomials is as lists of coefficients with the constant term first. For example, the polynomial $x^3 - 2x + 6$ is represented by the list $[6; -2; 0; 1]$. The corresponding 'evaluation' function is simply expressed as a primitive recursive definition over lists:

```
⊢ (poly [] x = &0) ∧
  (poly (CONS h t) x = h + x * poly t x)
```

This representation is used in a nested fashion to encode multivariate polynomials. A key point is that we can prove many analytical theorems such as special intermediate-value properties generically for all polynomials just by using `poly l` for a general list of reals `l` (the proofs usually proceed by induction over lists). Thus we can avoid proving many special cases for the actual polynomials we use: they are deduced by a single primitive inference step of variable instantiation from the generic versions.

4.2 Data Structures

The first difficult choice we encountered was how to represent our current knowledge about the state of the sign matrix. We were faced with the task of organizing the following information, for example, giving a partial sign matrix for the polynomials

$$p_0(x) = x^2$$
$$p_1(x) = x - 1$$

(Here, x_0 and x_1 correspond to the roots 0 of $\lambda x.\ x^2$ and 1 of $\lambda x.\ x - 1$ respectively.)

$$\forall x.\ x < x_0 \Rightarrow p_0(x) > 0$$
$$\forall x.\ x < x_0 \Rightarrow p_1(x) < 0$$
$$p_0(x_0) = 0$$
$$p_1(x_0) < 0$$
$$\forall x.\ x_0 < x < x_1 \Rightarrow p_0(x) > 0$$
$$\forall x.\ x_0 < x < x_1 \Rightarrow p_1(x) < 0$$
$$p_0(x_1) > 0$$
$$p_1(x_1) = 0$$
$$\forall x.\ x_1 < x \Rightarrow p_0(x) > 0$$
$$\forall x.\ x_1 < x \Rightarrow p_1(x) > 0$$

As the matrices become larger (which is immediate due to the exponential nature of the procedure), the task of managing these facts using simple data types like lists and products becomes daunting. Instead, we chose to define a series of predicates that allow us to organize this data in a succinct fashion. We begin by defining an enumerated type of signs with members Zero, Pos, Neg, Nonzero, Unknown. The additional element Unknown is useful in order to be able to make rigorous proven statements at intermediate steps, whereas in the original we could just say 'the setting of this sign may be wrong now but we'll fix it in the next step'. We then define a predicate which, given a domain and a polynomial, interprets the sign:

$$\text{interpsign } S\ p\ \text{Zero} := (\forall x.x \in S \Rightarrow (p(x) = 0))$$
$$\text{interpsign } S\ p\ \text{Pos} := (\forall x.x \in S \Rightarrow (p(x) > 0))$$
$$\text{interpsign } S\ p\ \text{Neg} := (\forall x.x \in S \Rightarrow (p(x) < 0))$$
$$\text{interpsign } S\ p\ \text{Nonzero} := (\forall x.x \in S \Rightarrow (p(x) \neq 0))$$
$$\text{interpsign } S\ p\ \text{Unknown} := (\forall x.x \in S \Rightarrow (p(x) = p(x)))$$

Now, the previous set of formulas can be written as follows:

$$\text{interpsign } (\lambda x.x < x_0)\ p_0\ \text{Pos}$$
$$\text{interpsign } (\lambda x.x < x_0)\ p_1\ \text{Neg}$$
$$\text{interpsign } (\lambda x.x = x_0)\ p_0\ \text{Zero}$$
$$\text{interpsign } (\lambda x.x = x_0)\ p_1\ \text{Neg}$$
$$\text{interpsign } (\lambda x.x_0 < x < x_1)\ p_0\ \text{Pos}$$
$$\text{interpsign } (\lambda x.x_0 < x < x_1)\ p_1\ \text{Neg}$$
$$\text{interpsign } (\lambda x.x = x_1)\ p_0\ \text{Pos}$$
$$\text{interpsign } (\lambda x.x = x_1)\ p_1\ \text{Zero}$$
$$\text{interpsign } (\lambda x.x_1 < x)\ p_0\ \text{Pos}$$
$$\text{interpsign } (\lambda x.x_1 < x)\ p_1\ \text{Pos}$$

These formulas are of a very regular form. We can thus use HOL Light's cababilities for making primitive recursive definitions:

ALL2 P [] $l2 \Leftrightarrow (l2 = [])$)
ALL2 P (CONS $h1\ t1$) $l2 \Leftrightarrow$
$\qquad$ if $l2 = []$ then F else $(P\ h1\ (\text{HD}\ l2)) \wedge (\text{ALL2}\ P\ t1\ (\text{TL}\ l2))$

This allows us to compact the formulas for a given set with another predicate:

interpsigns $polys\ S\ signs = $ ALL2 (interpsign S) $polys\ signs$

Our formulas can now be represented slightly more succinctly:

$\qquad$ interpsigns $(\lambda x.x < x_0)\ [p_0, p_1]\ [\text{Pos}, \text{Neg}]$
$\qquad$ interpsigns $(\lambda x.x = x_0)\ [p_0, p_1]\ [\text{Zero}, \text{Neg}]$
$\qquad$ interpsigns $(\lambda x.x_0 < x < x_1)\ [p_0, p_1]\ [\text{Pos}, \text{Neg}]$
$\qquad$ interpsigns $(\lambda x.x = x_1)\ [p_0, p_1]\ [\text{Pos}, \text{Zero}]$
$\qquad$ interpsign $(\lambda x.x_1 < x)\ [p_0, p_1]\ [\text{Pos}, \text{Pos}]$

Now, given a predicate OrderedList which indicates that the points in the list are sorted, and a function PartitionLine that breaks the real line into intervals based on the points of the list, we can represent the entire matrix with a final predicate.

interpmat $points\ polys\ signs =$
$\qquad$ OrderedList $points\ \wedge$
$\qquad$ ALL2 (interpsigns $polys$) (PartitionLine $points$) $signs$

Thus, our entire set of formulas is represented by the simple formula

interpmat $[x_0, x_1]\ [p_0, p_1]\ [[\text{Pos}, \text{Neg}], [\text{Zero}, \text{Neg}], [\text{Pos}, \text{Neg}], [\text{Pos}, \text{Zero}], [\text{Pos}, \text{Pos}]]$

As the sign matrix is the primary data structure in the algorithm, this succinct representation makes the implementation much smoother than dealing with the formulas individually.

One potential drawback to this approach is the time spent assembling and disassembling the representation to extract or add formulas. While in a high level programming language this would not be a concern, for large matrices this is a substantial amount of term rewriting.

On the other hand, we can take advantage of the representation to improve efficiency. For a simple example, suppose we've deduced that $p_2 = p_0$ and would like to replace p_0 by p_2 in the sign matrix. This is a trivial one step rewrite for our representation, while many rewrites would be required to achieve the same result if the matrix were represented by individual formulas. For a more interesting example, recall the part of the dedmatrix function where we add points at infinity. This complex step can be encapsulated by proving a number of general theorems such as:

```
|- !p ps r1 x s2 r2 r3.
interpmat [x]
  (CONS (\x. poly p x) (CONS (\x. poly (poly_diff p) x) ps))
  [CONS Unknown (CONS Neg r1);CONS s2 r2;CONS Unknown (CONS Pos r3)]   ==>
 nonconstant p ==>
?xminf xinf.
 interpmat [xminf; x; xinf]
  (CONS (\x. poly p x) (CONS (\x. poly (poly_diff p) x) ps))
    [CONS Pos (CONS Neg r1);
     CONS Pos (CONS Neg r1);
     CONS Unknown (CONS Neg r1);
     CONS s2 r2;
     CONS Unknown (CONS Pos r3);
     CONS Pos (CONS Pos r3);
     CONS Pos (CONS Pos r3)]
```

We originally coded the addition of these points by breaking up the sign matrix. The optimization using rewriting with theorems similar to this one runs 10 times faster than the orignial version. While these kinds of general theorems are much more difficult to prove in HOL Light , rewriting data structures directly has been the most important step in producing reasonable performance.

4.3   Effort Comparison

Generating proofs for each step of the code above was a significant challenge. To compare the procedures, we first consider the amount of code that was necessary to implement the procedure. The OCaml source code given above is somewhat less than 300 lines. In contrast, our version runs around 1300. This does not include the additional 8000 lines of proof scripts which prove lemmas needed at various points in the procedure.

5   Results

It is well-known that quantifier elimination for the reals is in general computationally intractable, both in theoretical complexity [31] and in the limited success on real applications. So we cannot start with high expectations of routinely solving really interesting problems. This applies with all the more force since proof production makes the algorithm considerably slower, apparently about 3 orders of magnitude for our current prototype.

However, in the field of interactive theorem proving, the algorithm could be an important tool. A great deal of time can be spent proving trivial facts of real arithmetic that do not fall into one of the well known decidable (and implemented) subsets, such as linear arithmetic, or linear programming. One author spent many hours proving simple lemmas in preparation for implementing this procedure. Some indicative examples are

$$\forall x \forall y. \, x \cdot y > 0 \Leftrightarrow (x > 0 \wedge y > 0) \vee (x < 0 \wedge y < 0)$$
$$\forall x \forall y. \, x \cdot y < 0 \Leftrightarrow (x > 0 \wedge y < 0) \vee (x < 0 \wedge y > 0)$$
$$\forall x \forall y. \, x < y \Rightarrow \exists z. \, x < z \wedge z < y$$

Our procedure easily dispenses with such problems, and compares favorably with the time it takes to prove such problems "by hand". Thus, it could be a potentially valuable tool in the day to day use of theorem provers like HOL Light.

5.1   Times

Without further ado, we give some of the problems the algorithm can solve and running times. The procedure is written in OCaml and was run uncompiled on a 3GHz Pentium 4 processor running Linux kernel 2.4. Times are in seconds unless otherwise indicated.

Univariate Examples. We've arranged the results into loose categories. We first consider some routine univariate examples, the results collected in Table 1. A final pair of examples demonstrates the key bound properties of the first two "Chebyshev Polynomials" which are useful in function approximation; these had running times of 4.7 and 13.7 seconds respectively.

$$\forall x. \; -1 \leq x \wedge x \leq 1 \Rightarrow -1 \leq 2x^2 - 1 \wedge 2x^2 - 1 \leq 1$$

$$\forall x. \; -1 \leq x \wedge x \leq 1 \Rightarrow -1 \leq 4x^3 - 3x \wedge 4x^3 - 3x \leq 1$$

Table 1. Runtimes on simple univariate examples

Category	Formula	Result	Running Time
Linear	$\exists x. \; x - 1 > 0$	T	0.2
Linear	$\exists x. \; 3 - x > 0 \wedge x - 1 > 0$	T	0.7
Quadratic	$\exists x. \; x^2 = 0$	T	0.4
Quadratic	$\exists x. \; x^2 + 1 = 0$	F	0.5
Quadratic	$\exists x. \; x^2 - 2x + 1 = 0$	T	0.5
Cubic	$\exists x. \; x^3 - 1 = 0$	T	1.1
Cubic	$\exists x. \; x^3 - 3x^2 + 3x - 1 > 0$	T	0.8
Cubic	$\exists x. \; x^3 - 4x^2 + 5x - 2 > 0$	T	2.0
Cubic	$\exists x. \; x^3 - 6x^2 + 11x - 6 = 0$	T	1.6
Quartic	$\exists x. x^4 - 1 > 0$	T	1.2
Quartic	$\exists x. x^4 + 1 < 0$	F	1.0
Quartic	$\exists x. x^4 - x^3 = 0$	T	0.8
Quartic	$\exists x. x^4 - 2 * x^2 + 2 = 0$	T	2.9
Quintic	$\exists x. x^5 - 15 * x^4 + 85 * x^3 - 225 * x^2 + 274 * x - 120 = 0$	T	19

Multivariate Examples

- Here is an instance of a more complicated quantifier structure. Our implementation returns the theorem in 11.6 seconds.

$$(\forall a \; f \; k. \; (\forall e. \; k < e \Rightarrow f < a \cdot e) \Rightarrow f \leq a \cdot k)$$

- Here is an example arising as a polynomial termination ordering for a rewrite system for group theory, which takes 23.5 seconds.

$$1 < 2 \wedge (\forall x. \; 1 < x \Rightarrow 1 < x^2) \wedge$$
$$(\forall x \; y. \; 1 < x \wedge 1 < y \Rightarrow 1 < x(1 + 2y))$$

- We can use open formulas to determine when polynomials have roots, as in the case mentioned above of a quadratic polynomial, $\exists x.\ ax^2 + bx + c = 0$ The following identity is established in 4.7 seconds.

```
val it : thm =
|- (?x. a * x pow 2 + b * x + c = &0) <=>
  (&0 + a * &1 = &0) /\
  ((&0 + b * &1 = &0) /\ (&0 + c * &1 = &0) \/
   ~(&0 + b * &1 = &0) /\ (&0 + b * &1 > &0 \/ &0 + b * &1 < &0)) \/
  ~(&0 + a * &1 = &0) /\
  (&0 + a * &1 > &0 /\
   ((&0 + a * ((&0 + b * (&0 + b * -- &1)) + a * (&0 + c * &4)) = &0) \/
    ~(&0 + a * ((&0 + b * (&0 + b * -- &1)) + a * (&0 + c * &4)) = &0) /\
    &0 + a * ((&0 + b * (&0 + b * -- &1)) + a * (&0 + c * &4)) < &0) \/
   &0 + a * &1 < &0 /\
   ((&0 + a * ((&0 + b * (&0 + b * -- &1)) + a * (&0 + c * &4)) = &0) \/
    ~(&0 + a * ((&0 + b * (&0 + b * -- &1)) + a * (&0 + c * &4)) = &0) /\
    &0 + a * ((&0 + b * (&0 + b * -- &1)) + a * (&0 + c * &4)) > &0))
```

While this is not particularly readable, it does give the necessary and sufficient conditions. We can express the answer in a somewhat nicer form by "guessing", and the proof takes 816 seconds:

$$\forall a \forall b \forall c. (\exists x. ax^2 + bx + c = 0) \Leftrightarrow$$
$$(((a = 0) \land ((b \neq 0) \lor (c = 0))) \lor$$
$$(a \neq 0) \land b^2 \geq 4ac)$$

- Robert Solovay has shown us a method by which formulas over general real vector spaces can be reduced to the present subset of reals. Consider the following formula, where x and y are vectors and u a real number, $x \cdot y$ is the inner (dot) product and $||x||$ is the norm (length) of x:

$$\forall x\ y.\ x \cdot y > 0 \Rightarrow \exists u.\ 0 < u \land ||uy - x|| < ||x||$$

Our implementation of Solovay's procedure returns the following formula over the reals that provably implies the original. (Note that the body can be subjected to some significant algebraic simplification, but this gets handled anyway by our transition to canonical polynomial form.) Our procedure proves this in 8.8 seconds.

$$\forall a\ b\ c.\ 0 \leq b \land 0 \leq c \land 0 < ac$$
$$\Rightarrow \exists u.\ 0 < u \land u(uc - ac) - (uac - (a^2c + b)) < a^2c + b$$

6   Future Work

The underlying algorithm is quite naive, and could be improved in many ways at relatively little cost in complexity. One very promising improvement is to directly exploit equations to substitute. At its simplest, if we are eliminating an existential quantifier from a conjunction containing an equation with the variable on one side, we can simply replace the variable with the other side of the equation. (At present, our algorithm uses the inefficient general sign-matrix process even when such obvious simplifications could be made.) Slightly more complicated methods can yield very good results for low-degree equations like quadratics [32]. More generally, even more complicated higher-degree equations can be used to substitute, and we can even try to factor. For example, consider the assertion that the logistic map $x \mapsto rx(1-x)$ has a cycle with period 2:

$$\exists x.\, 0 \leq x \wedge x \leq 1 \wedge r(rx(1-x))(1-rx(1-x)) = x \wedge \neg(rx(1-x) = x)$$

By factoring the equation and then using the remaining factor to substitute, we can reach the following formula, where the degree of x has been reduced, making the problem dramatically easier for the core algorithm:

$$\exists x.\, 0 \leq x \wedge x \leq 1 \wedge r^2 x^2 - r(1+r)x + (1+r) = 0 \wedge \neg(2rx = 1+r)$$

This can be solved by our algorithm in about 5 hours.

The translation to a proof-producing version was done quite directly, and there is probably considerable scope for improvement by making some of the inference steps more efficient. In the last example, the HOL Light implementation runs over 10^3 slower than the unchecked version. It seems that this gap can be significantly narrowed.

One reason for the large gap is our representation of the sign matrix. The use of lists made proving proforma theorems awkward. For an extreme example, to attempt to prove a general theorem regarding the `inferpsign` step, we proved the following lemma which states that if $p(x) = s(x) * q(x) + r(x)$, $q(x) = 0$, and $r(x) > 0$, then we can replace the unknown sign for p at x by *Pos*.

```
let INFERPSIGN_ZERO_EVEN = prove_by_refinement(
 `!a n p ps q qs pts r rs s s1 s2 s3 rest sgns.
  interpmat pts
    (APPEND (CONS p ps) (APPEND (CONS q qs) (CONS r rs)))
    (APPEND sgns
      (CONS (APPEND (CONS Unknown s1)
        (APPEND (CONS Zero s2) (CONS Zero s3))) rest)) ==>
  (LENGTH ps = LENGTH s1) ==>
  (LENGTH qs = LENGTH s2) ==>
  ODD (LENGTH sgns) ==>
  (!x. a pow n * p x = s x * q x + r x) ==>
  (a <> &0) ==>
  EVEN n ==>
  interpmat pts
    (APPEND (CONS p ps) (APPEND (CONS q qs) (CONS r rs)))
    (APPEND sgns
      (CONS (APPEND (CONS Zero s1)
        (APPEND (CONS Zero s2) (CONS Zero s3))) rest))`,
```

A number of arithmetic steps were necessary at each application of this lemma to prove the antecedents. It ended up being faster to expand all the definitions, do some simple real arithmetic, and reassemble the matrix. This is drastically inefficient and we are currently experimenting with new representations that will allow us to avoid such tedious and expensive logic.

One interesting continuation of the current work would be to see how easily our implementation could be translated to another theorem prover such as Isabelle [25]. Finally, there is the potential to use this procedure in a fully automated combined decision procedure environment such as CVC-Lite [1]. We have not explored these lines of research in any detail.

7   Conclusion

It is difficult to foresee the practical benefits of using general decision procedures such as this one in the field of interactive theorem proving. As this case study shows, even when they exist, it is not at all clear whether, due to complexity constraints, they will be applicable to even moderately difficult problems. Considering the examples given above, one might even dismiss such procedures outright as entirely too inefficient. For the *user* of such a system, however, it is procedures like this one which automate tedious low level tasks that make the process of theorem proving useful and enjoyable, or at least tolerable.

In conclusion, the work described above can be seen in two different lights. On the one hand, it is a rather inefficient implementation of an algorithm which, while mathematically and philosophically interesting, and theoretically applicable to an enormous range of difficult problems, is not yet practically useful for those problems. On the other hand, it can be viewed as another tool in the (human) theorem prover's tool chest. One that, given the wide range of applications of the real numbers in theorem proving, could be an important *practical* achievement.

Acknowledgments

The first author would like to thank Frank Pfenning and the National Science Foundation for supporting the Logosphere project[5] The second author is grateful to Loïc Pottier for first telling him about the Hörmander procedure.

References

1. C. Barrett and S. Berezin. CVC Lite: A new implementation of the cooperating validity checker. In R. Alur and D. A. Peled, editors, *Computer Aided Verification, 16th International Conference, CAV 2004*, volume 3114 of *Lecture Notes in Computer Science*, pages 515–518, Boston, MA, 2004. Springer-Verlag.

2. M. Blum. Program result checking: A new approach to making programs more reliable. In A. Lingas, R. Karlsson, and S. Carlsson, editors, *Automata, Languages and Programming, 20th International Colloquium, ICALP93, Proceedings*, volume 700 of *Lecture Notes in Computer Science*, pages 1–14, Lund, Sweden, 1993. Springer-Verlag.

3. J. Bochnak, M. Coste, and M.-F. Roy. *Real Algebraic Geometry*, volume 36 of *Ergebnisse der Mathematik und ihrer Grenzgebiete*. Springer-Verlag, 1998.

4. R. J. Boulton. Efficiency in a fully-expansive theorem prover. Technical Report 337, University of Cambridge Computer Laboratory, New Museums Site, Pembroke Street, Cambridge, CB2 3QG, UK, 1993. Author's PhD thesis.

5. B. F. Caviness and J. R. Johnson, editors. *Quantifier Elimination and Cylindrical Algebraic Decomposition*, Texts and monographs in symbolic computation. Springer-Verlag, 1998.

6. P. J. Cohen. Decision procedures for real and p-adic fields. *Communications in Pure and Applied Mathematics*, 22:131–151, 1969.

---

[5] Grant Number CCR-ITR-0325808.

7. G. E. Collins. Quantifier elimination for real closed fields by cylindrical algebraic decomposition. In H. Brakhage, editor, *Second GI Conference on Automata Theory and Formal Languages*, volume 33 of *Lecture Notes in Computer Science*, pages 134–183, Kaiserslautern, 1976. Springer-Verlag.

8. M. Davis. A computer program for Presburger's algorithm. In *Summaries of talks presented at the Summer Institute for Symbolic Logic, Cornell University*, pages 215–233. Institute for Defense Analyses, Princeton, NJ, 1957. Reprinted in [28], pp. 41–48.

9. E. Engeler. *Foundations of Mathematics: Questions of Analysis, Geometry and Algorithmics*. Springer-Verlag, 1993. Original German edition *Metamathematik der Elementarmathematik* in the *Series Hochschultext*.

10. L. Gårding. *Some Points of Analysis and Their History*, volume 11 of *University Lecture Series*. American Mathematical Society / Higher Education Press, 1997.

11. M. J. C. Gordon and T. F. Melham. *Introduction to HOL: a theorem proving environment for higher order logic*. Cambridge University Press, 1993.

12. M. J. C. Gordon, R. Milner, and C. P. Wadsworth. *Edinburgh LCF: A Mechanised Logic of Computation*, volume 78 of *Lecture Notes in Computer Science*. Springer-Verlag, 1979.

13. J. Harrison. Metatheory and reflection in theorem proving: A survey and critique. Technical Report CRC-053, SRI Cambridge, Millers Yard, Cambridge, UK, 1995. Available on the Web as http://www.cl.cam.ac.uk/users/jrh/papers/reflect.dvi.gz.

14. J. Harrison. HOL Light: A tutorial introduction. In M. Srivas and A. Camilleri, editors, *Proceedings of the First International Conference on Formal Methods in Computer-Aided Design (FMCAD'96)*, volume 1166 of *Lecture Notes in Computer Science*, pages 265–269. Springer-Verlag, 1996.

15. J. Harrison. *Theorem Proving with the Real Numbers*. Springer-Verlag, 1998. Revised version of author's PhD thesis.

16. J. Harrison. Complex quantifier elimination in HOL. In R. J. Boulton and P. B. Jackson, editors, *TPHOLs 2001: Supplemental Proceedings*, pages 159–174. Division of Informatics, University of Edinburgh, 2001. Published as Informatics Report Series EDI-INF-RR-0046. Available on the Web at http://www.informatics.ed.ac.uk/publications/report/0046.html.

17. L. Hörmander. *The Analysis of Linear Partial Differential Operators II*, volume 257 of *Grundlehren der mathematischen Wissenschaften*. Springer-Verlag, 1983.

18. W. A. Hunt, R. B. Krug, and J. Moore. Linear and nonlinear arithmetic in ACL2. In D. Geist, editor, *Proceedings of the 12th Advanced Research Working Conference on Correct Hardware Design and Verification Methods, CHARME 2003*, volume 2860 of *Lecture Notes in Computer Science*, pages 319–333. Springer-Verlag.

19. G. Kreisel and J.-L. Krivine. *Elements of mathematical logic: model theory*. Studies in Logic and the Foundations of Mathematics. North-Holland, revised second edition, 1971. First edition 1967. Translation of the French 'Eléments de logique mathématique, théorie des modeles' published by Dunod, Paris in 1964.

20. R. Kumar, T. Kropf, and K. Schneider. Integrating a first-order automatic prover in the HOL environment. In M. Archer, J. J. Joyce, K. N. Levitt, and P. J. Windley, editors, *Proceedings of the 1991 International Workshop on the HOL theorem proving system and its Applications*, pages 170–176, University of California at Davis, Davis CA, USA, 1991. IEEE Computer Society Press.

21. A. Mahboubi and L. Pottier. Elimination des quantificateurs sur les réels en Coq. In Journées Francophones des Langages Applicatifs (JFLA), available on the Web from http://pauillac.inria.fr/jfla/2002/actes/index.html08-mahboubi.ps, 2002.

22. C. Michaux and A. Ozturk. Quantifier elimination following Muchnik. Université de Mons-Hainaut, Institute de Mathématique, Preprint 10, `http://w3.umh.ac.be/math/preprints/src/Ozturk020411.pdf`, 2002.

23. G. C. Necula. Proof-carrying code. In *Conference record of POPL'97: the 24th ACM SIGPLAN-SIGACT Symposium on Principles of Programming Languages*, pages 106–119. ACM Press, 1997.

24. P. Parrilo. Semidefinite programming relaxations for semialgebraic problems. Available from the Web at `citeseer.nj.nec.com/parrilo01semidefinite.html`, 2001.

25. L. C. Paulson. *Isabelle: a generic theorem prover*, volume 828 of *Lecture Notes in Computer Science*. Springer-Verlag, 1994. With contributions by Tobias Nipkow.

26. H. Schoutens. Muchnik's proof of Tarski-Seidenberg. Notes available from `http://www.math.ohio-state.edu/~schoutens/PDF/Muchnik.pdf`, 2001.

27. A. Seidenberg. A new decision method for elementary algebra. *Annals of Mathematics*, 60:365–374, 1954.

28. J. Siekmann and G. Wrightson, editors. *Automation of Reasoning — Classical Papers on Computational Logic, Vol. I (1957-1966)*. Springer-Verlag, 1983.

29. C. Sturm. Mémoire sue la résolution des équations numériques. *Mémoire des Savants Etrangers*, 6:271–318, 1835.

30. A. Tarski. *A Decision Method for Elementary Algebra and Geometry*. University of California Press, 1951. Previous version published as a technical report by the RAND Corporation, 1948; prepared for publication by J. C. C. McKinsey. Reprinted in [5], pp. 24–84.

31. N. N. Vorobjov. Deciding consistency of systems of polynomial in exponent inequalities in subexponential time. In T. Mora and C. Traverso, editors, *Proceedings of the MEGA-90 Symposium on Effective Methods in Algebraic Geometry*, volume 94 of *Progress in Mathematics*, pages 491–500, Castiglioncello, Livorno, Italy, 1990. Birkhäuser.

32. V. Weispfenning. Quantifier elimination for real algebra — the quadratic case and beyond. *Applicable Algebra in Engineering Communications and Computing*, 8:85–101, 1997.

The MathSAT 3 System[*]

Marco Bozzano[1], Roberto Bruttomesso[1], Alessandro Cimatti[1], Tommi Junttila[2],
Peter van Rossum[3], Stephan Schulz[4], and Roberto Sebastiani[5]

[1] ITC-IRST, Via Sommarive 18, 38050 Povo, Trento, Italy
[2] Helsinki University of Technology, P.O.Box 5400, FI-02015 TKK, Finland
[3] Radboud University, Toernooiveld 1, 6525 ED Nijmegen, The Netherlands
[4] University of Verona, Strada le Grazie 15, 37134 Verona, Italy
[5] Università di Trento, Via Sommarive 14, 38050 Povo, Trento, Italy

1 Introduction

Satisfiability Modulo Theories (SMT) can be seen as an extended form of propositional
satisfiability, where propositions are either simple boolean propositions or quantifier-
free atomic constraints in a specific theory. In this paper we present MATHSAT version
3 [6,7,8], a DPLL-based decision procedure for the SMT problem for various theories,
including those of Equality and Uninterpreted Functions ($\mathcal{EUF}$)[1], of Separation Logic
($\mathcal{SEP}$), and of Linear Arithmetic on the Reals ($\mathcal{LA}(\mathbb{R})$) and on the integers ($\mathcal{LA}(\mathbb{Z})$).
MATHSAT is also able to solve the SMT problem for combined $\mathcal{EUF} + \mathcal{SEP}$, $\mathcal{EUF}$
$+\mathcal{LA}(\mathbb{R})$, and $\mathcal{EUF} +\mathcal{LA}(\mathbb{Z})$, either by means of Ackermann's reduction [1] or using
our new approach called Delayed Theory Combination (DTC) [6], which is alternative
to the classical Nelson-Oppen or Shostak integration schemata.

MATHSAT is based on the approach of integrating a state-of-the-art SAT solver with
a hierarchy of dedicated theory solvers. It is a re-implementation of an older version
of the same tool [3,4], supporting more extended theories, and their combination, and
implementing a number of important optimization techniques [2].

MATHSAT has been and is currently used in many projects, both as a platform
for experimenting new automated reasoning techniques, and as a workhorse reasoning
procedure onto which to develop formal verification tools. Our main target application
domains are those of formal verification of RTL circuit designs, and of timed and hybrid systems. MATHSAT has been and is currently widely used by many authors for
empirical tests on SMT problems (see, e.g., [11,12,2]).

For lack of space, in this paper we omit any description of empirical results, which
can be found in [6,7,8]. A Linux executable of the solver, together with the papers
[6,7,8] and the benchmarks used there, is available from http://mathsat.itc.it/.

---

[*] This work has been partly supported by ISAAC, an European sponsored project, contract no.
AST3-CT-2003-501848, by ORCHID, a project sponsored by Provincia Autonoma di Trento,
and by BOWLING, a project sponsored by a grant from Intel Corporation. The work of T.
Junttila has also been supported by the Academy of Finland, projects 53695 and 211025.
[1] More precisely, $\mathcal{EUF}$ with some extensions for arithmetic predicates and constants.
[2] We notice that some ideas related to the mathematical solver(s) presented in this paper (i.e.,
layering, stack-based interfaces, theory-deduction) are to some extent similar to ideas pio-
neered by Constraint Logic Programming (see, e.g., [14]).

R. Nieuwenhuis (Ed.): CADE 2005, LNAI 3632, pp. 315–321, 2005.

2   The Main Procedure

MATHSAT is built on top of the standard "online" lazy integration schema used in many SMT tools (see, e.g., [3,11,2]). In short: after some preprocessing to the input formula ϕ, a DPLL-based SAT solver is used as an enumerator of (possibly partial) truth assignments for (the boolean abstraction of) ϕ; the consistency in $\mathcal{T}$ of (the set of atomic constraints corresponding to) each assignment is checked by a solver $\mathcal{T}$-SOLVER. This is done until either one $\mathcal{T}$-consistent assignment is found, or all assignments have been checked.

MATHSAT 3 [6,7,8] is a complete re-implementation of the previous versions described in [3,4], supporting more theories (e.g., $\mathcal{EUF}$, $\mathcal{EUF}$ $+\mathcal{LA}(\mathbb{R})$, $\mathcal{EUF}$ $+\mathcal{LA}(\mathbb{Z})$) and a richer input language (e.g., non-clausal forms, if-then-else). It features a new preprocessor, a new SAT solver, and a much more sophisticate theory solver, which we describe in this section. It also features new optimization techniques, which we describe in the next sections. [3]

2.1   The Preprocessor

First, MATHSAT allows the input formula to be in non-clausal form and to include operations such as if-then-else's over non-boolean terms. The preprocessor translates the input formula into conjunctive normal form by using a standard linear-time satisfiability preserving translation. Second, the input formula may contain constraints that mix theories in a way that cannot be handled either by the $\mathcal{EUF}$ solver or the linear arithmetic solver alone (e.g., $f(x) + f(z) = 4$). To handle these constraints, the preprocessor either (i) eliminates them by applying Ackermann's reduction [1] or (ii) purifies them into a normal form if the Delayed Theory Combination scheme [6] is applied.

2.2   The SAT Solver

In MATHSAT, the boolean solver is built upon the MINISAT solver [10]. Thus it inherits for free conflict-driven learning and backjumping [15], restarts [13], optimized boolean constraint propagation based on the two-watched literal scheme [16], and an effective splitting heuristics VSIDS [16]. It communicates with $\mathcal{T}$-SOLVER through a stack-based interface that passes assigned literals, $\mathcal{T}$-consistency queries and backtracking commands to $\mathcal{T}$-SOLVER, and gets back answers to the queries, $\mathcal{T}$-inconsistent sets (theory conflict sets) and $\mathcal{T}$-implied literals.

2.3   The Theory Solver $\mathcal{T}$-SOLVER

$\mathcal{T}$-SOLVER gets in input a set of quantifier-free constraints μ and checks whether μ is $\mathcal{T}$-satisfiable or not. In the first case, it also tries to perform and return deductions in

---

[3] In order to distinguish what is new in MATHSAT 3 wrt. previous versions, in the next sections we label by "[3,4]" those techniques which were already present in previous versions, with "[7,8]" the new techniques, and with "[3,4,7,8]" those techniques which have been proposed in earlier versions and have been significantly improved or extended in MATHSAT 3.

the form $\mu' \models_T l$ s.t. l is a literal representing a truth assignment to a not-yet-assigned atom occurring in the input formula, and $\mu' \subseteq \mu$ in the (possibly minimal) set of literals entailing l. In the second case, it returns the (possibly minimal) sub-assignment $\mu' \subseteq \mu$ which caused the inconsistency (*conflict set*). Due to the early pruning step (see § 3), T-SOLVER is typically invoked on *incremental* assignments. When a conflict is found, the search backtracks to a previous point, and T-SOLVER then restarts from a previously visited state. Based on these considerations, T-SOLVER has a persistent state, and is *incremental* and *backtrackable*: incremental means that it avoids restarting the computation from scratch whenever it is given as input an assignment μ' such that $\mu' \supset \mu$ and μ has already been proved satisfiable; backtrackable means that it is possible to return to a previous state on the stack in a relatively efficient manner.

T-SOLVER consists mainly on two main layers: an *Equational Satisfiability Procedure* for $\mathcal{E}\mathcal{U}\mathcal{F}$ and a *Linear Arithmetic Procedure for $\mathcal{S}\mathcal{E}\mathcal{P}$, $\mathcal{L}\mathcal{A}(\mathbb{R})$ and $\mathcal{L}\mathcal{A}(\mathbb{Z})$*.

The Equational Satisfiability Procedure The first layer of T-SOLVER is provided by the equational solver, a satisfiability checker for $\mathcal{E}\mathcal{U}\mathcal{F}$ with minor extensions for arithmetic predicates and constants. The solver is based on the congruence closure algorithm presented in [17], and reuses some of the data structures of the theorem prover E [19] to store and process terms and atoms. It is incremental and supports efficient backtracking. The solver generates conflict sets and produces deductions for equational literals. It also implicitly knows that syntactically different numerical constants are semantically distinct, and efficiently detects and signals if a new equation forces the identification of distinct domain elements.

The Linear Arithmetic Procedure The second layer of T-SOLVER is given by a procedure for the satisfiability of sets of linear arithmetic constraints in the form $(\sum_i c_i v_i \bowtie c_j)$, with $\bowtie \in \{=, \neq, >, <, \geq, \leq\}$. The linear solver is *layered*, running faster, more general solvers first and using slower, more specialized solvers only if the early ones do not detect an inconsistency. The control flow through the linear solver is given in Fig. 1.

First, we consider only those constraints that are in the difference logic fragment, i.e., the subassignment of μ consisting of all constraints of the forms $v_i - v_j \bowtie c$ and $v_i \bowtie c$, with $\bowtie \in \{=, \neq, <, >, \leq, \geq\}$. Satisfiability checking for this subassignment is performed by an incremental version of the Bellman-Ford algorithm [9], which allows for deriving minimal conflict sets. Second, we try to determine if the current assignment μ is consistent over the reals, by means of the Cassowary constraint solver. Cassowary [5] is a simplex-based solver over the reals, using slack variables to efficiently allow the addition and removal of constraints and the generation of a minimal conflict

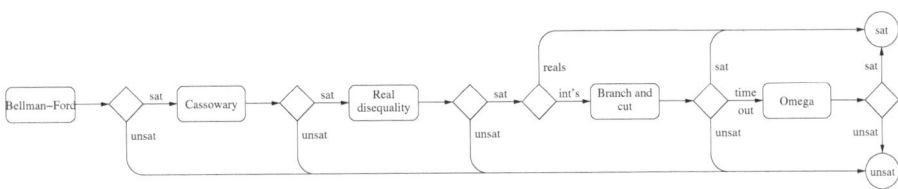

Fig. 1. The control flow of the linear solver

set. Cassowary has been extended by an ad-hoc technique to handle disequalities on $\mathbb{R}$ and with arbitrary precision arithmetic.

If the variables are interpreted over the integers, and the problem is unsatisfiable in the reals, then it is so in the integers. When the problem is satisfiable in the reals, a simple form of branch-and-cut is carried out, to search for solutions over the integers, using Cassowary's incremental and backtrackable machinery. If branch-and-cut does not find either an integer solution or a conflict within a small, predetermined amount of search, the Omega constraint solver [18] is called on the current assignment.

3   Tightly-Integrated SAT and Theory Solvers

In MATHSAT the naive DPLL+$\mathcal{T}$-SOLVER integration schema is enriched by the following optimization techniques [3,4,7,8]. Apart from theory-driven backjumping and learning, all these optimizations can be disabled/enabled by the user.

Early Pruning [3,4] Before every boolean decision step, $\mathcal{T}$-SOLVER is invoked on the current assignment μ. If this is found unsatisfiable, then there is no need to proceed, and the procedure backtracks.

Theory-Driven Backjumping [3,4,7,8] When $\mathcal{T}$-SOLVER finds the assignment μ to be $\mathcal{T}$-unsatisfiable, it also returns a conflict set η causing the unsatisfiability. This enables MINISAT to backjump in its search to the most recent branching point in which at least one literal $l \in \eta$ is not assigned a truth value, pruning the search space below.

Theory-Driven Learning [3,4,7,8] When $\mathcal{T}$-SOLVER returns a conflict set η, the clause $\neg\eta$ can be added in conjunction to φ: this will prevent MINISAT from generating again any branch containing η.

Theory-Driven Deduction (and Learning) [3,4,7,8] With early pruning, if a call to $\mathcal{T}$-SOLVER produces some deduction in the form $\mu' \models_{\mathcal{T}} l$ (e.g., by the $\mathcal{EUF}$ solver), then l is returned to the SAT solver, which uses it for boolean constraint propagation, triggering new boolean simplification. Moreover, the implication clause $\mu \rightarrow l$ can be learned and added to the main formula, pruning the remaining boolean search.

Static Learning [7,8] Before the main solver is invoked, short clauses valid in $\mathcal{T}$ like, e.g., $\neg(t = 1) \vee \neg(t = 2)$, $\neg(t_1 - t_2 \leq 3) \vee (t_2 - t_1 > -4)$, $(t_1 - t_2 \leq 3) \wedge (t_2 - t_3 < 5) \rightarrow (t_1 - t_3 < 9)$, are added off-line to the input formula ϕ if their atoms occur in ϕ. This helps pruning the search space at the boolean level.

Clause Discharge [7,8] MATHSAT inherits MINISAT's feature of periodically discarding some of the learned clauses to prevent explosion of the formula size. However, because the clauses generated by theory-driven learning and theory deduction mechanisms may have required a lot of work in $\mathcal{T}$-SOLVER, as a default option they are never discarded.

Control on Split Literals [3,4,7,8] In MATHSAT it is possible to initialize the weights of the literals in the VSIDS splitting heuristics so that either boolean or mathematical atoms are preferred as splitting choices early in the DPLL search.

4   An Optimized Theory Solver

In MATHSAT, $\mathcal{T}$-SOLVER benefits of the following optimization techniques [3,4,7,8]. All these optimizations can be disabled/enabled by the user.

Clustering [7,8] At the beginning of the search, the set of all atoms occurring in the formula is partitioned into disjoint *clusters* $L_1, ..., L_k$: intuitively, two atoms (literals) belong to the same cluster if they share a variable. Thus, instead of having a single, monolithic solver for linear arithmetic, k different solvers are constructed, each responsible for the handling of a single cluster. This allows for "dividing-and-conquering" the mathematical component of search, and for generating shorter conflict sets.

EQ-Layering [7,8] The equational solver can be used not only as a solver for $\mathcal{EUF}$, but also as a layer in the arithmetic reasoning process. In that case, *all* constraints, including those involving arithmetic operators, are passed to the equational solver. Arithmetic function symbols are treated as fully uninterpreted. However, the solver has a limited interpretation of the predicates $<$ and $\leq$, knowing only that $s < t$ implies $s \neq t$, and $s = t$ implies $s \leq t$ and $\neg(s < t)$. Thus, the equational interpretation is a (rough) approximation of the arithmetic interpretation, and all conflicts and deductions found by the equational solver under $\mathcal{EUF}$ semantics are valid under fully interpreted semantics. Hence, they can be used to prune the search. Thus, given the efficiency and the deduction capabilities of the equality solver, this process in many cases significantly improves the overall performances [8].

Filtering [3,4,7,8] MATHSAT simplifies the set of constraints passed to $\mathcal{T}$-SOLVER by "filtering" unnecessary literals. If an atom ψ which occurs only positively (resp., negatively) in the input formula ϕ is assigned to false (resp., true) in the current truth assignment μ, then it is dropped from μ without loosing correctness and completeness. If an atom ψ is assigned by unit propagation on clauses resulting from theory-driven learning, theory-driven deduction, or static learning, then it is dropped from the set of constraints μ to check because it is a consequence in $\mathcal{T}$ of other literals in μ.

Weakened Early Pruning [7,8] During early pruning calls, $\mathcal{T}$-SOLVER does not have to detect *all* inconsistencies; as long as calls to $\mathcal{T}$-SOLVER at the end of a search branch faithfully detect inconsistency, correctness is guaranteed. We exploit this by using a faster, but less powerful version of $\mathcal{T}$-SOLVER for early pruning calls. Specifically, as the theory of linear arithmetic on $\mathbb{Z}$ is much harder, in theory and in practice, than that on $\mathbb{R}$, during early pruning calls, $\mathcal{T}$-SOLVER looks for a solution on the reals only.

5   Delayed Theory Combination of $\mathcal{T}$-Solvers

In the standard Nelson-Oppen's (or Shostak's) approaches, the $\mathcal{T}$-solvers for two different theories T_1, T_2 interact with each other by deducing and exchanging (disjunctions of) *interface equalities* in the form $(v_i = v_j)$, v_i, v_j being variables labeling terms in the different theories; the SAT solver interacts with this integrated $T_1 \cup T_2$-solver according to the standard SMT paradigm.

In Delayed Theory Combination (DTC) [6], instead, the SAT solver is in charge of finding suitable truth value assignments not only to the atoms occurring in the formula, but also for their relative interface equalities. Each $\mathcal{T}$-solver works in isolation, without direct exchange of information, and interacts only with the SAT solver, which gives it as input not only the set of atomic constraints for its specific theory, but also the truth value assignment to the interface equalities. Under such conditions, two theory-specific models found by the two $\mathcal{T}$-solvers can be merged into a model for the input formula.

We notice the following facts [6]. DTC does not require the direct combination of $\mathcal{T}$-solvers for T_1 and T_2; the construction of conflict sets involving multiple theories is straightforward; with DTC the $\mathcal{T}$-solvers are not required to have deduction capabilities (though, the integration benefits from them); DTC extends naturally to the case of more than two theories; DTC does not suffer in the case of non-convex theories, like $\mathcal{LA}(\mathbb{Z})$.

References

1. W. Ackermann. *Solvable Cases of the Decision Problem*. North Holland Pub. Co., Amsterdam, 1954.
2. A. Armando, C. Castellini, E. Giunchiglia, and M. Maratea. A SAT-based Decision Procedure for the Boolean Combination of Difference Constraints, 2004. In SAT 2004 conference.
3. G. Audemard, P. Bertoli, A. Cimatti, A. Korniłowicz, and R. Sebastiani. A SAT Based Approach for Solving Formulas over Boolean and Linear Mathematical Propositions. In *CADE-18*, volume 2392 of *LNAI*, pages 195–210. Springer, 2002.
4. G. Audemard, P. Bertoli, A. Cimatti, A. Korniłowicz, and R. Sebastiani. Integrating boolean and mathematical solving: Foundations, basic algorithms and requirements. In *AISC 2002*, volume 2385 of *LNAI*, pages 231–245. Springer, 2002.
5. G. J. Badros, A. Borning, and P. J. Stuckey. The Cassowary Linear Arithmetic Constraint Solving Algorithm. *ACM Trans. on Computer-Human Interaction*, 8(4):267–306, 2001.
6. M. Bozzano, R. Bruttomesso, A. Cimatti, T. Junttila, S. Ranise, P. van Rossum, and R. Sebastiani. Efficient Satisfiability Modulo Theories via Delayed Theory Combination. In *CAV 2005*, LNCS. Springer, 2005.
7. M. Bozzano, R. Bruttomesso, A. Cimatti, T. Junttila, P. van Rossum, S. Schulz, and R. Sebastiani. An incremental and Layered Procedure for the Satisfiability of Linear Arithmetic Logic. In *TACAS 2005*, volume 3440 of *LNCS*, pages 317–333. Springer, 2005.
8. M. Bozzano, R. Bruttomesso, A. Cimatti, T. Junttila, P. van Rossum, S. Schulz, and R. Sebastiani. MathSAT: Tight Integration of SAT and Mathematical Decision Procedure. *Journal of Automated Reasoning*, to appear.
9. B. V. Cherkassky and A. V. Goldberg. Negative-Cycle Detection Algorithms. *Mathematical Programming*, 85(2):277–311, 1999.
10. N. Eén and N. Sörensson. An extensible SAT-solver. In *SAT 2003*, volume 2919 of *LNCS*, pages 502–518. Springer, 2004.
11. C. Flanagan, R. Joshi, X. Ou, and J.B. Saxe. Theorem Proving using Lazy Proof Explication. In *CAV 2003*, volume 2725 of *LNCS*, pages 355–367. Springer, 2003.
12. H. Ganzinger, G. Hagen, R. Nieuwenhuis, A. Oliveras, and C. Tinelli. DPLL(T): Fast decision procedures. In *CAV 2004*, volume 3114 of *LNCS*, pages 175–188. Springer, 2004.
13. C. P. Gomes, B. Selman, and H. A. Kautz. Boosting Combinatorial Search Through Randomization. In *AAAI/IAAI 1998*, pages 431–437. AAAI Press / The MIT Press, 1998.
14. J. Jaffar, S. Michaylov, P. J. Stuckey, and R. H. C. Yap. The CLP(R) Language and System. *ACM Trans. on Programming Languages and Systems*, 14(3):339–395, 1992.
15. J. P. Marques-Silva and K. A. Sakallah. GRASP: A Search Algorithm for Propositional Satisfiability. *IEEE Trans. on Computers*, 48(5):506–521, 1999.

16. M. W. Moskewicz, C. F. Madigan, Y. Zhao, L. Zhang, and S. Malik. Chaff: Engineering an Efficient SAT Solver. In *DAC 2001*, pages 530–535. ACM, 2001.
17. R. Nieuwenhuis and A. Oliveras. Congruence Closure with Integer Offsets. In *LPAR 2003*, volume 2850 of *LNAI*, pages 78–90. Springer, 2003.
18. Omega. `http://www.cs.umd.edu/projects/omega`.
19. S. Schulz. E – A Brainiac Theorem Prover. *AI Communications*, 15(2/3):111–126, 2002.

Deduction with XOR Constraints in Security API Modelling

Graham Steel

School of Informatics, University of Edinburgh,
Edinburgh, EH8 9LE, Scotland
graham.steel@ed.ac.uk
http://homepages.inf.ed.ac.uk/gsteel

Abstract. We introduce XOR constraints, and show how they enable a theorem prover to reason effectively about security critical subsystems which employ bitwise XOR. Our primary case study is the API of the IBM 4758 hardware security module. We also show how our technique can be applied to standard security protocols.

1 Introduction

The application program interfaces (APIs) of several secure cryptographic hardware modules, such as are used in cash machines and electronic payment devices, have been shown to have subtle flaws which could be used to mount financially lucrative attacks, [5,9]. These flaws were, however, only discovered after laborious hand analysis. This suggests them as promising candidates for formal analysis in a theorem prover. The APIs can be thought of as defining a set of two-party protocols, and modelled in a similar way to security protocols, which have attracted a large amount of attention from formal methods researchers in recent years. However, one key difference is that the specifications of the APIs are at the concrete bit by bit level, rather than at the abstract level of public or symmetric key cryptography. It has been by exploiting bit-level operations in the APIs that attacks have been found. One example of such an operation is bitwise exclusive-or (XOR). In the IBM 4758 Common Cryptographic Architecture API, XOR is used extensively. In an attack discovered by Bond, [5], the self-inverse property of XOR can be exploited, together with some other coincidences in the API transaction set, to reveal a customer's PIN. However, the combinatorial possibilities caused by the associative, commutative and self-inverse properties of XOR pose a significant challenge to formal analysis. It this challenge we address in the work described in this paper. We introduce XOR constraints, which state that two terms must be equal modulo XOR. We describe first-order theorem proving in a calculus modified to use these constraints, and show how this allows models of such APIs to be reasoned with effectively.

In the rest of this paper, we will first, in §2, give some more background in the field of secure hardware API analysis. In §3, we describe our approach using XOR constraints. Our results using this method are given in §4. We conclude in §5, with some evaluation and comparison to related work.

R. Nieuwenhuis (Ed.): CADE 2005, LNAI 3632, pp. 322–336, 2005.

2 Background

Hardware security modules (HSMs) are widely used in security critical systems such as electronic payment and automated teller machine (ATM) networks. They typically consist of a cryptoprocessor and a small amount of memory inside a tamper-proof enclosure. They are designed so that should an intruder open the casing, or insert probes to try to read the memory, it will auto-erase in a matter of nanoseconds. In a typical application, this tamper-proof memory will be used to store the master keys for the device. Further keys will then be stored on the hard drive of a computer connected to the HSM, encrypted under the HSM's master key. These keys can then only be used by sending them back into the HSM, together with some other data, under the application program interface (API). The API will typically consist of a several dozen key management and PIN processing commands, and their prescribed responses. Although carefully designed, various subtle flaws in these APIs have been found, for example by Bond, [5], and Clulow, [9].

One can think of the API as defining a number of two-party protocols between a user and the HSM. This suggests that we might profitably employ methods developed for the analysis of security protocols, e.g. [23,18,3]. However, there are some important differences in the scenario to be modelled. We are concerned only with two 'agents', the HSM itself and a malicious intruder. The HSM itself is (usually) 'stateless', in the sense that no information inside the HSM changes during transactions. This means we need only concern ourselves with the knowledge of the intruder, and not with any other information about the state of the protocol participants. Readers familiar with conventional security protocol analysis will appreciate that this constitutes a significant simplification of the general problem. However, there are other aspects which add complexity not normally considered in standard protocol analysis. In order to look for attacks, we must analyse the composition of several dozen short two-party protocols, that may be combined in any order. This is a significant increase in complexity compared to analysing key exchange protocols, which typically entail half a dozen messages at most. Additionally, it is not sufficient to model APIs in a 'perfect encryption' or free algebra model, where algebraic properties of the crypto functions are ignored. Most of the attacks that have been found exploit properties of these functions. The example we look at in this paper is the XOR function, which is widely used in the Common Cryptographic Architecture (CCA) API of the IBM 4758[1].

The attacks on the 4758 CCA were first discovered by Bond and Anderson, [5]. They exploit the way the 4758 constructs keys of different types. The CCA supports various key types, such as data keys, key encrypting keys, import keys and export keys. Each type has an associated 'control vector' (a public value), which is XORed against the master key, km, to produce the required key type. So if *data* is the control vector for data keys, then all data keys will be stored outside

---

[1] http://www-3.ibm.com/security/cryptocards/pcicc.shtml

the 4758 box encrypted under $km \oplus data^2$. Under the API, only particular types of keys will be accepted by the HSM for particular operations. For example, data keys may be used to encrypt given plaintext, but PIN derivation keys may not. Bond discovered an attack whereby the attacker changes the type of a key he has by exploiting the properties of XOR. In Bond's attack, the API's Key Part Import command is used to generate keys with a known difference. The Key Import command can then be used to change a key of one type to any other type. This allows a PIN derivation key to be converted to a data key, and then used to encrypt data, which is a critical flaw: a customer's PIN is just his account number encrypted under a PIN derivation key. So, the attack allows a criminal to generate a PIN for any card number. Recent versions of the CCA manual (2.41 onwards) give procedural restrictions on the key part import commands to prevent the attack.

After discovering the attack, Bond made an effort to formalise the problem and rediscover the attack, using the theorem prover SPASS, [26]. He produced a first-order model with one predicate, P, to indicate that a term was known outside the box (and therefore available to an intruder). The XOR operation was modelled by equational axioms giving its associative, commutative and self-inverse properties. However, SPASS was unable to rediscover the attack, even after days of run time, and with only the three relevant commands from the API included in the model. Bond was able to prove that the attack was in the search space modelled, by inserting various hints and proving intermediate conjectures, but the combinatorial blow-up caused by the AC and self-inverse properties, and the fact that the spy can XOR together any two terms he knows to create a new term, prevented a genuine rediscovery.

Ganapathy et al. have recently attacked the same rediscovery problem using bounded model checking, [12]. They successfully rediscovered the part of the attack where a key's type is changed. However, they included only two of the APIs commands in their model, so there was very little search involved. Again, associativity, commutativity and self-inverse were modelled by explicit axioms in the model.

The aim of the work described in this paper was to develop a framework for handling XOR that allows the whole of an API to be modelled and reasoned with effectively. Our approach involved adapting a theorem prover supporting AC unification (daTac, [25]) to handle constraints specifying equality between terms modulo XOR. Constraints have been used before in first-order theorem proving, for a variety of purposes. They provide a natural way of implementing ordering restrictions, [20], and the basicness restriction, [2], handling AC unification, [21,25], and incorporating the axioms of Abelian groups, [13]. As has been remarked e.g. in [22], they are an attractive solution for bringing domain knowledge into deduction systems, because they provide a clean separation between the general purpose logic, and the special purpose constraints. Our results using the XOR constraint framework include a rediscovery of Bond's attack in a model formalising the whole of the CCA key-management transaction set. Additionally,

---

[2] We use $\oplus$ as the infix XOR operator.

we rediscovered a second, stronger, and more complex attack which, although already discovered by Bond, had not been previously modelled. We have also used our XOR constraint technique to rediscover an attack on a variant of the Needham-Schroeder-Lowe protocol, using XOR, given in [7]. As far as we are aware, this attack had also not been rediscovered automatically before.

3   XOR Constraints

We will first explain how our work on API analysis suggested the development of XOR constraints. We then define deduction, redundancy checking and simplification with XOR constraints. Finally, we discuss some implementation details.

3.1   The Need for XOR Constraints

The attacks we are concerned with on the 4758 CCA all employ the Key Import command. This command is designed to be used, for example, to import a key from another 4758 unit. Keys are transported between 4758s under 'key encrypting keys', or *KEK*s. Here is the command as a two-message protocol:

Key Import:

1. User $\rightarrow$ HSM : $\{\!|$ KEY1 $|\!\}$ $_{\mathrm{KEK}\oplus\mathrm{TYPE}}$, TYPE, $\{\!|$ KEK $|\!\}$ $_{\mathrm{km}\oplus\mathrm{imp}}$
2. HSM $\rightarrow$ User : $\{\!|$ KEY1 $|\!\}$ $_{\mathrm{km}\oplus\mathrm{TYPE}}$

We write the master key km and the import control vector imp in lowercase to show these are values specific to a particular 4758, and other values in caps to show these are in effect variables; for these, any values (modulo a certain amount of error checking) can be used. In the above protocol, the HSM does not know in advance the key required to decrypt the first packet, $\{\!|$ KEY1 $|\!\}_{\mathrm{KEK}\oplus\mathrm{TYPE}}$. It must first decrypt the final packet to obtain KEK, and then XOR this against $TYPE$, which is given in the clear. The result can then be used as a key to decrypt the first packet. It is this XORing together of parts that is exploited in Bond's attacks, the first variant of which runs like this, [4, §7.3.4]:

> Suppose some corrupt bank insider has seen $\{\!|$ pp $|\!\}_{\mathrm{kek}\oplus\mathrm{pin}}$, a PIN derivation key ($pp$), encrypted under some key-encrypting key, kek, XORed against the appropriate control vector, pin. This is the form in which it would be sent to a 4758 in a bank. Additionally, we suppose that an attacker has access to, and can tamper with, a 'key part' for the key. Keys are divided into parts to allow two (or more) separate people to physically transfer key information to another 4758 module. The idea of dividing the key into parts is that attacks requiring collusion between multiple employees are considered infeasible. The key parts for the 4758 are combined using XOR. Because of the properties of XOR, no single officer, in possession of a single key part, has any information about the

value of the final key. Here is the sequence of commands required to import a two-part key $k1 = k1a \oplus k1b$:

Key Part Import(1):

1. Host $\rightarrow$ HSM : k1, TYPE
2. HSM $\rightarrow$ Host : $\{\!|$ k1a $|\!\}$ km⊕kp⊕TYPE

Key Part Import(2):

1. Host $\rightarrow$ HSM : $\{\!|$ k1a $|\!\}$ km⊕kp⊕TYPE, k1b, TYPE
2. HSM $\rightarrow$ Host : $\{\!|$ k1a $\oplus$ k1b$|\!\}$ km⊕TYPE

The kp control vector indicates that a key is still only partially complete. For Bond's attack, we suppose the attacker is a single corrupt insider responsible for adding the final key part. So, he has a partial key which looks like $\{\!|$ kek $\oplus$ k2 $|\!\}$ km⊕imp⊕kp, where $k2$ is his own key part. Before addng his own key part, he XORs it against *data* and *pin*. The final key part import command will then yield $\{\!|$ kek $\oplus$ data $\oplus$ pin$|\!\}$ km⊕imp, which he can use in the Key Import command in the following way:

1. User $\rightarrow$ HSM : $\{\!|$ pp $|\!\}$ kek⊕pin, data, $\{\!|$ kek $\oplus$ data $\oplus$ pin $|\!\}$ km⊕imp
2. HSM $\rightarrow$ User : $\{\!|$ pp $|\!\}$ km⊕data

On receiving message 1, the HSM forms $kek \oplus data \oplus pin \oplus data$, but with the self-inverse cancellation of bitwise XOR, this is the same as $kek \oplus pin$. So the encrypted key pp is output successfully, but instead of being under its correct control vector, as a PIN derivation key, it is now returned as a data key, allowing the attacker to generate customer's PINs using the Encrypt Data command[3]:

1. User $\rightarrow$ HSM : $\{\!|$ pp $|\!\}$ km⊕data, PAN
2. HSM $\rightarrow$ User : $\{\!|$ PAN $|\!\}$ pp

In order to capture these kinds of attacks, we must model the command in the way it is implemented, i.e. we must model the HSM constructing the decryption key for packet 1, $KEK \oplus TYPE$, from the other two packets. This seemingly requires us to abandon the so-called 'implicit decryption' assumption usually used in modelling protocols, [19]. Instead, we must model the decryption of packet 1 explicitly, like this[4]:

$$P(crypt(xkek1 \oplus xtype1, xk1)) \wedge$$
$$P(xtype2) \wedge P(crypt(km \oplus imp, xkek2))$$
$$\rightarrow P(crypt(km \oplus xtype2,$$
$$\quad decrypt(xkek2 \oplus xtype2,$$
$$\quad\quad crypt(xkek1 \oplus xtype1, xk1))))$$

and introduce the cancellation rule:

---

[3] PAN stands for 'personal account number'. Recall that a PIN is just a customer's account number encrypted under a PIN derivation key.

[4] We follow Bond's original model, with the P predicate signifying a term known outside the HSM.

$$\to decrypt(x1, crypt(x1, x2)) = x2$$

This brings the attacks into the search space, but only allows them to be found by forward search through the model. This is because the result of the key import command will not unify with the result required in the attack (which is $P(crypt(km * data, pp))$) until after the equations specifying the self-cancelling properties of XOR, and the encrypt/decrypt cancellation rule above, have been applied. To allow this to happen in backwards search, we would have to allow the cancellation rules to be applied in reverse, which would cause an enormous blow-up in the search space. When searching forwards, the branching rate is still very large, since for any terms x and y, if we have $P(x)$ and $P(y)$, we can always combine these with XOR to produce $P(x \oplus y)$. We have to search a large space of these possible combinations in order to derive the terms required to make attacks.

Our solution is to introduce XOR constraints to state that in the key import command, the key used to encrypt the first packet, and the key derived by the HSM from the other packets, must be equal modulo XOR. This allows us to use implicit encryption. The clause to model the command now looks like this:

$$P(crypt(x1, xk1)), P(xtype), P(crypt(km \oplus imp, xkek2))$$
$$\to P(crypt(km \oplus xtype, xk1))$$
$$\text{IF } xkek2 \oplus xtype =_{XOR} x1$$

Semantically, the constraints are a restriction on the universally quantified variables in the clauses. Rather than generating instantiations to solve the constraints directly, we allow the search to proceed normally, but check after each deduction that the constraints remain soluble. Note that the output of the new form of the command will now unify with terms in a backwards search from the goal. To encourage this, we employ the 'basic strategy', [2], which seems highly effective for these problems.

3.2 Deduction with XOR Constraints

We employ the constrained resolution/paramodulation calculus of [25], with the addition of XOR constraints. The XOR constraints of two resolving clauses are combined in the resolvent using logical AND, just like the pre-existing ordering constraints and substitution constraints. When generating new inferences we apply the substitution required for the inference to the constraints before checking for solubility. Because of the self-cancelling property of XOR, this amounts to checking whether the constraint is ground and has been solved, or whether there are still variables at positions which could solve the constraint. More formally, solubility of an XOR constraint $s_1 \oplus \ldots \oplus s_m =_{XOR} t_1 \oplus \ldots \oplus t_n$ is checked like this:

1. If any of $s_1, \ldots, s_m, t_1, \ldots, t_n$ contain variables, the constraint is regarded as soluble.
2. If all $s_1, \ldots, s_m, t_1, \ldots, t_n$ are ground, then first discard zeros, and then count up the number of occurrences of each term in the set $\{s_1, \ldots, s_m, t_1, \ldots, t_n\}$. If all terms occur an even number of times, the constraint is soluble. If not, it is insoluble.

Only inferences producing clauses with soluble constraints are permitted. These simple syntactic checks can be made very quickly. Note that condition 1 may allow us to keep some clauses with constraints which cannot be solved. We do not attempt to check that, in the theory in question, the variable may eventually be instantiated to a value which satisfies the constraint. However, we will certainly only discard genuinely insoluble constraints, thereby preserving completeness. We can eliminate some further insoluble sets of XOR constraints by a simple pairwise check for inconsistency. If two constraints, both attached to the same clause, equate the same variables to different ground terms, then they are inconsistent and the clause can be pruned away. Again, this check may allow some insoluble constraints through, but it is quick to perform, preserves completeness and seems useful in practice.

3.3   Subsumption with XOR Constraints

Simplification and checking for redundancy are of paramount importance in practical theorem proving. Modifications to the resolution/paramodulation calculus which prevent the use of subsumption checking rules are usually useless in practice, whatever their apparent theoretical advantages. The use of XOR constraints allows subsumption checking, though with the following restriction: the solutions of the constraints in a clause that is subsumed must be a subset of the solutions of the clause we retain. For XOR constraints, this occurs just when:

1. The more general clause has no XOR constraint, or
2. The XOR constraints of the subsumed clause and the subsuming clause are identical (modulo AC), *after* any substitution required to make the subsumption has been applied to the XOR constraint.

This is similar to the standard rule for subsumption in the constrained calculus, [25, §5.1]. To see why condition 2 is not only sufficient but necessary to ensure that all solutions of some constraint, T_1, are solutions of another, T_2, the reader is invited to write down two identical constraints, and then add a single variable or ground term to either. Immediately it is possible to construct solutions of the first which are not solutions of the second, and vice versa.

In practice, many checks for subsumption are ruled out because of incompatible XOR constraints. Fortunately, the check can be made quite quickly (see §3.5, below).

3.4   Simplification with XOR Constraints

Reduction of newly produced clauses by demodulation, clausal simplification, etc. is also permissible for XOR-constrained clauses. In our implementation, we only allow demodulation by clauses with no XOR constraint, since in our experiments, it seems to be only these clauses we would like to use.

3.5   Implementation

Our prototype implementation is in daTac, [25], which uses the constrained basic calculus we require. It also supports AC-unification, implemented via constraints. We store XOR constraints in a normal form, $v_1 \oplus \ldots \oplus v_m =_{XOR} t_1 \oplus \ldots \oplus t_n$, where the v_i are pure variables, and the t_j are atoms or complex terms. After an inference, we apply the required substitution to the XOR constraint, and then re-normalise it, by first moving any non-variables from the v_i to the right hand side of the equality, and then by removing from the t_j any occurrences of the identity element for XOR, and any pairs of identical ground terms. We also order the v_i and t_j according to some arbitrary total ordering. After the simplification, our checks for solubility described in §3.2 can be readily performed. Additionally, keeping the constraints in a normal form speeds up the check for identical constraints required in §3.3.

4   Results

We present first our results on the command set of the 4758 API. Then, we present results from experiments using XOR constraints in the modelling of a standard key-exchange protocol, which has an attack that can only be found when the properties of XOR are taken into account. All the model files and results are available via http at http://homepages.inf.ed.ac.uk/gsteel/xor/

4.1   The 4758 CCA Attacks

Our model for the first variant of Bond's attack includes all the symmetric key management commands in the 4758 CCA API, except those which do not output keys (MAC Generate, MAC Verify and Key Test). We employ XOR constraints to model any command in which the HSM must decrypt a packet using a key formed by XOR. In the 4758 CCA, the Key Import command and the Key Translate command require XOR constraints. The XOR operation is modelled by an operator (*) which is declared to be AC. We add the following two equations to the model to define an identity element for the XOR operation:

$$x1 * id = x1$$
$$x1 * x1 = id$$

These properties of the XOR function are also implicitly modelled by the way we check the solubility of XOR constraints, as described in §3.2. However, the XOR constraints will only account for these properties when the XOR cancellation happens during a command being executed by the HSM. We have to add these equations to allow for the possibility of the attacker doing his own manipulations outside the HSM.

The intruder in our model is given the initial knowledge specified by Bond (see §3.1), along with some other public terms such as the values of all the control vectors. We chose a precedence ordering which set the P predicate largest, the

XOR operator smallest, and put the key and control vector identifiers in a set of symbols of equal precedence in between. The conjecture is that a term of the form $\{\!|\ PAN |\!\}_{PDK}$, i.e. a PIN, may become public. daTac finds the attack in just under 1 second[5], generating 162 new clauses for a model with 21 initial clauses.

We also modelled the second, stronger variant of Bond's attack, the key import/export attack, described in [4, §7.3.5]. This attack does not require the intruder to have seen a PIN derivation key being sent to the 4758 under attack, nor does it require him to have access to a key part for that PDK. Instead, he converts a PDK already imported for use on the box into a data key by first creating a pair of import/export keys with a known difference. The cleartext value of these keys is unknown – they are generated by a process known as 'key conjuring', [4, §7.2.3], where random values are tried until one is accepted by the HSM. We approximate this by explicitly adding the conjured parts to the model. The idea is to export the PDK under the exporter key, then to re-import it under the importer key to turn it into a data key, using the same XOR trick as in the previous attack. In fact, in this attack, the 'trick' must be used twice: once to create the related import export pair from conjured key parts, and once to use it to change the type of the PDK. In our model for this experiment, in addition to the conjured key parts, we include in the intruder's initial knowledge a PIN derivation key encrypted under $km \oplus pin$ (instead of $kek \oplus pin$, as before). The attack is found after 1.47 seconds, and the generation of 601 clauses in a model with 23 initial clauses.

As a further experiment, we modelled the attack exhibited by IBM engineers in their response to Bond's discovery, described in [9, p. 54]. This attack requires that we take into account the fact that the true value of the 'data' control vector is 0. This we model with the additional rule:

$$P(crypt(xk * data, x)) \rightarrow P(crypt(xk, x))$$

1 The attack is quite simple, and is found in 0.2 seconds after generating 229 clauses from an initial theory of 21 clauses. Together, these results represent a qualitative improvement over the previous efforts to reason with XOR described in §2. We have found stronger, and more complex attacks on the whole command set, while Bond's original model, and the Ganapathy et al. model, both required hints and/or simplifications to discover even part of the original, shorter attack. We believe it is the XOR constraints that make the difference, and to investigate this, we tried a version of the model without them, modelling the key import command with explicit encryption and decryption, as described in §3.1. If we model only the commands required for the attack, daTac finds the attack after about 5 minutes, generating more than 10 000 clauses. If the whole command set is used, the attack is not found after more than 24 hours run-time. This supports the hypothesis that it is the XOR constraints that are making the difference.

---

[5] Running under Linux 2.4.20 on a 2GHz Pentium 4 machine.

4.2   Attacking a Variant of the Needham–Schroeder–Lowe Protocol

In [7], Chevalier et al. exhibit a variant of the well known Needham–Schroeder–Lowe protocol, [17], which uses XOR to bind Bob's identifier to Alice's nonce in message 2. They show that this protocol can be attacked, but only when the AC and self-inverse properties of XOR are taken into account. We were keen to see if our XOR constraints could be more generally applied to security problems, so we formalised this protocol. Our model for the protocol is based on the first-order model devised by Jacquemard, Rusinowitch and Vigneron, [14]. They also used daTac, since the model uses AC unification to deal with the set of intruder knowledge, and to model the set of states of agents. The protocol in question runs like this:

1. $A \rightarrow B : \{\!| N_A, A |\!\}_{pubK_B}$
2. $B \rightarrow A : \{\!| N_A \oplus B, N_B |\!\}_{pubK_A}$
3. $A \rightarrow B : \{\!| N_B |\!\}_{pubK_B}$

To attack the protocol, we assume Alice starts a session with our intruder, I, who is accepted as an honest agent. He forwards on the nonce and identifier to the victim, Bob, *after* XORing the nonce against B and I. He can now carry out Lowe's classic man-in-the middle attack:

1. $A \rightarrow I \ : \{\!| N_A, A |\!\}_{pubK_I}$
1'. $I_A \rightarrow B \ : \{\!| N_A \oplus B \oplus I, A |\!\}_{pubK_B}$
2'. $B \rightarrow I_A : \{\!| N_A \oplus B \oplus I \oplus B, N_B |\!\}_{pubK_A}$
2. $I \rightarrow A \ : \{\!| N_A \oplus B \oplus I \oplus B, N_B |\!\}_{pubK_A}$
3. $A \rightarrow I \ : \{\!| N_B |\!\}_{pubK_I}$
3.' $I_A \rightarrow B \ : \{\!| N_B |\!\}_{pubK_B}$

In our model, we use an XOR constraint to model the fact that A will accept message 2 from B only if, after XORing it against the identifier for B, it is equal to the nonce she sent to B in message 1. In the Jacquemard et. al model, agents move into a state of waiting for an appropriate response at the same time as they send a message. So in this protocol, a single clause models A sending a message 1 containing nonce N_A, and moving into a state of waiting for a response containing $N_A \oplus B$. We simply change the clause so that A is waiting for a response containing some fresh variable X, and add the constraint $X \oplus B =_{XOR} N_A$. We also allow the intruder to XOR arbitrary terms he knows against each other to generate new ones, and allow him to send these terms in messages.

These modifications bring the attack into the search space, but we are faced with the usual XOR problem: there are simply too many ways the spy can combine terms together. However, we can fix this problem using the same trick Jacquemard et al. used for modelling intruder knowledge: use AC unification. In all protocols, it will only pay the intruder to consider the self-inverse properties of XOR when he is trying to fool other agents. This we model with XOR constraints, which take into account self-inverse. Now XOR is just an

AC combinator, equivalent to the set combinator used in the model to collect together terms. When faking a message, the spy simply picks a term from his knowledge by unification, and we allow the XOR constraints mechanism to check whether this term is satisfactory when the combinator is considered to be XOR.

Using this model, daTac finds the attack presented in [7] in 12.35 seconds, deriving 1652 clauses. This is in the middle of the range of times reported in [8], where daTac was used to find flaws in a number of protocols in a free algebra model. One should bear in mind that the first-order model had by then evolved a little from the one reported in [14], and hardware used was probably slower (details are not given in the paper). Even so, we have succeeded in automatically discovering a protocol attack in a model using XOR. We compare this to related work in the next section.

5   Conclusions

In the experiments we have performed, the use of XOR constraints has greatly improved the performance of the theorem prover on the problems examined. In the analysis of the 4758 CCA API, the greatest benefit seems to come from the fact that XOR constraints allow us to use a goal-directed search, via the basic strategy, in a model where explicit encryption and decryption would otherwise prevent this. Experiments with using a positive strategy on the same models yielded no proofs after 24 hours run-time. For the protocol considered in §4.2, the benefit is that by considering the self-inverse properties of XOR only in the constraints, we can treat XOR as a standard AC combinator, and so use AC unification to build the terms we need to satisfy the constraints. Again, running daTac on a model where XOR terms must be built up explicitly yields no proofs after hours of run-time.

In the early nineties, Longley and Rigby were the first to look at the problem of automatically analysing the key-management schemes presented by HSM APIs, [16]. They searched for attacks by querying a PROLOG database of API rules with terms that an attacker might try to obtain. They did not consider any algebraic properties of the cryptofunctions used, however. In §2, we mentioned more recent related work by Ganapathy et al., which did consider the properties of XOR. In §4.1, we showed how our results have improved on theirs, by allowing attack discovery even when the whole symmetric-key management command set is modelled, and by discovering the more complex key import/export attack, which had not been modelled before.

In the field of security protocol analysis, several researchers have recently started to consider algebraic properties of crypto functions, such as XOR. The decidability of insecurity for protocols using XOR, provided the number of sessions is bounded, has been shown in [7] and [10]. Their decision procedures for insecurity are, however, purely theoretical. We can relate decidability of insecurity for APIs, specified in the way we have shown in this paper, to these decidability results: first, observe that if we bind the number of times each API command can be executed, there are a bound number of ways these commands can be chosen and ordered. Consider each command as a two message protocol, as we have suggested in this paper. Consider each choice and ordering of the commands as a complete protocol. Then a bound API specifies a set of bound protocols. We can now augment an NP decision procedure for XOR protocols by first guessing a protocol from our set. Unfortunately, this is not quite enough to settle the question of decidability, since we also have to change the formalism for protocols to allow for steps like the Key Import command, where a key used for decryption must be generated using XOR in the same step. Delaune and Jacquemard have proposed a decidability procedure for protocols with explicit encryption and decryption, which we could adapt to such API commands, [11]. Unfortunately, they have so far been unable to show whether the XOR operation can be covered by their results. We suspect that one of these works could indeed be adapted to show decidability of insecurity for bound API transaction sets. However, for the moment, it remains an open question.

In further work, we plan to improve the performance of the prover when handling XOR constraints. We have observed that in our proof searches, often very early a clause is derived containing just $P(x_1) \wedge \ldots \wedge P(x_m) \Rightarrow$, and an XOR constraint $x_1 \oplus \ldots \oplus x_m =_{XOR} t_1 \oplus \ldots \oplus t_n$, with all the terms t_i needed to solve the constraint available in the domain of P. However, it still takes a long time to finally resolve this to the empty clause, since there are still many ways the XOR combination rule, $P(x) \wedge P(y) \Rightarrow P(x \oplus y)$, can be applied to the clause. We discovered this by using *viz*, our tool for visualising first-order proof search, [24]. The output from *viz* for the first 4758 attack is given in Figure 1. Note that *viz* produces output in scalable vector graphics (SVG) format, which is designed to be viewed interactively in a viewer supporting zooming and panning. However, for small proofs, even a straight printout of the search space can be revealing, as in this case. The nodes on the graph are clauses, and the edges mark logical dependency. Nodes on the 'critical path', i.e. ones used in the proof, are diamond-shaped. The nodes are coloured to indicate when they were generated in the search. Darker nodes were generated first, and lighter nodes generated later. The circle we have added to the diagram surrounds Clause 50, a clause of the form we have just described, which appears like this in the daTac output:

```
Clause 50: P(x1), P(x2) => \
       IF   x1 * x2 =x data * pin * k2
```

Clause 50 is eventually solved to give the proof, but we can see from Figure 1 that most of the work in the proof search is just to find the terms requires to solve Clause 50's XOR constraint (this is what is happening in the large collection of nodes in the bottom portion of the diagram). Finding the combination of API commands required to effect the attack takes a relatively small amount of search (in the top of the diagram). To address this problem, we could perhaps employ our AC unification trick again somehow, or use a specialised 'terminator' tactic, [1], to try to finish off these clauses.

There are many other electronic payment and banking APIs waiting to be formalised. They pose more challenges, such as reasoning about the relative

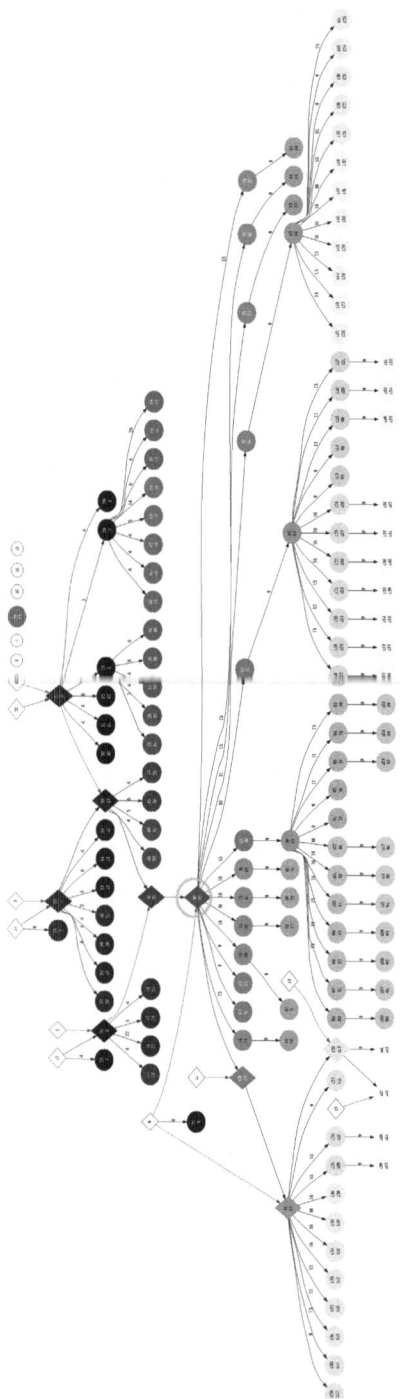

Fig. 1. The search for the first 4758 attack

complexity of breaking different crypto functions. We will be continuing our research in this area.

Acknowledgements

The API of the 4758 CCA is specified in a 500 page manual. Mike Bond's unpublished 4 page 'summary sheet' for the key management transaction set was therefore invaluable, and we are grateful to him for providing us with a copy. We are also grateful to Laurent Vigneron for providing the source code for daTac, and to the anonymous referees for their considered criticisms and suggestions for improvements.

References

1. G. Antoniou and H. J. Ohlbach. TERMINATOR. In *International Joint Conference on Artificial Intelligence*, pages 916–919, 1983.
2. L. Bachmair, H. Ganzinger, C. Lynch, and W. Snyder. Basic paramodulation and superposition. In D. Kapur, editor, *11th Conference on Automated Deduction*, number 607 in LNCS, pages 462–476, 1992.
3. D. Basin, S. Mödersheim, and L. Viganò. An on-the-fly model-checker for security protocol analysis. In *Proceedings of the 2003 European Symposium on Research in Computer Security*, pages 253–270, 2003. Extended version available as Technical Report 404, ETH Zurich.
4. M. Bond. *Understanding Security APIs*. PhD thesis, University of Cambridge, 2004.
5. M. Bond and R. Anderson. API level attacks on embedded systems. *IEEE Computer Magazine*, pages 67–75, October 2001.
6. A. Bundy, editor. *Automated Deduction - CADE-12, 12th International Conference on Automated Deduction, Nancy, France, June 26 - July 1, 1994, Proceedings*, volume 814 of *Lecture Notes in Computer Science*. Springer, 1994.
7. Y. Chevalier, R. Küsters, M. Rusinowitch, and M. Turuani. An NP decision procedure for protocol insecurity with XOR. In Kolaitis [15], pages 261–270.
8. Y. Chevalier and L. Vigneron. Automated unbounded verification of security protocols. In E. Brinksma and K. Larsen, editors, *Computer Aided Verification, 14th International Conference*, volume 2404 of *Lecture Notes in Computer Science*, pages 324–337, Copenhagen, Denmark, July 2002. Springer.
9. J. Clulow. The design and analysis of cryptographic APIs for security devices. Master's thesis, University of Natal, Durban, 2003.
10. H. Comon-Lundh and V. Shmatikov. Intruder deductions, constraint solving and insecurity decision in presence of exclusive or. In Kolaitis [15], pages 271–281.
11. S. Delaune and F. Jacquemard. A decision procedure for the verification of security protocols with explicit destructors. In *CCS '04: Proceedings of the 11th ACM conference on Computer and communications security*, pages 278–287. ACM Press, 2004.
12. V. Ganapathy, S. A. Seshia, S. Jha, T. W. Reps, and R. E. Bryant. Automatic discovery of API-level exploits. In *ICSE '05: Proceedings of the 27th International Conference on Software Engineering*, pages 312–321, New York, NY, USA, May 2005. ACM Press.

13. G. Godoy and R. Nieuwenhuis. Superposition with completely built-in abelian groups. *J. Symb. Comput.*, 37(1):1–33, 2004.

14. F. Jacquemard, M. Rusinowitch, and L. Vigneron. Compiling and verifying security protocols. In Michel Parigot and Andrei Voronkov, editors, *LPAR*, volume 1955 of *Lecture Notes in Computer Science*, pages 131–160. Springer, 2000.

15. P. G. Kolaitis, editor. *18th IEEE Symposium on Logic in Computer Science (LICS 2003), 22-25 June 2003, Ottawa, Canada, Proceedings.* IEEE Computer Society, 2003.

16. D. Longley and S. Rigby. An automatic search for security flaws in key management schemes. *Computers and Security*, 11(1):75–89, March 1992.

17. G. Lowe. An attack on the Needham-Schroeder public-key authentication protocol. *Information Processing Letters*, 56(3):131–133, 1995.

18. G. Lowe. Breaking and fixing the Needham Schroeder public-key protocol using FDR. In *Proceedings of TACAS*, volume 1055, pages 147–166. Springer Verlag, 1996.

19. J. K. Millen. On the freedom of decryption. *Inf. Process. Lett.*, 86(6):329–333, 2003.

20. R. Nieuwenhuis and A. Rubio. Theorem proving with ordering constrained clauses. In D. Kapur, editor, *11th Conference on Automated Deduction*, number 607 in LNCS, pages 477–491, 1992.

21. R. Nieuwenhuis and A. Rubio. AC-superposition with constraints: No AC-unifiers needed. In Bundy [6], pages 545–559.

22. R. Nieuwenhuis and A. Rubio. Paramodulation-based theorem proving. In John Alan Robinson and Andrei Voronkov, editors, *Handbook of Automated Reasoning*, pages 371–443. Elsevier and MIT Press, 2001.

23. L.C. Paulson. The Inductive Approach to Verifying Cryptographic Protocols. *Journal of Computer Security*, 6:85–128, 1998.

24. G. Steel. Visualising first-order proof search. In *Workshop on User Interfaces for Theorem Provers (UITP '05)*, pages 179–189, Edinburgh, Scotland, April 2005.

25. L. Vigneron. Associative-commutative deduction with constraints. In Bundy [6], pages 530–544.

26. C. Weidenbach et al. System description: SPASS version 1.0.0. In H. Ganzinger, editor, *Automated Deduction - CADE-16, 16th International Conference on Automated Deduction*, LNAI 1632, pages 378–382, Trento, Italy, July 1999. Springer-Verlag.

On the Complexity of Equational Horn Clauses

Kumar Neeraj Verma[1], Helmut Seidl[1], and Thomas Schwentick[2]

[1] Institut für Informatik, Technische Universität München, Germany
{verma, seidl}@in.tum.de
[2] Fachbereich Mathematik und Informatik, Philipps-Universität Marburg, Germany
tick@informatik.uni-marburg.de

Abstract. Security protocols employing cryptographic primitives with algebraic properties are conveniently modeled using Horn clauses modulo equational theories. We consider clauses corresponding to the class $\mathcal{H}3$ of Nielson, Nielson and Seidl. We show that modulo the theory ACU of an associative-commutative symbol with unit, as well as its variants like the theory XOR and the theory AG of Abelian groups, unsatisfiability is NP-complete. Also membership and intersection-non-emptiness problems for the closely related class of one-way as well as two-way tree automata modulo these equational theories are NP-complete. A key technical tool is a linear time construction of an existential Presburger formula corresponding to the Parikh image of a context-free language. Our algorithms require deterministic polynomial time using an oracle for existential Presburger formulas, suggesting efficient implementations are possible.

1 Introduction

In [1], Blanchet proposes to use first-order Horn clauses for verifying secrecy of cryptographic protocols. Among others, this approach has later-on also been advocated by Goubault-Larrecq and Parrennes [10], Comon-Lundh and Cortier [5] and Seidl and Verma [19] who consider rich decidable fragments of clauses which still allow us to model many useful protocols. While traditional methods for verifying cryptographic protocols have been based on the *perfect cryptography* assumption, a more accurate analysis of these protocols requires us to take into account algebraic properties of cryptographic primitives, modeled using equational theories. For example modeling of protocols based on modular exponentiation must account for properties like associativity and commutativity [11]. In general what we require most often in protocols are the associative and commutative theories ACU, XOR and AG (i.e., the theory of Abelian groups) [6]. While the case of protocols with bounded numbers of sessions has already received considerable attention [2], there exist very few decidability results in the case of unbounded number of sessions, in the presence of equational theories. Horn clauses modulo equational theories provide a suitable framework for modeling such protocols. A decidable class of clauses with the theory XOR is studied in [5] where a non-elementary upper bound is proposed. In [11,20], this problem has been attacked by forms of Horn clauses corresponding to two-way automata (see e.g. [4], Chapter 7), in the presence of several variants of the theory of associativity and commutativity. In this framework, automata-theoretic problems like membership and intersection-non-emptiness correspond to the unsatisfiability problem for clauses.

R. Nieuwenhuis (Ed.): CADE 2005, LNAI 3632, pp. 337–352, 2005.

Dealing with first order clauses in the presence of equational theories, in particular associative-commutative theories, is also of more general interest, and has received considerable interest in the past [16]. While most work has focused on obtaining sound and complete inference systems for general forms of clauses, very little work has been done on obtaining decidable fragments of clauses in the presence of such theories.

In this paper, we start from the class $\mathcal{H}3$ of Horn clauses which has been proposed in [13] for control-flow analysis and also is used by Goubault-Larrecq for cryptographic analysis of C programs [10]. This class is closely related to two-way automata [20,4] and has a polynomial unsatisfiability problem. We extend this class by operators satisfying associative-commutative theories. We show that unsatisfiability then becomes NP-complete for the theories XOR and AG. For the theory ACU of an associative-commutative symbol with unit, the same holds true under suitable restrictions.

Independently of the application to cryptographic protocols, related notions of tree automata have also been studied by others [14,3], notably for applications to XML document processing [17,18,12]. The languages accepted by unordered Presburger tree automata [17], for example, are essentially those accepted by our one-way ACU automata. While very general classes of these automata have been shown to be decidable, their complexity remains mostly unknown. A common idea underlying all these classes is their connection to *Parikh images* of context free languages [15], i.e. *semilinear* or Presburger-definable sets [9] which are closed under Boolean operations. For example in [20], to decide intersection-non-emptiness of two ACU automata, the product automaton is computed by first computing semilinear sets corresponding to the automata and then computing intersection of the semilinear sets. Both steps are expensive, and such ideas are unlikely to give us optimal algorithms.

In this paper we show how to obtain optimal algorithms for $\mathcal{H}3$ clause sets, or two-way tree automata, modulo associative-commutative theories, without computing product automata. A key technique we rely on is a linear time construction of an existential Presburger formula corresponding to the Parikh image of a context-free language, allowing us to show that membership and intersection-non-emptiness for one-way as well as two-way tree automata modulo the theories ACU, XOR and AG are NP-complete. NP-completeness of the membership problem for one-way ACU automata is also shown in [14], although the complexity of the intersection-non-emptiness problem is left open there. We resolve both questions for each of the theories ACU, XOR and AG, besides others like the theory XOR_p which contains the axiom $\sum_{i=1}^{p} x = 0$ besides the axioms of ACU (XOR is the special case $p = 2$). We further extend these complexity results to two-way automata as well. As a consequence the non-emptiness problem, which requires linear time in the one-way case, is also NP-complete in the two-way case. Further we obtain NP-completeness of the unsatisfiability problem for $\mathcal{H}3$ modulo ACU, XOR, AG and XOR_p. Note that the technique of [14] is not useful here since it is specific to the membership problem and to one-way automata.

Outline. We start in Section 2 by introducing our classes of automata and clauses, and demonstrate how they model cryptographic protocols. To deal with these classes, we give in Section 3 a linear time construction of existential Presburger formulas corresponding to Parikh images of context free languages. This is used to deal with one-way ACU automata in Section 4. The one-way XOR and AG cases are similarly dealt with in Section 5. These results are used to deal with two-way automata and $\mathcal{H}3$ in Section 6. The readers are also referred to [20] which uses similar techniques to show decidability of most of the problems which we show here to be NP-complete.

2   Clauses, Automata and Cryptographic Protocols

Fix a signature Σ of function symbols. Since we deal with variants of the ACU theory, we assume that Σ contains at least the symbols $+$ and 0, and additionally the symbol $-$ when dealing with the theories ACUD or AG, defined below. Symbols in $\Sigma_f = \Sigma \setminus \{+, -, 0\}$, are *free*. Free symbols of zero arity are *constants*. Terms of the form $f(t_1, ..., t_n)$ where f is free are *functional terms*. The equational theory ACU consists of the equations $x + (y + z) = (x + y) + z, x + y = y + x$ and $x + 0 = x$. The other theories we deal with are obtained by adding equations to this theory. The theory XOR is obtained by adding the equation $x + x = 0$. More generally, the theory XOR_p for $p \geq 2$ is obtained by the equation $\sum_{i=1}^{p} x = 0$. The theory AG is obtained by the equation $x + (-x) = 0$. The theory ACUD, obtained by the equations $-(x + y) = (-x) + (-y)$ and $-(-x) = x$, is weaker than AG and is introduced as a tool to deal with the theory AG which is of interest to us. The equation $x + x = x$ gives the theory ACUI of idempotent commutative monoids. Throughout this paper, if $s =_{\text{ACU}} t$ or $s =_{\text{ACUD}} t$ then s and t are treated as the same object.

A *clause* is a finite set of *literals* A (a *positive* literal) or $-A$ (a *negative* literal), where A is an atom $P(t_1, \ldots, t_n)$. A *Horn clause* contains at most one positive literal. The clause $A \vee -A_1 \vee \ldots \vee -A_n$ is written as $A \Leftarrow A_1 \wedge \ldots \wedge A_n$ and called a *definite* clause. The clause $-A_1 \vee \ldots \vee -A_n$ is written as $\bot \Leftarrow A_1 \wedge \ldots \wedge A_n$ and called a *goal* clause. A is *head* of the first clause, while $-A_1 \vee \ldots \vee -A_n$ is the *tail* of both clauses. Satisfiability of clauses modulo equational theories is defined as usual. To every clause C we can associate a variable dependence graph G_C whose nodes are the literals of C, and two literals are adjacent if they share a variable. A clause C is called H3 if

(1) every literal of C is linear, i.e. no variable occurs twice in the literal
(2) G_C is acyclic, and adjacent literals in G_C share at most one variable.
(3) the symbol $+$ does not occur in non-ground negative literals, except in case of theories XOR and AG.

The class $\mathcal{H}3$ consists of finite sets of H3 clauses. The first two conditions above are as in [13] in the non-equational case. In the equational case, we now also impose the third condition. Without this restriction, the unsatisfiability problem in the ACU case would subsume [7,22] the provability problem in MELL (Multiplicative Exponential Linear Logic) which itself subsumes the reachability problem in VASS (Vector Addition Systems with States). The latter is decidable and EXPSPACE-hard, while decidability of the former is still open. Examples of H3 clauses are one-way and two-way equational tree automata clauses defined below.

An *(equational) tree automaton* $\mathcal{A}$ is a finite set of definite clauses involving only unary predicates. We read an atom $P(t)$ as "term t is accepted at state P". We write $\mathcal{A}/\mathcal{E}$ to emphasize the equational theory $\mathcal{E}$ modulo which the automaton is considered. *Derivations* of ground atoms in the automaton are defined using the following two rules:

$$\frac{P_1(t_1\sigma)\dots P_n(t_n\sigma)}{P(t\sigma)} \; (P(t) \Leftarrow P_1(t_1) \wedge \dots \wedge P_n(t_n) \in \mathcal{A}) \qquad \frac{P(s)}{P(t)} \; (s =_{\mathcal{E}} t)$$

where substitution σ maps all variables to ground terms, and $=_{\mathcal{E}}$ is the congruence on terms induced by $\mathcal{E}$. Hence the derivable atoms are exactly the elements of the least Herbrand model modulo $\mathcal{E}$. We define the language $L_P(\mathcal{A}/\mathcal{E}) = \{t \mid P(t) \text{ is derivable}\}$. When $\mathcal{E}$ is the empty theory, we also write it as $L_P(\mathcal{A})$. If in addition some state P is designated as *final* then the language accepted by the automaton is $L_P(\mathcal{A}/\mathcal{E})$. For a language L we define $\mathcal{E}(L) = \{s \mid \exists t \in L \cdot s =_{\mathcal{E}} t\}$. Note that automata-theoretic problems are closely related to the unsatisfiability problem of Horn clauses:

Lemma 1. *Let $\mathcal{A}$ be a tree automaton and $\mathcal{E}$ an equational theory.*

(i) $L_P(\mathcal{A}/\mathcal{E}) \neq \emptyset$ iff $\mathcal{A} \cup \{\bot \Leftarrow P(x)\}$ is unsatisfiable modulo $\mathcal{E}$.
(ii) $t \in L_P(\mathcal{A}/\mathcal{E})$ iff $\mathcal{A} \cup \{\bot \Leftarrow P(t)\}$ is unsatisfiable modulo $\mathcal{E}$, where t is ground.
(iii) $L_P(\mathcal{A}/\mathcal{E}) \cap L_Q(\mathcal{A}/\mathcal{E}) \neq \emptyset$ iff $\mathcal{A} \cup \{\bot \Leftarrow P(x) \wedge Q(x)\}$ is unsatisfiable modulo $\mathcal{E}$.

Hence the results in this paper can be interpreted from an automata-theoretic viewpoint as well as from a logical viewpoint. *One-way automata* consist of clauses:

$$P(f(x_1, ..., x_n)) \Leftarrow P_1(x_1) \wedge \dots \wedge P_n(x_n) \quad (1) \qquad P(x) \Leftarrow P_1(x) \qquad (2)$$

which we call *pop clauses* and *ε-clauses* respectively. In (1), the variables $x_1, ..., x_n$ are mutually distinct. In the non-equational case, one-way automata are exactly the tree automata usually found in the literature, and which accept regular tree languages. We also recall from [20]:

Lemma 2. *We have $\mathcal{L}_P(\mathcal{A}/\mathcal{E}) = \mathcal{E}(\mathcal{L}_P(\mathcal{A}))$ for any one-way automaton $\mathcal{A}$ and equational theory $\mathcal{E}$. In particular, emptiness for one-way $\mathcal{E}$ tree-automata is decidable in linear time.*

Because of the form of signatures that we consider, the pop clauses in our automata are of the following form,

$$P(x+y) \Leftarrow P_1(x) \wedge P_2(y) \quad (3) \qquad P(a) \text{ where } a \text{ is a constant} \quad (5)$$
$$P(0) \quad (4) \qquad P(-x) \Leftarrow P_1(x) \quad (6)$$
$$P(f(x_1, ..., x_n)) \Leftarrow P_1(x_1) \wedge \dots \wedge P_n(x_n) \qquad (f \text{ is free}) \quad (7)$$

called *+-pop clauses*, *zero clauses*, *constant clauses*, *minus clauses* and *free pop clauses*. Clauses (5) are special cases of clauses (7). For $\mathcal{E} \in \{\text{ACU}, \text{XOR}, \text{AG}, \text{ACUD}\}$, one-way $\mathcal{E}$-tree automata are sets of clauses (2–7) (clause (6) is present only when $- \in \Sigma$). We define two-way automata by adding the following kind of clauses

$$Q(x_i) \Leftarrow P(f(x_1, ..., x_n)) \wedge \bigwedge_{j\in\{1,...,n\}\setminus\{i\}} Q_j(x_j) \qquad (f \text{ is free}, 1 \le i \le n) \quad (8)$$

called *push clauses*, to one-way automata (the variables $x_1, ..., x_n$ are mutually distinct.) Hence *two-way* automata are sets of clauses (2–8) (clause (6) is included only when $- \in \Sigma$). For a two-way automaton $\mathcal{A}$, we let $\mathcal{A}_{eq}$ denote the set of ϵ-clauses, +-pop clauses, zero clauses and $-$ clauses in $\mathcal{A}$. These are the *equational clauses* of $\mathcal{A}$. Let us discuss the side-conditions in the above push clause. We first prohibit the variable x_i to occur twice in the tail of the clause. Removing this restriction allows us to encode alternating tree automata. In the non-equational case, this leads to an EXPTIME-complete emptiness problem. In case of the theories ACU, AG and ACUD, the emptiness problem becomes undecidable [20]. The XOR case is decidable [21], though the complexity seems high. Secondly we have restricted f to be free. In the ACU case, the justification is as for $\mathcal{H}3$ clauses. In case of the theory XOR the clause $P(x) \Leftarrow Q(x+y) \land R(y)$ is equivalent to the clause $P(x + y) \Leftarrow Q(x) \land R(y)$. In the AG case, the former clause is equivalent to the clauses $P(x + y) \Leftarrow Q(x) \land R'(y)$ and $R'(-y) \Leftarrow R(y)$ for fresh R'. Push clauses $P(x) \Leftarrow Q(-x)$ involving the $-$ symbol are equivalent to the minus clause $P(-x) \Leftarrow Q(x)$ modulo our equational theories.

Note that we have restricted each $x_j \neq x_i$ to occur exactly twice in the tail. Our complexity results hold even if we allow more atoms of the form $Q_j^1(x_j), Q_j^2(x_j) \ldots$ in the tail provided the number of repetitions of each variable is bounded by a constant. Allowing the variables x_j to occur an arbitrarily large number of times in the tail makes the non-emptiness problem subsume the intersection-non-emptiness problem for a sequence of tree automata, leading to EXPTIME-hardness already in the non-equational case. In the equational case, the complexity is likely to be higher.

2.1 Modeling Cryptographic Protocols

To demonstrate the modeling of protocols using equational H3 clauses, we take the following variant of the Needham-Schroeder-Lowe protocol from [2], in standard notation. The operator + obeys XOR laws. Note that the analysis of [2] is for bounded number of sessions whereas our interest is in the analysis for unbounded number of sessions.

$$A \to B : \{N_A, A\}_{K_B}$$
$$B \to A : \{N_B, N_A + B\}_{K_A}$$
$$A \to B : \{N_B\}_{K_B}$$

We model it using Horn clauses as in [5]. We let the function symbols $\{_-\}_-$ and $\langle _-, _- \rangle$ denote encryption and pairing. Each protocol step is repeated arbitrarily many times, although only finitely many nonces are used. For every pair of distinct agents a and b we choose constants n_{ab}^1 and n_{ab}^2 representing the nonces N_A and N_B which are used in sessions between A and B. We choose a predicate known to represent messages known to the adversary. For every pair of agents a and b, we have the following three clauses corresponding to the three protocol steps:

$$\mathsf{known}(\{\langle n_{ab}^1, a \rangle\}_{K_b})$$
$$\mathsf{known}(\{\langle n_{ab}^2, x + b \rangle\}_{K_a}) \Leftarrow \mathsf{known}(\{\langle x, a \rangle\}_{K_b})$$
$$\mathsf{known}(\{x\}_{K_b}) \Leftarrow \mathsf{known}(\{\langle x, n_{ab}^1 + b \rangle\}_{K_a})$$

This is based on the well known assumption that the adversary has full control over the network. All messages sent by agents are sent to him and all messages received by agents are received from him. The second clause for example represents the fact that if b receives the message $\{\langle x, a \rangle\}_{K_B}$ for any x then he will send the message $\{\langle n_{ab}^2, x +$

$b\rangle\}_{K_a}$. In place of x, b expects some nonce generated by a, however the adversary can fool b by sending a message with something else in place of x. We need other clauses to represent the ability of the adversary to compute new messages from existing messages. We have the clause known($\{x\}_y$) $\Leftarrow$ known(x) $\wedge$ known(y) to represent the ability of the adversary to encrypt messages. His decryption ability is represented by clauses known(x) $\Leftarrow$ known($\{x\}_k$) $\wedge$ known(k^{-1}) for every key k, where k^{-1} is the inverse of k. The ability of the adversary to pair and unpair messages is represented by the clauses known($\langle x, y \rangle$) $\Leftarrow$ known(x) $\wedge$ known(y), known(x) $\Leftarrow$ known($\langle x, y \rangle$) and known(y) $\Leftarrow$ known($\langle x, y \rangle$). The clause known($x + y$) $\Leftarrow$ known(x) $\wedge$ known(y) represents the ability of the adversary to apply the $+$ operation on known messages. The adversary's knowledge of other messages m like identities of agents, public keys, private keys of dishonest agents, is represented by clauses known(m). Finally to check secrecy of a message S we add the clause $\bot \Leftarrow$ known(S) and check that the resulting clause set is satisfiable. All these clauses are H3. Our modeling used only finitely many nonces in infinitely many sessions. This is a safe abstraction: we detect all attacks against the protocols. However the secrecy problem is still undecidable. Indeed not all protocols can be modeled using H3 clauses without further safe abstractions.

As further examples, note that the modeling in [11] of the IKA.1 initial key agreement protocol requires only H3, or two-way automata clauses, modulo ACU. The verification in [11] is done using approximation techniques since the known algorithms for two-way ACU automata were too expensive. Our improved algorithms in this paper should let us dispense with approximation techniques in dealing with clauses required for such protocols. The clauses required for the modeling of the example protocol using XOR in [5] are also H3, whereas the upper bound provided for their class of clauses modulo XOR is non-elementary. The results in this paper are likely to provide efficient techniques for dealing with a large number of protocols in practice.

3   Parikh Images of Context-Free Grammars

Our equational tree automata are closely related to Parikh images of context free languages and Presburger formulas, as we show in this section. The *Parikh image* $\mathcal{P}(x)$ of a string x on some alphabet maps each symbol a to the number of occurrences of a in x. The Parikh image of a set of strings is the set of Parikh images of its members. It is well-known [15] that Parikh images of context-free languages are *semilinear* sets, which are exactly the sets definable by (existential) Presburger formulas. We first improve this result by showing that for every context-free grammar G one can compute in *linear time* an *existential* Presburger formula ϕ_G which characterizes the Parikh image of the language $L(G)$ generated by G. The proof combines a result from [8] with techniques from [18]. Recall that existential Presburger formulas ϕ are defined by the following grammar and interpreted over natural numbers:

$$t ::= 0 \mid 1 \mid x \mid t_1 + t_2 \qquad \phi ::= t_1 = t_2 \mid t_1 > t_2 \mid \phi_1 \wedge \phi_2 \mid \phi_1 \vee \phi_2 \mid \exists x \cdot \phi_1$$

The result in [8] is formulated in terms of communication-free Petri nets whose definition we recall next. A *Net* $N = (S, T, W)$ consists of a set S of *places*, a set T of *transitions* and a weight function $W : (S \times T) \cup (T \times S) \to \mathbb{N}$. If $W(x, y) > 0$ we say that there is an edge from x to y of weight $W(x, y)$. A net is *communication-free*, if for each transition t there is at most one place s with $W(s, t) > 0$ and furthermore $W(s, t) = 1$. A *marking* M associates a number of *tokens* which each place, formally it is simply a function $S \to \mathbb{N}$. A *communication-free Petri net* is a pair (N, M_0), where N is a communication-free net and M_0 is a marking. A marking M *enables* a transition t in a communication-free Petri net if $M(s) > 0$, for the place s with $W(s, t) > 0$. If a transition t is enabled for a marking M, then it can *occur* resulting in the marking M' defined by $M'(s) = M(s) + W(t, s) - W(s, t)$, for every place s. A marking M' is *reachable* from a marking M, if there is a sequence $\sigma = t_1 \cdots t_m$ of transitions and a sequence of markings $M_0 = M, M_1, \ldots, M_m = M'$ such that, for each i the occurrence of t_i in (N, M_{i-1}) results in M_i. We say also that σ *can occur at* M in that case. The following result was shown in [8] (Lemma 3.1 and Theorem 3.1).

Theorem 3. *Let (N, M_0) be a communication-free Petri net with transition set T and let X be a function from T to $\mathbb{N}$. There exists a sequence σ of transitions with $\mathcal{P}(\sigma) = X$ that can occur in (N, M_0) if and only if the following two conditions hold.*

(a) *For each place s, it holds $M_0(s) + \sum_{t \in T}[(W(t, s) - W(s, t))X(t)] \geq 0$, and*
(b) *in the subgraph of N which is induced by the transitions $t \in T$ with $X(t) > 0$, every place is reachable (in the graph-theoretical sense) from some place s with $M_0(s) > 0$.*

The intimate relationship between context-free grammars and communication-free Petri nets can be seen as follows. Let G be a grammar with non-terminal set V, terminal set U, start symbol A_0 and set P of productions. With G we associate a net $N_G = (V \cup U, P, W)$. If $A \to \alpha$ is a production p from P then $W(A, p) = 1$ and $W(p, B)$ is the number of times which B occurs in α, for each $B \in V \cup U$. The Petri net (N_G, M_G) is then obtained by setting $M_G(A_0) = 1$ and $M_G(A) = 0$ for all other A. Note that it is communication-free. An application of a production p now corresponds to the occurrence of the transition p in the net. Hence, it is not hard to see that X is is the Parikh image of a sequence that can occur in (N_G, M_G) if and only if there is a derivation of G in which each production p is used exactly $X(p)$ times.

Given a context-free grammar G on terminals $a_1, \ldots, a_p$, we now compute an existential Presburger formula $\phi_G(x_{a_1}, \ldots, x_{a_p})$ representing its Parikh image, i.e. such that $\phi_G(n_1, \ldots, n_p)$ holds iff some string in $L(G)$ contains each a_i exactly n_i times. For this it basically remains to express requirement (b) of Theorem 3. This can be done analogously as in [18].

Theorem 4. *Given a context-free grammar G, one can compute an existential Presburger formula ϕ_G for the Parikh image of $L(G)$ in linear time.*

Proof. Let $G = (V, U, P, A_0)$ be context-free and let N_G and M_G be defined as above. Let, for each $A \in U$, x_A be a variable, and for each $p \in P$, let y_p be a variable. Clearly, the free variables of ϕ_G will be the variables x_A with $A \in U$. We need three kinds of quantifier-free subformulas.

 - First, for each $A \in V$ there is one equation which is directly determined from requirement (a) of Theorem 3. To this end, let $p_1, \ldots, p_k$ be all productions with

A on the left-hand side and let, for each production p, $A(p)$ denote the number of occurrences of A on the right hand side of p. Then ϕ_G contains the equation $M_G(A) + \sum_{p \in P} A(p)y_p - \sum_{i=1}^{k} y_{p_i} = 0$. Note that we have $= 0$ instead of ≥ 0 here, as we have to make sure that the derivation under consideration is complete, i.e., there are no remaining non-terminals. Note further, that we do not need such subformulas for $A \in U$ as such A only occur on right-hand sides of productions.

- Next, we have to make sure that the values x_A are consistent with the y_p. To this end, we have, for each $A \in U$ an equation $x_A = \sum_{p \in P} A(p)y_p$.
- Finally, it remains to express requirement (b) of Theorem 3. For this purpose, we use additional variables z_A, for each $A \in U \cup V$. The idea is that the z_A reflect the distance of A from A_0 in a spanning tree on the subgraph of N_G induced by those p with $y_p > 0$. To this end, we use the following kinds of formulas.
  - We have $x_A = 0 \vee z_A > 0$, for each $A \in U$.
  - If $p_1, \ldots, p_l$ are the productions with A on the right-hand side and $B_1, \ldots, B_l$ are their corresponding left-hand sides then we have a formula $(z_A = 0) \vee \bigvee_{i=1}^{l}(z_A = z_{B_i} + 1 \wedge y_{p_i} > 0 \wedge z_{B_i} > 0)$. If one of the B_i is the start symbol A_0 the corresponding disjunct is replaced by $z_A = 1 \wedge y_{p_i} > 0$.

It is not hard to prove that the resulting formula ϕ_G, in which all variables, except the x_A with $A \in U$, are existentially quantified, characterizes exactly the Parikh image of $L(G)$. For the one direction, if a vector is in the Parikh image of $L(G)$, the variables can be chosen such that they correspond to a derivation of G. Otherwise, if a vector satisfies ϕ_G then the equations for the z_A make sure that condition (b) of Theorem 3 is fulfilled and the remaining equations verify that there is a derivation of a string with the corresponding numbers of symbols.

Finally, the size of ϕ_G is linear in the size of G, i.e., basically the size of P. Note that, in the sums over $p \in P$ only summands with $A(p) > 0$ are taken. Implemented thoroughly on a register machine, the construction of ϕ_G is possible in linear time. $\square$

Recall also that to check satisfiability of an existential Presburger formula, we can first move quantifiers to the top, then non-deterministically replace subformulas $\phi_1 \vee \phi_2$ by ϕ_1 or ϕ_2, and check satisfiability of the resulting formula $\exists x_1 \cdot \ldots \exists x_n \cdot \phi$ where ϕ is a conjunction of equations and inequations. As satisfiability for formulas in the latter form is NP-complete, we have:

Lemma 5. *Satisfiability of existential Presburger formulas is NP-complete.*

It remains to relate equational tree automata to context free grammars. Let $\mathcal{A}$ be a constants-only ACU automaton on constants $a_1, \ldots, a_p$. Modulo ACU, the terms are then of the form $n_1 a_1 + \ldots + n_p a_p$, equivalently tuples $(n_1, \ldots, n_p) \in \mathbb{N}^p$. We consider $\mathcal{A}$ as a context-free grammar. States of $\mathcal{A}$ are non-terminals and constants are terminals. Clauses $P(0)$, $P(a)$, $P(x) \Leftarrow Q(x)$ and $P(x+y) \Leftarrow P_1(x) \wedge P_2(y)$ are productions $P \to \lambda$, $P \to a$, $P \to Q$ and $P \to P_1 P_2$ respectively where λ is the empty string. The final state P is the start symbol. From Theorem 4 we obtain an existential Presburger formula, which we denote as $\phi_{\mathcal{A},P}(x_{a_1}, \ldots, x_{a_p})$, such that $\phi_{\mathcal{A},P}(n_1, \ldots, n_p)$ holds iff $n_1 a_1 + \ldots + n_p a_p \in L_P(\mathcal{A}/\mathrm{ACU})$.

Lemma 6. *For any constants-only ACU automaton $\mathcal{A}$ and state P we can compute in linear time an existential Presburger formula $\phi_{\mathcal{A},P}$ describing $L_P(\mathcal{A}/\mathrm{ACU})$.*

4 One-Way ACU Automata

We show in this section that membership and intersection-non-emptiness for one-way ACU automata are NP-complete. We first make the following observation from [20] about derivations in ACU automata, allowing us to reuse certain parts of derivations.

Lemma 7. *Let $\mathcal{E}$ be any set of equations containing ACU. Consider a derivation δ of an atom $P(t)$ modulo $\mathcal{E}$. Let $\delta_1, ..., \delta_n$ be non-overlapping subderivations of δ such that outside the δ_i's, the only equations used are ACU and the set S of clauses used contains only equational clauses (see Figure 1.) Suppose the conclusions of $\delta_1, ..., \delta_n$ are $P_1(t_1), ..., P_n(t_n)$. Then*

1. *$t =_{\text{ACU}} t_1 + ... + t_n$*
2. *If there are derivations $\delta'_1, ..., \delta'_n$ of atoms $P_1(s_1), ..., P_n(s_n)$ modulo $\mathcal{E}$ then there is a derivation δ' of $P(s_1 + ... + s_n)$ modulo $\mathcal{E}$, containing δ'_i's as subderivations, such that outside the δ'_i's, the only equations used are ACU, and all clauses used belong to S.*

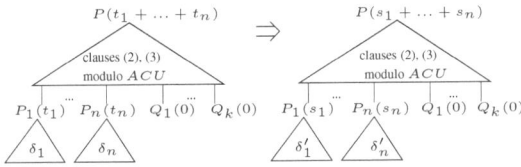

Fig. 1. Reuse of ACU derivations

The following definition from [20] gives one way of computing such δ_i's and $P_i(t_i)$'s:

Definition 1. *Consider a derivation δ of an atom $P(t)$ in a one-way automaton modulo ACU. Let $\delta_1, ..., \delta_n$ be the set of maximal subderivations of δ in which the last step used is an application of a free pop clause (or base clause). Suppose the conclusions of $\delta_1, ..., \delta_n$ are $P_1(t_1), ..., P_n(t_n)$ (in which case $t_1, ..., t_n$ must be functional). Then we will say that the (unordered) list of atoms $P_1(t_1), ..., P_n(t_n)$ is the* functional support *of the derivation δ. (From Lemma 7 we have $t =_{\text{ACU}} t_1 + ... + t_n$).*

Lemma 7 tells us how to reuse an arbitrarily large derivation involving only $+$-pop clauses, zero clauses and ϵ-clauses. Sets of such derivations can also be represented by Presburger-formulas, by using some constants to represent the effect of other clauses. Formally consider a set S of $+$-pop clauses, zero clauses, ϵ-clauses on some set of predicates $\mathbb{P}$. Introduce constants $a_{P,Q}$ for states P and Q. Intuitively $a_{P,Q}$ represents terms accepted by both P and Q. For a set $Z \subseteq \mathbb{P}^2$, define constants-only ACU automaton $S[Z] = S \cup \{P(a_{P,Q}), Q(a_{P,Q}) \mid (P,Q) \in Z\}$. Lemma 6 then allows us to represent the languages $L_P(S[Z]/\text{ACU})$ by existential Presburger formulas $\phi_{S[Z],P}$.

We now show how to decide intersection-non-emptiness of one-way ACU automata. The procedure can be thought of as marking of non-empty states in the product automaton computed in [20]. The relations $\Rightarrow_{\text{ACU}}$ and $\Rightarrow_{free}$ below allow us to mark new states. The proof of the NP upper bound then involves guessing an increasing sequence of marked states. Formally, consider one-way automata $\mathcal{A}$ and $\mathcal{B}$ on sets of states $\mathbb{P}$ and $\mathbb{Q}$. Given $Z \subseteq \mathbb{P} \times \mathbb{Q}$ and $(P, Q) \in \mathbb{P} \times \mathbb{Q}$, if $L_P(\mathcal{A}_{eq}[Z]/\text{ACU}) \cap L_Q(\mathcal{B}_{eq}[Z]/\text{ACU}) \neq \emptyset$ then we say that $Z \Rrightarrow_{\text{ACU}} (P, Q)$. Intuitively this means that if the pairs of states in Z have non-empty intersection, then on the basis of the equational clauses in $\mathcal{A}$ and $\mathcal{B}$ we can conclude that P and Q have non-empty intersection. A term in $L_P(\mathcal{A}_{eq}[Z]/\text{ACU})$ represents the effect of an arbitrarily large derivation using equational clauses of $\mathcal{A}$, starting from derivations of terms in intersections of pair of states of Z, and ending at P. Formally:

Lemma 8. *If* $L_{P'}(\mathcal{A}/\text{ACU}) \cap L_{Q'}(\mathcal{B}/\text{ACU}) \neq \emptyset$ *for all* $(P', Q') \in Z$ *and* $Z \Rrightarrow_{\text{ACU}}$ (P, Q) *then* $L_P(\mathcal{A}/\text{ACU}) \cap L_Q(\mathcal{B}/\text{ACU}) \neq \emptyset$.

Proof. For each $(P', Q') \in Z$ we have terms $t_{P',Q'}$ such that $P'(t_{P',Q'})$ and $Q'(t_{P',Q'})$ are derivable in $\mathcal{A}/\text{ACU}$ and $\mathcal{B}/\text{ACU}$ respectively. As $Z \Rrightarrow_{\text{ACU}} (P, Q)$ hence we have some $a_{P_1,Q_1} + \ldots + a_{P_n,Q_n} \in L_P(\mathcal{A}_{eq}[Z]/\text{ACU}) \cap L_Q(\mathcal{B}_{eq}[Z]/\text{ACU})$. From the definition of $\mathcal{A}_{eq}[Z]$ the derivation of $P(a_{P_1,Q_1} + \ldots + a_{P_n,Q_n})$ in $\mathcal{A}_{eq}[Z]/\text{ACU}$ has a functional support $P_1(a_{P_1,Q_1}), \ldots, P_n(a_{P_n,Q_n})$. Also $P_i(t_{P_i,Q_i})$ are derivable in $\mathcal{A}/\text{ACU}$. By Lemma 7 $P(t_{P_1,Q_1} + \ldots + t_{P_n,Q_n})$ is derivable in $\mathcal{A}/\text{ACU}$. Similarly $Q(t_{P_1,Q_1} + \ldots + t_{P_n,Q_n})$ is derivable in $\mathcal{B}/\text{ACU}$. Hence $L_P(\mathcal{A}/\text{ACU}) \cap L_Q(\mathcal{B}/\text{ACU}) \neq \emptyset$. $\square$

We write $Z \Rrightarrow_{free} (P, Q)$ to mean that $\mathcal{A}$ has some free pop clause $P(f(x_1, \ldots, x_n)) \Leftarrow P_1(x_1) \wedge \ldots \wedge P_n(x_n)$ and $\mathcal{B}$ has some free pop clause $Q(f(x_1, \ldots, x_n)) \Leftarrow Q_1(x_1) \wedge \ldots \wedge P_n(x_n)$ such that $\{(P_i, Q_i)\} \in Z$ for $1 \leq i \leq n$. Intuitively this means that if the pairs of states in Z have non-empty intersection, then on the basis of the free pop clauses in $\mathcal{A}$ and $\mathcal{B}$ we can conclude that P and Q have non-empty intersection.

We write $Z \Rightarrow (P, Q)$ to say that there are some $(P_1, Q_1), \ldots, (P_n, Q_n) \in \mathbb{P} \times \mathbb{Q}$ such that $(P_n, Q_n) = (P, Q)$ and for $1 \leq i \leq n$ we have

$$Z \cup \{(P_1, Q_1), \ldots, (P_{i-1}, Q_{i-1})\} \, (\Rrightarrow_{\text{ACU}} \cup \Rrightarrow_{free}) \, (P_i, Q_i).$$

Intuitively this represents the effect of a sequence of conclusions using the $\Rrightarrow_{\text{ACU}}$ and $\Rrightarrow_{free}$ rules. This rule suffices for detecting all pairs having non-empty intersection:

Lemma 9. *If* $L_P(\mathcal{A}/\text{ACU}) \cap L_Q(\mathcal{B}/\text{ACU}) \neq \emptyset$ *then* $\emptyset \Rightarrow (P, Q)$.

Proof. We do induction on the size of the given term $t \in L_P(\mathcal{A}/\text{ACU}) \cap L_Q(\mathcal{B}/\text{ACU})$. Let $t = t_1 + \ldots + t_m$ where $t_i = f_i(t_i^1, \ldots, t_i^{k_i})$ is functional for $1 \leq i \leq m$. The derivation of $P(t)$ has some functional support $P_1(t_1), \ldots, P_m(t_m)$ where for $1 \leq i \leq m$ the derivation of $P_i(t_i)$ uses a clause $P_i(f_i(x_1, \ldots, x_{k_i})) \Leftarrow P_i^1(x_1) \wedge \ldots \wedge P_i^{k_i}(x_{k_i})$ and the derivations of $P_i^1(t_i^1), \ldots, P_i^{k_i}(t_i^{k_i})$. Similarly the derivation of $Q(t)$ has some functional support $Q_1(t_1), \ldots, Q_m(t_m)$ where for $1 \leq i \leq m$ the derivation of $Q_i(t_i)$ uses a clause $Q_i(f_i(x_1, \ldots, x_{k_i})) \Leftarrow Q_i^1(x_1) \wedge \ldots \wedge Q_i^{k_i}(x_{k_i})$ and the derivations of $Q_i^1(t_i^1), \ldots, Q_i^{k_i}(t_i^{k_i})$. Hence $t_i^j \in L_{P_i^j}(\mathcal{A}/\text{ACU}) \cap L_{Q_{j_i}}(\mathcal{B}/\text{ACU})$ for $1 \leq i \leq m, 1 \leq j \leq k_i$. By induction hypothesis we have $\emptyset \Rightarrow (P_i^j, Q_i^j)$ for $1 \leq i \leq m, 1 \leq j \leq k_i$. For $1 \leq i \leq m$, $\{(P_i^j, Q_i^j) \mid 1 \leq j \leq k_i\} \Rrightarrow_{free} (P_i, Q_i)$. Hence $\emptyset \Rightarrow (P_i, Q_i)$

for $1 \leq i \leq m$. Also because of the above two functional supports we know from Lemma 7 that $a_{(P_1,Q_1)} + \ldots + a_{(P_m,Q_m)} \in L_P(\mathcal{A}_{eq}[Z]/\text{ACU}) \cap L_Q(\mathcal{B}_{eq}[Z]/\text{ACU})$ where $Z = \{(P_i, Q_i) \mid 1 \leq i \leq m\}$. Hence $Z \Rightarrow_{\text{ACU}} (P, Q)$. Hence $\emptyset \Rightarrow (P, Q)$. □

Lemma 10. *Intersection-non-emptiness for one-way* ACU *automata is in NP.*

Proof. Let P and Q be the final states of $\mathcal{A}$ and $\mathcal{B}$ respectively. From Lemmas 8 and 9 $L_P(\mathcal{A}/\text{ACU}) \cap L_Q(\mathcal{B}/\text{ACU}) \neq \emptyset$ iff $\emptyset \Rightarrow (P, Q)$. The latter is equivalent to existence of (mutually distinct) pairs $(P_1, Q_1), \ldots, (P_n, Q_n) \in \mathbb{P} \times \mathbb{Q}$ such that $(P_n, Q_n) = (P, Q)$ and for $1 \leq i \leq n$, we have

$$Z_i = \{(P_1, Q_1), \ldots, (P_{i-1}, Q_{i-1})\}(\Rightarrow_{\text{ACU}} \cup \Rightarrow_{free})(P_i, Q_i).$$

There are polynomially many such pairs. Checking $Z_i \Rightarrow_{free} (P_i, Q_i)$ requires polynomial time. To check that $Z_i \Rightarrow_{\text{ACU}} (P_i, Q_i)$ we check satisfiability (Lemma 5) of

$$\phi_{\mathcal{A}_{eq}[Z_i], P_i}(x_{a_{P_1,Q_1}}, \ldots, x_{a_{P_{i-1},Q_{i-1}}}) \wedge \phi_{\mathcal{B}_{eq}[Z_i], Q_i}(x_{a_{P_1,Q_1}}, \ldots, x_{a_{P_{i-1},Q_{i-1}}})$$

which can be computed in polynomial time using Lemma 6. □

Hence the membership problem is also in NP. It is in fact NP-complete since the membership problem for Parikh images of languages generated by context free grammars is NP-hard [8]. This is also shown in [14], but the complexity of the intersection-non-emptiness problem is left open there. Indeed the technique of [14] is too specific to the membership problem. We now have:

Theorem 11. *The membership and intersection-non-emptiness problems for one-way* ACU *automata are NP-complete.*

5 Theories XOR and AG

We now show NP-completeness of membership and intersection-non-emptiness for one-way automata modulo other theories including XOR and AG. First we describe the XOR case. For $n \in \mathbb{N}$ if n is odd then define $n^* = 1$ otherwise define $n^* = 0$. If $a_1, \ldots, a_k$ are mutually distinct constants then define $(n_1 a_1 + \ldots + n_k a_k)^* = n_1^* a_1 + \ldots + n_k^* a_k$. For a set L of such terms define $L^* = \{t^* \mid t \in L\}$. Because of the cancellation axiom, we also need to decide intersection-non-emptiness of all pairs of states in the same automaton. Hence we consider a single one-way automaton $\mathcal{A}$ on set of states $\mathbb{P}$, instead of two different automata as in the ACU case. Given $Z \subseteq \mathbb{P}^2$ and $(P, Q) \in \mathbb{P}^2$, we say $Z \Rightarrow_{\text{XOR}} (P, Q)$ to mean that $(L_P(\mathcal{A}_{eq}[Z]/\text{ACU}))^* \cap (L_Q(\mathcal{A}_{eq}[Z]/\text{ACU}))^* \neq \emptyset$. This is the counterpart of the $\Rightarrow_{\text{ACU}}$ relation in the ACU case. Intuitively, because of the cancellation axiom of XOR, we only need to check whether a constant occurs an even or odd number of times. The relations $\Rightarrow_{free}$ and $\Rightarrow$ are redefined as expected. As in the ACU case, states P and Q have non-empty intersection iff $\emptyset \Rightarrow (P, Q)$. The construction is easily generalized to the XOR_p theory for any (fixed) $p \geq 2$: we need to consider intersection-non-emptiness for p-tuples of states.

Lemma 12. *Intersection-non-emptiness for one-way* XOR_p *automata is in NP. This holds in particular for the theory* XOR.

Proof. (Sketch:) The algorithm works as in the ACU case. To check that $Z = \{(P_1, Q_1),$ $\ldots, (P_n, Q_n)\} \Rightarrow_{\text{XOR}} (P, Q)$ we check that the formula

$$\phi_{\mathcal{A}_{eq}[Z], P}(x_{a_{P_1, Q_1}}, \ldots, x_{a_{P_n, Q_n}}) \wedge \phi_{\mathcal{B}_{eq}[Z], Q}(y_{a_{P_1, Q_1}}, \ldots, y_{a_{P_n, Q_n}})$$

$$\wedge \bigwedge_{1 \leq i \leq n}(even(x_{a_{P_i, Q_i}}) \wedge even(y_{a_{P_i, Q_i}}) \vee odd(x_{a_{P_i, Q_i}}) \wedge odd(y_{a_{P_i, Q_i}}))$$

is satisfiable where $even(x) \equiv \exists y \cdot x = 2y$ and $odd(x) \equiv \exists y \cdot x = 2y + 1$.   □

Next we prove NP-hardness of membership.

Lemma 13. *Membership for one-way* XOR_p *automata is NP-hard for* $p \geq 2$.

Proof. (Sketch:) We use a reduction from the 3-colorability problem for undirected graphs. For a graph $G = (V, E)$ we define the automaton $\mathcal{A}_G$ as follows. We have the set $\{r, b, g\}$ of three colors. For every vertex v, color c and edge e adjacent on v, introduce a fresh constant $a_{e,v,c}$, representing the assignment of color c to v. For every edge e joining u and v, introduce a fresh predicate P_e. For every pair (c_1, c_2) of distinct colors, add clause $P_e(a_{e,u,c_1} + a_{e,v,c_2})$ to $\mathcal{A}_G$. For every vertex v introduce a fresh predicate P_v. For every color c, add the clause $P_v(\sum_{e \in E, e \text{ adjacent on } v}(p - 1)a_{e,v,c})$ to $\mathcal{A}_G$. Finally add the clause $P(\sum_{v \in V} x_v + \sum_{e \in E} x_e) \Leftarrow \bigwedge_{v \in V} P_v(x_v) \wedge \bigwedge_{e \in E} P_e(x_e)$ to $\mathcal{A}_G$. Then 3-colorability of G is equivalent to $0 \in L_P(\mathcal{A}_G / \text{XOR}_p)$.   □

Theorem 14. *The membership and intersection non-emptiness problems for one-way* XOR_p *automata are NP-complete for* $p \geq 2$. *This holds in particular for XOR.*

To deal with the AG case, we use the ACUD theory as a tool. This is similar to the way the XOR case was dealt with using the ACU theory. The ACUD theory allows us to normalize terms by pushing the $-$ symbol downwards to functional terms. First we consider constant only ACUD automata. Σ_f is the set of constants in our signature. Let $\overline{\Sigma}_f$ be a set of fresh constants $\{\bar{a} \mid a \in \Sigma_f\}$. Terms built from $\Sigma_f \cup \{+, -, 0\}$ modulo ACUD are of the form $a_1 + \ldots + a_m - b_1 - \ldots - b_n$ $(m, n \geq 0)$ while those built from $\Sigma_f \cup \overline{\Sigma}_f \cup \{+, 0\}$ modulo ACU are of the form $a_1 + \ldots + a_m + \bar{b}_1 + \ldots + \bar{b}_n$ $(m, n \geq 0)$. Hence there is a natural 1-1 correspondence between terms (languages) on $\Sigma_f \cup \{+, -, 0\}$ modulo ACUD and terms (languages) on $\Sigma \cup \overline{\Sigma}_f \cup \{+, 0\}$ modulo ACU. Then modulo this correspondence of languages we have [20]:

Lemma 15. *The language accepted by a constants-only ACUD automaton $\mathcal{A}$ with constants from Σ_f is also accepted by a constants-only ACU automaton $\mathcal{B}$ with constants from $\Sigma_f \cup \overline{\Sigma}_f$. $\mathcal{B}$ is computable in linear time from $\mathcal{A}$.*

For a set S of +-pop clauses, minus clauses, zero clauses and ϵ-clauses, and a set Z of pairs of states, the automaton $S[Z]$ is defined as before. If the elements of Z are $(P_1, Q_1), \ldots, (P_k, Q_k)$ then using Lemmas 6 and 15 we can compute in polynomial time an existential Presburger formula $\phi_{S[Z], P}(x_{a_{P_1, Q_1}}, \ldots, x_{a_{P_k, Q_k}}, x'_{a_{P_1, Q_1}},$ $\ldots, x'_{a_{P_k, Q_k}})$ such that $\phi_{S[Z], P}(n_1, \ldots, n_k, m_1, \ldots, m_k)$ holds iff $n_1 a_{P_1, Q_1} + \ldots + n_k a_{P_k, Q_k} - m_1 a_{P_1, Q_1} - \ldots - m_k a_{P_k, Q_k} \in L_P(S[Z] / \text{ACUD})$.

We now show how to decide intersection-non-emptiness of one-way AG automata. Consider a one-way automaton $\mathcal{A}$ on set of states $\mathbb{P}$. Given $Z \subseteq \mathbb{P}^2$ and $(P, Q) \in \mathbb{P}^2$, we say $Z \Rightarrow_{AG} (P, Q)$ to mean that some $n_1 a_{P_1,Q_1} + \ldots + n_p a_{P_p,Q_p} - m_1 a_{P_1,Q_1} - \ldots - m_p a_{P_p,Q_p} \in L_P(\mathcal{A}/\text{ACUD})$ and some $n'_1 a_{P_1,Q_1} + \ldots + n'_p a_{P_p,Q_p} - m'_1 a_{P_1,Q_1} - \ldots - m'_p a_{P_p,Q_p} \in L_Q(\mathcal{A}/\text{ACUD})$ where $(P_1, Q_1), \ldots, (P_p, Q_p)$ are the mutually distinct elements of Z and $n_i - m_i = n'_i - m'_i$ for $1 \leq i \leq p$. This corresponds to the relations $\Rightarrow_{\text{ACU}}$ and $\Rightarrow_{\text{XOR}}$ in the ACU and XOR cases respectively, and takes care of possible cancellations using the AG axioms. Note the use of ACUD above instead of ACU. The relations $\Rightarrow_{free}$ and $\Rightarrow$ are defined as expected. We use a generalization of Lemma 7 to the ACUD case. The rest works as in the previous cases, and we have:

Theorem 16. *The membership and intersection-non-emptiness problems for one-way AG automata are NP-complete.*

Note that the above lower bound is inherited from the ACU case. Also, while we introduced the ACUD theory only as a tool to deal with the AG theory, these techniques clearly also work in the ACUD case. In fact the proofs become simpler because we do not have to deal with cancellations. The lower bound is also inherited from the ACU case. We merely state the result:

Theorem 17. *The membership and intersection-non-emptiness problems for one-way ACUD automata are NP-complete.*

6 $\mathcal{H}3$ and Two-Way Automata

We now show how to deal with two-way automata and $\mathcal{H}3$. First note that membership and intersection-non-emptiness modulo our equational theories are NP-hard, as for the one-way case. While non-emptiness in the one-way case is decidable in linear time, in the two-way case it becomes NP-hard because the intersection-non-emptiness problem reduces to the non-emptiness problem. To decide intersection-non-emptiness of states P_1 and P_2 we create fresh states P_3, P_4 and P_5 and add clauses $P_3(0)$, $P_4(f(x, y)) \Leftarrow P_1(x) \wedge P_3(y)$ and $P_5(x) \Leftarrow P_4(f(x, y)) \wedge P_2(y)$. Then non-emptiness of P_5 is equivalent to intersection-non-emptiness of P_1 and P_2. We now first show that the NP-upper bound holds also for two-way automata modulo our equational theories. We illustrate the techniques for the XOR case, the other cases are similar.

The key idea is to add new ϵ-clauses to the automata till the push clauses become redundant. For example the push clause $P(x) \Leftarrow Q(f(x))$ and the pop clause $Q(f(x)) \Leftarrow R(x)$ can be "short-cut" to produce the ϵ-clause $P(x) \Leftarrow R(x)$. This is the main idea in the non-equational case. In the XOR case, there can be arbitrarily many applications of $+$-pop clauses in between the applications of the free pop clause and the push clause. However after cancellations using the XOR axioms, only a functional term should be left before application of the push clause. This is formalized as follows, as in [20]. Consider a two-way automaton $\mathcal{A}$. For any two-way automaton $\mathcal{A}'$, $\mathcal{A}'_{ow}$ denotes the subset of $\mathcal{A}$ without the push clauses. If there is some $Z \subseteq \mathbb{P}^2$ such that

(1) $R(x_i) \Leftarrow P(f(x_1, \ldots, x_n)) \wedge \bigwedge_{j \in \{1, \ldots, n\} \setminus \{i\}} R_j(x_j) \in \mathcal{A}$
(2) $Q(f(x_1, \ldots, x_n)) \Leftarrow Q_1(x_1) \wedge \ldots \wedge Q_n(x_n) \in \mathcal{A}$
(3) $a_{Q,Q} + 2n_1 a_{P_1,P'_1} + \ldots + 2n_m a_{P_m,P'_m} \in L_P(\mathcal{A}_{eq}[Z \cup \{(Q, Q)\}]/\text{ACU})$

(4) $L_{P_j}(\mathcal{A}_{ow}/\text{XOR}) \cap L_{P'_j}(\mathcal{A}_{ow}/\text{XOR}) \neq \emptyset$ for $1 \leq j \leq m$

(5) for $j \in \{1, \ldots, n\} \setminus \{i\}, L_{Q_j}(\mathcal{A}_{ow}/\text{XOR}) \cap L_{R_j}(\mathcal{A}_{ow}/\text{XOR}) \neq \emptyset$

then we write $\mathcal{A} \triangleright \mathcal{A} \cup \{R(x_i) \Leftarrow Q_i(x_i)\}$ provided $R(x_i) \Leftarrow Q_i(x_i) \notin \mathcal{A}$. Note that the pairs (P_j, P'_j) in (3) are necessarily from $Z \cup \{(Q, Q)\}$. The relation $\triangleright$ is one-step of our saturation procedure. It does not affect the set of derivable atoms modulo XOR[20]. From Theorem 4 we now know that the validity of a saturation step is in NP.

Note that there are only polynomially many ϵ-clauses possible. Given a two-way automaton $\mathcal{A}$, we can keep on adding new ϵ-clauses as long as possible. In the end we have an automaton $\mathcal{B}$. We then remove all push clauses to get one-way automaton $\mathcal{B}_{ow}$. This last step does not affect the set of derivable atoms modulo XOR[20]. This gives us a way of deciding intersection-non-emptiness in NP. We guess the saturated automaton by choosing a sequence of saturation steps. We then remove all push clauses and check intersection-non-emptiness on the resulting one-way automaton. The same techniques work in the case of other theories, and we have:

Theorem 18. *Let* $\mathcal{E} \in \{\text{ACU}, \text{XOR}, \text{AG}, \text{ACUD}, \text{XOR}_p \mid p \geq 2\}$. *The membership, non-emptiness and intersection-non-emptiness problems for two-way* $\mathcal{E}$ *tree automata are NP-complete.*

This also allows us to deal with the class $\mathcal{H}3$. For this we apply the transformation given in [13] which takes polynomial time and produces a set of H3 clauses of the form

(1) $P(x_i) \Leftarrow Q(f(x_1, \ldots, x_n)) \wedge \bigwedge_{j \in \{1,\ldots,n\} \setminus \{i\}} P_j(x_j)$

(2) $P(x_1, \ldots, x_n) \Leftarrow P_1(x_1) \wedge \ldots \wedge P_n(x_n) \wedge \bigwedge_{1 \leq i \leq m} Q_i()$

(3) $P(x_i) \Leftarrow Q(x_1, \ldots, x_n) \wedge \bigwedge_{j \in \{1,\ldots,n\} \setminus \{i\}} P_j(x_j)$

(4) $P() \Leftarrow Q(x_1, \ldots, x_n) \wedge \bigwedge_{1 \leq i \leq n} P_i(x_i)$

besides one-way automata clauses. The transformation preserves satisfiability. Also condition (3) of the definition of H3 clauses ensures that $+$ does not occur in the tail of clauses except in case of theories XOR or AG. In these cases, $+$ can be removed from tails as in Section 2. The symbol $-$ is also removed from tails as in Section 2. Next to get rid of predicates with arbitrary arities, we encode atoms $P(t_1, \ldots, t_n)$ as $P'(f_n(t_1, \ldots, t_n))$. Then we are left with just two-way automata clauses, except for the presence of nullary predicates. The saturation procedure above is easily generalized to take care of nullary predicates: it can now generate clauses of the form $P()$. Clauses $\bot \Leftarrow A_1 \wedge \ldots \wedge A_n$ are also treated as definite clauses by treating $\bot$ as a nullary predicate. To check unsatisfiability we check that the clause $\bot$ can be generated using saturation steps.

Theorem 19. *Unsatisfiability for* $\mathcal{H}3$ *modulo each of the theories* ACU, XOR, AG, ACUD *and* XOR$_p$ *for* $p \geq 2$ *is NP-complete.*

In case of the related theory ACUI of idempotent commutative monoids, it is known that two-way automata are powerful enough to encode alternation [20]. Hence the associated problems become EXPTIME-hard, though they are decidable using the techniques of [21]. This is similar to the XOR case, whereas alternating automata are undecidable in the case of the theories ACU, AG and ACUD.

7   Conclusion

We have shown that the polynomially decidable class $\mathcal{H}3$ of Horn clauses can be extended with the theories $\mathrm{ACU, XOR, AG, ACUD}$ and XOR_p for $p \geq 2$, to obtain NP-complete unsatisfiability problems. This improves known algorithms for one-way as well as two-way tree automata modulo these theories: essentially all problems are NP-complete. Our algorithms require deterministic polynomial time using an oracle for existential Presburger formulas, suggesting efficient implementations are possible.

While these decidability results can also be extended to the theory ACUI of idempotent commutative monoids, the complexity is EXPTIME-hard and probably higher. In the ACU case, we have forbidden the symbol $+$ to appear in non-ground negative literals. Without this restriction, the decidability question is still open. However the problems involved subsume other problems which are known to be EXPSPACE-hard and whose decidability is still open.

References

1. B. Blanchet. An efficient cryptographic protocol verifier based on Prolog rules. In *CSFW'01*, pages 82–96. IEEE Computer Society Press, 2001.
2. Y. Chevalier, R. Küsters, M. Rusinowitch, and M. Turuani. An NP decision procedure for protocol insecurity with XOR. In *LICS'03*, pages 261–270, 2003.
3. T. Colcombet. Rewriting in the partial algebra of typed terms modulo AC. In *Electronic Notes in Theoretical Computer Science*, volume 68. Elsevier Science Publishers, 2002.
4. H. Comon, M. Dauchet, R. Gilleron, F. Jacquemard, D. Lugiez, S. Tison, and M. Tommasi. Tree automata techniques and applications. http://www.grappa.univ-lille3.fr/tata, 1997.
5. H. Comon-Lundh and V. Cortier. New decidability results for fragments of first-order logic and application to cryptographic protocols. In *RTA'03*, pages 148–164. Springer-Verlag LNCS 2706, 2003.
6. V. Cortier, S. Delaune, and P. Lafourcade. A survey of algebraic properties used in cryptographic protocols. *Journal of Computer Security*, 2005. To appear.
7. P. de Groote, B. Guillaume, and S. Salvati. Vector addition tree automata. In *LICS'04*, pages 64–73. IEEE Computer Society Press, 2004.
8. J. Esparza. Petri nets, commutative context-free grammars, and basic parallel processes. *Fundam. Inform.*, 31(1):13–25, 1997.
9. S. Ginsburg and E. H. Spanier. Semigroups, Presburger formulas and languages. *Pacific Journal of Mathematic*, 16(2):285–296, 1966.
10. J. Goubault-Larrecq and F. Parrennes. Cryptographic protocol analysis on real C code. In *VMCAI'05*, pages 363–379. Springer-Verlag LNCS 3385, 2005.
11. J. Goubault-Larrecq, M. Roger, and K. N. Verma. Abstraction and resolution modulo AC: How to verify Diffie-Hellman-like protocols automatically. *Journal of Logic and Algebraic Programming*, 2005. To Appear. Available as Research Report LSV-04-7, LSV, ENS Cachan.
12. D. Lugiez. Counting and equality constraints for multitree automata. In *FOSSACS'03*, pages 328–342. Springer-Verlag LNCS 2620, 2003.
13. F. Nielson, H. R. Nielson, and H. Seidl. Normalizable Horn clauses, strongly recognizable relations and Spi. In *SAS'02*, pages 20–35. Springer-Verlag LNCS 2477, 2002.
14. H. Ohsaki and T. Takai. Decidability and closure properties of equational tree languages. In *RTA'02*, pages 114–128. Springer-Verlag LNCS 2378, 2002.

15. R. J. Parikh. On context-free languages. *Journal of the ACM*, 13(4):570–581, October 1966.
16. M. Rusinowitch and L. Vigneron. Automated deduction with associative-commutative operators. *Applicable Algebra in Engineering, Communication and Computation*, 6:23–56, 1995.
17. H. Seidl, T. Schwentick, and A. Muscholl. Numerical document queries. In *PODS'03*, pages 155–166, 2003.
18. H. Seidl, T. Schwentick, A. Muscholl, and P. Habermehl. Counting in trees for free. In *ICALP'04*, pages 1136–1149. Springer-Verlag LNCS 3142, 2004.
19. H. Seidl and K. N. Verma. Flat and one-variable clauses: Complexity of verifying cryptographic protocols with single blind copying. In *LPAR'04*, pages 79–94. Springer-Verlag LNCS 3452, 2004.
20. K. N. Verma. Two-way equational tree automata for AC-like theories: Decidability and closure properties. In *RTA'03*, pages 180–196. Springer-Verlag LNCS 2706, 2003.
21. K. N. Verma. Alternation in equational tree automata modulo XOR. In *FSTTCS'04*, pages 518–530. Springer-Verlag LNCS 3328, 2004.
22. K. N. Verma and J. Goubault-Larrecq. Karp-Miller trees for a branching extension of VASS. Research Report LSV-04-3, LSV, ENS Cachan, France, January 2004.

A Combination Method for Generating Interpolants

Greta Yorsh[1] and Madanlal Musuvathi[2]

[1] School of Comp. Sci., Tel Aviv Univ.
gretay@post.tau.ac.il
[2] Microsoft Research, Redmond
madanm@microsoft.com

Abstract. We present a *combination* method for generating interpolants for a class of first-order theories. Using interpolant-generation procedures for individual theories as black-boxes, our method modularly generates interpolants for the combined theory. Our combination method applies for a broad class of first-order theories, which we characterize as *equality-interpolating* Nelson-Oppen theories. This class includes many useful theories such as the quantifier-free theories of uninterpreted functions, linear inequalities over reals, and Lisp structures. The combination method can be implemented within existing Nelson-Oppen-style decision procedures (such as Simplify, Verifun, ICS, CVC-Lite, and Zap).

1   Introduction

Given two logical formulas A and B such that $A \wedge B$ is unsatisfiable, an interpolant I is a formula such that (i) A implies I, (ii) $I \wedge B$ is unsatisfiable, and (iii) every non-logical symbol that appears in I appears in both A and B. Craig interpolation theorem [2], a classic result in logic, proves the existence of interpolants for all first-order formulas A and B.

Motivation. While interpolation theorems are of great theoretical significance, our interest in interpolants is particularly motivated by their use in program analysis and model checking [12,11,13]. When A represents the current state of a system, and B represents the error state of the system, an interpolant I for A and B can be used as a *goal-directed* over-approximation of A. [12] uses this technique to achieve faster termination while model checking finite state systems, and [13] explores the possibility of using interpolants for model checking infinite systems. Such applications typically require an efficient procedure to generate an interpolant of A and B from the proof of unsatisfiability of $A \wedge B$. Existing procedures are either too expensive or work only for specific first-order theories (see Sec. 6).

In this paper, we provide a novel *combination* method for generating interpolants for a class of first-order theories. Using interpolant-generation procedures for component theories as black-boxes, this method generates interpolants for formulas in the combined theory. Provided the individual procedures for the

R. Nieuwenhuis (Ed.): CADE 2005, LNAI 3632, pp. 353–368, 2005.

component theories can generate interpolants in polynomial time, our method generates interpolants for the combined theory in polynomial time.

Our combination method relies on the Nelson-Oppen [14] framework for combining decision procedures. In this framework, the decision procedures for component theories communicate by propagating entailed equalities. The crucial idea behind our combination method is to associate a *partial interpolant* (Sec. 3.1) with each propagated equality. Whenever a component theory propagates an equality, the combination method uses the interpolant-generation procedure for that theory to generate the partial interpolant for the equality. When a theory detects a contradiction, the combination method uses the partial interpolants of all propagated equalities to compute the interpolant for the input formulas.

Our method places some restrictions on the theories that can be combined. The Nelson-Oppen combination method requires that the component theories have disjoint signatures and be *stably-infinite* [14,16]. Our method naturally inherits these restrictions. Additionally, our combination method restricts the form of equalities that can be shared by the component theories. Specifically, if a propagated equality contains a symbol that appears only in the input formula A, then it does not contain symbols that appear only in B, and vice versa. We show that this restricted form of equality propagation is sufficient for a class of theories, which we characterize as *equality-interpolating* theories (Sec. 4). Many useful theories including the quantifier-free theories of uninterpreted functions, linear arithmetic, and Lisp structures are equality-interpolating, and thus can be combined with our method.

Our method handles arbitrary quantifier-free input formulas. To handle Boolean structure in the input formula and to extend the combination to non-convex theories [14], we use an extended version of Pudlák's algorithm for generating interpolants for the propositional part. To show correctness of interpolants generated this way, we give an alternative explanation of Pudlák's algorithm based on partial interpolants.

The combination method has definite advantages over an interpolant generation procedure that is specific to a particular theory [17] or specific to a particular combination of theories [13]. First, by being modular, this method greatly simplifies the exposition, the proof of correctness, and the implementation of interpolant-generation procedures. More importantly, the combination method makes it easy to incrementally extend interpolant generation for additional theories. From a practical perspective, our combination method can be easily integrated with existing Nelson-Oppen-style decision procedures, such as [3,1,4,5], greatly enhancing their utility for program analysis.

In summary, this paper makes the following contributions. First, the paper presents an efficient method to generate interpolants for a general class of first-order theories, namely the union of equality-interpolating theories. Second, this paper shows that the classic Nelson-Oppen framework for combining decision procedures can be extended in a novel way to combine interpolant-generation procedures. Finally, we show that the basic combination algorithm can be generalized to work for both convex and non-convex theories. Due to a lack of space, this version does not contain proofs and contains only a partial discussion of the results. For a full version, the reader is referred to our technical report [19].

2   Preliminaries

Throughout this paper we use A and B to denote logical formulas of interest, restricting the syntax of A and B in different sections. Free variables in all formulas are implicitly existentially quantified, as we are checking satisfiability. We use *symbol* to refer to non-logical symbols and variables.[1]

Let $\mathcal{V}(\phi)$ be the set of symbols that appear in a formula or a term ϕ. Given formulas A and B, a symbol is A-local if it is in $\mathcal{V}(A) - \mathcal{V}(B)$. Similarly, a symbol is B-local if it is in $\mathcal{V}(B) - \mathcal{V}(A)$. A symbol is AB-common if it is in $\mathcal{V}(A) \cap \mathcal{V}(B)$. A formula or a term ϕ is AB-pure when either $\mathcal{V}(\phi) \subseteq \mathcal{V}(A)$ or $\mathcal{V}(\phi) \subseteq \mathcal{V}(B)$. Otherwise, ϕ is AB-mixed. Note that an AB-mixed formula or term contains at least one A-local symbol and at least one B-local symbol. Throughout this paper, we use a to refer to an A-local variable, b to refer to a B-local variable, and c, x and y to refer to a AB-common variables.

We use the standard notations $\vdash$, $\bot$, and $\top$ for entailment, contradiction, and tautology.

Definition 1. (Craig interpolant) *Given two first-order logical formulas A and B such that $A \wedge B \vdash \bot$, an interpolant for $\langle A, B \rangle$ is a first-order formula I such that (i) $A \vdash I$, (ii) $I \wedge B \vdash \bot$, and (iii) I refers only to AB-common symbols.*

Example 1. A is $(a = c) \wedge (f(c) = a)$ and B is $(c = b) \wedge \neg(b = f(c))$. The variables a, b, c are respectively A-local, B-local, AB-common variables, and f is an AB-common function symbol. The interpolant $f(c) = c$ for $\langle A, B \rangle$ involves only AB-common symbols f and c. By definition, A and B contain only AB-pure terms and formulas, however, AB-mixed terms and formulas may appear in the proof of unsatisfiability of $A \wedge B$. In this example, an AB-mixed equality $a = b$ can be generated in a proof of unsatisfiability of $A \wedge B$, as discussed later.

2.1   Theory-Specific Interpolants

As opposed to a well-known definition of Craig interpolants (Def. 1), interpolants for a specific first-order theory can be defined in several ways. In this section, we provide the definition used in this paper, and discuss alternative definitions in [19]. We adapt Def. 1 to the context of a specific first-order theory T. We use $\vdash_T$ to denote entailment in theory T. A theory T can contain uninterpreted

---

[1] Like much literature on decision procedures and model-checking, we use the term "variables" to refer to what in logic and theorem proving are "free constants." Similarly, "variables" in quantifier-free formulas refer to the corresponding Skolem-constants, as the formulas are implicitly existentially quantified.

symbols (e.g., uninterpreted functions), as well as a designated set of *interpreted* symbols (with intended meaning in T, or implicitly defined by T). [2]

Definition 2. (theory-specific interpolant) *Let T be a first-order theory of a signature Σ and let $\mathcal{L}$ be the class of quantifier-free Σ-formulas. Let $\Sigma_T \subseteq \Sigma$ denote a designated set of interpreted symbols in T. Let A and B be formulas in $\mathcal{L}$ such that $A \wedge B \vdash_T \perp$, i.e., the formula $A \wedge B$ is unsatisfiable in the theory T. We define a theory-specific interpolant for $\langle A, B \rangle$ in T to be a formula I in $\mathcal{L}$ such that (i) $A \vdash_T I$, (ii) $I \wedge B \vdash_T \perp$, and (iii) I refers only to AB-common symbols, and symbols in Σ_T (interpreted by the theory T).*

This definition differs from the traditional notion of an interpolant in two important ways. First, we require a theory-specific interpolant to be a quantifier-free formula. Second, a theory-specific interpolant can contain a symbol interpreted by the theory, even when if the symbol is A-local, or B-local or does not appear at all in A and B. The following example and discussion motivates these differences.

Example 2. Let A be $c_2 = \mathtt{car}(c_1) \wedge c_3 = \mathtt{cdr}(c_1) \wedge \neg\mathtt{atom}(c_1)$ and B be $\neg c_1 = \mathtt{cons}(c_2, c_3)$ in the theory of Lisp structures [15]. This theory interprets all function symbols that appear in this example, i.e., $\Sigma_T = \{\mathtt{car}, \mathtt{cdr}, \mathtt{cons}, \mathtt{atom}\}$. The variables c_1, c_2, and c_3 are AB-common, $\mathtt{cons}$ is B-local, and other function symbols are A-local. $A \wedge B$ is unsatisfiable in the theory of Lisp structures because A entails $c_1 = \mathtt{cons}(c_2, c_3)$ using the axiom: $\forall x, y, z : \neg\mathtt{atom}(x) \Rightarrow \mathtt{cons}(\mathtt{car}(x), \mathtt{cdr}(x)) = x$. According to Def. 2, $c_1 = \mathtt{cons}(c_2, c_3)$ is a theory-specific interpolant for $\langle A, B \rangle$. Note that if we do not allow $\mathtt{cons}, \mathtt{car}, \mathtt{cdr}$, and $\mathtt{atom}$ to appear in the interpolant, then there is no first-order formula that is an interpolant for $\langle A, B \rangle$.

A theory T has *no* interpolants (or does not have the interpolation theorem), if there exists a pair of input formulas A and B in $\mathcal{L}$ such that $A \wedge B \vdash_T \perp$ but there exists no formula in $\mathcal{L}$ that satisfies conditions (i)-(iii). If we allow any first-order Σ-formula as a theory-specific interpolant (instead of a quantifier-free Σ-formula), then every theory has theory-specific interpolants, as follows from Craig interpolation theorem. In practice however, we are interested in quantifier-free interpolants to guarantee that the satisfiability checks involving interpolants are complete, say in the subsequent stages of a program analysis. [3]

If we strengthen requirement (iii) from Def. 2 to eliminate interpreted symbols as well, then many interesting theories do not have interpolants (even if the generated interpolants are not restricted to quantifier-free formulas). Example 2 demonstrates that the theory of Lisp structures does not have interpolants under

---

[2] The details of how T is defined (e.g., set of axioms, set of models) are not essential to our method.

[3] In general, the language $\mathcal{L}$ of input formulas A and B is not necessarily the same as the language of the generated interpolants. For example, one could require quantifier-free formulas as input, but allow that generated interpolants contain quantifiers. Our method applies to this generalized definition of theory-specific interpolants [19].

this stronger requirement. However, a weaker requirement, as stated in Def. 2, is sufficient for our purposes, because in program analysis interpolants are used to eliminate state information, encoded by uninterpreted symbols, whereas interpreted symbols encode persistent semantics of statements, such as arithmetic operations and memory manipulations.

2.2   Interpolants for Combined Theories

In this paper, we address the problem of computing interpolants for a combined theory T. Without loss of generality, let T be a combination of two theories T_1 and T_2.[4] Let T_i be a first-order theory of signature Σ_i, with a set of interpreted symbols $\Sigma_{T_i} \subseteq \Sigma_i$, and $\mathcal{L}_i$ be a class of Σ_i-formulas, for $i = 1, 2$. The signature Σ of the combined theory T is a union of Σ_1 and Σ_2; also, the set of interpreted symbols of T is $\Sigma_T = \Sigma_{T_1} \cup \Sigma_{T_2}$. Let $\mathcal{L}$ be a class of Σ-formulas.

The input of the combination method consists of two $\mathcal{L}$-formulas A and B. Note that the input may contain *mixed* terms from both Σ_1 and Σ_2, but it does not contain AB-mixed terms, by definition. An interpolant for $\langle A, B \rangle$ in the combined theory T is an $\mathcal{L}$-formula which may contain mixed terms from both Σ_1 and Σ_2, but contains only AB-common uninterpreted symbols, or symbols interpreted by T_1 and T_2.

Example 3. Consider a combination of the theory of uninterpreted functions and the theory of linear inequalities (where the symbols $\{+, <, \leq\}$ have the standard interpretation over reals) with the input formulas:

$$A \stackrel{\text{def}}{=} (f(x_1) + x_2 = x_3) \wedge (f(y_1) + y_2 = y_3) \wedge (y_1 \leq x_1)$$
$$B \stackrel{\text{def}}{=} (x_2 = g(b)) \wedge (y_2 = g(b)) \wedge (x_1 \leq y_1) \wedge (x_3 < y_3)$$

The first subformula of A contains a mixed term, with both f from the theory of uninterpreted functions and $+$ from the theory of linear inequalities.

We assume that T_1 and T_2 are stably-infinite theories with disjoint signatures, i.e., the only common symbol for Σ_1 and Σ_2 is equality. Each T_i has a decision procedure for satisfiability of a (quantifier-free) conjunction of Σ_i-literals (a literal is an atomic formula or its negation). These are standard requirements of the component theories in the Nelson-Oppen framework. In addition, we assume that each T_i has an efficient interpolant generation procedure that takes as input a pair of conjunctions of Σ_i-literals A_i and B_i, (i.e., A_i and B_i are pure Σ_i formulas). It returns an $\mathcal{L}_i$-formula as a theory-specific interpolant for A_i and B_i. Finally, we make the assumption that each T_i is an equality-interpolating theory; we explain and justify it in the next sections.

3   The Combination Method

This section deals with the simple case in which (i) input formulas are conjunctions of pure literals, and (ii) all theories are convex, i.e., if a disjunction

---

[4] As usual, combination of theories means union of axioms or, equivalently, intersection of sets of models.

of equalities between variables is entailed, then at least one of the disjuncts is entailed [14]. These restrictions greatly simplify our description, while capturing the intuition behind our algorithm; they are relaxed in the following sections.

Purification. Our method uses the Nelson-Oppen framework [14] to compute an interpolant for the combined theory. We assume that the unsatisfiability of input formula ψ was proved by a Nelson-Oppen procedure. The first step of the Nelson-Oppen procedure is *purification*. Given a mixed formula ψ, it constructs an equisatisfiable formula $\psi_1 \wedge \psi_2$, where ψ_i consists only of pure Σ_i-literals. Purification introduces new variables to replace terms of one signature that appear as sub-terms of terms in the other signature. Equalities defining these variables are added to the input formulas.

In our setting, the input ψ is a conjunction $A \wedge B$. We purify A and B separately. This guarantees that the new variables generated by the purification of A do not appear in the interpolant, because these variables are A-local. The result of purification of A is $A_1 \wedge A_2$ such that A_i contains only symbols from Σ_i and $A_1 \wedge A_2$ is satisfiable if and only if A is satisfiable; similarly, for B.

Example 4. After purifying A and B from Example 3 separately, we have that $A = A_{UIF} \wedge A_{LI}$ and $B = B_{UIF} \wedge B_{LI}$, where

$$A_{UIF} \overset{\text{def}}{=} a_1 = f(x_1) \wedge a_2 = f(y_1)$$
$$A_{LI} \overset{\text{def}}{=} a_1 + x_2 = x_3 \wedge a_2 + y_2 = y_3 \wedge y_1 \le x_1$$
$$B_{UIF} \overset{\text{def}}{=} x_2 = g(b) \wedge y_2 = g(b)$$
$$B_{LI} \overset{\text{def}}{=} x_1 \le y_1 \wedge x_3 < y_3$$

a_1 and a_2 are A-local variables, b is a B-local variables, and f and g are A-local and B-local function symbols, respectively. We will use this as a running example.

Equality Propagation. Let A and B be conjunctions of pure literals in the signature Σ of the combined theory T. A is $A_1 \wedge A_2$ such that A_i contains only symbols from Σ_i; similarly, for B. Let $\psi_i \overset{\text{def}}{=} A_i \wedge B_i$, for $i = 1, 2$. Note that ψ_i is a pure formula in Σ_i, but it is not AB-pure. Suppose that a Nelson-Oppen procedure shows the unsatisfiability of $\psi_1 \wedge \psi_2$ in T. It generates the set of equalities between variables, denoted by Eq.[5] Eq is sufficient to show the unsatisfiability of $A \wedge B$ using only one of the theories; assume, without loss of generality, that this theory is T_1. That is, $A \wedge B \vdash_T \bot$ follows from the fact that $Eq \wedge A_1 \wedge B_1 \vdash_{T_1} \bot$.

Example 5. The input $A \wedge B$ from Example 4 is not satisfiable, because $A \wedge B \vdash_T$ $x_1 = y_1 \wedge a_1 = a_2 \wedge x_2 = y_2 \wedge x_3 = y_3$, which contradicts $x_3 < y_3$ from B. The set of equalities $Eq = \{x_1 = y_1, a_1 = a_2, x_2 = y_2\}$ is sufficient to derive a contradiction using only the theory of linear inequalities: $A_{LI} \wedge B_{LI} \wedge$

---

[5] [14] shows that it is sufficient to propagate only equalities between variables.

$Eq \vdash_{LI} \bot$. An interpolant for $\langle A, B \rangle$ is $y_1 < x_1 \lor x_2 - y_2 = x_3 - y_3$. Note that the symbols g and f are eliminated because they denote local uninterpreted functions, but theory-specific interpreted functions $+$ and $<$ are not eliminated, recall the discussion of Def. 2(iii).

Overview of Our Combination Method. The idea is to use an interpolant generated by T_1 from a proof of unsatisfiability of $Eq \land A_1 \land B_1$, to generate an interpolant for $\langle A, B \rangle$ in the combined theory T.

The interpolant-generation procedure for T_1 takes as input two formulas A' and B' for which the conjunction $A' \land B'$ is not satisfiable. In our case, the unsatisfiable conjunction is $Eq \land A_1 \land B_1$. The question is how to split it into two formulas, A' and B'. The condition for splitting is that the common symbols for $\langle A', B' \rangle$ should be (a subset of) AB-common symbols, because we would like to use the resultant interpolant for $\langle A', B' \rangle$ as a part of an interpolant for the original A and B.

Suppose that Eq contains only AB-pure equalities. We split Eq into an A-part and a B-part: all the equalities from Eq that involve A-local symbols are added to the A-part; the B-part contains the rest of Eq.[6] We define A' to be a conjunction of A_1 and the A-part of Eq, and similarly for B'. Now, we can generate an interpolant for $\langle A', B' \rangle$ in theory T_1, using the interpolant generation procedure for T_1, as we planned.

It is important to note that the theory-specific interpolant for $\langle A', B' \rangle$ in theory T_1 is not an interpolant for the input formula $\langle A, B \rangle$ in the combined theory T. It uses only AB-common symbols, i.e., satisfies property (iii) of Def. 2, but it need not satisfy the properties (i) and (ii). The reason, intuitively, is that A does not imply A', because A' contains equalities which cannot be derived without information from B, as shown in the following example:

Example 6. In Example 5, the theory of linear inequalities derives a contradiction from $A' = A_{LI} \land a_1 = a_2$ and $B' = B_{LI} \land x_1 = y_1 \land x_2 = y_2$. The theory-specific partial interpolant for $\langle A', B' \rangle$ is $x_2 - y_2 = x_3 - y_3$. It is *not* an interpolant for the input $\langle A, B \rangle$, because A does not entail $x_2 - y_2 = x_3 - y_3$ (in the combined theory), as A alone does not entail $a_1 = a_2$.

To address this problem, we attach, for each propagated equality, additional information, called "partial interpolant". This notion is formally defined in Sec. 3.1, and used along with a theory-specific interpolant for $\langle A', B' \rangle$ in theory T_1 to generate an interpolant for $\langle A, B \rangle$ in T.

Finally, if Eq contains AB-mixed equalities, we construct an equivalent set of AB-pure equalities, as explained in Sec. 4. This is the reason for restricting our method only to equality-interpolating theories.

---

[6] For equalities with only AB-common variables, the question as to whether to add them "to B or not to B" can be answered in either way, as long as the answer is always the same for a given variable, and consistent for all equalities that appear in the proof of unsatisfiability. This gives us some control over how precise the interpolant is (how "close" it is to A or B), but we have not explored this direction yet. In all our examples, we add AB-common equalities to the B-part.

3.1 Partial Interpolants

In the following definitions we assume that all equalities generated in the Nelson-Oppen framework are AB-pure equalities, as guaranteed by Sec. 4.

A *partial interpolant* is a formula associated with each formula derived by the theories in the proof of unsatisfiability. To simplify the definitions, we associate a (trivial) partial interpolant with each input formula as well. A partial interpolant for $\perp$ is the interpolant for the input formula $\langle A, B \rangle$. The crucial part of our combination method is a way to associate a partial interpolant with each propagated equality. Whenever a decision procedure for a component theory T_i generates an equality e that needs to be propagated, our method provides a partial interpolant for that equality. Our method does not require a special interface for generating partial interpolants, but uses the interpolant-generation procedures of the component theories. A partial interpolant is a boolean combination of the partial interpolants for the inputs and a *theory-specific partial interpolant*. A theory-specific partial interpolant is generated by T_1 using only the input formulas and equalities generated for other theories, without using their partial interpolants.

Definition 3. (projection) *Let* Θ *be a conjunction of AB-pure literals. Let* $\Theta|_A$ *be a conjunction of A-local literals of* Θ, *and* $\Theta|_B$ *be a conjunction of B-local and AB-common literals of* Θ. *Note that* $\Theta = \Theta|_A \wedge \Theta|_B$.

Definition 4. (theory-specific partial interpolant) *Let A' and B' be conjunctions of pure literals in Σ_i, and let e be an AB-pure atomic formula generated by the decision procedure for the theory T_i, i.e.,* $A' \wedge B' \vdash_{T_i} e$ *(thus,* $A' \wedge B' \wedge \neg e \vdash_{T_i} \perp$).

An interpolant generated for $\langle A' \wedge \neg(e|_{A'}), B' \wedge \neg(e|_{B'}) \rangle$ *by T_i's procedure is a theory-specific partial interpolant for e w.r.t.* $\langle A', B' \rangle$, *denoted by* $\phi^i_{A',B'}(e)$.

Intuitively, we add $\neg e$ to A' if e contains an A-local symbol, otherwise we add it to B', using the assumption that e is AB-pure, i.e., e cannot contain both A-local and B-local symbols. Thus, any theory-specific partial interpolant for e contains only AB-common symbols.

Example 7. The first step of a proof of unsatisfiability in Example 4 uses the theory of linear inequalities to derive the equality $x_1 = y_1$ from $A' \stackrel{\text{def}}{=} A_{LI}$, which contains the literal $(y_1 \leq x_1)$, and $B' \stackrel{\text{def}}{=} B_{LI}$, which contains the literal $(x_1 \leq y_1)$. We use interpolant generation of the theory of linear inequalities to derive an interpolant for $(y_1 \leq x_1)$ and $(x_1 \leq y_1) \wedge \neg(x_1 = y_1)$. Trivially, the interpolant is $y_1 \leq x_1$, which is the theory-specific partial interpolant for $x_1 = y_1$, denoted by $\phi^{LI}_{A',B'}(x_1 = y_1)$.

Let e be an AB-pure equality such that $A \wedge B \vdash e$. We define a *partial interpolant* for e w.r.t. $\langle A, B \rangle$ as follows.

Definition 5. (partial interpolant) *Suppose that e is derived from $A \wedge B$ in the Nelson-Oppen framework by a theory T_i. Suppose that T_i derives e from two* *conjunctions of pure literals in Σ_i, denoted by A_i and B_i, and a set of AB-pure equalities Eq: $A_i \wedge B_i \wedge Eq \vdash e$.*

A partial interpolant for e w.r.t. $\langle A, B \rangle$ denoted by $\phi_{A,B}(e)$ is defined inductively. The base cases: If $e \in A_i$ then $\phi_{A,B}(e)$ is $\bot$. If $e \in B_i$ then $\phi_{A,B}(e)$ is $\top$. The inductive definition: Let $A' \stackrel{\text{def}}{=} A_i \wedge Eq|_A$ and $B' \stackrel{\text{def}}{=} B_i \wedge Eq|_B$.

$$\phi_{A,B}(e) = (\phi^i_{A',B'}(e) \vee \bigvee_{a \in A'} \phi_{A,B}(a)) \wedge \bigwedge_{b \in B'} \phi_{A,B}(b) \tag{1}$$

Note that this definition includes the special case when e is $\bot$.

Example 8. Table 1 shows the partial interpolants generated by the combination method for the input formulas from Example 4. In the second step of the proof, the decision procedure for the theory of uninterpreted functions generates the equality $a_1 = a_2$ from the input A_{UIF} and B_{UIF} defined in Example 4, and the equality $(x_1 = y_1)$. First, we compute a theory-specific partial interpolant for $a_1 = a_2$, denoted by $\phi^{UIF}_{A',B'}(a_1 = a_2)$, where $A' = A_{UIF}$ and $B' = B_{UIF} \wedge (x_1 = y_1)$, because x_1 and y_1 are AB-common. By Def. 4, we run the interpolant-generation procedure of the theory of uninterpreted functions with the input $A' \wedge \neg(a_1 = a_2)$ and B', and we get $\neg(x_1 = y_1)$, which is $\phi^{UIF}_{A',B'}(a_1 = a_2)$.

We compute a partial interpolant $\phi_{A,B}(a_1 = a_2)$ using $\phi^{UIF}_{A',B'}(a_1 = a_2) = \neg(x_1 = y_1)$ and $\phi_{A,B}(x_1 = y_1) = (y_1 \leq x_1)$ (and the partial interpolant for the input equality $x_1 = y_1$, generated in the previous step). The result $\phi_{A,B}(a_1 = a_2)$ is $(y_1 \leq x_1) \wedge \neg(x_1 = y_1)$.

Table 1. Partial interpolants generated by the combination method for the input formulas from Example 4. In each step of the process, the decision procedure for the component theory T generates a formula e and the corresponding partial interpolant

Theory T	e	$\phi^{\mathbf{T}}_{\mathbf{A'},\mathbf{B'}}(\mathbf{a_1 = a_2})$	$\phi_{\mathbf{A,B}}(e)$
LI	$x_1 = y_1$	$y_1 \leq x_1$	$y_1 \leq x_1$
UIF	$a_1 = a_2$	$\neg(x_1 = y_1)$	$y_1 < x_1$
UIF	$x_2 = y_2$	$\top$	$\top$
LI	$\bot$	$x_2 - y_2 = x_3 - y_3$	$x_2 - y_2 = x_3 - y_3 \vee y_1 < x_1$

If a theory proves unsatisfiability, the partial interpolant $\phi_{A,B}(\bot)$ is an interpolant for $\langle A, B \rangle$, as shown in the following lemma.

Lemma 1. *The partial interpolant $\phi_{A,B}(e)$ from Def. 5 is an interpolant for $A \wedge \neg(e|_A)$ and $B \wedge \neg(e|_B)$ in the combined theory T. In the special case when e is $\bot$, $\phi_{A,B}(\bot)$ is an interpolant for $\langle A, B \rangle$.*

Example 9. In the last step, $\phi_{A,B}(\bot)$ is $x_2 - y_2 = x_3 - y_3 \vee y_1 < x_1$. It is easy to verify that $\phi_{A,B}(\bot)$ is an interpolant for the input formulas A and B from Example 3.

4   Equality-Interpolating Theories

The combination method in Sec. 3 requires that each component theory only propagates AB-pure equalities. This restriction arose as partial interpolants are not defined for AB-mixed equalities. This section justifies the restriction by showing that for a class of first-order theories, defined as *equality-interpolating* theories, it is sufficient to share AB-pure equalities. Also, this section shows that many interesting theories including the quantifier-free theories of uninterpreted functions, linear arithmetic, and Lisp structures are equality-interpolating.

The basic idea behind equality-interpolating theories is the following. Whenever a decision procedure for a component theory generates an equality $a = b$, where a is an A-local variable and b is a B-local variable the combination method requires that the theory also produce an AB-common term t, such that $a = t$ and $t = b$. Instead of propagating $a = b$, the theory now propagates these two AB-pure equalities separately. For an *equality-interpolating* theory, such an AB-common term t exists for all entailed AB-mixed equalities:

Definition 6. (equality-interpolating theory) *Theory T is equality-interpolating when for all A and B in T, and for all AB-mixed equalities between variables $a = b$ such that $A \wedge B \vdash_T a = b$, there exists a term t such that $A \wedge B \vdash a = t \wedge t = b$ and t contains only AB-common symbols. We say that t is an equality-interpolating term for $a = b$ w.r.t. $\langle A, B \rangle$.*

Example 10. This example shows that not all theories are equality-interpolating. Consider a theory with two relation symbols P and Q, and the axiom $\forall abc\, P(a,c) \wedge Q(c,b) \Rightarrow a = b$. When A contains $P(a,c)$ and B contains $Q(c,b)$, $A \wedge B \vdash a = b$. However, there is no equality-interpolating term for $a = b$.

The proof of correctness of the combination method for equality-interpolating theories follows from Def. 6 and Lem. 1. Whenever a decision procedure for a component theory generates an AB-mixed equality $a = b$, the combination method propagates two equalities $a = v_t$ and $v_t = b$, where v_t is a previously unseen variable representing the equality-interpolating term t. The combination method treats these new variables v_t as AB-common symbols, thus the two equalities propagated instead of $a = b$ are AB-pure and so the correctness of the combination method described in Sec. 3 applies. After the interpolant is generated, occurrences of v_t in it are replaced by the associated term t. By Def. 6, t contains only AB-common symbols, thus the interpolant is an AB-common formula, as required.

The modification mentioned above does not affect the Nelson-Oppen framework in terms of complexity, termination or soundness. All component theories contain equality axioms, thus each theory can infer $a = b$ from the equalities $a = v_t$ and $v_t = b$. Moreover, as the variable v_t is previously unseen, this is the only inference the theories can make. Note, the number of new variables v_t generated in this process is bounded by the number of AB-mixed equalities used in the proof of unsatisfiability of $A \wedge B$.

In the remainder of this section, we prove that some useful theories are equality-interpolating.

The Theory of Uninterpreted Functions. A decision procedure for the theory of uninterpreted functions can be easily modified to generate only AB-pure equalities. The idea is to modify the implementation of the congruence closure algorithm [15] to choose a representative for an equivalence class to be an AB-common term, whenever an equivalence class contains at least one such term. When an equivalence class contains both A-local and B-local terms, it also contains an AB-common term, as follows from:

Lemma 2. *The theory of uninterpreted functions is equality-interpolating.*

Sketch of Proof: We prove a stronger claim: there exists an interpolating term t for all equalities of the form $t_a = t_b$ entailed by $A \wedge B$, where t_a is an A-local term (involves at least one A-local symbols or variable) and t_b is a B-local term.

Note that $t_a = t_b$ is an AB-mixed equality, but it does not contain AB-mixed terms. Every proof of $t_a = t_b$ that uses AB-mixed terms can be transformed into a proof of $t_a = t_b$, in which all derivations involve only AB-pure terms. Assume that a proof of $t_a = t_b$ from $A \wedge B$ uses only AB-pure terms in all derivations. The proof proceeds by induction on the proof tree of $t_a = t_b$ from $A \wedge B$.

The Theory of Lisp Structures. A decision procedure for the theory of Lisp structures based on the congruence closure algorithm is described in [15]. A proof generated by the decision procedure for the theory of Lisp structures can contain proof rules of the theory of uninterpreted functions and an additional rule:

$$\frac{z = cons(x, y)}{x = car(z) \wedge y = cdr(z) \wedge \neg atom(z)} \tag{2}$$

The interpolant-generation procedure for the theory of uninterpreted functions [13] can be adapted to the theory of Lisp structures, as follows. Given a proof P of unsatisfiability of $A \wedge B$ in the theory of Lisp structures, we replace each derivation in P that uses proof rule (2) by the formula it derives, which is treated as an axiom. The result is a proof of unsatisfiability P_{UIF} in the theory of uninterpreted functions, where the symbols car, cdr, cons, and atom are treated as uninterpreted function symbols.

Let H be the set of new axioms added to P. Using Lem. 2, we can ensure that all formulas in H are AB-pure. Thus, the interpolant generated for $A \wedge (H|_A)$ and $B \wedge (H|_B)$ by the theory of uninterpreted functions using the proof P_{UIF}, is the interpolant for $\langle A, B \rangle$ in the theory of Lisp structures.

The Theory of Linear Inequalities. To show that the theory of linear inequalities is equality-interpolating, we first show that there is an *inequality-interpolating* term for every AB-mixed inequality between variables, entailed by $A \wedge B$.

Definition 7. (inequality-interpolant) *For an AB-mixed inequality $a \leq b$ such that $A \wedge B \vdash a \leq b$, an inequality-interpolant is an AB-common term t such that $A \wedge B \vdash a \leq t \leq b$.*

If $A \wedge B \vdash (a \leq b) \wedge (b \leq a)$ and t_1, t_2 are the inequality-interpolating terms for $a \leq b$ and $b \leq a$ respectively, it follows that $a = b = t_1 = t_2$. Thus, t_1 (or t_2) is an equality-interpolating term for $a = b$.

Lemma 3. *Given conjunctions of linear arithmetic constraints A, B, an inequality-interpolant exists for every AB-mixed inequality $a \leq b$ entailed by $A \wedge B$.*

Proof: Consider an AB-mixed inequality $a \leq b$ derived from $A \wedge B$. Let A and B respectively contain m and n linear constraints. These constraints are of the following form

$$A \equiv \bigwedge_{1 \leq i \leq m} 0 \leq s_i + t_i \quad \text{and} \quad B \equiv \bigwedge_{1 \leq j \leq n} 0 \leq s'_j + t'_j$$

where the terms s_i are A-local, the terms s'_j are B-local, and the terms t_i and t'_j are AB-common. Any linear inequality that can be derived from A and B can be obtained by a linear combination of the constraints in A and B. As $A \wedge B \vdash a \leq b$ (or equivalently $0 \leq -a + b$), there exist non-negative constants $d_1, d_2, \ldots, d_m, d'_1, d'_2, \ldots, d'_n$ such that

$$0 \leq \sum_{i=1}^{m} d_i(s_i + t_i) + \sum_{j=1}^{n} d'_j(s'_j + t'_j) = -a + b \tag{3}$$

Consider the linear combination in Equation 3 restricted to the terms in A. As the terms s'_j and t'_j contain no A-local variables, we have $0 \leq \sum_{i=1}^{m} d_i(s_i + t_i) = -a + t$ for some AB-common term t. Similarly, considering the linear combination restricted to terms in B, we have $0 \leq \sum_{j=1}^{n} d'_j(s'_j + t'_j) = t' + b$ for some AB-common term t'. From Equation 3 it follows that $t = -t'$. Thus, $A \wedge B \vdash (a \leq t) \wedge (t \leq b)$, and t is the inequality-interpolating term for $a \leq b$.

Lemma 4. *Theory of linear arithmetic is equality-interpolating.*

Proof: Directly follows from Lem. 3 and the discussion following Def. 7.

5   Interpolants for Arbitrary Quantifier-Free Formulas

Sec. 3 describes the combination of interpolant-generation procedures for convex theories with disjoint signatures, when the input formulas A and B are conjunctions of literals. This section relaxes these constraints. First, we describe Pudlák's algorithm for generating propositional interpolants [17]. Then, we integrate our method with an extended version of Pudlák's algorithm in the *lazy-proof-explication* framework for checking satisfiability of quantifier free first-order formulas with arbitrary boolean structure.

Pudlák's Algorithm. We describe Pudlák's algorithm for generating propositional interpolants and give an alternative correctness argument based on partial interpolants. This algorithm takes as input a proof of unsatisfiability of a propositional formula $A \wedge B$ and generates a propositional interpolant I for $\langle A, B \rangle$. The algorithm generates a partial interpolant $p(c)$ for each clause c derived in the proof, as described below. The partial interpolant generated for the empty clause $p(\bot)$ is an interpolant I for $\langle A, B \rangle$.

Definition 8. *Given two clauses of the form $c_1 \overset{\text{def}}{=} x \vee c_1'$ and $c_2 \overset{\text{def}}{=} \neg x \vee c_2'$, the resolution of c_1 and c_2 is a clause $c \overset{\text{def}}{=} c_1' \vee c_2'$, denoted by $c = resolve_x(c_1, c_2)$, where x is called the pivot variable.*

Let $\langle A, B \rangle$ be a pair of clause sets such that $A \wedge B \vdash \bot$. Let $\mathcal{T}$ be a proof of unsatisfiability of $A \wedge B$. The propositional formula $p(c)$ for a clause c in $\mathcal{T}$ is defined by induction on the proof structure:

(i) if c is one of the input clauses then
  (a) if $c \in A$, then $p(c) := \bot$;
  (b) if $c \in B$, then $p(c) := \top$.
(ii) otherwise, c is a result of resolution, i.e., $c = resolve_x(c_1, c_2)$
  (a) if $x \in A$ and $x \notin B$ (x is A-local), then $p(c) := p(c_1) \vee p(c_2)$
  (b) if $x \notin A$ and $x \in B$ (x is B-local), then $p(c) := p(c_1) \wedge p(c_2)$
  (c) otherwise (x is AB-common), $p(c) := (x \vee p(c_1)) \wedge (\neg x \vee p(c_2))$.

The correctness of the algorithm is guaranteed by the following invariant: for each clause $c \in \mathcal{T}$, the partial interpolant $p(c)$ is an interpolant for $\langle g_A(c), g_B(c) \rangle$ where $g_A(c) \overset{\text{def}}{=} A \wedge ((\neg c)|_A)$ and $g_B(c) \overset{\text{def}}{=} B \wedge ((\neg c)|_B)$ When c is an empty clause $\bot$, we get that $\langle g_A(\bot), g_B(\bot) \rangle$ is $\langle A, B \rangle$, and the formula $p(\bot)$ is the result.

Lazy Proof-Explication Framework. In order to leverage advances in SAT solving, state-of-the-art decision procedures [1,4,5] based on the Nelson-Oppen framework use a SAT solver to perform propositional reasoning. Given an input formula, the SAT solver treats all atomic formulas occurring in the input formula as free boolean variables. Suppose that the SAT solver finds an assignment to the boolean variables that satisfies the input formula propositionally. This assignment is a conjunction of (first-order) literals. It is passed to a Nelson-Oppen decision procedure.

The decision procedure attempts to derive a contradiction from this conjunction of literals. If it cannot derive a contradiction, the input formula is declared as satisfiable. If a contradiction is detected, the negation of the current assignment is added to the SAT solver as a new conflict clause. Because this new clause is in conflict with the current assignment, the SAT solver backtracks, searching for a new assignment. If it cannot find another assignment, it has proved that the propositional abstraction is unsatisfiable. Thus, the input formula is unsatisfiable.

Integration with an Extended Pudlák's Algorithm. We assume that the unsatisfiability of A and B in theory T is proved by a lazy proof-explication framework. That is, a SAT solver proved propositional unsatisfiability of A and B using a set of conflict clauses C. For each conflict clause c in C, $\neg c$ is a conjunction of (first-order) literals. By construction, we have a proof of unsatisfiability of $\neg c$. Also, it is guaranteed that $\neg c$ contains only AB-pure literals, as it contains only the original literals from A or B (each of which is AB-pure by definition). Therefore, we can use the method described in Sec. 3 to generate an interpolant between the A-part of $\neg c$ and the B-part of $\neg c$, which is also called a partial interpolant for the conflict clause c.

Definition 9. (partial interpolants for clauses) *Let $A \wedge B \vdash_T \bot$ be proved by a decision procedure for T using a corresponding set of conflict clauses C, such that $A \wedge B \wedge C$ is propositionally unsatisfiable. A partial interpolant for a clause c is denoted by $\phi_{A,B}(c)$. If $c \in A$, then $\phi_{A,B}(c) = \bot$, if $c \in B$, then $\phi_{A,B}(c) = \top$, otherwise, for a conflict clause $c \in C$, a partial interpolant $\phi_{A,B}(c)$ is the interpolant for $\langle (\neg c)|_A, (\neg c)|_B \rangle$ in theory T, where the projection operation is given in Def. 3.*

We use partial interpolants $\phi_{A,B}(c)$ defined above as initial values for $p(c)$ in the extended version of Pudlák's algorithm, instead of using the standard initialization of Pudlák's algorithm from Def. 8(i). (To see why this change is necessary, recall that a conflict clause may involve both A-local and B-local literals.) Partial interpolants in Def. 9 satisfy the invariant of Pudlak's algorithm.

There is no change in phase (ii) of Def. 8. The input for the extended algorithm consists of three clause sets, denoted by $\langle A, B; C \rangle$, all three of them are necessary for an unsatisfiability proof. However, in each resolution step, the pivot is guarantee to be in A or B, because all literals in conflict clauses C appear in the original formulas A and B. Note that the result is a first-order interpolant, which is a combination of the original clauses and the interpolants generated for the conflict clauses. The correctness of the interpolant generated by the extended Pudlák's algorithm follows from the correctness of theory-specific interpolants for conflict clauses.

Lemma 5. *The interpolant for $\langle A, B; C \rangle$ generated by the extended Pudlak's algorithm using partial interpolants for clauses as in Def. 9 is an interpolant for the input $\langle A, B \rangle$ in theory T.*

6   Related Work

Interpolants are of great theoretical and practical significance. Our interest in interpolants is particularly motivated by their use in program analysis and model checking. [12] uses interpolants to achieve faster termination while model checking finite state systems, and [13] explores the possibility of using interpolants for model checking infinite systems.

Craig interpolation theorem for first-order logic [2] shows the existence of a first-order interpolant for any pair of formulas in first-order logic. While constructive proofs of Craig interpolation theorem exist [6,18,9], these proofs are (to the best of our knowledge) based on cut elimination and result in very expensive interplation-generation procedures. (See [8] for references on the complexity of cut elimination.)

On the other hand, interpolants can be generated efficiently for formulas in a restricted subclass of first-order logic. When the input formulas A and B are propositional, or when they are both conjunctions of linear constraints, Pudlák [17] provides interpolant-generation procedures that are linear in the proof of unsatisfiability of $A \wedge B$. McMillan [13] extends these procedures to compute interpolants for quantifier-free formulas in the combined theory of uninterpreted functions and linear arithmetic.

Finally, the Nelson-Oppen framework is being constantly improved. For example, a recent work by [7] defines sufficient conditions for extending the Nelson-Oppen framework to theories with non-disjoint signatures, e.g., the theory of bit-vectors, presburger arithmetic, a theory of Lists with length operator, or theories of many-sorted logics. In the extended framework, the theories can exchange atomic formulas over the intersection of their signatures, and not only equalities between variables. Our method, by being modular, is well-suited to support such advances in the Nelson-Oppen framework. Given the unsatisfiability proof with only AB-pure equalities, our method can generate interpolants for non-disjoint theories, because partial interpolants in Def. 4 and Def. 5 do not assume that the theories exchange only equalities.

7   Conclusions and Future Work

The combination method for equality-interpolating theories presented in this paper proves existence of quantifier-free interpolants for a combined theory, if all the component theories have quantifier-free interpolants. If some of the theories has quantified interpolants, our method produces correct, quantified interpolant for the combined theory. Currently, our method applies only to quantifier-free input formulas. We believe the method can be extended to handle quantified formulas, because the proof of unsatisfiability contains a finite instantiation of the quantified variables.

Our method shows how to integrate interpolant generation for various theories within the existing satisfiability-checking tools, adding only a small overhead. This provides a practical way to use interpolants for speeding up termination of software model checking and real-time model checking.

Finally, the combination of interpolant-generation procedures demonstrates that *equality propagation* in the Nelson-Oppen framework can be used to combine operations other than satisfiability checking. Recently, [10] have used a similar approach to combine abstract domains. We believe that similar combination methods for operations would enhance program analysis tools while retaining the flexibility of the Nelson-Oppen framework to extend with additional theories.

References

1. C. Barrett and S. Berezin. CVC Lite: A new implementation of the cooperating validity checker. In *CAV*, pages 515–518, 2004.
2. W. Craig. Linear reasoning. a new form of the herbrand-gentzen theorem. *J. Symbolic Logic*, 22:250–268, 1957.
3. D. Detlefs, G. Nelson, and J. Saxe. Simplify theorem prover. http://research.compaq.com/SRC/esc/Simplify.html.
4. J.-C. Filliâtre, S. Owre, H. Rueß, and N. Shankar. ICS: integrated canonizer and solver. In *CAV*, pages 246–249, 2001.
5. C. Flanagan, R. Joshi, X. Ou, and J. B. Saxe. Theorem proving using lazy proof explication. In *CAV*, pages 355–367, 2003.
6. J. H. Gallier. *Logic for Computer Science: Foundations of Automatic Theorem Proving*. John Wiley & Sons, New York, 1987.
7. V. Ganesh, S. Berezin, C. Tinelli, and D. L. Dill. Combination results for many-sorted theories with overlapping signatures. Technical report, Department of Computer Science, Stanford University, 2004.
8. P. Gerhardy. Refined Complexity Analysis of Cut Elimination. In *CSL*, pages 212–225, 2003.
9. G.Takeuti. *Studies in Logic*, volume 81. Elsevier, Amsterdam, North Holland, 1975.
10. Sumit Gulwani and Ashish Tiwari. Unpublished manuscript.
11. T. A. Henzinger, R. Jhala, R. Majumdar, and K. L. McMillan. Abstractions from proofs. In *POPL*, pages 232–244, 2004.
12. K.L. McMillan. Interpolation and sat-based model checking. In *CAV*, pages 1–13, 2003.
13. K.L. McMillan. An interpolating theorem prover. In *TACAS*, pages 16–30, 2004.
14. G. Nelson and D. C. Oppen. Simplification by cooperating decision procedures. *ACM Transactions on Programming Languages and Systems*, 1(2):245–257, October 1979.
15. G. Nelson and D. C. Oppen. Fast decision procedures based on congruence closure. *J. ACM*, 27(2):356–364, 1980.
16. Derek C. Oppen. Complexity, convexity and combinations of theories. In *Theoretical Computer Science*, volume 12, pages 291–302, 1980.
17. P. Pudlák. Lower bounds for resolution and cutting planes proofs and monotone computations. *J. of Symbolic Logic*, 62(3):981–998, 1995.
18. S.C.Kleene. *Mathematical Logic*. Wiley Interscience, New York, 1967.
19. G. Yorsh and M. Musuvathi. A combination method for generating interpolants. Technical Report MSR-TR-2004-108, Microsoft Research, October 2004.

sKizzo: A Suite to Evaluate and Certify QBFs

Marco Benedetti

Istituto per la Ricerca Scientifica e Tecnologica (IRST),
Via Sommarive 18, 38055 Povo, Trento, Italy
benedetti@itc.it

Abstract. We present sKizzo, a system designed to evaluate and certify QBFs by means of propositional skolemization and symbolic reasoning.

1 Introduction

We present sKizzo [2,3], a software suite for dealing with *Quantified Boolean Formulas* (QBFs). sKizzo is mainly aimed at evaluating prenex CNF formulas by means of a novel *symbolic skolemization* technique[4]. In addition, it enables the user to (A) experiment with *quantifier trees*[6], (B) certify the (un)satisfiability of formulas[5] and (possibly) extract *unsatisfiable cores*, and (C) compute, manage, and query stand-alone certificates of satisfiability for QBFs. Both quantifier tree extraction and answer certification have never been attempted so far on QBFs.

At the hearth of sKizzo stays a new kind of symbolic representation for clauses and formulas, based on *Binary Decision Diagrams* (BDDs). As opposed to previous BDD-based approaches to propositional logic, sKizzo's one employs a two-level data structure [2] designed to take advantage of the distinguishing features of QBFs. Besides allowing for a novel style of (complete/incomplete) *symbolic* reasoning, such representation makes it possible to unify within a coherent framework all the other approaches to QBF-satisfiability implemented so far. Namely: DPLL-like branching reasoning, q-resolution based algorithms, and compilation-to-SAT techniques.

2 Representation of QBF Instances

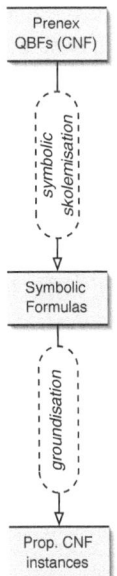

Three representation spaces for QBFs coexist within sKizzo. They are interconnected by two satisfiability-preserving trasformations (applied one-way), as reported in the picture aside. The first transformation leverages *outer skolemization* to map any (prenex CNF) instance $F \in QBFs$ onto a *symbolic* formula $\mathcal{F} = SymbSk(F)$, which is said to be *symbolic* as it couples list-based and BDD-based data structures to compactly represent a (possibly) exponentially less succinct propositional formula. The sentence $\mathcal{F}$ encodes the definability of a set of Skolem functions that capture a model (if any) of the original instance, according to the *symbolic skolemization* technique presented in [4]. A formal semantics is associated to symbolic formulas in such a way that $F \overset{sat}{\equiv} SymbSk(F)$ for every F. The other transformation—called *groundization*—translates a

R. Nieuwenhuis (Ed.): CADE 2005, LNAI 3632, pp. 369–376, 2005.
© Springer-Verlag Berlin Heidelberg 2005

symbolic formula $\mathcal{F}$ into a purely existential CNF propositional instance $Prop(\mathcal{F})$ (a SAT problem) such that $F \stackrel{sat}{\equiv} SymbSk(F) \stackrel{sat}{\equiv} Prop(SymbSk(F))$.

The role of these representations is as follows: Plain QBFs are handled in a preprocessing phase. Then, sKizzo moves to the symbolic representation and performs its work thereon. Zero or more CNF instances are generated/solved during the whole process.

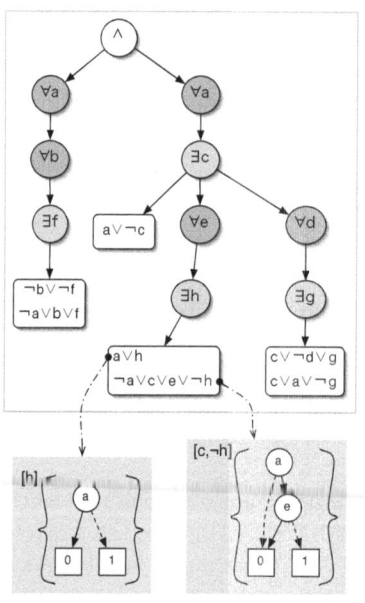

Symbolic skolemization (and most of the processes described below) relies on the existence of a *quantifier tree* stating which existential variables are in the scope of which universal variables. Such tree-shaped structures are extracted out of the flat prenex input according to [6]. They replace linear prefixes so to more closely reflect the intrinsic dependencies in the matrix. A sample quantifier tree for the QBF $\forall a \forall b \exists c \forall d \forall e \exists f \exists g \exists h.(a\vee\neg c) \wedge (a\vee h) \wedge (c\vee\neg d\vee g) \wedge (\neg a\vee b\vee f) \wedge (\neg a\vee c\vee e\vee\neg h) \wedge (\neg b\vee\neg f) \wedge (a\vee c\vee\neg g)$ is depicted aside. The symbolic representation is designed to allow for efficient forms of *symbolic reasoning* (Section 2), where universal reasoning is taken apart form existential reasoning (ROBDDs conveniently deal with the former, list-based representations with the latter). A symbolic formula is made up by symbolic clauses. During symbolic skolemization, one symbolic clause is extracted out of each QBF clause. The two major components of a symbolic clause $\Gamma_{\mathcal{I}}$ are a list Γ of existential literals and an index-set $\mathcal{I}$ represented via a ROBDD whose support set is the set of universals dominating the existential node at which the clause is attached in the quantifier tree. For example, the symbolic clauses $[h]_{\{00,01\}}$ and $[c,\neg h]_{\{10\}}$ are extracted out of $a \vee h$ and $\neg a \vee c \vee e \vee \neg h$ respectively (see the picture). Each symbolic clause $\Gamma_{\mathcal{I}}$ compactly represents a set $Prop(\Gamma_{\mathcal{I}})$ of $|\mathcal{I}|$ propositional clauses, in such a way that F is **sat** iff $Prop(\mathcal{F})$ is **sat**. For example, $Prop([c,g]_{\{01,10\}}) = \{c_0 \vee g_{01}, c_1 \vee g_{10}\}$. The *symbolic size* of $\mathcal{F}$ is $|\mathcal{F}|$, its *ground size* is $|Prop(\mathcal{F})|$: The initial symbolic size of $\mathcal{F}$ is thus linear in $|F|$. For details, see [4].

3 Inference Strategy

The *inference strategy* followed by sKizzo changes accordingly to a finite state machine whose inference states $S^{inf} = \{G, S, R, B, G\}$ are traversed.

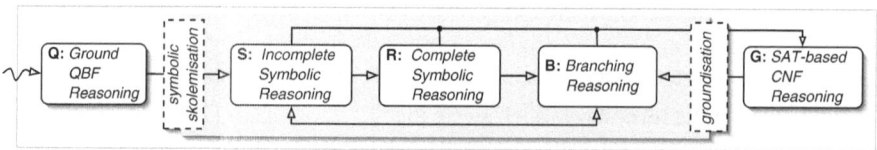

Each state in S^{inf} is associated to the application of an *inference style*. Each transition $x \to y$ in the picture, $x, y \in S^{inf}$, is labeled by a condition that triggers the shift from the style x to y (possibly requiring a satisfiability-preserving transformation).We now describe each state and transition.

Q: Ground QBF Reasoning.   In the Q-state `sKizzo` works in the original QBF space, as represented aside. The step Q_1 amounts to apply the quantified form of three simple (incomplete) inference rules: *unit clause propagation*, *pure literal elimination*, and *forall-reduction*. The transition $Q_1 \to Q_2$ is triggered when all these rules reach their fixpoint. Bounded variable elimination (Q_3) applies *q-resolution* to eliminate a selected existentially quantified variable v in the deepest existential scope of some branch of the quantifier tree. This is done by substituting all the clauses containing v with the set of all the resolvents over v. As repeated applications of variable elimination often lead to an unmanageable explosion of the number of clauses, a *bounded* form of elimination is employed: Only variables whose elimination shrink the overall number of literals or clauses are eligible for elimination. The transition $Q_3 \to Q_1$ is selected when at least one variable has been eliminated during the last round, $Q_3 \to S$ is followed otherwise.

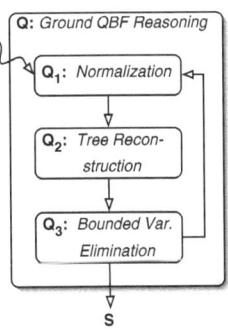

S: Incomplete Symbolic Reasoning. The instance is attacked by means of a set of (incomplete) *symbolic inference rules*, designed after their ground counterparts to achieve in one single application on symbolic clauses the same result they would obtain if applied to each ground clause separately.

SUCP (Symbolic Unit Clause Propagation). This rule builds on top of the observation that each symbolic unit clause $[\gamma]_{\mathcal{I}}$ in the formula represents a set $\{\gamma_i | i \in \mathcal{I}\}$ of ground unit literals. All of them are symbolically assigned at once to avoid an immediate contradiction.

SPLE (Symbolic Pure Literal Elimination). This rule computes a symbolic representation for the set of pure literals, then simplify the formula by assigning all of them at once. It comes in two flavors: a monolitic (one variable per step) and an incremental (one clause per step) version.

SSUB (Symbolic SUBsumption). This rules removes all the symbolic clauses that are subsumed by other clauses (*forward subsumption*). It employs scheduling heuristics, lazy computations, and a signature-based mechanism to minimize the overall effort. This rule complements the *backward subsumption* mechanism which is applied on-the-fly at each clause insertion.

SHBR (Symbolic Hyper Binary Resolution). This rules enumerates all the resolution chains of binary symbolic clauses in the formula, looking for contradictions. Each such contradiction determines a necessary consequence of the formula, compactly represented as a unit symbolic clause which is added to the instance (SUCP then draws all the entailed consequences).

SER (Symbolic Equivalency Reasoning). This rules look for non-empty strongly connected components in the *symbolic binary implication graph*[2] of the formula, and applies the resulting symbolic equivalences to simplify the formula.

A carefully designed application schedule is necessary to profit from the above set of rules as a whole. sKizzo implements the following dynamic scheduling policy.

1. The inference process is divided into subsequent *inference rounds*. At each round, the rules that have the rights to do so (see below) are sequentially executed.
2. The rule currently working is monitored during its execution. When certain resource limits are exceeded (inference steps undertaken, time elapsed, memory allocated, etc.), the rule is preemptively stopped (the rule's context is saved to re-start working from the interruption point).
3. When all the rules in the inference round have been executed, they are ranked according to their *relative efficiency*. The resource limits for the next rounds are redistributed on a meritocratic basis: the better a rule has proved to be, the larger the resources it will be granted next.
4. Rules failing to be effective loose the right to execute for a number of inference steps that enlarges with the number of rounds they have been performing poorly. The longer they keep on being ineffective, the more sparingly they are given a try.

The assessment of rules' efficiency is a major issue in the above policy. As all the rules reduce the ground size of the instance at each application (conversely, the symbolic size might be enlarged), the ground-size-shrink-percentage-per-resource-unit is assumed as a measure of efficiency. This measure needs itself resources to be computed. When BDD primitives and lazy evaluation do not suffice to keep the cost of assessment within pre-established limits, sKizzo resorts to approximated measures. The transition S → G is triggered if the ground size of the current problem becomes *affordable* via SAT-based reasoning (see the G-style), unless the symbolic reasoning is behaving so efficiently that ground reasoning is estimated not to pay back. The transition S → R is activated when the rules adopted come out to be unable to solve the problem. This happens under two circumstances: (1) the overall fixpoint is reached but no decision is obtained, or (2) the rate at which the problem is being shrunk has been staying below a certain threshold since a given number of inference rounds.

R: Complete Symbolic Reasoning. This state is similar to S, with one major exception: a refutationally complete rule is inserted in the pool of symbolic rules exercised at each inference round.

SDR (Symbolic Directional Resolution). This rules eliminates one symbolic variable per step by substituting the set of resolving clauses with the set of their symbolically computed resolvents.

Efficiency as size-shrinking measurement is unfair for SDR. This rule may need to pass through intermediate clause-sets that are much larger than the originating instance to come to a solution. So, SDR is given the change to consume more and more resources regardless of the size of the formula it is constructing. The other rules are still applied/evaluated in a round robin way (SSUB is especially useful here to reduce the redundancy SDR generates). Two outcomes are possible: (1) the largest intermediate result fits within the physical memory of the machine on which sKizzo is running—so the instance is solved, or (2) an out-of-memory condition occurs. As sKizzo keeps on monitoring its own resource consumption, he is able to detect the latter occurrence and give up resolution-based reasoning. The transition R $\rightarrow$ B is triggered. As usual, the transition R $\rightarrow$ G is followed if (and as soon as) the current problem becomes *affordable* via SAT-based reasoning (see the G-style).

A checkpointing mechanism is implemented against the unlucky possibilities that no consistent formula representation exists when mem-out occurs, or the formula yielded by SDR is so larger than the input formula that we would prefer to restart working on the original version: Symbolic formulas have to be explicitly committed or rolled-back depending on their eventual characteristics. This ensures that blow-up phenomena do not negatively affect the rest of the inference process.

B: Branching Reasoning. In this status, a search-based branching decision procedures extending the DPLL approach to the quantified case is applied. Models are searched following the left-to-right prefix order of variables during a depth-first visit of the semantic evaluation tree of the formula. Existential variables generate *or* nodes, universal quantifiers are associated to *and* nodes. Distinguishing features of sKizzo:

- Both universal and existential splits are performed symbolically.
- The partial order induced by the internal structure of the quantifier tree is substituted for the left-to-right order of variables in the prefix. The main advantage is that nodes with more than one child induce sets of *disjoint* sub-instances that are solved in isolation of one another.
- After each existential split, the cofactored matrix undergoes further incomplete symbolic normalization (transition B $\rightarrow$ S and back). This mechanism extends the unit-clause-propagation based form of look-ahead used in purely branching solvers.
- The base case of the recursion does not deal with trivial sub-formulas. Well in advance, either symbolic reasoning (transition B $\rightarrow$ S, whenever the current instance falls within its deductive power) or ground reasoning (transition B $\rightarrow$ G, whenever the ground version of the problem is affordable) decide every sub-instance, acting as powerful look-ahead tools.

Many enhancements to the basic DPLL procedure are implemented. A conflict-analysis machinery is employed in the event of inconsistent partial assignment to isolate the branching steps responsible for the contradiction to arise. This information is used to perform a conflict-directed backjumping. As contradictions follow in general from a mix of branching steps, symbolic reasoning, and SAT-based reasoning, the three of these inference styles share a common conflict-analysis engine. A symbolic learning mechanism extracts symbolic clauses out of contradictions (prune the rest of the search). Size-bounded and relevance-bounded heuristics are used to constraint the required amount of memory. Branching heuristics are also enrolled: MOMS and VSDIS are implemented.

G: SAT-Based Ground Reasoning. We explicitly construct $Prop(SymbSk(F))$ and solve it via state-of-the-art SAT solvers (they come out to be very efficient on such instances). This amounts to (1) build an encoding from the structured namespace of symbolic literals/clauses onto a flat propositional space, (2) generate all the necessary clauses, (3) make the SAT solver handle the resulting instance: Quite some "almost-existential" families of instances are successfully dealt with in the G status (hash-table based mechanism are implemented to make the translation fast). A transition $x \rightarrow$ G, $x \in \{S, R, B\}$, is triggered as soon as the groundization of the current formula becomes *affordable*. At the beginning, this notion is simply given in terms of memory require-ments: The ground version of the instance fits into the memory and leaves enough space for the SAT solver to work. By construction, the transitions $x \rightarrow$ G, $x \in \{S, R\}$ are triggered at most once, yielding an instance SAT-equivalent to the original QBF prob-lem. Conversely, B generates a (long) chain of SAT instances, each one encoding the outcome of the exploration of an entire sub-tree of the QBF semantic evaluation tree. Along this chain, the notion of affordability is adjusted by a learning algorithm that tries to guess a good switch size between B and G. Furthermore, for the G-state to ac-tively participate in conflict analysis, we map unsatisfiable ground cores (extracted by analyzing the ground inference trace) onto symbolic cores, then onto branching choices.

4   Certification

sKizzo implements a mechanism to certify its claims of (un)satisfiability. Evaluation and certification are completely decoupled, with almost no overhead for the former. The two meshes of the chain are connected through an *inference log*, produced by the solver and subsequently red by an external *certifier*. The log contains information about (1) the context switches between inference styles, (2) the sequence of the (symbolic) instantiations of inference rules undertaken (resolutions, substitutions, assignments), (3) entries for rollback/commit points and other control information.

By reading the log forward, the cer-tifier is able to reproduce the deriva-tion of the empty clause (unsat in-stances) and its graph of dependen-cies, thus extracting an unsatisfiable core. On sat instances, the certi-fier applies an *inductive model re-construction*[5] procedure while pars-ing the log backward. It constructs

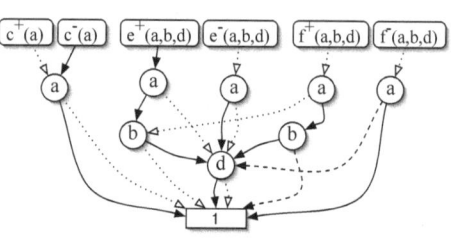

a stand-alone, BDD-based sat-certificate encoding a QBF model. As an exam-ple, the picture aside depicts the sat-certificate produced for $\forall a \forall b \exists c \forall d \exists e \exists f$. $(\neg b \vee e \vee f) \wedge (a \vee c \vee f) \wedge (a \vee d \vee e) \wedge (\neg a \vee \neg b \vee \neg d \vee e) \wedge (\neg a \vee b \vee \neg c) \wedge (\neg a \vee \neg c \vee \neg f) \wedge (a \vee \neg d \vee \neg e) \wedge (\neg a \vee d \vee \neg e) \wedge (a \vee \neg e \vee \neg f)$. A model is encoded into such certificate: By assigning the existential variable e (similarly for c and f) to TRUE when $e^+(a, b, d) = 1$ and to FALSE when $e^-(a, b, d) = 1$ the matrix is always satisfied.

5   Implementation and Experimentation

sKizzo is a 50k-line piece of code written in C using an object-oriented program-ming style. It has been developed on a PowerPC/MacOS X platform, then migrated to

Table 1. On the left: some *2004-hard* instances (remained unsolved during the SAT04 evaluation) solved by sKizzo. On the right: evaluation compared to SAT-certification. We report: number of existential/universal variables ($\exists/\forall$), quantifier alternations ($Al.$), time to solve/reconstruct/verify (T_s, T_r, T_v), number of steps in the log ($|\mathcal{L}|$), and number of nodes in the certificate ($|\mathcal{C}|$)

Instance	Al.	T_s	Instance	Al.	T_s	instance	$\forall$	$\exists$	Al.	T_s	T_r	T_v	$|\mathcal{L}|$	$|\mathcal{C}|$
adder12	3	190.0	cnt15	30	33.2	adder-16	1672	3096	3	1200.0	2025.2	1.1	$3.5 \cdot 10^3$	$2.4 \cdot 10^5$
adder14	3	670.0	cnt16	32	44.7	Adder2-10	545	7424	5	360.0	97.8	0.1	$8.5 \cdot 10^3$	$6.1 \cdot 10^4$
adder16	3	1200.0	s713_d3_s	2	384.7	cnt09re	9	609	18	280.0	0.4	0.1	$5.8 \cdot 10^3$	$9.0 \cdot 10^1$
cnt09re	18	259.6	s499_d7_s	2	107.5	cnt16	16	1650	32	45.0	36.0	0.1	$3.4 \cdot 10^5$	$5.9 \cdot 10^2$
cnt10r	20	67.5	s499_d10_s	2	493.1	k_grz_n18	24	767	16	55.0	0.8	0.1	$1.8 \cdot 10^3$	$3.2 \cdot 10^3$
cnt10e	20	923.1	s386_d3_s	2	23.9	k_poly_n18	110	1354	112	4.8	11.6	0.1	$4.5 \cdot 10^3$	$1.1 \cdot 10^3$
cnt11r	22	190.38	s386_d4_s	2	631.0	k_ph_n15	10	4833	4	86.0	1.5	0.4	$1.1 \cdot 10^4$	$2.8 \cdot 10^3$
cnt12r	24	548.68	s386_d5_s	2	795.3	k_d4_n16	69	1368	40	11.0	149.0	0.3	$1.1 \cdot 10^4$	$4.8 \cdot 10^4$

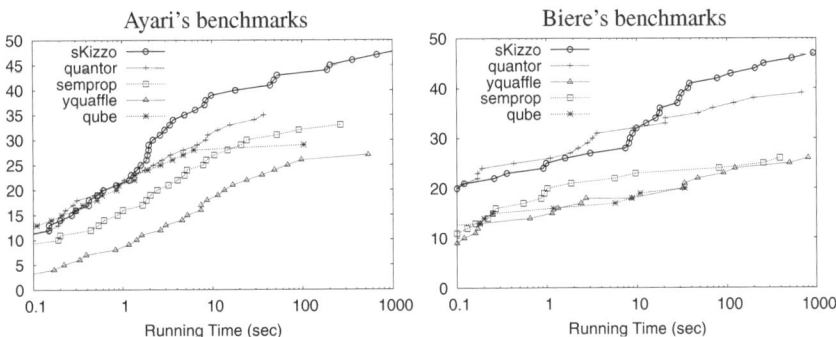

Fig. 1. Number of solved instances (Y) for each timeout up to 1000s (X). Solvers: QuBE-LRN [8], v. 1.3, a search-based solver featuring lazy data structures and conflict/solution learning. Quantor [7], v. 2004.01.25, a solver employing q-resolution and expansion to eliminate quantifiers. SEMPROP [10], v. 24.02.02, a search-based solver featuring directed backtracking and lemma/model caching. yQuaffle [11], v. 09.30.04, a search-based solver featuring conflict-driven learning, inversion of quantifiers, solution-based backtracking

i386/Linux. It relies on the CUDD package 2.4.0 and DDDMP 2.0 for BDD manipulations, and on zChaff 2004.5.13 and siege v4 for SAT solving. Command-line options allow the user to individually (de)activate inference rules, and to construct *solving personalities* by forbidding the visit of some states of the inference FSM. Syntactic trees, CNF instances and certificates may be dumped to secondary memory.

The experimental evaluation of our suite yields a large amount of data, for which we refer the reader to [3]. Here we limit our presentation to a performance comparison (shown in Figure 1 and performed on a 2.6 GHz P4, 1Gb main memory, running Linux v2.4) between sKizzo and the best publically available state-of-the-art QBF solvers [9] over two challenging groups of QBF instances extracted from the QBFLIB's archive [8]: Biere's benchmarks [7], made up of 64 instances divided into 4 families, where the n-th instance in each family refers to a model checking problem on a n-bit counter, and Ayari's benchmarks [1], made up of 72 instances divided into 5 families, obtained from real-world verification problems on circuits and protocol descriptions.

Table 1 presents a few more results showing that sKizzo has solved instances never solved before. In addition, the results concerning SAT-certificate extraction suggest that certification is actually feasible. Although QBF verification cannot be in general accomplished in polynomial time, the task of building and verifying a certificate comes out not to be overwhelming on application-related instances: sKizzo has been able to certify all the satisfiable formulas it has solved.

References

1. A. Ayari and D. Basin. Bounded Model Construction for Monadic Second-order Logics. In *Proc. of CAV'00*, 2000.
2. M. Benedetti. sKizzo: a QBF Decision Procedure based on Propositional Skolemization and Symbolic Reasoning, Tech.Rep. 04-11-03, ITC-irst, 2004.
3. M. Benedetti. sKizzo's web site, sra.itc.it/people/benedetti/sKizzo, 2004.
4. M. Benedetti. Evaluating QBFs via Symbolic Skolemization. In *Proc. of the 11th International Conference on Logic for Programming, Artificial Intelligence, and Reasoning (LPAR04)*, number 3452 in LNCS. Springer, 2005.
5. M. Benedetti. Extracting Certificates from Quantified Boolean Formulas. In *Proc. of IJCAI05*, 2005.
6. M. Benedetti. Quantifier Trees for QBFs. In *Proc. of SAT05*, 2005.
7. A. Biere. Resolve and Expand. In *Proc. of SAT'04*, pages 238–246, 2004.
8. E. Giunchiglia, M. Narizzano, and A. Tacchella. QuBE: A system for deciding Quantified Boolean Formulas Satisfiability. In *Proc. of the International Joint Conference on Automated Reasoning (IJCAR'2001)*, 2001.
9. D. Le Berre, M. Narizzano, L. Simon, and A. Tacchella. Second QBF solvers evaluation, avaliable on-line at www.qbflib.org, 2004.
10. R. Letz. Advances in Decision Procedures for Quantified Boolean Formulas. In *Proceedings of the First International Workshop on Quantified Boolean Formulae (QBF'01)*, pages 55–64, 2001.
11. L. Zhang and S. Malik. Towards Symmetric Treatment of Conflicts And Satisfaction in Quantified Boolean Satisfiability Solver. In *Proc. of CP'02*, 2002.

Regular Protocols and Attacks with Regular Knowledge

Tomasz Truderung[*]

LORIA-INRIA-Lorraine, France,
Institute of Computer Science, Wrocław University, Poland

Abstract. We prove that, if the initial knowledge of the intruder is given by a deterministic bottom-up tree automaton, then the insecurity problem for cryptographic protocols with atomic keys for a bounded number of sessions is NP-complete. We prove also that if regural languages (given by tree automata) are used in protocol descriptions to restrict the form of messages, then the insecurity problem is NEXPTIME-complete.

Furthermore, we define a class of cryptographic protocols, called *regular protocols*, such that the knowledge which the intruder can gain during an unlimited number of sessions of a protocol is a regular language.

1 Introduction

Formal verification of cryptographic protocols has been attracting much attention in the recent years (see [10,4] for an overview). It has been very succesful in finding flaws in cryptographic protocols. Althout the general verification problem is undecidable [6,1,7], there are interesting and important decidable variants [5,6,13,2]. One of them is the insecurity problem of protocols analyzed w.r.t. a bounded number of sessions, in presence of the so-called Dolev-Yao intruder, which is NP-complete [13]. In this case, one assumes that the initial knowledge of the intruder is a finite set of terms.

In this paper, we prove the decidability of security for bounded number of sessions, when the initial knowledge of the intruder is a regular language, with the assumption that keys used in protocols are atomic. We show that if the initial knowledge of the intruder is given by a deterministic bottom-up tree automaton, then the existence of an attack remains NP-complete.

A regular language which represents the initial knowledge of the intruder can be an approximation or an exact representation of the set of messages which could have been intercepted during an unbounded number of prior executions of some protocols. In fact, approximating the knowledge of the intruder by means of finite tree automata or similar formalisms has been studied by several authors (see e.g. [9,11]). As a complementary result we define also a class of cryptographic protocols, called *regular protocols*, such that the exact knowledge which the intruder can gain during an unbounded number of sessions of a protocol is a

---

[*] Partially supported by the RNTL project PROUVE-03V360 and by IST-2001-39252 AVISPA.

R. Nieuwenhuis (Ed.): CADE 2005, LNAI 3632, pp. 377–391, 2005.

regular language given by an alternating tree automaton of polynomial size w.r.t. the size of the protocol. The security problem for such protocols is DEXPTIME-complete. As an immediate consequence we obtain also an NEXPTIME algorithm for deciding protocols which consist of two phases: in the first one, only regular rules can be used, these rules can be, however, executed an unbounded number of times. Then, in the second phase, non-regular rules (i.e. rules of an arbitrary form as long as they have atomic keys) can be used a fixed number of times.

We extend also our decidability result to *protocols which regular constraints*, i.e. protocols which impose some *well-formness* constraints on messages that can be sent. In [13], a receive-send action of a principal is described by a rewrite rule $t \rightarrow s$ (where t, s are terms). The meaning of such a rule is that a principal after receiving any ground instance $t\theta$ of t (for any ground substitution θ), replies $s\theta$. It is impossible to model the behaviour of a principal who replies only if the term $t\theta$ has some form which cannot be expressed by patern matching (e.g. if $t\theta$ is list of encrypted messages). Protocols with regular constraints allow us to express required form of messages by constraints of the form $x \in L$, where L is a regular language (given by some tree automaton). Such constraints may express some *integrity* requirements. For instance, a checksum for a message m can be simulated by a term $f(m)$ (where f is a new function symbol), which can be adequate, if checksums are collision-free. This approach can be however inadequate, when *weak* checksums (in which, given a checksum of a message, it is possible to produce another message that evaluates to the same checksum) are considered. Modeling the set $\{\langle m, c \rangle \mid c$ is the checksum of $m\}$ by a regular language, and using regular constraints can give more precise results.

We show that the insecurity problem of protocols with regular constraints, and with the initial knowledge of the intruder given by finite tree automata is NEXPTIME-complete (it is NEXPTIME-hard, when the used automata are deterministic, and remains in NEXPTIME even for alternating tree automata).

In this paper, we make the following abstractions. We use the Dolev-Yao model of the intruder [5], which is a standard practice in formal verification of cryptographic protocols. We formulate protocols in the rule-based model used in [2,3,13]. In the case of regular protocols, where unbounded number of sessions is considered, this model cannot express fresh nonces which have to be replaced by constants (or some other terms). It implies that some false attacks can be found. It should be mentioned, that this approach is also quite standard, since verification of protocols with nonces is undecidable even in the restricted case, where the size of messages is bounded.

Related Work. The security problem of protocols when the initial knowledge of the intruder is given by finite automata has not been considered so far. Similarly, there are no previous decidability results for protocols with regular constraints.

There are, however, many results related to regular protocols defined in Sect. 5. Regular protocols are a generalization of regular unary-predicate programs proposed in [8]. They are also closely related to a class of monadic Horn theories defined in [15]. Our regular protocols are more general than the class $\mathcal{H}1$ defined in [12]. In [1], the authors specify a class of protocols (without nonces, and satisfying so called *independence* condition) which is DEXPTIME-hard. Regular protocols are also more general than this class.

Structure of the Paper. Sect. 2 contains some basic definitions. In Sect. 3, we prove that the insecurity problem for bounded number of sessions, when the initial intruder knowledge is given by a deterministic tree automaton is NP-complete. Sect. 4 contains complexity results for protocols with regular constraints. In Sect. 5, we define regular protocols, and prove their properties.

2   Preliminaries

Terms and Term-DAGs. Let $T(\Sigma, V)$ denote the set of terms over the signature Σ and the set of variables V. If $V = \emptyset$, then we can write $T(\Sigma)$ instead of $T(\Sigma, V)$. A term is *ground*, if it does not contain variables. A (ground) *substitution* is a mapping from variables to (ground) terms, which, in a natural way, is extended to a mapping from term to terms.

For a given signature Σ, a *term*-DAG D is a labelled directed acyclic ordered graph such that, if a node v is labelled with a function symbol f of arity n, then it has n ordered immediate successors $v_1, \ldots, v_n$. In such a case we write $v =_D f(v_1, \ldots, v_n)$. For a term-DAG D, and a vertex $v =_D f(v_1, \ldots, v_n)$, we recursively define the *term $t(v, D)$ represented by v in D* by the equation $t(v, D) = f(t(v_1, D), \ldots, t(v_n, D))$.

Unary Definite Logic Programs. Let Σ be a signature, V be a set of variables, and P be a set of predicate symbols (we assume here that all predicates are unary). If $p \in P$, and $t \in T(\Sigma, V)$, then $p(t)$ is an *atomic formula*. An atomic formula $p(t)$ is *ground*, if t is ground. A *unary definite logic program* is a finite set of *clauses* of the form $a_0 \leftarrow a_1, \ldots, a_n$, where $a_0, \ldots, a_n$ are atomic formulas.

We will use the following notation. Let P be a unary definite logic program, let A, B be sets of ground atomic formulas. We write $A \vdash_P B$, if there exists *a proof of B with respect to P assuming A*, i.e. a sequence $a_1, \ldots, a_n$ of atomic formulas such that each element of B occurs in $a_1, \ldots, a_n$, and, for each $i = 1, \ldots, n$, we have either (i) $a_i \in A$, or (ii) there exists a clause $b_0 \leftarrow b_1, \ldots, b_m$ in P, and a substitution θ such that $a_i = b_0\theta$, and each of $b_1\theta, \ldots, b_m\theta$ occurs in $a_1, \ldots, a_{i-1}$.

For a set of atomic formulas A, and an atomic formula a, we write $A \vdash_P a$ for $A \vdash_P \{a\}$. We write also $\vdash_P B$ for $\emptyset \vdash_P B$. It is easy to show that $A \vdash_P a$, if and only if a is in the least Herbrand Model of $P \cup A$.

Messages, Protocols, and Intruder. *Messages* are ground terms over the signature Σ consisting of constants (*atomic messages* such as principal names, nonces, keys), and the following binary function symbols: $\langle \cdot, \cdot \rangle$ (*pairing*) $\{\cdot\}$. (*symmetric encryption*), and $\{\cdot\}^p$ (*public key encryption*), with the restriction that keys used in public key encryption are constants, i.e. that a term of the form $\{t\}^p_s$ is valid only if s is a constant. We assume that there is a bijection $\cdot^{-1}$ on atomic messages which maps every public (private) key k to its corresponding

$$I(\langle x,y \rangle) \leftarrow I(x), I(y), \qquad I(x) \leftarrow I(\langle x,y \rangle), \qquad\qquad I(y) \leftarrow I(\langle x,y \rangle), \qquad (1)$$

$$I(\{x\}_y) \leftarrow I(x), I(y), \qquad I(x) \leftarrow I(\{x\}_y), I(y), \qquad\qquad\qquad\qquad (2)$$

$$I(\{x\}_k^p) \leftarrow I(x), I(k), \qquad I(x) \leftarrow I(\{x\}_k^p), I(k^{-1}) \qquad \text{(for each key } k) \qquad (3)$$

Fig. 1. T_I — Intruder Rules

private (public) key k^{-1}. We assume that Σ contains special constant Sec (a secret). We will sometimes omit $\langle \cdot, \cdot \rangle$, and write, for instance, $\{t, s\}_k$ instead of $\{\langle t, s \rangle\}_k$.

A *principal* Π is a sequence $(r_i \to s_i)_{i=1}^n$ of *rules*, where, for each $i = 1, \ldots, n$, we have $r_i, s_i \in T(\Sigma, V)$, for a set of variables V, and every variable in s_i occurs in $r_1, \ldots, r_i$. By $|\Pi|$ we denote the number of rules of Π (i.e. the length of the sequence Π). A rule $(r \to s)$ is intended to specify receive-send action of a principal who after receiving $r\theta$, for a ground substitution θ, replies $s\theta$. A *protocol* is a finite set of principals. This method of representing principals and protocols follows [13,3,2], where examples of modeling protocols in this framework can be found. The set of variables occurring in a protocol P will be denoted by $Var(P)$.

In the Dolev-Yao model [5], the intruder has the entire control over the network. He can intercept and memorize messages, generate new messages and send them to participants with a false identity. We express the ability of the intruder to generate (derive) new messages from a given set of messages by the program T_I in Figure 1, where the predicate symbol I is intended to describe the intruder knowledge. The rules (1) express his ability to construct new messages by pairing known messages, and by deconstructing them. The rules (2) and (3) express his ability to crypt and decrypt messages, when he has appropriate keys. For a set A of messages, let $I(A) = \{I(t) \mid t \in A\}$. We will say that the intruder can *derive a message t from messages A*, if $I(A) \vdash_{T_I} I(t)$.

Now we give a definition of an *attack for a bounded number of sessions*. In an attack, the intruder chooses some execution order of the rules of the given protocol and then produces input messages for these rules. These input messages have to be derived from the intruder's initial knowledge and the output messages obtained so far. The aim of the intruder is to derive the secret message Sec. Note that in this definition of an attack, only security (or more precisely *secrecy*) is the concern. We do not study here properties like for instance authentication or liveness. If some number of interleaving sessions of a protocol is to be analyzed, then these sessions have to be encoded into the protocol, which is the standard approach when protocols are analyzed w.r.t. a bounded number of sessions (see, for instance [13,2]).

Formally, given a protocol $P = \{\Pi_1, \ldots, \Pi_l\}$, a *protocol execution scheme* is a sequence of rules $\pi = \pi_1, \ldots, \pi_n$ such that each element of π can be assigned to one of the participants $\Pi_1, \ldots, \Pi_l$, and, for each participant Π_k ($k = 1, \ldots, l$), the subsequence of the elements of π assigned to Π_k is $\Pi_k^1, \ldots, \Pi_k^m$, for some

$m \leq |\Pi|$, where Π_k^i is the i-th rule of Π_k.[1] An *attack* with an initial knowledge A_0 is a pair (π, σ), where π is a protocol execution scheme, and σ is a ground substitution such that, for all $i = 1, \ldots, n$, we have

$$I(A_0), I(s_1\sigma), \ldots, I(s_{i-1}\sigma) \vdash_{T_I} I(r_i\sigma), \text{ and} \tag{4}$$

$$I(A_0), I(s_1\sigma), \ldots, I(s_n\sigma) \vdash_{T_I} I(Sec). \tag{5}$$

A protocol is *insecure*, if there exists an attack on it.

Finite Tree Automata. We will express finite tree automata by means of unary logic programs. We say that a logic program T with a set of accepting predicate symbols Q_F is an *alternating finite tree automaton*, if each rule of T has the form

$$p_0(f(x_1, \ldots, x_n)) \leftarrow p_1(y_1), \ldots, p_m(y_m) \tag{6}$$

where $x_1, \ldots, x_n$ are distinct variables, and for each $i = 1, \ldots, m$, the variable $y_i \in \{x_1, \ldots, x_n\}$. A program is a *nondeterministic finite tree automaton*, if each its rule has the form

$$p_0(f(x_1, \ldots, x_n)) \leftarrow p_1(x_1), \ldots, p_n(x_n). \tag{7}$$

where $x_1, \ldots, x_n$ are distinct variables. A program T is a *deterministic bottom-up finite tree automaton*, if each its rule has the form (7), and for each function symbol f and each sequence of predicate symbols $p_1, \ldots, p_n$, the program contains at most one clause of the form (7). It is easy to see that, in this case, for each term t, there exists at most one predicate symbol p such that $\vdash_T p(t)$.

   Let T with Q_F be an automaton. A term t is *accepted by* (T, Q_F), if $\vdash_T q(t)$, for some $q \in Q_F$. The set of terms accepted by (T, Q_F) will be denoted by $L(T, Q_F)$.

3   Attacks with Regular Knowledge

In this section we consider the insecurity problem of protocols analyzed w.r.t. a bounded number of sessions, assuming that the initial knowledge of the intruder is a regular language given by a finite tree automaton. We assume that keys (both in symmetric and public key encryption) are atomic, which is the only assumption not made in [13], where only keys used in public key encryption were assumed to be atomic (in other respects, the result presented here subsumes the decidability result from [13]).

   The rest of this section is devoted to prove that, when the initial knowledge of the intruder is given by a deterministic bottom-up tree automaton, then the insecurity problem is NP-complete. The proof proceeds in two steps. First, in

---

[1] More formally, a sequence $\pi_1, \ldots, \pi_n$ of rules is a protocol execution scheme, if there is a function $f : \{1, \ldots, n\} \to \{1, \ldots, l\}$ such that, for each $k = 1, \ldots, l$, assuming that integers $i_1 < \cdots < i_m$ are all the elements of $f^{-1}(k)$, we have $\pi_{i_j} = \Pi_k^j$, for each $j = 1, \ldots, m$.

Section 3.1, we introduce *stage theories of protocols* which allow us to represent attacks in more uniform way. Next, in Section 3.2, we introduce the notion of ADAGs which are labelled term-DAGs suitable to represent attacks. We show that if an ADAG exists, then there exists an ADAG of polynomial size. It gives rise to the nondeterministic polynomial-time algorithm for the insecurity problem.

3.1   Stage Theories

In this subsection we express the existence of an attack, using *a stage theory of a protocol* which takes into account the fact that A_0 is a regular language represented by a logic program (and hence A_0 and the intruder inference rules can be represented in a uniform way). Second, instead of representing the knowledge of the intruder by the predicate I, the family of predicate symbols $I_0, \ldots, I_m$ is used to represent his knowledge at different stages of an attack.

Let P be a protocol, and A_0 be the initial knowledge of the intruder, represented by a finite tree automaton (T, Q_F). Let $\mathcal{K}$ be the set consisting of the constant *Sec*, and all the keys of the given protocol. We can assume without loss of generality that no rule of P have the form $a \to s$, for $a \in \mathcal{K}$ (if it is the case, we can replace it by e.g. $\langle a, a \rangle \to s$, obtaining a protocols which is equivalent w.r.t. the existence of an attack).

Let $\pi = (r_i \to s_i)_{i=1}^n$ be a protocol execution scheme, and $\Omega = \mathcal{K} \cup \{1, \ldots, n\}$. A sequence $e = e_1, \ldots, e_m$ of elements of Ω is called a *stage sequence for π*, if e contains all the elements $Sec, 1, \ldots, n$, and whenever $e_i = k$ and $e_j = l$, for $i < j$, then $k < l$.

For $e \in \Omega$, let us define e^r, and e^s by the equations $e^r = r_e$, $e^s = s_e$, if $e \in \{1, \ldots, n\}$, and $e^r = e^s = e$, otherwise. Let $E_i^r = \{e_1^r, \ldots, e_i^r\}$, and $E_i^s = \{e_1^s, \ldots, e_i^s\}$. The set E_i^s represents keys and terms of the form $s_j \sigma$ available to the intruder at the i-th stage of an attack. The set E_i^r represents keys and terms of the form $r_j \sigma$ which should be known to the intruder before the i-th stage. Let T_e denote the program T extended with the stage theory for T and e (Figure 2), where $Q_I = \{I_0, \ldots, I_m\}$ are fresh predicate symbols. The predicate symbol I_k is intended to describe the intruder knowledge at the k-th stage of an attack with a substitution σ, where the terms from $\{t\sigma \mid t \in E_k^s\}$ are available to him.

$$I_i(f(x_1, \ldots, x_n)) \leftarrow q_1(x_1), \ldots, q_n(x_n) \tag{8}$$

whenever $q_0(f(x_1, \ldots, x_n)) \leftarrow q_1(x_1), \ldots, q_n(x_n)$ is a rule of T, and $q_0 \in Q_F$;

$$I_i(\langle x, y \rangle) \leftarrow I_j(x), I_k(y) \qquad\qquad \text{if } i \geq j, k \tag{9}$$

$$I_i(x) \leftarrow I_j(\langle x, y \rangle) \qquad I_i(y) \leftarrow I_j(\langle x, y \rangle) \qquad \text{if } i \geq j \tag{10}$$

$$I_i(x) \leftarrow I_j(\{x\}_a) \qquad I_i(x) \leftarrow I_j(\{x\}_{a-1}^p) \qquad \text{if } i \geq j, \text{ and } a \in E_i^s, \tag{11}$$

$$I_i(\{x\}_a) \leftarrow I_j(x) \qquad I_i(\{x\}_a^p) \leftarrow I_j(x) \qquad \text{if } i \geq j, \text{ and } a \in E_i^s. \tag{12}$$

Fig. 2. The Stage Theory for T and e

Lemma 1. *Let π be a protocol execution scheme, and σ be a ground substitution. The pair (π, σ) is an attack iff there is a stage sequence e for π such that*

$$I_1(e_1^s\sigma), \ldots, I_m(e_m^s\sigma) \vdash_{T_e} I_0(e_1^r\sigma), \ldots, I_{m-1}(e_m^r\sigma). \tag{13}$$

Proof. First, suppose that (13) holds, for some π, e, and σ, and that Δ is a proof of it. Without loss of generality, we can assume that $I_k(t)$ occurs in Δ before $I_l(s)$, if $k < l$. Let Δ_i denote the subsequence of Δ containing only facts of the form $I_i(t)$, and let $\Delta_{\leq i}$ be the concatenation of $\Delta_1, \ldots, \Delta_i$. Let $\Delta_{\leq i}^*$ be the sequence obtained from $\Delta_{\leq i}$ by substituting I_k by I. One can show, by induction on i, that $\Delta_{\leq i}^*$ is a proof w.r.t. T_I which uses only assumptions from $I(A_0) \cup \{I(s_j\sigma) : s_j \in E_i^s\}$ (i.e. $\Delta_{\leq i}^*$ is a proof of $I(A_0) \cup \{I(s_j\sigma) : s_j \in E_i^s\} \vdash_{T_I} \emptyset$). Now, let k be any integer from $\{1, \ldots, n\}$. There exists i such that $e_i = k$. By the definition of E_i^s, we have $e_i^s = s_k \notin E_{i-1}^s$. Moreover, if $s_l \in E_{i-1}^s$, then $l < k$. So, $\Delta_{\leq i-1}^*$ is a proof w.r.t. T_I which uses only assumption from $I(A_0), I(s_1\sigma), \ldots, I(s_{k-1}\sigma)$. By the definition of Δ, we have $I_{i-1}(r_k\sigma) \in \Delta_{\leq i-1}$ (because $r_k = e_i^r$), hence $\Delta_{\leq i-1}^*$ is a proof of $I(A_0), I(s_1\sigma), \ldots, I(s_{k-1}\sigma) \vdash I(r_k\sigma)$. Similarly, we show that (5) holds. So, we can conclude that (π, σ) is an attack.

Now, suppose that we have an attack (π, σ). Let Π_i be a proof of (4), for $i = 1, \ldots, n$, and let Π_{n+1} be a proof of (5). We split each Π_k (for $k = 1, \ldots, (n+1)$) into the maximal (w.r.t. its length) sequence $\Pi_k^1, \ldots, \Pi_k^{m_k}$ such that the last element of Π_k^i, for $1 \leq i < m_k$, is of the form $I(a)$ for $a \in \mathcal{K}$, and this occurrence of $I(a)$ is the only one in $\Pi_1, \ldots, \Pi_{k-1}, \Pi_k^1, \ldots, \Pi_k^i$. We want to re-index the obtained sequence of Π_k^i, so let $\hat{\Pi}_1, \ldots, \hat{\Pi}_N = \Pi_1^1, \ldots, \Pi_1^{m_1}, \ldots, \Pi_{n+1}^1, \ldots, \Pi_{n+1}^{m_{n+1}}$.

For $i = 1, \ldots, N$, let Δ_i be the sequence of facts obtained from $\hat{\Pi}_i$ by substituting each $I(t)$ by $I_{i-1}(t)$, and let e_i be equal to k, if $\hat{\Pi}_i = \Pi_k^{m_k}$, for some k, and, otherwise, let e_i be a, where $I(a)$ is the last element of $\hat{\Pi}_i$. Finally, let $S = \{t \in A_0 \mid I(t) \text{ occurs in } \Pi_1, \ldots, \Pi_{n+1}\}$, and let Δ_0 be a proof of $\vdash_{T_e} I_0(S)$. One can prove that the concatenation of $\Delta_0, \ldots, \Delta_N$ is a proof of (13).[2]   □

A proof is *normal*, if for each term t, it contains at most one fact of the form $I_k(t)$ (for some k). The following fact is easy to prove.

Lemma 2. *It holds (13) iff there is a normal proof Δ of*

$$I_1(e_1^s\sigma), \ldots, I_m(e_m^s\sigma) \vdash_{T_e} I_{i_1}(e_1^r\sigma), \ldots, I_{i_m}(e_m^r\sigma), \tag{14}$$

where, for each $k = 1, \ldots, m$, we have $0 \leq i_k < k$.

3.2   DAG of the Attack

Suppose that we have a protocol P, a protocol execution scheme $\pi = (r_i \to s_i)_{i=1}^n$, and a stage sequence e for π. We denote by $\mathcal{T}(P)$ the set of subterms of $\{r_i, s_i\}_{i=1}^n \cup \mathcal{K}$. Suppose that the initial knowledge of the intruder is given by a deterministic bottom-up automaton (T, Q_F) with the set of predicate symbols Q, and the set of accepting predicate symbols Q_F. Let Z be the set of elements of the form ϵ, and $I_k^\downarrow$, $I_k^\uparrow$ (for $0 \leq k \leq |e|$).

---

[2] We use here the assumption that no rule of P is of the form $a \to s$, for $a \in \mathcal{K}$.

Definition 1. A DAG *of the attack* (an ADAG for short) for P, e is a tuple $\langle D, \alpha, \delta_1, \delta_2 \rangle$ where D is a term-DAG over Σ with the set of vertices V, $\delta_1 : V \to Q$, $\delta_2 : V \to Z$, and α is a mapping from $\mathcal{T}(P)$ to V such that

(i) if $\alpha(f(t_1, \ldots, t_n)) = v$, then $v =_D f(v_1, \ldots, v_n)$, and $\alpha(t_i) = v_i$, for $i = 1, \ldots, n$,

(ii) if $v_0 =_D f(v_1, \ldots, v_n)$, and $\delta_1(v_i) = q_i$, for $i = 0, \ldots, n$, then T contains the rule $q_0(f(x_1, \ldots, x_n)) \leftarrow q_1(x_1), \ldots, q_n(x_n)$,

(iii) if $\delta_2(v) = I_i^\uparrow$, then we have either (a) $\delta_1(v) \in Q_F$, or (b) for each child v' of v, $\delta_2(v') = I_j^\downarrow$ or $\delta_2(v') = I_j^\uparrow$, for some $j \leq i$, and if $v =_D \{v'\}_a$ or $v =_D \{v'\}_a^p$ then $a \in E_i^s$,

(iv) if $\delta_2(v) = I_i^\downarrow$, then either (a) $v = \alpha(s_k)$, for $s_k = e_i^s$, or (b) for some parent v' of v, $\delta_2(v') = I_j^\downarrow$, for some $j \leq i$, and if $v' =_D \{v\}_a$ or $v' =_D \{v\}_{a-1}^p$, then $a \in E_i^s$,

(v) if $v = \alpha(e_i^r)$, then $\delta_2(v) = I_j^\downarrow$ or $\delta_2(v) = I_j^\uparrow$, for some $j < i$.

The following lemma links the existence of an attack and the existence of an ADAG for a given protocol and stage sequence.

Lemma 3. *Let P be a protocol. There exists an attack on P iff there exists a stage sequence e and an ADAG for P, e.*

Proof. Suppose that there is an attack (π, σ). By Lemma 1 and Lemma 2, there is a sequence e, and a normal proof Δ of (14). Let D be the DAG representing all the terms of the form $t\sigma$, where $t \in \mathcal{T}(P)$ (i.e. for each term s of the form $t\sigma$, D contains a vertex v representing s). For $t \in \mathcal{T}(P)$, let $\alpha(t)$ be the vertex v such that $t(v, D) = t\sigma$. For a vertex v of D, let $\delta_1(v)$ be (the only) state which T assigns to $t_v = t(v, D)$. Let $\delta_2(v) = \epsilon$, if Δ does not contain $I_j(t_v)$, for any j. If $I_j(t_v)$ occurs in Δ, then let $\delta_2(v)$ be $I_j^\uparrow$, if $I_j(t_v)$ is obtained using (9) or (12), and let $\delta_2(v)$ be $I_j^\downarrow$, otherwise (in this case either $t_v = s_k\sigma$, for $s_k = e_j^s$, or $I_j(t_v)$ is obtained in Δ using (10) or (11)). One can show that $\langle D, \alpha, \delta_1, \delta_2 \rangle$ is an ADAG.

Now, suppose that $\langle D, \alpha, \delta_1, \delta_2 \rangle$ is an ADAG for P, e. Let $\sigma(x) = t(\alpha(x), D)$. We produce the following sequence of facts: First, we put all the fact of the form $I_k(t)$, where $\delta_2(v) = I_k^\downarrow$ (for some k), and $t = t(v, D)$, in such a way that $q(t)$ is before $q'(t')$, if $t > t'$. Second, we put all the facts of the form $p(t)$, where $\delta_1(v) = p$, for $t = t(v, D)$, and all the facts of the form $I_k(t)$ (for some k), where $\delta_2(v) = I_k^\uparrow$, for $t = t(v, D)$, in such a way that $q(t)$ is before $q'(t')$, if $t < t'$. One can prove that this sequence is a normal proof of (14), which by Lemma 1 and Lemma 2, implies that there exists an attack. □

Lemma 3 is a crucial step of our construction, because it characterizes the existence of an attack by a structure which is defined by some local properties ((i)–(v) of Definition 1). As we will see, it allows us to minimize ADAGs, roughly speaking, by merging vertices which are indistinguishable from the point of view of this local properties.

Let D be an ADAG. We say that $v \in V$ is *free*, if $v \neq \alpha(t)$, for each $t \in \mathcal{T}(P)$. Let $\delta(v) = (\delta_1(v), \delta_2(v))$. A vertex v is said to be a *push vertex*, if $\delta_2(v) = I_k^\downarrow$, for some k; otherwise it is a *non-push vertex*. A vertex v is a *top* vertex, if $\delta_2(v) = I_i^\downarrow$ (and so it is a push vertex), and $v = \alpha(s_k)$, for $s_k = e_i^5$ (and so we do not have to use its parents in order to ensure that (iv) of Definition 1 is met).

Now, we will show that if there exists an ADAG (for some P, e) then there exists an ADAG of polynomial size. The proof proceeds in two steps. First, in Lemma 4, we minimize the number of non-push vertices. It is a simple step which resembles the proof of pumping lemma for regular (tree) languages. In the second step (Lemma 5), we show how to minimize the number of push vertices. To explain this step, it is convenient to think that Item (iv) of Definition 1 allows us to transfer labels of the form $I_k^\downarrow$ down the ADAG, so that it can be used by pop vertices (Item (iii)). Now, roughly speaking, if a number of pop vertices is polynomially bounded, then a polynomially bonded number of push vertices is sufficient to transfer the necessary information from top vertices to pop-vertices (which is expressed by the *pushing relation* in the proof of Lemma 5).

Lemma 4. *If there is an* ADAG D, *then there is an* ADAG D' *with the same number of push vertices, and with the set of non-push free vertices of the size at most $c = m \cdot (2n + 1)$, where n is the length of e, and m is the number of predicate symbols of T.*

Proof. Let v, v' be free non-push vertices of D with $\delta(v) = \delta(v')$. We can assume that $v \not< v'$ (if it is not the case, we can switch them). Let us remove v and replace it by v' (i.e. whenever v was a child of u, we make v' a child of u instead). One can show that in this way we obtain an ADAG. We repeat this step until there are no two distinct free non-push vertices with the same value of δ.   □

Lemma 5. *If there is an* ADAG D, *then there is an* ADAG D^* *of polynomial size w.r.t. the size of the given protocol, and the program T.*

Proof. Suppose that D is an ADAG Let D' be the ADAG obtained from D using Lemma 4. Let W be the set of all the push vertices of D' which either are not free, or are children of some non-push vertices. Note that $|W| \leq 2c + |P|$, where c is the constant from Lemma 4.

For each non-top vertex v with $\delta_2(v) = I_k^\downarrow$, we chose one of its parents $h(v)$ such that $\delta_2(h(v)) = I_{k'}^\downarrow$, for some $k' \leq k$ (so $h(v)$ can be used to verify the point (iv) of Definition 1). We will write $v' \mapsto_h v$, if $v' = h(v)$, and denote the transitive closure of $\mapsto_h$ by $\mapsto_h^*$. We will call $\mapsto_h$ a *pushing relation* of D. Note that $\mapsto_h$ defines a forest such that the roots of its trees are top vertices, and every push vertex is a node of this forest. Let us denote this forest by T_h. For a push vertex v, let $G(v)$ be the set $\{w \in W \mid v \mapsto_h^* w\}$ (note that if $v \in W$, then $v \in G(v)$).

Now, we perform the following changes in D'. Let us set δ_2 to ϵ in each free push vertex v such that $G(v) = \emptyset$. One can show that in this way we obtain an ADAG Next, suppose that v, v' are distinct free vertices such that

$\delta(v) = \delta(v') = (q, I_k^\downarrow)$ with $G(v) = G(v') \neq \emptyset$. Note that $v \mapsto_h^* v'$ or $v' \mapsto_h^* v$. We assume the former case. Let us remove v and replace it by v', and put $h(v') = h(v)$. Let $\delta_2(u)$ be set to ϵ in each push vertex u such that $v \mapsto_h^* u$, and $v' \not\mapsto_h^* u$. One can prove that what we have obtained is an ADAG Note that no vertex from W has been removed, and moreover, for $v \in W$, the value of $\delta(v)$ has not been changed.

We repeat this step until there are no two distinct free push vertices v, v' with $\delta(v) = \delta(v')$ and $G(v) = G(v')$. Note that each time we modify the ADAG we modify also its pushing relation. Let D'' be the ADAG obtained in this way, and let $\mapsto_{h''}$ be its pushing relation. Because W is polynomial, $T_{h''}$ is polynomial as well: this forest has at most $|W|$ leafs (each leaf is an element of W), and each its path is not longer than $|W| \cdot c$ (note that c is the number of distinct values of δ, and $|W|$ is the maximal number of distinct values of the function G on each path). Each push vertex of D'' is in $T_{h''}$, so the number of push vertices in D'' is polynomial. Let us apply Lemma 4 to D'' obtaining D^*. The number of push vertices is unchanged, and the number of free non-push vertices is polynomial. Thus D^* has polynomial size. □

Theorem 1. *Protocol insecurity for a bounded number of sessions, with the initial knowledge of the intruder given by a deterministic bottom-up tree automaton is* NP-*complete.*

Proof. For deciding a protocol, we guess a protocol execution scheme, a sequence e for it, then we guess an ADAG of polynomial size (verifying whether such a guessed structure is an ADAG can be easily done in polynomial time). NP-hardness follows from NP-hardness of deciding protocols without composed keys, with the initial knowledge of the intruder given as a finite set [13]. □

4   Protocols with Regular Constraints

Definition 2. *A protocol with regular constraints is a tuple* $(P, \mathcal{D})$, *where* P *is a protocol, and* $\mathcal{D}$ *is a domain assignment which assigns a regular language* $\mathcal{D}_x$ *(the domain of* x*) to each variable* $x \in Var(P)$.

For a protocol with regular constraint $(P, \mathcal{D})$, *a pair* (π, σ) *is an attack on* $(P, \mathcal{D})$, *if it is an attack on* P, *and furthermore, for each* $x \in Var(P)$, *we have* $x\sigma \in \mathcal{D}_x$.

We consider the problem of deciding protocols with regular constraints, where both the initial knowledge of the intruder, and languages $\mathcal{D}_x$ are given by finite tree automata. As we will see the choice of the type of automata (deterministic, nondeterministic, alternating) does not have any impact on the complexity of the problem: in all these cases the problem turns out to be NEXPTIME-complete.

Proposition 1. *The problem of deciding a protocol with constraints* $(P, \mathcal{D})$, *where the initial knowledge of the intruder and the languages* $\mathcal{D}_x$, *for* $x \in Var(P)$, *are given by alternating tree automata can be reduced to the problem of deciding a protocol (without constraints) with a regular initial knowledge of the intruder given by an alternating automaton.*

Proof. Suppose that $(P, \mathcal{D})$ is a protocol with regular constraints, and that $Var(V) = \{x_1, \ldots, x_m\}$. Let A_0 and $\{A_i\}_{i=1}^m$ be alternating tree automata which describe the initial knowledge of the intruder and the languages $\mathcal{D}_{x_i}$, respectively. We assume that these automata have disjoint sets of states, and that the accepting state of A_i is q_i (for $0 \leq i \leq m$). Let A denote the union of $A_0, \ldots, A_m$ with the accepting state q_0 (recall that it is the accepting state of A_0).

Let P' be the protocol P with one additional principal having the only rule

$$\mathsf{Sec}, \{x_1\}_{k_1}, \ldots, \{x_m\}_{k_m} \rightarrow \mathsf{Sec'},$$

where $k_1, \ldots, k_m$ and $\mathsf{Sec'}$ are fresh constants. Let A' be the automaton A with additional transitions that assign the state q_0 to a term $\{t\}_{k_i}$ only if t can be assigned the state q_i. One can show that the intruder with the initial knowledge given by A_0 can derive Sec in the protocol $(P, \mathcal{D})$, if and only if the intruder with the initial knowledge given by A' can derive $\mathsf{Sec'}$ in the protocol P'. $\square$

It is known that, for an alternating tree automaton, one can construct an equivalent deterministic bottom-up tree automaton of exponential size. Hence, Proposition 1, and Theorem 1 have the following consequence.

Theorem 2. *The insecurity of a protocol $(P, \mathcal{D})$ with the initial knowledge of the intruder and the languages $\mathcal{D}_x$ given by alternating tree automata is in* NEXP-PTIME.

One can show that the exponential bounded tiling problem (which is NEXP-TIME-hard) can be reduced to the problem of deciding a protocol with regular constraints which use deterministic automata only. Thus we have the following result (the proof is given the extended version of this paper [14]).

Theorem 3. *The insecurity of a protocol $(P, \mathcal{D})$ with regular constraints is* NEXPTIME-*hard, even if the initial knowledge of the intruder and languages $\mathcal{D}_x$ are given by bottom-up deterministic tree automata.*

Let us note that the reduction given in the proof of Proposition 1 has the following property: if the initial knowledge of the intruder and the languages $\mathcal{D}_x$ are given by nondeterministic (but not alternating) tree automata, then the resulting automaton A' is also nondeterministic (does not use alternations). We can use this fact and Theorem 3 to obtain the following result, which shows that the assumption about the determinism of the automaton in Theorem 1 is essential.

Corollary 1. *The insecurity problem of protocols (without constraints) with the initial intruder knowledge given by nondeterministic tree automata is* NEXPTIME-*hard.*

5   Regular Protocols

The aim of this section is to specify a (possibly general) class of protocols such that each protocol P in this class has the following property: the knowledge which the intruder can gain during an unbounded number of sessions of P is a regular language. The class defined here is closely related to regular unary-predicate programs defined in [8], and to a class of monadic Horn theories defined in [15].

In this section we consider the analysis w.r.t. unbounded number of sessions. We should note that in this case, the formalism used do describe protocols does not model nonces (in the case of a bounded number of sessions nonces can be modeled by constants). Hence, we can assume without loss of generality, that a protocol is just a set of (independent) rules[3], and that each of its rules $r \rightarrow s$ says that if the intruder knows a term $r\theta$, than it can also know $s\theta$, for any ground substitution θ.

Definition 3. A term s *covers* x *in a term* t, if either $s = x$, or $s = f(s_1, \ldots, s_n)$, for some $f \in \Sigma$, and each occurrence of x in t is in the context of one of $s_1, \ldots, s_n$.

For instance, $s = \langle \{x\}_b, y \rangle$ covers x in $t = \{\{x\}_b, \{y, \{x\}_b\}_a\}_a$ (because each occurrence of x in t is in the context of $\{x\}_b$), but s does not cover x in $\{\{x\}_c\}_b$. Note also that any term covers x in $\{y\}_a$.

Definition 4. Let φ be the function, which assigns a set of terms to a term, defined by the equations $\varphi(t) = \varphi(t_1) \cup \varphi(t_2)$, if $t = \langle t_1, t_2 \rangle$, and $\varphi(t) = \{t\}$, otherwise.

For instance $\varphi(\langle \{b\}_k, \langle \{b, c\}_k, d \rangle \rangle) = \{\{b\}_k, \{b, c\}_k, d\}$.

Definition 5. A rule $r \rightarrow s$ is *regular*, if for each $s' \in \varphi(s)$ the following conditions hold: s' is linear, and each term $r' \in \varphi(r)$ can be assign a subterm $\gamma_{s'}(r')$ of s', such that:

(i) for each $r' \in \varphi(r)$ and each $x \in Var(s')$, the term $\gamma_{s'}(r')$ covers x in r',
(ii) for each $r', r'' \in \varphi(r)$, if a variable $y \notin Var(s')$ occurs in both r' and r'', then $\gamma_{s'}(r') = \gamma_{s'}(r'')$.

A *protocol is regular*, if it consists of regular rules only.

Example 1. The rule $r \rightarrow s$, where $r = \{N_A, x, B, \{x, A\}_{K_B}^p\}_{K_A}^p$ and $s = \{x, A\}_{K_B}^p$ is regular. In fact, for $\gamma_s(r) = x$, the conditions of Definition 5 hold (it is because x covers x in any term, and $Var(s) = \{x\}$; note also that $\varphi(r) = \{r\}$ and $\varphi(s) = \{s\}$). Similarly, one can easily check that each rule which has only one occurrence of a variable on the right-hand side, is regular.

Example 2. The rule $\{\{x, y\}_a, z\}_b, \{z, z\}_c \rightarrow \{\{y, x\}_b, d\}_c, \{z, \{x, y\}_a\}_c$ is regular. To show it, let us denote the left hand side by r, and the right-hand side by s. Note that $\varphi(r) = \{r_1, r_2\}$, where $r_1 = \{\{x, y\}_a, z\}_b$ and $r_2 = \{z, z\}_c$, and

---

[3] If it is not the case, each principal $\{r_i \rightarrow s_i\}_{i=1}^n$ can be transformed to n principals with rules $r_1, \ldots, r_i \rightarrow s_i$, for each $i = 1, \ldots, n$. It is easy to check that this transformation is correct in the following sense: the sets of messages the intruder can gain during an unbounded number of sessions of the original protocol and the protocol after the transformation are the same.

$\varphi(s) = \{s_1, s_2\}$, where $s_1 = \{\{y, x\}_b, d\}_c$ and $s_2 = \{z, \{x, y\}_a\}_c$. Clearly, terms s_1 and s_2 are linear. So, let $\gamma_{s_1}(r_1) = \gamma_{s_1}(r_2) = \langle y, x \rangle$ (note that $\langle y, x \rangle$ is a subterm of s_1, because $\{y, x\}_b$ is a shorthand for $\{\langle y, x \rangle\}_b$), and $\gamma_{s_2}(r_1) = \gamma_{s_2}(r_2) = \langle z, \{x, y\}_a \rangle$. One can see that $\langle y, x \rangle$ covers x and y in r_1 and r_2. One can also see that $\langle z, \{x, y\}_a \rangle$ covers x, y, and z in r_1 and r_2.

Similarly, we can show that the rule $\{z, \{\{y\}_a, x\}_b\}_a \rightarrow \{\{x, \{y\}_a\}_b, z'\}_c$ is regular. The rule $\{\{x\}_b, y\}_a \rightarrow \{x, \{y\}_b\}_a$ is not regular.

Theorem 4. *The knowledge which the intruder can gain during an unbounded number of sessions of a regular protocol, can be described by an alternating tree automaton with the polynomial number of states w.r.t. the size of the protocol. Moreover, such an automaton can be computed in exponential time.*

Proof (sketch). First, we translate a given regular protocol to a logic program: for each rule $r \rightarrow s$, we produce clauses of the form $I(s') \leftarrow I(r_1), \ldots, I(r_n)$, where $s' \in \varphi(s)$, and $\{r_1, \ldots, r_n\} = \varphi(r)$. Suppose that T is a logic program obtained in this way Let $T' = T \cup T_I$. One can show that the knowledge that the intruder can gain during the protocol execution is the interpretation of I in the least Herbrand model of T'. Moreover one can show that each clause $s \leftarrow r_1 \ldots r_n$ of T' meets the following conditions: s is linear, and each term r_i (for $i = 1, \ldots, n$) can be assign a subterm $\gamma(r_i)$ of s, such that: (i) for each $i = 1, \ldots, n$, and each $x \in Var(s)$, the term $\gamma(r_i)$ covers x in r_i, (ii) for each $i, j = 1, \ldots, n$, if a variable $y \notin Var(s)$ occurs in both r_i and r_j, then $\gamma(r_i) = \gamma(r_j)$. We will call clauses of this form *regular*.

Now, T' can be translated to equivalent program T'' which consists of rules of the following form only:

$$p(f(x_1, \ldots, x_n)) \leftarrow p_1(t_1), \ldots, p_n(t_n), \quad \text{where } f(x_1, \ldots, x_n) \text{ is linear.} \quad (15)$$

In order to obtain T'', one can first eliminate clauses with the head of the form $p(x)$ (we assume that we have a fixed signature). Now, suppose that a clause has the form $p(\langle s_1, s_2 \rangle) \leftarrow p_1(t_1), \ldots, p_n(t_n)$ (for other function symbols the proof proceeds similarly). Let γ be as in the definition of regular clauses. We divide the literals $p_1(t_1), \ldots, p_n(t_n)$ into three groups A, B, C such that $p(t_i) \in A$ iff $\gamma(t_i) = \langle s_1, s_2 \rangle$, $t_i \in B$ iff $\gamma(t_i) \leq s_1$, and $t_i \in C$ iff $\gamma(t_i) \leq s_2$. We remove the rule, and add the following ones:

$$p(\langle x, y \rangle) \leftarrow A[s_1/x, s_2/y], p'(x), p''(y), \qquad p'(s_1) \leftarrow B, \qquad p''(s_2) \leftarrow C,$$

where p' and p'' are fresh predicate symbols. We recursively repeat this procedure for p' and p''. One can show that the size of T'' is polynomial w.r.t. the size of T.

Monadic Horn theories consisting of clauses of the form (15) are considered in [15], where it is shown that they can be finitely saturated by a sort resolution[4]. We can proceed similarly. Roughly speaking, we saturate P', successively adding simpler clauses, and finally, we remove all the clauses which are not of the form

---

[4] One can also show that the program T'' is in the class $\mathcal{H}1$ defined in [12], and so, by Theorem 1 of [12], is normalizable.

(6) (see page 381). Thus the obtained program is just an alternating automaton. We show that the saturation process stops after at most exponential number of steps, and that the obtained program is equivalent to P. The detailed proof can be found in the extended version of the paper [14].

Theorem 5. *Secrecy of a regular protocol is* DEXPTIME-*complete.*

Proof. To decide a secrecy of a regular protocol, we build (in exponential time) an alternating tree automaton A of polynomial number of states which describes the knowledge of the intruder, and check whether $Sec \in L(A)$, which can be done in exponential time.

We prove DEXPTIME-hardness by reduction of the emptiness of the intersection of regular tree languages given by n finite automata. We build a protocol that encode all these automata in such a way that the i-th automaton recognizes a term t iff the intruder knows the term $\{t\}_{k_i}$. We add the rule $\{x\}_{k_1}, \ldots, \{x\}_{k_n} \rightarrow Sec$ to the protocol. One can see that the protocol is insecure, iff the intersection of the given automata is not empty. $\square$

By a very similar technique, regular protocols can be extended to work with regular constraints: we can encode a finite state automaton A by some regular rules so that $t \in L(A)$ iff $I(\{t\}_{k_A})$, and add terms of the form $\{x\}_{k_A}$ to the left-hand side of rules.

The results of this section and Sections 3 can be easily combined to achieve decidability of secrecy of the following two-phases protocols. Suppose that a protocol, which uses only atomic keys, consists of some regular rules P_1, and some rules P_2 of arbitrary form. The intruder can execute rules from P_1 unbounded number of times (building a knowledge which is a regular language), and then he can execute the rules of P_2 at most once. Because, for an alternating tree automaton, one can construct an equivalent deterministic bottom-up tree automaton of exponential size, by Theorems 4 and 1, the insecurity problem of such a protocol can be decided in NEXPTIME.

6   Conclusions

We have extended the decidability result for protocols analyzed w.r.t. a bounded number of sessions to the case when the initial knowledge of the intruder is a regular language. We have shown that if this language is given by a deterministic bottom-up automaton, then the insecurity problem of a protocol is NP-complete, assuming that complex keys are not allowed. We have showed also that if we add to protocols regular constraints which guarantee that messages have a required form, then the problem of deciding protocols is NEXPTIME-complete. These results can be a starting point for developing practical algorithms for detecting attacks with regular initial knowledge.

We have also defined a family of protocols such that the set of messages that the intruder can gain during unbounded number of sessions is exactly a regular language.

An open problem is decidability of the security of protocols with *complex keys* against attacks with regular initial knowledge.

References

1. R. M. AMADIO AND W. CHARATONIK, *On name generation and set-based analysis in the Dolev-Yao model.*, in CONCUR, vol. 2421 of Lecture Notes in Computer Science, Springer, 2002, pp. 499–514.
2. Y. CHEVALIER, R. KÜSTERS, M. RUSINOWITCH, AND M. TURUANI, *An NP decision procedure for protocol insecurity with XOR.*, in LICS, IEEE Computer Society, 2003, pp. 261–270.
3. Y. CHEVALIER, R. KÜSTERS, M. RUSINOWITCH, M. TURUANI, AND L. VIGNERON, *Extending the Dolev-Yao intruder for analyzing an unbounded number of sessions.*, in CSL, vol. 2803 of Lecture Notes in Computer Science, Springer, 2003, pp. 128–141.
4. H. COMON AND V. SHMATIKOV, *Is it possible to decide whether a cryptographic protocol is secure or not?*, Journal of Telecommunications and Information Technology, special issue on cryptographic protocol verification, 4 (2002), pp. 5–15.
5. D. DOLEV AND A. YAO, *On the security of public-key protocols*, IEEE Transactions on Information Theory, 29 (1983), pp. 198–208.
6. N. DURGIN, P. LINCOLN, J. MITCHELL, AND A. SCEDROV, *Undecidability of bounded security protocols*, in Workshop on Formal Methods and Security Protocols (FMSP'99), 1999.
7. S. EVEN AND O. GOLDREICH, *On the security of multi-party ping-pong protocols*, in Technical Report 285, Israel Institute of Technology, 1983.
8. T. W. FRÜHWIRTH, E. Y. SHAPIRO, M. Y. VARDI, AND E. YARDENI, *Logic programs as types for logic programs*, in LICS, 1991, pp. 300–309.
9. T. GENET AND F. KLAY, *Rewriting for cryptographic protocol verification.*, in CADE, vol. 1831 of Lecture Notes in Computer Science, Springer, 2000, pp. 271–290.
10. C. MEADOWS, *Formal methods for cryptographic protocol analysis: Emerging issues and trends*, IEEE Journal on Selected Areas in Communication, 21 (2003), pp. 44–54.
11. D. MONNIAUX, *Abstracting cryptographic protocols with tree automata.*, in SAS, A. Cortesi and G. Filé, eds., vol. 1694 of Lecture Notes in Computer Science, Springer, 1999, pp. 149–163.
12. F. NIELSON, H. R. NIELSON, AND H. SEIDL, *Normalizable horn clauses, strongly recognizable relations, and spi.*, in SAS, vol. 2477 of Lecture Notes in Computer Science, Springer, 2002, pp. 20–35.
13. M. RUSINOWITCH AND M. TURUANI, *Protocol insecurity with a finite number of sessions, composed keys is NP-complete.*, Theor. Comput. Sci., 1-3 (2003), pp. 451–475.
14. T. TRUDERUNG, *Regular protocols and attacks with regular knowledge. Extended version*, 2005. Available at http://www.ii.uni.wroc.pl/~tt/papers/.
15. C. WEIDENBACH, *Towards an automatic analysis of security protocols in first-order logic.*, in CADE, vol. 1632 of Lecture Notes in Computer Science, Springer, 1999, pp. 314–328.

The Model Evolution Calculus with Equality

Peter Baumgartner[1] and Cesare Tinelli[2]

[1] Max-Planck Institute for Computer Science, Saarbrücken
baumgart@mpi-sb.mpg.de
[2] Department of Computer Science, The University of Iowa
tinelli@cs.uiowa.edu

Abstract. In many theorem proving applications, a proper treatment of equational theories or equality is mandatory. In this paper we show how to integrate a modern treatment of equality in the Model Evolution calculus ($\mathcal{ME}$), a first-order version of the propositional DPLL procedure. The new calculus, $\mathcal{ME}_E$, is a proper extension of the $\mathcal{ME}$ calculus without equality. Like $\mathcal{ME}$ it maintains an explicit *candidate model*, which is searched for by DPLL-style splitting. For equational reasoning $\mathcal{ME}_E$ uses an adapted version of the ordered paramodulation inference rule, where equations used for paramodulation are drawn (only) from the candidate model. The calculus also features a generic, semantically justified simplification rule which covers many simplification techniques known from superposition-style theorem proving. Our main result is the correctness of the $\mathcal{ME}_E$ calculus in the presence of very general redundancy elimination criteria.

1 Introduction

The Model Evolution ($\mathcal{ME}$) Calculus [4] has recently been introduced by the authors of this paper as a first-order version of the propositional DPLL procedure [7]. Compared to its predecessor, the FDPLL calculus [2], it lifts to the first-order case not only the core of the DPLL procedure, the splitting rule, but also DPLL's simplification rules, which are crucial for effectiveness in practice.

Our implementation of the $\mathcal{ME}$ calculus, the Darwin system [3], performs well in some domains, but, unsurprisingly, it generally performs poorly in domains with equality. In this paper we address this issue and propose an extension of the $\mathcal{ME}$ calculus with dedicated inference rules for equality reasoning. These rules are centered around a version the *ordered paramodulation* inference rule adapted to the $\mathcal{ME}$ calculus. The new calculus, $\mathcal{ME}_E$, is a proper extension of the $\mathcal{ME}$ calculus without equality. Like $\mathcal{ME}$, it searches for a model of the input clause set by maintaining and incrementally modifying a finite representation, called a *context*, of a *candidate model* for the clause set. In $\mathcal{ME}_E$, equations from the context, and only those, are used for ordered paramodulation inferences into the current clause set. The used equations are kept together with the clause paramodulated into and act as passive constraints in the search for a model.

In this paper we present the calculus and discuss its soundness and completeness. The completeness proof is obtained as an extension of the completeness proof of the $\mathcal{ME}$ calculus (without equality) by adapting techniques from the Bachmair/Ganzinger framework developed for proving the completeness of the superposition calculus [1,11,

R. Nieuwenhuis (Ed.): CADE 2005, LNAI 3632, pp. 392–408, 2005.
© Springer-Verlag Berlin Heidelberg 2005

e.g.]. The underlying model construction technique allows us to justify a rather general simplification rule on semantic grounds. The simplification rule is based on a general redundancy criterion that covers many simplification techniques known from superposition-style theorem proving.

Related Work. Like $\mathcal{ME}$, the $\mathcal{ME}_E$ calculus is related to *instance based methods (IMs)*, a family of calculi and proof procedures developed over the last ten years. What has been said in [4] about $\mathcal{ME}$ in relation to IMs also applies to $\mathcal{ME}_E$ when equality is not an issue, and the points made there will not be repeated here in detail. Instead, we focus on instance based methods that natively support equality reasoning.

Among them is Ordered Semantic Hyperlinking (OSHL) [12]. OSHL uses rewriting and narrowing (paramodulation) with unit equations, but requires some other mechanism such as Brand's transformation to handle equations that appear in nonunit clauses.

To our knowledge there are only two instance-based methods that have been extended with dedicated equality inference rules for full equational clausal logic. One is called disconnection tableaux, which is a successor of the disconnection method [6].[1] The other is the IM described in [9]. Both methods are conceptually rather different from $\mathcal{ME}$ in that the main derivation rules there are based on resolving *pairs* of complementary literals (connections) from *two* clauses, whereas $\mathcal{ME}$'s splitting rule is based on evaluating *all* literals of a *single* clause against a current candidate model.

The article [10] discusses various ways of integrating equality reasoning in disconnection tableaux. It includes a variant based on ordered paramodulation, where paramodulation inferences are determined by inspecting connections between literals of two clauses. Only comparably weak redundancy criteria are available.

The instance based method in [9] has been extended with equality in [8]. Beyond what has been said above there is one more conceptual difference, in that the inference step for equality reasoning is based on refuting, as a subtask, a set of unit clauses (which is obtained by picking clause literals).

Paper Organization. We start with an informal explanation of the main ideas behind the $\mathcal{ME}_E$ calculus in Section 2, followed by a more formal treatment of *contexts* and their associated interpretations in Section 3. Then, in Section 4, we present what we call *constrained clauses* and a way to perform equality reasoning on them. We describe the $\mathcal{ME}_E$ calculus over constrained clauses in Section 6, and discuss its correctness in Section 7. For space constraints we cannot provide proofs of the results presented in this paper. Also, we must assume that the reader has already some familiarity with the $\mathcal{ME}$ calculus. All proofs as well as a more detailed exposition the calculus can be found in the paper's extended version [5].

2 Main Ideas

The $\mathcal{ME}$ calculus of [4], and by extension the $\mathcal{ME}_E$ calculus, is informally best described with an eye to the propositional DPLL procedure, of which $\mathcal{ME}$ is a first-order

---

[1] Even in that early paper a paramodulation-like inference rule was considered, however a rather weak one.

lifting. DPLL can be viewed as a procedure that searches the space of possible interpretations for a given clause set until it finds one that satisfies the clause set, if it exists. This can be done by keeping a current candidate model and *repairing* it as needed until it satisfies every input clause. The repairs are done incrementally by changing the truth value of one clause literal at a time, and involve a non-deterministic guess (a "split") on whether the value of a selected literal should be changed or kept as it is. The number of guesses is limited by a constraint propagation process ("unit propagation") that is able to deduce deterministically the value of some input literals.

Both $\mathcal{ME}$ and $\mathcal{ME}_E$ lift this idea to first-order logic by maintaining a *first-order* candidate model, by identifying *instances* of input clauses that are falsified by the model, and by repairing the model incrementally until it satisfies all of these instances. The difference between the two calculi is that $\mathcal{ME}_E$ works with *equational* models, or *E-interpretations*, that is, Herbrand interpretations in which the equality symbol is the only predicate symbols and always denotes a congruence relation.

The current E-interpretation is represented (or more precisely, induced) by a *context*, a finite set of non-ground equations and disequations directly processed by the calculus. Context literals can be built over two kinds of variables: *universal* and *parametric* variables. The difference between the two lies in how they constrain the possible additions of further literals to a context and, as a consequence, the possible repairs to its induced E-interpretation. As far as the induced E-interpretation is concerned, however, the two types of variables are interchangeable. The construction of this E-interpretation is best explained in two stages, each based on an ordering on terms/atoms: the usual instantiation preordering $\gtrsim$ with its strict subset $\gtrsim_{\sim}$, and an arbitrary reduction ordering $\succ$ total on ground terms. Using the first we associate to a context Λ, similarly to the $\mathcal{ME}$ calculus, a (non-equational) interpretation I_Λ. Roughly, and modulo symmetry of $\approx$, this interpretation satisfies a ground equation $s'' \approx t''$, over an underlying signature Σ, iff $s'' \approx t''$ is an instance of an equation $s \approx t$ in Λ without being an instance of any equation $s' \approx t'$ such that $s \approx t \gtrsim_{\sim} s' \approx t'$ and $s' \not\approx t' \in \Lambda$. For instance, if $\Lambda = \{f(u) \approx u, f(a) \not\approx a\}$ where u is a (parametric) variable and the signature Σ consists of the unary function symbol f and the constant symbols a and b, then I_Λ is the symmetric closure of $\{f^{n+1}(b) \approx f^n(b) \mid n \geq 0\} \cup \{f^{n+1}(a) \approx f^n(a) \mid n \geq 1\}$.

In general I_Λ is not an E-interpretation. Its purpose is merely to supply a set of candidate equations that determine the final E-interpretation induced by Λ. This E-interpretation, denoted by R_Λ^E, is defined as the smallest congruence on ground Σ-terms that includes a specific set R_Λ of ordered equations selected from I_Λ. The set R_Λ is constructed inductively on the reduction ordering $\succ$ by adding to it an ordered equation $s \to t$ iff $s \approx t$ or $t \approx s$ is in I_Λ, $s \succ t$ and both s and t are irreducible wrt. the equations of R_Λ that are smaller than $s \to t$. This construction guarantees that R_Λ is a convergent rewrite system. In the example above, R_Λ is $\{f(b) \to b, f(f(a)) \to f(a)\}$ for any reduction ordering $\succ$; the E-interpretation R_Λ^E induced by Λ is the congruence closure of $\{f(b) \approx b, f(f(a)) \approx f(a)\}$. Since R_Λ is convergent by construction for any context Λ, any two ground Σ-terms are equal in R_Λ^E iff they have the same R_Λ-normal form.

Now that we have sketched how the E-interpretation is constructed, we can explain how the calculus detects the need to repair the current *E*-interpretation and how it goes about repairing it. To simplify the exposition we consider here only ground input clauses. A repair involves conceptually two steps: (i) determining whether a given clause C is false in the E-interpretation R_Λ^E, and (ii) if so, modifying Λ so that the new R_Λ^E satisfies it.

For step (i), by congruence it suffices to rewrite the literals of C with the rewrite rules R_Λ to normal form. If $C{\downarrow}_{R_\Lambda}$ denotes that normal form, then R_Λ^E falsifies C iff all equations in $C{\downarrow}_{R_\Lambda}$ are of the form $s \approx t$ with $s \neq t$, and all disequations are of the form $s \not\approx s$. In the earlier example, if $C = f(a) \approx a \vee f(f(a)) \approx b \vee f(b) \not\approx b$ then $C{\downarrow}_{R_\Lambda} = f(a) \approx a \vee f(a) \approx b \vee b \not\approx b$, meaning that R_Λ^E indeed falsifies C.

For step (ii), we first point out that the actual repair needs to be carried out only on the literals of $C{\downarrow}_{R_\Lambda}$, not on the literals of C. More precisely, the calculus considers only the positive equations of $C{\downarrow}_{R_\Lambda}$, as the trivial disequations $s \not\approx s$ in it do not provide any usable information. To repair the E-interpretation it is enough to modify Λ so that R_Λ contains one of the positive equations $s \approx t$ of $C {\downarrow}_{R_\Lambda}$. Then, by congruence, R_Λ^E will also satisfy C, as desired. Concretely, Λ is modified by creating a choice point and adding to Λ one of the literals L of $C{\downarrow}_{R_\Lambda}$ or its complement. Adding L—which is possible only provided that neither L not its complement are contradictory, in a precise sense defined later, with Λ—will make sure that the new R_Λ^E satisfies C. Adding the complement of L instead will not make C satisfiable in the new candidate E-model. However, it is necessary for soundness and marks some progress in the derivation because it will force the calculus to consider other literals of $C {\downarrow}_{R_\Lambda}$ for addition to the context.

Referring again to our running example, of the two positive literals of $C{\downarrow}_{R_\Lambda} = f(a) \approx a \vee f(a) \approx b \vee b \not\approx b$, only $f(a) \approx b$ can be added to the context $\Lambda = \{f(u) \approx u, f(a) \not\approx a\}$ because neither it nor its complement is contradictory with Λ (by contrast $f(a) \approx a$ is contradictory with Λ). With $\Lambda = \{f(u) \approx u, f(a) \not\approx a, f(a) \approx b\}$, now $R_\Lambda = \{f(b) \rightarrow b, f(a) \rightarrow b\}$ and $C{\downarrow}_{R_\Lambda}$ becomes $b \approx a \vee b \approx b \vee b \not\approx b$, which means that C is satisfied by R_Λ^E.

We point out that adding positive equations to the context is not always enough. Sometimes it is necessary to add negative equations, whose effect is to eliminate from R_Λ rewrite rules that cause the disequations of C to rewrite to trivial disequations. The calculus takes care of this possibility as well. To achieve that we found it convenient to have $\mathcal{ME}_E$ work with a slightly generalized data structure. More precisely, instead of clauses C we consider *constrained clauses* $C \cdot \Gamma$, where Γ is a set of rewrite rules. The constraint Γ consists just of those (instances of) *equations* from a context Λ that were used to obtain C from some input clause (whose constraint is empty).

Reusing our example, the clause C would be represented as the constraint clause $C \cdot \Gamma = f(a) \approx a \vee f(f(a)) \approx b \vee f(b) \not\approx b \cdot \emptyset$, with its R_Λ-normal form being $C {\downarrow}_{R_\Lambda} \cdot \Gamma = f(a) \approx a \vee f(a) \approx b \vee b \not\approx b \cdot f(f(a) \rightarrow f(a), f(b) \rightarrow b$ for $\Lambda = \{f(u) \approx u, f(a) \not\approx a\}$. Now, the rewrite rule $f(b) \rightarrow b$ used to obtain the normal form is available in the constraint part, as written. The calculus may add its negation $f(b) \not\approx b$ to Λ, with the effect of removing $f(b) \rightarrow b$ from R_Λ. The resulting context and rewrite system would be, respectively, $\Lambda'' = \{f(u) \approx u, f(a) \not\approx a, f(b) \not\approx b\}$, and $R_{\Lambda''} = \{f(f(b)) \rightarrow f(b), f(f(a)) \rightarrow f(a)\}$. It is easy to see that the new $I_{R_\Lambda}^E$ satisfies C as well, as desired.

While the above informal description illustrates the main ideas behind $\mathcal{ME}_E$, it is not entirely faithful to the actual calculus as defined later in the paper. Perhaps the most significant differences to mention here are that (i) the calculus works with non-ground clauses as well (by treating them, as usual in refutation-based calculi, as schematic for their ground instances and relying heavily on unification), and (ii) the normal form of a constrained clause is not derived in one sweep, as presented above. Instead the calculus, when equipped with a fair strategy, derives all intermediate constrained clauses as well. It does so by a suitably defined paramodulation rule, where the equations paramodulating (only) into the clause part of a constrained clause are drawn from the current context Λ. The rationale is that the rewrite system R_Λ is in general not available to the calculus. Hence rewriting (ground) clause literals with rules from R_Λ, which would theoretically suffice to obtain a complete calculus at the ground level, is approximated by ordered paramodulation with equations from Λ instead.

3   Contexts and Induced Interpretations

We start with some formal preliminaries. We will use two disjoint, infinite sets of variables: a set X of *universal* variables, which we will refer to just as variables, and another set V, which we will always refer to as *parameters*. We will use u and v to denote elements of V and x and y to denote elements of X. We fix a signature Σ throughout the paper and denote by Σ^{sko} the expansion of Σ obtained by adding to Σ an infinite number of fresh (Skolem) constants. If t is a term we denote by $\mathcal{V}ar(t)$ the set of t's variables and by $\mathcal{P}ar(t)$ the set of t's parameters. A term t is *ground* iff $\mathcal{V}ar(t) = \mathcal{P}ar(t) = \emptyset$.

A substitution ρ is a *renaming on* $W \subseteq (V \cup X)$ iff its restriction to W is a bijection of W onto itself; ρ is simply a *renaming* if it is a renaming on $V \cup X$. A substitution σ is *p-preserving* (short for parameter preserving) if it is a renaming on V. If s and t are two terms, we write $s \gtrsim t$, iff there is a substitution σ such that $s\sigma = t$.[2] We say that s is *a variant of* t, and write $s \sim t$, iff $s \gtrsim t$ and $t \gtrsim s$ or, equivalently, iff there is a renaming ρ such that $s\rho = t$. We write $s \gtrsim\!\!\!\!_\sim t$ if $s \gtrsim t$ but $s \not\sim t$. We write $s \geq t$ and say that t is *a p-instance of* s iff there is a p-preserving substitution σ such that $s\sigma = t$. We say that s is *a p-variant of* t, and write $s \simeq t$, iff $s \geq t$ and $t \geq s$; equivalently, iff there is a p-preserving renaming ρ such that $s\rho = t$. The notation $s[t]_p$ means that the term t occurs in the term s at position p, as usual.

All of the above is extended from terms to literals in the obvious way.

In this paper we restrict to equational clause logic. Therefore, and essentially without loss of generality, we assume that the only predicate symbol in Σ is $\approx$. An atom then is always an equation, and a literal then is always an equation or the negation of an equation. Literals of the latter kind, i.e., literals of the form $\neg(s \approx t)$ are also called *negative equations* and generally written $s \not\approx t$ instead. We call a literal *trivial* if it is of the form $t \approx t$ or $t \not\approx t$. We denote literals by the letters K and L. We denote by $\overline{L}$ the complement of a literal L, and by L^{sko} the result of replacing each variable of L by a fresh Skolem constant in $\Sigma^{\text{sko}} \setminus \Sigma$. We denote clauses by the letters C and D, and the empty clause by $\square$. We will write $L \vee C$ to denote a clause obtained as the disjunction of a (possibly empty) clause C and a literal L.

A *(Herbrand) interpretation* I is a set of ground Σ-equations—those that are true in the interpretation. Satisfiability/validity of ground Σ-literals, Σ-clauses, and clause sets

---

[2] Note that many authors would write $s \lesssim t$ in this case.

in a Herbrand interpretation is defined as usual. We write $I \models F$ to denote the fact that I satisfies F, where F is a ground Σ-literal or a Σ-clause (set). An *E-interpretation* is an interpretation that is also a congruence relation on the Σ-terms. If I is an interpretation, we denote by I^E the smallest congruence relation on the Σ-terms that includes I, which is an E-interpretation. We say that I *E-satisfies* F iff $I^E \models F$. Instead of $I^E \models F$ we generally write $I \models_E F$. We say that F *E-entails* F', written $F \models_E F'$, iff every E-interpretation that satisfies F also satisfies F'. We say that F and F' are *E-equivalent* iff $F \models_E F'$ and $F' \models_E F$.

The Model Evolution calculus, with and without equality, works with sequents of the form $\Lambda \vdash \Phi$, where Λ is a finite set of literals possibly with variables or with parameters called a context, and Φ is a finite set of clauses possibly with variables. As in [4], we impose for simplicity that literals in a context can contain parameters or variables but not both, but this limitation can be overcome.

Definition 3.1 (Context [4]). *A context is a set of the form $\{\neg v\} \cup S$ where $v \in V$ and S is a finite set of literals each of which is parameter-free or variable-free.*

Differently from [4], we implicitly treat any context Λ as if it contained the symmetric version of each of its literals. For instance, if $\Lambda = \{\neg v, f(u) \approx a, f(x) \not\approx x\}$ then $a \approx f(u), f(u) \approx a, x \not\approx f(x), f(x) \not\approx x$ are all considered to be literals of Λ, and we write, for instance, $a \approx f(u) \in \Lambda$.

Where L is a literal and Λ a context, we write $L \in_\sim \Lambda$ if L is a variant of a literal in Λ, write $L \in_\simeq \Lambda$ if L is a p-variant of a literal in Λ, and write $L \in_\geq \Lambda$ if L is a p-instance of a literal in Λ. A literal L is *contradictory with* a context Λ iff $L\sigma = \overline{K}\sigma$ for some $K \in_\sim \Lambda$ and some p-preserving substitution σ. A context Λ is *contradictory* iff it contains a literal that is contradictory with Λ. Referring to the context Λ above, $f(v) \not\approx a$, $a \not\approx f(v), a \approx f(a), f(a) \approx a$ all are contradictory with Λ. Notice that an equation $s \approx t$ is contradictory with a context Λ if and only if $t \approx s$ is so. The same applies to negative equations.

We will work only with non-contradictory contexts. Thanks to the next two notions, such contexts can be used as finite denotations of (certain) Herbrand interpretations. Let L be a literal and Λ a context. A literal K is *a most specific generalization (msg) of L in Λ* iff $K \gtrsim L$ and there is no $K' \in \Lambda$ such that $K \gtrsim_{\not\sim} K' \gtrsim L$.

Definition 3.2 (Productivity [4]). *Let L be a literal, C a clause, and Λ a context. A literal K produces L in Λ iff (i) K is an msg of L in Λ, and (ii) there is no $K' \in_\geq \Lambda$ such that $K \gtrsim_{\not\sim} \overline{K'} \gtrsim L$. The context Λ produces L iff it contains a literal K that produces L in Λ.*

Notice that a literal K produces a literal L in a context Λ if and only if K produces the symmetric version of L in Λ. For instance, the context Λ above produces $f(b) \approx a$ and $a \approx f(b)$ but Λ produces neither $f(a) \approx a$ nor $a \approx f(a)$. Instead it produces both $a \not\approx f(a)$ and $f(a) \not\approx a$.

A non-contradictory context Λ uniquely induces a (Herbrand) Σ-interpretation I_Λ, defined as follows:

$$I_\Lambda := \{l \approx r \mid l \approx r \text{ is a positive ground } \Sigma\text{-equation and } \Lambda \text{ produces } l \approx r\}$$

For instance, if $\Lambda = \{x \approx f(x)\}$ and Σ consists of a constant a and the unary function symbol f then $I_\Lambda = \{a \approx f(a), f(a) \approx a, f(a) \approx f(f(a)), f(f(a)) \approx f(a), \ldots\}$.

A consequence of the presence of the pseudo-literal $\neg v$ in every context Λ is that Λ produces L or $\overline{L}$ for every literal L. Moreover, it can be easily shown that whenever $I_\Lambda \models L$ then Λ produces L, even when L is a negative literal. This fact provides a "syntactic" handle on literals satisfied by I_Λ. The induced interpretation I_Λ is not an E-interpretation in general.[3] But we will use it to define a unique E-interpretation associated to Λ.

4   Equality Reasoning on Constrained Clauses

The $\mathcal{ME}_E$ calculus operates with *constrained clauses*, defined below. In this section we will introduce derivation rules for equality reasoning on constrained clauses. These derivation rules will be used by the $\mathcal{ME}_E$ calculus in a modular way. The section concludes with a first soundness and completeness result, which will serve as a lemma for the completeness proof of the $\mathcal{ME}_E$ calculus.

As an important preliminary remark, whenever the choice of the signature makes a difference in this section, e.g. in the definition of grounding substitution, we always implicitly meant the signature Σ, not the signature Σ^{sko}.

Constrained Clauses. A *(rewrite) rule* is an expression of the form $l \to r$ where l and r are Σ-terms. Given a parameter-free Σ-clause $C = L_1 \vee \cdots \vee L_n$ and a set of parameter-free Σ-rewrite rules $\Gamma = \{A_1, \ldots, A_m\}$, the expression $C \cdot \Gamma$ is called a *constrained clause (with constraint Γ)*. Instead of $C \cdot \{A_1, \ldots, A_m\}$ we generally write $C \cdot A_1, \ldots, A_m$. The notation $C \cdot \Gamma, A$ means $C \cdot \Gamma \cup \{A\}$.

A constrained clause $C \cdot \Gamma$ is a *constrained clause without expansion constraints* iff Γ contains no *expansion rules*, i.e., rules of the form $x \to t$, where x is a variable and t is a term. A *constrained clause set without expansion constraints* is a constrained clause set that consists of constrained clauses without expansion constraints. The $\mathcal{ME}_E$ calculus works only with such constrained clause sets.[4]

Applying a substitution σ to $C \cdot \Gamma$, written as $(C \cdot \Gamma)\sigma$, means to apply σ to C and all rewrite rules in Γ. A constrained clause $C \cdot \Gamma$ is *ground* iff both C and Γ are ground. If γ is a substitution such that $(C \cdot \Gamma)\gamma$ is ground, then $(C \cdot \Gamma)\gamma$ is called a *ground instance* of $C \cdot \Gamma$, and γ is called a *grounding substitution* for $C \cdot \Gamma$. We say that $C \cdot \Gamma$ *properly subsumes* $C' \cdot \Gamma'$ iff there is a substitution σ such that $C\sigma \subset C'$ and $\Gamma\sigma \subseteq \Gamma'$ or $C\sigma \subseteq C'$ and $\Gamma\sigma \subset \Gamma'$. We say that $C \cdot \Gamma$ *non-properly subsumes* $C' \cdot \Gamma'$ iff there is a substitution σ such that $C\sigma = C'$ and $\Gamma\sigma = \Gamma'$. The constrained clauses $C \cdot \Gamma$ and $C' \cdot \Gamma'$ are *variants* iff $C \cdot \Gamma$ non-properly subsumes $C' \cdot \Gamma'$ and vice versa. For a set of constrained clauses Φ, Φ^{gr} denotes the set of all ground Σ-instances of all constrained clauses in Φ.

In principle, a constraint clause $C \cdot \Gamma = L_1 \vee \cdots \vee L_m \cdot l_{m+1} \to r_{m+1}, \ldots, l_n \to r_n$ could be understood as standing for the ordinary clause $L_1 \vee \cdots \vee L_m \vee l_{m+1} \not\approx r_{m+1} \vee \cdots \vee l_n \not\approx r_n$, which we call the *clausal form of $C \cdot \Gamma$* and denote by $(C \cdot \Gamma)^c$. In effect, however, constrained clauses and their clausal forms are rather different from an operational point

---

[3] In fact, in the earlier example $a \approx f(f(a)) \notin I_\Lambda$.

[4] As will become clear later, disallowing expansion constraints comes from the fact that paramodulation into variables is unnecessary in $\mathcal{ME}_E$ as well.

of view. The derivation rules for equality reasoning below, in particular paramodulation, are *never* applied to constraints—as a consequence, the calculus cannot be said to be a resolution calculus.

Orderings. We suppose as given a reduction ordering $\succ$ that is total on ground Σ-terms. It has to be extended to rewrite rules, equations and constrained clauses. Following usual techniques [1,11, e.g.], rewrite rules and equations are compared by comparing the multisets of their top-level terms with the multiset extension of the base ordering $\succ$. There is no need in our framework to distinguish between positive and negative equations. It is important, though, that when comparing constrained clauses the clause part is given precedence over the constraint part. This can be achieved by defining $C \cdot \Gamma \succ C' \cdot \Gamma'$ iff (C, Γ) is strictly greater than (C', Γ') in the lexicographical ordering over the multiset extension of the above ordering on equations and rewrite rules. (See [5] for an alternative definition.) This way, the calculus' derivation rules $\mathsf{Ref}_{\mathcal{ME}}$ and $\mathsf{Para}_{\mathcal{ME}}$ for equality reasoning defined in Section 5 work in an order-decreasing way.

Derivation Rules. We first define two auxiliary derivation rules for equality reasoning on constrained clauses. The rules will be used later in the $\mathcal{ME}_E$ calculus.

$$\mathsf{Ref}(\sigma) \quad \frac{s \not\approx t \vee C \cdot \Gamma}{(C \cdot \Gamma)\sigma} \quad \text{if } \sigma \text{ is a mgu of } s \text{ and } t.$$

We write $s \not\approx t \vee C \cdot \Gamma \Rightarrow_{\mathsf{Ref}(\sigma)} (C \cdot \Gamma)\sigma$ to denote a Ref inference.[5]

$$\mathsf{Para}(l \approx r, \sigma) \quad \frac{L[t]_p \vee C \cdot \Gamma}{(L[r]_p \vee C \cdot \Gamma, l \to r)\sigma} \quad \text{if } \begin{cases} t \text{ is not a variable,} \\ \sigma \text{ is a mgu of } t \text{ and } l, \text{ and} \\ l\sigma \not\preceq r\sigma. \end{cases}$$

We write $L[t]_p \vee C \cdot \Gamma \Rightarrow_{\mathsf{Para}(l \approx r, \sigma)} (L[r]_p \vee C \cdot \Gamma, l \to r)\sigma$ to denote a Para inference.

A Ref or Para inference is *ground* if both its premise and conclusion are ground and as well as the equation $l \approx r$ in the Para case. If from a given Ref or Para inference a ground inference results by applying a substitution γ to the premise, the conclusion and the used equation $l \approx r$ in case of Para, we call the resulting ground inference a *ground instance via γ (of the inference).*

As in the superposition calculus, *model construction*, *redundancy* and *saturation* are core concepts for the understanding of the $\mathcal{ME}_E$ calculus.

Model Construction. A rewrite system is a set of Σ-rewrite rules. A ground rewrite system R is *ordered by* $\succ$ iff $l \succ r$, for every rule $l \to r \in R$. As a non-standard notion, we define a *rewrite system without overlaps* to be a ground rewrite system R that is ordered by $\succ$, and whenever $l \to r \in R$ then there is no other rule in R of the form $s[l] \to t$ or $s \to t[l]$. In other words, no rule can be reduced by another rule, neither the left hand side *nor the right hand side*. Any rewrite system without overlaps is a convergent ground rewrite system. In the sequel, the letter R will always denote a (ground) rewrite system without overlaps.

---

[5] An *inference* is an instance of a derivation rule that satisfies the rule's side condition.

We show how every non-contradictory context Λ induces a ground rewrite system R_Λ without overlaps. The general technique is taken from the completeness proof of the superposition calculus [1,11] but adapted to our needs.

First, for a given non-contradictory context Λ and positive ground Σ-equation $s \approx t$ we define by induction on the literal ordering $\succ$ sets of rewrite rules $\varepsilon^\Lambda_{s \approx t}$ and $R^\Lambda_{s \approx t}$ as follows. Assume that $\varepsilon^\Lambda_{s' \approx t'}$ has already been defined for all ground Σ-equations $s' \approx t'$ with $s \approx t \succ s' \approx t'$. Where $R^\Lambda_{s \approx t} = \bigcup_{s \approx t \succ s' \approx t'} \varepsilon^\Lambda_{s' \approx t'}$, define

$$
\varepsilon^\Lambda_{s \approx t} = \begin{cases} \{s \to t\} & \text{if } I_\Lambda \models s \approx t,\, s \succ t,\, \text{and } s \text{ and } t \text{ are irreducible wrt. } R^\Lambda_{s \approx t} \\ \emptyset & \text{otherwise} \end{cases}
$$

Then, $R_\Lambda = \bigcup_{s \approx t} \varepsilon^\Lambda_{s \approx t}$ where s and t range over all ground Σ-terms.

By construction, R_Λ has no critical pairs, neither with left hand sides nor with right hand sides, and thus is a rewrite system without overlaps. Since $\succ$ is a well-founded ordering, R_Λ is a convergent rewrite system by construction. The given context Λ comes into play as stated in the first condition of the definition of $\varepsilon^\Lambda_{s \approx t}$, which says, in other words, that Λ must produce $s \approx t$ as a necessary condition for $s \to t$ to be contained in R_Λ. An important detail is that whenever Λ is non-contradictory and produces $s \approx t$, then it will also produce $t \approx s$. Thus, if $s \prec t$ then $s \approx t$ may still be turned into the rewrite rule $t \to s$ in R_Λ by means of its symmetric version $t \approx s$.

Where the $\mathcal{ME}$ calculus would associate to a sequent $\Lambda \vdash \Phi$ the interpretation I_Λ as a candidate model of Φ, the $\mathcal{ME}_E$ calculus will instead associate to it the E-interpretation R^E_Λ, the congruence closure of R_Λ (or, more correctly, of the interpretation containing the same equations as R_Λ). There is an interesting connection between the two interpretations: if L is a ground literal and $L{\downarrow}_{R_\Lambda}$ is the normal form of L wrt. R_Λ then $R^E_\Lambda \models L$ (or, equivalently, $R_\Lambda \models_E L$) iff $I_\Lambda \models L{\downarrow}_{R_\Lambda}$ or $L{\downarrow}_{R_\Lambda}$ is a trivial equation. This connection is fundamental to $\mathcal{ME}_E$, as it makes it possible to reduce satisfiability in the intended E-interpretation R^E_Λ to satisfiability in I_Λ.

For an example for the model construction let $\Lambda = \{a \approx u, b \approx c, a \not\approx c\}$ a non-contradictory context. With the ordering $a \succ b \succ c$ the induced rewrite system R_Λ is again $\{b \to c\}$. To see why, observe that the candidate rule $a \to c$ is assigned false by I_Λ, as Λ does not produce $a \approx c$, and that the other candidate $a \to b$ is reducible by the smaller rule $b \to c$. Had we chosen to omit in the definition of ε the condition "t is irreducible wrt $R^\Lambda_{s \approx t}$" [6] the construction would have given $R_\Lambda = \{a \to b, b \to c\}$. This leads to the undesirable situation that a constrained clause, say, $a \not\approx c \cdot \emptyset$ is falsified by R^E_Λ. But the $\mathcal{ME}_E$ calculus cannot modify Λ to revert this situation, and to detect the inconsistency (ordered) paramodulation into variables would be needed.

Semantics of Constrained Clauses. Let $C \cdot \Gamma$ be a ground constrained clause and R a ground rewrite system. We say that R is an *E-model* of $C \cdot \Gamma$ and write $R \models_E C \cdot \Gamma$ iff $\Gamma \not\subseteq R$ or $R \models_E C$ (in the sense of Section 3, by treating R as an interpretation). We write $R \models_E \Phi$ for a set Φ of constrained clauses iff $R \models_E C \cdot \Gamma$ for all $C \cdot \Gamma \in \Phi$. If F is a non-ground constrained clause (set) we write $R \models_E F$ iff $R \models_E F^{gr}$.

---

[6] This condition is absent in the model construction for the superposition calculus. Its presence in the end explains why paramodulation into smaller sides of equations is necessary.

The general intuition for this notion of satisfiability for constrained clauses is that ground constrained clauses whose constraint is not a subset of a rewrite system R are considered to be trivially satisfied by R, while the other constrained clauses are considered to be satisfied by R exactly when their non-constraint part is E-satisfied by R. Note that for constrained clauses $C \cdot \emptyset$ with an empty constraint, $R \models_E C \cdot \emptyset$ iff $R \models_E C$.

If Φ and Φ' are sets of constrained clauses, we say that Φ *entails* Φ' *wrt.* R, written as $\Phi \models_R \Phi'$, iff $R \models_E \Phi$ implies $R \models_E \Phi'$.

Redundancy. Let Φ be a set of constrained clauses and $C \cdot \Gamma$ a ground constrained clause. Define $\Phi_{C \cdot \Gamma} = \{C' \cdot \Gamma' \in \Phi^{gr} \mid C' \cdot \Gamma' \prec C \cdot \Gamma\}$ as the set of ground instances of clauses from Φ that are smaller than $C \cdot \Gamma$.

Let R be a rewrite system without overlaps. We say that the ground constrained clause $C \cdot \Gamma$ is *redundant wrt.* Φ *and* R iff $\Phi_{C \cdot \Gamma} \models_R C \cdot \Gamma$, that is, iff $C \cdot \Gamma$ is entailed wrt. R by smaller ground instances of clauses from Φ. Notice that if $\Gamma \not\subseteq R$ then $C \cdot \Gamma$ is trivially redundant wrt. every constrained clause set and R (as R is ordered by $\succ$). For a (possibly non-ground) constrained clause $C \cdot \Gamma$ we say that $C \cdot \Gamma$ is *redundant wrt.* Φ *and* R iff all ground instances of $C \cdot \Gamma$ are redundant wrt. Φ and R.

Suppose $C \cdot \Gamma \Rightarrow_D C' \cdot \Gamma'$ is a ground inference, for some constrained clause $C' \cdot \Gamma'$, where D stands for $\text{Ref}(\varepsilon)$ or $\text{Para}(l \approx r, \varepsilon)$ (with $l \approx r$ ground). The ground inference is called *redundant wrt.* Φ *and* R iff $\Phi_{C \cdot \Gamma} \models_R C' \cdot \Gamma'$. We say that a Ref or Para inference is *redundant wrt.* Φ *and* R iff every ground instance of it is redundant wrt. Φ and R.

Saturation. Let Λ be a context. Let $R^\Lambda_{s \approx t} = \bigcup_{s \approx t \succ s' \approx t'} \varepsilon^\Lambda_{s' \approx t'}$ be the rewrite system defined earlier and consisting of those ground rules true in I_Λ that are smaller than $s \approx t$.

Definition 4.1 (Productive Constrained Clause). *Let* $C \cdot \Gamma = A_1 \vee \cdots \vee A_m \cdot \Gamma$ *be a ground constrained clause, for some* $m \geq 0$, *where* A_i *is a positive non-trivial equation for all* $i = 1, \ldots, m$. *We say that* $C \cdot \Gamma$ *is* productive *wrt.* Λ *iff* $\Gamma \subseteq R_\Lambda$ *and* A_i *is irreducible wrt.* $R^\Lambda_{A_i}$ *for all* $i = 1, \ldots, m$. *A (possibly non-ground) constrained clause* $C \cdot \Gamma$ *is* productive *wrt.* Λ *iff some ground instance of* $C \cdot \Gamma$ *is productive wrt.* Λ.

Intuitively, if $C \cdot \Gamma$ is a productive ground constrained clauses wrt. Λ then C provides positive equations, all irreducible in the sense as stated, at least one of which must be satisfied by I_Λ, so that in consequence R^E_Λ satisfies $C \cdot \Gamma$. The following definition turns this intuition into a demand on Λ (in its second item).

Definition 4.2 (Saturation up to Redundancy). *A sequent* $\Lambda \vdash \Phi$ *is* saturated up to redundancy *iff for all* $C \cdot \Gamma \in \Phi$ *such that* $C \cdot \Gamma$ *is not redundant wrt.* Φ *and* R_Λ, *the following hold:*

1. *For every inference* $C \cdot \Gamma \Rightarrow_D C' \cdot \Gamma'$, *where* D *stands for* $\text{Ref}(\sigma)$ *or* $\text{Para}(l \approx r, \sigma)$ *with a parameter-free* $l \approx r \in_\sim \Lambda$, *the clause* $(C \cdot \Gamma)\sigma$ *is redundant wrt.* Φ *and* R_Λ *or the inference* $C \cdot \Gamma \Rightarrow_D C' \cdot \Gamma'$ *is redundant wrt.* Φ *and* R_Λ.
2. *For every grounding substitution* γ *for* $C \cdot \Gamma$, *if* $C \neq \square$ *and* $(C \cdot \Gamma)\gamma$ *is productive wrt.* Λ *and non-redundant wrt.* Φ *and* R_Λ, *then* $I_\Lambda \models C\gamma$.

Referring back to our informal explanation of the calculus, and ignoring the redundancy concepts in Definition 4.2, ground instances of constrained clauses that are not productive wrt. Λ are subject to the first condition. It requires a sufficient number of applications of the Ref and Para rules to reduce (lifted versions of) such constrained clauses to constrained clauses productive wrt. Λ. The equality reasoning rules in $\mathcal{ME_E}$, which are based on Ref and Para, together with the Split rule, all defined in the next section, make sure that both conditions will be met in the limit of a derivation.

The next proposition clarifies under what conditions R_Λ^E is a model for all constrained clauses Φ in a sequent $\Lambda \vdash \Phi$ saturated up to redundancy.

Proposition 4.3. *Let $\Lambda \vdash \Phi$ be a sequent saturated up to redundancy and suppose Φ is a constrained clause set without expansion constraints. Then, $R_\Lambda \models_E \Phi$ if and only if Φ contains no constrained clause of the form $\square \cdot \Gamma$ that is productive wrt. Λ and non-redundant wrt. Φ and R_Λ.*

Notice that Proposition 4.3 applies to a *statically* given sequent $\Lambda \vdash \Phi$. The connection to the *dynamic* derivation process of the $\mathcal{ME_E}$ calculus will be given later, and Proposition 4.3 will be essential then in proving the correctness of the $\mathcal{ME_E}$ calculus.

5 $\mathcal{ME_E}$ Calculus

Like its predecessor, the $\mathcal{ME_E}$ calculus consists of a few basic derivation rules and a number of optional ones meant to improve the performance of implementations of the calculus. The basic derivation rules include rules for equality reasoning and two rules, namely Split and Close, which are not specific to the theory of equality. We start with a description of the basic rules.

Derivation Rules for Equality Reasoning. The following rules $\text{Ref}_{\mathcal{ME}}$ and $\text{Para}_{\mathcal{ME}}$, the only mandatory ones for equational reasoning, extend the derivation rules of Section 4 to sequents.

$$\text{Ref}_{\mathcal{ME}}(\sigma) \quad \frac{\Lambda \vdash \Phi, C \cdot \Gamma}{\Lambda \vdash \Phi, C \cdot \Gamma, C' \cdot \Gamma'} \quad \text{if} \begin{cases} C \cdot \Gamma \Rightarrow_{\text{Ref}(\sigma)} C' \cdot \Gamma', \text{ and} \\ \Phi \cup \{C \cdot \Gamma\} \text{ contains no variant of } C' \cdot \Gamma'. \end{cases}$$

$$\text{Para}_{\mathcal{ME}}(l \approx r, \sigma) \quad \frac{\Lambda \vdash \Phi, C \cdot \Gamma}{\Lambda \vdash \Phi, C \cdot \Gamma, C' \cdot \Gamma'} \quad \text{if} \begin{cases} l \approx r \text{ is a parameter-free fresh variant} \\ \text{of a } \Sigma\text{-equation in } \Lambda, \\ C \cdot \Gamma \Rightarrow_{\text{Para}(l \approx r, \sigma)} C' \cdot \Gamma', \text{ and} \\ \text{no variant of } C' \cdot \Gamma' \text{ is in } \Phi \cup \{C \cdot \Gamma\}. \end{cases}$$

The purpose of both the $\text{Ref}_{\mathcal{ME}}$ and $\text{Para}_{\mathcal{ME}}$ rules is to reduce the question of satisfiability of a constrained clause in the intended E-interpretation $R_{\Lambda_B}^E$, where Λ_B is a certain limit context (cf. Section 6), to deriving a smaller one and answering the question wrt. that one. Notice that constraints have a rather passive rôle in both derivation rules. In particular, Para is not applicable to constraints. The requirement in $\text{Para}_{\mathcal{ME}}$ that $l \approx r$ be a *parameter-free variant* of an equation in the context guarantees that all constrained clause sets derivable by the calculus are parameter-free.

Basic Derivation Rules. The mandatory rules Split and Close below are taken with only minor modifications from the $\mathcal{ME}$ calculus without equality [4]. This is possible because the equality reasoning is done *only* by the Ref$_{\mathcal{ME}}$ and Para$_{\mathcal{ME}}$ rules above. Both the Split and Close rule are based on the concept of a *context unifier*.

Definition 5.1 (Context Unifier). *Let Λ be a context and $C = L_1 \vee \cdots \vee L_m$ an ordinary clause. A substitution σ is a context unifier of C against Λ iff there are fresh p-variants $K_1, \ldots, K_m \in_{\simeq} \Lambda$ such that σ is a most general simultaneous unifier of the sets $\{K_1, \overline{L_1}\}, \ldots, \{K_m, \overline{L_m}\}$.*

For each $i = 1, \ldots, m$, we say that a literal $K_i' \in \Lambda$ is a context literal of σ if $K_i' \simeq K_i$, and that $L_i \sigma$ is a remainder literal of σ if $(\mathcal{P}ar(K_i))\sigma \not\subseteq V$. We say that σ is productive iff K_i produces $\overline{L_i}\sigma$ in Λ for all $i = 1, \ldots, m$.

A context unifier σ of C against Λ is *admissible (for* Split*)* iff every remainder literal L of σ is parameter- or variable-free and for all distinct remainder literals L and K of σ $\mathcal{V}ar(L) \cap \mathcal{V}ar(K) = \emptyset$.

$$\text{Split}(L, \sigma) \quad \frac{\Lambda \vdash \Phi, C \cdot \Gamma}{\Lambda, L \vdash \Phi, C \cdot \Gamma \quad \Lambda, \overline{L}^{\text{sko}} \vdash \Phi, C \cdot \Gamma} \quad \text{if} \quad \begin{cases} C = A_1 \vee \cdots \vee A_m \text{ with } m \geq 0 \\ \text{and for all } i = 1, \ldots, m, A_i \text{ is a} \\ \text{positive non-trivial equation, } \sigma \\ \text{is an admissible context unifier} \\ \text{of } (C \cdot \Gamma)^c \text{ against } \Lambda \text{ with} \\ \text{remainder literal } L, \text{ and neither} \\ L \text{ nor } \overline{L}^{\text{sko}} \text{ is contradictory with} \\ \Lambda. \end{cases}$$

A Split inference is *productive* iff σ is a *productive* context unifier of $(C \cdot \Gamma)^c$ against Λ.

To obtain a complete calculus Split needs to be applied only when $C \cdot \Gamma$ has an R_Λ-irreducible ground instance that is falsified by the E-interpretation R_Λ^E. Technically, these ground instances are approximated by the productive ones, in terms of Definition 4.1, and a productive context unifier is guaranteed to exist then. Applying a Split inference then will modify the context so that it E-satisfies such a ground instance afterwards, which marks some progress in the derivation.

$$\text{Close}(\sigma) \quad \frac{\Lambda \vdash \Phi, C \cdot \Gamma}{\Lambda \vdash \Box \cdot \emptyset} \quad \text{if} \quad \begin{cases} \Phi \neq \emptyset \text{ or } C \cdot \Gamma \neq \Box \cdot \emptyset, \text{ and} \\ \sigma \text{ is a context unifier of } (C \cdot \Gamma)^c \text{ against } \Lambda \\ \text{with no remainder literals.} \end{cases}$$

The purpose of the Close rule is to detect a trivial inconsistency between the context and a constrained clause.

Optional Derivation Rules. Like DPLL, the $\mathcal{ME}$ calculus includes an optional derivation rule, called Assert, to insert a literal into a context without causing branching. In $\mathcal{ME}$ this rule bears close resemblance to the unit-resulting resolution rule. The $\mathcal{ME}_E$ calculus has a suitable version of the Assert rule which is also more general than the one in $\mathcal{ME}$. To define it we need some more preliminaries first.

Let us fix a constant a from the signature $\Sigma^{\text{sko}} \setminus \Sigma$ and consider the substitution $\alpha := \{v \mapsto a \mid v \in V\}$. Given a literal L, we denote by L^a the literal $L\alpha$. Note that L^a is ground if, and only if, L is variable-free. Similarly, given a context Λ, we denote by

Λ^a the set of *unit clauses* obtained from Λ by removing the pseudo-literal $\neg v$, replacing each literal L of Λ with L^a, and considering it as a unit clause.[7]

$$\mathsf{Assert}(L) \quad \frac{\Lambda \vdash \Phi}{\Lambda, L \vdash \Phi} \quad \text{if} \quad \begin{cases} \Lambda^a \cup \Phi^c \models_{\mathrm{E}} L^a, \\ L \text{ is non-contradictory with } \Lambda, \text{ and} \\ \text{there is no } K \in_{\sim} \Lambda \text{ such that } K \geq L. \end{cases}$$

As an example, Assert is applicable to the sequent $\neg v, P(u,b) \approx \mathbf{t}, b \approx c \vdash P(x,y) \not\approx \mathbf{t} \vee f(x) \approx y \cdot \mathbf{0}$ to yield the new context equation $f(u) \approx c$.

The third condition of Assert avoids the introduction of superfluous literals in the context. The first condition is needed for soundness. This condition is not decidable in its full generality and so can only be approximated. This, however, is not a problem given that Assert is an optional rule in $\mathcal{ME}_{\mathrm{E}}$. See [5] for an explanation of how the Assert rule of $\mathcal{ME}$ (with its concrete preconditions) can be seen as a special case of Assert above.

Simplification. The purpose of simplification is to replace a constrained clause by a *simpler* one. The optional Simp rule below is general enough to accomodate the simplification rules of $\mathcal{ME}$ [8] and also various new simplification rules connected with equality. To formulate it we need one more notion.

For any context Λ, a (ground) rewrite system R without overlaps is *compatible with* Λ iff there is no $l \rightarrow r \in R$ and no parameter-free $s \not\approx t \in \Lambda$ such that $s \approx t \gtrsim l \approx r$.

$$\mathsf{Simp} \; \frac{\Lambda \vdash \Phi, C \cdot \Gamma}{\Lambda \vdash \Phi, C' \cdot \Gamma'} \quad \text{if} \quad \begin{cases} \text{(i)} \; C' \cdot \Gamma' \in \Phi \text{ and } C' \cdot \Gamma' \text{ non-properly subsumes } C \cdot \Gamma, \text{ or} \\ \text{(ii) for every rewrite system } R \text{ compatible with } \Lambda\text{:} \\ \quad C \cdot \Gamma \text{ is redundant wrt. } \Phi \cup \{C' \cdot \Gamma'\} \text{ and } R, \\ \quad C' \cdot \Gamma' \text{ is a constrained clause over } \Sigma \text{ without} \\ \quad \text{expansion constraints, and} \\ \quad \Lambda^a \cup (\Phi \cup \{C \cdot \Gamma\})^c \models_{\mathrm{E}} (C' \cdot \Gamma')^c. \end{cases}$$

The last condition in the definition of the Simp rule guarantees soundness.

As a simple instance of the Simp rule, any constrained clause $C \cdot \Gamma$ of the form $s \approx s \vee D \cdot \Gamma$ can be simplified to $\mathbf{t} \approx \mathbf{t} \cdot \mathbf{0}$. This simplification step actually yields the same effect as if $C \cdot \Gamma$ were deleted. Dually, any constrained clause $C \cdot \Gamma$ of the form $s \not\approx s \vee D \cdot \Gamma$ can be simplified to $D \cdot \Gamma$. Also, as observed previously, when the constraint Γ of a constrained clause $C \cdot \Gamma$ contains a rule $l \rightarrow r$ such that $l \prec r$ then this rule is trivially redundant wrt. any rewrite system ordered by $\succ$ and so the clause can be simplified to $\mathbf{t} \approx \mathbf{t} \cdot \mathbf{0}$. As a simple example that takes the context into account, consider the sequent $f(x) \not\approx x \vdash a \approx b \cdot f(a) \rightarrow a$. Now, no rewrite system compatible with $\{f(x) \not\approx x\}$ can contain $f(a) \rightarrow a$. The constrained clause can therefore again be simplified to $\mathbf{t} \approx \mathbf{t} \cdot \mathbf{0}$. Dually, in the sequent $f(x) \approx x \vdash a \approx b \cdot f(a) \rightarrow a$ the constrained clause can be simplified to $a \approx b \cdot \mathbf{0}$. (Notice in particular that this simplification is indeed sound.)

---

[7] Here and below Φ^c denotes the set of clausal forms of all constrained clauses in Φ.
[8] Except for the Subsume rule.

As illustrated by the last two examples, the practically important unit-resolution like rule of $\mathcal{ME}$, Resolve, is covered by the Simp rule.

Derivation Example. The following excerpt from an $\mathcal{ME}_E$ derivation demonstrates Para, Simp and Split in combination. It follows the example in Section 2 by taking the same context $\Lambda = \{f(u) \approx u, f(a) \not\approx a\}$. However, to be more instructive, it uses a lifted version $f(x) \approx x \vee f(f(x)) \approx b \vee f(b) \not\approx b$ of the ground clause there.

$$\cdots$$

$$\neg v, \underline{f(u) \approx u}, f(a) \not\approx a \vdash \ldots, f(x) \approx x \vee \underline{f(f(x))} \approx b \vee f(b) \not\approx b \cdot \emptyset$$

$$\neg v, \underline{f(u) \approx u}, f(a) \not\approx a \vdash \ldots, \frac{f(x) \approx x \vee f(f(x)) \approx b \vee f(b) \not\approx b \cdot \emptyset,}{f(x) \approx x \vee \ f(x) \approx b \ \vee \underline{f(b)} \not\approx b \cdot f(f(x)) \to f(x)} \quad \text{(By Para)}$$

$$\neg v, \underline{f(u) \approx u}, f(a) \not\approx a \vdash \ldots, \begin{array}{l} f(x) \approx x \vee f(f(x)) \approx b \vee f(b) \not\approx b \cdot \emptyset, \\ f(x) \approx x \vee \ f(x) \approx b \ \vee f(b) \not\approx b \cdot f(f(x)) \to f(x) \\ f(x) \approx x \vee \ f(x) \approx b \ \vee \ \underline{b \not\approx b} \ \cdot f(f(x)) \to f(x), f(b) \to b \end{array}$$

$$\text{(By Para)}$$

$$\neg v, f(u) \approx u, f(a) \not\approx a \vdash \ldots, \begin{array}{l} f(x) \approx x \vee f(f(x)) \approx b \vee f(b) \not\approx b \cdot \emptyset, \\ f(x) \approx x \vee \ f(x) \approx b \ \vee f(b) \not\approx b \cdot f(f(x)) \to f(x) \\ f(x) \approx x \vee \ f(x) \approx b \qquad \qquad \cdot f(f(x)) \to f(x), f(b) \to b \end{array}$$

$$\text{(By Simp)}$$

Among the alternatives to proceed now we focus on possible Split inferences. Consider the last sequent with the constrained clause $f(x) \approx x \vee f(x) \approx b \cdot f(f(x)) \to f(x), f(b) \to b$ and its clausal form $f(x) \approx x \vee f(x) \approx b \vee f(f(x)) \not\approx f(x) \vee f(b) \not\approx b$. Simultaneous unification of that clause literals with fresh variants of the context literals $f(a) \not\approx a, \neg v, f(u) \approx u, f(u) \approx u$, respectively, gives the (productive and admissible) context unifier $\sigma = \{x \mapsto a, \ldots\}$. The remainder literals of σ are $f(a) \approx b$, $f(f(a)) \not\approx f(a)$ and $f(b) \not\approx b$ (notice that the clause instance literal $f(a) \approx a$ is contradictory with the context and hence is a non-remainder literal). Each of them can be selected for Split. The effect of selecting $f(a) \approx b$ or $f(b) \not\approx b$ was already described in Section 2.

6 Correctness of the $\mathcal{ME}_E$ Calculus

Similarly to the $\mathcal{ME}$ calculus, derivations in $\mathcal{ME}_E$ are formally defined in terms of derivation trees. The purpose of the calculus is to build for a given clause set a derivation tree all of whose branches are failed iff the clause set is unsatisfiable. The soundness argument for the calculus is relatively straightforward and analogous to the one for the $\mathcal{ME}$ calculus. Therefore, in this section we concentrate just on completeness. A detailed soundness proof can be found in [5].

A *derivation tree* of a set $\{C_1, \ldots, C_n\}$ of Σ-clauses is a finite tree over sequents in which the root node is the sequent $\neg v \vdash C_1 \cdot \emptyset, \ldots, C_n \cdot \emptyset$, and each non-root node is the result of applying one of the derivation rules to the node's parent.

Let $\mathbf{T}$ be a derivation tree presented as a pair $(\mathbf{N}, \mathbf{E})$, where $\mathbf{N}$ is the set of the nodes of $\mathbf{T}$ and $\mathbf{E}$ is the set of the edges of $\mathbf{T}$. A derivation $\mathcal{D} = (\mathbf{T}_i)_{i < \kappa}$ in $\mathcal{ME}_E$ is a possibly infinite sequence of derivation trees defined in the obvious way. Each *derivation* $\mathcal{D} =$

$((\mathbf{N}_i, \mathbf{E}_i))_{i<\kappa}$ determines a *limit tree* $\mathbf{T} := (\bigcup_{i<\kappa}\mathbf{N}_i, \bigcup_{i<\kappa}\mathbf{E}_i)$. It is easy to show that a limit tree of a derivation $\mathcal{D}$ is indeed a tree. But note that it will not be a derivation tree unless $\mathcal{D}$ is finite.

Now let $\mathbf{T}$ be the limit tree of some derivation, let $\mathbf{B} = (N_i)_{i<\kappa}$ be a branch in $\mathbf{T}$ with κ nodes, and let $\Lambda_i \vdash \Phi_i$ be the sequent labeling node N_i, for all $i < \kappa$. Define $\Lambda_{\mathbf{B}} = \bigcup_{i<\kappa}\bigcap_{i\le j<\kappa}\Lambda_j$ and $\Phi_{\mathbf{B}} = \bigcup_{i<\kappa}\bigcap_{i\le j<\kappa}\Phi_j$, the sets of *persistent context literals* and *persistent clauses*, respectively. These two sets can be combined to obtain the *limit sequent* $\Lambda_{\mathbf{B}} \vdash \Phi_{\mathbf{B}}$ (of $\mathbf{T}$).

As usual, the completeness of $\mathcal{ME}_E$ relies on a suitable notion of fairness.

Definition 6.1 (Exhausted Branch). *Let $\mathbf{T}$ be a limit tree, and let $\mathbf{B} = (N_i)_{i<\kappa}$ be a branch in $\mathbf{T}$ with κ nodes. For all $i < \kappa$, let $\Lambda_i \vdash \Phi_i$ be the sequent labeling node N_i. The branch $\mathbf{B}$ is* exhausted *iff for each constrained clause $C \cdot \Gamma \in \Phi_{\mathbf{B}}$ that is not redundant wrt. Φ_j and $R_{\Lambda_{\mathbf{B}}}$, for some $j < \kappa$, all of the following hold, for all $i < \kappa$ such that $C \cdot \Gamma \in \Phi_i$:*

(i) *if $\mathsf{Ref}_{\mathcal{ME}}$ is applicable to $\Lambda_i \vdash \Phi_i$ with selected constrained clause $C \cdot \Gamma$ and underlying Ref inference $C \cdot \Gamma \Rightarrow_{\mathsf{Ref}(\sigma)} C' \cdot \Gamma'$, and $(C \cdot \Gamma)\sigma$ is not redundant wrt. Φ_i and $R_{\Lambda_{\mathbf{B}}}$, then there is a $j < \kappa$ such that the inference $C \cdot \Gamma \Rightarrow_{\mathsf{Ref}(\sigma)} C' \cdot \Gamma'$ is redundant wrt. Φ_j and $R_{\Lambda_{\mathbf{B}}}$.*

(ii) *if $\mathsf{Para}_{\mathcal{ME}}$ is applicable to $\Lambda_i \vdash \Phi_i$ with selected constrained clause $C \cdot \Gamma$ and underlying Para inference $C \cdot \Gamma \Rightarrow_{\mathsf{Para}(l \approx r, \sigma)} C' \cdot \Gamma'$, where $l \approx r \in_{\sim} \Lambda_{\mathbf{B}}$ and $\Lambda_{\mathbf{B}}$ produces $(l \approx r)\sigma$, and $(C \cdot \Gamma)\sigma$ is not redundant wrt. Φ_i and $R_{\Lambda_{\mathbf{B}}}$, then there is a $j < \kappa$ such that the inference $C \cdot \Gamma \Rightarrow_{\mathsf{Para}(l \approx r, \sigma)} C' \cdot \Gamma'$ is redundant wrt. Φ_j and $R_{\Lambda_{\mathbf{B}}}$.*

(iii) *if Split is applicable to $\Lambda_i \vdash \Phi_i$ with selected constrained clause $C \cdot \Gamma$ and productive context unifier σ such that every context literal K of σ is a Σ-literal[9] and $K \in_{\sim} \Lambda_{\mathbf{B}}$, and $(C \cdot \Gamma)\sigma$ is productive wrt. $\Lambda_{\mathbf{B}}$, then there is a $j < \kappa$ such that $(C \cdot \Gamma)\sigma$ is redundant wrt. Φ_j and $R_{\Lambda_{\mathbf{B}}}$ or there is a remainder literal L of σ and a $j \ge i$ with $j < \kappa$ such that Λ_j produces L but not $\overline{L}$.*

(iv) Close *is not applicable to $\Lambda_i \vdash \Phi_i$ with selected constrained clause $C \cdot \Gamma$ and any context unifier σ such that $K \in_{\sim} \Lambda_{\mathbf{B}}$ for every context literal K of σ.*

(v) $\Phi_i \ne \{\Box \cdot \emptyset\}$.

A limit tree of a derivation is *fair* iff it is a refutation tree that is, a finite tree all of whose leafs are conclusions of the Close rule, or it has an exhausted branch. A derivation is *fair* iff its limit tree is fair.

It is not hard to see that actually carrying out a $\mathsf{Ref}_{\mathcal{ME}}$ or $\mathsf{Para}_{\mathcal{ME}}$ inference renders the underlying Ref or Para inference redundant *wrt. any rewrite system ordered by $\succ$*. Concerning Split, like in the $\mathcal{ME}$ calculus carrying out a Split inference also achieves what fairness demands for. These considerations indicate that a fair proof procedure indeed exists. It should not be too difficult to modify the proof procedure (and implementation) for the Model Evolution calculus described in [3] accordingly.

Definition 6.1 provides a framework for fair derivations based on redundant clauses and redundant inferences. The redundancy criteria are formulated wrt. $R_{\Lambda_{\mathbf{B}}}$, an object

---

[9] Note the restriction to Σ-literals; it is *not* possible to restrict condition (iv) in the same way.

not available during a derivation. The redundancy tests are therefore impossible to effectively realize in their full strength. Nethertheless, there are some effective and inexpensive redundancy tests similar to those discussed in conjunction with the Simp rule.

Proposition 6.2 (Exhausted Branches are Saturated up to Redundancy). *If* **B** *is an exhausted branch of a limit tree of some fair derivation then (i)* $\Lambda_\mathbf{B} \vdash \Phi_\mathbf{B}$ *is saturated up to redundancy, (ii)* $\Phi_\mathbf{B}$ *is a constrained clause set without expansion constraints, and (iii)* $\Phi_\mathbf{B}$ *contains no constrained clause of the form* $\square \cdot \Gamma$ *that is productive wrt.* $\Lambda_\mathbf{B}$ *and that is not redundant wrt.* $\Phi_\mathbf{B}$ *and* $R_{\Lambda_\mathbf{B}}$.

Propositions 6.2 and 4.3 together entail our main result:

Theorem 6.3 (Completeness of $\mathcal{ME}_E$). *Let* Ψ *be a parameter-free Σ-clause set, and* **T** *be the limit tree of a fair derivation of* Ψ. *If* **T** *is not a refutation tree, then* Ψ *is satisfiable; more specifically, for every exhausted branch* **B** *of* **T**, $R_{\Lambda_\mathbf{B}} \models_E \Psi$.

7   Conclusions

We have presented the $\mathcal{ME}_E$ calculus, an extension of the Model Evolution calculus by paramodulation-based inference rules for equality. Our main result is its correctness, in particular the completeness in combination with redundancy criteria. As for future work, we will extend the implementation of the model evolution calculus, the Darwin system [3] to the $\mathcal{ME}_E$ calculus.

There are also some theoretical issues to be addressed. The perhaps most pressing theoretical question is if or when paramodulation into smaller sides of equations can be avoided. It is clear that the current completeness proof breaks down when such inferences are no longer subject to fairness. Other questions concern further, useful instantiations of our simplification rule.

Acknowledgements. We would like to thank the reviewers for their valuable comments.

References

1. L. Bachmair and H. Ganzinger. Equational Reasoning in Saturation-Based Theorem Proving. In W. Bibel and P. H. Schmitt, ed., *Automated Deduction. A Basis for Applications*, Volume I: Foundations. Calculi and Refinements, pp. 353–398. Kluwer, 1998.
2. P. Baumgartner. FDPLL – A First-Order Davis-Putnam-Logeman-Loveland Procedure. In D. McAllester, ed., Proc. *CADE-17*, LNAI 1831, pp. 200–219. Springer, 2000.
3. P. Baumgartner, A. Fuchs, and C. Tinelli. Implementing the Model Evolution Calculus. *International Journal on Artificial Intelligence Tools (IJAIT)*, 2005. To appear.
4. P. Baumgartner and C. Tinelli. The Model Evolution Calculus. In F. Baader, ed., Proc. *CADE-19*, LNAI 2741, pp. 350–364. Springer, 2003.
5. P. Baumgartner and C. Tinelli. The Model Evolution Calculus with Equality, 2005. http://www.mpi-sb.mpg.de/~baumgart/publications/MEE.pdf.
6. J.-P. Billon. The Disconnection Method. In P. Miglioli, U. Moscato, D. Mundici, and M. Ornaghi, eds., Proc. *TABLEAUX*, LNAI 1071, pp. 110–126. Springer, 1996.
7. M. Davis, G. Logemann, and D. Loveland. A Machine Program for Theorem Proving. *Communications of the ACM*, 5(7):394–397, July 1962.

8. H. Ganzinger and K. Korovin. Integrating Equational Reasoning into Instantiation-Based Theorem Proving. In Proc. *CSL'04*, LNCS 3210, pp. 71–84. Springer, 2004.

9. H. Ganzinger and K. Korovin. New Directions in Instance-Based Theorem Proving. In Proc. *LICS*, 2003.

10. R. Letz and G. Stenz. Integration of Equality Reasoning into the Disconnection Calculus. In U. Egly and C. G. Fermüller, eds., *TABLEAUX*, LNCS 2381, pp. 176–190. Springer, 2002.

11. R. Nieuwenhuis and A. Rubio. Paramodulation-Based Theorem Proving. In J. A. Robinson and A. Voronkov, eds., *Handbook of Automated Reasoning*, pp. 371–443. Elsevier, 2001.

12. D. A. Plaisted and Y. Zhu. Ordered Semantic Hyper Linking. *Journal of Automated Reasoning*, 25(3):167–217, 2000.

Model Representation via Contexts and Implicit Generalizations

Christian G. Fermüller and Reinhard Pichler

Technische Universität Wien, A-1040 Vienna, Austria
{chrisf, reini}@logic.at

Abstract. Some results on expressibility and complexity issues in model representation are presented. In particular, the relation between so-called 'contexts' as recently introduced in [4] for the model evolution calculus and the more traditional 'disjunctions of implicit generalizations' (DIGs) [21] is clarified: contexts are as expressible as DIGs, but DIGs may represent the same model exponentially more succinctly. The clause evaluation problem and the equivalence problem for DIGs and contexts, respectively, are all shown to be coNP-complete.

Keywords: Model computation, model representation, clause evaluation.

1 Introduction

Computations that refer to model structures arise in many areas of computer science. Even when the attention is restricted to first-order clause logic and corresponding Herbrand models (term models), a rich research field emerges. Obviously, since Herbrand models are infinite objects, one has to devise formalisms for the finite *representation of models* to render computations with models effective. *Automated Model Building* studies algorithms that extract such representations of models from satisfiable sets of clause. Much of the work along this line of research is summarized in the forthcoming monograph [8]. A somewhat more general point of view on the relevant research issues is expressed by the term '*Model Computation*'. Vivid interest in this topic is documented, e.g., by workshops on Model Computation at CADE-17 and CADE-19.[1]

Besides the general interest in Model Computation, we are motivated by the fact that the calculus of *model evolution*, which has recently been introduced by Baumgartner and Tinelli [4,5], relies heavily on a particular model representation formalism. Moreover, a very similar form of model representation is also used in disconnection calculi (see, e.g., [6,23,24]).

Model evolution amounts to a sophisticated generalization to first-order clause logic of the well-known DPLL procedure for propositional logic and

---

[1] See the proceedings at http://www.uni-koblenz.de/~peter/CADE17-WS-MODELS/ http://www.uni-koblenz.de/~peter/models03/, respectively.

R. Nieuwenhuis (Ed.): CADE 2005, LNAI 3632, pp. 409–423, 2005.

promises to lift at least some of the properties that make DPLL so successful to the first-order level. We will not study the deduction mechanism of model
evolution directly, but investigate in detail a crucial component of it: so-called
contexts, which, being a computationally motivated mechanism for model representation, are also of independent interest.

Contexts are related to *disjunctions of implicit generalizations (DIGs)*, a
more traditional formalism for representing sets of ground terms and therefore
also Herbrand models (cf. [21]). We prove that the 'expressive power' of contexts
is equivalent to that of DIGs, in the sense that for every context there is a DIG
representing the same model, and vice versa. However whereas, given a context,
one can construct an equivalent DIG in polynomial time, we show that DIGs
can be exponentially more succinct than shortest equivalent contexts.

In reference to general principles of Model Representation, we investigate the
computational complexity of the following decision problems:

- Given a clause C and a DIG Δ (a context Λ), is C true in the model represented by Δ (by Λ)?
- Given two DIGs (contexts), are they representing they same model?

We show that all four decision problems are coNP-complete.

A number of related mechanisms for model representation have been studied
in the literature. An overview can be found in [29] and [8]. We will comment
on related work in Section 5, but it is important to take note right away of an
essential difference between our scenario and that of most other forms of model
representation: Whereas usually the underlying signature is assumed to be *finite*,
contexts (and therefore also the corresponding DIGs) refer to a signature that
contains *infinitely* many different constants.

Because of space restrictions some (parts of) proofs are not reproduced here.
They will be included in a forthcoming journal version of this paper.

2   Basic Concepts of Model Representation

A *model representation* is a syntactic structure D associated with a unique
model $\mathcal{M}(D)$ over a given signature. Throughout this work we restrict our attention to first-order clause logic without equality. For the intended applications,
a model representation D should satisfy the following properties (cf. [14]):

(1) For ground atoms A, it can be checked efficiently whether A is true in $\mathcal{M}(D)$.
(2) Given a clause C, it is decidable whether C is true in $\mathcal{M}(D)$.
(3) Given another structure D' (of the same type as D), it is decidable whether
    $\mathcal{M}(D') = \mathcal{M}(D)$.

A paradigmatic example of model representation is the explicit specification
of a *finite* model by tables ('diagrams') for each predicate and function symbol,
that record the value for each tuple of domain elements. However, in our case,
representations of *Herbrand models* (term models) are more important. We will
only deal with Herbrand models and therefore simply speak of a *model*. Since

there are uncountably many such models over a given signature in general, only a subset of all corresponding models can be represented by a given syntactic formalism. Thus, in contrast to finite models, the issue of 'expressive power' of a model representation formalism arises. Clearly, a trade-off between the expressiveness of a formalism and the corresponding complexity of the above mentioned decision problems is to be expected.

A model is identified with the set of ground atoms that are true in it. In the light of condition (1) model representation formalisms seek to specify this set as explicitly as possible. The simplest form of a (Herbrand) model representation consists in a finite set of ground atoms. Clearly, the range of models in which only finitely many ground atoms are true is very limited. The expressive power is easily enhanced by stipulating *general atoms* to represent the set of all its ground instances. Such *atomic representations of models (ARMs)* were used and investigated, e.g., in [13,14,15,8]. We only remark in passing that hyperresolution directly generates an atomic representation of a model of a set of Horn clauses S whenever it terminates on the input clauses S without deriving the empty clause. Generalizations of this model building mechanism to classes of clause sets that are not necessarily Horn have been studied in [13,14].

2.1   Contexts and Model Evolution

Recently Baumgartner and Tinelli [4,5,1] have introduced a calculus for clause logic, called *model evolution*, that relies heavily on a particular form of model representation. Model evolution can be seen as belonging to the family of calculi known as 'model elimination' and are related to 'connection tableaux' (see, e.g., [22]) and 'disconnection tableaux' (see, e.g., [6,23,24]), that also integrate deduction and model building. We will not directly investigate deduction in model evolution, which is a rather sophisticated form of lifting the well-known DPLL-procedure for propositional logic to the first-order level. To motivate our investigation of the underlying model representation mechanism it suffices to point out that each inference step in model evolution refers to models that are represented by so-called 'contexts'. We will investigate the expressiveness of contexts and the complexity of clause evaluation and of testing equivalence for contexts.

Formally a context is simply a finite set of literals. However, an important ingredient is the distinction between 'universal variables' and 'parameters'. The latter can be viewed as a form of variables, that do not necessarily admit instantiation by arbitrary ground terms; whereas universal variables can be understood as placeholders for arbitrary terms of the Herbrand universe. (A corresponding distinction appears in [24], where parameters are called 'shared variables' and universal variables are called 'local variables').

Throughout this paper we fix a signature Σ that contains infinitely many constants. This is motivated by the intended use of contexts and amounts to an important difference to representation formalisms based on finite signatures. We will use 0, 1, a, b, c, and d (possibly with subscripts) to denote constants; f will denote a function symbol. There is an infinite set of *(universal) variables*, as well as an infinite set of *parameters*. We will use x, y, z to denote variables, and u,

v, w for parameters. *Terms* $(s, t, \ldots)$ and *atoms* $(A, B, \ldots)$ are built up from constants, variables, and parameters, using function and predicate symbols, as usual. *Literals* (denoted by K, L, M) are either atoms, called *positive literals*, or negated atoms, called *negative literals*. We write $\overline{L}$ for the literal that is dual to L; i.e., $\overline{A} = \neg A$ and $\overline{\neg A} = A$. A *clause* $C = L_1 \vee \cdots \vee L_k$ is a disjunction of literals. Terms, literals and clauses are called *ground* if no variables or parameters occur in them.

Substitutions are mappings from variables and parameters to terms that have fixpoints almost everywhere. A substitution is called a *renaming* if it is a permutation that maps variables to variables and parameters to parameters. If the restriction of a substitution σ to parameters is a renaming, then we call σ *p-preserving*.

We write $s \precsim t$ if s is an *instance* of t; i.e.: if there is a substitution σ such that $s = t\sigma$. The term t is called a *generalization* of s. If $s = t\sigma$ is a ground term then s is called *ground instance* of t and σ is called *ground substitution*. In case σ is a renaming then we call s a *variant* of t. A term $t \in T$ is called *most specific* among a set of terms T if T does not contain a proper instance of t. In particular, we say that t is a *most specific generalization* (for short, an *msg*) of s in S, if t is most specific in $T = \{s' \in S \mid s \precsim s'\}$. These definitions are generalized from terms to atoms and literals in the obvious way.

In addition to ordinary literals, the parameter v is used as a *pseudo-atom*, potentially representing all ground atoms. The dual of v is the *pseudo-literal* $\neg v$. Every positive (negative) literal is considered a proper instance of the pseudo-literal v ($\neg v$). The *pseudo-literal* $\neg v$ is used to guarantee that all atoms that are not explicitly specified to be true are false by default in contexts.

We assume the reader is familiar with unification and most general unifiers. We write $\vartheta = mgu\{E_1, \ldots, E_n\}$ to denote that ϑ is the most general unifier of the terms or atoms E_i $(1 \leq i \leq n)$.

From now on, we will assume that each atom only contains either variables or parameters, but not both. Correspondingly, we speak of *universal atoms (literals)* and *parameter atoms (literals)*, respectively.[2] (Only ground atoms are parameter atoms and universal at the same time.)

Definition 1. *A* context Λ *is a finite set of literals containing the pseudo-literal* $\neg v$. Λ *is contradictory iff* $L\sigma = \overline{K}\sigma$ *for some variants* L, K *of elements in* Λ *and a p-preserving substitution* σ.

Example 1. The context $\Lambda_1 = \{\neg v, P(x, f(y)), \neg P(a, x)\}$ is contradictory, since $P(x, f(y))$ and $P(a, x')$ are unifiable, and the corresponding unifier is p-preserving (since no parameters occur in the atoms). However the context $\Lambda_2 = \{\neg v, P(x, f(y)), \neg P(a, u)\}$ is non-contradictory since, for all substitutions σ such that $P(x, f(y))\sigma = P(a, u)\sigma$ holds, the parameter u has to be instantiated. Likewise the (pseudo-)parameter v has to be instantiated when unified with $P(x, f(y))$.

---

[2] In [4,5], the simultaneous occurrence of parameters and variables in an atom is allowed, but not used in the formalism. In fact the restriction does not affect the expressive power of the formalism.

Important Remark. From now on we only consider *non-contradictory* contexts and thus drop the adjective, unless we want to emphasize this property.

Below, we present a definition for the production of ground atoms, that differs from the corresponding one in [4,5], but makes the different roles of universal literals and parameter literals more transparent. Moreover, our Definition 2 can be easily shown to be equivalent to the corresponding definition in [4,5].

Definition 2. *A ground atom A is produced by a context Λ iff one the following conditions holds:*

1. *A is an instance of some universal literal in Λ, or*
2. *A is an instance of a parameter literal $L \in \Lambda$, but $\neg A$ is not an instance of a literal $\neg B \in \Lambda$, where B is either universal or a proper instance of L.*

Note that a literal is a (proper) instance of another literal L iff it is a (proper) instance of a variant of L. Consequently the literals of a context can always be assumed to be variable and parameter disjoint.

Informally, Definition 2 can be paraphrased as follows: All instances of universal atoms are produced; in contrast, the (potential) production of an instance A of a parameter atom L is blocked if $\neg A$ is an instance of a more specific literal or of a universal literal in the context.

Example 2. Consider the context $\Lambda = \{\neg v, \ P(c,x,y), \ P(u,a,v), \ \neg P(u,u,b), \ P(u,v,b)\}$. Since all instances of $P(c,x,y)$ are produced, Λ produces all atoms of the form $P(c,s,t)$, for arbitrary ground terms s and t. $P(u,a,v)$ and therefore Λ produces, for instance, the atom $P(a,a,a)$. On the other hand, $\neg P(u,u,b)$ prevents $P(u,v,b)$ from producing $P(d,d,b)$. Thus $P(d,d,b)$ is not produced by Λ. Note that $\neg P(u,u,b)$ also prevents $P(u,v,b)$ from producing $P(a,a,b)$ and $P(c,c,b)$. Nevertheless, both $P(a,a,b)$ and $P(c,c,b)$ are produced by Λ, namely by $P(u,a,v)$ and $P(c,x,y)$, respectively.

Definition 3. $\mathcal{M}(\Lambda)$, *the model induced by the context Λ, is the set of ground atoms that are produced by a context Λ. We call a model $\mathcal{N}$ context representable if $\mathcal{N} = \mathcal{M}(\Lambda)$ for some context Λ.*

By the above definition of $\mathcal{M}(\Lambda)$ and Definition 2, it is clear that the truth value of a ground atom A in a model represented by a context Λ can be computed efficiently. In fact, this can be easily done in deterministic polynomial time, since it only requires linearly many matching tests.

Example 3. The reader is invited to check that the context Λ of Example 2 induces the model $\mathcal{M}(\Lambda) = \{P(r,s,t) \mid r = c \lor s = a \lor (r \neq s \land t = b)\}$.

2.2   Explicit and Implicit Generalizations

As already mentioned, a simple form of model representation is constituted by a finite set of atoms, stipulated to represent the set of all their ground instances. In fact, this amounts to a special case of contexts: Every set $\Lambda =$

$\{\neg v\} \cup \{A_1, \ldots, A_n\}$, where the A_i are (universal or parameter) atoms, is a context. The represented model $\mathcal{M}(\Lambda)$ is the set of all ground instances of the A_i ($1 \leq i \leq n$). In this respect $\{A_1, \ldots, A_n\}$ is also called an *atomic representation* of $\mathcal{M}(\Lambda)$ (see, e.g., [14]).

An atom is sometimes understood as the *explicit generalization* of all its ground instances. Consequently, an atomic representation can alternatively be viewed as a *disjunction of explicit generalizations*.

Following ideas in [21], these notions can be generalized as follows:

Definition 4. *An* implicit generalization Γ *is an expression of the form* $A/\mathcal{B}$, *where* A *is an atom and* $\mathcal{B}$ *is a finite set of atoms. We simply write* A *for* $A/\{\}$. *Every ground atom that is an instance of* A, *but not an instance of any* $B \in \mathcal{B}$ *is said to be* contained *in* $A/\mathcal{B}$.

A disjunction of implicit generalizations Δ *(shortly: DIG) is defined as an expression of the form* $A_1/\mathcal{B}_1 \sqcup \ldots \sqcup A_n/\mathcal{B}_n$, *also written as* $\bigsqcup_{1 \leq i \leq n} A_i/\mathcal{B}_i$. *A ground atom is said to be* contained *in* Δ *if it is contained in* $A_i/\mathcal{B}_i$ *for some* $i \in \{1, \ldots, n\}$.

The model $\mathcal{M}(\Delta)$ *represented by a DIG* Δ *is the set of all ground atoms that are contained in* Δ. *We call a model* $\mathcal{N}$ DIG representable *if* $\mathcal{N} = \mathcal{M}(\Delta)$ *for some DIG* Δ.

Remark. Note that our notation differs slightly from the one in [21], where implicit generalizations are written in the form $t/\{t_1 \vee \ldots \vee t_n\}$. In particular, "disjunctions" are indeed denoted by "$\vee$". In order to avoid confusion with clauses of the form $C = L_1 \vee \ldots \vee L_n$, we have decided to write DIGs in the way described above. In particular, "disjunctions" are denoted by "$\sqcup$". Of course, this is only syntactic sugar.

That DIGs denote a natural and useful concept is highlighted by the fact that they seem to have been (re)discovered several times. In particular, Baumgartner et. al. [3,2] introduce them as 'atoms with exceptions (AWEs)' and describe their use in tableau based inference in the context of their 'Living Book' technology. Further applications of implicit generalizations are mentioned in Section 5.

Although the difference between universal variables and parameters is, in principle, irrelevant for DIGs, we will from now on assume that DIGs consist of parameter atoms only. This is in accordance with the intended semantics of parameters as 'variables that may not be instantiated arbitrarily'.

The truth value of a ground atom A in a model represented by a DIG Δ can be computed efficiently. As in the case of models represented by a context, checking whether $A \in \mathcal{M}(\Delta)$ reduces to linearly many matching tests.

Note that for any implicit generalization $A/\mathcal{B}$, every $B \in \mathcal{B}$ can be replaced by $B\sigma$, where σ is the *mgu* of (parameter disjoint copies of) A and B, without affecting the set of contained ground atoms. If $B\sigma$ is a variant of A (i.e., if A is an instance of B) then $\mathcal{M}(A/\mathcal{B})$ is empty. In other words: for any implicit generalization $A/\mathcal{B}$ one may assume without loss of generality that the atoms in $\mathcal{B}$ are *proper instances* of A. We call a DIG *normalized* if, for all implicit generalizations $A/\mathcal{B}$ in it, $\mathcal{B}$ consists only of proper instances of A and, moreover, all atoms occurring in the DIG are pairwise parameter disjoint.

Note that a single normalized implicit generalization can be considered as a special form of contexts. Indeed, the following equivalence follows immediately from the Definitions 2, 3, and 4:

Proposition 1. *For any normalized implicit generalization* $\Gamma = A/\mathcal{B}$ *the set* $\Lambda_\Gamma = \{\neg v\} \cup \{A\} \cup \{\neg B \mid B \in \mathcal{B}\}$ *is a context with* $\mathcal{M}(\Lambda_\Gamma) = \mathcal{M}(\Gamma)$.

3 Expressive Power of DIGs and Contexts

In this section, we show that contexts and DIGs have the same *expressive power*, i.e., the set of models representable by one formalism coincides with the models representable by the other formalism (see Theorem 1). However, if we also take into account the actual *cost of transforming one model representation into the other*, then a significant difference between contexts and DIGs emerges, namely: DIGs can represent models significantly more succinctly than contexts (see Theorems 2 and 3).

Theorem 1. *Contexts and DIGs have the same expressive power; i.e., a model* $\mathcal{N}$ *is context representable* $\Leftrightarrow$ $\mathcal{N}$ *is DIG representable.*

Proof.
"$\Rightarrow$": Let $\Lambda = \{\neg v\} \cup \Lambda_u^+ \cup \Lambda_p^+ \cup \Lambda_u^- \cup \Lambda_p^-$ be a context, where Λ_u^+ are the positive universal literals, Λ_p^+ are the positive parameter literals, Λ_u^- are the negative universal literals, and Λ_p^- are the negative parameter literals of Λ, respectively. Moreover, let the DIG Δ_Λ be defined by:

$$\Delta_\Lambda = \bigsqcup_{K \in \Lambda_u^+} K \ \sqcup \ \bigsqcup_{K \in \Lambda_p^+} K/(\{\overline{L} \mid L \in \Lambda_p^-, \overline{L} \not\precsim K\} \cup \{\overline{L} \mid L \in \Lambda_u^-\})$$

The definition of Δ_Λ above makes immediate use of Definition 2, i.e., an atom $K \in \Lambda_u^+$ produces all its ground instances, while an atom $K \in \Lambda_p^+$ produces a ground instance only if this production is not prevented by the negative literals in Λ_u^- and Λ_p^-. It is straightforward to check that $\mathcal{M}(\Delta_\Lambda) = \mathcal{M}(\Lambda)$, i.e., Δ_Λ and Λ represent the same model.

"$\Leftarrow$": Let $\Delta = A_1/\mathcal{B}_1 \sqcup \ldots \sqcup A_n/\mathcal{B}_n$ be a normalized DIG. Moreover let Λ_Δ be the context defined as follows:[3]

$$\Lambda_\Delta = \{\neg v\} \ \cup \ \{A_i \mid 1 \le i \le n\}$$
$$\cup \ \{\neg B_1 \vartheta [= \ldots = \neg B_k \vartheta] \mid \vartheta = mgu\{B_1, \ldots, B_k\} \wedge \text{cond}_1 \wedge \text{cond}_2\},$$

---

[3] Recall from Section 2.2 that we assume that DIGs contain only parameters (and no universal variables). Of course, this assumption is not necessary; but it makes the definition of Λ_Δ much simpler (and more readable).

where

$$\text{cond}_1 = \{B_1, \ldots, B_k\} \subseteq \bigcup_{1 \le i \le n} \mathcal{B}_i$$
$$\text{cond}_2 = (\forall 1 \le \ell \le n)\{B_1, \ldots, B_k\} \cap \mathcal{B}_\ell = \emptyset \text{ implies } B_1 \vartheta \not\precsim A_\ell.$$

The idea of cond_1 is obvious: For all negative literals $\neg B \in \Lambda_\Delta$, the dual B must either occur on the right-hand side of some implicit generalization $A_\ell / \mathcal{B}_\ell$ or B must be the mgu of such atoms. The purpose of cond_2 is to make sure that a negative literal $\neg B \in \Lambda_\Delta$ does not prevent A_ℓ from producing any atom unless B is obtained via unification with some atom $B_j \in \mathcal{B}_\ell$. Note that cond_2 also guarantees that the resulting context Λ_Δ is non-contradictory. In fact, since Λ_Δ contains only parameter literals, the only possibility of a contradiction is that Λ_Δ contains two literals $\neg B_1 \vartheta$ and A_ℓ where $B_1 \vartheta$ is a variant of A_ℓ. However, by cond_2, $B_1 \vartheta$ is obtained via unification with some atom $B_j \in \mathcal{B}_\ell$. Moreover, since Δ is normalized, B_j (and thus also $B_1 \vartheta$) is a proper instance of A_ℓ.

Again, it is rather straightforward to check that $\mathcal{M}(\Delta) = \mathcal{M}(\Lambda_\Delta)$. ◇

Example 4. Recall the context $\Lambda = \{\neg v, P(c, x, y), P(u, a, v), \neg P(u, u, b), P(u, v, b)\}$ from Example 2. By the proof of Theorem 1, Λ represents the same model as $\Delta_\Lambda = P(c, x, y) \sqcup P(u, a, v) \sqcup P(u, v, b)/\{P(u, u, b)\}$. It is now obvious that the description of $\mathcal{M}(\Lambda)$ in Example 3 is correct.

Example 5. Consider the DIG $\Delta = P(u_1, u_2)/\{P(w_1, b)\} \sqcup P(u_3, u_4)/\{P(a, w_2)\}$. Clearly, Δ represents all atoms $P(s, t)$, s.t. either $s \ne a$ or $t \ne b$. In other words, Δ is equivalent to the implicit generalization $P(u_1, u_2)/\{P(a, b)\}$.

Analogously to the notation in the proof of Theorem 1, let $A_1 = P(u_1, u_2)$, $\mathcal{B}_1 = \{P(w_1, b)\}$, $A_2 = P(u_3, u_4)$, and $\mathcal{B}_2 = \{P(a, w_2)\}$. The context Λ_Δ is computed as follows: $\{\neg v\} \cup \{A_i \mid 1 \le i \le 2\} = \{\neg v, P(u_1, u_2), P(u_3, u_4)\}$. Moreover, $\{\neg B_1 \vartheta \mid \vartheta = mgu\{B_1, \ldots, B_k\} \wedge \text{cond}_1\} = \{\neg P(w_1, b), \neg P(a, w_2), \neg P(a, b)\}$. It remains to check which of the negative literals in this set also fulfill cond_2:

$\{P(w_1, b)\} \cap \mathcal{B}_2 = \emptyset$ but $P(w_1, b) \precsim A_2$. Likewise, $\{P(a, w_2)\} \cap \mathcal{B}_1 = \emptyset$ but $P(a, w_2) \precsim A_1$. Hence, in either case, cond_2 is violated. On the other hand, $P(a, b) = P(w_1, b)\vartheta$ with $\vartheta = mgu\{P(w_1, b), P(a, w_2)\}$. Hence, $P(a, b)$ fulfills cond_2 and, therefore, $\Lambda_\Delta = \{\neg v, P(u_1, u_2), P(u_3, u_4), \neg P(a, b)\}$. Of course, the atom $P(u_3, u_4)$ is redundant, but it does no harm. It is easy to see that Λ_Δ is indeed equivalent to $P(u_1, u_2)/\{P(a, b)\}$ and hence also to Δ.

Theorem 2. *Given a context Λ, a normalized DIG Δ with $\mathcal{M}(\Lambda) = \mathcal{M}(\Delta)$ can be computed in polynomial time.*

Proof. The construction of Δ_Λ in the proof of Theorem 1 only involves some instance checks and thus clearly is polynomial. However, note that in general Δ_Λ is not normalized. As described in Section 2.2, normalization requires the computation of (linearly many) most general unifiers. Using appropriate data structures, the corresponding instances can be generated in polynomial time. ◇

Theorem 3. *There exists a sequence Δ_n ($n > 1$) of DIGs, where the size of Δ_n is polynomial (in n), such that all contexts representing the same model as Δ_n are of exponential size (in n).*

Proof. Let $\Delta_n = \bigsqcup_{1 \leq i \leq n} P(u_1, \ldots, u_n)/\{P(u_1, \ldots, u_{i-1}, 0, u_{i+1}, \ldots, u_n),$
$$P(u_1, \ldots, u_{i-1}, 1, u_{i+1}, \ldots, u_n)\},$$
where the u_i ($1 \leq i \leq n$) are pairwise distinct parameters.

Note that the represented model $\mathcal{M}(\Delta_n)$ consists in all ground instances of $P(u_1, \ldots, u_n)$ except those where all parameters u_i are replaced by either 0 or 1. Let us denote the corresponding set of negative literals by Λ_n^-; i.e.:

$$\Lambda_n^- = \{\neg P(d_1, \ldots, d_n) \mid d_i \in \{0, 1\}, 1 \leq i \leq n\}$$

Clearly $|\Lambda_n^-| = 2^n$. It remains to show that for every context Λ_n with $\mathcal{M}(\Lambda_n) = \mathcal{M}(\Delta_n)$ we have $\Lambda_n^- \subseteq \Lambda_n$.

Let $c_1, \ldots, c_n$ be *fresh*, pairwise distinct constants, i.e., the c_i do not occur in Λ_n or Δ_n. In particular, the c_i are all different from 0 and 1. Clearly, the atom $P(c_1, \ldots, c_n)$ is in $\mathcal{M}(\Delta_n)$. But then Λ_n must contain an atom $P(u_1, \ldots, u_n)$, where the u_i are pairwise distinct parameters. Note that the u_i cannot be variables since this would mean that every ground instance of $P(u_1, \ldots, u_n)$ is in $\mathcal{M}(\Lambda_n)$; including those, where the u_i are instantiated by 0 or 1.

Let $\neg P(d_1, \ldots, d_n) \in \Lambda_n^-$, i.e., $d_i \in \{0, 1\}$ for all i. By $P(u_1, \ldots, u_n) \in \Lambda_n$, the context Λ_n has to contain some negative literal $\neg P(s_1, \ldots, s_n)$ with

$$P(d_1, \ldots, d_n) \precsim P(s_1, \ldots, s_n) \not\precsim P(u_1, \ldots, u_n)$$

in order to prevent Λ_n from producing $P(d_1, \ldots, d_n)$ (cf. Definition 2). Let $\neg P(s_1, \ldots, s_n)$ be most specific among the literals in Λ_n with this property. It can be shown that $P(s_1, \ldots, s_n)$ neither contains variables nor parameters since otherwise it would also prevent the production of atoms that are indeed contained in the model $\mathcal{M}(\Delta_n)$. Thus, we have $s_i = d_i$ for all $1 \leq i \leq n$. $\diamond$

4   Clause Evaluation and Testing Equivalence

Recall from Section 2 the conditions (1), (2), and (3) that any model representation formalism should fulfill. The fact that condition (1) (i.e., the efficient evaluation of ground atoms) holds for both contexts and DIGs, follows directly from the observations of Section 2. In this section we are going to show that also conditions (2) and (3) (i.e., the decidability of clause evaluation and of equivalence) hold. More formally, we investigate the following four decision problems:

Context-Clause-Evaluation		Context-Equivalence	
Input:	A clause C and a context Λ.	*Input:*	Two contexts Λ and Λ'.
Question:	Is C true in the model $\mathcal{M}(\Lambda)$?	*Question:*	Does $\mathcal{M}(\Lambda) = \mathcal{M}(\Lambda')$ hold?

DIG-Clause-Evaluation		DIG-Equivalence	
Input:	A clause C and a DIG Δ.	*Input:*	Two DIGs Δ and Δ'.
Question:	Is C true in the model $\mathcal{M}(\Delta)$?	*Question:*	Does $\mathcal{M}(\Delta) = \mathcal{M}(\Delta')$ hold?

Below we not only provide algorithms for these decision problems but we also prove upper and lower bounds for these problems.

For the upper bounds, the following proposition is helpful:

Proposition 2. *Let $A_1, \ldots, A_n, A'_1, \ldots, A'_m$ be parameter atoms that do not share parameters with any of the parameter atoms in $\{B_1, \ldots, B_n, B'_1, \ldots, B'_m\}$. It can be checked in deterministic polynomial time whether*

$$\bigwedge_{1 \leq i \leq n} A_i \gamma \precsim B_i \wedge \bigwedge_{1 \leq j \leq m} A'_j \gamma \not\precsim B'_j$$

for some ground substitution γ.

Proof. W.l.o.g., we may assume that the atoms $B_1, \ldots, B_n, B'_1, \ldots, B'_m$ are pairwise parameter disjoint. We claim that γ exists $\Leftrightarrow$ the following conditions hold:

1. A simultaneous *mgu* σ of $(\{A_1, B_1\}, \ldots, \{A_n, B_n\})$ exists and
2. $A'_j \sigma \not\precsim B'_j$ for all $j \in \{1, \ldots, m\}$.

Obviously these conditions can be tested in polynomial time with an efficient unification algorithm (see [28]). We show the two directions of the equivalence separately:

"$\Rightarrow$": Suppose that the desired ground substitution γ exists. Then, since $A_i \gamma \precsim B_i$ and the B_i are pairwise parameter disjoint, there is a most general simultaneous unifier σ of all pairs $\{A_i, B_i\}$; i.e., condition 1 holds. As for condition 2, it suffices to note that $\gamma = \sigma \vartheta$ for some substitution ϑ and consequently $A'_j \gamma \not\precsim B'_j$ implies $A'_j \sigma \not\precsim B'_j$.

"$\Leftarrow$": Suppose conditions 1 and 2 hold. Let ν be a substitution that maps each parameter that occurs in one of the atoms $A_i \sigma$ or $A'_j \sigma$ into a distinct, fresh constant. (By 'fresh' we mean that the constant does not occur in any of the A_i, B_i, A'_j, or B'_j.) We claim that $\gamma = \sigma \nu$ is the desired ground substitution. Since $A_i \gamma$ is an instance of $A_i \sigma = B_i \sigma$, we have $A_i \gamma \precsim B_i$. On the other hand, by the particular form of ν, $A'_j \sigma \not\precsim B'_j$ implies $A'_j \gamma = A'_j \sigma \nu \not\precsim B'_j$.   ◇

Note that the "$\Leftarrow$"-direction in the above proof (and hence, the PTIME-upper bound in Proposition 2) only holds because we are considering models over an *infinite* signature Σ here. Indeed, when considering a *finite* signature, we end up with an NP-lower bound for the analogous decision problem:

Proposition 3. *Let Σ be a finite signature with at least two distinct constant or function symbols. Moreover, let $A_1, \ldots, A_n, A'_1, \ldots, A'_m$ be parameter atoms that do not share parameters with any parameter atoms in $\{B_1, \ldots, B_n, B'_1, \ldots, B'_m\}$. Then it is NP-hard to check whether there exists a ground substitution γ, s.t. the following condition holds:*

$$\bigwedge_{1 \leq i \leq n} A_i \gamma \precsim B_i \wedge \bigwedge_{1 \leq j \leq m} A'_j \gamma \not\precsim B'_j .$$

Proof. The proof is by reduction from the non-emptiness problem of implicit generalizations whose NP-hardness was shown independently in [19] and [18], i.e.: Given an implicit generalization $\Gamma = A / \{\bar{A}_1, \ldots, \bar{A}_m\}$ over some *finite* signature

Σ with at least 2 distinct function symbols, does there exist a ground instance of A that is not an instance of any atom $\bar{A}_i$ on the right-hand side?

W.l.o.g., the atom A has no parameters in common with the atoms $\bar{A}_j$. Now let $A'_1 = A'_2 = \cdots = A'_m = A$ and $B'_j = \bar{A}_j$ for all $j \in \{1, \dots, m\}$. Then the above non-emptiness problem of implicit generalizations is equivalent to the condition that there exists a ground substitution γ with $\bigwedge\limits_{1 \leq j \leq m} A'_j \gamma \not\precsim B'_j$. $\diamond$

Lemma 1. DIG-Clause-Evaluation *is in coNP.*

Proof. Let $C = A_1 \vee \cdots \vee A_k \vee \neg A'_1 \vee \cdots \vee \neg A'_\ell$ and $\Delta = \bigsqcup_{1 \leq i \leq n} B_i / \mathcal{B}_i$. By definition, C is false in $\mathcal{M}(\Delta)$ iff for some ground instance $C\gamma$ none of the positive literals in $C\gamma$, but all negative literals in $C\gamma$ are contained in Δ. In other words, C evaluates to false in $\mathcal{M}(\Delta)$ iff there exists a ground substitution γ such that

$$\bigwedge_{j=1}^{k} \bigwedge_{i=1}^{n} \left(A_j \gamma \not\precsim B_i \vee \bigvee_{B \in \mathcal{B}_i} A_j \gamma \precsim B \right) \wedge \bigwedge_{j=1}^{\ell} \bigvee_{i=1}^{n} \left(A'_j \gamma \precsim B_i \wedge \bigwedge_{B \in \mathcal{B}_i} A'_j \gamma \not\precsim B \right)$$

Note that, w.l.o.g., we may assume that the atoms A_i, A'_j do not share parameters with the (parameter) atoms B_i and those in $\mathcal{B}_i$. Therefore, the problem is reduced to one which has the form exhibited in Proposition 2 by correctly guessing a disjunct (of the form $A_j \gamma \not\precsim B_i$ or $A_j \gamma \precsim B$) for each of the conjuncts of the conjunction on the left-hand side of the formula, as well as a disjunct (of the form $A'_j \gamma \precsim B_i \wedge \bigwedge_{B \in \mathcal{B}_i} A'_j \gamma \not\precsim B$) for each j ($1 \leq j \leq \ell$) at the right-hand side of the formula. Hence it follows from Proposition 2 that clauses can be checked to be false in $\mathcal{M}(\Delta)$ in non-deterministic polynomial time. In other words: checking whether a clause is true in such a model is in coNP. $\diamond$

Lemma 2. DIG-Equivalence *is in coNP.*

Proof. It suffices to show that, given DIGs Δ and Δ', it can be tested by a coNP-algorithm whether $\mathcal{M}(\Delta) \subseteq \mathcal{M}(\Delta')$. Let $\Delta = \bigsqcup_{1 \leq i \leq n} B_i / \mathcal{B}_i$. Note that $\mathcal{M}(B_i / \mathcal{B}_i) \subseteq \mathcal{M}(\Delta')$ iff $\mathcal{M}(B_i) \subseteq \mathcal{M}(\Delta''_i)$, where $\Delta''_i = \Delta' \sqcup \bigsqcup_{B \in \mathcal{B}_i} B$. In other words, we have: $\mathcal{M}(\Delta) \subseteq \mathcal{M}(\Delta')$ iff for all $i \in \{1, \dots, n\}$ the positive unit clause B_i is true in the model $\mathcal{M}(\Delta''_i)$. We have thus polynomially reduced the equivalence test to (linearly many) clause evaluations. Therefore the coNP-membership follows from Lemma 1. $\diamond$

Lemma 3. Context-Clause-Evaluation *is coNP-hard. It remains coNP-hard if the clauses to be evaluated are simply (non-ground) atoms.*

Proof. We prove the coNP-hardness by reducing the well-known NP-complete problem 3SAT to the co-problem of Context-Clause-Evaluation.

An instance of the 3SAT problem is given through a set $X = \{x_1, \dots, x_k\}$ of propositional variables and a Boolean formula $E = (l_{11} \vee l_{12} \vee l_{13}) \wedge \cdots \wedge (l_{n1} \vee l_{n2} \vee l_{n3})$, s.t. the l_{ij} are literals over the variables in X, i.e.: every l_{ij} is either of the form x_γ or $\neg x_\gamma$ for some $\gamma \in \{1, \dots, k\}$. Correspondingly, we define $A = P(a, \dots, a, y_1, \dots, y_k)$, where P is a predicate symbol of arity $n + k$, a is a constant and the y_i are pairwise distinct variables. Moreover, let $u_1, \ldots, u_n, v_1, \ldots, v_k$ be pairwise distinct parameters. We set $\Lambda_E = \{\neg v\} \cup \Lambda_E^+ \cup \Lambda_E^-$ where Λ_E^+ and Λ_E^- are defined as follows:

$$\Lambda_E^+ = \{P(a, u_2, \ldots, u_n, v_1, \ldots, v_k), P(u_1, a, u_3, \ldots, u_n, v_1, \ldots, v_k), \ldots, P(u_1, \ldots, u_{n-1}, a, v_1, \ldots, v_k)\},$$ i.e., each atom in Λ_E^+ has the constant a as one of the first n arguments. The remaining $n + k - 1$ arguments are (pairwise distinct) parameters.

$$\Lambda_E^- = \{\neg B_{11}, \neg B_{12}, \neg B_{13}, \neg B_{21}, \neg B_{22}, \neg B_{23}, \ldots, \neg B_{n1}, \neg B_{n2}, \neg B_{n3}\} \text{ with}$$

$$B_{ij} = \begin{cases} P(u_1, \ldots, u_{i-1}, a, u_{i+1}, \ldots, u_n, v_1, \ldots, v_{\gamma-1}, 1, v_{\gamma+1}, \ldots, v_k) & \text{if } l_{ij} = x_\gamma \\ P(u_1, \ldots, u_{i-1}, a, u_{i+1}, \ldots, u_n, v_1, \ldots, v_{\gamma-1}, 0, v_{\gamma+1}, \ldots, v_k) & \text{if } l_{ij} = \neg x_\gamma \end{cases}$$

It can be easily checked that the resulting context Λ is non-contradictory. Indeed, since Λ contains only parameter literals, a contradiction is only possible if Λ_E contains some atom plus its dual, which is obviously not the case.

This transformation can clearly be done in polynomial time. Moreover, it is straightforward to check that E is satisfiable iff A evaluates to false in $\mathcal{M}(\Lambda)$. $\diamond$

Lemma 4. Context-Equivalence *is coNP-hard.*

Proof. The coNP-hardness of the equivalence problem can be shown by reducing the co-3SAT problem to it. Let the Boolean formula E as well as the atom A and the context Λ be defined as in the proof of Lemma 3. We claim that the Boolean formula E is unsatisfiable $\Leftrightarrow$ the contexts Λ and $\Lambda' = \Lambda \cup \{A\}$ are equivalent. Of course, all ground atoms true in $\mathcal{M}(\Lambda)$ are also true in $\mathcal{M}(\Lambda')$. It thus remains to show that E is unsatisfiable $\Leftrightarrow$ all ground atoms produced by Λ' are also produced by Λ. This is straightforward, but omitted here for lack of space. $\diamond$

By the polynomial time reduction from contexts to DIGs shown in Theorem 2, the upper bounds on DIGs clearly carry over to contexts and, likewise, the lower bounds on contexts carry over to DIGs. We thus immediately get:

Theorem 4. *The four decision problems* Context-Clause-Evaluation, DIG-Clause-Evaluation, Context-Equivalence, *and* DIG-Equivalence *are coNP-complete.*

5   Related Work

We have shown that DIGs and contexts have the same expressive power for representing models. But DIGs are known to be equivalent to yet another important model representation formalism, namely atoms with equational constraints ('*constrained atoms*', for short). Models represented by constrained atoms have been extensively studied by Caferra, Peltier et al. (see [9,10,8]). Translating DIGs into sets of constrained atoms is straightforward; the translation of constrained atoms into DIGs can be done effectively via results in [25]. However, the latter translation has non-elementary complexity in the worst-case, if arbitrary quantifier alternations are allowed in the constraining formula (see [31]). The equivalence between DIGs and constrained atoms implies that the set of expressible models is closed under union, intersection and complement (using results presented, e.g., in [11]). Similar formalisms are investigated, e.g., in [7,11]. The dissertation [29] and, more recently, chapter 5 of [8] provides an overview of alternative model representation formalisms.

Besides their intended use in Automated Deduction, constrained atoms and related representations of Herbrand models have also received attention in Logic Programming and in Nonmonotonic Reasoning. In particular, [12] and [16] use certain forms of constrained atoms to represent stable models for logic programs with negation. Likewise, implicit generalizations (or equivalent notions) have been applied to many fields of Computer Science, like Machine Learning and the design of logic programming languages (cf. [21]), Program Verification and Program Transformation (cf. [26]), Functional Programming (cf. [20]), etc. However, the decision problems studied so far have been primarily the following ones:

1. The *emptiness problem*: Given an implicit generalization $\Gamma = A/\mathcal{B}$ over some fixed signature Σ, is every ground instance of A also a ground instance of at least one element $B \in \mathcal{B}$ (i.e., is Γ empty)?
2. The *explicit representability problem*: Given an implicit generalization $\Gamma = A/\mathcal{B}$ over some fixed signature Σ, does there exist an equivalent explicit generalization $\mathcal{A} = \{A_1, \ldots, A_n\}$, i.e., every ground atom contained in Γ is a ground instance of some atom $A_i \in \mathcal{A}$ and vice versa?

Related problems arise in the area of equational algebraic specifications (where the "sufficient-completeness" problem is of interest, see [18]) and in automating inductive proofs in equational theories (where the "ground reducibility" problem is fundamental, see [17]). Actually, both the emptiness problem and the explicit representability problem were shown to be coNP-complete for a *finite signature Σ* (see [18,19,26,27]). These coNP-completeness results still hold if we generalize these problems to disjunctions of implicit generalizations (see [30]).

Note that the situation changes completely if we consider implicit generalizations over an *infinite signature*, as is done here. It is shown in [21] that then the above two decision problems are in fact trivial and hence, clearly solvable in deterministic polynomial time. Conversely, if clause evaluation and the equivalence problem were studied over a *finite signature*, then the proof of Proposition 2 would be wrong. Indeed, the PTIME-upper bound shown in Proposition 2 for an infinite signature contrasts with the NP-lower bound shown in Proposition 3 for a finite signature. Hence, also the upper bounds shown in Lemma 1 and Lemma 2 would not necessarily hold any longer if we considered a finite signature here.

6   Conclusion

In the light of our results, DIGs may be considered superior to contexts as a representation formalism: as we have seen, they may allow one to represent models considerably more succinctly without sacrificing expressive power. However, contexts have been specifically designed for the stepwise, systematic construction of model descriptions in a calculus which supports efficient proof search [4,5]. It remains unclear whether DIGs can be used alternatively. (See [3,2] for a promising approach.) Moreover, one may seek to characterize the set of those models for which optimal DIG representations are indeed exponentially more succinct than all context representations. The fact that clause evaluation and the equivalence test are not harder for DIGs than for contexts (cf. Theorem 4) suggests that there are many models for which the representation as a context is not significantly longer than the representation as a DIG. In particular, this is the case for the family of contexts constructed in the NP-hardness proof in Lemma 3.

Another problem for future investigation is whether even more succinct representations of models (of the same class) are possible by, e.g., nesting the exception operator '/' used for DIGs. Moreover it might be interesting to identify fragments of contexts, DIGs, and other mechanisms that allow for polynomial-time clause evaluation.

Finally, recall that we have considered the case of an infinite signature here (motivated by the intended use of contexts). As has already been mentioned, the complexity results derived in Section 4 (in particular the upper bounds) may no longer hold if we switch to a finite signature. The precise complexity in the latter case has yet to be determined.

References

1. P. Baumgartner, A. Fuchs, and C. Tinelli. Darwin: A theorem prover for the model evolution calculus. In *Proceedings of ESFOR'04*. Elsevier, 2004.
2. P. Baumgartner, U. Furbach, M. Gross-Hardt, and A. Sinner. Living Book – Deduction, Slicing, and Interaction. *Journal of Automated Reasoning*, 32(3):259–286, 2004.
3. P. Baumgartner, M. Gross-Hardt, and A. Sinner. Living Book – Deduction, Slicing, and Interaction. Fachberichte Informatik 2/2003, Universität Koblenz Landau.
4. P. Baumgartner and C. Tinelli. The model evolution calculus. In *Proceedings of CADE-19*, LNCS 2741, pages 350–364. Springer, 2003.
5. P. Baumgartner and C. Tinelli. The model evolution calculus. Fachberichte Informatik, 1/2003, Universität Koblenz Landau, 2003. Extended Version of [4].
6. J.-Paul. Billon. The disconnection method – a confluent integration of unification in the analytic framework. In *Proceedings of TABLEAUX'96*, LNCS 1071, pages 110–126. Springer, 1996.
7. H.-J. Bürckert. Solving disequations in equational theories. In *Proceedings of CADE-9*, LNCS 310, pages 517–526. Springer, 1988.
8. R. Caferra, A. Leitsch, and N. Peltier. *Automated Model Building*, volume 31 of *Applied Logic Series*. Kluwer Academic Publishers, 2004.
9. R. Caferra and N. Peltier. Extending semantic resolution via automated model building: applications. In *Proceedings of IJCAI'95*, pages 328–334, 1995.
10. R. Caferra and N. Zabel. Extending resolution for model construction. In *Proceedings of JELIA'90*, LNAI 478, pages 153–169, 1991. Springer Verlag.
11. H. Comon. Disunification: a survey. In *Computational Logic: Essays in Honor of Alan Robinson*. MIT Press, 1991.

12. T. Eiter, G. Gottlob, and H. Veith. Modular logic programming and generalized quantifiers. In *Proceedings of LPNMR'97*, LNCS 1265, pages 290–309. 1997.
13. C.G. Fermüller and A. Leitsch. Model building by resolution. In *Proceedings of CSL'92*, LNCS 702, pages 134–148. Springer, 1993.
14. C.G. Fermüller and A. Leitsch. Hyperresolution and automated model building. *Journal of Logic and Computation*, 6(2):173–203, 1996.
15. G. Gottlob and R. Pichler. Working with ARMs: Complexity results on atomic representations of Herbrand models. *Information and Computation*, 165:183–207, 2001.
16. G. Gottlob, S. Marcus, A. Nerode, G. Salzer, and V.S. Subrahmanian. A non-ground realization of the stable and well-founded semantics. *Theoretical Computer Science*, 166(1&2):221–262, 1996.
17. J.-P. Jouannaud and E. Kounalis. Automatic proofs by induction in theories without constructors. *Information and Computation*, 82(1):1–33, 1989.
18. D. Kapur, P. Narendran, D. Rosenkrantz, and H. Zhang. Sufficient-completeness, ground-reducibility and their complexity. *Acta Informatica*, 28(4):311–350, 1991.
19. K. Kunen. Answer sets and negation as failure. In *Proceedings of ICLP'87*, pages 219–228, 1987. MIT Press.
20. J.-L. Lassez, M. Maher, and K. Marriott. Elimination of negation in term algebras. In *Proceedings of MFCS'91*, LNCS 520, pages 1–16, 1991. Springer Verlag.
21. J.-L. Lassez and K. Marriott. Explicit representation of terms defined by counter examples. *Journal of Automated Reasoning*, 3(3):301–317, 1987.
22. R. Letz and G. Stenz. Model Elimination and Connection Tableau Procedures. In *Handbook of Automated Reasoning*, volume II, pages 2015–2114. June 2001.
23. R. Letz and G. Stenz. Proof and Model Generation with Disconnection Tableaux. In *Proceedings of LPAR 2001*, LNAI 2250, pages 142–156. Springer, 2001.
24. R. Letz and G. Stenz. Generalised handling of variables in disconnection tableaux. In *Proceedings of IJCAR 2004*, LNAI 3097, pages 289–306. Springer, 2004.
25. M. Maher. Complete axiomatizations of the algebras of finite, rational and infinite trees. In *Proceedings of LICS'88*, pages 348–357, 1988. IEEE Computer Society.
26. M. Maher and P. Stuckey. On inductive inference of cyclic structures. *Annals of Mathematics and Artificial Intelligence*, 15(2):167–208, 1995.
27. K. Marriott. *Finding Explicit Representations for Subsets of the Herbrand Universe*. PhD thesis, University of Melbourne, Australia, 1988.
28. A. Martelli and U. Montanari. An efficient unification algorithm. *ACM Transactions on Programming Languages and Systems*, 4(2):258–282, 1982.
29. R. Matzinger. Computational representations of models in first-order logic. Technische Universität Wien, 2000. Dissertation (Ph.D. thesis).
30. R. Pichler. Explicit versus implicit representations of subsets of the Herbrand universe. *Theoretical Computer Science*, 290(1):1021–1056, 2003.
31. S. Vorobyov. An improved lower bound for the elementary theories of trees. In *Proceedings of CADE-13*, LNAI 690, pages 316–327, 1996. Springer Verlag.

Proving Properties of Incremental Merkle Trees

Mizuhito Ogawa[1], Eiichi Horita[2], and Satoshi Ono[3]

[1] Japan Advanced Institute of Science and Technology, Ishikawa, 923-1292 Japan
`mizuhito@jaist.ac.jp`
[2] NTT Information Sharing Platform Laboratories, Tokyo, 180-8585, Japan
`horita.eiichi@lab.ntt.co.jp`
[3] Kogakuin University, Tokyo, 163-8677 Japan
`ono@cpd.kogakuin.ac.jp`

Abstract. This paper proves two basic properties of the model of a single attack point-free event ordering system, developed by NTT. This model is based on an incremental construction of Merkle trees, and we show the correctness of (1) completion and (2) an incremental sanity check. These are mainly proved using the theorem prover MONA; especially, this paper gives the first proof of the correctness of the incremental sanity check.

Keywords: Merkle tree, theorem prover, temporal authentication.

1 Introduction

With the growth of the Internet, resilient temporal authentication for system failure and/or malicious attacks becomes important. The standard method is to use a *timestamp based on a public-key cryptosystem*. However, it has relatively short time span (most public keys are renewed each 5 years), and once the cryptography is compromised, all certificates become invalid.

With the aim for long-term validity (say, 20-30 years), NTT developed the event ordering system [5] based on a Merkle tree [8], which is a labeled binary tree such that a label of a node is recursively computed from labels of its child nodes using a hash function. Although an event ordering has relatively rough precision on a time scale, it relies only on the collision-resistance (and one-wayness) of a hash function, which is believed to be much harder than public key cryptosystems. Thus, this system is complementary to a timestamp system based on a public key cryptosystem; with its supplementary use, we can obtain both precision and long-term validity.

The event ordering system receives a hash value of a timestamp, and constructs a Merkle tree in an incremental manner. It issues a certificate immediately after a hash value is registered to a leaf label of a partially constructed Merkle tree. A newly registered hash value is recursively propagated in this bottom-up way, and a certificate is the known part of minimum information to compute the hash value at the root of a Merkle tree. When a whole Merkle tree has been constructed, the hash value at its root is released as a public witness.

R. Nieuwenhuis (Ed.): CADE 2005, LNAI 3632, pp. 424–440, 2005.

Such an incremental construction has been proposed in literature [3,2,12,9,7], and these studies support

- for long-term validity, the systems relay on only collision-resistance (and one-wayness) of a hash function, and
- each transaction message is kept in $O(log\ n)$ wrt the number of events; thus, they are scalable.

Our system further enhances these systems:

- single attack point free.
- even if the system halts, an intermediate snapshop supports relative correctness of temporal authentication.

This paper proves two basic properties of the incremental construction of a Merkle tree: (1) correctness of completion and (2) correctness of an incremental sanity check. They are mainly proved using theorem prover MONA [1]; especially, this paper first prove (2) correctness of incremental sanity check.

Sections 2 and 3 explain what Merkle trees and MONA are, respectively. Section 4 briefly introduces the event ordering system; terminology for an incremental Merkle forest and our protocol design are explained. Section 6 gives the proof of correctness of completion of an incremental Merkle forest. The proof is performed both by manual induction and by MONA for comparison. Section 7 gives the proof of correctness of the incremental sanity check proposed in [5]. This property has been checked by experiments with large-scale data, but without the proof. Although the proof is not fully formal, the main lemmata are verified by MONA.

2   Merkle Tree

$T = (V(T), E(T))$ is a *directed graph* if $E(T) \subseteq V(T) \times V(T)$. We call an element in $V(T)$ a *node*, and an element in $E(T)$ an *edge*. A *path* is a sequence $(t_0, \cdots, t_n)$ of nodes such that for each $1 \le i \le n$, $(t_{i-1}, t_i) \in E(T)$. A directed graph T is *acyclic* if there are no paths that visit the same node twice.

We say that an acyclic directed graph $T = (V(T), E(T))$ is a binary tree with the (unique) root denoted by $root(T)$ if

- $root(T) \in V(T)$,
- for each node $t \in V(T)$, there exists the unique path $(t_0, \cdots, t_n)$ such that $t_0 = root(T)$ and $t_n = t$, and
- for each node $t \in V(T)$, if $(\{t\} \times V(T)) \cap E(T) \neq \emptyset$, there are exactly two edges (t, t') and (t, t'') in $E(T)$ (we call t' and t'' the *child nodes* of t).

To distinguish the child nodes of t, we will give the explicit ordering denoted by $t.0$ (left-child) and $t.1$ (right-child). $s \le s'$ is equivalent to $\bar{v} \in \{0,1\}^*$ being a prefix of $\bar{w} \in \{0,1\}^*$ where $s = t.\bar{v}$ and $s' = t.\bar{w}$. We also say that $t.0$ is the *brother* of $t.1$, and vice versa. Note that the brother relation is symmetric, but not reflexive, i.e., $t.0$ and $t.1$ are not brothers of themselves, respectively. We say that a node t is a *leaf* if t has no child nodes, and the set of leaves in T is denoted by $leaves(T)$.

The *position* of $t \in V(T)$ is the sequence of 0's and 1's such that the corresponding the sequence of choice of left- and right-child from the root $root(T)$ results the path to t.

In the following, $T = (V(T), E(T))$ is always a binary tree with a root. $T' = (V(T'), E(T'))$ is a subtree of $T = (V(T), E(T))$, if $V(T') \subseteq V(T)$ and there exists $t' \in V(T')$ such that T' is a binary tree with the root t' (i.e., $root(T') = t'$).

As convention, we will denote binary trees by $T, T_1, T_2, ...$, nodes by $s, t, u, v, ...$ and $t_0, t_1, ...$, sets of nodes by $X, Y, Z, ...$, the set of labels by L, and descriptions in MONA by type writer fonts.

Definition 1. *Let $g : L \times L \to L$ be a binary function where L is the set of labels. Let $T = (V(T), E(T))$ be a binary tree with the root $root(T)$, where $V(T)$ and $E(T)$ are the sets of nodes and edges, respectively. A Merkle tree $MT = (V(T), E(T), \alpha)$ is a L-labeled binary tree with a labeling function $\alpha : leaves(T) \to L$. The label for non-leaf nodes is recursively defined by $\alpha(t) = g(\alpha(t.0), \alpha(t.1))$.*

We will often overload a tree T and a Merkle tree MT when it is clear from the context.

Remark 1. Originally, a Merkle tree was defined such that each path to a leaf from $root(T)$ has the same length [8]. We generalize a *Merkle tree*, such that paths to leaves may have different lengths. This generalization makes proof of the target properties easier and expressible in $WS2S$.

In our design of the event-ordering system, we assume that g *is a collision-resistant one-way hash function*. Although theoretically it may be difficult to guarantee a collision-resistant and one-way function, in practice we can set an appropriate function.

An event sequence corresponds to the set of leaves of a Merkle tree (in which each path to a leaf has the same length); as default, we consider that time proceeds in a left-to-right manner. Thus, if the level (the length from the root to a leaf) of a Merkle tree T is n, the start leaf is $root(T).\underbrace{0 \cdots 0}_{n}$ and the end leaf is $root(T).\underbrace{1 \cdots 1}_{n}$. At each time unit, the referred leaf shifts to the next (i.e., right neighborhood) leaf. When an event occurs, it put a label (e.g., the hash value of the transaction to be certificated) at the currently referred leaf. The label of each node is computed recursively when the labels of its both children are computed. Thus, if a referred leaf reaches to the end leaf, the label of $root(T)$ is computed.

If g is a collision-resistant one-way hash function, bottom-up computation of hash values (i.e., hash values of both child nodes are concatenated by suitable injective binary operation, and a hash function computes the hash value of their parent node) is easy; but topdown computation (i.e., from the label of a parent, guess labels of its child nodes) is infeasible. In other words, to interpolate the labels of children is expected to be practically impossible.

In the following definition, the authentication path at a node t is the minimum information to compute the label at $root(T)$ from the label of t, the left authentication path at a node t is the set of labels that were computed before t, and the right authentication path at a node t is the set of labels that will be computed after t. Note that our definition of an authentication path is not restricted to a leaf, but is also for a node.

Definition 2. *Let $t \in V(T)$ and let $(root(T), t_1, \cdots, t_{n-1}, t)$ be a path from $root(t)$ to t.*

- *The authentication path of t, denoted by $CA_T(t)$, is the set of brothers of $t_1, \cdots, t_{n-1}, t$.*
- *The left authentication path of t, denoted by $LA_T(t)$, is the intersection of $CA_T(t)$ and $\{root(T).0, t_1.0, \cdots, t_{n-1}.0\}$ (i.e., left brothers).*
- *The right authentication path of t, denoted by $RA_T(t)$, is the intersection of $CA_T(t)$ and $\{root(T).1, t_1.1, \cdots, t_{n-1}.1\}$ (i.e., right brothers).*
- *The root path of t, denoted by $path_T(t)$, is $\{root(T), t_1, \cdots, t_{n-1}, t\}$.*
- *The path closure of t, denoted by $pCls(t)$, is $path_T(t) \cup CA_T(t)$.*

We often omit T in $CA_T(t)$, $LA_T(t)$, $RA_T(t)$, $path_T(t)$, and $pCls_T(t)$ as $CA(t)$, $LA(t)$, $RA(t)$, $path(t)$, and $pCls(t)$, if T is clear from the context.

Remark 2. A left (resp. right) authentication path is called a *freshness* (resp. an *existential*) *token*.

3   Monadic Second Order Logic

3.1   (W)S2S

Monadic second order logic SnS is a logic on n-ary (possibly infinite) trees. We focus on $S2S$, a logic on binary trees, consisting of

- First order variable, $s, t, u, \cdots$
- Second order variable, $X, Y, Z, \cdots$
- Quantifiers, $\forall$, $\exists$
- Logical connectives, $\wedge$, $\vee$, $\neg$, $\Rightarrow$
- Set operations, $\in$, $\subseteq$, $\cup$, $\cap$, $\setminus$
- Function symbols, $root$, s_0, s_1
- Position relation, $<$, $\leq$

These are interpreted as logical operations on nodes of a binary tree. $root$ is the unique constant that represents the root of a binary tree. The order $s < t$ on nodes means that s is placed between $root$ and t, i.e., s is nearer to the root than t. Note that the satisfiability of an $S2S$-formula is decidable; that is, the satisfiability of an $S2S$-formula is equivalent to the emptiness problem of a Büchi tree automata [11].

$WS2S$ (weak $S2S$) is the restricted logic of $S2S$ such that the range of set variables (second-order variables) runs on sets of finite trees. The satisfiability of a $WS2S$-formula corresponds to the emptiness problem of a tree automata.

3.2 MONA

MONA is a batch-style satisfiability checker for $WS2S$ [1]. Although complexity of the satisfiability is non-elementary, MONA is efficiently implemented and practically quite usable. MONA's syntax for $WS2S$ formulae consists of

- First order variable, s, t, u, $\cdots$
- Second order variable, X, Y, Z, $\cdots$
- *Variable declaration*, var1, var2
- Quantifiers, all1, ex1, all2, ex2
- Logical connectives, &, |, $\sim$, =>
- Set operations, in, notin, sub, union, inter, \

```
# s is properly lefter than t
pred lefter(var1 s,t) = ex1 u: (u.0 <= s & u.1 <= t);
# u = glb(s,t)
pred glb(var1 s,t,u) =
  u <= s & u <= t & all1 v: ((v <= s & v <= t) => v <= u);
# s.1...1 = t
pred rightmost(var1 s,t)=s < t & all1 u: ((s <= u & u < t) => u.1 <= t);
# Each pair of nodes in A is incomparable
pred incomparable(var2 A) =
  all1 s,t : ((s in A & t in A) => (s = t | (~(s < t) & ~(t < s))));
# t is the last node in A
pred last(var1 t, var2 A) =
  t in A & all1 s: ((s in A & s ~= t) => lefter(s,t));
# t is the next node of s in A
pred next(var1 s,t, var2 A) =
  s in A & t in A & lefter(s,t) &
  all1 u : ((u in A & lefter(u,t)) => (u = s | lefter(u,s)));
# Y is the lower bound node set of X
pred lower_bound(var2 X,Y) =
  incomparable(Y) & all1 s: (s in X => ex1 t: (t in Y & t <= s));
# X is a (sub)bintree rooted at node s
pred bintree_at(var1 s, var2 X) =
  s in X & all1 t:((s <= t =>((t notin X =>(t.0 notin X & t.1 notin X)) &
                             (t in X => (t.0 in X <=> t.1 in X)))) &
              (~(s <= t) => t notin X));
# Y is the subtree of X below s
pred below(var1 s, var2 X,Y) = all1 t: ((s <= t & t in X) <=> t in Y);
```

Fig. 1. Library for proofs by MONA

- Function symbols, root, t.0, t.1, t^
- Position relation, <, <=

The difference with $WS2S$ is:

- Quantifiers are explicitly classified for first- or second-order variables.
- Free variables used in a formula need the variable declarations var1, var2 depending on whether they are first- or second-order free variables.
- Since negation (complement of tree automata) is an exponentially heavy operation, notin is prepared.
- t^ is prepared for the ancestor node of t.

Note that <= is a prefix relation between positions and => is a logical implication. The library used in the paper is shown in Fig. 1.

Example 1. The example below shows predicate definitions and a $WS2S$-formula that means *the closure operation is idempotent.*

```
ws2s;
var2 X,Y,Z;
pred closed(var2 Y) = all1 t: ((t.0 in Y & t.1 in Y) => t in Y);
pred closure(var2 X,Y) =
  closed(Y) & X sub Y & all2 Z:((closed(Z) & X sub Z) => Y sub Z);
(closure(X,Y) & closure(Y,Z)) => closure(X,Z);
```

The predicate closed(Y) means that for each node t, if both children t.0 (left child) and t.1 (right child) are in Y, then t is in Y. The predicate closure(X,Y) means that Y is the minimum set such that Y is closed and includes X. The last line describes the formula to be checked. When these lines are saved as, say, example.mona, type "mona example.mona"; then it is computed to be VALID.

4 Scalable Event-Ordering System

4.1 Incremental Merkle Forest

Definition 3. *Let T be a Merkle tree and let $t \in V(T)$. The* incremental Merkle *forest $IMF_T(t)$ is the union of binary sub-trees T' of T satisfying either*

- $s = root(T')$ *where s is the minimum node such that $s.1 \cdots 1 = t$, or*
- $s.0 = root(T')$ *where $s.1 \leq t$ and $s.\underbrace{1 \cdots 1}_{m} \neq t$ for $\forall m$.*

We will often omit T in $IMF_T(t)$ as $IMF(t)$ when T is clear from the context. Note that $IMF(t)$ is the set of subtrees in which the label (hash value) of each node is defined. In MONA, "Z is the set of roots of subtrees in $IMF(t)$" is described as IMFroot(t,Z) below.

```
pred defined(var1 s,t) = lefter(s,t) | rightmost(s,t) | s = t;
pred preIMF(var1 t, var2 Z) = all1 s: (s in Z => defined(s,t));
pred IMFroot(var1 t, var2 Z) =
  preIMF(t,Z) & all2 Y: (preIMF(t,Y) => lower_bound(Y,Z));
```

Remark 3. It is tempting to directly define $IMF(t, Z)$ as

```
pred IMF(var1 t, var2 Z) = all1 s: (s in Z <=> defined(s,t));
```

but, this makes Z in `IMF(t,Z)` run on infinite sets, i.e., beyond the scope of $WS2S$. For instance, $WS2S$ assumes that Z in `preIMF(t,Z)` runs on finite sets.

Note that the restriction to nodes in an incremental Merkle forest does not affect a left authentication path; however, it *does* affect a right authentication path. We say that for $s, t \in V(T)$, s is *lefter* than t (in T) if there exists $u \in V(T)$ such that $u.0 \leq s$ and $u.1 \leq t$ (which corresponds to `lefter(s,t` in Fig. 1).

Definition 4. *Let $s, t \in V(T)$ and let s be lefter than t. A The relative right authentication path $RA_{T,t}(s)$ of s wrt t is $RA_T(t) \cap IMF_T(t)$.*

We often omit T in $RA_{T,t}(s)$, if T is clear from the context. In MONA, $CA(t)$, $LA(t)$, $RA_t(s)$, and $pCls_T(t)$ are described as

```
# Authentication path
pred preCA(var1 t, var2 Y) = (all1 s : s.0 <= t => s.1 in Y) &
                             (all1 s : s.1 <= t => s.0 in Y);
pred CA(var1 t, var2 Y) =
  preCA(t,Y) & all2 Z : (preCA(t,Z) => Y sub Z);
# Left authentication path
pred preLA(var1 t, var2 Y) = all1 s : (s.1 <= t => s.0 in Y);
pred LA(var1 t, var2 Y) =
  preLA(t,Y) & all2 Z : (preLA(t,Z) => Y sub Z);
# Right authentication path
pred preRA(var1 s,t, var2 Y) =
  ex2 Z: (IMFroot(t,Z) &
          all1 u: ((u.0 <= s & ex1 v: (v in Z & v <= u.1))
                   => u.1 in Y));
pred RA(var1 s,t, var2 Y) =
  preRA(s,t,Y) & (all2 Z : preRA(s,t,Z) => Y sub Z);
# Path closure
pred pCls(var1 t, var2 X) =
  all1 s: (s in X <=> (s <= t | (ex2 Y: (CA(t,Y) & s in Y))));
```

Note that $PCls(t) = Cls(CA(t) \cup \{t\})$ where $Cls(X)$ is a *closure* operator is defined below.

Definition 5. *Let T be a Merkle tree. For $X \subseteq V(T)$, the closure $Cls(X)$ is the minimum set satisfying*

- $X \subseteq Cls(X)$, *and*
- *if both child nodes of t is in $Cls(X)$, then t is in $Cls(X)$.*

In MONA, "$Y = Cls(X)$" is described as `closure(X,Y)` (see Example 1).

For notational convenience, we define $LS(t) = LA(t) \cup \{t\}$ and $LSR_t(s) = LS(s) \cup RA_t(s)$, where $s, t \in V(T)$ such that s is lefter than t. In MONA, they are described as

```
pred LS(var1 t, var2 X) = ex2 Y: (LA(t,Y) & X = Y union {t});
pred LSR(var1 s,t, var2 Z) =
   ex2 X,Y: (LS(s,X) & RA(s,t,Y) & Z = X union Y);
```

4.2   Incremental Scheme for Optimal Slice Replication

Let $A = \{t_1, \cdots, t_k\}$ be a set of leaf nodes of T where one user requests to register events. An incremental Merkle forest $IMF(t_k)$ is also called a *temporal slice at t_k*. A *spatial slice of A* is the union of path closures of nodes in A (i.e., $\cup_{t_i \in A} pCls(t_i)$), and an *optimal slice* of A is the intersection of the temporal slice at t_k and the spatial slice of A (i.e., $(\cup_{t_i \in A} pCls(t_i)) \cap IMF(t_k)$).

A *path slice of t_i at t_j* (for $i < j$) is the intersection of the root path of t_i and the temporal slice at t_j, i.e., the fragment of the root path of t_i in which each hash value is known at t_j.

Fig. 2 shows the spatial/temporal/optimal slices of of $A = \{t_1, t_2, t_3, t_4, t_5\}$. The area surrounded by the dotted line is the spatial slice of A, the area surrounded by the thin line is the temporal slice at t_5, and their intersection is the optimal slice of A. The set of circled nodes is the left authentication path $LA(t_4)$ at t_4, the set of two boxed nodes is the right authentication path $RA(t_4)$, and $RA_{t_5}(t_4)$ consists of the node boxed with the line. The thick line that stems from t_4 shows the path slice of t_4 at t_5.

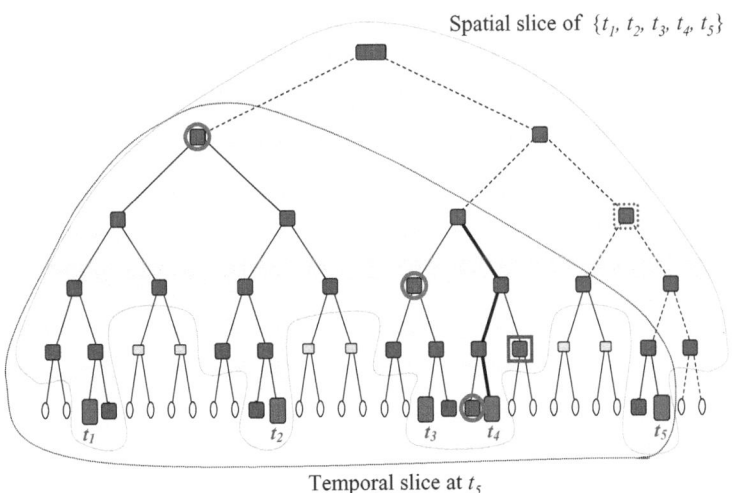

Spatial slice of $\{t_1, t_2, t_3, t_4, t_5\}$

Temporal slice at t_5

Fig. 2. Incremental Merkle forest

The protocol of our event-ordering system proceeds with the following transaction at each request from a user. The detailed algorithm is described in [5].

- For the request at t_1, return a pair $(\emptyset, LS(t_1))$.
- For a request at t_i with $1 < i \leq k$, return a pair $(RA_{t_i}(t_{i-1}), LS(t_i))$.

Theorem 1 guarantees that one can recover the optimal slice of A only from $LSR_{t_{i+1}}(t_i)$'s (for $1 \leq i < k$) and $LS(t_k)$, which are obtained by transactions of the protocol. This is called *completion*. Each message at a transaction is logarithmically small, and this gives an efficient optimal slice replication.

The event ordering system is designed to use this protocol between an auditor and a server, as well as an user and a server. One auditor is assumed to periodically register events $a_1, a_2, \cdots$. If these are sufficiently frequent, there will be an auditor's request a_j between one user's requests t_i and t_{i+1}. In such situation, the left authentication path $LS(a_j)$ has an overlap with the path slice of t_i at t_{i+1}. Then, the auditor can confirm that t_i occurs before a_j by comparing a hash value of an overlapping node at the user side and that at the auditor side.

The systems studied in [3,2,12,9,7] can perform similar event ordering, but only after a whole Merkle tree has been constructed; users need to collect information on right authentication paths from a server. Our system can perform the same thing *without* information from a server, because each participant keeps its own optimal slice replication. Thus, our system is safe from a server clash.

The optimal slice replication also enables participants to check each other without making inquiries to a server. We also assume that there are multiple auditors; this enables them to detect a malicious auditor even if a server halts. This setting guarantees single attack point free.

Theorem 2 guarantees the correctness of an efficient *incremental sanity check*, i.e., consistency among labels of nodes in $\{LSR_{t_{i+1}}(t_i) \mid 1 \leq i < k\} \cup \{LS(t_k)\}$ can be incrementally verified by weak consistency between each pair of neighbors $(LSR_{t_{i+1}}(t_i), LS(t_{i+1}))$ for $1 \leq i < k$.

During an incremental optimal slice replication, hash values may be computed at different moments even for the same node. The consistency among multiple definitions enables us (including a server itself) early detection of server errors and/or malicious attacks.

The proof of Theorem 1 (in Section 6) is fully performed by MONA, because it can be described in terms of nodes in a binary tree T. However, the proof of Theorem 2 (in Section 7) is only partially performed by MONA; the use of MONA is restricted to proofs of the main lemmata, which are essential for inductive steps in the full proof. MONA is fully automatic, thus its scope and ability are restricted. The main limitations here are:

- MONA lacks induction, and
- MONA lacks a description for equality between labels.

5   Characterization as a Pivoted Forest

Although the characterization given in this section is more than that needed in later sections (what we need in the proof of Lemma 8 is the fact that the union and the intersection of a forest of binary trees are again a forest of binary trees), this will clarify the perspective.

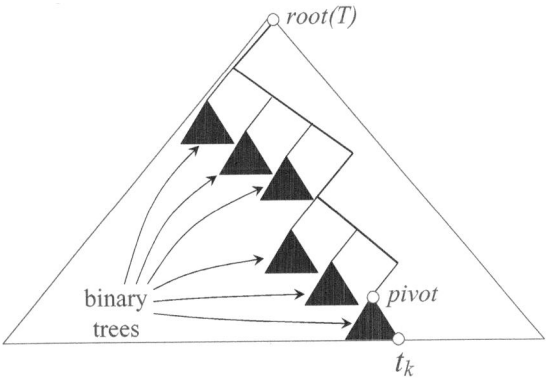

Fig. 3. $IMF(t_k)$ as a pivoted forest

Definition 6. *Let T be a Merkle tree. A node $t \in V(T)$ is a pivot if either*

- $t = root(T)$, or
- t is the left child of a node.

A forest $X \subseteq V(T)$ of binary trees is a pivoted forest (wrt a pivot t) if $X = \bigcup_{s \in LS(t)} X_s$ where X_s is a binary tree with $root(X_s) = s$.

In MONA, "X is a pivoted forest wrt t" is given as `pivoted_forest(t, X)`.

```
pred pivoted_forest(var1 t, var2 X) =
  all2 Y: (LS(t,Y) =>
          (lower_bound(X,Y) & Y sub X &
          all1 s: (s in Y =>
                  (all2 Z: (below(s,X,Z) => bintree_at(s,Z))))));
```

Let $A = \{t_1, \cdots, t_k\}$ be a set of leaf nodes of T such that t_i is lefter than t_{i+1}. We first show that an incremental Merkle forest $IMF(t_k)$ is a pivoted forest, as described in Fig. 3.

Lemma 1. *An incremental Merkle forest $IMF(t)$ is a pivoted tree where its pivot is the root of the rightmost component of $IMF(t)$.*

In MONA, Lemma 1 is described below and verified as `VALID`.

```
(IMFroot(t,X) & last(s,X)) => pivoted_forest(s,X);
```

Second, $Cls_T(LSR_{t_{i+1}}(t_i))$ for $1 \le i < k$, $Cls_T(LS(t_k))$, and their union are also pivoted forests.

Lemma 2.   *1. $Cls_T(LS(s))$ is a pivoted forest wrt u where u is the minimum node with $s = u.1 \cdots 1$.*
2. Let $s, t \in V(T)$ such that s is lefter than t and let $v = glb(s,t)$. Then, $Cls_T(LSR(s,t))$ is a pivoted forest wrt u where u is

- *the minimum node with* $t = u.1 \cdots 1$ *if* $t = v.1 \cdots 1$, *and*
- $v.0$ *otherwise.*

To describe Lemma 2, we prepare predicates that describe

- "$X = Cls_T(LS(s))$" as LSclosure(s,X),
- "$X = Cls_T(LSR_t(s))$" as LSRclosure(s,t,X),
- "u is a pivot of $Cls_T(LS(s))$" as LSpivot(s,u), and
- "u is a pivot of $Cls_T(LSR_t(s,u))$" as LSRpivot(s,t,u).

```
pred LSclosure(var1 t, var2 X) = ex2 Y: (LS(t,Y) & closure(Y,X));
pred LSRclosure(var1 s,t, var2 X) =
                              ex2 Y: (LSR(s,t,Y) & closure(Y,X));
pred LSpivot(var1 s,t) =
          (t = s | rightmost(t,s)) &
          (all1 u: ((u = s | rightmost(u,s)) => t <= u));
pred LSRpivot(var1 s,t,u) =
   all1 v: (((glb(s,t,v) & rightmost(v,t)) => LSpivot(t,u)) &
           ((glb(s,t,v) & ~rightmost(v,t)) => u = v.0));
```

In MONA, the statement of Lemma 2 is described as

```
(LS(s,X) & closure(X,Y) & LSpivot(s,t)) => pivoted_forest(t,Y);
(lefter(s,t) & LSR(s,t,X) & LSRpivot(s,t,u) & closure(X,Y))
                                => pivoted_forest(u,Y);
```

and is verified as VALID.

Lemma 3. *Let* X, Y *be pivoted forests wrt to pivots* s, t, *respectively. If* $s \in Y$, *then* $X \cup Y$ *(resp.* $X \cap Y$) *is a pivoted forest wrt* t *(resp.* s).

In MONA, this statement is described as

```
(pivoted_forest(s,X) & pivoted_forest(t,Y) & s in Y) =>
   ((pivoted_forest(t, X union Y) & pivoted_forest(s, X inter Y)));
```

and is verified as VALID.

Lemma 4. *If* s *is lefter than* t, *the pivot of* $Cls(LSR_t(s))$ *is in* $Cls(LS(t))$.

This is described as

```
(lefter(s,t) & LSRpivot(s,t,u) & LSclosure(t,X)) => t in X;
```

is verified as VALID by MONA. Thus, next Corollary is immediate.

Corollary 1. $(\cup_{1 \leq i < k} Cls_T(LSR_{t_{i+1}}(t_i))) \cup Cls_T(LS(t_k))$ *is a pivoted forest wrt* u *where* u *is the minimum node satisfying* $u.1 \cdots 1 = t_k$.

6   Completion

6.1   Completion in Incremental Merkle Forest

Intuitively, completion is a process to collect all nodes in an incremental Merkle forest such that their hash values can be computed only from issued certificates. Its correctness is, whether an optimal slice at the moment can be computed (Theorem 1) at a user side well a a server side.

Theorem 1. *(Theorem 1 in [5]) Let $A = \{t_1, t_2, \cdots, t_k\}$ be leaves in a Merkle tree T such that t_i is lefter than t_{i+1} for $1 \leq i < k$. Then,*

$$(\cup_{1 \leq i \leq k} pCls(t_i)) \cap IMF(t_k) = Cls(\cup_{1 \leq i < k} LSR_{t_{i+1}}(t_i) \cup LS(t_k)).$$

In the system, completion can be done efficiently by a right-to-left incremental closure operations. For notational convenience, we define a *path closure slice* $pClsSlc_t(s) = pCls(s) \cap IMF(t)$ where s is lefter than t. In MONA, $pClsSlc_t(s)$ is described as pClsSlc(s,t,X).

```
pred pClsSlc(var1 s,t, var2 X) =
  ex2 Y,Z: (pCls(s,Y) & IMFroot(t,Z) &
            all1 u:(u in X<=>(u in Y & ex1 v:(v in Z & v <= u))));
```

By the distributive law

$$(\cup_{1 \leq i \leq k} pCls(t_i)) \cap IMF(t_k) = \cup_{1 \leq i \leq k} (pCls(t_i) \cap IMF(t_k)),$$

the completion is enough to compute $pClsSlc_{t_k}(t_i)$ for $1 \leq i \leq k$. Then, the completion algorithm (guaranteed by Lemma 5) is:

1. Compute $pClsSlc_{t_k}(t_k)$, which is $Cls(LS(t_k))$.
2. When $pClsSlc_{t_k}(t_{i+1})$ (for $1 \leq i < k$) is computed, compute $pClsSlc_{t_k}(t_i)$, which is contained in $Cls(pClsSlc_{t_k}(t_{i+1}) \cup LSR_{t_{i+1}}(t_i))$.

Note that during computation, each step requires only logarithmic time.

Lemma 5. *Let $A = \{t_1, t_2, \cdots, t_k\}$ be leaves in a Merkle tree T such that t_i is lefter than t_{i+1} for $1 \leq i < k$. Then,*

- $pClsSlc_{t_k}(t_k) = Cls(LS(t_k))$.
- $pClsSlc_{t_k}(t_i) \subseteq Cls(pClsSlc_{t_k}(t_{i+1}) \cup LSR_{t_{i+1}}(t_i))$.

Section 6.2 will show a manual proof of Theorem 1, and Section 6.3 will show a formal proof by MONA for comparison of proofs by human and machine.

6.2   Proving Theorem 1 by Induction

Let $A = \{t_1, t_2, \cdots, t_k\}$ such that for t_i is lefter than t_{i+1} for $1 \leq i < k$.

Proof of Theorem 1 by Induction on k.   By induction on k. If $k = 1$, obvious. For $k > 1$, by induction hypothesis,

$$(\cup_{2 \leq i \leq k} pCls(t_i)) \cap IMF(t_k) = Cls(\cup_{2 \leq i < k} LSR_{t_{i+1}}(t_i) \cup LS(t_k)).$$

Let $u = glb(t_1, t_2)$. We denote the left child node of u by $u.0$ and the right child node by $u.1$, respectively. Since t_1 is lefter than t_2, $u.0 \leq t_1$ and $u.1 \leq t_2$. Since $u.0 \in LS(t_2)$, $u.0 \in IMF(t_k)$. Thus, $u.0 \in pClsSlc_{t_k}(t_1)$.

Let $X_1 = \{t \in pClsSlc_{t_k}(t_1) \mid u.0 \leq t\}$ and $X_2 = pClsSlc_{t_k}(t_1) \setminus X_1$. Then, $X_1 \subseteq Cls(LSR_{t_2}(t_1))$ and $X_2 \subseteq pClsSlc_{t_k}(u.0) \subseteq pClsSlc_{t_k}(t_2)$. Therefore

$$(\cup_{1 \leq i \leq k} pCls(t_i)) \cap IMF(t_k) \subseteq Cls(\cup_{1 \leq i < k} LSR_{t_{i+1}}(t_i) \cup LS(t_k)).$$

The opposite direction is obvious.   ∎

The proof of Lemma 5 can be performed similarly to that of Theorem 1.

6.3   Proving Theorem 1 by MONA

Let $A = \{t_1, t_2, \cdots, t_k\}$ such that t_i is lefter than t_{i+1} for $1 \leq i < k$. Define:

- "$X = \cup_{1 \leq i < k} LSR_{t_{i+1}}(t_i) \cup LS(t_k)$" is denoted by LSRunion(A,X).
- "$X = \cup_{1 \leq j \leq k} pCls(t_j)$" is denoted by spatial_slice(A,X).
- "$X = (\cup_{1 \leq j \leq k} pCls(t_j)) \cap IMF(t_k)$" is denoted by opt_slice(A,X).

```
pred LSRunion(var2 A,X) =
  all1 s: (s in X <=>
      ex1 t: ((ex1 u: ex2 Y: (next(t,u,A)&LSR(t,u,Y)&s in Y)) |
              (ex2 Z: (last(t,A) & LS(t,Z) & s in Z))));
pred spatial_slice(var2 A,X) =
  all1 s: (s in X <=>
          ex1 t: ex2 Y: (t in A & pCls(t,Y) & s in Y));
pred opt_slice(var2 A,X) =
  ex1 t: ex2 Y,Z: (last(t,A) & IMFroot(t,Y) & spatial_slice(A,Z) &
              all1 u: (u in X <=>
                      (u in Z & ex1 s: (s in Y & s <= u)))));
```

The statement of Theorem 1 is described as

```
(incomparable(A) & opt_slice(A,X) & LSRunion(A,Y) & closure(Y,Z))
                                                    => X = Z;
```

and is verified as VALID by MONA. The statement of Lemma 5 is described as

```
(LS(t,X) & pClsSlc(t,t,Y) & closure(X,Z)) => Y = Z;
(lefter(s,t) & (lefter(t,u) | t = u) & LSR(s,t,X) &
 pClsSlc(s,u,Y) & pClsSlc(t,u,Z) & closure(X union Z,C))=>Y sub C;
```

and also verified as VALID.

7   Incremental Sanity Check

7.1   Consistency

Let $A = \{t_1, t_2, \cdots, t_k\}$ such that for each pair (t_i, t_{i+1}) with $1 \leq i < k$, t_i is lefter than t_{i+1}. Upon completion, there may be nodes in a Merkle tree such that their labels are computed from different $LSR_{t_{i+1}}(t_i)$'s. If multiple computations of the label of a node coincide, this will be an indication of no system failures and/or no malicious attacks. This check of a server can be also performed by users and auditors, as well as self check by a server itself. This is called a *sanity check*; however, the naive way will be too expensive. We will show that an *incremental sanity check* that verifies weak consistency between each pair of neighbors $(LSR_{t_{i+1}}(t_i), LS(t_{i+1}))$ is enough.

To formalize the sanity check, we need to distinguish generated labels at each transaction; we associate a labeling (partial) function $\alpha_i : leaves(T) \rightarrow L$ to each pair $(LSR_{t_{i+1}}(t_i), LS(t_i))$. Note that during the sanity check, $g : L \times L \rightarrow L$ is fixed.

Definition 7. *Let $U_i \subseteq V(T)$ be a set of incomparable nodes in a Merkle tree T and let $\alpha_i : U_i \rightarrow L$ be a labeling (partial) function such that α_i is extended by $\alpha_i(t) = g(\alpha_i(t.0), \alpha_i(t.1))$ for $t \in cCLS_T(U_i)$.*

- *$\{(U_i, \alpha_i)\}$ is weakly consistent if $t \in cCLS_T(U_i) \cap cCLS_T(U_j)$ implies $\alpha_i(t) = \alpha_j(t)$.*
- *$\{(U_i, \alpha_i)\}$ is consistent if for each $t \in Cls_T(\cup U_i)$, $\alpha(t)$ is well-defined where*

$$\alpha(t) = \begin{cases} \alpha_i(t) & when\ t \in leaves(U_i) \\ g(\alpha(t.0), \alpha(t.1)) & when\ t \notin leaves(\cup U_i) \end{cases}$$

Note that $\alpha(t)$ may have multiple definitions, i.e., t may be a leaf node of some U_i and simultaneously t may be a non-leaf node of some U_j.

Theorem 2. *If $(LSR_{t_{i+1}}(t_i), \alpha_i)$ and $(LS(t_{i+1}), \alpha_{i+1})$ are weakly consistent for $1 \leq i < k$, $\{(LSR_{t_{i+1}}(t_i), \alpha_i) \mid 1 \leq i < k\} \cup \{(LS(t_k), \alpha_k)\}$ is consistent.*

Note that weak consistency between $(LSR_{t_{i+1}}(t_i), \alpha_i)$ and $(LS(t_{i+1}), \alpha_{i+1})$ can be checked quite efficiently. That is, by Lemma 2, 3, and 4, the set of minimum nodes in $Cls_T(LSR_{t_{i+1}}(t_i)) \cap Cls_T(LS(t_{i+1}))$ is $LS(u)$ where u is the pivot of $Cls_T(LS(t_{i+1}))$. In practice, we assume a collision-resistant one-way hash function g; thus, it is enough to check whether each hash value by α_i at a node in $LS(u)$ coincides with that by α_{i+1}.

7.2   Proving Weak Consistency

For the former half of the proof of Theorem 2, we will prove that *if $(LSR_{t_{i+1}}(t_i), \alpha_i)$ and $(LS(t_{i+1}), \alpha_{i+1})$ are weakly consistent for each i with $1 \leq i < k$, then $\{(LSR_{t_{i+1}}(t_i), \alpha_i) \mid 1 \leq i < k\} \cup \{(LSR_{t_k}(t_k), \alpha_k)\}$ is weakly consistent.*

Lemma 6. *Let* $s, t, u, v, w \in V(T)$ *such that* s *is lefter than* t, t *is lefter than or equal to* u, u *is lefter than or equal to* v, *and* v *is lefter than* w. *Then,*

1. $LSR_t(s) \cap LSR_w(v) \subseteq LS(u)$, *and*
2. $Cls_T(LSR_t(s)) \cap Cls_T(LSR_w(v)) \subseteq Cls_T(LS(u))$.

In MONA, the statement of Lemma 6 is described as

```
(lefter(s,t) & (t = u | lefter(t,u)) & (u = v | lefter(u,v)) &
 lefter(v,w) & LSR(s,t,X) & LS(u,Y) & LSR(v,w,Z))
                                          => X inter Z sub Y;
(lefter(s,t) & (t = u | lefter(t,u)) & (u = v | lefter(u,v)) &
 lefter(v,w) & LSRclosure(s,t,X) & LSclosure(u,Y)
                         & LSRclosure(v,w,Z)) => X inter Z sub Y;
```

and is verified as VALID.

Lemma 7. *If* $(LSR_{t_{i+1}}(t_i), \alpha_i)$ *and* $(LS(t_{i+1}), \alpha_{i+1})$ *are weakly consistent for each* i *with* $1 \le i < k$, *then* $\{(LSR_{t_{i+1}}(t_i), \alpha_i) \mid 1 \le i < k\} \cup \{(LS(t_k), \alpha_k)\}$ *is weakly consistent.*

Proof. By induction on k. If $k = 1$, obvious. Assume $k > 1$ and the statement holds for $k - 1$. Let $X = (\bigcup_{1 \le i < k-1} Cls_T(LSR_{t_{i+1}}(t_i)) \cup Cls_T(LS(t_{k-1}))$.

It is enough to consider the intersection $X_1 = X \cap Cls_T(LSR_{t_k}(t_{k-1}))$ and $X_2 = X \cap Cls_T(LS(t_k))$.

From Lemma 6, $X_1, X_2 \subseteq Cls_T(LS(t_{k-1}))$. Since $(LSR_{t_k}(t_{k-1}), \alpha_{k-1})$ and $(LS(t_k), \alpha_k)$ are weakly consistent, Lemma is proved. ∎

Note that MONA cannot verify Lemma 7, because it cannot describe the equality between labels.

7.3   Proving Consistency

For the latter half of the proof of Theorem 2, we will prove that *if*

$$\{(LSR_{t_{i+1}}(t_i), \alpha_i) \mid 1 \le i < k\} \cup \{(LS(t_k), \alpha_k)\}$$

is weakly consistent, they are consistent. This complete the proof of Theorem 2.

Lemma 8. *If* $\{(LSR_{t_{i+1}}(t_i), \alpha_i) \mid 1 \le i < k\} \cup \{(LS(t_k), \alpha_k)\}$ *is weakly consistent, they are consistent.*

For notational convenience, we define

$$LSR_A(t_i) = \begin{cases} LSR_{t_{i+1}}(t_i) & \text{for } 1 \le i < k \\ LS(t_k) & \text{for } i = k \end{cases}$$

for $A = \{t_1, \cdots, t_k\}$.

Proof. Let $X = (\cup_{1 \leq i < k} Cls_T(LSR_{t_{i+1}}(t_i))) \cup Cls_T(LS(t_k))$. From Corollary 1, X is a pivoted forest; thus, for each $t \in X$, $t.0 \in X$ if and only if $t.1 \in X$.

For each $t \in X$, we will prove that the labeling function $\alpha : V(X) \to L$ is well-defined by induction on the size of $X \cap T_t$ where $T_t = \{s \in V(T) \mid t \leq s\}$. If $|X \cap T_t| = 1$, this means $t \in LSR_A(t_i)$ or $t \notin Cls_T(LSR_A(t_i))$ for each i. Since $LSR_A(t_i)$'s are weakly consistent, $\alpha(t)$ is well-defined.

Assume $|X \cap T_t| > 1$. Since $X \cap T_t$ is a forest of binary trees, $t.0, t.1 \in X \cap T_t$. If $t \in Cls_T(LSR_A(t_i))$, either $t.0$, $t.1 \in Cls_T(LSR_A(t_i))$ or $t \in LSR_A(t_i)$.

Since $|X \cap T_{t.0}|$, $|X \cap T_{t.1}| < |X \cap T_t|$, induction hypothesis implies that $\alpha(t.0)$ and $\alpha(t.1)$ are well-defined. Let $I_0 = \{i \mid t.0, t.1 \in Cls_T(LSR_A(t_i))\}$ and $I_1 = \{i \mid t \in LSR_A(t_i)\}$. Since $|X \cap T_t| > 1$, $I_0 \neq \emptyset$.

Let $j \in I_0$; then $\alpha_j(t) = g(\alpha(t.0), \alpha(t.1))$. Weakly consistency of $LSR_A(t_i)$'s implies that $\alpha_j(t) = \alpha_i(t)$ for each $i \in I_1$. Thus, $\alpha(t) = \alpha_j(t)$ is well-defined. ∎

Theorem 2 is immediate from Lemma 7 and 8. Note that MONA cannot verify Lemma 8, because it cannot describe the equality between labels.

8   Conclusion

This paper proved two basic properties

1. correctness of completion
2. correctness of incremental sanity check

of an incremental Merkle forest, which is used in the event ordering system [5] developed by NTT. Especially, this paper is the first to prove (2) the correctness of an incremental sanity check.

During the proofs, we mainly used the automata-based theorem prover MONA [1]. Although MONA can treat only decidable properties, this does not mean that its use is easy. We need to find suitable formalization and key lemmata, which are essential in the whole proof and still provable by MONA. For instance, during the use of MONA, we have also simplified the manual proof of Theorem 1 (the original proof, by induction on the homogeneous depth of a Merkle tree, takes more than 1 page in two-column style).

Another notable example of $WSnS$ is an optimal reduction strategy of a strongly sequential term rewriting system [6]. This is known to be intricate; however it was clearly re-described in terms of $WSnS$ [4].

A drawback is that an automata-based prover does not give a deductive proof. Thus, incomplete descriptions may be easily neglected; instead, we often found them by test data. On the other hand, it is extremely powerful for detecting oversights and gaps in a proof draft, which are often found in tentative proof goals. At the moment, support for theorem prover is not enough; but, we feel it is possible for theorem provers to be an assistance even for constructing a new proof.

For future work, we are planning:

- full formal proof of Theorem 2 by combining MONA and an induction-based prover *Isabelle/HOL* [10].
- proofs for more detailed properties of the event ordering system.

Acknowledgments

This research is partially supported by Special Coordination Funds for Promoting Science and Technology and Scientific Research on Priority Area (No. 16016241) by Ministry of Education, Culture, Sports, Science and Technology, PRESTO by Japan Science and Technology Agency, and Kayamori Foundation of Informational Science Advancement.

References

1. MONA project. http://www.brics.dk/mona/.
2. C. Adams and et al. RFC3161, internet X.509 public key infrastructure time-stamp protocol (TSP). Technical report, IETF, 2001.
3. A. Buldas, H. Lipmaa, and B. Schoenmakers. Optimally efficient accountable time-stamping. In *Proc. 3rd International Workshop on Practice and Theory in Public Key Cryptography (PKC 2000)*, pages 293–305. Springer-Verlag, 2000. Lecture Notes in Computer Science, Vol.1751.
4. H. Comon. Sequentiality, monadic second-order logic and tree automata. *Information and Computation*, 157(1 & 2):25–51, 2000. Previously presented in *Proc. 10th IEEE Symposium on Logic in Computer Science*, pages 508–517, 1995.
5. E. Horita, S. Ono, and H. Ishimoto. Implementation mechanisms of scalable event-ordering system without single point of attack. Technical report, IEICE SIG-ISEC, 11 2004. *in Japanese.*
6. G. Huet and J.-J. Lévy. Computations in orthogonal rewriting systems I,II. In *Computational Logic: Essays in Honor of Alan Robinson*, pages 395–443. MIT Press, 1991. Previous version: Report 359, INRIA, 1979.
7. M. Jakobsson, F.T. Leighton, S. Micali, and M. Szydlo. Fractal Merkle tree representation and traversal. In *Proc. Topics in Cryptology - CT-RSA 2003, The Cryptographers' Track at the RSA Conference 2003*, pages 314–326. Springer-Verlag, 2003. Lecture Notes in Computer Science, Vol.2612.
8. R.C. Merkle. *Secrecy, Authentication, and Public Key Systems*. UMI Research Press, 1982. also appears as Ph.D thesis at Stanford University, 1979.
9. S. Michael. Merkle tree traversal in log space and time. In *Proc. International Conference on the Theory and Applications of Cryptographic Techniques, Advances in Cryptology - EUROCRYPT 2004*, pages 541–554. Springer-Verlag, 2004. Lecture Notes in Computer Science, Vol.3027.
10. T. Nipkow, L.C. Paulson, and M. Wenzel. *Isabelle/HOL, A proof assistant for higher-order logic*. Springer-Verlag, 2002. Lecture Notes in Computer Science, Vol.2283.
11. W. Thomas. Automata on infinite objects. In J. van Leeuwen, editor, *Handbook of Theoretical Computer Science*, volume B, chapter 4, pages 133–192. Elsevier, 1990.
12. J. Villemson. *Size-efficient interval time stamps*. PhD thesis, University of Tartu, Estonia, 2002.

Computer Search for Counterexamples to Wilkie's Identity*

Jian Zhang

Laboratory of Computer Science, Institute of Software,
Chinese Academy of Sciences, P.R. China
zj@ios.ac.cn

Abstract. Tarski raised the High School Problem, i.e., whether a set of 11 identities (denoted by HSI) serves as a basis for all the identities which hold for the natural numbers. It was answered by Wilkie in the negative, who gave an identity which holds for the natural numbers but cannot be derived from HSI. This paper describes some computer searching efforts which try to find a small model of HSI rejecting Wilkie's identity. The experimental results show that such a model has at least 11 elements. Some experiences are reported, and some issues are discussed.

1 Introduction

One learns in high school that the following identities are true in the set $\mathbf{N}$ of positive integers:

$$x + y = y + x$$
$$x + (y + z) = (x + y) + z$$
$$x * 1 = x$$
$$x * y = y * x$$
$$x * (y * z) = (x * y) * z$$
$$x * (y + z) = (x * y) + (x * z)$$
$$1^x = 1$$
$$x^1 = x$$
$$x^{y+z} = x^y * x^z$$
$$(x * y)^z = x^z * y^z$$
$$(x^y)^z = x^{y*z}$$

The above set of identities is denoted by HSI. Tarski's *High School Problem* is: does HSI serve as a basis for all the identities of $\mathbf{N}$? In other words, can any identity which holds for the natural numbers be derived from HSI using equational reasoning? For more details about this problem, see [2].

In 1981, Wilkie gave the following identity, denoted by $W(x, y)$, and showed that it holds in $\mathbf{N}$, but cannot be derived from HSI:

* Supported by the National Science Fund for Distinguished Young Scholars of China (grant No. 60125207).

R. Nieuwenhuis (Ed.): CADE 2005, LNAI 3632, pp. 441–451, 2005.

$$(P^x + Q^x)^y * (R^y + S^y)^x = (P^y + Q^y)^x * (R^x + S^x)^y$$

where $P = 1+x$, $Q = 1+x+x*x$, $R = 1+x*x*x$, and $S = 1+x*x+x*x*x*x$.

Wilkie's original proof was purely syntactic. Alternatively we can try to build a finite model of HSI in which $W(x,y)$ does not hold. In fact, Gurevič [4] constructed such a model which has 59 elements. Models like this are called *G-algebras* later in [1]. In this paper, such a model will be called a Gurevič-Burris Algebra (GBA). A GBA serves as a counterexample showing that the Wilkie identity does not follow from HSI. In the sequel, we use the constants a and b to denote a pair of elements in a GBA such that $W(a,b)$ does not hold.

A question naturally arises: what is the smallest GBA? It is proved in [1] that the size of any GBA is at least 7 and that a 15-element GBA exists. In [5], the lower bound is increased to 8, and the upper bound is decreased to 14. More recently, Burris and Yeats found a 12-element GBA [2].

All the above lower bounds are proved mathematically. Alternatively, we may also try to find a small GBA using computers. This sounds quite interesting, and also feasible, given that there have been significant advances in computer hardware and search algorithms. Over the past 10 years, we have attempted to search for a small GBA for several times. But it is a bit disappointing that no GBA was found. However, since the search algorithm is complete, the experimental results still tell us something (i.e., GBAs of certain sizes do not exist). This paper gives a brief summary of our experiments.

2   Search Programs

In the early 1990's, we developed a general-purpose search program for finding finite models, called FALCON [11,13]. Typically its input consists of a set of equations E and a positive integer n. The input may also include an inequation of the form s != t, where s and t are terms. The output of FALCON is an n-element model of E, if there is such a model. Usually we assume that the domain of the model is $D_n = \{0, 1, \ldots, n-1\}$. Such a model is an interpretation of the function symbols in D_n, such that every equation (inequation) holds.

SEM [14] can be regarded as a successor of FALCON. It is more efficient, and it can accept non-unit first-order clauses. To use SEM to find a 4-element GBA, we may give it the following input file:

```
4.

s(x,y) = s(y,x).
s(x,s(y,z)) = s(s(x,y),z).
p(x,1) = x.
p(x,y) = p(y,x).
p(x,p(y,z)) = p(p(x,y),z).
p(x,s(y,z)) = s(p(x,y),p(x,z)).
e(1,x) = 1.
e(x,1) = x.
```

```
e(x,s(y,z)) = p(e(x,y),e(x,z)).
e(p(x,y),z) = p(e(x,z),e(y,z)).
e(e(x,y),z) = e(x,p(y,z)).

c = p(a,a).
P = s(1,a).
Q = s(P,c).
R = s(1,p(a,c)).
S = s(s(1,c),p(c,c)).
p(e(s(e(P,a),e(Q,a)),b),e(s(e(R,b),e(S,b)),a)) !=
p(e(s(e(P,b),e(Q,b)),a),e(s(e(R,a),e(S,a)),b)).
```

Here the first line denotes the size of the model, and the rest of the file gives the formulas (in clausal form). The function symbols s, p and e denote summation, product and exponentiation, respectively. The variable x, y and z are assumed to be universally quantified.

In our recent experiments, we also used Mace4 [7], another general-purpose model searching program.

All of these programs work by backtracking search, which is exhaustive in nature. If such a program terminates without finding a model, it means that there is no model of the given size (assuming that the program is correct).

Let us briefly explain the search process. Suppose $n = 4$. Then we need to find suitable values for all the constants and these terms:

$$s(0,0), \ s(0,1), \ s(0,2), \ s(0,3), \ s(1,0), \ \ldots, \ p(0,0), \ p(0,1), \ \ldots, \ e(3,3).$$

Each of them can take a value from the domain D_4, i.e., $\{ 0, 1, 2, 3 \}$. Some of the terms should get certain values, e.g., $e(1,0) = e(1,1) = 1$. But for other terms, we can only "guess" their values. For instance, $s(0,0)$ may take either of the four values. So we assign each value to $s(0,0)$ and check if there is any contradiction. Similarly for other terms and constants.

3   Useful Mathematical Results

To show that a GBA has at least seven elements, Burris and Lee [1] first establish some lemmas. They are basically properties of elements x and y in an HSI-algebra which guarantee $W(x,y)$ holds in the algebra. Some of the lemmas (and theorems) are also helpful to computer search.

Following Burris and Lee (page 154 of [1]), we use "HSI $\vdash \Sigma \to W(x,y)$" to denote that

$$\text{HSI} \ \vdash \ \forall x \forall y \, [\Sigma \to W(x,y)].$$

Here Σ is an identity. We also use "$u \mid v$" to denote that $\exists w \, (v = u * w)$.

It is shown by Lee (Lemma 8.7 of [1]) that   HSI $\vdash x \mid y \to W(x,y)$. Thus in a GBA, $a \mid b$ does not hold. In other words, for any x,

$$(L1) \qquad b \neq a * x.$$

It is also shown by Lee (Lemma 8.13 of [1]) that, if Σ is one of the following conditions:

$$P \mid Q$$
$$Q \mid P$$
$$R \mid S$$
$$S \mid R$$

then we have   HSI $\vdash \Sigma \to W(x, y)$.

Similarly, we can conclude from the first 4 conditions that, for any x,

$$(L2) \quad Q \neq P * x$$
$$(L3) \quad P \neq Q * x$$
$$(L4) \quad S \neq R * x$$
$$(L5) \quad R \neq S * x$$

As in [1,2], we call the following elements of an HSI-algebra *integers*:

$$1, \quad 2 = 1 + 1, \quad 3 = 2 + 1, \quad \ldots$$

In [1], it is proved that a GBA must have at least 3 integers. Moreover, if $W(x, y)$ fails at a, b, then a and b should be distinct non-integer elements. Thus in a GBA, the elements $1, 2, 3, a, b$ are different from each other.

In addition, Lemma 8.20 of [1] tells us that, if Σ is one of the following conditions:

$$1 + x = 1$$
$$2 + x = 1$$
$$x + x = 1$$
$$x * x = 1$$
$$1 + x * x = 1$$
$$x * x * x = 1$$
$$1 + x = x$$
$$2 + x = x$$
$$x + x = x$$
$$x * x = x$$
$$1 + x * x = x$$
$$2 + x = 1 + x$$
$$x * x = 1 + x$$
$$x * x * x = 1 + x$$
$$x * x = 2 + x$$
$$x * x = x + x$$
$$1 + x * x = x * x$$

then we have HSI $\vdash$ $\Sigma \to W(x,y)$. Thus in a GBA,

$$
\begin{array}{ll}
(M01) & 1 + a \neq 1 \\
(M02) & 2 + a \neq 1 \\
(M03) & a + a \neq 1 \\
(M04) & a * a \neq 1 \\
(M05) & 1 + a * a \neq 1 \\
(M06) & a * a * a \neq 1 \\
(M07) & 1 + a \neq a \\
(M08) & 2 + a \neq a \\
(M09) & a + a \neq a \\
(M10) & a * a \neq a \\
(M11) & 1 + a * a \neq a \\
(M12) & 2 + a \neq 1 + a \\
(M13) & a * a \neq 1 + a \\
(M14) & a * a * a \neq 1 + a \\
(M15) & a * a \neq 2 + a \\
(M16) & a * a \neq a + a \\
(M17) & 1 + a * a \neq a * a
\end{array}
$$

Finally, let us quote Jackson (Lemma 1 in [5]):

Let $y = n + n_1 x + n_2 x^2 + \ldots + n_m x^m$, where $n, n_1, n_2, \ldots, n_m$ are integers. Then HSI $\vdash W(x,y)$.

Thus, to search for a GBA, we may add the following lemmas:

$$
\begin{array}{ll}
(J11) & b \neq 1 + 1 * a \\
(J12) & b \neq 1 + 2 * a \\
(J21) & b \neq 2 + 1 * a \\
& \ldots
\end{array}
$$

In mathematical terms, b can not be in the core generated by a.

4   Search for 7-Element and 8-Element GBAs

4.1   Size 7 by FALCON

In 1993 and 1994, while developing the finite algebra search program FALCON, the author chose the HSI problem as an "exercise" for the program. Since the program deals mainly with equations, we started from some partial models of HSI, and tried to extend each of them to a GBA.

For the 7-element case, there are 3 partial models initially:

(1) The integers are 0, 1, 2, with $s(0,0) = 1$, $s(0,1) = s(1,0) = 2$; and $a = 3$, $b = 4$.
(2) The integers are 0, 1, 2, 3, with $s(0,0) = 1$, $s(0,1) = s(1,0) = 2$, $s(1,1) = s(0,2) = s(2,0) = 3$; and $a = 4$, $b = 5$.

(3) The integers are 0, 1, 2, 3, 4, with $s(0,0) = 1$, $s(0,1) = s(1,0) = 2$, $s(0,2) = s(1,1) = s(2,0) = 3$, and $s(0,3) = s(1,2) = s(2,1) = s(3,0) = 4$; and $a = 5$, $b = 6$.

Several lemmas in $(M01)$–$(M17)$ are added to the input of FALCON. In addition, we established our own lemmas. When a partial model satisfies certain conditions, it will not be extended, because these conditions guarantee that Wilkie's identity holds. Some heuristics are also used in the search. For example, when choosing an unknown, we not only consider the number of possible values for it, but also consider the number of new assignments which may be generated when the unknown is assigned a value. More details are given in [11].

The search for a 7-element GBA was completed on SUN SPARCstations. It lasted for several days.

4.2   Size 8 by SEM

The search for an 8-element GBA was completed on HP workstations running UNIX, using SEM [14]. It also lasted for several days. As in the case of size 7, we started from many partial models and tried to extend each of them using SEM. The initial partial models were obtained by hand. The search was performed in 1995, during the author's visit to the University of Iowa, U.S.A.

5   Recent Experimental Results

In August 2004, we completed the search for 9-element and 10-element GBAs. No model was found. The searches were mainly performed on a desktop (Dell Optiplex GX270: Pentium 4, 2.8 GHz CPU, 2G memory) running RedHat Linux. Other similar machines were also used when searching for an 11-element GBA. We used both SEM [14] and Mace4 [7] (mace4-2004-C).

5.1   Problem Formulation

In addition to the lemmas $(M01)$–$(M17)$, we also added the lemmas $(L1)$, $(L2)$, $(L3)$, $(L4)$, $(L5)$. The basic formulation is given in the Appendix. But in individual searches, we used more formulas representing different partial models.

5.2   The Case of Size 9

This case is divided further into several subcases:

(1) There are 3 integers: 1, 2, 3.
(2) There are 4 integers: 1, 2, 3, 4.
    (a) $s(4,1) = 4$.
    (b) $s(4,1) = 3$.
    (c) ...
(3) There are 5 integers: 1, 2, 3, 4, 5.
(4) There are 6 integers: 1, 2, 3, 4, 5, 6.
(5) There are 7 integers: 1, 2, 3, 4, 5, 6, 7.

5.3   The Case of Size 10

The "search tree" is roughly like the following:

(1) There are 3 integers: 1, 2, 3.
(2) There are 4 integers: 1, 2, 3, 4.
    (a) s(4,1) = 4.
        i. s(0,0) = 2.
        ii. s(0,0) = 6.
    (b) s(4,1) = 3.
    (c) ...
(3) There are 5 integers: 1, 2, 3, 4, 5.
(4) There are 6 integers: 1, 2, 3, 4, 5, 6.
(5) There are 7 integers: 1, 2, 3, 4, 5, 6, 7.
(6) There are 8 integers: 1, 2, 3, 4, 5, 6, 7, 8.

5.4   The Case of Size 11

The search for an 11-element counterexample has not been completed yet. So far we have only eliminated some subcases. Our conclusion is that, if a GBA of size 11 exists, it should have 3 or 4 integers.

This case is much more difficult than the search for smaller models. For example, when the size is 10, the subcase of 6 integers takes SEM a little more than 10 minutes; but when the size is 11, the subcase of 6 integers takes SEM more than 150 hours.

As previously, we examine many partial models and try to extend each of them. The longest completed single run of Mace4 lasted for about 11 days. For some subcases, Mace4 did not complete the search after running for two weeks, and the process was killed.

6   Some Experiences

In addition to SEM and Mace4, we have tried other programs such as Paradox [3] and Gandalf [10]. A problem with Gandalf is that it starts the search from 1-element model to 2-element model and so on. There is no convenient way to ask it to search for 9-element model only, for example.

The GBA problem is highly "first-order". It seems that SAT-based tools are not so advantageous as on other problems. The running times of MACE2 [6] are not satisfactory. This may be due to the inefficiency of its internal SAT solver. The performance of Paradox (version 1.0-casc) is similar to that of SEM.

The problem is also highly "equational", since most of the (ground) clauses are equations/identities. We found that it is better to turn off the negative propagation rules in SEM and Mace4, as shown in Table 1. Informally speaking, such rules try to deduce from negated equations or try to generate negated equations, while ordinary (positive) propagation rules deduce equations from equations.

Table 1. Negative Propagation

size	program	neg_prop	time
7	SEM	Yes	0.24
7	SEM	No	0.18
8	SEM	Yes	7.96
8	SEM	No	5.32
7	Mace4	Yes	0.64
7	Mace4	No	0.39
8	Mace4	Yes	25.14
8	Mace4	No	11.41

Table 2. Using Lemmas

size	program	lemma	time
7	SEM	Yes	0.24
7	SEM	No	19.36
7	Mace4	Yes	0.64
7	Mace4	No	43.76

The lemmas (esp. short lemmas) are quite useful in general. Table 2 compares the performances of SEM and Mace4, when the input has or does not have the lemmas $(L1)$–$(L5)$. In both cases, the lemmas $(M01)$–$(M17)$ are included in the input file, and negative propagation is not used.

In Table 1 and Table 2, "size" denotes the size of the model. The running times ("time") are given in seconds. The data were obtained on a Dell Optiplex GX270 (Pentium 4, 2.8 GHz CPU, 2G memory, RedHat Linux).

From Table 2 we see that the five lemmas can greatly reduce the search time. We briefly described the search process at the end of Sec. 2. During the search, some terms get values, while others do not. Let us look at lemma $(L1)$ which says, for any x, $b \neq a * x$. Usually we can assume that $a = 0$ and the integers are $1, 2, 3, \ldots$ With the above lemma, if we know the value of b, we can safely exclude this value from consideration when trying to find a value for $p(0,0)$. Similarly for $p(0,1)$, $p(0,2)$ and so on. Thus the number of choices is reduced, and the search time is reduced too. We note that, about ten years ago, Slaney et $al.$ [9] also used some extra constraints when solving the quasigroup problems.

Although the lemmas are generally helpful, there are also some exceptions. For example, consider the subsubcase in 11-element GBA search where there are 7 integers and $7 + 1 = 7$. If we add the Jackson lemmas $(J11, \ldots, J77)$, the running time of Mace4 is 257 seconds. But without these lemmas, the running time is 144 seconds.

Electricity outage occurred once (in a weekend). Fortunately, SEM did not run for too long before it was interrupted. We think it wise to set a time limit on each run. The '-t' and '-f' options of SEM can be helpful. When the time limit is reached, SEM saves the current path of the search tree in a file called "_UF". From this file, we can restart the search later. We can also see the (approximately maximum) depth of the search tree. The deepest search path in our _UF files has 61 branches/decisions (i.e., assignments to cells).

7   Concluding Remarks

Search for a small GBA is an interesting problem for mathematicians. In this paper, we summarize some computer search results, which show that the smallest counterexample to Wilkie's identity has at least 11 elements.

Of course, this conclusion is not proved mathematically. It is possible that the programs have some bugs, or the user (myself) made some errors. Like Slaney [8], we believe that it is valuable to double check the results using other programs. This may increase their reliability.

The following picture shows some recent attempts to increase the lower bound and decrease the upper bound on the size of the smallest GBA:

		[5]								
Refs.		[1] [11] [14]	*	*	[2]		[5] [1]			
bound		7 8 9	10	11	12	13	14	15		

In the picture, related papers and documents are given above the line. For example, it was established in [14] that the lower bound is 9 (because there is no 8-element model). The "*" denotes this paper. When the lower bound and the upper bound meet, the problem will be solved.

In addition to its implications for mathematics, GBA search should also stimulate research in automated reasoning. It can be a challenging problem for first-order model searching programs, SAT solvers and more general constraint solvers. It was suggested as a benchmark problem more than 10 years ago [12]. Due to the advancement in hardware, improvements on algorithms and data structures, as well as useful mathematical results, we can now easily solve problem instances that were very difficult in the early 1990's. We hope that they will become even easier in the near future.

There are still some issues which deserve further investigation. First of all, are there more "efficient" specifications of the problem? Of course, this depends on which tool you are using. But it appears that certain short lemmas are always helpful. Secondly, can we prove the correctness of the search results (rather than the correctness of the search programs, which is very difficult)? Finally, we should pay more attention to the "engineering" aspects of model searching, when the problem is very difficult.

Acknowledgements

The author is very grateful to the anonymous referees for their critical comments and constructive suggestions.

References

1. S. Burris and S. Lee, Small models of the high school identities, *Int'l J. of Algebra and Computation* 2: 139–178, 1992.
2. S. Burris and K. Yeats, The saga of the high school identities, *Algebra Universalis*, to appear.
3. K. Claessen and N. Sörensson, New techniques that improve MACE-style finite model finding. In: *Model Computation – Principles, Algorithms, Applications*, CADE-19 Workshop W4, Miami, Florida, USA, 2003.
4. R. Gurevič, Equational theory of positive numbers with exponentiation. *Proc. Amer. Math. Soc.* 94: 135–141, 1985.
5. M.G. Jackson, A note on HSI-algebras and counterexamples to Wilkie's identity, *Algebra Universalis* 36: 528–535, 1996.
6. W. McCune, MACE 2.0 reference manual and guide, Technical Memorandum ANL/MCS-TM-249, Argonne National Laboratory, Argonne, IL, USA, May 2001.
7. W. McCune, *Mace4 Reference Manual and Guide*, Technical Memorandum No. 264, Argonne National Laboratory, Argonne, IL, USA, 2003.
8. J.K. Slaney, The crisis in finite mathematics: Automated reasoning as cause and cure, *Proc. 12th Int'l Conf. on Automated Deduction (CADE-12)*, 1–13, 1994.
9. J. Slaney, M. Fujita and M. Stickel, Automated reasoning and exhaustive search: Quasigroup existence problems, *Computers and Mathematics with Applications* 29(2): 115–132, 1995.
10. T. Tammet, Finite model building: improvements and comparisons In: *Model Computation – Principles, Algorithms, Applications*, CADE-19 Workshop W4, Miami, Florida, USA, 2003.
11. J. Zhang, *The Generation and Applications of Finite Models*, PhD thesis, Institute of Software, Chinese Academy of Sciences, Beijing, 1994.
12. J. Zhang, Problems on the generation of finite models, *Proc. of the 12th Int'l Conf. on Automated Deduction (CADE-12)*, LNAI Vol. 814, 753–757, 1994.
13. J. Zhang, Constructing finite algebras with FALCON, *J. Automated Reasoning* 17(1): 1–22, 1996.
14. J. Zhang and H. Zhang, SEM: a System for Enumerating Models. *Proc. of the 14th Int'l Joint Conf. on Artif. Intel. (IJCAI-95)*, 298–303, 1995.

Appendix

The SEM input for a 7-element GBA search can be like the following:

```
7.

s(x,y) = s(y,x).
s(x,s(y,z)) = s(s(x,y),z).
p(x,1) = x.
p(x,y) = p(y,x).
p(x,p(y,z)) = p(p(x,y),z).
p(x,s(y,z)) = s(p(x,y),p(x,z)).
e(1,x) = 1.
e(x,1) = x.
```

```
e(x,s(y,z)) = p(e(x,y),e(x,z)).
e(p(x,y),z) = p(e(x,z),e(y,z)).
e(e(x,y),z) = e(x,p(y,z)).
c = p(0,0).
P = s(1,0).
Q = s(P,c).
R = s(1,p(0,c)).
S = s(s(1,c),p(c,c)).
p(e(s(e(P,0),e(Q,0)),b),e(s(e(R,b),e(S,b)),0)) !=
p(e(s(e(P,b),e(Q,b)),0),e(s(e(R,0),e(S,0)),b)).

2 = s(1,1).
b != 0.
b != 1.
b != 2.
b != p(0,x).
P != p(Q,x).
Q != p(P,x).
R != p(S,x).
S != p(R,x).
s(1,0) != 1.
s(2,0) != 1.
s(0,0) != 1.
     c != 1.
s(1,c) != 1.
p(c,0) != 1.
s(1,0) != 0.
s(2,0) != 0.
s(0,0) != 0.
     c != 0.
s(1,c) != 0.
s(2,0) != s(1,0).
     c != s(1,0).
p(c,0) != s(1,0).
     c != s(2,0).
     c != s(0,0).
s(1,c) != c.
```

Here we assume that a is the element 0. Given the above input, SEM and Mace4 can complete the search very quickly when the size is 7 or 8. See Table 1.

KRHyper - In Your Pocket
System Description

Alex Sinner and Thomas Kleemann

University of Koblenz-Landau, Department of Computer Science,
D-56070 Koblenz, Germany
{sinner, tomkl}@uni-koblenz.de

Abstract. Pocket KRHyper is a reasoning system for Java-enabled mobile devices. The core of the system is a first order theorem prover and model generator based on the hyper tableau calculus.

The development of Pocket KRHyper was motivated by the arising need for reasoning on mobile devices for mobile semantic web applications. To satisfy this need, a Description Logics (DL) interface is provided, which allows DL reasoning by transforming DL Expressions into first order clausal logic.

1 Introduction

Pocket KRHyper is a Java 2 Mobile Edition (J2ME[1]) software library for automated reasoning. It can be embedded in applications on mobile devices like PDAs, smartphones, and cell phones, but can also be used in Java 2 Standard Edition (J2SE[2]) applications.

The reasoning engine is based on the hyper tableau calculus [5] and can be considered as a resource optimized version of the KRHyper [16] system. The original KRHyper system has been used successfully in a multitude of knowledge management applications (See [4,6,7]), but is not designed for running on mobile devices.

Pocket KRHyper was developed explicitly for use in mobile devices with modest resources, and is actually the first reasoner for mobile devices able to tackle useful first order logic problems.

1.1 The Need for Mobile Reasoning

The need for mobile reasoning has arisen through the development of mobile semantic web services (See [13,15]). To provide reasoning support on a mobile device, the easiest solution was to connect to a special reasoning server on the Internet. Unfortunately, mobile Internet connections are usually quite expensive, and even if bandwidth is expected to become cheaper for mobile devices, there are some other good reasons to do the reasoning on your own device.

---

[1] http://java.sun.com/j2me
[2] http://java.sun.com/j2se

R. Nieuwenhuis (Ed.): CADE 2005, LNAI 3632, pp. 452–457, 2005.

First of all, the computing power of modern mobile devices is quite impressive considering their size. Pocket KRHyper may not solve open TPTP problems, but it tackles many small and useful semantic web or knowledge management problems in acceptable time.

Second, there are many places where a cellular Internet connection is not available or possible, like, for example, some basement pubs or restaurants. Mobile applications may rely on their bluetooth interface or the built-in camera to receive semantically annotated information. In these cases the independence from an Internet connection is a big advantage for such mobile reasoning applications.

Another problem of centralized Internet reasoning servers is that they don't scale well. Many mobile devices accessing the same reasoning server on the Internet cause trouble. If the reasoning tasks are distributed to the devices that use them, scalability is no longer an issue.

Finally, there is a privacy issue. The reasoning data may be confidential like, for example, a user profile. Many people will not feel comfortable if their private data is sent to an Internet server to be processed and might prefer not to use such services at all.

These points show that indeed, there is a need for mobile reasoning. Pocket KRHyper satisfies this need without relying on reasoning servers on the Internet.

2 Features of Pocket KRHyper

2.1 Hyper Tableaux Calculus

Pocket KRHyper is an implementation of the hyper tableau calculus [5,2]. Here we will only present some of the features of Pocket KRHyper, but restrain from describing the calculus.

In comparison to the desktop KRHyper system [16], Pocket KRHyper lacks some features like default negation and term indexing, but it is also very resource friendly so it can run on mobile devices. It still provides all the main features that made the original KRHyper a useful tool.

Pocket KRHyper can tackle first order logic problems in clausal form. The heads of clauses may contain disjunctions, but the literals in the head may not share variables. If they do, an exception is thrown. The reasoning process is very similar to the one of the original KRHyper system and is described in more detail in [16].

2.2 Description Logics Transformation

Common to Semantic Web applications is the use of ontologies and queries corresponding to DL formalisms. The transformation of these DL formula into sets of clauses enables the Pocket KRHyper to be used in this context. Matchmaking requires both the annotation as well as the profile to be represented by a DL concept. A match is detected by the subsumption or the satisfiability of the

```
// import ontology and transform to set of clauses
// take snapshot of knowledgebase
// import profile and transform to set of clauses
kb.setProfileBoundary ();              // remember profile snapshot
Reasoner krh = new KrHyper();          // instanciate the reasoner
krh.setKnowledgeBase(knowledgeBase);
int currentMatch = MatchMaker.MATCH_NOMATCH;
try { // test the satifiability of intersection
        // add message annotation to knowledgebase here
    kb.startQuery ();
    kb.addClause(LogicFactory.newClause("message(a)."));
    kb.addClause(LogicFactory.newClause("profile(a)."));
    if ( krh.reason(minTermWeight, maxTermWeight, timeout) )
        kb.removeQuery (); // model found, satisfiable intersection
        currentMatch = MatchMaker.MATCH_INTERSECT;
        kb.startQuery ();    // test for subsumption
        kb.addClause(LogicFactory.newClause("message(a)."));
        kb.addClause(LogicFactory.newClause(":- profile(a)."));
        if ( !krh.reason(minTermWeight, maxTermWeight, timeout) )
            currentMatch = MatchMaker.MATCH_PLUGIN;
        // refutation indicates message subsumed by profile
    }
}
kb.removeQuery ();
} catch ... // handle all kinds of exceptions
kb.truncateToProfile (); // drop the messageannotation
return (currentMatch );
```

Fig. 1. Matchmaking of *profile* and *message*

intersection of these concepts with respect to a terminology (see [11,13].) The terminology is considered to be a finite set of axioms $C \sqsubseteq D$ and $C \equiv D$, where C, D are concepts of the DL $\mathcal{ALC}$ extended by inverse roles and role hierarchies (see [1].) The syntax follows the lisp-like KRSS [12] and is a subset of the RACER-syntax [10]. Thus the development of the terminology may use standard desktop applications.

The transformation into sets of clauses does not enforce blocking [9] of generated role-successors. The termination of the proof is limited to acyclic terminologies. A more sophisticated transformation, that is compatible with the hyper tableau calculus is given in [4], but will be used in the resource restricted environment only if the terminology requires blocking.

Because KRHyper does not enforce the clauses to be Horn clauses, this approach supports terminologies beyond DLP [8] and enables the addition of rules and constraints.

To reduce the resource consumption, the knowledgeBase (a set of clauses) is partitioned into the terminological, profile, and query sets of clauses. Methods to manage these parts reset the knowledge base to its state before the addition of a query or profile. This allows for a single transformation of the terminological part.

The code snippet in Fig. 1 details the matchmaking of a given profile with the annotation of a message. Both are represented by a DL concept. At most two invocations of the reasoner check the satisfiability of the intersection *profile* $\sqcap$ *message* and the subsumption *message* $\sqsubseteq$ *profile* with respect to the ontology. The clauses derived from the annotation and the tests are removed afterwards.

```java
public static void main(String[] args){
    int minTermWeight = 5; //initial search depth is a term weight of 5
    int maxTermWeight = 0; //no maximum term weight
    int timeout = 1000;    //stop after 1000 ms
    boolean model = false;
    KnowledgeBase kb = new KnowledgeBase();
    kb.addClause(LogicFactory.newClause("ufo(enterprise)."));
    kb.addClause(LogicFactory.newClause("false:-ufo(X)."));
    Reasoner krhyper = new KrHyper();
    krhyper.setKnowledgeBase(kb);
    try {
        model = krhyper.reason(minTermWeight, maxTermWeight, timeout);
        if (!model){
            // Refutation Found
        } else {
            // Model Found
            Vector model = krhyper.getModel();
        }
    } catch (ProofNotFoundException ex){
        //Timeout reached
    } catch (OutOfMemoryError err){
        //Out of Memory
    }
}
```

Fig. 2. Pocket KRHyper Code Example

3   Using Pocket KRHyper

Pocket KRHyper is a Java library designed for the Java 2 Platform, Micro Edition (J2ME), supporting both Connected Limited Device Configuration (CLDC[3]) version ≥ 1.0 and Mobile Information Device Profile (MIDP[4]) version ≥ 1.0. The library can however also be used within Standard or Enterprise Edition Java applications.

3.1   Using the First Order Reasoner

To use the first order reasoner, you need to create an instance of KnowledgeBase, fill it with some Clause objects and pass it to the Reasoner instance. Figure 2 shows a small example program using Pocket KRHyper.

The code example shows the basic usage of the Pocket KRHyper library. First, a KnowledgeBase instance is created. Then, using the special LogicFactory static class, some Clause instances are created by giving a string in Protein [3] syntax as a parameter. After the knowledge base contains all the clauses for your reasoning problem, you pass it to the reasoner using Reasoner.setKnowledgeBase(KnowledgeBase kb).

To start the reasoner, use Reasoner.reason(int mintermweight, int maxtermweight, int timeout). The reasoning algorithm performs an iterative deepening search for a model, expanding the proof tree in each iteration up to

---

[3] http://java.sun.com/products/cldc
[4] http://java.sun.com/products/midp

a certain term weight. The term weight parameters control with which bound the search is started and after which bound the search is completely abandoned. The timeout parameter sets the maximum time allowed for performing a search. If `maxtermweight` or `timeout` are set to 0, they are ignored.

The reasoning process may terminate in several different ways. Optimally, it has either found a refutation or a model. A model can be retrieved using the `Reasoner.getModel()` method. There are some other reasons why the reasoning process can stop. Either it is interrupted by the user (`Reasoner.interruptReasoner()`), a timeout has occurred or the maximum term weight bound is exceeded. In these cases, the reasoner throws a `ProofNotFound Exception`. If the virtual machine runs out of memory, an `OutOfMemoryError` is thrown. Exceptions and errors can be caught safely (See the example code in Fig. 2) without causing the whole application to crash.

In the case that the programmer chose not to set a timeout or maximum term weight bound, the calculus may not terminate (e.g. if the problem has an infinite model). The programmer should be aware of this pitfall and always provide a means to manually interrupt the reasoning process.

4   Performance Evaluation

Pocket KRHyper was evaluated mainly on a Sony-Ericsson P910i smartphone with four different subsets of the TPTP library v.3.0.0 [14], namely Horn satisfiable and unsatisfiable problems, and non-Horn satisfiable and unsatisfiable problems. Problems with status open or unknown are excluded, as are those with equality, which is not supported. Also, only range restricted non-Horn problems have been evaluated, since Pocket KRHyper can't handle clauses with shared variables in the head without preprocessing.

With a timeout setting of 10 seconds and no term weight upper bound, 35% of Horn satisfiable problems, 29% of Horn unsatisfiable problems, 54% non Horn satisfiable problems, and 39% of non Horn unsatisfiable problems have been solved. Memory was not an issue.

4.1   Conclusions

Pocket KRHyper, a first order theorem prover and model generator system for mobile devices, has been presented. It runs on most Java platforms, including J2ME, J2SE, J2EE. The system is intended to be used in mobile semantic web applications, to which end a description logics interface is also provided.

The reasoner has been tested with a subset of TPTP problems with satisfying results. As future work, we will evaluate the reasoner with more real-world related problems. A sample mobile application using Pocket KRHyper can be downloaded from `http://www.uni-koblenz.de/~{}iason/downloads`. It requires a MIDP 2.0 and CLDC 1.0 compliant mobile device to run.

Acknowledgments. This work is part of the IASON[5] project and is funded by the "Stiftung Rheinland-Pfalz für Innovation".

References

1. F. Baader, D. Calvanese, D. McGuinness, D. Nardi, and P. Patel-Schneider, editors. *Description Logic Handbook.* Cambridge University Press, 2002.
2. P. Baumgartner. Hyper Tableaux — The Next Generation. Technical Report 32–97, Universität Koblenz-Landau, 1997.
3. P. Baumgartner and U. Furbach. PROTEIN: A PROver with a theory extension INterface. In *Conference on Automated Deduction*, pages 769–773, 1994.
4. P. Baumgartner, U. Furbach, M. Gross-Hardt, and T. Kleemann. Model Based Deduction for Database Schema Reasoning. In S. Biundo, T. Frühwirth, and G. Palm, editors, *KI 2004: Advances in Artificial Intelligence*, volume 3238, pages 168–182. Springer Verlag, Berlin, Heidelberg, New-York, 2004.
5. P. Baumgartner, U. Furbach, and I. Niemelä. Hyper Tableaux. Technical Report 8–96, Universität Koblenz-Landau, 1996.
6. P. Baumgartner, U. Furbach, and A. H. Yahya. Automated reasoning, knowledge representation and management. *KI*, 1:5–11, 2005.
7. P. Baumgartner and F. M. Suchanek. Automated Reasoning Support for SUMO/KIF. Submitted, 2005.
8. B. Grosof, I. Horrocks, R. Volz, and S. Decker. Description Logic Programs: Combining Logic Programs with Description Logic. In *Proceedings of the Twelfth International World Wide Web Conference (WWW'2003)*. ACM, 2003.
9. V. Haarslev and R. Möller. Expressive ABox Reasoning with Number Restrictions, Role Hierarchies, and Transitively Closed Roles. In *KR2000: Principles of Knowledge Representation and Reasoning*, pages 273–284. Morgan Kaufmann, 2000.
10. V. Haarslev and R. Möller. RACER system description. In *IJCAR '01: Proceedings of the First International Joint Conference on Automated Reasoning*, volume 2083, pages 701–706. Springer-Verlag, 2001.
11. L. Li and I. Horrocks. A software framework for matchmaking based on semantic web technology. In *Proceedings of the Twelfth International World Wide Web Conference (WWW'2003)*, pages 331–339. ACM, 2003.
12. P. F. Patel-Schneider and B. Swartout. Description-logic knowledge representation system specification, Nov. 1993.
13. A. Sinner, T. Kleemann, and A. von Hessling. Semantic user profiles and their applications in a mobile environment. *Artificial Intelligence in Mobile Systems 2004*, 2004.
14. G. Sutcliffe and C. Suttner. The TPTP Problem Library: CNF Release v1.2.1. *Journal of Automated Reasoning*, 21(2):177–203, 1998.
15. W. Wahlster. Smartweb: Mobile applications of the semantic web. In *GI Jahrestagung (1)*, pages 26–27, 2004.
16. C. Wernhard. System Description: KRHyper. Fachberichte Informatik 14–2003, Universität Koblenz-Landau, 2003.

---

[5] http://www.uni-koblenz.de/~{}iason

Author Index

Autexier, Serge   84

Baader, Franz   278
Baumgartner, Peter   392
Bechhofer, Sean   177
Benedetti, Marco   369
Bozzano, Marco   315
Brown, Chad E.   23
Bruttomesso, Roberto   315
Bryant, Randal E.   255

Castellini, Claudio   235
Chaudhuri, Kaustuv   69
Cimatti, Alessandro   315
Contejean, Evelyne   7
Corbineau, Pierre   7

Dowek, Gilles   1
Dufay, Guillaume   116

Felty, Amy   116
Fermüller, Christian G.   409

Ghilardi, Silvio   278
Godoy, Guillem   164

Harrison, John   295
Horita, Eiichi   424
Horrocks, Ian   177
Hustadt, Ullrich   204

Immerman, N.   99

Junttila, Tommi   315

Kleemann, Thomas   452
Konev, Boris   182, 204
Kuncak, Viktor   260

Lev-Ami, T.   99
Levy, Jordi   149

Manna, Zohar   131
Matwin, Stan   116
McLaughlin, Sean   295
Meier, Andreas   250
Melis, Erica   250
Musuvathi, Madanlal   353

Nguyen, Huu Hai   260
Niehren, Joachim   149

Ogawa, Mizuhito   424
Ono, Satoshi   424

Pfenning, Frank   69
Pichler, Reinhard   409
Pientka, Brigitte   54

Reps, T.   99
Rinard, Martin   260

Sagiv, M.   99
Schmidt, Renate A.   204
Schulz, Stephan   315
Schwentick, Thomas   337
Sebastiani, Roberto   315
Seidl, Helmut   337
Seshia, Sanjit A.   255
Sinner, Alex   452
Sipma, Henny B.   131
Smaill, Alan   235
Sofronie-Stokkermans, Viorica   219
Srivastava, S.   99
Steel, Graham   322

Tasson, Christine   38
Tinelli, Cesare   392
Tiwari, Ashish   164
Truderung, Tomasz   377
Turi, Daniele   177

Urban, Christian   38

van Rossum, Peter   315
Verma, Kumar Neeraj   337
Villaret, Mateu   149

Wolter, Frank   182

Yorsh, Greta   99, 353

Zakharyaschev, Michael   182
Zhang, Jian   441
Zhang, Ting   131

Lecture Notes in Artificial Intelligence (LNAI)

Vol. 3632: R. Nieuwenhuis (Ed.), Automated Deduction – CADE-20. XIII, 459 pages. 2005.

Vol. 3626: B. Ganter, G. Stumme, R. Wille (Eds.), Formal Concept Analysis. X, 349 pages. 2005.

Vol. 3607: J.-D. Zucker, L. Saitta (Eds.), Abstraction, Reformulation and Approximation. XII, 376 pages. 2005.

Vol. 3596: F. Dau, M.-L. Mugnier, G. Stumme (Eds.), Conceptual Structures: Common Semantics for Sharing Knowledge. XI, 467 pages. 2005.

Vol. 3587: P. Perner, A. Imiya (Eds.), Machine Learning and Data Mining in Pattern Recognition. XVII, 695 pages. 2005.

Vol. 3584: X. Li, S. Wang, Z.Y. Dong (Eds.), Advanced Data Mining and Applications. XIX, 835 pages. 2005.

Vol. 3575: S. Wermter, G. Palm, M. Elshaw (Eds.), Biomimetic Neural Learning for Intelligent Robots. IX, 383 pages. 2005.

Vol. 3571: L. Godo (Ed.), Symbolic and Quantitative Approaches to Reasoning with Uncertainty. XVI, 1028 pages. 2005.

Vol. 3559: P. Auer, R. Meir (Eds.), Learning Theory. XI, 692 pages. 2005.

Vol. 3558: V. Torra, Y. Narukawa, S. Miyamoto (Eds.), Modeling Decisions for Artificial Intelligence. XII, 470 pages. 2005.

Vol. 3554: A. Dey, B. Kokinov, D. Leake, R. Turner (Eds.), Modeling and Using Context. XIV, 572 pages. 2005.

Vol. 3539: K. Morik, J.-F. Boulicaut, A. Siebes (Eds.), Local Pattern Detection. XI, 233 pages. 2005.

Vol. 3538: L. Ardissono, P. Brna, A. Mitrovic (Eds.), User Modeling 2005. XVI, 533 pages. 2005.

Vol. 3533: M. Ali, F. Esposito (Eds.), Innovations in Applied Artificial Intelligence. XX, 858 pages. 2005.

Vol. 3528: P.S. Szczepaniak, J. Kacprzyk, A. Niewiadomski (Eds.), Advances in Web Intelligence. XVII, 513 pages. 2005.

Vol. 3518: T.B. Ho, D. Cheung, H. Liu (Eds.), Advances in Knowledge Discovery and Data Mining. XXI, 864 pages. 2005.

Vol. 3508: P. Bresciani, P. Giorgini, B. Henderson-Sellers, G. Low, M. Winikoff (Eds.), Agent-Oriented Information Systems II. X, 227 pages. 2005.

Vol. 3505: V. Gorodetsky, J. Liu, V. A. Skormin (Eds.), Autonomous Intelligent Systems: Agents and Data Mining. XIII, 303 pages. 2005.

Vol. 3501: B. Kégl, G. Lapalme (Eds.), Advances in Artificial Intelligence. XV, 458 pages. 2005.

Vol. 3492: P. Blache, E. Stabler, J. Busquets, R. Moot (Eds.), Logical Aspects of Computational Linguistics. X, 363 pages. 2005.

Vol. 3488: M.-S. Hacid, N.V. Murray, Z.W. Raś, S. Tsumoto (Eds.), Foundations of Intelligent Systems. XIII, 700 pages. 2005.

Vol. 3476: J. Leite, A. Omicini, P. Torroni, P. Yolum (Eds.), Declarative Agent Languages and Technologies II. XII, 289 pages. 2005.

Vol. 3464: S.A. Brueckner, G.D.M. Serugendo, A. Karageorgos, R. Nagpal (Eds.), Engineering Self-Organising Systems. XIII, 299 pages. 2005.

Vol. 3452: F. Baader, A. Voronkov (Eds.), Logic for Programming, Artificial Intelligence, and Reasoning. XI, 562 pages. 2005.

Vol. 3451: M.-P. Gleizes, A. Omicini, F. Zambonelli (Eds.), Engineering Societies in the Agents World. XIII, 349 pages. 2005.

Vol. 3446: T. Ishida, L. Gasser, H. Nakashima (Eds.), Massively Multi-Agent Systems I. XI, 349 pages. 2005.

Vol. 3445: G. Chollet, A. Esposito, M. Faundez-Zanuy, M. Marinaro (Eds.), Nonlinear Speech Modeling and Applications. XIII, 433 pages. 2005.

Vol. 3438: H. Christiansen, P.R. Skadhauge, J. Villadsen (Eds.), Constraint Solving and Language Processing. VIII, 205 pages. 2005.

Vol. 3430: S. Tsumoto, T. Yamaguchi, M. Numao, H. Motoda (Eds.), Active Mining. XII, 349 pages. 2005.

Vol. 3419: B. Faltings, A. Petcu, F. Fages, F. Rossi (Eds.), Constraint Satisfaction and Constraint Logic Programming. X, 217 pages. 2005.

Vol. 3416: M. Böhlen, J. Gamper, W. Polasek, M.A. Wimmer (Eds.), E-Government: Towards Electronic Democracy. XIII, 311 pages. 2005.

Vol. 3415: P. Davidsson, B. Logan, K. Takadama (Eds.), Multi-Agent and Multi-Agent-Based Simulation. X, 265 pages. 2005.

Vol. 3403: B. Ganter, R. Godin (Eds.), Formal Concept Analysis. XI, 419 pages. 2005.

Vol. 3398: D.-K. Baik (Ed.), Systems Modeling and Simulation: Theory and Applications. XIV, 733 pages. 2005.

Vol. 3397: T.G. Kim (Ed.), Artificial Intelligence and Simulation. XV, 711 pages. 2005.

Vol. 3396: R.M. van Eijk, M.-P. Huget, F. Dignum (Eds.), Agent Communication. X, 261 pages. 2005.

Vol. 3394: D. Kudenko, D. Kazakov, E. Alonso (Eds.), Adaptive Agents and Multi-Agent Systems II. VIII, 313 pages. 2005.

Vol. 3392: D. Seipel, M. Hanus, U. Geske, O. Barten-stein (Eds.), Applications of Declarative Programming and Knowledge Management. X, 309 pages. 2005.

Vol. 3374: D. Weyns, H. V.D. Parunak, F. Michel (Eds.), Environments for Multi-Agent Systems. X, 279 pages. 2005.

Vol. 3371: M.W. Barley, N. Kasabov (Eds.), Intelligent Agents and Multi-Agent Systems. X, 329 pages. 2005.

Vol. 3369: V. R. Benjamins, P. Casanovas, J. Breuker, A. Gangemi (Eds.), Law and the Semantic Web. XII, 249 pages. 2005.

Vol. 3366: I. Rahwan, P. Moraitis, C. Reed (Eds.), Argu-mentation in Multi-Agent Systems. XII, 263 pages. 2005.

Vol. 3359: G. Grieser, Y. Tanaka (Eds.), Intuitive Human Interfaces for Organizing and Accessing Intellectual As-sets. XIV, 257 pages. 2005.

Vol. 3346: R.H. Bordini, M. Dastani, J. Dix, A.E.F. Seghrouchni (Eds.), Programming Multi-Agent Systems. XIV, 249 pages. 2005.

Vol. 3345: Y. Cai (Ed.), Ambient Intelligence for Scientific Discovery. XII, 311 pages. 2005.

Vol. 3343: C. Freksa, M. Knauff, B. Krieg-Brückner, B. Nebel, T. Barkowsky (Eds.), Spatial Cognition IV. XIII, 519 pages. 2005.

Vol. 3339: G.I. Webb, X. Yu (Eds.), AI 2004: Advances in Artificial Intelligence. XXII, 1272 pages. 2004.

Vol. 3336: D. Karagiannis, U. Reimer (Eds.), Practical Aspects of Knowledge Management. X, 523 pages. 2004.

Vol. 3327: Y. Shi, W. Xu, Z. Chen (Eds.), Data Mining and Knowledge Management. XIII, 263 pages. 2005.

Vol. 3315: C. Lemaître, C.A. Reyes, J.A. González (Eds.), Advances in Artificial Intelligence – IBERAMIA 2004. XX, 987 pages. 2004.

Vol. 3303: J.A. López, E. Benfenati, W. Dubitzky (Eds.), Knowledge Exploration in Life Science Informatics. X, 249 pages. 2004.

Vol. 3301: G. Kern-Isberner, W. Rödder, F. Kulmann (Eds.), Conditionals, Information, and Inference. XII, 219 pages. 2005.

Vol. 3276: D. Nardi, M. Riedmiller, C. Sammut, J. Santos-Victor (Eds.), RoboCup 2004: Robot Soccer World Cup VIII. XVIII, 678 pages. 2005.

Vol. 3275: P. Perner (Ed.), Advances in Data Mining. VIII, 173 pages. 2004.

Vol. 3265: R.E. Frederking, K.B. Taylor (Eds.), Machine Translation: From Real Users to Research. XI, 392 pages. 2004.

Vol. 3264: G. Paliouras, Y. Sakakibara (Eds.), Gram-matical Inference: Algorithms and Applications. XI, 291 pages. 2004.

Vol. 3259: J. Dix, J. Leite (Eds.), Computational Logic in Multi-Agent Systems. XII, 251 pages. 2004.

Vol. 3257: E. Motta, N.R. Shadbolt, A. Stutt, N. Gibbins (Eds.), Engineering Knowledge in the Age of the Semantic Web. XVII, 517 pages. 2004.

Vol. 3249: B. Buchberger, J.A. Campbell (Eds.), Artificial Intelligence and Symbolic Computation. X, 285 pages. 2004.

Vol. 3248: K.-Y. Su, J. Tsujii, J.-H. Lee, O.Y. Kwong (Eds.), Natural Language Processing – IJCNLP 2004. XVIII, 817 pages. 2005.

Vol. 3245: E. Suzuki, S. Arikawa (Eds.), Discovery Sci-ence. XIV, 430 pages. 2004.

Vol. 3244: S. Ben-David, J. Case, A. Maruoka (Eds.), Al-gorithmic Learning Theory. XIV, 505 pages. 2004.

Vol. 3238: S. Biundo, T. Frühwirth, G. Palm (Eds.), KI 2004: Advances in Artificial Intelligence. XI, 467 pages. 2004.

Vol. 3230: J.L. Vicedo, P. Martínez-Barco, R. Muñoz, M. Saiz Noeda (Eds.), Advances in Natural Language Pro-cessing. XII, 488 pages. 2004.

Vol. 3229: J.J. Alferes, J. Leite (Eds.), Logics in Artificial Intelligence. XIV, 744 pages. 2004.

Vol. 3228: M.G. Hinchey, J.L. Rash, W.F. Truszkowski, C.A. Rouff (Eds.), Formal Approaches to Agent-Based Systems. VIII, 290 pages. 2004.

Vol. 3215: M.G.. Negoita, R.J. Howlett, L.C. Jain (Eds.), Knowledge-Based Intelligent Information and Engineer-ing Systems, Part III. LVII, 906 pages. 2004.

Vol. 3214: M.G.. Negoita, R.J. Howlett, L.C. Jain (Eds.), Knowledge-Based Intelligent Information and Engineer-ing Systems, Part II. LVIII, 1302 pages. 2004.

Vol. 3213: M.G.. Negoita, R.J. Howlett, L.C. Jain (Eds.), Knowledge-Based Intelligent Information and Engineer-ing Systems, Part I. LVIII, 1280 pages. 2004.

Vol. 3209: B. Berendt, A. Hotho, D. Mladenic, M. van Someren, M. Spiliopoulou, G. Stumme (Eds.), Web Min-ing: From Web to Semantic Web. IX, 201 pages. 2004.

Vol. 3206: P. Sojka, I. Kopecek, K. Pala (Eds.), Text, Speech and Dialogue. XIII, 667 pages. 2004.

Vol. 3202: J.-F. Boulicaut, F. Esposito, F. Giannotti, D. Pedreschi (Eds.), Knowledge Discovery in Databases: PKDD 2004. XIX, 560 pages. 2004.

Vol. 3201: J.-F. Boulicaut, F. Esposito, F. Giannotti, D. Pedreschi (Eds.), Machine Learning: ECML 2004. XVIII, 580 pages. 2004.

Vol. 3194: R. Camacho, R. King, A. Srinivasan (Eds.), Inductive Logic Programming. XI, 361 pages. 2004.

Vol. 3192: C. Bussler, D. Fensel (Eds.), Artificial Intel-ligence: Methodology, Systems, and Applications. XIII, 522 pages. 2004.

Vol. 3191: M. Klusch, S. Ossowski, V. Kashyap, R. Un-land (Eds.), Cooperative Information Agents VIII. XI, 303 pages. 2004.

Vol. 3187: G. Lindemann, J. Denzinger, I.J. Timm, R. Un-land (Eds.), Multiagent System Technologies. XIII, 341 pages. 2004.

Vol. 3176: O. Bousquet, U. von Luxburg, G. Rätsch (Eds.), Advanced Lectures on Machine Learning. IX, 241 pages. 2004.

Vol. 3171: A.L. C. Bazzan, S. Labidi (Eds.), Advances in Artificial Intelligence – SBIA 2004. XVII, 548 pages. 2004.

Vol. 3159: U. Visser, Intelligent Information Integration for the Semantic Web. XIV, 150 pages. 2004.